# The Catch

This definitive environmental history of medieval fish and fisheries provides a comprehensive examination of European engagement with aquatic systems between *c.* 500 and 1500 CE. Using textual, zooarchaeological, and natural records, Richard C. Hoffmann's unique study spans marine and freshwater fisheries across western Christendom, discusses effects of human–nature relations and presents a deeper understanding of evolving European aquatic ecosystems. Changing climates, landscapes, and fishing pressures affected local stocks enough to shift values of fish, fishing rights, and dietary expectations. Readers learn what the abbess Waldetrudis in seventh-century Hainault, King Ramiro II (d.1157) of Aragon, and thirteenth-century physician Aldebrandin of Siena shared with English antiquarian William Worcester (d.1482), and the young Martin Luther growing up in Germany soon thereafter. Sturgeon and herring, carp, cod, and tuna played distinctive roles. Hoffmann highlights how encounters between medieval Europeans and fish had consequences for society and the environment – then and now.

**Richard C. Hoffmann** is Professor Emeritus in History at York University, Toronto, and author of the acclaimed *An Environmental History of Medieval Europe* (Cambridge, 2014).

## Studies in Environment and History

**Editors**

J. R. McNeill, *Georgetown University*
Ling Zhang, *Boston College*

**Editors Emeriti**

Alfred W. Crosby, *University of Texas at Austin*
Edmund P. Russell, *Carnegie Mellon University*
Donald Worster, *University of Kansas*

**Other Books in the Series**

Samuel Dolbee *Locusts of Power: Borders, Empire, and Environment in the Modern Middle East*
Andy Bruno *Tunguska: A Siberian Mystery and Its Environmental Legacy*
Lionel Frost et al. *Cities in a Sunburnt Country: Water and the Making of Urban Australia*
Adam Sundberg *Natural Disaster at the Closing of the Dutch Golden Age: Floods, Worms, and Cattle Plague*
Germán Vergara *Fueling Mexico: Energy and Environment, 1850–1950*
Peder Anker *The Power of the Periphery: How Norway Became an Environmental Pioneer for the World*
David Moon *The American Steppes: The Unexpected Russian Roots of Great Plains Agriculture, 1870s–1930s*
James L. A. Webb, Jr. *The Guts of the Matter: A Global Environmental History of Human Waste and Infectious Intestinal Disease*
Maya K. Peterson *Pipe Dreams: Water and Empire in Central Asia's Aral Sea Basin*
Thomas M. Wickman *Snowshoe Country: An Environmental and Cultural History of Winter in the Early American Northeast*
Debjani Bhattacharyya *Empire and Ecology in the Bengal Delta: The Making of Calcutta*
Chris Courtney *The Nature of Disaster in China: The 1931 Yangzi River Flood*
Dagomar Degroot *The Frigid Golden Age: Climate Change, the Little Ice Age, and the Dutch Republic, 1560–1720*
Edmund Russell *Greyhound Nation: A Coevolutionary History of England, 1200–1900*
Timothy J. LeCain *The Matter of History: How Things Create the Past*
Ling Zhang *The River, the Plain, and the State: An Environmental Drama in Northern Song China, 1048–1128*
Abraham H. Gibson *Feral Animals in the American South: An Evolutionary History*
Andy Bruno *The Nature of Soviet Power: An Arctic Environmental History*
David A. Bello *Across Forest, Steppe, and Mountain: Environment, Identity, and Empire in Qing China's Borderlands*
Erik Loomis *Empire of Timber: Labor Unions and the Pacific Northwest Forests*
Peter Thorsheim *Waste into Weapons: Recycling in Britain during the Second World War*
Kieko Matteson *Forests in Revolutionary France: Conservation, Community, and Conflict, 1669–1848*
Micah S. Muscolino *The Ecology of War in China: Henan Province, the Yellow River, and Beyond, 1938–950*

George Colpitts *Pemmican Empire: Food, Trade, and the Last Bison Hunts in the North American Plains, 1780–1882*
John L. Brooke *Climate Change and the Course of Global History: A Rough Journey*
Paul Josephson et al. *An Environmental History of Russia*
Emmanuel Kreike *Environmental Infrastructure in African History: Examining the Myth of Natural Resource Management*
Gregory T. Cushman *Guano and the Opening of the Pacific World: A Global Ecological History*
Sam White *The Climate of Rebellion in the Early Modern Ottoman Empire*
Edmund Russell *Evolutionary History: Uniting History and Biology to Understand Life on Earth*
Alan Mikhail *Nature and Empire in Ottoman Egypt: An Environmental History*
Richard W. Judd *The Untilled Garden: Natural History and the Spirit of Conservation in America, 1740–1840*
James L. A. Webb, Jr. *Humanity's Burden: A Global History of Malaria*
Myrna I. Santiago *The Ecology of Oil: Environment, Labor, and the Mexican Revolution, 1900–1938*
Frank Uekoetter *The Green and the Brown: A History of Conservation in Nazi Germany*
Matthew D. Evenden *Fish versus Power: An Environmental History of the Fraser River*
Alfred W. Crosby *Ecological Imperialism: The Biological Expansion of Europe, 900–1900, second edition*
Nancy J. Jacobs *Environment, Power, and Injustice: A South African History*
Edmund Russell *War and Nature: Fighting Humans and Insects with Chemicals from World War I to Silent Spring*
Adam Rome *The Bulldozer in the Countryside: Suburban Sprawl and the Rise of American Environmentalism*
Judith Shapiro *Mao's War against Nature: Politics and the Environment in Revolutionary China*
Andrew Isenberg *The Destruction of the Bison: An Environmental History*
Thomas Dunlap *Nature and the English Diaspora*
Robert B. Marks *Tigers, Rice, Silk, and Silt: Environment and Economy in Late Imperial South China*
Mark Elvin and Tsui'jung Liu *Sediments of Time: Environment and Society in Chinese History*
Richard H. Grove *Green Imperialism: Colonial Expansion, Tropical Island Edens and the Origins of Environmentalism, 1600–1860*
Thorkild Kjærgaard *The Danish Revolution, 1500–1800: An Ecohistorical Interpretation*
Donald Worster *Nature's Economy: A History of Ecological Ideas, second edition*
Elinor G. K. Melville *A Plague of Sheep: Environmental Consequences of the Conquest of Mexico*
J. R. McNeill *The Mountains of the Mediterranean World: An Environmental History*
Theodore Steinberg *Nature Incorporated: Industrialization and the Waters of New England*
Timothy Silver *A New Face on the Countryside: Indians, Colonists, and Slaves in the South Atlantic Forests, 1500–1800*
Michael Williams *Americans and Their Forests: A Historical Geography*
Donald Worster *The Ends of the Earth: Perspectives on Modern Environmental History*
Robert Harms *Games against Nature: An Eco-Cultural History of the Nunu of Equatorial Africa*

Warren Dean *Brazil and the Struggle for Rubber: A Study in Environmental History*
Samuel P. Hays *Beauty, Health, and Permanence: Environmental Politics in the United States, 1955–1985*
Arthur F. McEvoy *The Fisherman's Problem: Ecology and Law in the California Fisheries, 1850–1980*
Kenneth F. Kiple *The Caribbean Slave: A Biological History*
Richard C. Hoffmann *The Catch: An Environmental History of Medieval European Fisheries*

# The Catch

## *An Environmental History of Medieval European Fisheries*

Richard C. Hoffmann

*York University*

CAMBRIDGE
UNIVERSITY PRESS

CAMBRIDGE
UNIVERSITY PRESS

Shaftesbury Road, Cambridge CB2 8EA, United Kingdom

One Liberty Plaza, 20th Floor, New York, NY 10006, USA

477 Williamstown Road, Port Melbourne, VIC 3207, Australia

314–321, 3rd Floor, Plot 3, Splendor Forum, Jasola District Centre, New Delhi – 110025, India

103 Penang Road, #05–06/07, Visioncrest Commercial, Singapore 238467

Cambridge University Press is part of Cambridge University Press & Assessment, a department of the University of Cambridge.

We share the University's mission to contribute to society through the pursuit of education, learning and research at the highest international levels of excellence.

www.cambridge.org
Information on this title: www.cambridge.org/9781108845465

DOI: 10.1017/9781108955898



First published 2023

*A catalogue record for this publication is available from the British Library.*

*Library of Congress Cataloging-in-Publication Data*
Names: Hoffmann, Richard C. (Richard Charles), 1943- author.
Title: The catch : an environmental history of medieval European fisheries / Richard C. Hoffmann, York University, Toronto.
Description: Cambridge, United Kingdom ; New York, NY : Cambridge University Press, 2023. | Includes bibliographical references and index.
Identifiers: LCCN 2022043228 (print) | LCCN 2022043229 (ebook) | ISBN 9781108845465 (hardback) | ISBN 9781108958202 (paperback) | ISBN 9781108955898 (epub)
Subjects: LCSH: Fisheries–Europe, Western–History–To 1500. | Fishing–Europe, Western–History–To 1500. | Fishers–Europe, Western–Social life and customs–To 1500. | Fisheries–Environmental aspects–Europe, Western–History–To 1500. | Fisheries–Economic aspects–Europe, Western–History–To 1500.
Classification: LCC SH254.E85 H64 2023 (print) | LCC SH254.E85 (ebook) | DDC 639.2094–dc23/eng/20220928
LC record available at https://lccn.loc.gov/2022043228
LC ebook record available at https://lccn.loc.gov/2022043229

ISBN 978-1-108-84546-5 Hardback
ISBN 978-1-108-95820-2 Paperback

for Ellen,
finally and always

Catch, noun

*Oxford English Dictionary* 1989

2a. The catching of fish

b. The number of fish caught at one time, or during one season. 1465

d. A stretch of water in which fish may be caught.

6. A trick. Obs. 1430

7a. A catching or entangling question. Obs. 1674

c. Catch 22 … a set of circumstances in which one requirement etc., is dependent upon another, which is, in turn dependent upon the first. Joseph Heller, *Catch-22. A novel* 1961.

10. concr. That by which anything is caught and held; any contrivance for checking the motion of a piece of mechanism, a door, etc. 1420

*Webster's New Collegiate Dictionary* 1975

3. Something that checks or holds immoveable.

7. A concealed difficulty

# Contents

Additional resources can be found online at www.cambridge.org/TheCatch

# List of Figures

# List of Maps



TABLE

# Preface and Acknowledgements

The past does not change. Our perceptions and understanding of it, which we call history, do change with new questions, new awareness of what constitutes evidence of a particular past, and new paradigms for provisional apprehension of that past. Environmental history is barely more than a scholarly generation old. The questions it poses to known traditional sources have altered much understanding of fairly recent pasts. Its liaisons with various branches of natural science have revealed hitherto unknown traces of past phenomena in natural and in human cultural realms. Of preindustrial Europe, for instance, we now know (or think we know) more about diseases (plague, rinderpest) and atmospheric events (drought, storms, cold spells) and thus of their places in human lives than did anyone in the late 1960s when humans first left their planet and safely returned. What has been learned of past people's experiences with pandemics, storm floods, and changing regional weather patterns ('climate') can help guide responses to present-day socio-natural events.

And what about human, or just one culture's, historic relations with dominant forms of aquatic life, fishes? That is and was not the first question to come to mind for an interdisciplinary medievalist agrarian historian, even after a childhood filled with experiences in a quasi-rural landscape beside a great inland sea. Nor even with early and later revived pleasure in recreational angling, due at least in part to its function as release from professional scholarly routines. Even after taking professional note of some fishing communities amidst the grain-growing landscape of medieval Silesia, it took an unexpected and unexplained invitation to write a brief encyclopedia article on medieval fishponds to prompt initially cursory interest. The necessary deep dive into a literature mostly ill-informed about fish, history, or both did pique idle curiosity into still peripheral investigation of historic aquaculture, angling, and ever less peripheral attention to medieval uses of inland waters as resources for transport, power, and food. Although the latter motivated initial contact with archaeozoologists who studied remains of fishes found in medieval trash heaps, my approaches long remained via economic and social avenues of enquiry, even as my late and sorely missed office neighbour Elinor Melville admonished "Richard, you are really doing environmental history!," i.e. history as if nature mattered.

So what environmental history might a medievalist of materialist bent do in the mid-1990s? Self-identified environmental historians, then and still deservedly well known, wrote articles and books on how human activities affected their natural surroundings. That was not a topic much pursued by medievalists. Even important research on woodland clearance and wetland drainage was cast in terms of economic development, not environmental impact. But I did know enough of that literature and its sources to turn it on its head and infer, even describe, the effects those and other medieval economic activities had on aquatic habitats, organisms, and ecosystems. A conference paper prepared to help Canadian representation at the International Congress of Historical Sciences in Montreal, 1995, turned into the first environmental history article to grace the cover of the *American Historical Review* in 1996, and then underwent metamorphosis into a book proposal and sabbatical/leave fellowship in 1997. That manuscript (Catch I?) was two-thirds complete when events at York University stalled my ability to think and write creatively for something more than a year. Subsequent service obligations to my department and colleagues in an agitated institution while simultaneously helping to establish a graduate field in preindustrial environmental history then consumed another two years. Happily my collegial friends in the Fish Remains Working Group continued to enjoy and encourage my efforts to turn their site reports into small works of history.

Teaching, and publishing articles and book chapters, on the general theme of medieval European environmental history led me to doubt the efficacy of a primarily economic approach to the field in general and to historic fisheries in particular. Models then prevalent for relations among economy, culture, and nature (to mention one of several paradigms) left me dissatisfied and unable to move from abstractions to operational propositions and testable relationships. Was catching and eating fish economy or culture? If a fish in the water is nature, does it become something else in a net? Or in a pan? Or as skeletal remains in a midden? While it became clear that the approach I had followed in the articles and manuscript of 1996–1997 had fatal flaws, it was only after time spent with Viennese colleagues at the Zentrum für Umweltgeschichte of the then Faculty for Interdisciplinary Research and Education (IFF) and more encounters with aquatic scientists that I found in the interaction model treating human society and its works as hybrids of autonomous natural and cultural forces a heuristic that seemed to work. By my next sabbatical in 2006–2007, I was well into a new book manuscript (Catch II?) with a more dynamic role for natural forces interacting with medieval human culture bearers, while also exploiting more fully especially new research findings of archaeozoologist colleagues. Then an invitation to write the first wide-ranging environmental history of medieval Europe (not just episodes of clearances, pollution, St. Francis, and plagues) offered an opportunity to apply and disseminate the interactive metabolic approach to broader audiences. With the fishes set aside more willing this time, the *Environmental History of Medieval Europe* was drafted by 2011, revised in 2012–2013, and published in 2014.

Returning to 'The Catch' in 2013 meant newly appeared literature in several relevant fields had to be reviewed, assessed, and integrated. Experience with the medieval environmental history book and burgeoning research by climate historians and marine biologists together opened exciting new ecological perspectives on several key fisheries. A decade of new findings from stable isotope analysis of human remains, a source just gaining recognition in 2007, cast essential new light on the evolving place of fish in medieval diets. Despite other commitments, for the first time a complete manuscript (Catch III?) was drafted by spring 2019, only to fall for the rest of that year into a prospective publisher's sadly black hole. It finally went out to readers for Cambridge at the end of that year and was accepted subject to the anticipated beneficial revisions at the very moment the COVID-19 virus embarked on its global rampage. Ensuing cultural precautions, restrictions, and lockdowns on both sides of the Atlantic slowed by months the processes of revision (inaccessible libraries, inability to replace obsolete technology, etc.) and the acquisition and management of illustrations. These latest hurdles have, however, neither changed a past millennium of relations among Europeans and fishes, nor the value of understanding, as best we can for now, the implications this view of medieval engagement with the natural world had for lives and ecosystems then, and may have for present-day grasp of our own planetary situation.

What have three decades in pursuit of long-dead fish, fishers, and piscivores shown me? An initial surprise was the mutual ignorance of fisheries scientists and writers regarding history and its methods on the one hand and of medievalists, even economic historians, regarding fish (among other things of nature) on the other. Over time my surprise dwindled to almost cynical expectations of inattention, while gradually some medievalists and fish people did come to show genuine respect and interest for each other's methods and findings. I'd like to hope that environmental historians and collegial archaeozoologists have shown the way to this common objective. Data assembled by both bone specialists and archival historians must be read with full contextual awareness to make sense of the past. Training as an interdisciplinary medievalist may ease a path to consilience, even if traditional medievalists were rarely alert to such mundane matters as fishes. As an agrarian historian, field naturalist, fly-fisher, and stream rehab volunteer, I was less surprised than most to find medieval sources replete with (always skewed and partial) references to fish and fishing. More recently historians of climate have emerged in similar manner and to like effect. History where nature matters.

*The Catch* turned out to be a story of similarities and patterns and of specific and local differences alike. It repeatedly demonstrates the ubiquity of fisheries across medieval European Christendom, not just or even primarily by the piscaterian monks of present-day popular myth. Almost as widespread were the adaptations regional medieval human communities made to the particulars of their neighbouring aquatic ecosystems with distinctive habitats, species mixes, seasonalities, and opportunities to take fishes in small or quite large numbers. Yet I remain slightly puzzled about how much, if anything, dominant medieval elites actually knew about how their subjects really caught fish, even when obliged to do so.

One key aspect of fisheries technology was (and is) keeping a highly perishable material from rapid decay into uselessness. Early medieval preservation techniques, like those for capture, involved diverse applications of general procedures, in this case drying, smoking, and dry salting. Then by the high Middle Ages and often under mercantile influence, certain methods, notably calling for barrels and specific treatments with salt or brine, spread with, and in essential support of, new consumer markets. Dissemination of innovative capture gears came later.

Finally, and to the satisfaction of believers in the alert practical intelligence of ordinary medieval people, I have been mildly surprised and pleased to uncover the widespread concern of medieval communities and authorities to maintain their supplies of fish through measures ranging across spectra from preservation of catches to provisioning markets, defending access to local stocks, and eventually to practices meant to ensure continuity of fish reproduction and habitats with what might now be called their sustainable use. Medieval Europeans, like other indigenous cultures, were aware of their natural world and could, in pursuit of their own survival values, put that knowledge to protective purposes, whether literate elites recognized that or not. Thus pursuing the exploitation of fisheries renders a mildly hopeful tale of traditional ecological knowledge and its superior ability to support continual careful resource use. Less prudent approaches set stocks and entire ecosystems at risk.

No major research project, and especially not one engaging several disciplines and sub-disciplines, multiple methodologies, more than a dozen present-day countries (and more medieval polities), and as many languages modern and medieval over three decades, can be accomplished without essential aid from colleagues, librarians, archivists, various technical specialists, and the help and questions of students. I rejoice finally to acknowledge with public thanks the critical collegial support and key contributions 'medfish' has gratefully received over the years from archaeozoologists, aquatic ecologists, historians of many stripes, and successive former graduate students and research assistants, now significant scholars in their own right (all in alphabetical order). I could not have done this without you.

Ellen Arnold
Cristina Arrigoni Martelli
Janos Bak†
Eugene Balon†
James Barrett
László Bartosiewicz
Steven Bednarski
Renaud Benarrous
Rosalind Berlow
Donna Bilak
Scott Bruce
Michael Bürkner
Markus Cerman†
Benoît Clavel
Thomas V. Cohen
William Crossgrove†
E. J. Crossman†
Dagomar Degroot
Jonathan Edmondson
Anton Ervynck
James Galloway
Aleksander Gieysztor†
Simone Häberle
Gertrud Haidvogl
Dirk Heinrich
Heide Hüster Plogmann
Jack Imhof
Stuart Jenks
Leif Jonsson
Danielle Joyner

Terryl Kinder
Rachel Koopmans
Maryanne Kowaleski
T. J. A. LeGoff
Heike Lotze
Daniel Makowiecki
Marcel Martel
Gundula Müldner
Hans Hermann Müller†
Arturo Morales Muniz
Tim Newfield
David Orton
Duane Osheim
Didier Pont
Pierre Reynard
Alasdair Ross†
Richard Schneider†
Geneviève Séguin
Jean Sergeantson
Christoph Sonnlechner
Paolo Squatriti
William TeBrake
Richard Unger
Petra van Dam
Wim Van Neer
Ryan Whibbs
Verena Winiwarter

Innumerable listeners and questioners of papers delivered at meetings of the annual Congress of Medieval Studies in Kalamazoo; biennial assemblies of the Fish Remains Working Group of ICAZ; more than one meeting of the Medieval Academy of America, American Historical Society, American Society for Environmental History, and European Society for Environmental History; and guest lectures and workshops in Canada, the United States, Scotland, England, the Netherlands, Belgium, Germany, Poland, Austria, Switzerland, France, Norway, Italy, and probably elsewhere also deserve my gratitude. Good audiences show an author what did not work.

A host of librarians at York University and custodians in dozens of collections elsewhere located and lent obscure books and periodicals; archivists in England, Scotland, France, Germany, Austria, Switzerland, Sweden, and Poland made available a (to them) puzzling array of manuscript materials.

Two anonymous readers for Cambridge University Press identified areas needing further work, followed by the editorial advice of John McNeill, Lucy Rhymer, and Rachel Blaifeder.

Of course none of these generous and hardworking folk bear any responsibilities for my errors of omission, commission, and general misunderstanding of matters medieval, ecological, or typographical. Like every book *The Catch* is an invitation for present and future scholars and scientists to correct the author's mistakes and misconceptions.

At various times aspects of my research and writing on medieval fish, fishing, and fisheries have been supported by

- Faculty fellowships and research grants from the Faculty of Arts (now Liberal Studies), York University
- Sabbatical Leave Fellowships from York University
- Major Research Grant from the Social Science and Humanities Research Council of Canada
- several grants from the Travel and Small Grants Committee, SSHRC
- research grant from the American Philosophical Society
- research travel grant from the International Research and Exchanges Board
- Senior Fellowship from the Shelby Cullom Davis Center for Historical Studies, Princeton University
- Visiting Professorship, Department of Medieval Studies, Central European University, Budapest, Hungary
- Visiting Professorship, Institute for Hydrobiology and Aquatic Ecosystem Management, Department of Water, Atmosphere and Environment, University of Natural Resources and Life Sciences, Vienna, Austria
- Leverhulme Trust Visiting Professorship, Research Centre for Environmental History and Policy, University of Stirling, Scotland.

This is not a simple story. There's many a catch here.

medfish@yorku.ca
King City, Ontario

# Abbreviations Used in Notes and Bibliography

| | |
|---|---|
| AW | Österreichischen Akademie der Wissenschaften, Vienna |
| BHSA | Bayerische Hauptstaatsarchiv, Munich |
| BL | British Library, London |
| BNF | Bibliothèque Nationale de France, Paris |
| BSB | Bayerische Staatsbibliothek, Munich |
| *CdMP* | *Codex diplomaticus Maioris Poloniae* |
| Fishbase | FishBase Froese, R. and D. Pauly, eds., 2020. *FishBase.* World Wide Web electronic publication. www.fishbase .org, version (12/2020). Last viewed February 17, 2021. |
| GHSA | Geheime Staatsarchiv, Berlin |
| KL | Klosterliteralien |
| MGH | Monumenta Germaniae Historica |
| OBA | Ordensbriefarchiv |
| OÖLA | Ober-Österreichische Landesarchiv, Linz |
| PAN | Polska Akademia Nauk |
| PIMS | Pontifical Institute of Mediaeval Studies, Toronto |
| PKB | Preussische Kulturbesitz |
| *PL* | *Patrologia Latina*, ed. Migne |
| PROME | Parliamentary Rolls of Medieval England http://www.sd-editions.com/PROME/ |
| SrM | Scriptores rerum Merovingiacarum |
| SS | Scriptores |
| SUB | Appelt and Irgang, eds. *Schlesisches Urkundenbuch* |
| TNA | The National Archive, London (formerly PRO, Public Record Office) |
| ZBZ | Zentralbibliothek Zürich ZBZ |

# Introduction
## Considering Fisheries: Medieval Europe and Its Legacies

*The Catch* explores historic interactions between two dynamic communities, namely the human societies and the aquatic ecosystems of Europe during western Christendom's medieval millennium, roughly 500–1500 CE. The book traces common patterns, variants, and changes over time in medieval experiences exploiting Europe's diverse fish populations. Human desires, abilities, and impacts played off in sometimes unexpected ways against changing aquatic habitats and fish stocks. Unsatisfied wants and newly available opportunities transformed social and environmental relations, affecting people's experiences of their natural surroundings then and in a much longer term. The Middle Ages look somewhat different when viewed from the water and so do present-day issues in fisheries science and management when seen in perspective of a thousand years long past. One catch for author and readers alike is the difference between discourses of interdisciplinary natural science and of interdisciplinary medieval studies. But that is inherent in environmental history, history studied as if nature matters. This enterprise must aspire to mediate between the two conversations.

### I.1 Fish Tales

Humans and fishes, creatures terrestrial and aquatic, inhabit what medieval Europeans conceived as two different elements, earth and water. That they were and are intimately engaged some observers then and now remained wholly oblivious. Others can tell stories, in some way testifying to how fish and fisheries were thought to matter across a millennium and more of Europeans' encounters with life forms in and beyond their own waters. An opening sampler of three tales alerts readers of various backgrounds to the confluence of methods and data sets and to some of the diverse perspectives and issues to be encountered in this realm of interaction between European natures and the medieval cultures that thrived among them.

### I.1.1 Observations and Inferences: The Long Demise of the Atlantic Sturgeon

The sad yet curiously complicated tale of the European Atlantic (or sea) sturgeon[1] begins in distant times, for this family displays many primitive physiological features. Reaching four to six meters and living sixty to a hundred years these creatures of coasts, estuaries, and large rivers consume bottom organisms and small fishes. Spawned over river gravels, the young live for a time in estuarine shallows before slowly growing up to 400 kilograms, making them the largest fish in western European fresh waters. Sturgeons first reproduce after age twelve and thereafter only every few years. They lack bony skeletons, but their backs and sides bear horny plates called scutes which endure in archaeological settings. And sturgeon flesh was historically prized. Roman luxury demand had by the first century CE seriously depleted once abundant stocks in the Rhône estuary. They took centuries to recover.[2]

Early medieval elites continued to enjoy eating sturgeon. A Byzantine physician at the court of Frankish king Theodoric (511–534) recommended it for diners in good health. Centuries later Prüm abbey in Luxembourg required serfs from its Rhine delta manors annually to haul an eleven-foot (three-metre) specimen some 200 kilometers overland or upriver for the monks to eat.[3] Finds from seventh- to ninth-century high-status sites along the southern Baltic coast suggest that sturgeon there provided up to 70 percent of the weight of fish consumed, results corroborated from more recently and precisely excavated tenth-century Gdańsk.[4]

Already by the twelfth century surviving records indicate fishing pressure and drainage projects were reducing the frequency, relative share, and size of harvested sturgeons. In the same south Baltic sites just mentioned, their share fell below 20, even to 10, percent of recovered fish remains. The average size of sturgeon consumed at Gdańsk fell from

1 Each species of European fishes here mentioned is identified in text by its common English name, where such exists. See the Appendix for the scientific name and some aspects of the organism's habitat. Only in cases of potential ambiguity or unusual need for distinction do scientific or dialect names appear in the text. On European sea sturgeon see www.fishbase.de/summary/Acipenser-sturio.html (last consulted 8 June 2020) and Kottelat and Freyhof, *Handbook*, 57–58.

2 On Roman taste see Pliny, *Natural History*, lib. 9, §§17 and 27–29. Desse and Desse-Berset, "Pêche et surpêche," 335–337; Desse-Berset, "Sturgeons of the Rhône," 81–90; Sternberg, "Rôle des fleuves dans la pêche," 190–195.

3 Anthimus, *De observatione ciborum*, ed. Lichtenhahn, 41–42; Schwab, ed., *Prümer Urbar*, 232–33 and 244. Loveluck, *Northwest Europe*, 69–70, 131–132, and 139, reports sturgeon remains at sixth–eighth-century Frisian sites in the Rhine delta and from elite Frankish sites near Lille and on the Marne just upstream of Paris.

4 Benecke, "Remarks on sturgeon," 9–17; Makowiecki, "Usefulness," 108–109.

the tenth to thirteenth century.[5] Declines in the North Sea drainage were similar. From the river Ems the sturgeon present up to the thirteenth century had vanished by the next. Written records from the Low Countries mark further losses among coastal stocks from the eleventh through the fourteenth century. The genus then disappears from food waste at the castle in Namur and also from archaeological finds at Schaffhausen, the probable upper limit of its range in the Rhine.[6] Thirteenth-century French and English kings claimed exclusive rights to all sturgeons taken in their realms. Throughout coastal Atlantic waters, by late and post-medieval centuries these mythic fish had become a rare and sometimes newsworthy catch. Some chefs even then developed recipes "to make sturgeon from veal."[7] Surviving wall paintings at several English parish churches further reflect the continued cultural resonance of sturgeon, as its distinctive form stands out among the fishes commonly depicted in waters through which St. Christopher carries the Christ Child.[8]

Medieval sturgeon populations in the western Mediterranean, notably Italy, may have been more resilient – or just longer adapted to humanized landscapes and human predation. Tenth- and eleventh-century prelates and local potentates laid claim to specimens pulled from nearby rivers: the archbishop of Ravenna, for instance, had preemptive rights over any *storionem* greater than six feet taken from the Padoreno, a now obliterated tributary of the Po.[9] The fish were deliberately pursued in the late medieval Po, Arno, and Tiber. The latter seems the best studied. Rights to take sturgeon in permanent weirs and traps at Rome's bridges were distinguished from rights to shad and other migratory fishes. Catches from the delta were valued as well.[10] Regulations issued for

[5] *Ibid.*; and Susłowska and Urbanowicz, "Szczątki kostne ryb," 53–65. For like data from other urban and elite sites in northern Poland see Makowiecki, *Historia Ryb*, 110–112 and 115, and Makowiecki, *Hodowla*, 41–42.

[6] Lampen, *Fischerei und Fischhandel*, 208; Boddeke, *Vissen*, 169–174; Pigière et al., "Status as reflected in food refuse," 238–241; Huster-Plogmann and Rehazek, "Historical record versus archaeological data."

[7] *Fleta*, I: 44, ed. Richardson and Sayles, p. 100; Clavel, *L'Animal*, 144–145; *Ménagier*, ed. Brereton and Ferrier, 239–240; Plouvier, "La gastronomie," 154; Hieatt and Butler, eds., *Curye on Inglysch*, 155–156; Hieatt, ed., *Ordinance*, item 93. Some fifteenth-century English menus do show sturgeon being served at royal banquets, but other recipes suggest their possible availability on the (black?) market (Austin, ed., *Fifteenth-Century Cookery*, 10–59, 104, and 115–117; Sutton and Hammond, eds., *Coronation of Richard III*, 292–295; Hieatt, ed., *Ordinance*, item 95).

[8] As discovered by Frederick Buller, *Fish & Fishermen*, 15–16, 55, 74, 97, 105–106, and 135–136. The same scenes in the Krems collection of central European images contain no identifiable sturgeon but I have been unable to learn if the same is true further west.

[9] Montanari, *L'alimentazione contadina*, 281 and 289.

[10] Vendittelli, "Diritti ed impianti," 394–397 and 409–422. Elsewhere see Fumagalli, *Landscapes of Fear*, 137, and Nada Patrone, *Il cibo*, 319.

Rome's fish market in 1405 make special mention of selling portions cut from a sturgeon, about the same time as quantities appropriate for retail sales appear in financial accounts of small merchants in Prato.[11] A few decades later the master chef to the duke of Savoy named sturgeon as a major import to the court and a dish appropriate for luxury dining. Maestro Martino de Como, a Lombard cook in the following generation, included in his recipe collection two for sturgeon which famous Renaissance chef Platina later copied.[12]

Even in the south yields hint at limited availability. Moralist preacher San Bernardino of Siena in 1425 ranked eating sturgeon among the worst signs of a luxurious path to perdition, and humanist Aeneas Silvius Piccolomini become Pope Pius II (1458–1464) celebrated in his *Commentaries* the seven "much prized" sturgeons up to 250 'pounds' from the Tiber delta given to him on a visit to Ostia.[13] Papal courtier and humanist Paolo Giovio's erudite but observant 1524 booklet on the Roman fish market, *De romanis piscibus libellus*, also celebrated at least the literary reputation and memory of sturgeon from the Tiber.[14] Perhaps under long-standing preindustrial conditions Italian sturgeon stocks had stabilized at a level still sufficient for consistent elite use.

The medieval tale of sturgeon thus rests on verbal and material evidence of luxury demand and consumption, which led fishers to seek this large and impressive trophy. Actual medieval capture methods and biological knowledge are unrecorded, though plainly effective enough when accompanied by damage to habitat that stocks everywhere diminished. The downward trend appears across the medieval record. This is not to assert local, much less general, extirpation, but that quantitative evidence from the nineteenth century was unlikely to reflect a pristine stock. Present-day palaoscience, however, reveals diversity previous consumers or fishers could not know.

Giovio and Piccolomini both referred to *sturione*, employing the vulgar Latin and cognate old Germanic root *stur/stor* rather than Pliny's elite Greek-derived *acipenser*. Their usage followed learned and vernacular nomenclature for all sea sturgeons across western Europe. But Latin

[11] Lanconelli, "Gli *Statuta pescivendulorum urbis*," 116–117 (article 36); Marshall, *Local Merchants*, 19.

[12] Scully, ed., "*Du fait de cuysine*," 135 and 164; Faccioli, ed., *Arte*, vol. I, p. 184; Platina, *De honesta voluptate et valetudine* and *Il piacere onesto*, ed Faccioli.

[13] Welch, *Shopping*, 33; Piccolomini, *Commentarii*, lib. XI, cap. 19 (ed. Heck, vol. 2, pp. 694–697, tr. Gragg and Gabel, *Commentaries*, abridged ed., pp. 306 and 308–309). By ancient or nineteenth-century standards this was not a remarkably large sturgeon.

[14] Giovio, "*De romanis piscibus*," cap. iiii (Travi and Penio, eds., pp. 16–26). Giovio's career is studied in Zimmermann, *Paolo Giovio*, which treats this early work on pp. 16 and 64.

writers around the late medieval Baltic applied an originally Celtic (Gallic) vernacular for salmon, *isox/esox*, to several unrelated but notably large fishes. In Poland, for instance, these included the sturgeon documented prior to 1400 in the native Slavic vernacular as *jesiotr*.[15] Until the present century this taxonomic peculiarity remained a mere antiquarian anecdote, as naturalists and after them fish scientists long understood the entire western European sturgeon population from the Baltic to Italy as a single species, European Atlantic sturgeon. In 1758 Swedish biologist Linnaeus gave it the scientific name *Acipenser sturio*.[16]

Early twenty-first-century techniques to extract ancient DNA, however, tell a different story: genetic markers indicate that even by the start of the Neolithic the American Atlantic sturgeon, *Acipenser oxyrinchus*, had colonized northernmost European coasts. The closely related American species is now thought to have arrived through North Atlantic drift currents during one or more of the cold periods which punctuated transition from the Pleistocene to the Holocene, roughly 12,000 years ago. *A. oxyrinchus* spawns successfully in temperatures some degrees colder than does *A. sturio*, and by the Middle Ages was dominant in populations from the North and Baltic Seas east to Lake Ladoga. More balanced sympatric (multi-species) stocks formed along the French Atlantic coast.[17] Most of the sturgeon eaten by early medieval Slavic princes and some of those claimed by French and English rulers were thus of long present but invasive origin. Both human and natural forces could shape the evolution of medieval fish and fisheries. Might what looks like the greater adaptability of Mediterranean sturgeons to human predation even have a genetic basis?

Genetic diversity could not, however, ward off post-medieval pressures on sea sturgeons in European waters. Ongoing human predation, pollution, and physical modifications to rivers have since the nineteenth century continued to suppress sturgeon numbers and the size and age of individuals across its range, resulting in incremental extirpation

[15] Rostafiński, *Średniowieczna historia naturalna*, pt. 1, p. 402.

[16] East-flowing rivers, notably the Danube system, contain several other species of sturgeons, including the migratory beluga (*Huso huso*), famous for its roe and the best caviar (Bartosiewicz et al., "Sturgeon fishing in the middle and lower Danube region").

[17] Thieren et al., "The Holocene occurrence of *Acipenser* spp.," brings together DNA studies of more than 7,000 dated sturgeon remains as well as earlier findings including those of Ludwig et al., "When the American sea sturgeon swam east," 447–448; Ludwig et al., "Further evidence"; Tiedemann et al., "Atlantic sturgeons (*Acipenser sturio, Acipenser oxyrinchus*)"; Desse-Berset and Williot, "Emerging questions"; Chassaing et al., "Palaeogenetics of western French sturgeons"; and Galimova et al., "Fish Bones … Staraya Ladoga," 63. Additional DNA results from German rivers are in Nikulina and Schmölcke, "Reconstruction of the historical distribution of sturgeons."

everywhere. In 1991 people took, killed, and proudly ate the last known specimens of *A. sturio* from both the Black Sea and the Mediterranean. The last recorded on the Atlantic coast of Iberia was served in a restaurant in 1992 and the last from the North Sea met a like end the following year. One spawning population may still survive in the Gironde, where, however, the last known success was in 1994, although a juvenile with that genetic marker was taken and released in 2007. Nor do wild American sea sturgeon survive in Europe, as the last in the Baltic was taken in 1996 and in the British Isles in 2005. Of the twenty-five species in the sturgeon family (*Acipenseridae*) worldwide, the International Union for the Conservation of Nature (IUCN ) rates nineteen as endangered or worse and only one as "least concern" for extinction.[18]

The separate views of medieval commentators, modern archaeozoologists, and recent geneticists trace the fading trail of a fish of impressive scale, once a key component in estuarine ecosystems, but long coveted by the rich and powerful, so fatally vulnerable.

### *I.1.2 Communities, Culture, and Sustained Fishing on Lake Constance*

While sturgeons never passed the Rhine falls at Schaffhausen, a millennial tale of fishing in medieval Lake Constance (German *Bodensee*) traces evolving sociocultural approaches to diverse freshwater ecosystems in western and central Europe's third largest natural lake.[19] Now bordered by Germany, Switzerland, and Austria, and thus once near the geographical centre of medieval Latin Christendom, the lake's two principal basins, Obersee and Untersee, are joined at Constance by a two-kilometer strait sometimes called the 'Lake Rhine' (*Seerhein*). With cold depths to 252 metres plus extensive shallows, Lake Constance now supports at least fifty fish species, while two endemics, a whitefish and a char, are extinct. Yet this tale is more about humans learning to live with their aquatic neighbours. The evolving significance of the lake's fisheries for medieval contemporaries and for present-day observers

[18] Williot et al., eds., *Biology and conservation of the European sturgeon*, 3–13 *et passim*; Kottelat and Freyhof, *Handbook*, 53–58; www.iucn.org/74928 and http://cmsdata.iucn.org/downloads/species_status_and_population_trend_of_sturgeon_on_the_red_list.pdf (both consulted 10 August 2015).

[19] Lake Geneva holds more water and Balaton covers a greater area. At 536 km$^2$ Lake Constance is almost twice the size of Lough Neagh, largest body of water in the British Isles, but only 10% that of Sweden's Lake Vänern, largest in medieval western Christendom. From another perspective it is barely 2% the area of Lake Ontario, smallest of the Laurentian Great Lakes of North America.

exemplifies key source genres and issues for wide-ranging enquiry into fisheries throughout western Christendom.

Remains of Neolithic pile dwellings along the lake shore and a passing comment from Pliny (*Nat. Hist.*, IX, 29) confirm pre-medieval fishing in Lake Constance, but the story really begins with sub- and post-Roman Alemannic (Swabian) settlement and arrival of Christian missionaries and monastic foundations during the seventh to ninth centuries. By around 800, traditional accounts made much of the fishing by Gall, a companion 200 years earlier to the wandering Irish monk Columbanus. While the party of exiles lived in the dilapidated Roman town of Bregenz, Gall made nets and fished the lake to supply monks, visitors, and local people alike. When the group left for Italy, illness kept Gall behind. After Gall's recovery he and a skilled local outdoorsman, the deacon Hiltibod, traveled up the little river Steinach to establish a hermitage and there, too, Gall's skill with his nets and some miraculous assistance again delivered the necessary rations.[20]

From local oral tradition later scribes at two great monasteries beside the lake wrote down the *Life* of St. Gall. A century or more after the holy man's death a Benedictine house had been founded on the presumed site of his hermitage. By the mid-ninth century royal Carolingian patronage made this abbey of St. Gallen[21] a major cultural centre and rival with the bishop of Constance over fishing rights on much of the south shore of the Obersee. Meanwhile since 724 what soon became another great Frankish royal abbey had been established on the island of Reichenau in the Untersee. Monks from the communities of St. Gallen, Reichenau, and the twelfth-century Cistercian foundation of Salem on the north shore of the Obersee provide most of the surviving information about use of the lake's fisheries into the thirteenth century.[22]

Knowing that St. Gallen possessed extensive fishing rights in the lake barely ten kilometers away at Rörschach, mid-tenth-century prelates arriving for an inspection assumed the kitchen had ample supplies of fish. This the monks denied, asserting the lake produced few fish at great expense, and hoped thereby to excuse their eating meat.[23] Two generations later the learned Ekkehard IV (*c.* 980–*c.* 1060), who retrospectively reported that tale, was himself well aware of local/regional fishes and their at least occasional presence on monastic tables. Although Ekkehard's

[20] *Vita Galli*, cap. 6–12 and 28, ed. Krusch pp. 252–253, 260–272, and 289–307; tr. Joynt, pp. 72–83 and 103–104.

[21] To avoid confusion with the original hermit I use the German name for the abbey and its town.

[22] Mayer, "Konstanz und St. Gallen"; Rösener, *Reichsabtei Salem*, 51–52.

[23] Ekkehard, *Casus Sancti Galli*, § 105 (Haefele, ed., pp. 212–213).

*Benedictiones ad mensam* ("Blessings at table") surely emulated not an actual menu but the fish list Isidore of Seville had provided in his sixth-century encyclopedic *Etymologiae*,[24] the Swiss monk named twenty-four 'fishes' (including four synonyms) in a mix of Latin, Latinized German, and local dialect.[25] Ekkehard thus provides the oldest known verbal sign that people around the lake recognized distinct fish varieties, from pike, trout, and char to gudgeon, perch, and chub. Most of them were indigenous to Lake Constance, including its distinctive char ("*Rubelga*", dialect *Rötel*), and one of the whitefishes ("*Illanch allemannia*" reflecting dialect "*rinanke*", most likely modern so-called Blaufelchen, *Coregonus wartmanni*[26]).

About the same time as Ekkehard at St. Gallen was cataloging and at least thinking to eat many fishes from the lake,[27] over at Reichenau a contingent of fishers, specialized abbey servants, were among the few non-monks permitted to live on the holy island. Their situation, and thus that of people who actually worked at the fishery, becomes clear at some point in the 1140s–50s, when Reichenau set out procedures for its cellarer's office, the agency responsible for the abbey economy.[28] Growing hemp to make rope for fishing nets was an obligation of all peasant tenures, while manors along streams tributary to the Untersee were to build boats and provide annual cartloads of stakes and willow branches for making fish traps. Four fisher's tenures located on the mainland where the modern causeway leads to the island were to fish

[24] Isidore, *Etymologiae*, lib. 12, cap. 6 (ed. and tr. André, pp. 180–223; tr. Barney et al., pp. 259–263, although the latter's translations are untrustworthy) mentions more than fifty aquatic taxa, but of those only the grayling (*thymallus*) and trout (*trocta*) are to be found in the fresh waters of the upper Rhine basin and only the eel (*anguilla*) would likely migrate into them.

[25] Ekkehard, *Benedictiones*, ed. Egli, 285–289. Duft, *Bodensee*, 21–23 and 90–91, Schulz, *Essen und Trinken*, 586–593 (which includes a reprint of Egli's text), and works there cited correct the reading and contest the interpretation of Egli, "Küchenzettel." To call this a 'menu' is hyperbole, but while Isidore had cobbled together fanciful etymologies of classical Latin names for Mediterranean and mythic aquatic creatures, Ekkehard plainly knew his local fishes (including the beaver (*fiber piscis*), understood as dwelling in water with other 'fishes').

[26] For one authoritative view of the 'species swarm' or sympatry among at least four, possibly six, distinct populations of whitefishes (genus *Coregonus*) in Lake Constance, see Kottelat and Freyhof, *Handbook*, 361–364, and more generally 350. Other fish taxonomists may differ.

[27] Sadly, while the monastic library at St. Gallen is rich beyond imagination, the administrative archive normally so useful for environmental and economic historians is notably thin, medieval records reportedly having been discarded in the mid-eighteenth century to free up storage space.

[28] Rösener, *Grundherrschaft*, 223–227 and 234–235, uses the text from Schulte, "Die Urkunde," 352–353, who showed this so-called *Kelleramtordnung* was falsely backdated to the ninth century. No fish are named.

each morning for the abbey and receive their midday meal when they delivered the catch. In cold seasons they ran the seine through the mouth of the Rhine channel (where prized whitefish then spawned) but after Easter they worked shoreline reed beds. The labour of servants and subjects with traditional local knowledge served the culturally determined dietary needs of the monks

But the economic autarky of the old abbeys was fading. While Cistercians were established in 1138 at Salem, about as far inland from the northwestern arm of the Obersee as St. Gallen was from the south shore, only a century later do the newer abbey's archives exhibit active interests in the lake fishery. In 1260 a share in a site to pull a seine net near Constance was yielding annual dues of 2,000 whitefish. When the monks acquired from a local knightly family full control over this site in 1290, it alone annually gave them 15,000 "dried fish which are commonly called gangfish" (*piscium aridorum, qui vulgariter dicuntur gangvische*).[29] These were not, however, the product of specialized servants, but caught and processed by independent leaseholders (legally termed 'fief-holders').

While voices of the religious dominate the lake's early record, less articulate, literate, or durable lay lords and lakeside communities also acquired fishing rights from the crown or by simple undisputed exercise. Common rights of use in the Obersee extended over areas not in the recognized private possession of individuals or urban communities. This comprised all the offshore waters of the main lake basin. While some ordinary citizens of Constance possessed their own private fisheries, the city limited use of its municipal waters to recognized fishers, who had also the right to fish elsewhere in the lake.[30] From at latest some time in the 1200s these artisans sold their catch at a public market on the shore, where archaeologists have turned up remains of pike, carp, and unidentified members of the carp family (*Cyprinidae* sp.) older than 1350. Material from this site was not sieved, so whereas large bones of large fishes are identifiable, not so the smaller and less durable remains of whitefishes and other species of potential interest.[31] Both the documents of traditional history and the material remains of archaeology are

[29] Early generations of Cistercians were almost entirely vegetarian, eating only tiny amounts of fish. Mone, "Ueber die Flußfischerei," 72 n. 4; Staiger, *Salem*, 95–97; Rösener, *Reichsabtei Salem*, 51–52. *Gangfisch* were and are a local delicacy, small egg-filled whitefish taken before the spawn, salted down, and then smoked.

[30] Cahn, *Recht der Binnenfischerei*, 86–87.

[31] Prillof, *Tierknochen*, 18 and 211–212. The carp themselves may here be thought an invasive species, having arrived from the lower Danube into the Rhine basin only since around 1000.

somewhat skewed samples calling for critical assessment to provide a partial view of people and environments in past time. By the late thirteenth century, pressures from concentrated urban demand called up local measures to regulate fishing and fish sales in the interest of consumers and subsequent formation of self-governing fishers' guilds under town supervision. Guilds in Constance and Lindau emerged as leading voices respectively in the west and east ends of the lake. Ever more people and interests were engaged with the lake's fisheries.

Several late medieval developments and events impinged on human relations with the lake. As elsewhere, growth of even fairly small urban populations concentrated flows of organic effluents into nearby waters. Between the thirteenth century and the sixteenth a lagoon-like shoreline at Constance gradually filled with town wastes. From layers of those deposits the remains of plants native to clean waters gradually disappear, replaced by a filamentous algae closely associated with strongly eutrophic or polluted conditions.[32] Consumer doubts about sharp business practices on the waterfront surface in a satirical dialect poem from about 1400. In *Des Teufels Netz* the devil explains how he drags off to hell in his net sinners of all sorts: fishers and fishmongers end up there for taking both small and large fish, keeping the catch too long in their boats, and splashing dead fish with water to appear fresh on the sale table. Honest dealers, though, evade the devil by smoking the dead ones to make the dried fish good to eat.[33]

Then in 1414 a council of the western church assembled in Constance to resolve, among other matters, the competitive presence of not two but three rival popes. Some 600 prelates, princes, professors, and ambassadors plus untold servants and hangers-on streamed into a town with probably closer to three than five thousand inhabitants. Many visitors stayed for almost four years. Their dietary needs likely tested even the honest fishmongers and the entire lake-based ecosystem. According to eyewitness burger Ulrich Richental, a joint commission of church and town councillors so managed markets and prices of victuals, notably including fish, that neither inflation nor supply shortage marred the council sessions. When bread went for a pfennig the loaf, and beef at 3pf the pound, the lake supplied pike, carp, tench, bream and the larger whitefish called *felchen* at 12–20 pf the pound and several small species such as gudgeon or dace at 24–27pf the 'measure'. But the *ganckvisch*,

[32] Küster, "Mittelalterliche Eingriffe," 72.

[33] Barack, ed., *Des Teufels Netz*, ll. 9593–9685. Ehlers, *Des Teufels Netz*, 9–15, notes the survival of five different and independent manuscript redactions and places the anonymous author in the vicinity of Lake Constance about the turn of the century.

salt or fresh, had no set price. Imports were necessary: salt sturgeon from Lombardy; large and small *stockvisch*, meaning air- or salt-dried headless cod from Norway, Iceland, or shores of the North Sea; brined herring; and from Verona the big Lake Garda trout preserved in oil.[34] If not earlier the case, the council brought fishing and fish consumption around Lake Constance into a western European network of trade in preserved fishes from widely scattered aquatic ecosystems.

Whether triggered by the council or not, ensuing human generations around the lake undertook precocious vernacular writings on local fishes, both when to eat them and how to catch them. Already in 1416 Johannes Schenklin, a cleric at Wil, managerial centre of the St. Gallen abbot's estates, glossed into a liturgical manuscript a *Nota de piscibus*, calendaring the two best to eat in each month. Of the two dozen varieties, all named in dialect, one cannot now be identified, three (huchen, salmon, and eel) were not native or present in Lake Constance but available nearby, and the remaining twenty all indigenous, five in genus *Coregonus*. In ensuing decades Schenklin's list was at least twice recopied into other St. Gallen manuscripts.[35] A choir monk, Gall Kemly, was more original, writing up in 1469 and again shortly before his death in 1477 a catalog "On varieties of fishes and the time for eating them" (*De piscium generibus et tempore comestionis ipsorum*) with seasonal and culinary advice about twenty-seven locally available fishes.[36] Compared to Ekkehard some four centuries earlier these independent texts share much greater familiarity with the lake's diverse taxa and local vernacular taxonomy.

An anonymous mid-century townsman whose dialect and references indicate he lived near the lake's outlet considered the other end of the cultural processes which transformed fishes from swimming organisms to a diner's plate. This fellow assembled a memorandum of twenty-nine recipes of baits "*Vische zů vahen*" ("to catch fish"). Into a loose collection

34 Richental, *Konzilschronik*, facsimile fols. 22b–26a; Loomis tr., pp. 98–101, misidentifies several of the fishes.

35 Stiftsbibliothek St. Gallen, Codex Csg 1050, pp. 74–75, is Schenklin's autograph, and in Csg 26, pp. 1–12, his information plus some medicinal remarks is glossed into the calendar of a liturgical manuscript done elsewhere but appropriate to the liturgy of St. Gallen. For commentary, not necessarily well informed of fishes or other early writings on them, see Wickersheimer, "Zur Fischdiätetik," 414–415, and Duft, *Bodensee*, 23 and 91.

36 Kemly's texts appear in Csg 919, pp. 220–221, and Zentralbibliothek Zürich, cod. C 150, fols. 46r–47r, and receive cursory treatment in Wickersheimer, "Zur Fischdiätetik," 412–414, and Duft, *Bodensee*, 23 and 91. Kemly's catalog would resurface in 1490 as "*Hie merck was zeyt und monat im iar ein iegelicher visch am besten sey*" in the printing of *Der Pfaffe von Kalenberg* by Heinrich Knoblochtzer at Heidelberg and in several later publications. See for the latter Hoffmann, *Fishers' Craft*, 31–52 and 92–95.

of household information he included instructions for baits to put on hooks or in traps, for attracting fish by olfactory or quasi-magical means, and for making a snorkel.[37] Like similar fifteenth-century textual objects elsewhere, this manuscript brought into a literate cultural sphere knowledge hitherto transmitted orally among professional and household fishers. At the end of our period Gregor Mangolt, a Protestant bookseller exiled from Habsburg Constance in 1548, assembled both the capture and the cooking of the lake's fishes into a single *Fischbouch*.[38] This illustrated and soon printed handbook brought the regional fishery into the sixteenth-century discourse of an emerging scientific ichthyology.

Intensified contemporary concern for regional fish supplies manifested itself in several ways. Ulrich Rösch, abbot of St. Gallen 1463–1491, is credited with restoring the abbey's long-neglected governance and economy, so laying the foundation for its early modern survival as a religious community and lordship. His regard for fisheries was reflected in acquisition and revindication of fishing rights on the lake and close supervision of the fishing 'fiefs' let out for annual rents in fish or cash. Sadly, the internal management records which would reveal the operation of abbey fisheries during his tenure as abbot or earlier as cellarer (1453–1463) survive only in fragments. The kitchen ordinance Rösch instituted about 1480 called for fish in the regular weekly round of meals on Wednesdays, Fridays, and Saturdays. His kitchen master was to buy herring and stockfish as needed and serve the latter at least once a week.[39] Seeking more secure and ample supplies of fresh fish, in 1483 Rösch undertook to move the entire abbey to lakeside Rörschach (for barely five years until angry townsmen tore down the still half-built structure).[40]

The abbot further sought to increase fish supplies by constructing ponds (*Weiher*) on many abbey properties, some of them meant to rear

[37] Hoffmeister, "Fischer- und Tauchertexte," provides an introduction and transcription of Donaueschingen Schlossbibliothek cod. 792, fols. 48r–55v. The only varieties named are eel and crayfish. For discussion of this form of recipe collection see Hoffmann, *Fishers' Craft*, 322–325, and compare a distinctly more ordered manuscript tract produced about the same time as transcribed and discussed in Hoffmann, "Haslinger Breviary," 9–12 and 16.

[38] Mangolt, *Fischbouch*. For the author's biography and the unusual relationship between the original manuscript, the pirated printing of 1557, and the author's revision now Zürich Stadtbibliothek Ms Simmler 425, see both Mayer's edition of 1905, 121–126, and the commentary in Ribi, *Fischbenennung des Unterseegebietes*, 59–79.

[39] Schwarz, "Küchenordnung," 271–276.

[40] Vogler, "Wirtschafts- und Finanzpolitik," notably 138–141 and 153–164. Duft, *Bodensee*, 91–92, quotes the contemporary chronicle for Rösch's reasoning: *"Item nützlich vnd fruchtpar den gaistlichen vne trostlichen zü ewigen zitten mitt fürbindigen guten vischen, nach jettlichem zit, vnd allweg frisch vnd vmb ain rechten pfening, daran der gaistlichait vil gelegen ist."*

carp. Other Swiss landowners since at latest the 1460s were obtaining juvenile carp from Lake Constance to stock their own ponds.[41] Financial records from Salem covering 1489–1530 likewise document not only extensive fishponds but regular purchases of dried cod and brined herring as well as heavy Lenten consumption of the lake's own *gangvisch* and *felchen*.[42]

Persistent and widespread concern for fish supplies in the growing and prosperous region around fifteenth-century Lake Constance provided context for an accelerating shift from open exploitation or local guild and municipal control of fisheries to collective joint management of the lake for sake of sustainability.[43] Having established their own regimes to protect consumers and reduce disruptive conflicts among fishers, towns and guilds began in mid-century to impose rules protecting young fish, bleak, perch, and whitefish, by seasonal closures and bans on specific kinds of nets. Minimum sizes and maximum legitimate catches ensued. With so much of the main lake open to common use – though effectively accessible only to well-equipped professionals – individual jurisdictions had, however, limited effect. Multilateral agreements among urban and seigneurial powers curbed competition and eased enforcement. Constance, Überlingen, and the principalities of Salem, Mainau, and Heiligenberg entered into a regulatory treaty in 1481. The parties' enlarged agreement of 1536 was emulated for the lake's eastern end under Lindau's leadership (1537). Abbey lordships did likewise, with Reichenau setting rules for the Untersee about 1450 and refining their terms at decadal or shorter intervals. St. Gallen turned its own regulations into a treaty for the entire south shore in 1544. This regulatory regime lasted into the nineteenth century and then only gave way to an international joint commission.

As early as 1531 the master of the Constance fishers' guild clearly articulated in a letter to his counterpart in Überlingen the rationale underlying all the regulatory measures: "Were all other fishers to hold to such an ordinance and spare the young fishes, this lake would again be as rich in fish as it is otherwise almost emptied [of them]." Pointing out that one winter's protection of young perch yielded fishers of Thurgau many more adults the next summer, he continued, "A like idea also

41 Amacher, "Teichwirtschaft," 83. See also Häberle and Marti-Grädel, "Teichwirtschaft," 151–157, and Hoffmann, "Der Karpfen," 163–166.

42 Ammann, "Untersuchungen … Kloster Salem," 286–287.

43 An environmental understanding of late medieval and early modern fisheries regulation on Lake Constance is the achievement of Michael Zeheter. For what follows see his "To prevent a disorder" and *Ordnung*, notably 82–116.

applies to the juvenile whitefish and other fishes. A pail of juveniles left uncaught would a year later give a heap of *gangvisch*."[44]

For more than 500 years medieval residents around Lake Constance saw ideological and economic value in its fishes. This motivated their cultural interest and institutional structures to exploit this resource. People of the lake were familiar with various fish varieties, knew how to catch them, and ways to get that catch from the water to consumers both promptly and in preserved form also for later use. By the last medieval centuries, if not before, they were aware of their own effect on the resource. As well as considering substitutes reared in artificial ponds or imported from afar, these medieval Europeans devised methods to sustain natural wild stocks. The latter strategy even transcended competition between different groups among fishers and among consumers. While challenged by human impacts in subsequent ages, Lake Constance remains a viable fishery a half-millennium later still.

### *I.1.3 From Fish to Commodity*

Unlike sturgeon or whitefishes, cod are probably familiar, at least on the dining table, to many present consumers – or would be if more fisheries still survived and marketers less pirated the well-known name to apply to more or less distantly related white-fleshed fishes from elsewhere. Atlantic cod is a marine creature, so fitting the modernist misapprehension of most consumers and writers that fish of economic significance always come from the sea. Atlantic cod are bottom-oriented open ocean apex predators in a range along shorelines and continental shelves spanning cold North Atlantic waters.[45] Without here going into detail, historically important stocks include those off arctic Norway, around Iceland, in the North Sea, and on the oceanic banks between Newfoundland and Cape Cod. Where abundant, cod are considered keystone species, as their predation shapes relationships among other animals in their ecosystem. Recent research has shown cod to be highly sensitive to temperature fluctuations and other environmental stresses. Diners can enjoy the delicate white flesh of cod for the same reason it can easily be preserved by drying or salting: fish of the cod family

[44] *Wiewol ich achten wo all ander vischer wer wie ain ordnung hielten und volab der jugent vischenten dieser [See] werd wider vischrich so er sunst vast erlärt ist … Ain gliche maindg hats ouch mit den selen und andern vischen. Ain pfe____er [pfeimern?] selen (deren wir kainr niender fahen wrent) gäb aber ain Jar ain hufen gangkfisch.* Transcription from Stadtarchiv Überlingen C976/ thanks to Michael Zeheter. Translation by the present author.

[45] www.fishbase.org/summary/Gadus-morhua.html (consulted 8 August 2015) and references there provide information not otherwise cited below.

(*Gadidae*, gadids or gadoids) store energy as oil in their livers rather than muscles (unlike, for example, tunas, mackerels, and salmonids).

The deepest history of cod fishing and consumption belongs to arctic Norway, where seasonal nearshore spawning concentrations of large cods and frigid but dry late winter conditions enabled coast dwellers since the Bronze Age to produce stockfish, the board-like slabs of cod dried without need for salt. Up to modern times fishers took cod with baited hooks sunk to the bottom, whether on individual hand lines or long lines with many hooks. Processing into stockfish leaves middens of skeletal remains (heads, anterior vertebrae) which identify a key subsistence good and the store of capital behind Iron Age chieftaincies there. Ninth-century Norse settlers in Orkney and Iceland carried with them the same pattern of resource use.[46] Europeans elsewhere then made no significant efforts to exploit inshore stocks of smaller cod.

The Viking Age (*c.* 800–1080) initiated a southwards expansion of taste for cod, followed by export of northern stockfish to societies in northwestern Europe.[47] Tenth-century trash heaps where Scandinavians settled in eastern England and northern Scotland held remains of headless cods (i.e. stockfish), but ensuing generations at York and London substituted local catches of cods and other gadids from the North Sea.[48] By sometime in the twelfth century fishing for cod had become ubiquitous in those waters, with at least some of the take salted down for future consumption or commerce.[49] Norwegian exports grew simultaneously. During the thirteenth and fourteenth centuries large shipments of northern stockfish passed regularly through Bergen to consumers in England, north Germany, and the Baltic. Traders from the Hanse towns of northern Germany handled the greater share, followed by the English, but as the one legal point of exchange Bergen grew into Norway's chief urban and commercial centre. All of these fisheries and trades achieved their highest volume in decades around 1300, when human numbers also reached their medieval zenith. Even so, fishing and eating cod were less

[46] Perdikaris, "Scaly heads and tales," and "Chiefly provisioning"; Barrett, "Farming and fishing"; Barrett et al., "Archaeo-ichthyological evidence"; Barrett, *Being an Islander*; Perdikaris and McGovern, "Codfish and kings" and "Viking Age economics."

[47] Barrett et al., "'Dark Age economics' revisited," the now-classic statement on origins of marine fishing for cod and other species in northwestern Europe, is more broadly nuanced and grounded in Barrett, "Medieval sea fishing" and the essays in Barrett and Orton, eds., *Cod and Herring*, some of which are cited individually in what follows.

[48] Harland et al., "Fishing and fish trade"; Orton et al., "Fish for London."

[49] For much of what follows see the concluding overview in Barrett and Orton, eds., *Cod and Herring*, namely Barrett's own "Medieval sea fishing, AD 500–1550," notably 253–262. Details for England are in Kowaleski, "Commercialization" and "Early documentary evidence."

common in the high and late medieval Low Countries, where inshore waters supported good concentrations of other gadids (whiting) and flatfishes.[50] On more southerly tables cod would be hard to find before roughly 1350.

During late medieval centuries consumption of cod deepened in some western regions and spread haltingly eastwards and southwards into places where no such fish ever lived. Origins and features of the product changed as well. Consumption of Nordic cods at London and York peaked in the early 1300s, when stockfish from Iceland first entered the market. After some post-plague generations of diminished demand, English consumption of northern catches, now mainly of Icelandic origin, revived only briefly in the mid-fifteenth century. Much more salt-dried or even wet-salted cod from stocks around the British Isles more than took up any slack, since about 1400 this species surpassed herring as the principal source of fish flesh in English diets. Boat owners in both eastern and southwestern England replaced crews working on shares with individually hired hands. As DNA analysis of cod remains from the *Mary Rose* (sank 1545) has brought home, after 1500 large cod of Atlantic origin supplemented regional sources of supply. Across the northern European mainland, too, fifteenth-century markets and tables came to hold cod in different preserved forms from various distant and local sources.[51]

After centuries of lag, from around 1400 cod did begin to penetrate more southerly markets and diets. This growth relied as much on use of salt to preserve catches from coastal stocks as it did on expansion of the dried stockfish industry. In interior central Europe, stockfish were among the victuals with managed prices during the Council of Constance, but the Alsatian chef there hired by the council's elected pope reported they were for Thuringians, Hessians, and Swabians, not the French or Italians.[52] An occasional stockfish recipe crops up in fifteenth-century

50 On Norwegian exports, see Nedkvitne, *Hansa and Bergen*, 25–96, and summarized in his "Development." For the Baltic the vast historiography on Hanseatic trade receives material content in Holm, "Commercial sea fisheries," 18–19; Makowiecki et al., "Cod and herring," 123–129; and Lõugas, "Fishing and fish trade." For sparse representation of cod on Flemish markets and sites until the very late 1400s see Van Neer and Ervynck, "Rise of sea-fish consumption," 159–167.

51 Nedkvitne, *Hansa and Bergen*, 96–249; Barrett, "Medieval sea fishing," 262–264; Harland et al., "Fish and fish trade," 191–196 and tables 15.2–15.3; Orton et al., "Fish for the city"; Orton et al., "Fish for London"; Locker, *Role of Stored Fish*, 282; Hamilton-Dyer, "Fish in Tudor naval diet"; Hutchinson et al., "Globalization of naval provisioning"; and Clavel, *L'Animal*, 162–163.

52 Richental, *Konzilschronik*, fol. 25b (tr. Loomis, p. 101), where reading of *stockvisch* as 'salt cod' seems to me dubious. On Germans' taste for these fish see Laurioux, ed., "Le 'Registre de cuisine'," pp. 741–742 (recipe nr. 69). Prillof, *Tierknochen*, 18 and 212, reports cod vertebrae from the Constance market site dating *c.* 1450.

Bavarian and Austrian cookbooks, but no such traces in their kitchen middens or even their kitchen accounts. Further to the south and west the prevalent preserved white-fleshed fish was hake, the resident bottom-oriented predatory gadid in the Channel Approaches and Bay of Biscay. Historical confusion reigns when incautious modern writers translate *merluz*, *merluccio*, and like words as 'stockfish', not the distinctly smaller hake taken and salt-dried by fishers from Cornwall and Brittany south to Galicia and Portugal.[53] In Spain and Italy, as in France, salt-dried cod were a new product characteristic of the sixteenth century. Papal courtier Paolo Giovio, whom we met praising the sturgeons of Rome, may epitomize the ambiguity and transition to a stiff and headless commodity:

> Also there are brought from the bounds of Sweden and Norway arm-long *merlucciae*, so hard and stretched out as to be like boards, which we see to be the choice of the Germans who inhabit the city.[54]

Even in 1524 the fish of far northern origin remained alien to native Romans. An age of *baccalà* was yet to come. In any case, this commodity was far far away from any living fish elsewhere called cod.

Even as Giovio wrote the cod tale was changing. In a story less widely known than it deserves, a Venetian navigator financed in part by Bristol merchants whom Hanseatic competition had squeezed out of the Bergen market and Icelandic fishery found his way to a bonanza of cod in the western Atlantic and back to England to spread the news. One of Zuan Caboto's English crewmen in 1497 told a Spanish intelligence agent of unimaginable abundance off the "New found land," which would free English mariners and markets from dependence on Iceland. The Cabot circle would later speak of "infinite fish."[55] First to exploit the hitherto untouched Grand Banks were Portuguese (1506), Normans (1508), Basques and Bretons (1511). Yet within a generation relations between humans and fish in the northwest Atlantic were transformed.[56] Soon French, English, Spanish, and eventually American boats and fleets would compete for those rich schools through a half-millennium. By the late sixteenth century European markets, diets, coastal settlements, and

[53] See for instance Kowaleski, "Expansion"; Hocquet, "Pêcheries médiévales," 86–87, 96–98, and 111–112; Kerhervé, *L'État breton*, 686–692; and Ferreira Priegue, "Pesca y economia."

[54] *Adducuntur quoque ex finibus Gothiae atque Norvegiae cubitales merlucciae, adeo durae et extentae, ut fustibus assimilentur, quas in delitiis Germanorum, qui Urbem incolunt, esse videmus.* Giovio, *De romanis piscibus*, cap. xlii (p. 60).

[55] Williamson, ed., *Cabot Voyages*, 207–214, 266–268, and 297. Later writers, notably Zuan's son Sebastian, asserted fish so abundant they "stayed the passage of his ships."

[56] Holm et al., "North Atlantic fish revolution," puts the early modern case in a progressive economic mode.

fisheries politics felt waves of the revolutionary disruption radiating from the epicentre off Newfoundland. Salt-dried cod went to markets in Spain, Portugal, Italy, southern France, and eventually to Caribbean plantations as rations for slaves. The English colony become Commonwealth of Massachusetts would make a sculptured golden cod a symbol of its autonomy, but what reached distant consumers were hard, headless, salt-encrusted slabs.

Fast forward five centuries: environmentally aware consumers now think of cod fisheries in terms of competition and collapse. While fisheries experts and contemporary historians continue to debate the relative roles in the cod debacle of the late twentieth century of overfishing, of mismanagement, and of environmental change (regional fluctuations in water temperatures),[57] in a longer view all this was 'new' only in scale. Since documented European discovery of the banks between New England and Newfoundland, the prevalent cultural regime has sought to satisfy as much demand and to fish as intensively, as available technologies for capture and preservation would allow. Perhaps already by 1600 and surely in the late seventeenth century and again from the mid-nineteenth, shortfalls in catches caused some local observers to worry about disruption of aquatic life and future depletion. But each time new technologies and practices (long multi-hook lines, nets and trawls, steam power, factory ships), new fishing frontiers (further offshore, further north, longer seasons), and a shared desire for wealth and prosperity pushed the catches up, changed the sense of normal abundance (what ecologists call the 'shifting baseline syndrome'), and let both decision-makers and fishers set aside all doubts.[58] The ever greater distance of investors, managers, and consumers from the waters and from the live animals made decisions easier, too.

The sorry tale of late twentieth-century cod is a reasonably clear progression from the late fifteenth-century arrival of medieval Europeans in the western Atlantic and its 'infinite fish'. But what lay behind this evolution? Had it anything to do with European experience of the cod and, by reasonable extension, other fisheries during the previous medieval millennium? How and why the parallels between the long demise of the sturgeon and the collapse of the cod in one-third the time? What had fishers of Lake Constance been able to learn and others not?

[57] Rose, *Cod: The Ecological History*; Bavington, "From hunting fish to managing populations" and *Managed Annihilation*. Further online in Supplement 0.1.2.

[58] A narrative developed in evocative detail by Bolster, *Mortal Sea*, and sustained with documentary precision in Pope, "Early estimates," 12 and 27.

### I.1.4 *Telling Tales in Time and Space*

Whether centred on a natural organism, human groups, or the natural and historic settings of their encounters, our three fish tales offer glimpses of medieval Europeans interacting with surrounding aquatic nature. Sea sturgeons swam at the centre of a story of human desire to possess and consume an impressive, very distinctive fish. Common people worked in the natural world so medieval elites of wealth and power could enjoy this increasingly rare trophy. Attributes of the animal itself framed human efforts and impacts on aquatic ecosystems as well as the response of sturgeon populations to those pressures. The tale of Lake Constance revolved around medieval communities living beside a particular water body and turning to it for important corporeal and cultural needs. They successively adapted techniques and skills to exploit diverse fish stocks within their evolving sociopolitical structures – subsistence, lordships, urban markets – and their cultural understandings. Aware of pressures on resources, leading elements in regional society responded in several practical ways.

Even in the cursory form given here the three-act story of cod covered the longest time and widest stage. Millennia of local subsistence use anchored the narrative and even some of its durable technical features off northern coastal Norway. Across five subsequent medieval centuries northern fishing evolved into commercial production of a standardized commodity for long-distance trade even as its methods came to be emulated by regional societies elsewhere in the fish's range. Better access to resources (salt) let late adopters develop superior (or at least cheaper) techniques for preservation and thus commerce. Fishers but especially traders in cod early helped build a European network of interregional trade later joined by, for instance, the Lake Constance fishery. The end of the Middle Ages initiated a new five-century global expansion of cod which widened the gap – spatial, economic, perceptual – separating the animal and its fishers from the commodity and its consumers. Competitive conflicts meant dominant interest groups (nations, investors, traders) responded to successive waves of depletion with separate, even contradictory, technological and finally reluctant regulatory measures.

Every historic aspect, attribute, and evolution in these fish tales will recur in the wider exploration of western Christendom's encounters with its aquatic environments. Different habitats affect biodiversity and the resources accessible to local human communities. Economic institutions evolved differently across Europe's many regions. Species characteristics and consumption patterns shaped potential for innovation in catching, preserving, and marketing fish. Some themes can be seen as shared across broad areas, but the particulars will differ. Medieval fisheries are not all one story. The tales simply indicate some patterns and variations.

Each tale has further been informed by a confluence of methods and data sets. These few pages have drawn upon ichthyology and ecology of present experience with fishes as well as the methods and findings of archaeozoology which can make sense of recovered remains of fishes long dead and (mostly) eaten. In the traditional historians' written record critical readings of various documentary types and genres indicated something of past writers' purposes, perceptions, and unintentional assumptions, even ignorance, regarding the natural world in general and fishes in particular. Integrated with these anthropogenic remnants of the medieval European past are the findings of palaeoscience, the present-day recovery of natural features such as ancient DNA, climate proxies, trace elements, and stable isotopes unimagined by medieval people. The interdisciplinary approaches deployed as opportune in the fairly simple fish tales will help call up a more diverse but interwoven understanding of people and fishes across western Christendom's medieval millennium and something of the legacy their experiences, actions, and consequences left for their future, which is our present.

From northern Norway to Sicily and Ireland to the Danube, European habitats held a multiplicity of aquatic creatures. Medieval regional vernacular cultures knew equally variegated vocabularies for those animals and for the means by which people caught them. Though acknowledging both the biogeographical integrity of European waters and the cultural unities of western Latin Christendom, most studies of fish and fisheries and nearly all of those with historical or antiquarian intent have (understandably) been local, regional, or at most national in scope. Yet only rare highly localized species are confined to a single country, and many European watersheds and coasts encompass several modern (and medieval) polities.[59] National, regional or single-species studies – Danish

[59] Robinson and Starkey, "Sea fisheries of the British Isles," 121–122, in 1996 rightly criticized their predecessors for lack of context and narrowly national perspectives. Their essay and the ongoing research and publications such as Roberts, *Unnatural History of the Sea* (2007) and works of the History of Marine Animals Project (HMAP) actively and successfully build toward broader and more ecologically informed approaches (see, for example, Holm et al., "Marine animal populations," 3–23). Further progress is evident in Rick and Erlandson, eds, *Human Impacts on Ancient Marine Ecosystems*, an important collective effort from mainly archaeological perspectives; in the essay "Of seascapes and people" wherein historian-editors Kathleen Schwerdtner Mánez and Bo Poulsen introduce their innovative multidisciplinary *Perspectives on Oceans Past*; and in the collective prolegomonon of Holm et al., "North Atlantic Fish Revolution." Yet little of this excellent work in marine environmental history or historical ecology thinks to link with freshwater fisheries science, history, or other scholarship.

herrings, Norwegian cod, English fishers, etc. – neglect competitors on the water and in broader markets as well as other species of some regional importance. Customary separation of fisheries studies into freshwater and marine pools simply violates the commonalities of aquatic life and of terrestrial humans engaging with creatures from another medium. Most medieval Europeans lived far from the sea and, as will appear below, medieval fishing long remained a primarily inland and coastal activity carried out within sight of land. For all these reasons, this book attempts European, not territorial or salinity-based, coverage. As a work of environmental history it aims to explore and tell tales of a thousand years of encounters between the cultural community of western Christendom and the animals and ecosystems of waters flowing through and washing the bounds of that community.

## I.2 Doing Environmental History

### *I.2.1 Interactions: Nature and Human Culture, People and Places*

An effective environmental history can avoid a priori assumptions of environmental, material, or cultural determinism by starting from the palpably human constructs of 'Culture' and 'Nature' as identifying two autonomous realms of identity and causation.[60] The former comprises the means by which humans comprehend and communicate their own being and the world around them. Such communication creates human knowledge, beliefs, institutions, and purposes. Everything from divinely ordained dietary rules to seigneurial authority and markets to metaphysics is a learned element of culture. The latter, 'Nature', refers to the material universe, living and non-living, and its operations free of human (or evident supernatural) wishes or cultural preferences. 'Laws of Nature' describe, among other things, how flows of energy and matter (carbon, nitrogen, oxygen, etc.) link living organisms to one another and

[60] Hoffmann, *Environmental History*, 3–16, more fully discusses methods for environmental history. For a similar way of thinking with different conceptual vocabulary see the multi-authored essays "Historical ecology and the longue durée" and "Concepts for integrated research in historical ecology" in Crumley et al., eds., *Issues and Concepts in Historical Ecology*, 13–40 and 145–181 respectively. These approaches have evolved from and beyond the mutual connection of 'ecology', 'economy', and 'mentality' which now classic pioneering American environmental historians postulated in the 1980s (McEvoy, "Interactive theory of nature and culture," and Worster, "Doing environmental history" in *Ends of the Earth*, 289–307).

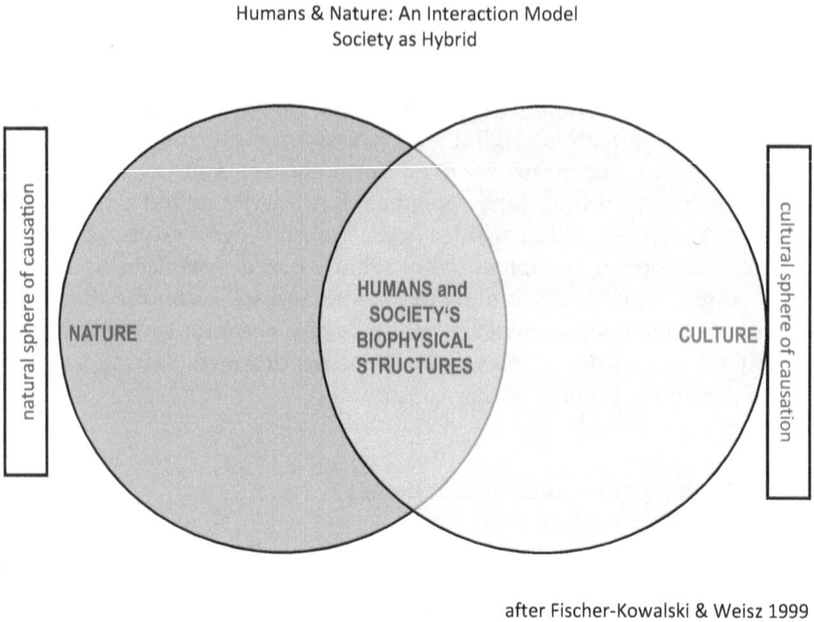


Figure I.1 Humans and Nature: an interaction model – society as hybrid.

to their surroundings in networks now called ecosystems. For each living thing, those flows constitute its metabolism. Humans and their societies have metabolisms, too.

In terms articulated by a group of social ecologists in Vienna, the realms of culture and nature *interact* in the actual world. Human bodies, artifacts, and societies necessarily function as *hybrids* of nature and culture, possessing both symbolic and material attributes. Figure I.1 attempts to visualize this condition. Material actions of humans are subject to natural forces, while the realm of nature enters human perceptions only through cultural forms, notably language. When, however, humans interact with the natural world, they modify it, intentionally, knowingly, or not. Changed flows of materials and energy result in new ecosystems, not purely natural but 'colonized' for human cultural purposes (Figure I.2). Yet in even the most humanized of circumstances, natural forces remain, sometimes giving human interventions *un*intended results. Our fish tales have already illustrated this.

Two further aspects of this interactive or metabolic perspective help organize data and thinking in environmental history in general and about

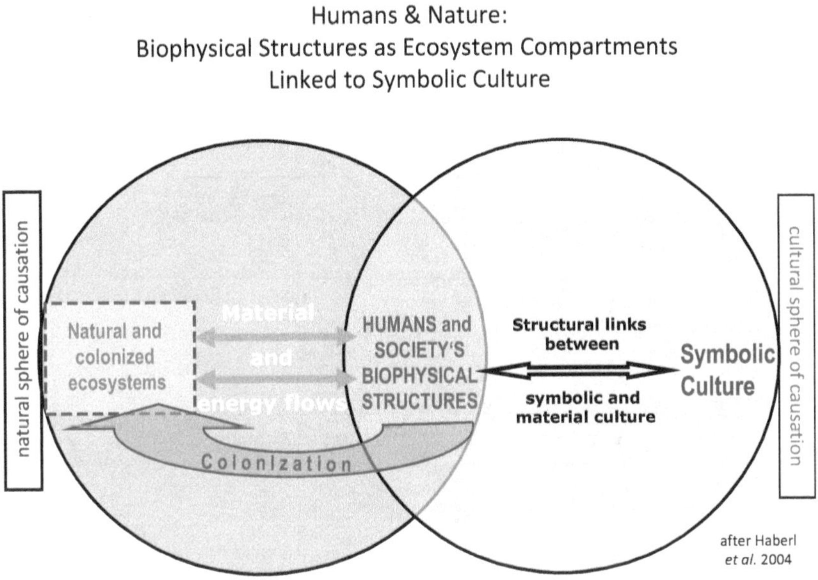


Figure I.2 Humans and Nature: biophysical structures as ecosystem compartments linked to symbolic culture.

medieval fisheries in particular. First, relations between the natural and the cultural spheres are *reciprocal* and cumulative, driving but not determining change in both realms (Figure I.3). When people experience their physical environment, they make sense of this experience as a cultural representation, which then becomes an object of processes within the cultural sphere. So long as those processes remain symbolic, even if they become coherent programs to do something in material life (such as make a net to catch fish or erect a temple), they lack environmental effect. Whether driven by moral values or what might be called economic concerns, cultural programs enter the natural sphere through human physical work, which both changes and is itself subject to the natural sphere. Subsequent human experience with this new colonized nature (the net catches fish but destroys habitat, so driving other fish away; the new cathedral removes large trees for miles around) results in new representations. Over time nature and culture co-adapt, resulting in the *co-evolution* of both.

Second, it is soon apparent that the heuristics of 'Nature' and 'Culture' are themselves cultural constructs, generalized concepts free from time and space, by which we try to make sense of our experiences. What rather occurs in a 'real' world of human behaviour is that (potentially) identifiable

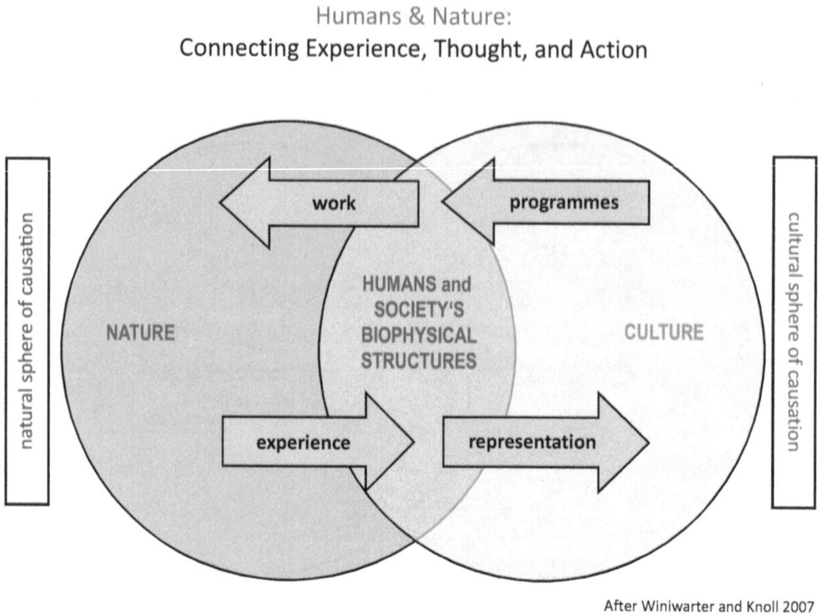


Figure I.3 Humans and Nature: connecting experience, thought, and action.

individuals or groups with specific culturally defined purposes and practices pursue these in specific places with their own natural features: a Genoese fisher setting out to catch sardines in the Tyrrhenian Sea is not the same as a Swedish miller with salmon traps in the mill leet. The 'hybridity' and 'co-adaptation' of theory are thus but labels for what people and some things of nature do at particular 'socio-natural sites'. Historians' generalizations about, for instance, the construction of weirs to take migratory salmon, shad, or eel and the effects of these structures on the runs of fish are therefore assembled from what specific men and women did at particular times on (then) identifiable reaches of the Blackwater, the Humber, the Dordogne, or Rhine, each understandable as a socio-natural site. The same applies at larger scale to the waters and shores of the Danish Øresund, where thousands of men and women assembled each summer in late medieval centuries to catch and cure herring. The concept of socio-natural sites reminds present-day historians and their readers that actions in the natural world are by individual, named, people at places with distinctive features. Medievalists can rarely see the personal side of ordinary peasants, fishers, or even seigneurs. Environmental histories often treat only policymakers or collectivities. So individuals who do appear in relevant sources here will often be remarked. So will the specific places engaged by those people. The abbey of Cluny and its priory of Paray le Monial are but fifty kilometers apart, each on a river in southern Burgundy draining from the Massif Central. The former sat beside the Grosne, a tributary of the Saône and thus of the Rhône, the latter on a minor affluent of the Loire. Monks or their servants at Paray could exploit migratory Atlantic salmon at their doorstep, but not so those at the mother house.

Fishing is simultaneously an ecological and a cultural (economic, symbolic, social) act. This book attempts to ascertain the people, practices, and installations intended to take fish under certain conditions from identifiable medieval waters and put those fish to human use. Even fragmentary knowledge can help establish human experiences, programs, and work with consequences for natural and social systems alike. If and when such actions and impacts reach a scale sufficient to transform an entire socio-natural site (or collectivity of such sites) into something quite different we can speak of an 'ecological revolution' or a 'regime shift'.[61] The ability of a defined system (social and ecological) to persist under such pressures is a good working definition of 'resilience'.[62] In shifting scale from site to socio-economic structures to ecological revolutions we also go from individual human experience to large collective generalizations which may lie beyond contemporary awareness.

### *I.2.2 Interrogating What Remains*

Historians habitually borrow concepts and theories from other disciplines – supply and demand from economics, queering from sex and gender studies – to test if these ideas help them understand conditions, events, and developments in the past. Medievalists are likewise familiar with employing multiple tools to squeeze meaning from scattered and skewed fragments from a largely illiterate and long-gone culture. The inherent interdisciplinarity of environmental history was already manifest in introductory fish tales, where use of saint's legends might surprise fisheries ecologists, references to ancient DNA and water chemistry may have caused some medievalists to blink, and ecological concepts lent meaning to data. Environmental history often ranges widely not only for concepts but for empirical methods to obtain critically credible data about the past.

[61] Merchant, "Theoretical Structure of Ecological Revolutions" and Scheffer, *Critical Transitions*, 104–105, 121–122, 186–216, and 357.

[62] Holling, "Resilience and stability," is the classic formulation.

Despite different media and sometimes analytic vocabularies, traditional history, conventional archaeology, and the zooarchaeology heavily used in this book have similar needs and ways to establish the identity of an object, its likely relationship to a particular past, and (the limits of) its ability to provide information about that past. The textual and [icono]-graphic objects [sources] commonly examined by historians and other humanist scholars were once created with some kind of communication (or memory) in mind, though sometimes a new viewpoint can reveal information other than that intended. For instance, that it took a miracle to provide Thomas Aquinas, dying in 1274 at a monastery on the road from Naples to Rome, with his desired dish of herring (transformed from sardines)[63] unknowingly provides a good sign that the widely traded Atlantic fish had not reached Italy before 1300. The hybrid quality necessarily imparted to fish from their capture through eventual consumption is obvious, and more so if marketed in the process. Records of prices thus indicate points of plausible consensus between supplier and purchaser/consumer on the value of this object at a particular time and place. At several points in this book historically retrieved fish prices serve to shed light on the relative value of different fish varieties, both biological species and forms of processing; on the value placed on fish compared to other foodstuffs, wages, or other quantitative social markers; and eventually on some changes over time in price series. Supplement I.2.2 online treats critical distinctions among sources of price citations and the constraints this places on inferences from them.

Traditional archaeologists recover material artifacts and structures, objects rarely created or deposited with communicative intent. Having, like a historian, established the character and date of the object, the archaeologist can infer the purpose and evident abilities of its makers. Patterned finds of large stones and beams in what once were river gravels can indicate a fish trap and, if enough of the structure is recovered, its orientation in the flow shows whether upstream migrant salmon and shad or downstream migrant adult eel were the quarry. The structure alone neither identifies the builders or operators nor reveals when it was used (but see below).

Zooarchaeologists work with what some call 'ecofacts', pieces of the natural world modified and commonly deposited by humans. In the present context this means fish bones which a skilled analyst can identify to the level of family, genus, or species. An individual bone may indicate

[63] As reported by the first hagiographer of Thomas and principal investigator for his canonization in 1323, Guillaume de Tocco, in *Ystoria sancti Thome de Aquino*, 192–194 (ch. LVI).

the animal's presence and likely ecological origin – preferred habitat, range, etc. – but say little of its sociocultural significance. An assemblage produces a calculation of relative taxonomic abundance, and perhaps a pattern in the size of fish or in the presence or absence of certain skeletal elements indicative of butchering techniques.[64] But now there comes into play what archaeozoologists call 'taphonomy', considering what happens to organisms after they die and leading to reconstructing site formation and recovery. The process is always somehow selective. While all fish bones are less durable than those of domestic mammals, those of some varieties, the salmonid family in particular, decay more rapidly and completely than do others. Large bones of large fishes are sturdier and more easily recognized than the small bones of small species. Excavation strategy becomes critical: presence of pike-perch or cod in a sample only hand-picked is convincing, but absence or low representation of eel, whitefish, or sardine says little if the context was not sieved. Yet similar critical reasoning applies to the most traditional historical sources: monastic cartularies retain the charters where the house gained, not those recording an opponent's victory. An artist's observation or convention may or may not replicate distinctive features of a fish or of a capture device. Basket traps (pot gears) are commonplace in medieval illustrations, nets shown deployed from boats rarely comprehensible, and major barrier traps simply invisible.

In words of archaeozoologist László Bartosiewicz, "It [taphonomy] may be seen as the common denominator between archaeological, iconographic and textual data within the context of reconstructing life during the Middle Ages."[65] Taphonomy provides a paradigm for systematic analysis of selective and differential loss and preservation/recovery of remains from the medieval past and hence, whatever the discipline associated with a particular medium, the variety of source materials comprising the historical record for medieval Europe. This book follows such a holistic critical approach to evidence of people and fish.

Finally, the practice of environmental history in the early twenty-first century draws increasingly on findings from palaeosciences, recovering information about the past from materials and features outside any realm of past human awareness or purpose. In certain circumstances those recovered logs from fish traps are datable by analysis of tree rings and/or radiocarbon. Certain stable isotopes extracted from collagen in fish

[64] Orton, "Archaeology as a tool for understanding," provides a useful summary catalog of zooarchaeological evidence for historical ecology, including some of the palaeoscientific methods discussed below.

[65] Bartosiewicz, "Like a headless chicken," 21.

remains can indicate the region where the fish were taken – so far notably to discriminate among cod bones of Baltic, North Sea, Norwegian, or New World origin.[66] Other stable isotopes in protein from human skeletal remains can signal significant consumption of marine organisms during at least the last decades of the individual's life. Marine sediments provide proxy indicators for water temperature and datable river deposits for pollution by heavy metals. Most of what is now known about climate and its changes in Europe's medieval millennium comes not only from textual references to weather but from tree rings and from isotopes and volcanic ash in ice cores from glaciers and sediment cores from deep seas and lakes.

The operative mode of enquiry underlying this book is shared among disciplines, namely critical assessment of relations between specific relics of the past and the way each has reached the present. Historical texts call for source criticism, archaeological materials for taphonomic analysis, and palaeosciences for the processes and calibrations which make certain observations convincing proxies for past environmental conditions. All procedures ask what statements regarding a specific past this data can support, which it would negate, and where it is mute. Just as failure to sieve means no information about consumption of small fish, lack of literate interest in techniques of capture does not mean none were being used. The goal is consilience, a concordance of evidence from independent, unrelated sources, or good reasons for its absence.[67]

A further epistemological problem is negotiating matters of scale. How to get from an individual set of market regulations, the archaeological context of a single bone-filled kitchen midden, or an individual socio-natural site to infer cultural norms, trends, or the historic range of a fish species? Truly quantitative or quantifiable source materials and comparative methods have limited applicability for the Middle Ages. In what follows, this issue shall call for documenting multiple cases and variants across spaces and/or times: authorities labeled fish traps hazards to navigation in the sixth-century Po, thirteenth-century Thames, and fifteenth-century Danube; in one early medieval site after another across Europe all the fish identified, though of different species, ranges, and habitats, were recognizably native to nearby waters. Beyond hopefully representative examples in the text, more often appear in the online Supplement. *The Catch* thus aims to piece together fragments of information to reveal what

[66] Barrett et al., "Interpreting the expansion."

[67] Izdebski et al., "Realising consilience"; Newfield and Labuhn, "Realizing consilience"; and Haldon et al., "History meets palaeoscience," apply Wilson's classic *Consilience* to research on medieval environments.

medieval people said, what they seemed to know and do without telling, and what those contemporary witnesses could neither know nor comprehend about their encounters with Europe's aquatic life. Of course when even present-day environmental consequences seem to hover at the edge of human awareness, none should be surprised to find here occasionally hypothetical reconstruction from an exiguous and commonly unplanned medieval record. Scraps of historical information are patterned by known ecological relationships and processes or by known cultural mechanisms linking extraction of fish from their habitats and distribution to people who ate them.

A well-conceived environmental history offers perspective in time for current dilemmas and responses. It can sustain this first comprehensive interpretation of preindustrial European fisheries from an environmental and not simply economic point of view. It further recognizes the dynamism and resilience of the non-human in shaping the long-term encounter of aquatic nature and medieval Europeans. This permits historically and biologically knowledgeable tracing of antecedents for a present environmental crisis back before the most recent centuries and to roots of dominant paradigms for market-based development and resistance to it.

## I.3 Then and Now and Then

Tales of medieval sturgeon and cod eventually fit into a present narrative of crises in aquatic systems and thus the human communities entwined with them. A generation and more of scientific research has broadly identified a global 'fisheries crisis' and richly exemplified its diverse local aspects, only some of which our fish tales have highlighted. More than four in five stocks are now assessed as fully exploited, overexploited, or depleted. From an ecological perspective, especially noteworthy is the depletion by an estimated 90 percent of large predatory fish populations, the top of the food chain, in all the world's oceans.[68] Effects cascade down entire ecosystems. Biodiversity and natural ecosystemic interactions are put at risk by depletion, pollution, habitat destruction, and invasive species. Rising global temperatures draw attention to strong scientific research findings that climatic fluctuations and changes do cause and have caused large-scale changes in fisheries at both short and

[68] McNeill, *Something New under the Sun*, 243–252 and 257–260, surveys the present crisis. For quantitative perspectives see Worm and Branch, "The future of fish," 594–599, and FAO, *State of World Fisheries 2014*, p. 3. Smil, *Harvesting the Biosphere*, 206–210 and 212–214, critiques such estimates. An adeptly contextualized historical overview of destructive modern commercial fisheries is Simmons, *Global Environmental History*, 139–41 and 193–94.

long term, especially under conditions of heavy human exploitation and/or for species near the limits of their range.[69]

Proposed remedies for particular and global fisheries crises are not lacking, though all pose proven or potential economic, cultural, and ecological risks. The very regulatory regimes meant to conserve wild stocks and protect their habitats have been blamed for the collapse of bottom fisheries in both the Canadian and American Atlantic. The relative conservation value of community control or privatization of aquatic resources is fiercely debated. Environmentalists, local fishing communities, and large fisheries corporations hold different views on the value and effects of establishing large protected areas or aquacultures. Environmental manipulation always seems to have a 'catch', a collateral or unanticipated consequence. Today's acute and pressing problems of humans and fish around the world frame this book's investigation. The millennial history of medieval European fisheries offers long and well-documented parallels, examples, and warnings for present-day concerns. But "the past is a foreign country, they do things differently there."[70] Fishers, fisheries scientists, managers, and conservationists need translators to understand what interdisciplinary visitors might there see, hear, or read. For a millennium of the European experience, medievalists must assume that role – if, when it comes to fisheries, they know the terrain and can read the water.

[69] Sample case studies and big data in Aarts et al., "Habitat loss"; Airoldi and Beck, "Coastal marine habitats"; Tockner et al., *Rivers of Europe, passim*; von der Ohe et al., "Water quality indices across Europe," 970–978; Humphries and Winemiller, "Historical impacts on river fauna"; Lassalle et al., "Diadromous fish conservation"; Allison et al., "Vulnerability of national economies to the impacts of climate change on fisheries," 173–196; Dulvy et al., "Climate change and deepening of the North Sea fish assemblage"; Brander, "Impacts of climate change"; Smil, *Harvesting the Biosphere*, 246.

[70] A phrase shameless historians borrow from novelist Leslie P. Hartley, *The Go-Between*, 1.

# 1 "Natural" Aquatic Ecosystems around Late Holocene Europe

Discussing historic fisheries calls for a common store of information, vocabularies, and ideas. This includes working acquaintance with the waters and fishes of post-Roman Europe and the conceptual framework of freshwater and marine ecology used to describe them. For readers not so equipped this chapter lays out that basic knowledge, beginning with characteristic regional waters, going on to their distinctive fishes, and then to flows of energy and materials through those ecosystems and organisms. Streams and lakes are familiar to most people. Runoff of rain and snow-melt is the point in the hydrological cycle when liquid water gains biological significance on the Earth's surface. Historically seminal medieval fisheries took shape in Europe's streams and lakes. For these reasons, this discussion starts in fresh water and flows to the sea, then sketches the communities of life forms, especially fishes, living in those habitats. The aim is to identify patterned relationships among aquatic organisms and their surroundings, including humans, in the earlier Middle Ages. The web of life goes around in interconnections, as does the interaction of natural and cultural spheres, so any starting juncture will include links explicable only from other perspectives. With loose ends eventually tied back together, this chapter and the next two establish a baseline – not of humans entering a pristine state of nature but of where medieval Europeans and their fishes started – to recognize and assess later changes.

## 1.1 Watersheds and Seas[1]

Natural aquatic ecosystems in Europe result from the physical geography and Pleistocene glaciation of the subcontinent. Europe forms a peninsula

[1] On congruent cultural and natural grounds this book focuses on the lands of medieval western Christendom from Mediterranean coasts to Scandinavia and the eastern Baltic. It ranges little into the southern Balkans or eastern Slavic territories, which belonged in the Middle Ages to the distinctive sibling culture of orthodox Christendom, sometimes called the 'Byzantine Commonwealth'. These regions hold vastly more complex and varied fish fauna in unglaciated (Balkans) or truly continental (Ukraine, Russia) habitats. For

from the larger land mass of Eurasia, which is a situation ecologists find commonly reducing the variety of all resident life forms (species). Still speaking in the broadest terms, from the north side of the mountains bordering the Mediterranean – Pyrenees, Alps, Balkan ranges – most of the subcontinent slopes to the north and west through uneven (but not abrupt) relief and becomes wetter. How did this topography shape fish life?

Water from rain or snow falling on western and central Europe eventually reaches the sea through one of Europe's watersheds (river basins), which are the fundamental territorial units of freshwater habitats (see Map 1.1). Those of the north European plain are long and broad, gathering waters from part of the Alps and from many lesser interior mountains and draining northward to the Baltic and North Sea or, west of the Rhine–Maas delta, to the English Channel and Bay of Biscay. Most of the Iberian peninsula also comprises large, west-flowing river basins. Scandinavian waters drain from the central mountain ridge and other uplands westwards to the North Atlantic or south and east to the Baltic, often pausing in extensive post-glacial lakes. From almost at the Rhine and then east through the centre of the continent runs the Danube basin, largest in the west and the only major system to flow from western and central Europe into the Black Sea. The Mediterranean basin has characteristically shorter, and so more numerous, watersheds. Only the Ebro, Rhône, and Po drain extensive areas.[2] Thus Europe's terrain and humid temperate climate make for many flowing waters, but fewer and localized permanent natural standing waters. Local and regional variation in gradient, substrate (river bed materials), and seasonal runoff result in reaches of different character, some more dynamic than others and subject to change over time.

The political authority of Rome over western and southern Europe broke down in the course of the fourth through sixth centuries. So did human numbers and the intensity of land use there. For the next half-millennium or so, native plant communities blanketed much of the irregular and damp landscape north of the Alps. Vegetative cover on the drier and more abrupt Italian and Iberian peninsulas continued to show effects from intensive human use during ancient times.[3] Some

orientation see Baron, "Approach to Byzantine environmental history," 180–184. Similarly, the fisheries of the Iberian peninsula will gain in interest as Christian Spaniards and Portuguese supplanted Muslim al-Andalus.

2 For present-day overviews see Tockner et al., eds., *Rivers of Europe* .

3 Fumagalli, *Landscapes of Fear*, 99–103; Vita-Finzi, *Mediterranean Valleys*, 65–82 and 91–120; McNeill, *Mountains of the Mediterranean World*, 16–18, 72–74, and 84–86; Hughes, *Pan's Travail*, 16–18, 77–84, and 190; Grove and Rackham, *Nature of Mediterranean Europe*, 141–150 and 167–189; Hoffmann, *Environmental History*, 21–50. See also note 10 below.

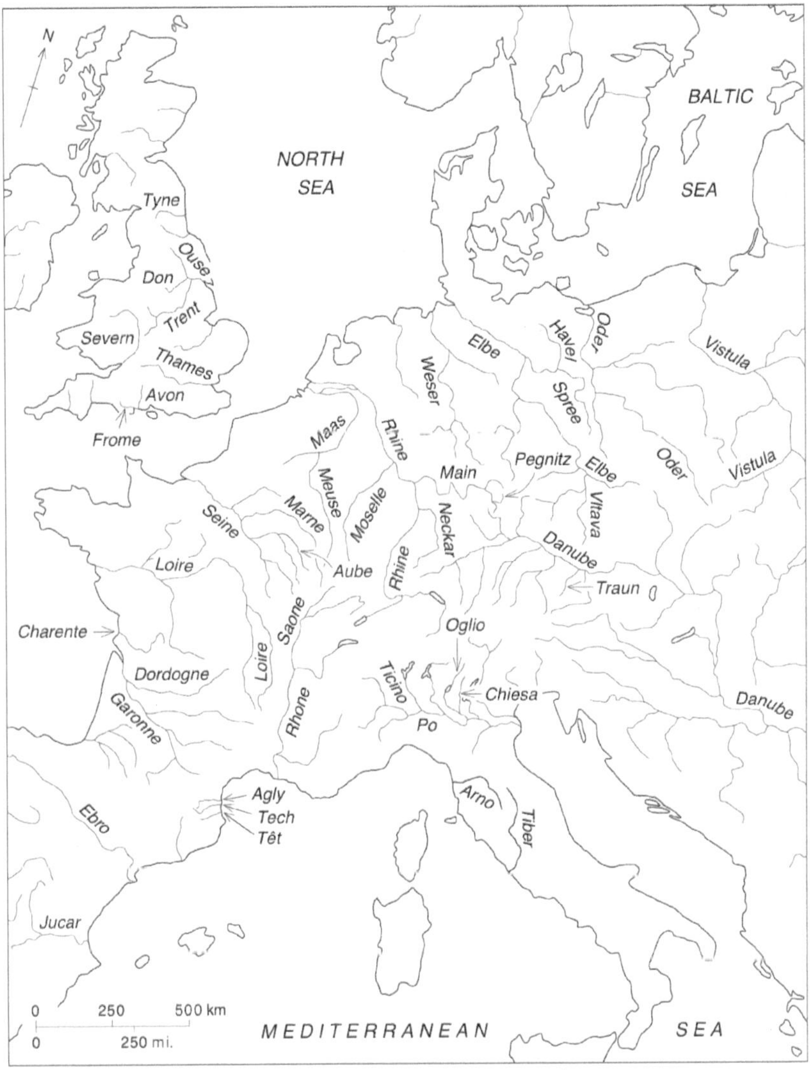


Map 1.1 Some major European rivers (mentioned in this book).

good-sized areas throughout Europe were still under dense wild woodland, where human activity was little to be noticed. More early medieval terrain held open woods and brush lands subject to intermittent and extensive use by scattered human communities and their livestock. Tight growths of woody and large herbaceous plants especially covered the well-irrigated margins of watercourses.[4] So protected, the land absorbed precipitation and retained both its soils and the mineral plant nutrients in them. Fed by well-filtered and slowly released surface runoff and ground water, many European streams ran clear, cool, and stable. Seasonal high waters inundated only natural floodplains. Even in wetlands, rooted aquatic vegetation absorbed soluble nutrients and kept the water clear. In the late fourth century Gallo-Roman poet Ausonius saw the Moselle "bright as water in crystal goblets," and his "vision, when it pierces this deep stream, finds the open secrets of the bottom."[5] Limpid flows tumble through other poetic Gallic landscapes all the way to 1200.[6] In Thuringian Wolfram von Eschenbach's romance *Titurel* (*c.* 1210/20), the young hero Schionatulander angled for trout and grayling wading "through the cool and clear-running brook."[7] A real-life mid-twelfth-century sportsman, French cleric Gui of Bazoches, contrasted the clear streams still issuing from the Ardennes with those of more agricultural Champagne.[8]

[4] Darby, "Clearing of the woodland in Europe," 183–216, is a classic, but for the importance of early medieval woodland see the pioneering essay of Higounet, "Les forêts de l'Europe," 343–398, and the more recent examination of its physical and economic attributes by Wickham, "European forests," 479–545. Stable and well-vegetated riparian zones are described in Petts, "Forested river corridors," 12–34, and other medieval examples appear in Fumagalli, *Landscapes of Fear*, 99–103; Derville, "Rivières et canaux," 19; Arnold, *Negotiating the Landscape*, 57–60 and 75–76; Hoffmann, *Land, Liberties, and Lordship*, 13–14, 100–104, and 255–256; and in the exhaustive coverage of riverine landscapes of ancient and medieval Brenne by Benarrous, *Grande Brenne*, 258–342. Lambrick, "Alluvial archaeology," 222, notes a decline in the rate of alluviation at the start of the Saxon period in the Upper Thames Basin, as for the immediately post-Roman Paris basin does Pastre et al., "Variations paléoenvironnementales," 29–44.

[5] Ausonius, *Mosella*, ed. Green, ll. 22–74, notably 27–28, "*devexas pronus in undas ut fluvius, vitreoque lacus imitate profundo*," and 61–62, "*sic demersa procul durante per intima visu cernimus arcanique patet penetrale profundi.*" A century later Sidonius Apollinaris described another transparent river and lake in *Epistolae*, II, 2 and 9 (Anderson, ed. and tr.). So, too, did sixth-century poet and bishop Venantius Fortunatus wax eloquent about the waters of the Moselle and other streams issuing from the Ardennes and Vosges (Arnold, *Negotiating the Landscape*, 57–60).

[6] Guillerme, *Age of Water*, 76–77, provides several examples. Another is the description of convalescent monks watching fish play where the Aube entered the gardens of early thirteenth-century Clairvaux in "*Descriptio positionis seu situationis monasterii Claraevallensis*," *PL* vol. 185, cols. 569–572. Though translated as "Description of the position and site of the Abbey of Clairvaux" in Bernard of Clairvaux, *Life and Works*, tr. Eales, vol. 2: 460–465, the text postdates Bernard's death. Fossier, "L'essor économique de Clairvaux," 110–112, treats pond building at Clairvaux.

[7] *Titurel*, strophe 159, 3, ed. Schröder and Hollandt, 617: "*dâ er stuont ûf blozen blanken beinen durh die küele in lütersnellem bache.*"

[8] Gui, Ep. 21, ll. 70–88 (Adolfsson, ed.), *Liber epistularum*, 97–98.

Such evidence of limited human impacts should not, however, be taken to imply static natural riverscapes. Multichannel rivers in early medieval England and high gradient sections of the Danube, for example, continually reworked their beds. Reaches of the Loire and Rhône shifted from incised meanders to braided form with the wetter post-Roman climate, which also brought frequent flooding and alterations to the Po basin during especially the sixth and seventh centuries.[9] Figure 1.1, though created much later, still depicts the generic complexity of most medieval rivers. Freshwater fish populations confronted changing local conditions and habitats.

Europe, subcontinental peninsula of peninsulas, is deeply penetrated by the sea into which its rivers flow: through salt ponds and marshes as the Po, Rhône, and Tiber; through channels among sand bars and peat bogs as the Rhine in Frisia and Great Ouse in the English fens; or through broad bays with barrier islands as the Weser and Wisła or rocky headlands as the Tejo and Seine. Great streams fed sediments from high mountains; Rhine, Rhône, Po, and Danube, formed extensive deltas. In those estuaries fresh waters transition into marine environments.

Along Europe's southern margin the Mediterranean and Black Seas have mountainous shorelines and deep waters close to shore.[10] In their nearly three million square kilometers of water surface the only extensive areas shallower than 200 meters (600 feet, 100 fathoms) are the upper Adriatic and the northwestern quarter of the Black Sea, both lying off large sediment-rich rivers. The Mediterranean itself has a warm, dry climate, and few major sources of fresh water. On balance, more water evaporates from its surface than feeds it, which leaves Mediterranean water salty and dense. Lighter, less salty seawater flows steadily into the Mediterranean from the Black Sea and, at the surface, from the Atlantic. Down below, a strong current carries cold and salty Mediterranean water west past Gibraltar into the Atlantic. With few nutrients from surrounding landscapes, little upwelling, and almost no tidal action, the Mediterranean, especially along its northern shores, is famous for

[9] Generally on Holocene variability, especially regarding river deltas and estuaries, see Dickinson, "Changing times," 483–502. Hoffmann, *Environmental History*, 67–71, treats climatic fluctuations in the earliest Middle Ages. Particular cases are discussed in Lewin, "Medieval environmental impacts," 267–311; Hohensinner et al., "Two steps back, one step forward," 132–133; Hohensinner and Jungwirth, "Hydromorphological characteristics," 33–38; Galinié et al., "Quelques aspects documentés," 127–136; Burnouf et al., "Fluvial metamorphosis of the Loire," 163–171; Straffin and Blum, "Late and post-glacial fluvial dynamics," 85–99; Bravard et al., "La diversité spatiale," 209; and Squatriti, *Water and Society*, 67–76.

[10] For physical geography, climate, and hydrography of the Mediterranean see Rohling et al., "The marine environment: present and past."

Figure 1.1 A major European river, the Upper Rhine, not yet 'regulated'.
The view from the Isteiner Klotz looking upriver toward Basel as painted by Peter Birmann about 1800, before the famous engineering of the river begun by Johann Gottfried Tulla in 1817. Though used by small boats and the village on its bank, the Rhine is dotted with islands, sand bars, and other irregular features.
Kunstmuseum Basel, Birmann-Sammlung 1859, Inv. 755. Open access and public use (*Gemeinfrei*).

its clear – because naturally infertile – waters. The Black Sea, meanwhile, circulates much surface water from large rivers toward the strong outflow to the Mediterranean, but the shape of its basin and prevailing winds cause little vertical mixing. Below 100 meters low oxygen levels leave the Black Sea lifeless.

Unlike the south, from the inner angle of the Bay of Biscay north to Shetland and the central Norwegian coast western and northern Europe front on the wide continental shelf of the eastern Atlantic Ocean and its extensions, the North Sea and Baltic. Only in a narrow trench off south-western Norway do soundings greater than 200 meters come nearer to shore than hundreds of kilometers. Hence, the million or so square kilometers of more or less enclosed waters in the European Atlantic, though less than half the area of the Mediterranean and Black Sea, provide two or three times more continental shelf. Even along the Cantabrian and Portuguese coasts, where great depths do reach closer to land, the shallow coastal zone is wider than in most of the western and central Mediterranean. The eastern Atlantic is swept, even into the outer reaches of the North Sea, by the great oceanic circulation of the Gulf Stream; it is agitated twice daily by tides and fed by large temperate rivers. Fresh water flowing into the Baltic, where long cool seasons limit evaporation, lowers salinity there and feeds a steady current running out the Danish straits.

By late Roman times humans had crisscrossed the Mediterranean for millennia, explored all its shores, and settled those they found useful – though winter broke their sailing season and skippers always preferred coasting routes. Comparable long-distance knowledge of movement across northern European seas was acquired only during the Early Middle Ages, as exhibited by settlement of Irish monks, then Norse farming families, on the previously uninhabited Shetland, Faeroes, and Iceland, not to mention movements by other Celtic and Germanic peoples across the protected narrow seas between Scandinavia, the British Isles, and Brittany. Of course none of this hard-gained experience meant systematic knowledge of the oceanic systems, water chemistry, or bottom profiles just described. It is always essential to distinguish between what we can now say or reconstruct of aquatic environments, with its necessary consequences for human activities, and what medievals were then aware of, even when their ignorance in no way impeded, for instance, their ability to catch and eat tuna in the Mediterranean, herring in the Baltic, pike-perch in the Danube, or sturgeon in the Loire.[11]

[11] Medieval people had participant knowledge, in technical terms an 'emic' perspective, on their world, and present-day oceanographers an observer's view, called 'etic.' Both approaches run as threads throughout this book.

## 1.2 Which Fishes Lived Where?

Fishes are and were no more uniformly distributed through Europe's waters than are other animals on land. Most fish cannot survive in the air long enough to get overland from one water body to another. Nor can most fish even move safely between salt and fresh water, for the distinct relations of these to the body chemistry of an aquatic animal can be deadly. Seawater has a higher concentration of salts than is in the blood of fish. To maintain their own more dilute and particular osmotic balance, marine organisms must actually drink water and excrete salts. Those in fresh water face the opposite problem: their kidneys must extract surplus water and pump it out in copious urine. While most fishes can tolerate some range of salinity, only a few can reverse kidney function to live in both fresh and salt water. These are notably certain inshore marine species (called euryhaline) such as flounder and mullet, and diadromous fishes, such as salmon, shad, or eel, whose life cycle stages are adapted to alternate between the two. Most fishes, however, are either marine or freshwater.[12]

### *1.2.1 In Freshwater Europe*[13]

Freshwater fishes can naturally move from one habitat to another only through a temporary or permanent freshwater link where both temperature and dissolved oxygen levels fall within the tolerance of the species. That meant, until humans built the first major inter-basin canals around 1700, European fishes of fresh water could of their own accord move between watersheds only when geological processes slowly changed the land itself. For instance, the barbel, a resident of moderately flowing rivers on the continent, naturally occurs in Britain only in those eastern watersheds which once joined the Pleistocene Rhine in its route across the then dry or ice-covered floor of what later rising sea levels have made into the North Sea and English Channel.

[12] As noted in the Appendix.

[13] A classic overview of the zoogeography of European fishes is Lepiksaar, *Faunengeschichte der Süsswasserfische*, rev. ed. by Heinrich. Still valuable older authorities include Varley, *British Freshwater Fishes*, and Thienemann, *Verbreitungsgeschichte der Süsswassertierwelt*. For further updates see Bănărescu, "Zoogeography," 88–107, and Evans, "Conservation of European freshwater fishes," 107–119, as well as treatments of individual species and families by Kottelat and Freyhof, *Handbook*, which employs more molecular genetics than could the earlier works. The biologically comfortable can consult www.fishbase.de/search.php.

Given that a fish species could reach a watershed in the past, its presence and distribution there depends on its adaptability to certain environmental factors. Water temperature and dissolved oxygen are critical for cold-blooded, oxygen-respiring animals. Cold water holds more oxygen than warm. As with salinity, each species has a range of tolerance for both variables. Current speed is important in the lives of fishes, too, for it makes them expend energy and through erosion it modifies beds, banks, and water clarity. Suitable food supplies and spawning sites further determine the presence and abundance of a species in a watershed. When any of these factors violates threshold values, the species will be extirpated.

Factors of water temperature are thought to be the greatest single cause for the gross distribution of fishes among European watersheds (and, as we'll see shortly, in marine environments, too). Cool-water varieties like salmon, trout, whitefishes, and pike were naturally absent from most Mediterranean watersheds or there confined to high-altitude headwaters. Especially in Spain and the Balkans their niches are filled by more tolerant local fishes.

The number of resident freshwater fish species is low throughout Europe and lowest in the northwest.[14] Many thousands of years ago, a continental ice sheet spread over most of northern and western Europe. As Pleistocene waters chilled below the tolerance level of some, then all, aquatic organisms, those unable to move into salt water died or withdrew to the southeast, the only direction where warm fresh water was accessible. Only cold- and salt-tolerant varieties like salmon, trout, and their kin (collectively called *Salmonidae*) could survive in oceanic coastal waters and frigid ice-front lakes. Fishes needing warm and fresh water, notably Eurasia's many representatives of the carp family (*Cyprinidae*), survived the Ice Age only in the refuge of the lower Danube and other rivers running from the Balkans into the Black Sea and Caspian. From there they could only very slowly recolonize the greater part of peninsular western Europe.[15] Hence the ancestors of all life forms in early medieval waters had survived or repopulated the frozen antithesis of a freshwater

[14] As traditionally defined, European freshwater fishes number only 118 species in but 6 families. "Western Europe is exceptional in that it has a very limited freshwater fish fauna, probably because of the Pleistocene glaciations. There the Alps prevented both retreat and recolonization, and many species must have been eliminated ..." (Hynes, *Ecology of Running Waters*, 337–338). See also Bănărescu, "Zoogeography," 105, and Evans, "Conservation of European freshwater fishes," 107. Compare Kottelat and Freyhof, *Handbook*, *passim.*

[15] A case study of retreat and recolonization by two warm-water species, catfish and pike-perch, is Heinrich, "Nordwestlichen Verbreitung des Welses," 303–320.

habitat, and the numbers of fish species in European watersheds fall sharply along an east–west gradient.

The clear, cool inland waters of post-glacial western Europe thus held relatively simple fish communities with a low diversity of native species. Most were well adapted to cold, well-oxygenated, alkaline, running-water habitats with low levels of dissolved nutrients. Many fishes in drainages running to the Atlantic, North Sea, and Baltic show adaptations to marine access and to conditions in the ice-front lakes that had joined these basins while the glacier withdrew slowly downslope to the north.[16] Notably important as fish biomass and for human use in watersheds from the Garonne to the Wisła were sturgeon, shads, and closely related families of salmon, trout, and whitefish. Most of these fishes are anadromous, meaning they spend their adult lives more or less at sea and return to fresh water to reproduce. Two of western Europe's three species of lampreys are also anadromous. Eel, however, a species of large and growing early medieval economic importance, have the reverse habit of 'catadromy', for they enter freshwater as larvae to grow and mature, and return to the sea to spawn and die. Eel, lampreys, sturgeon, and shads were also important natives of western Mediterranean watersheds, but marine migratory salmonids do not occur there. Other ubiquitous (apart from peninsular Italy and Spain), numerous, and oft-eaten natives were the European representatives of two circumpolar genera, the perch, and the pike. About 1080 the monk Ulrich compiled a dictionary of the sign language used at Cluny, then western Christendom's most respected monastery, during times of compulsory silence. His section on foods named only six fishes, but five of them were from fresh water: salmon, trout, pike, eel, and lamprey.[17]

Other parts of the Old World have many representatives of the huge *Cyprinidae* family – they comprise 80 percent of China's 600 freshwater species – which commonly favour warm still waters. By comparison, however, rather few members of this family live in western Europe, and those naturally present, like the barbel, bream, roach, minnow, or gudgeon, are unusually adapted to cool and/or running water. Only barbel and gudgeon,

Spain and southern Italy have species inventories still more abbreviated than elsewhere in Europe and a high proportion of distinctive autochthonous taxa. Several historically significant native inland fisheries in Italy exploited migratory fishes which also use the Mediterranean. See Kottelat and Freyhof, *Handbook*, 25; Dill, *Inland Fisheries*, 255; and general remarks in Prat et al., "Llobregat," 542, and de Sostoa and Lobon-Cervia, "Fish and fisheries of the River Ebro."

16 Besides items cited above, for what follows see Kottelat and Freyhof, *Handbook*; Wheeler, *Fishes of the British Isles*; and Blanc et al., *European Inland Water Fish.*

17 Ulrich of Cluny, "Consuetudines Cluniacensis," II.4 (*PL*, 149, col. 703). Note the absence of species preferring warm still waters. Cluny, on a tributary of the Saône in southern Burgundy, is 300 km from salt water.

two cyprinids especially suited to swiftly flowing middle reaches of streams, joined perch, pike, salmon, and eel in the catches twelfth-century sporting cleric Gui of Bazoches made in northern French waters.[18]

The post-glacial ice-front lakes for a time united what would become distinct watersheds. As the ice retreated, fishes from the lakes spread downstream to the northwest while migratory species moved along the coasts. (Both groups left relict populations in habitats like cold Alpine lakes.) At the same time, warm-water varieties could work up the Danube system and, using occasional ice lake or swampland inter-basin connections in poorly drained post-glacial Europe, gradually reenter the Atlantic drainage from east to west. The roach made it all the way to central Scotland and to rivers running into the Gulf of Bothnia, and the barbel to the Thames and north Germany. The European catfish or wels spread only so far as the upper Rhine, Elbe, and a small beachhead in southern Sweden. Carp had under wild conditions by the early Middle Ages reached only the middle Danube.

Finally, and quite apart from issues of water temperature, Europe's peninsular character has held up the westward spread of freshwater fishes with more easterly distributions. The migratory Atlantic salmon is absent from Mediterranean and Black Sea watersheds. In the latter its niche is filled by the huchen or 'Danubian salmon', a non-migratory form with congeners in central Asia. Likewise neither cold nor heat barred from the Rhine, Rhône, and other western watersheds the pike-perch or zander, close relative of the North American walleye and occupant of a broad Pontic, Russian, Scandinavian, and east-central European range.

One side of that peninsular Europe, the Baltic Sea, can better be understood as one of the world's largest estuaries, a body of water where fresh and salt mix. This makes for a hard environment, but for fishes which can adapt, also a fecund one. The Baltic at least once held major populations of some cyprinids and pike in its freshest eastern reaches, salmon, sea trout, and sturgeon throughout, and in the saltiest west, at times legendary numbers of herring and cod.

### *1.2.2 Marine Zoogeography*

The global distribution of marine fishes is less well understood than are freshwater varieties, if only because more species cover larger areas with less visible boundaries among them. In place of explanatory principles marine zoogeographers offer descriptive regions set by oceanographic

[18] Gui, Ep. 21, ll. 70–88 (Adolfsson, ed., *Liber epistularum*, 97–98). Gui's fishing is discussed in Hoffmann, "Fishing for sport in medieval Europe," 886–887.

conditions and continental shelves, where currents and especially water temperature seem influential.

Fishes in historically relevant nearshore European waters chiefly form a 'Mediterranean–Eastern Atlantic' temperate region, bounded fuzzily on the north along a line from Iceland to northern Norway by a distinctive circumpolar Arctic or boreal region. In the south, the Mediterranean fauna, with many (540) species, has much in common with the Atlantic south of the English Channel, where summer water temperatures also stay above 12°C. Fishes from typically tropical families, the mullets and sea breams or porgies, for instance, are common inshore, and so are the sardine/pilchard and anchovy. Low salinity and restricted marine access let only some of the Mediterranean fishes populate the Black Sea. More northerly Atlantic waters are consistently colder, averaging between 12° and 6°C, and containing important representatives of coldwater fish families, the codfishes, flatfishes, herrings, and salmonids. Only some of these range into the Bay of Biscay. Colder waters typically contain fewer species and very large numbers of individuals in those species.

Two other marine regions need mention. Open offshore surface waters provide a different kind of migratory habitat to distinctive, sometimes world-girdling fish families. South of the British Isles and in the Mediterranean the tuna or tunny, mackerels, and swordfish, are familiar, at least in warmer months. Further north this role is more that of adult salmon. Finally, deep-sea fishes, those which live on and below the continental slopes, seem relatively undiversified – and remained all but unknown to human fishers before the industrial age.

## 1.3 Ecosystems, Habitats, and Communities

Even in native waters few fishes live everywhere. Variables of current speed, light penetration, and nutrient levels separate different habitat zones to which various species are more or less adapted. Inland, the basic division is between moving and still waters; the former, where constant movement and mixing dominate, contain what ecologists call 'lotic' habitats, and the latter, more influenced by depth and water chemistry, 'lentic' habitats. In such terms marine environments might be conceived as oversized and chemically distinctive lakes, with their own analogous habitat zones and nomenclatures (pp. 44–45 below).[19]

[19] See also Appendix. Fishes themselves can be classed as rheophilic (liking currents), eurytopic (a wide range of habitats), or limnophilic (preferring still waters). For a more elaborate discussion of diversity and bioproductivity see Rohde, "Latitudinal diversity gradients," 155.

In the lotic habitats of running waters European biologists commonly distinguish four zones of decreasing current speed, each named after a typical resident fish.[20] The *trout zone* is in mountainous headwaters where steep gradients cause very fast currents. These make for much dissolved oxygen but so erode the banks and bottom as to leave them mostly rock and gravel. Trout and young salmon are the most common fish species. As the bed flattens and currents slow slightly, the stream enters the *grayling zone*. Oxygen levels remain high, but more gravel is found, and sheltered places hold enough sand or silt for rooted aquatic plants. Trout or grayling are the most abundant fishes, but commonly joined by cyprinids like barbel, dace, and chub. Third, the *barbel zone* is found where the gradient becomes gentle and the current moderate. The bottom is predominantly soft, and rooted plants are common. In a mixed fish fauna the cyprinids dominate; if water temperatures remain low, trout and grayling may still occur, but piscivorous pike and perch are also important. Finally, the *bream zone* is named for the slow-water cyprinid native to western Europe. In quiet, warm, and weedy waters they are accompanied by roach, rudd, tench, predatory fishes, and now the introduced carp. Not until this last zone in the lowest and gentlest reaches of rivers are depth or circulation likely to impede constant temperature, light, and nutrient levels throughout the water column. Streams rising in bogs or having high gradients in middle or lower reaches will depart from the normal zonal pattern and can accordingly show unusual distributions of fishes.

Biologists commonly stratify the lentic habitats of still waters (lakes, ponds) by depth and light penetration into three major zones, which they label littoral, limnetic, and benthic. The littoral zone extends from the shore to the depth limit of locally prevalent rooted aquatic plants. It contains many nutrients, the warmest water of the lake, and the greatest variety of fish species. The limnetic or pelagic zone is the lakeward extension of the littoral. Its open surface waters are well-lighted and thus can hold large populations of drifting microscopic plants, plankton, on which certain fishes, such as some whitefishes and cyprinids, can feed. Like littoral waters, the upper layer of the limnetic is subject to rapid changes through wave action or air temperature. Beneath the limnetic zone the dark benthic (profundal) zone goes to the bottom of the lake. It receives nutrients from above, but is commonly cold and often short of oxygen. The profundal zone is chiefly inhabited by bottom-feeding fishes and their predators, such as European catfish or burbot, the freshwater

[20] Classically formulated in Huet, "Profiles and biology of Western European streams," with more specific detail in Logez et al., "Modelling habitat requirements."

cod with a circumpolar range. The relative importance of each zone depends on the predominant physical features of each lake or lake region. Deep and steep-sided lakes may have no littoral zone and shallow ones nothing else.

Unlike rivers, ponds and lakes are ephemeral features of geologically 'young' landscapes. As the land erodes, all still waters accumulate nutrients, become more fertile, and slowly fill in. This natural process, called eutrophication, moves lakes through three conventional stages. In Europe lakes at different stages are naturally characteristic of certain physiographic regions. Mountain lakes are commonly deep and narrow, hence typically cold habitats with low levels of dissolved nutrients and rooted plants. Biologists call infertile habitats oligotrophic. They are suited for salmonids, plankton-eating whitefishes, and associated predators. Unexploited oligotrophic lakes can contain quite large biomass concentrated in large, old, and slow-growing adult fishes. Where the land is more rolling and fertile, as around the post-glacial lakes of northern Germany and Poland, the still waters hold more nutrients and more shoreline vegetation. This middle group, called mesotrophic lakes, supports a mix of salmonids and cyprinids. Finally there are the highly fertile eutrophic lakes, relatively shallow, soft-bottomed, and full of plant life. In Europe they are naturally associated with slow-moving reaches of rivers (the bream zone) and inhabited by the same fishes.

At a scale smaller than whole watersheds, then, European regions and localities contain characteristic freshwater habitat zones and characteristically abundant or rare fish species. Comparable differences occur in marine ecosystems, where a general classification of habitats as coastal (neritic), offshore surface (called variously pelagic, epipelagic, or oceanic), and bottom-oriented continental shelf (sublittoral, benthic) seems to organize the environments historically engaged by human activities.[21]

The sea is a moving and changing environment with patterned uniformities, which exhibit more variation where seawater contacts air or earth. Hence marine ecosystems classify into habitat zones oriented to the surface (neritic and epipelagic) and to the bottom (benthic). The neritic zone is the water over the continental shelf, so extending downward to about 200 meters. This water is well supplied with light, oxygen, nutrients, and wave and current action, all subject to seasonal variation. Such variety supports many habitat niches and hence the greatest

[21] Marine biologists have evolved far more complex schemes, especially for deep-water (pelagic) and bottom (benthic) environments. Brief examples are in Lagler et al., *Ichthyology*, 424–429, and Moyle and Cech, *Fishes: An Introduction to Ichthyology*, 585–596.

diversity of marine fishes. Herrings, eels, tunas, and mackerels are familiar inhabitants of the neritic zone. Continuation of this same 200-meter-thick, well-lighted and well-oxygenated layer out beyond the continental shelf produces the epipelagic zone, a huge and featureless expanse of open ocean waters. Most fishes – think of tunas, mackerels, and swordfishes, but also of sardines and anchovies – concentrate in nutrient-rich areas of upwelling or near coasts. Beneath the neritic zone are the bottom-oriented habitats of the benthic, from the intertidal area, exposed to the air up to half of every day, down to the edges of the continental slope, where even seasonal changes are barely perceptible. Across most of this submarine expanse, wave action and the makeup of the bottom (substrate) rank behind only temperature in distinguishing habitats for bottom-oriented (demersal) fishes. Important in European waters are the flatfishes – flounder in estuaries, plaice inshore, halibut well out on the shelf, etc. – and codfishes – whiting along shore, often with small cod and pollack not much further out, adult cod and hake in deeper waters, the former northwards in the Atlantic, the latter more common south of the British Isles, and ling favouring the deepest zones on the slope. Waters below 200 meters lack light and seasonal changes, while temperature declines and pressure increases with depth. Numbers of fish species and individuals diminish, too. All organisms there must eat (once-)living objects fallen from above, whatever they can catch on brief feeding forays upwards, or one another.

Finally, a transitional and fluctuating quality dominates estuarine ecosystems, where fresh and salt waters meet in a turbid, sandy river mouth, a deep, clear, rocky fjord, or perhaps a maze of channels, bars, and weed or shellfish beds. Diadromous fishes use estuaries to move between spawning and feeding areas. The layering of dense salt water under lighter fresh water and their eventual mixing by currents and tides cause high biological productivity, so many fishes use estuaries as nurseries for their young. Consequently, estuaries often hold large seasonal concentrations of certain species.

## 1.4 Food Webs and Other Relationships

Like all of Earth's widespread ecosystems, those in water depend on energy from sunlight, which powers the winds and currents circulating air and water and which green plants, large and small, use to produce organic material (a store of chemical energy) and oxygen from water and carbon dioxide. Only plants, from single-celled drifting plankton to well-anchored kelp or massive oaks dropping leaves into a brook, are 'producer organisms', and that only when exposed to sunlight. The 'euphotic' zone where light is available penetrates at most some 200 meters into clear seawater, and much less into water clouded or stained with suspended sediment, soluble chemicals, ice, or drifting plankton. In lentic fresh waters and in the sea, then, plants can live only in the littoral or neritic zones along the shore or in the (epi)pelagic surface layer; elsewhere they stop making food and oxygen and first temporarily (in nighttime respiration), then permanently (in decomposition) absorb oxygen to break down organic hydrocarbons. Turbid waters may limit the productive euphotic zone to only a few meters.

Survival of green plants, the basis of all life on Earth, demands air, water, and also access to the relatively rare soluble forms of nitrogen and phosphorus, mineral salts or ions of nitrates and phosphates. These nutrient chemicals are obtained in solution from terrestrial runoff and concentrated in the bodies of plants and animals. Solid organic wastes, including dead bodies, gradually sink to the floor of a still water body and thus out of the euphotic zone, where plants can reuse their constituent minerals. Inshore and nearshore marine regions where soluble nutrients are continually recharged with the runoff of large rivers – like, for instance, the North Sea, Bay of Biscay, or Gulf of St Lawrence – can support richer rooted and planktonic flora than can more sterile waters – the Mediterranean or the epipelagic areas of mid-ocean. Aquatic life also flourishes where natural mechanisms – upwellings of deep water like the Labrador Current or the annual 'turnover' of water in temperate-zone lakes before freezeup in fall and after breakup of ice in the spring[22] – stir mineral nutrients back up from the sterile and Stygian depths.

All animals consume and transform plant materials. The plant-eating herbivores (primary consumers) are followed by succeeding trophic levels of animal-eating predators (secondary, tertiary, etc. consumers). Most fishes occupy an identifiable niche or level in this interlocking web of eating and being eaten. Besides the invertebrate animals which do much primary consumption in many aquatic ecosystems, European fishes filling that role include the minnow in the trout zone, whitefishes in the limnetic zone of oligotrophic lakes, and the roach in warm and fertile waters. Top-level predators, the uneaten eaters, in the same habitats include large trout in cold streams and lakes, pike, and catfish. Cyprinids of large adult size such as barbel, carp, or bream commonly function as herbivores (and important prey for predators) when young,

[22] Because water reaches its maximum density at 4°C, cooling surface waters sink and mix until that temperature, then stratify before freezing to a floating layer of ice. Melting surface water also mixes at 4°C, but further warming creates a surface layer of lighter and warmer water. The deepest ocean water is always about 4°C.

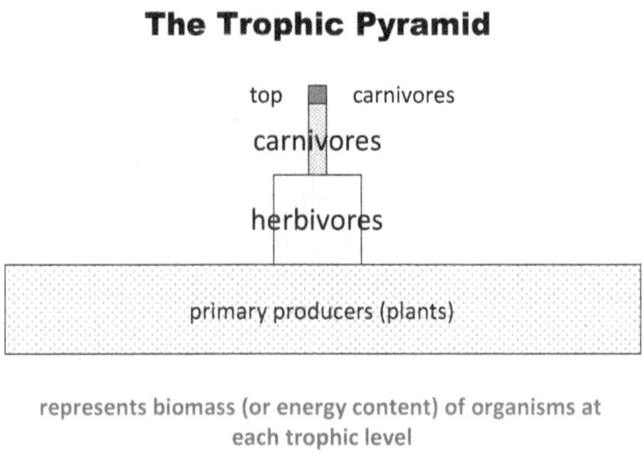


Figure 1.2 The trophic pyramid.

and when larger, as low-level predators consuming many invertebrates along with their diet of plant materials. Along coasts from Brittany to Scandinavia tiny young cod near bottom inshore and tiny young sand eels offshore eat zooplankton, so are secondary consumers; adult cod eating adult sand eels are at a higher trophic level. Western Mediterranean pelagic tuna often eat Spanish mackerel, which commonly eat anchovies and sardines, which as adults eat mainly zooplankton, which eat phytoplankton.

Because living things convert only about 10 percent of the energy from one trophic level to the next, organisms nearest the base of the food web (producers or primary consumers) have necessarily the greatest total biomass and numbers of individuals; those with the most intervening stages are reciprocally low in biomass and numbers. The relationship can be visualized as a trophic pyramid (Figure 1.2)[23] Predators are characteristically larger in individual size but fewer in number and less in total mass than their prey. Predators and prey coexist in nature – sometimes the interaction controls the numbers of both – but the relationship is not necessarily stable and environmental changes of any origin can cause great dislocation.

In natural ecosystems the food web is completed through the activities of detritivores (waste-consuming organisms) and the process

[23] The valuable heuristic of the pyramid obscures observed reality that most consuming organisms do feed at several levels, at least across their life cycle. Studies of fish diets result in average trophic levels, such as roach at 3.0, herring at 3.4, sturgeon at 3.5, pike and Atlantic cod at 4.1, and both bluefin tuna and Atlantic salmon at 4.5. Ratings for each species are found in FishBase (www.fishbase.de/Summary/SpeciesSummary.php).

of decomposition, which break down complex tissues and recycle the material components of life (minerals, carbon, water) back into food for plants. Organisms removed from a watershed take the embodied nutrients with them. Lotic (running-water) ecosystems continually export nutrients downstream and so must receive regular inputs from the surrounding terrestrial environment.

Food webs are not the only historically significant patterned relationship among different species or individuals of the same species. Different animals compete for access to resources which their ecosystem holds in limited supply: think of food sources, spawning sites, even, in some situations, oxygen. Generally speaking, two species with identical habitat requirements cannot coexist; one or both must adapt or die out.[24] Among fishes, patterns of competition commonly shift with phases of a species' life cycle. The small fishes vying for zooplankton or insect food in the shallows of a sluggish English river may include young pike which as adults will eat their one-time competitor roach or loach. Adult salmon and trout choose gravels of different size to make spawning sites in the same reach of a stream, but for a year or two after hatching their offspring will compete to live from the same drifting or bottom-dwelling invertebrates.

Competition also occurs within a single species and population – over food, cover, nesting sites, and so on. Young carp eat much the same bottom invertebrates as do adults, and spread the total carp biomass across many small individuals. Young pike eat insects and smaller fish and so convert abundant but small-sized food into a more efficient meal for large pike – whose cannibalistic proclivities were familiar to medieval observers.[25] The hazard of competitive and predatory interaction between age classes is one reason many fishes have breeding grounds and nursery habitat distinct from adult feeding areas – which, in turn, makes those species dependent on more than one habitat. Note further that such patterns of interaction across the species's life cycle can relate to the structure of fish populations. Schools of pelagic herrings in the North Sea, for example, are successively made up of particular dominant year classes, great numbers of fish all born in the same especially good year for survival and growth of the larvae – rather like the baby-boomer cohort among late twentieth-century North Americans and Europeans. Among inland lake fishes – charr are a prime example – good conditions for small fish but limited food supplies will spread the biomass across

[24] When two closely related and similar species do coexist, as once with American and European sea sturgeons along French Atlantic coasts, the situation is called sympatry.

[25] Vincent of Beauvais, *Speculum naturale*, I:17:64; Thomas of Cantimpré, *Liber de natura rerum* (ed. Boese 1973), 7:48; Albert the Great, *De animalibus*, 26:73.

large numbers at the lower end of the size range; limited reproductive success, on the other hand, will let large, even possibly slow-growing older fish dominate. The latter is also true of sturgeons, which take decades to reach sexual maturity.

From a broader perspective, considering flows of energy and competition over them is just one way to think about the structure, preferences, and vulnerabilities of aquatic communities. Especially recent work in community and restoration ecology groups species, irrespective of their evolutionary relationships, in terms of overlapping habitat requirements into 'ecological guilds'. Trophic levels may thus be cross-cut by such shared preferences as reproduction conditions (spawning on gravel, vegetation, sand, or open water with or without current or nest building), current velocity, migratory or schooling behaviours, and relative tolerance or sensitivity to water quality, temperature, etc. 'Guild' membership means certain environmental changes can have similar effects on unrelated fish species, resulting in alterations to an entire community.[26]

Another ecological principle emphasizes the importance of 'edge' between two adjoining habitats as affording access to resources in both. This helps explain the rich concentration of fish life near the surface, where light and oxygen come from the air, in the open sea and open lakes; in the well-mixed and ever-changing tidal zones, with their many closely defined habitat niches; and in churned-up, nutrient-rich waters of estuaries. Plant life and herbivorous organisms attract an ascending pyramid of predators. The attraction works in other directions, too. Rich populations of fishes in Europe's streams and coastal inlets attracted many air-breathing predators, birds, mammals, and humans, to what were in the early Middle Ages still relatively natural boundaries between water and land.[27]

Humans who insert themselves into aquatic ecosystems by taking fish for food become a new apex predator atop the trophic pyramid, but one with potentially distinctive impacts. Aquatic food webs, especially those exploited by humans, are characteristically longer than terrestrial ones, as people in most cultures eat few aquatic plants or herbivores but rather the larger carnivorous fishes at the top of the food chain. Human choices of

[26] Noble et al., "Assessing the health of European rivers using functional ecological guilds of fish communities," 381–392, and as applied to river fishes, Logez et al. "Modelling the habitat requirement of riverine fish species," 266–282. In the form articulated by Austen et al., "Importance of the guild concept," this proves a more flexible frame for historical reference than the sensitivity scale of 'white', 'grey', and 'black' fish communities introduced by Welcomme and further developed in Regier et al., "Rehabilitation of degraded river ecosystems."

[27] Compare Schalk, "Structure of an anadromous fish resource."

prey species may be set by biological variables (prey numbers, habits, accessibility), or be determined by such cultural aspects as taste, prestige, or technical knowledge. Human fishers can anticipate future concentrations of favoured fishes or wait and select even rare varieties yielding fewer calories than the work necessary to catch them.[28] Different methods of catching fishes may be more or less selective of species, habitat, fish size, or habits in a given aquatic community.[29] The chosen bait can strongly affect which species are taken with hook and line: medieval fishers knew well that cyprinids liked bread paste with strong herbal flavours and pike a shiny little fish. The season and habitat fished have more influence on the catches made with basket traps or a trawl: among the oldest written records of fishing in lakes near Salzburg is a mid-ninth-century agreement which distinguished the autumn run of lake-dwelling trout from the spring fishery for *albuli pisces* (probably cyprinids, possibly Coregonid whitefishes).[30] Size of mesh in a net determines the share of small individuals taken, but fish of all sizes are available for gathering after being stunned with a submarine explosion or a poison. Hence different fisheries can have highly variable effects on an aquatic ecosystem.

Two effects of predation may be especially connected to humans. Human preferences for larger fish, whether cultural or technical in origin, often have distinctive and detectable results. In part because fish, unlike other vertebrate animals, grow continually throughout their lives, and in part from human cultural and device biases, even moderate fishing pressure removes the largest individuals and reduces the probability of others surviving to achieve great age and size. This commonly affects the naturally larger and less numerous predator species, with the further consequence of triggering a 'trophic cascade': as natural apex predators are removed, numbers of lower-ranking carnivores increase and can, in turn, consume more of their own (still smaller) prey … ultimately even altering any limits the feeding of herbivores may have imposed on plant biomass. Secondly, when human predators (fishers and/or consumers)

[28] Tudge, *Time before History*, 198, 258, 308. The ostentatious serving of the 'last' sturgeons from European habitats mentioned in the Introduction fits this attitude all too well.

[29] In treating capture methods I attempt to follow the analytical system of Brandt, *Fish Catching Methods*, 389–394. For the inherent selectivity of certain methods see Hubert, "Passive capture techniques," and Hayes, "Active fish capture methods," in Nielsen and Johnson, *Fisheries Techniques*, 95–146, or, from an archaeological perspective more congruent with concerns of the present work, Colley, "Fishing for facts," 17–18.

Worth noting here is the lack of medieval records quantifying catches and of even proxies for fishing effort (ships, fishers, days fished, etc.) and thus any meaningful use of the Catch per Unit Effort (CPUE) so important in present-day fisheries science.

[30] *Salzburger Urkundenbuch*, I: 914–915.

find themselves deprived of once favoured large carnivorous fishes, they may respond by 'fishing down the food chain', that is, targeting and developing a taste for more numerous species from lower trophic levels (perch or roach in place of pike or pike-perch; sardine or herring instead of tuna or cod).[31] Precise local balances might then determine the effects of a fishery on its supporting ecosystem. Consequences also follow from ways humans finally consume the fishes they catch and allow both waste and consumed nutrients eventually to re-enter an aquatic ecosystem. From a metabolic perspective it matters if the nutrients are being restored to the system from which they had been removed or added to some other waters. It is rare for effluents from small and scattered human populations to have impacts that are more than local or short-term. From this historical distance, at least, none are readily visible in early medieval Europe.

But in contemporary view and in historical retrospect the study of fisheries abounds with observational hazards deeper still than that between one elemental medium and another. Whether held by present-day scientific observers, past participants, or environmental historians looking to reconstruct past conditions, human cultural representations are susceptible to at least two diametrically opposed delusions. One, the shifting baseline syndrome, arises from the short professional lifespan of experiential observers, fishers or scientists, for whom conditions of their youth become personal benchmarks for subsequent assessments of change. The syndrome tends to suppress perception of trends occurring over longer periods of time than the subject could witness. A possible shifting baseline mindset calls for critical care in using both recent scientists' assertions of no meaningful changes to a fish stock or ecosystem prior to a certain recent event and past narrators' anecdotes of stability 'since time immemorial'.[32]

The other misapprehension assumes the great and stable abundance of a mythic past before personal experience. According to a tale well-reported in European and North American sources since at latest the seventeenth century, for instance, not so long ago Atlantic salmon used to be so abundant that household servants demanded contractual or even

[31] See Pauly et al., "Fishing down marine food webs."

[32] Famously formulated by Daniel Pauly, "Anecdotes and the shifting baseline syndrome of fisheries" in 1995, and a decade later brought explicitly into global and a long-term views also of freshwater systems by Pinnegar and Engelhard, "The 'shifting baseline' phenomenon," and Humphries and Winemiller, "Historical impacts on river fauna." Wider implications for historical study are developed in Klein and Thurstan, "Acknowledging long-term ecological change" and Hilding-Rydevik et al., "Baselines and the shifting baseline syndrome."

legislative protection against being served salmon more than once (or twice or thrice) a week. Yet no contemporary testimony or evidence of such documents has ever verified this retrospective dream.[33] So, too, back in the early twelfth century Cosmas of Prague, first Czech historiographer, wrote down how the Czech people first entered Bohemia (fifth/sixth century?) and there found "tasty fishes, healthy to eat ... from waters full of fishes beyond measure."[34] While the myth of hyperabundance imagines a pristine past and degenerate present, shifting baselines limit change to what the subject could her- or himself observe. Both attitudes and assumptions formulate an unchanging ideal past and a human-driven present of one-way decline. Unchecked, both discourses are inimical to good science and to good history. Evidence of the past itself, that of palaeoscience and of the historical written and material record, better reflects and echoes both stabilities and changes.

Humans are not the sole drivers of massive change to aquatic or other ecosystems. Events as local and abrupt as entry of a new species or as large and perhaps slow as a new pattern of runoff or general climate change can, did, and do alter basic ecological relations to the point of what ecologists call a 'regime' change or shift.[35] Species once keystones can be eliminated or so reduced in biomass that changes cascade up and down the trophic pyramid. In the post-glacial long run, for instance, both the Baltic and the Mediterranean experienced periods disconnected from direct access to the Atlantic, becoming respectively freshwater or highly saline lakes with fish assemblages unlike those of late antiquity or today.

## 1.5 Rome and After

Early medieval fish populations are not to be thought as a pristine nature in stable habitats unsullied by human activity. By about 500 CE, Europeans had been taking and eating fish for millennia, with or without written records. The classical Greco-Roman Mediterranean is being explored by marine environmental historians despite how literary proclivities of elite Roman writers, lack of any operational records for ancient

[33] Schwarz, "Der Weserlachs und die bremischen Dienstboten" and "Nochmals: Der Lachs und die Dienstboten"; Thibault and Garçon, "Un problème d'écohistoire"; Danker-Carstensen, "Wiefiel Störfleisch" and "Stör oder Lachs."

[34] "*Aque illic nimis perspicue et ad humanos usus sane, similiter et pisces suaves et ad comedendum salubres. ... Aque ex omni parte copiose et ultra modum piscose*" (Cosmas, *Chronicon bohemorum*, lib. 1, cap. 1, ed. Bretholz, pp. 6–7, tr. Wolverton, *Chronicle of the Czechs*, pp. 34–36).

[35] Classic discussion of theory and empirical evidence in Scheffer, *Critical Transitions in Nature and Society*, 104–105, 121–122, 186–216, and 265–284; more recently Seddon, "What do we mean by regime shift?."

fisheries enterprises, and relative paucity of domestic or market refuse yielding large samples of consumption together curtail opportunities to establish genuine ecological perspectives. Recent research has found fish a more important dietary element than was previously thought. Fresh fish were an elite dish, supplied by local artisanal fishers from nearby waters. Processed fish products had wider distribution in salted form, especially in the various fermented sauces generically called 'garum'. Sustained by urban demand scores of industrial scale processing sites around and beyond the western Mediterranean made sauce out of the tiniest specimens, bycatches, and at least scraps of tuna and other large varieties. Both segments of the industry relied on ecological knowledge of local fishers, some of it possibly even passed on into modern times. The best-documented Roman fisheries were exclusively a marine activity which in the West faded together with Roman naval superiority.[36]

At least two ecological impacts mark the historical record before the end of Roman times.[37] While archaeological evidence from previously uninhabited early Neolithic Cyprus and of cods taken in the early sixteenth century from the formerly unexploited Grand Banks both show initially remarkable abundance of very large old specimens, even pre-Roman fish remains from Europe lack any such features. But regionally confined populations of especially desirable fishes did incur further more particular and long-lasting damages during Greco-Roman antiquity. Roman writers blamed high prices on depleted coastal fisheries.[38] As remarked in the Introduction, long-term archaeozoological study of sturgeon remains from the Rhône estuary shows heavy commercial fishing reducing first the average size, age, and then the numbers caught of this large and long-lived but slow-growing favourite of Roman gourmets. By late antiquity local overfishing had essentially removed sturgeon from the Provençal catch and dining table. Subsequent release of pressure still meant that stock needed centuries to recover its pre-Roman abundance.

In general, then, human influence on aquatic life in Europe during the post-Roman, early medieval centuries up to about 1000 CE was diffuse and marginal, limited with respect to habitat even in the most-impacted Mediterranean to the irreversible hydrological effects of classical land-use patterns, and otherwise to local or superficial results of long-standing

[36] Multiple examples and overviews appear in Gertwagen et al., eds., *Le interazioni tra uomo ed ambiente*; Marzano, *Harvesting the Sea*; Bekker-Nielsen, and Gertwagen, eds., *The Inland Seas*; Mylona and Nicholson, eds., *The Bountiful Sea*. Squatriti, *Water and Society*, 98–103, and Salvadori, "Transition from late antiquity," document the contraction of marine fishing.

[37] Desse and Desse-Berset, "Pêche et surpêche," 332–337.

[38] Trakadas, "Exhausted by fishermen's nets."

but rarely intensive fisheries. Ecologies in and around Europe's fresh and coastal waters were largely consequences of natural drivers, processes, and their long-term changes. These had shaped a subcontinent of cool running waters with stable seasonal cycles in diverse riverine habitats, a sparsely differentiated fish fauna much adapted to marine access, and extensive tracts of marine coastal and inshore habitats both in the warm but nutrient-poor Mediterranean and the nutrient-rich but cold western and northern seas. But grounds for human pursuit of fish were shifting.

# 2 Protein, Penance, and Prestige
## Medieval Demand for Fish

Fish as food, an object of human consumption, is the primary human colonization of aquatic nature, seeking to turn one element of an ecosystem into a cultural object. Fish and human consumer alike enter the hybrid zone where culturally defined needs and wants impact upon natural processes. Human needs and wants generate what economists call demand, which in turn drives the human work of catching and preparing fish to eat. This chapter explores the special place of fish in medieval European consumption patterns and several ways in which cultural understandings of what people ate shaped a socially stratified demand structure for fish.[1]

### 2.1 Dietary Protein

Most medieval Europeans obtained most of their calories from grain products. Cereal eating gained importance between the ninth and fourteenth centuries and then in the fourteenth and fifteenth shrank before more consumption of animal products in many areas and social groups. The particular organisms or forms which people preferably or actually took as food commonly followed regional and social lines, with some tendency over the medieval millennium for distinctions of rank to supplant those of place. While a seventh-century Swabian chief and peasant

[1] For all the proliferation of food history over recent decades, the broad picture remains little changed. See such exemplary studies of medieval diets and perspectives on regional cuisines as Stouff, *Ravitaillement en Provence*; Nada Patrone, *Il cibo*; Menjot, ed., *Manger et boire*; Dyer, "Consumer and the market," 305–327; Santich, "Éléments distinctifs"; Adamson, ed., *Food in the Middle Ages* and *Regional Cuisines*; Cavaciocchi, ed. *Alimentazione e nutrizione*; Carlin and Rosenthal, eds., *Food and Eating*; Rippmann and Neumeister-Taroni, eds., *Les mangeurs de l'an 1000*; Dam and Winter, "Theorie en praktijk"; Laurioux, *Une histoire culinaire*; Schubert, *Essen und Trinken*; Woolgar, Serjeantson, and Waldron, eds., *Food in Medieval England*; Ravoire and Dietrich, eds, *La cuisine et la table*; Schulz, *Essen und Trinken*; Van Molle and Segers, eds., *The Agro-Food Market*; Hoffmann, *Environmental History*, 114–119; Broadberry et al., *British Economic Growth*, 280–295; Woolgar, *Culture of Food*, 112–119.

both ate spelt, a locally predominant primitive wheat, and their Pictish contemporaries rye or oats, for instance, by about 1400 dukes of Württemberg and of Albany more likely shared with one another tables of white wheaten bread, sweetened wine, and jellied meats than either did the dishes of farmers on their lands.

### 2.1.1 *Fish on Medieval Menus*

Within the broad dietary pattern, medieval western Christians, who seem to have lacked anything resembling the modern concept of 'protein', were well aware that fish functioned as a meat substitute or supplement. In their consistent differentiation between a 'flesh day' and a 'fish day' when meat foods were culturally taboo ('lean day,' 'fast day'), people continually voiced this understanding.[2] Tenth- and eleventh-century monastic reformers, aiming to restore St. Benedict's ban on meat for monks, compensated with larger rations of cheese, eggs, and fish.[3] Later Latin and vernacular literary battles pitted allegorical armies of fleshy foods – roasts, lard, pheasants, sausages, hams – under the gluttonous 'Lord Carnival' against gaunt 'Lord [sometimes Lady] Lent's battalions of herrings, trouts, crayfish, carps, haddocks, eels, oysters, porpoises, and minnows.[4] Market ordinances from fourteenth-century Kraków and popular ditties from Paris drew the same distinction.[5] So did medieval cookbooks: the oldest surviving Franco-Latin manuscripts from around 1300 and the first printed German *Kuchenmeisterei* from 1490 grouped recipes and menus for flesh days and for fish days; an Italian tradition gave alternate ingredients for each dish.[6]

[2] Scully, *Art of Cookery*, 72–79; Laurioux, *Une histoire culinaire*, 351–352. Medieval dietary concepts are explored below, as is the ideology of abstinence. The point here is the polarity of flesh and fish.

[3] Zimmermann, *Ordensleben*, 81–87. Where ninth-century Benedictines did closely follow the *Rule*, Rouché, "A faim à l'époque carolingienne," found them suffering too little protein consumption and symptoms of gastric distress and malnutrition.

[4] Grinberg and Kinser, "Combats de Carnaval et Carême," survey variations on the theme. Compare Scully, *Art of Cookery*, 62–64, and the elaborated catalogs of foodstuffs in Lozinski, ed., *Le Bataille*, and in Juan Ruiz's *Libro de Buen Amor*, ed. Zahares, stanzas 1067–1127.

[5] Piekosiński, ed., *Kodeks dyplomatyczne*, #336; Bucher, ed., *Alten Zunft- und Verkehrs-Ordnungen*; Nystrom, *Poèmes français*, 122–135. In the 1526 colloquy "A Fish Diet" Erasmus of Rotterdam poked fun at the whole issue, neglecting to mention that his "insuperable distaste" for fish had gained him full dispensation (*Colloques*, tr. Thompson, pp. 312–357).

[6] Scully *Art of Cookery*, 124–126, is an overview. See examples in Mulon, ed., "Deux traités inédits"; Lozinski, ed., *Le Bataille*, 181–190; *Kuchenmeisterei 1490* and related texts as in Ehlert, ed. *Küchenmeisterei*, 222–235; and Grieco, "From the cookbook to the table." Similar characteristics of English cookbooks in Hieatt and Butler, eds., *Curye on Inglysch*;

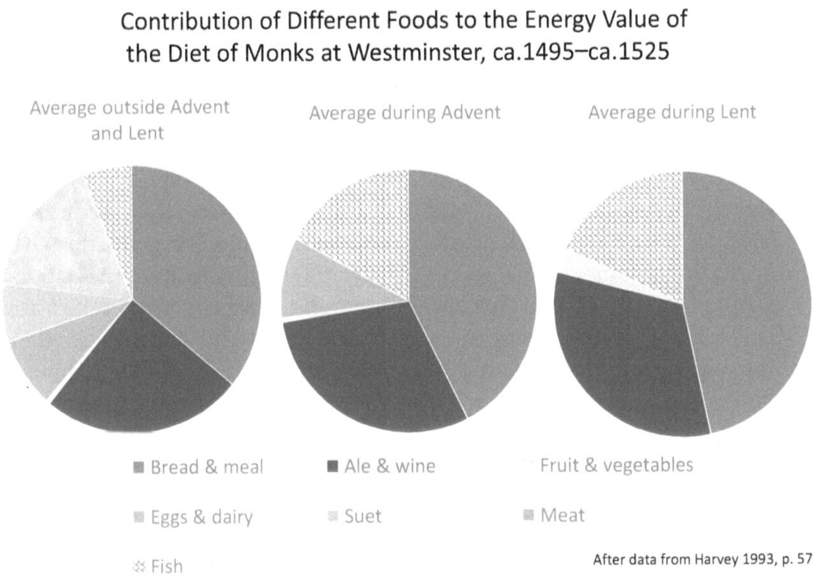


Figure 2.1 Contribution of different foods to the energy value of the diet of monks at Westminster, *c.* 1495–*c.* 1525.

Dietary practice carried through the idea of substitution, too. In minutely itemized kitchen accounts at late medieval Westminster abbey – where the *Rule* was once again bent – the share of calories provided by fish rose in direct proportion to staged restrictions on meat, dairy, and eggs (see Figure 2.1). At Carpentras in Provence, when Lent forbade meat eating the business of butchers collapsed and they closed their shops, while fishmongers' sales multiplied.[7] From a modern dietary point of view, the equivalence is not complete, for compared to an equal weight of red meat (beef) fish provide about 15–20 percent more protein but about a third as many calories and a tenth as much fat.[8] What difference that makes to an individual eater depends on the rest of the diet.

Fish were not just a theoretical or occasional dietary option, but are repeatedly verifiable in different settings and types of evidence for what

Hieatt and Jones, "Two Anglo-Norman culinary collections"; and Hieat, "Sorting through the titles"; and of French counterparts in Scully, ed., "*Du fait de cuysine*"; Scully, ed. and tr., *Chiquart's On Cookery*; and Lambert, ed., *Le recueil de Riom*.

7 Harvey, *Living and Dying*, 57; Stouff, *Ravitaillement*, 182, 192–193, and 211.

8 Might this help explain the attraction of such oily fishes as herring, sardines, salmon, and tuna?

medieval people actually ate. In remains of human food from numerous well-excavated archaeological sites the significant presence of fish in actual medieval diets is plain, though not susceptible to measured comparison with meat consumption.[9] A few examples might be selected for wide geographic, temporal, and social variety. In cisterns filled with soil from late Roman Carthage some 3,000 fishbones indicate human food waste of thirteen marine fishes and one from freshwater, all taxa common to Mediterranean in- and nearshore habitats.[10] Just a few centuries later both the inhabitants of Eketorp village on Öland in the central Baltic and others south of that sea at a fort and *suburbium* later called the 'Mecklenburg' were tossing onto garbage heaps the leavings of their tables: each group had eaten a dozen and more kinds of fishes, half of those the same local north European varieties, but where the coastal site favoured herring, cod, flatfishes, sturgeon, and pike, the more inland Mecklenburgers ate mostly pike and cyprinids.[11] Three well-excavated communities have large collections of discarded fish remains of mainly tenth- and eleventh-century date: in the 13,842 identified fishbones from the Viking entrepôt at Haithabu, herring, perch, and pike are most common; at the Slavic–Magyar village of Zalavár on a wetland beside Lake Balaton people ate catfish, carp, and pike-perch; late Anglo-Saxon *Hamwic*, the seaport precursor of Southampton, consumed mainly marine and migratory fishes, notably herring, flatfishes, and eel.[12] At York the many contexts from before about the year 1000 contain almost exclusively the bones of freshwater varieties, while those dating to the eleventh century reveal a great increase in marine species, especially herring, and by the years around 1100 growing numbers of cod.[13] About the same time inhabitants of the Castello di Manzano, a fortified

[9] On the differential survival and recovery rates of fish and mammalian remains in archaeological contexts see, for instance, Clason and Prummel, "Collecting, sieving and archaeozoological research"; Jones, "Bulk-sieving"; Benecke, "Sozialökonomische Interpretation"; Benecke, "Untersuchungen zum Einfluß der Bergungsmethode"; and Jones, "Survival of fish remains."

[10] Reese, "Faunal remains from three cisterns." In contrast, early medieval sites from northern Italy have only local freshwater fishes (Baker, "Fauna"; Baker, "Subsistence"; Baker, "Le rôle de la chasse").

[11] Hallström. "Die Fischknochen," 424; Benecke, "Fischreste"; Enghoff, "Fishing in the Baltic region," 56–58 and 74–75.

[12] Lepiksaar and Heinrich, *Fischresten aus Haithabu*, 17 and 119; Bökönyi, "Wirbeltierfauna"; Coy, "Provision of fowls and fish," tables 1 and 2.

[13] Barrett et al., "'Dark Age economics' revisited"; Barrett et al., "Origins of intensive marine fishing"; Harland et al., "A case study"; and Harland et al., "Fishing and fish trade." Loveluck, *Northwest Europe*, 186, 198–200, 202, and 211, assembles evidence of marine fish consumption well before 1000 at coastal settlements on both sides of the Channel.

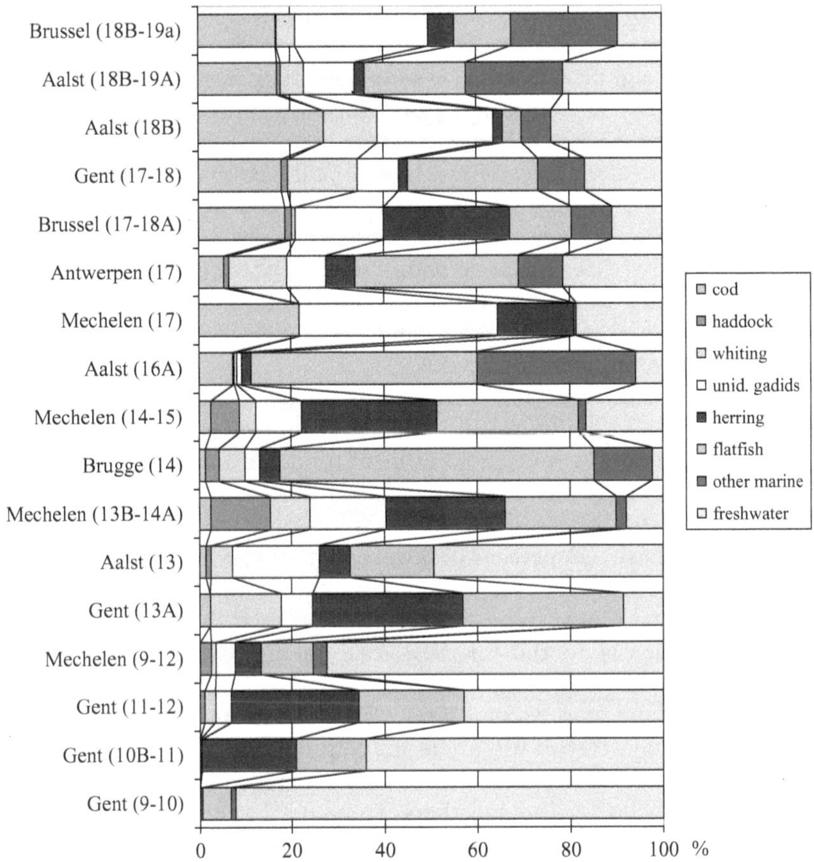


Figure 2.2 Fish varieties in remains recovered from medieval and early modern urban sites in the Scheldt basin of Flanders (with thanks to Dr. Anton Ervynck, Flemish Heritage Agency).

village built shortly before 1014 in the northwestern Italian piedmont and abandoned in 1243, who ate many kinds of wild and domestic animals, included tench, chub, and perch in that diet.[14]

Large-scale archaeozoological studies compile fish remains from several regional centres. As illustrated in Figure 2.2, at the emerging urban sites of medieval Flanders, into the thirteenth century freshwater fishes commonly made up the largest proportion of consumption remains, although by about 1000 inshore marine varieties had a significant presence ($\geq$20 percent). The latter share further increased by and after 1200.

[14] Bedini, "Osservazioni preliminari."

During later and post-medieval times (roughly 1300–1550) inshore marine species dominate and remains from possibly more distant seas (i.e. cod) gain significance equal to or greater than those from fresh water (≈20 percent).[15] In contrast, at the collegiate church of Saint-Pierre-Lent in Orleans 15,004 fish remains were identified from a latrine filled with food waste dated to *c.* 1490–1568: here fifteen marine taxa contributed about 40 percent of those bones, with cod at 23 percent, and herring, flatfishes, and rays each at about 5 percent; fifteen diadromous and freshwater species made up about 60 percent of remains, but bones of common carp were just over half of the total, followed by 5 percent other cyprinids, 2 percent perch, and mere traces of all other taxa.[16] (More cases in Supplement 2.1.1.)

Detailed financial records compiled by medieval authorities – and so chiefly artifacts of the fourteenth and fifteenth centuries – do show the share of fish in food a household or institution consumed or purchased. In an exceptionally early example, during 1296–1297 the household of Joan de Valence, countess of Pembroke, spent on staple fish (herring, salt fish, and stockfish) 39 percent of the kitchen budget or 9 percent of the total for food and drink.[17] A leading burgher of late fourteenth-century Köln, Hermann Goch, more market-oriented than the countess, spent on fish as much as he did for meat, each taking 15–20 percent of his household's outlay on food.[18] During 1403–1405 fish also comprised 17 percent of the food expenditures of the Polish royal court.[19] (More household accounts with fish are in the Supplement.)

Written sources also confirm fish as a distinct element in the ordinary diets of people who were not members of the elite. Twelfth-century Parisian teacher Alan of Lille said herring "relieved the hunger of the poor," which was exactly how King João II of Portugal (1481–95) spoke of sardines.[20] Some public artistic representations of calendars in twelfth- and thirteenth-century France and illuminations in contemporary English manuscripts

[15] Van Neer and Ervynck, "Rise of sea-fish consumption," 162–164.

[16] Marinval-Vigne, "Consommation d'animaux sauvages," 478–482. Throughout the medieval period, however, Italian monastic sites, more often contain at least some marine fish (Salvadori, "Transition from late antiquity").

[17] These numbers may inflate the share of fish, as the accounts omit products of the countess's estate, notably meats, cereals, and the take from its river fisheries. But the proportions spent on fish by the household of John de Vere, Earl of Oxford, in 1431–1432 were just about the same (Woolgar, *Great Household*, 112–121 with table 7).

[18] Irsigler, "Ein großbürgerlicher Kölner Haushalt."

[19] Piekosiński, ed., *Rachunki dworu*, 286–319.

[20] Alan, "De planctu naturae," tr. Sheridan, 94–98; da Cruz, "Apontamentos sobre a comida," 95.

Figure 2.3 Fourteenth-century English commoners grilling fish. As depicted in the Holkham Bible Picture Book, BL Ms Add. 47682, fol. 37r. Garbed as English commoners, Christ's disciples grill fish they just caught and serve them to the risen Christ. Reference is John 21: 1–14, the miraculous catch of 153 fish from the Sea of Galilee following instructions of the risen Christ. Sharp-eyed observers will notice that the fish look very like Atlantic mackerel, a species not known to be resident in the Sea of Galilee or the eastern Mediterranean, but common along English coasts. Reproduced with permission of the British Library.

(Figure 2.3) show peasants eating fish.[21] Residents in the papal almshouse in Rome were served fish on 117 days during accounting year 1285–1286.[22] By that time, and on into the 1430s, peasants who received meals for compulsory work in the Norfolk harvest were getting 3–5 percent of their calories from fish worth 7–25 percent of the total cost of their feeding.[23] Household servants at the royal Polish castle in Korzyn got fish on every meatless day in 1405–1408. Two generations later this was also the experience of municipal workers at Basel and servants of the duke of Bavaria.[24]

[21] Mane, "L'alimentation des paysans," 323.
[22] Cortonesi, "Le spese in *victualibus*," 203–204.
[23] Dyer, *Standards of Living*, 158. Harvest workers got fish in Portugal and Catalonia, too (da Cruz, "Apontamentos sobre a comida," 98–99; Altisent, *Les Granges de Poblet* 33–34).
[24] Piekosiński, ed., *Rachunki dworu*, 320–391; Hundsbichler, "Nahrung," 201–202.

### 2.1.2 *We Are What We Ate, and So Are Our Remains*

Written and archaeozoological sources convincingly imply medieval Europeans eating fish, only recent advances in human bioarchaeology have begun to offer some quantitative indications of the fish protein these people actually ingested. The technique analyzes stable isotopes in protein recovered from human skeletal remains. The chemical characteristics of foods we (and other organisms) eat leave traces in the composition of the consuming body. Many chemical elements naturally occur in molecules of more than one weight. The relative proportions of stable isotopes of certain elements taken in over a decade or a lifetime are reflected in the consumer's own tissues.[25]

In particular the isotopic ratios of carbon ($^{13}C$ vs $^{12}C$) and of nitrogen ($^{15}N$ vs $^{14}N$) in the body's proteins reflect those ratios in the foods consumed during life.[26] The ratios can be measured from the bone protein (collagen) which often survives in even millennia-old human and animal remains. Carbon isotopic ratios differentiate between terrestrial and marine-based food webs: in temperate climates like medieval Europe the bone collagen of humans who ate foods of terrestrial origin shows $\delta^{13}C$ values around −20 permil, and that of individuals getting their protein from marine sources has $\delta^{13}C$ of −12 permil (so less negative).[27] Nitrogen isotopic ratios signify the trophic level from which protein is consumed: the share of $^{15}N$ rises by 3–5 permil at each step up a trophic pyramid. The longer food chains in aquatic systems (see Chapter 1) result in higher $\delta^{15}N$ values for individuals with water-based diets. Typically persons with purely terrestrial diets display $\delta^{15}N$ values around 8–10 permil and those getting protein from aquatic sources go as high as 20 permil. As depicted in Figure 2.4, then, low (large negative) $\delta^{13}C$ signals terrestrial, not marine-based, dietary protein, and high

[25] Stable isotopes are not to be confused with radioactive (unstable) isotopes such as carbon 14 ($^{14}C$), well known as a means of dating organic materials from the past, or strontium 90 ($^{90}Sr$), a hazardous byproduct of nuclear testing.

[26] Stable carbon and nitrogen isotopic values are reported as the ratio of the heavier isotope to the lighter isotope relative to an internationally defined scale, VPDB for carbon and AIR for nitrogen. Isotopic results are reported as δ values ($\delta^{13}C$ and $\delta^{15}N$) in parts per thousand or 'permil' (‰), where $\delta^{15}N_{AIR} = [(^{15/14}N\ \text{sample}/^{15/14}N_{AIR}) - 1] \times 1000$. The more positive the δ value, the more enriched the sample is with the heavier isotope.

[27] Most terrestrial plants photosynthesize their carbon from the air with its standard ratio of the heavier and the lighter isotope. Marine organisms get their carbon from soluble bicarbonates, which are enriched in $^{13}C$. Only in some parts of southern Europe did medieval people possibly raise and eat millet, which belongs to a class of so-called C4 plants which adapted to warm arid conditions by evolving a different photosynthetic system which took more $^{13}C$ from the air. Careful archaeological science in the case studies referenced below takes into account the likelihood of C4 influences on dietary outcomes.

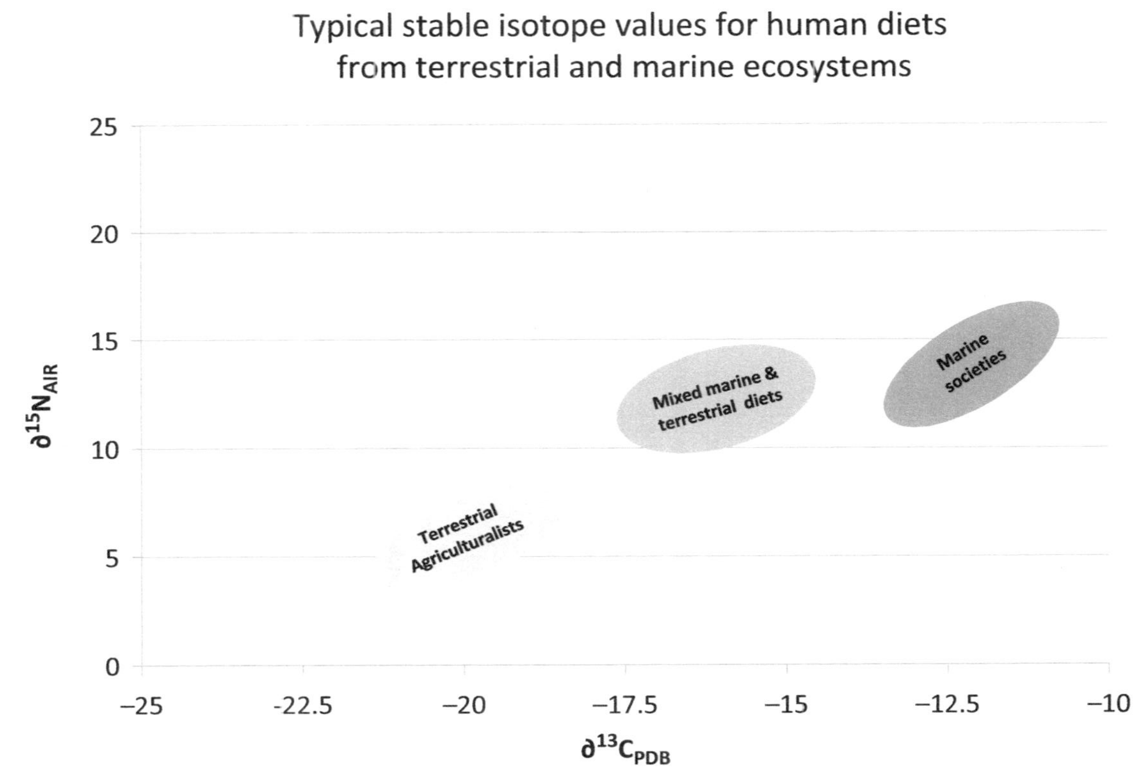


A simplified diagram redrawn by R. Hoffmann after Müldner & Richards, "Diet in Medieval England," fig. 16.1.

Figure 2.4 Typical stable isotope values for human diets from terrestrial and marine ecosystems.

$\delta^{13}C$ (though still negative) a significant proportion of seafoods. ($\delta^{13}C$ values for freshwater fishes may be lower because much of the nutrients in fresh waters comes from terrestrial sources.) High (positive) nitrogen values ($\delta^{15}N$) register consumers at the top of a (long) aquatic food chain and low those individuals whose protein was built from that of plants or herbivorous animals.[28] Ideally the stable isotope values of human remains should be compared to those of plants, herbivores, and several levels of carnivores in the same ecosystem, but this benchmark is not always provided.[29]

Certain caveats must be kept in mind. The threshold for either $\delta^{13}C$ or $\delta^{15}N$ values to reflect significant consumption of 'fish' is about 20–25 percent of the individual's intake of protein. Even a regular weekly meal of fish will likely not detectably affect that individual's isotopic signature. Nor are the fish species consumed or a mix of freshwater and marine organisms to be determined by this means. Protein obtained from diadromous fishes registers where the animals ate (adult eel in freshwater; adult salmon or shad in salt), not where they were captured during migrations. Nevertheless, this analytical method will identify consuming individuals and populations for whom marine protein sources were a regular dietary component and, perhaps less surely, those eating significant amounts of freshwater fish. Lacking corroborative evidence, however, the absence of an isotopic signature cannot be taken to mean people ate no fish at all. Further caution is advised because so far most publicly available results of such isotopic tests remain limited to relatively small samples of individuals from geographically biased regions of medieval western Christendom. Nevertheless the findings do point to some regional chronologies and sociocultural groups with distinctive patterns of fish consumption.

[28] For the scientific context and methods see Müldner and Richards, "Fast or feast"; Müldner and Richards, "Diet in medieval England"; Müldner, "Investigating medieval diet and society"; Hull and O'Connell, "Diet: recent evidence"; Lightfoot, Šlaus, and O'Connell, "Changing cultures, changing cuisines"; and Müldner, "Marine fish consumption," 239–249. Larger predatory fish and people who eat them show higher $\delta^{15}N$ than do smaller fish (Ervynck et al., "Assessing radiocarbon freshwater reservoir effect," 12–14).

[29] See. however, Müldner and Richards, "Diet in medieval England," fig. 16.3; Herrscher, "Inferring diet by stable isotope analysis," 146; Fuller et al., "Carbon and nitrogen stable isotope ratio analysis"; and Reitsema et al., "Human–environmental interactions." A baseline for isotopic values of freshwater fishes in medieval Switzerland is provided by Häberle et al., "Inter- and intraspecies variability." Ervynck et al., "Assessing radiocarbon freshwater reservoir effect," 17–18, question the temporal consistency of species values when remains have been subject to anthropogenic influence through farming and fertilizing. Use of some other stable isotopes to identify regions of origin for archaeological fish remains is discussed in Chapters 5 and 8 below.

Where, then, can physiologically significant medieval consumption of fish protein be confirmed? Answer: only in certain groups, localities, and periods. Into the eleventh century few Europeans ate enough fish of any kind to register in their remains,[30] but exceptional people and communities did. Corporeal relics of St. Waldetrudis (d.688), a noble monastic founder and abbess in Hainaut, clearly signal her consumption of freshwater, not marine, organisms. So do bones of two eleventh-century bishops of Tournai.[31] Among early to middle Anglo-Saxon villagers with terrestrial diets certain high-status burials stand out for high $\delta^{15}N$ values from aquatic food chains, those on the coast have high $\delta^{13}C$ from seafoods and those at riverine sites the low $\delta^{13}C$ indicative of fresh water.[32] By 800 or so, inhabitants of a manor house in Uppland, Sweden, and some in coastal Dalmatia also ate enough fresh- or brackish-water fishes to register relatively high $\delta^{15}N$ against low $\delta^{13}C$.[33] More pervasive use of seafishes is indicated at scattered coastal sites in eastern Britain and in the Mediterranean on Balearic and Greek islands then under Byzantine rule. Norse coastal societies, however, more generally consumed marine protein, as is visible at eighth- through eleventh-century Haithabu, on ninth- through twelfth-century Gotland, and in the remains of a band of Viking raiders who perished in tenth-century Wessex.[34]

Detectable marine dietary components increase in frequency during and after the eleventh and twelfth centuries. The most robust and so far

30 See, for example, Polet and Katzenberg, "Comportements alimentaires de trois populations médiévales belges"; Salamon et al., "Consilience of historical and isotopic approaches," 1667–1672; Wiedemann and Bocherens, "Spurenelement- und Isotopenanalyse"; Hakenbeck, et al., "Diet and mobility"; Privat et al., "Stable isotope analysis of human and faunal remains at Berinsfield"; Hull and O'Connell, "Diet: recent evidence"; Curtis-Summers et al., "Stable isotope evidence at Portmahomack."

Later medieval communities with no detectable fish consumption are reported in Ervynck, et al., "Dating human remains," 785; Reitsema et al., "Preliminary evidence for medieval Polish diet"; Yoder, "Diet in medieval Denmark"; Reitsema and Vercellotti, "Stable isotope evidence at medieval Trino Vercellese"; Herrscher, "Inferring diet"; and Ciaffi et al., "Palaeobiology of Albano."

31 Boudin et al. "An archaeological mystery," 607–608 and 611–615.

32 Hull and O'Connell, "Diet: recent evidence from analytical chemical techniques"; Mays and Beavan, "Investigations of diet in early Anglo-Saxon England." While early medieval Italian sites consistently lack evident marine consumption, a few show the high $\delta^{15}N$ typical of fresh water. An example is the Tiberside site of San Pancrazio in Rome ($\delta^{15}N$ = 11.5‰, $\delta^{13}C$ = 18.3‰) as discussed in Varano et al., "Edge of the empire," table 3 and supplemental figure 1.

33 Olsson and Isaksson, "Molecular and isotopic traces of cooking"; Lightfoot, "Changing cultures, changing cuisines."

34 Grupe, Heinrich, and Peters, "A brackish water aquatic foodweb"; Becker and Grupe, "Teamplayer oder Gegenspieler?" and Becker and Grupe, "Archaeometry meets archaeozoology"; Kosiba, Tykot, and Carlsson, "Stable isotopes as indicators of change"; Pollard et al., "Sprouting like cockle amongst the wheat."

earliest indications of transition come from Anglo-Norse and early Anglo-Norman York, where approximately one in three burials in the eleventh and twelfth centuries had consumed significant amounts of protein from marine sources. By and after the thirteenth century food from the sea was a regular part of most diets in York, though some men ate a lot more than did their neighbours and women in the community averaged somewhat less.[35] In other northern English sites remains of late medieval people across spectra of gender, social rank, and settlement types all show $\delta^{13}C$ around $-19.5$ permil and $\delta^{15}N$ over 12 permil, which the analysts interpret as signaling diets with aquatic foods both freshwater and marine.[36] Equally cogent findings of general large use of fish are rarer elsewhere and mainly associated with coastal sites. Late medieval inhabitants of Portmahomack on Scotland's east coast clearly did eat fish from the sea, and so too most at Gandia in Valencia. Their contemporaries buried in and beside the Cistercian abbey church of Ter Duinen in coastal Flanders did so as well, perhaps along with freshwater fishes.[37]

Elsewhere isotopic evidence again identifies in late medieval populations minorities who obtained protein from long aquatic food chains but ambiguity or diversity between its freshwater and marine origins. At the Christian cult centre of Whithorn in southwestern Scotland, six twelfth- to fourteenth-century bishops, senior clerics, and high-status lay benefactors buried inside the cathedral had consumed significantly more fish than the lower-status laity buried on its periphery. Just three men among the twelve male and twelve female skeletons tested from a mainly peasant cemetery near Poznań show traces of marine protein. In contemporary Tuscany certain castle dwellers and townspeople also stand out for high $\delta^{15}N$ despite $\delta^{13}C$ values indicating low marine intake.[38] The same

35 Müldner and Richards, "Stable isotope evidence for 1500 years"; Müldner and Richards, "Diet and diversity." The earliest noteworthy eaters of marine fishes at York seem to have been young men buried in the neighbourhood of Fishergate, maybe employed in maritime trades.

36 Müldner and Richards, "Diet in medieval England," fig. 16.2 and pp. 232–234, suggest commoners ate eel.

37 Curtis-Summer et al., "Stable isotope evidence at Portmahomack" and "From Picts to parish," find a change from the terrestrial diets of Pictish forebears, earliest monks included. Alexander et al., "Diet, society, and economy"; Polet and Katzenberg, "Reconstruction of the diet."

38 Buonincontri et al., "Multiproxy approach," notably figure 3; Müldner, et al., "Isotopes and individuals"; Reitsema et al., "Preliminary evidence for medieval Polish diet." Ayers, *German Ocean*, 88–89, calls attention to individual variation among inhabitants of Holbaek, Denmark, while Martínez-Jarreta et al., "Stable isotope and radiocarbon dating of the remains of the medieval royal House of Aragon," can identify those twelfth- and fourteenth-century individuals who had eaten more fish than had their kin as specific men and women known for their religious callings.

pattern occurs in seven of the thirty-six bodies of plague victims buried in a Roman neighbourhood about 1480.[39] On the other hand, the isolated individual men with high-status burials at riverine and urban locations among eight cemeteries of twelfth- and thirteenth-century Asturias more likely had eaten freshwater fishes, and so, too, the few people with signs of significant fish consumption from twelfth- through fifteenth-century burials at the poorhouse in Regensburg.[40]

From the perspective of dietary protein, then, the contribution of fish in general to human metabolisms in medieval Europe was limited, not absent, but surpassing the threshold of detectability only in certain social and regional circumstances. Elites as opposed to commoners, men in contrast to women, the religious compared to lay, and urban more than country people were at least marginally more likely to eat a significant amount of fish. Where so signaled, an increased consumption of marine fishes plausibly spread from Scandinavian and coastal British and continental areas toward the east and south, starting already around 1000 and continuing into early modern times. Stable isotope analysis of human remains cannot, however, itself now discriminate between consumption based on northern or on Mediterranean habitats, nor can it alone date the process or chronology which resulted in significant consumption of marine protein at sites some distance from salt water.[41] Throughout the period people at places across Europe continued to eat organisms from fresh waters. For the present, however, despite the limitations of stable isotopes, use of this method confirms that, from the earliest to the latest Middle Ages, at least some Europeans ate enough fish to provide 20–25 percent or more of their protein intake.

[39] The seven average $\delta^{15}$N 13.9 ± 0.6‰ and the others 10.2 ± 0.9‰. Salamon et al., "Consilience of historical and isotopic approaches," 1669–1670, would here see marine fish from the Atlantic. The stratification in this evidence is clear, but its enriched $\delta^{15}$N signifying a long aquatic food chain accompanies $\delta^{13}$C in the range of −19‰, so far from a convincing signal of marine rather than freshwater sources, much less any distant sea.

[40] MacKinnon, "Dietary reconstruction"; Olsen, "A multi-isotope investigation of two medieval German populations," 90–94. Zechini et al., "Diachronic changes in diet," find no sign of marine consumption in sixty-six adult skeletons dating 1100–1717 from the oldest cemetery in Berlin, but likely detectable eating of freshwater fish after the Black Death.

[41] From the perspective of fisheries, the greatest desideratum is for stable isotope analysis ($\delta^{13}$C and $\delta^{15}$N ) of good-sized samples of medieval human remains from inland France and central Europe, urban northern Italy, and the medieval Crown of Castile.

## 2.2 Prerogatives of Culture

There can be no doubt that medieval Europeans ate fish with some overall regularity and so generally recognized and used fish as food. But students of food acknowledge that what people eat often bears symbolic weight at least as great as the simple functional intake of bodily fuel. At different times or even all at the same time foods and their consumption may serve to entertain or comfort the eater or to display and validate group membership and social rank. Biology may set physiological requirements and limits on the intake of the organism but what humans define and consume as food is a distinct prerogative of culture, so learned, not inherited.[42] Medieval European consumption of fish was significantly shaped by religious taboos, by medical theory, and by the public display of status.

### *2.2.1 Religious Taboos on Meat*

The dominant religious orthodoxy of the medieval west, Latin Catholic Christianity, imposed severe restrictions on eating meat. The prohibitions normally applied on all Fridays and Saturdays (and/or, in some dioceses, Wednesdays), the days before major religious festivals, four quarterly 'ember days', the three days before the feast of Ascension, four weeks before Christmas, and the forty days of Lent during late winter and early spring before the high feast of Easter.[43] Abstinence from the eating of animal flesh was rooted in ancient Jewish traditions of purity and in ascetic practices of dietary deprivation which developed among late Roman Christians. Periodic observance had come to define normative Christian behaviour by the time of Charlemagne, whose Capitulary of Paderborn in 785 ordered the forcibly baptized Saxons under pain of death to observe the Lenten fast.[44] Lenten and Friday 'fasts' were the essential marks of membership in Latin Christendom during the Middle

[42] Teuteberg, "Agenda for a comparative European history of diet," 4–5, is a model formulation, given more discursive form in, for example, Fernández-Armesto, *Near a Thousand Tables*; Sarti, *Europe at Home*, 148–191; Fernández-Armesto and Smail, "Shared substance: food"; and Frantzen, *Food, Eating, and Identity*.

[43] Generally and from several perspectives see Zug-Tucci, "Il mondo medievale dei pesci," 294–301; Bérard, "La consommation du poisson," 171–173; Bynum, *Holy Feast and Holy Fast*, 32–42; Scully, *Art of Cookery*, 58–62; Nigro, "Mangiare di grasso," 113–146; Albala, *Food in Early Modern Europe*, 193–199; Winter, "Obligatory fasts and voluntary asceticism," and "Fasting and abstinence"; Grumett and Muers, *Theology on the Menu*, 22–24. Note that for lay people only, not clerics, the Lenten fast was lifted on Sundays.

[44] Boretius and Krause, eds., *Capitularia Regum Francorum*, vol. 2, #26 (Loyn and Percival, tr., *Reign of Charlemagne*, 51–53).

Ages. Monastic communities, for whom symbolic separation of spirit from flesh had fundamental importance, practiced abstinence still more regularly, and even permanently.

The rules and distinctions of dietary abstinence had considerable consistency in medieval cultural terms, if not in those of today. Though commonly called a fast (*jejunium*), the practice was rather to refrain from eating certain foods. Whether for monks and nuns or ordinary laymen and -women the operative ban was on the flesh of terrestrial quadrupeds, a classification not wholly congruent with the modern class of mammals, warm-blooded animals without feathers and with glands to provide milk for their young. The terms were classically set in the sixth-century monastic *Rule* of St. Benedict (chapters 36 and 39), which absolutely barred the meat of four-footed land animals to all monks but the sick. Several centuries ensued before Benedictine practice became the accepted standard across the Christian west.

Several ramifications gradually followed. 'Meat' itself had to be defined, and the practice gradually evolved to include a prohibition during Lent of all so-called fleshy foods, animal products such as cheese, eggs, and lard.[45] On the other hand the rules of abstinence were more than once modified in practice, both as a pattern of 'abuse' in socially too well-integrated monasteries of the post-Carolingian period, and by stepwise relaxation of the letter of the law during the later Middle Ages. In the latter case, what was banned from the refectory might be taken in the infirmary or a midway 'misericord,' and a papal decree from 1366 let this pass so long as half the monks ate in the refectory for each meal and all obeyed the letter of the *Rule* on Wednesdays, Fridays, Saturdays, four weeks of Advent, and eight weeks of Lent. Pope Alexander VI loosened the Lenten prohibition of butter in 1491 and encouraged purchase of indulgences exempting the buyer from other restrictions. Layfolk had long before learned to break the fast with a 'light' evening meal or to turn penitent austerity into a feast of elaborate dishes from permitted ingredients.[46]

45 Zimmermann, *Ordensleben*, 60–61, is augmented and refined, but not superseded, by Serjeantson and Woolgar, "Fish consumption," 102–104; Winter, "Fasting and abstinence"; and Fritsch, *Das Refektorium*, 45–53.

46 Bynum, *Holy Feast and Holy Fast*, 41; Harvey, *Living and Dying*, 38–41; Dam and Winter, "Theorie en praktijk," 393–398; and Albala, *Food in Early Modern Europe*, 199–207. Heinrich, "Untersuchungen an Fischresten aus St. Irminen," found vastly fewer and less diverse fish being eaten by nuns at an elite Trier convent in the 'degenerate' tenth century than by the founding generation back in the eighth. For subsequent debates over religious abstinence see Albala, "Ideology of fasting," 41–58.

Nevertheless and most importantly, the ban on meat actively encouraged the eating of fish. As Ekkehard IV at mid-eleventh-century St. Gallen cleverly phrased his scholar's Latin, *Omne natans trinus licitum benedicat et unus* ("The Three and One blesses as permitted all swimming things")[47] – having just imagined dining on a list of 'fishes' including whale, crayfish, and *fiber piscis*, 'beaver fish'. Rooted in Hebrew scriptural classification of animals by habitat, definition of all aquatic creatures as 'fish' was the (entirely sensible) standard throughout medieval times, from the *Etymologies* of sixth-century Isidore of Seville, through the scholastic encyclopedists of the mid-thirteenth century, up to and including the sixteenth-century pioneers of scientific ichthyology, Pierre Belon, Gillaume Rondelet, Conrad Gessner, and Ulisse Aldrovandi.[48] Early Greek-speaking Christians had associated fish with Christ, seeing ΙΧΘΟΣ as an anagram for the initials of "Jesus Christ, God, Saviour," and the earliest evolution from ascetic fasting to penitential abstinence more or less willingly accepted fish in lieu of flesh.[49]

Charlemagne considered fish the typical 'Lenten fare'.[50] Later more reform-minded ascetics in ninth-century Frankland, tenth-century England, and among early twelfth-century Cistercians were less enthusiastic, eschewing any hint of expensive quasi-carnivorous luxury for sparse almost vegan diets. But, in a pattern common to many revolutionary movements, over time occasional modest 'treats' ('pittances') were

[47] Schulz, *Essen und Trinken*, 598, l. 71, reprints the Latin text of Ekkehard's "Benedictiones ad mensas" from the Egli edition and (pp. 586–593) establishes a clear critical perspective for reading the monk's learned rhetoric.

[48] On the scaly tail of beaver (and at times the entire animal) as 'fish', see Blaschitz, "Der Biber im Topf," 416–425; Mänd, *Urban Carnival*, 205, 226, and works there cited.

[49] Less persuasive, however, is any direct positive causal link running from Christian monastic enchantment with fish as a spiritual metaphor – living in the water of baptism, reflecting by their agility in the water the same quality of the soul, safe in their pond like the monk in his cloister, etc. – to greater monastic consumption of fish as food. Indeed, the connection, if any, seems more plausibly reversed: to fish-eating and, to some degree, fish-catching, fish-keeping, and perhaps even fish-rearing monks these creatures were familiar enough to support pharmaceutical applications and didactic allegories. For instances see Zug Tucci, "Il mondo medievale dei pesci," 322–360; Moulinier, "L'abbesse et les poissons," 462–464, and the famous examples of Hildegard's *Physica*, lib. 5 (ed. Hildebrandt and Gloning, vol. 1: 259–285, tr. Throop, 159–175); Bernard of Clairvaux's sermon for the feast of St. Andreas (*PL* 183, 503–509); or the elaborate allegory of fish as Christ and Christ-like martyr propagated by Thomas of Monmouth in his martyrology of William of Norwich (Blurton, "Language of the liturgy," 1056–1058).

[50] Boretius and Krause, eds., *Capitularia Regum Francorum*, vol. 2, #32 (Capitulary de villis, cap. 44), "*quadragesimale*" (Loyn and Percival, tr., *Reign of Charlemagne*, 70).

rationalized as harmless.[51] Abbots hoped to wean perhaps reluctant brethren from clandestine defiance or to celebrate special events in their community. Donors reinforced annual remembrance with an extra but licit dish. By the twelfth century old-line reformed Benedictines (Cluniacs and their English and German counterparts) were defending their piscivorous practices against censorious Cistercians ... whose houses themselves were a century later everywhere accepting analogous gifts of fish.[52] Ordinary laypeople were more likely to grumble about seasonal loss of meat, butter, and eggs than be troubled by the acceptance of fish.

Much later what had become standard western Christian practice was given systematic *ex post* theological grounding by, for instance, Thomas Aquinas.[53] Fish was less 'fleshy' and so thought less to incite sexual passions, a point which Isidore had made some 600 years before.[54] Fish plainly did not copulate, so were thought not to reproduce by seed (a sexual act) in the manner of humans, terrestrial animals, and birds, giving them a 'pure' quality already noted by St. Augustine but later also by dualist heretics. Avoiding the meat of animals was a reminder of the Edenic state of sinlessness and peace between humans and animals. Indeed as God had spared fish in the Flood, they could be thought free of sin. Lacking the fatty mouth-feel of meat or butter, fish fit the penitent attitude of denying luxury.[55] And, finally, fish provided a practical alternative for meat.

51 Zimmermann, *Ordensleben*, 81–87 and 262; Harvey, "Monastic pittances," 216–227. Frantzen, *Food, Eating, and Identity*, 238–258, correctly documents the coolness of Anglo-Saxon monastic reformers towards fish consumption on the part of monks. For twelfth-century coastal princes endowing monastic houses with annuities of thousands of herrings see Bernard and Bruel, eds., *Recueil des chartes*, vol. 1, #4132 and 4143; Coopland, *St. Bertin*, 47; and Bougard and Delmaire, eds., *Cartulaire d'Avesnes*, #33.

52 For Cluniac pittances see Zimmermann, *Ordensleben*, vol. II: 262. Not a decade after the death of Bernard of Clairvaux, in 1157 the Cistercian General Chapter was setting guidelines for the purchase of fish, and by the twelfth century's end it had explicitly allowed monks to eat herring during Advent and Lent, while reserving for its support staff at Citeaux all purchases of trout from the Lake of Lausanne during the season of the annual meeting (Canivez, ed., *Statuta capitulorum*, vol. I, *sub anni* 1157, cap. 44; 1195, cap. 26; and 1199, cap. 9). An almost random survey of chartularies reveals Cistercian fish pittances at Eberbach in Hessen in 1212 (Schneider, "Lebensverhältnsse bei den Zisterziensern," 54), Le Pin in Poitou in 1225 (Brien, "Dévelopement de l'ordre cistercien," 43–44), and Austrian Wilherling in 1237 (Jungwirth, "Fischereigeschichte," 306–312). In later centuries both the Benedictine abbey in Abingdon and the Hospitallers house in Haarlem gave the pittancer charge of the corporate budget for fish (Kirk, ed., *Accounts of the Obedientiars*, 3 and 16–22; Winter, "Low Countries," 201).

53 Thomas, *Summa theologica*, IIa, IIae, q. 147 (Blackfriars ed., vol. 43: 90–117); see also Grumett and Muers, *Theology on the Menu*, 74–85.

54 Isidore, *Etymologiae*, XX: 2, 22.

55 Dam, "Fish for feast and fast," 313.

Indeed fish did. Medieval Christians were never obliged to eat fish (no, not even monks), but for a third and more of the days every year, fish were the best substitute for taboo flesh, and the eating of fish openly demonstrated adhesion to the norms of Christian society. A review of the liturgical calendar shows about 135 'fish days'. This could vary. Edward IV's household managers specified fish service on 168 days in the 1470s, but the contemporary bishop of Speyer observed only 130. At one extreme, the ostentatiously orthodox crusading knights of the Teutonic Order in thirteenth- and fourteenth-century Prussia observed abstinence on about 250 days of the year and the monks of Westminster on 215, at least in their refectory.[56] At another extreme, a poor widow eating barley gruel or pease porridge added no more fish on Friday than she did meat on Thursday. The kitchen office of French king Charles VI (1380–1422) replaced all purchases of meat or poultry during Lent with a long and diverse list of fishes. In 1469 Siena's communal government contracted with that of Perugia for annual shipment of 200,000 pounds of fish from Lake Trasimeno, 120,000 pounds during Lent and 80,000 pounds over the rest of the year.[57] Religious culture thus strongly encouraged medieval Europeans to consume fish in lieu of meat, it did this throughout the year and more so at certain seasons, but ideology itself never compelled anyone to eat fish.[58]

### 2.2.2 *Medieval Concepts of Health and Diet*

Both learned and popular ideas of how food affected well-being likely did motivate some medieval Europeans to adjust their eating habits. Such concepts certainly sustained an extensive dietary literature. Learned medieval physicians, their well-to-do patients, and the chief cooks of at least the greatest of the latter understood physiology and nutrition in terms of the ancient Greco-Roman theory of humours (Figure 2.5), given medical substance especially by the Roman Galen (129–200/216) and his later Arab followers. Humoural theory conceived the physical world as a mixture of the elemental polarities hot–cold, wet–dry,

[56] Myers, ed., *Household of Edward IV*, 108–110 *et passim*; Fouquet, "Wie die kuchenspise sin solle," 19; Willam, "Fischerei des Deutschen Ordens," 75–77; Harvey, *Living and Dying*, 46.

[57] Moirez, "L'approvisionnement de l'Hotel du roi"; Lanconelli, "La pesca nelle acque interne," 14.

[58] Zimmermann, *Ordensleben*, 81–87, points out that not even Cistercians required individuals to eat all the dishes served. Nor did any known rules or practices distinguish between organisms from fresh or salt water with respect to abstinent consumption. Kowaleski, "Seasonality," treats the interplay of natural and cultural seasons in English marine fishing.

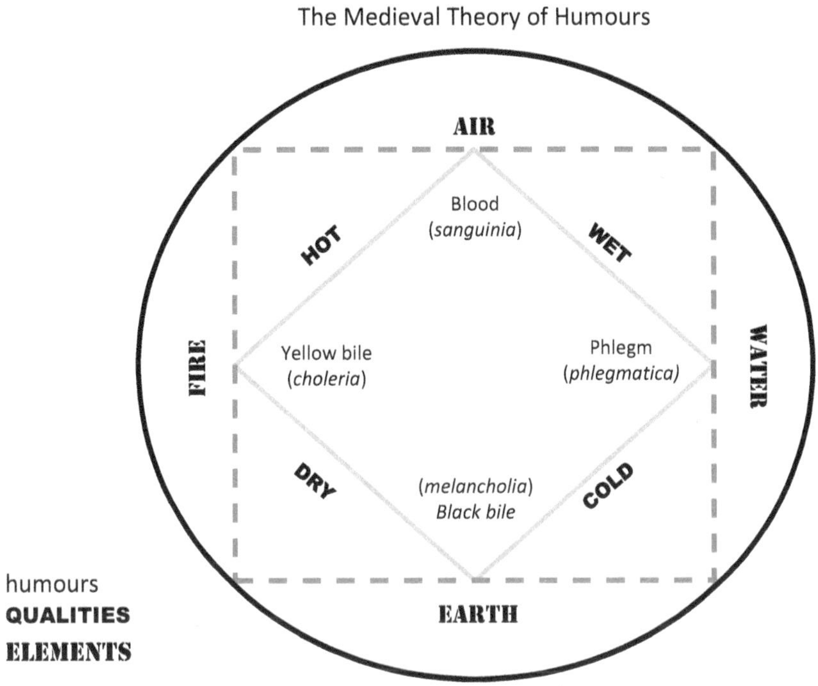


Figure 2.5 The medieval theory of humours.

expressed in the human body as the humours blood, phlegm, black bile, and yellow bile. Personal character and health were determined by the balance of these qualities, while illness, conceived as the imbalance of humours, was to be avoided or cured by readjusting the balance, often by consuming foods of appropriate humoural quality.

Not surprisingly, medico-dietary theory identified fish as cold and wet. These qualities of a food interacted with an individual consumer's basic character (temperament) and momentary medical condition but could be balanced by cooking methods and by serving the food with appropriate sauces.[59] Fish were thus by preference to be fried (cooked hot and dry), and to be followed in the menu by a service of nuts (also hot and

[59] Adamson, *Medieval Dietetics*, 10–18; Scully, *Art of Cookery*, 40–58 and 127–136; Scully, "Tempering medieval food"; Rippmann and Neumeister-Taroni, eds., *Les manguers de l'an 1000*, 114–129; Albala, *Eating Right*, 47–78; Winter, "Fish recipes," 271–274; Woolgar, "Feasting and fasting," 167–172; Rippmann, "Un aliment saine dans un corps sain," 39–64; Schulz, *Essen und Trinken*, 346–350. For Galenic teachings on fish see Powell, ed. and tr., *Galen. On the Properties of Foodstuffs*, book III, §§23–31, pp. 135–145.

dry). The more 'gross' or 'bestial' the flesh of the fish, the more it needed a 'warm' and 'sharp' dressing: relatively 'gross' salmon and trout – fishes with oily, often coloured flesh – demanded pepper and saffron sauce but red mullet or gurnard – white and dry meat – only a simple and mild one. Some Italian physicians simply advised against eating tench, carp, lamprey, and eel as being too 'glutinous', that is, slimy.[60]

Hence dietetic handbooks or sections in a guide to health (*Regimen sanitatis*) were among the earlier texts to draw practical distinctions among fish varieties.[61] A late but notably clear example comes from the cookbook prepared in the 1430s by Eberhard, chef for the duke of Bavaria-Landshut:

> The fish which swim in stony and flowing water and have many scales and are neither too large nor too small and not too fat such as eel and salmon, which are sweet and do not taste bad, they are the best if fresh and not spoiled. But you should know that all green [i.e. fresh] fish are cold and damp and difficult to digest and make for thirst and bad blood and make a bad stomach and much phlegm in the stomach and harm all men who are sick from excessive cold. Still they are useful for people who are warm and dry.
>
> All fish which go in muddy or stagnant waters, they are bad. All salt fish are unhealthy, and one should eat them little. Still, fish which are freshly salted are the best among them.[62]

Eberhard took his ideas and many of his very words from early fourteenth-century physicians Conrad of Eichstätt and Maino de'Maineri, a Milanese who worked in the Low Countries, and from the wise twelfth-century abbess Hildegard of Bingen.[63] But the same concepts of fish, of diet, and of the different needs of people had been equally present in the advice on

[60] Scully, "*Opusculum de saporibus*"; Thorndike, "Medieval sauce-book"; Naso, "Il pesce che nuoce, il pesce che sana," 96–99.

[61] Adamson, *Medieval Dietetics*, covers the genre, but for detailed discussion see Nicoud, *Les régimes de santé au moyen âge*.

[62] Eberhard, fol. 63 in Feyl, ed., "Das Kochbuch des Eberhard" and her dissertation, "Das Kochbuch Meister Eberhards," although both versions omit passages Eberhard borrowed from earlier writers.

[63] *Ibid.*; Weiss-Amer, "Food and drink," "Die 'Physica' Hildegards von Bingen als Quelle für das 'Kochbuch'," and "Role of medieval physicians," all treat Eberhard in his cultural context. Compare Scully, "*Opusculum de saporibus*," *Viandier of Taillevent*, and *Art of Cookery*, 44; Thorndike, "Medieval sauce-book"; Adamson *Medieval Dietetics*, 142–149; and Hagenmeyer, ed., *Regimen Sanitatis Konrads*. A thirteenth-century precursor, written in Old French, is *Le régime du corps*, of Aldobrandin of Siena, who died in Troyes, Champagne, in 1287 (Aldebrandin, *Le régime du corps*, ed. Landouzy and Pépin, 174–177); compare Nicoud, *Les régimes de santé au moyen âge*, II: 987–988). Because fish engender "fatty and viscous humours," Aldobrandin thought them better suited for people of warm, thin complexion than cool and phlegmatic ones. He recommended marine and freshwater varieties over the "less nutritious" migratory ones, and strongly urged selection of fishes from running water (rivers or ocean currents) rather than still bays or ponds, where filth and "gross humours" might accumulate.

eleven fish varieties which Greek physician Anthimus gave his sixth-century royal Frankish patient Theodoric, son of Clovis. The doctor recommended fresh salmon, but after a few days it became bad for the stomach, while salted salmon caused bad humours.[64] This kind of advice eventually went back to Galen himself. Hildegard, however, showed more influence from scriptural ideology when she argued against eating eels on grounds their flesh resembled pork.[65]

Doctor Anthimus further advised that sole, which he would cook in oil with a little salt, was notably good for sick people, and sturgeon for the healthy.[66] Comparable distinctions, if less theoretically integrated than the choices of species, preparation, and sauce, also reverberate through the medieval medico-dietary literature. One of the oldest recipe collections now known to survive, the late thirteenth-century *Tractatus de modo preparandi et condiendi omnia cibaria* ("Tract on means of preparing and seasoning all foods") alludes to but fails to detail the special needs of the sick.[67] While Hildegard had been especially concerned that the ill not consume eel, at the same time she found its gall bladder useful for treating those with failing vision. By the fifteenth century a more popular German-language tradition of "fish doctoring" taught that

> Pike are good for the sick and the healthy; one who eats the fat of the liver makes a healthy stomach. The kidney, roe, and milt of barbel are unhealthy, but the rest of the fish can properly be eaten by people in good health. Carp are good for the healthy but sick people should not eat either the roe or the milt of carp.[68]

Sir Henry Stafford, son of the Duke of Buckingham, was likely ignorant of that German text, but when he suffered a long illness in summer 1501,

[64] Anthimus, *De observatione ciborum* (Liechtenhahn, ed. and tr., 18–20; more modernized in Grant, ed. and tr.., §41 (pp. 66–67). The useful discussion in Adamson, *Medieval Dietetics*, 31–32, incorrectly reads *isoce* as sturgeon.

[65] Weiss-Amer, "Physica Hildegards"; Moulinier, "L'abbesse et les poissons," 467–471; Hildegard, *Physica*, lib. 5:33 (ed. Hildebrandt and Gloning, vol. 1: 283–284; tr. Throop, p. 174). Newman, "Sibyl of the Rhine," provides biographical background and Glaze, "Medical writer," 132–133, an appreciation of the *Physica* as a whole.

[66] Anthimus, *De observatione ciborum* (Liechtenhahn, ed. and tr., 18–20; Grant, tr. and ed., §41, pp. 66–67).

[67] Mulon, ed., "Deux traités inédits," 380; the general treatment in Scully, *Art of Cookery*, 185–195, ignores fish.

[68] "*Von visch artzenie. Der hechte ist gesunt siechen und gesunden der die leber dick ißet die machet einen gesunden magen &c. Des barben hirne und der rogen und die millich sint ungesunt daz ander daz ist gut geßen den gesunden &c. Der karpffe ist des gesunden gut. Die siechen sollent sin nit eßen bede dez rogen und der milch under den karpffen.*" is in Salzburg Universitätsbibliothek Codex M III 3, fol. 356v, a collection of medical advice dated 1439. Closely parallel statements and more also occur in Paris Bibliothèque National Ms. lat. 6952, fol. 238v, a fifteenth-century manuscript from a Speyer apothecary reported by Wickersheimer, "Zur spätmittelalterlichen Fischdiätetik," 415–416 (and worth revisiting in light of what has now been found at Salzburg).

pike were a feature of his therapeutic regime. Contemporary peasants in Upper Austria even complained that private and commercial possession of what had been open fisheries deprived them of fish they needed in times of sickness.[69]

As remarked by the Bavarian chef Eberhard, preserved fish had a bad reputation. Physicians thought them hazardous food for humans. The regimen by Thomas of Wrocław, a mid-fourteenth-century English-born and French-trained clerical physician, strongly advised fresh over brined or salted fish.[70] Thomas thus echoed other medico-dietary writers from the influential school of Arnald of Villanova and his follower Conrad of Eichstätt, whose manuscript treatises circulated widely in central and eastern Europe. Teaching that "truly, in no way are salted fishes for eating" ultimately derived from such influential Arab physicians as the learned eleventh-century Baghdadis, Avicenna (Ibn Sina) and Ibn Butlan.[71] Such articulate medical laymen as sixteenth-century Swedish cleric Olaus Magnus likewise favoured the health and food value, as well as the taste, of air-dried fishes over salted ones, but ranked all preserved forms below fresh fish.[72]

Similar criteria also took seasonal form. Thirteenth-century physician Aldobrandin of Siena thought cold, wet fish more healthful to eat in summer than in winter, but more than 100 years later a Parisian who compiled a housekeeping manual for his young wife drew closer distinctions: round fish from the sea were better in winter and flatfish in the summer, but no marine fish were good in rainy and damp weather.[73] Later advisories often specified varieties best for each month, always with much local diversity. For instance during April: St. Gallen cleric Johannes Schencklin (see Introduction) liked trout, dace, and minnow;[74]

69 Serjeantson and Woolgar, "Fish consumption," 129; Zauner, "Beschwerden der oberösterrreichischen Bauern," 114–115.

70 Thomas, *Higiena wedłe Tomasza*, ed. Burchardt, 80.

71 "*pisces vero saliti nullo modo sunt comedendi*," Hagenmeyer, ed., *Regimen Sanitatis Konrads*, 99–101, 211, and 245, with wider context and examples in Adamson, *Medieval Dietetics*, 142–149, 180–190, *et passim*, and Thorndike, "Three tracts on food," 361–364. Naso, "Il pesce che nuoce," 101–102, provides Italian examples.

72 Magnus, *Historia*, ed. Granlund, lib. 20, cap. 9 (p. 705) regarding pike in particular, and cap. 26–27 (pp. 722–723) on fresh and preserved fishes in general; tr. Fisher and Higgens, 1039 and 1058–1060, respectively.

73 Aldobrandin, *Le régime du corps*, ed. Landouzy and Pépin, 174–177; *Ménagier*, ed. Brereton and Ferrier, 236.

74 Wickersheimer, "Zur spätmittelalterlichen Fischdiätetik," 414–415. The same approach later appeared in the *Fischbouch* which Andreas Gessner printed at Zürich in 1557 under the name of the surprised author Gregor Mangolt, whose authentic holographs lack calendrical interpolations (Mangolt, *Fischbouch*, 13–43, 2d ed. Meyer, 136–145, is compared with ZBZ Mss A 83, fols. 211r–214r, and S 425, fols. 197r–208r, in Hoffmann, *Fishers' Craft*, 56–58).

a sixteenth-century English report on "the Seson and Seu(er)alte of all ffysses in tymes that they ben fresshe" recommended lampern and lamprey;[75] and variant texts along the upper Rhine advised salmon (adult and parr!), dace, trout, burbot, and "The gudgeons are good in February and March and April until May. But the young gudgeons are good with parsley all the time."[76]

For a time early Renaissance physicians and dieticians reverted to a strict reading of Galen himself. Extreme aversion to fish and dairy foods is reported to start with Marsilio Ficino's *De vita triplici* of 1489 and remained fashionable into the mid-1500s.[77] Advocates mainly repeated medieval worries about cold and watery humours, sliminess, and 'cold' pathologies of phlegm and respiratory inflammation, and again stressed the importance of water quality as a criterion for selecting fish to eat. But the medical consensus remained that food preparation would correct the harmful properties. Reformation debates over abstinence resulted in Protestant regimes banning fish days as a Catholic practice, but some later restored them to encourage maritime employment or mitigate meat shortages. Diet was but one consideration.

Plainly medieval ideas about food and about health so intertwined that recipes and advice are often ill-served by modern distinctions between the culinary and the medicinal.[78] To untangle them here is moot. Nutritional assessments were certainly not uniform but the variations of time, place, and cultural background remain to be mapped out. More important is the presence of learned theory – and surely folk beliefs, too – to guide medieval Europeans in deciding which fishes were preferred food and, if natural constraints as well as access to socio-economic power allowed, in choosing what fishes they would eat.

## 2.3 Social Display

Much medieval social behaviour revolved around public display of wealth and rank. Costume and rituals demonstrated superiority, subordination, and membership in groups. Men at arms accompanied their

[75] BL Additional manuscript 25238, fol. 56r–v.

[76] "*Die grundelen sind gut im hornung mertzen vnd apprillen vntz meyen. Aber dye iungen grundelen sind alzeyt gut mit peterlin.*" Hoffmann, *Fishers' Craft*, 94–95, where the various redactions of "The Seasons" are discussed 30–35 and 50–52. English king Richard III did enjoy gudgeons with parsley at a dinner on 5 July 1483 to celebrate new knights made during his coronation (Sutton and Hammond, eds., *Coronation of Richard III*, 203).

[77] Albala, *Eating Right*, 95–96, 102–103, and 122; Naso, "Il pesce che nuoce," 99–108; Woolgar, "Feasting and fasting," 195.

[78] A point worth repeating from Weiss-Amer, "Die 'Physica' Hildegards," 95–96.

Figure 2.6 French nobles dine in luxury on fish.
An elite fish banquet as depicted in an eleventh/twelfth-century French manuscript.
Redrawn by D. Bilak for R. Hoffmann from a reproduction in Alfred Gottschalk, *Histoire de l'alimentation* …, vol. 1, p. 341 (Paris: Editions Hippocrate, 1948), who provides no further reference than "Festin princier sous les premiers Capétiens (d'après une miniature de l'époque)."

lord in his livery, and the mounted king in rich attire received the keys to the city gates from municipal councillors who stood together in their robes of office. Feasting, the conspicuous and festive consumption of food, was equally part of these social relations, important to the high and the low alike.[79] Here the eating of fishes served distinct and symbolic roles. As in other times and places, the desire to be seen consuming what social scientists call 'positional goods' drove demand from the rich, powerful, and aspiring.[80]

### 2.3.1 *Fish to Mark High Status and Honour*

Several sorts of evidence and incidents suggest medieval awareness that fish on the table, indeed even certain kinds or qualities of fish, were a social marker (Figure 2.6). Much-prized sturgeon are already familiar. Early medieval Italian elites, whether warriors at a castle on Lake Como or the

[79] Bynum, *Holy Feast and Holy Fast*, 60; Grieco, "Food and social classes"; Müller, "Food, hierarchy and class conflict"; Freedman, "Lamprey and herring," 196–197; and Mänd, *Urban Carnival*; with further ample anecdotes in Henisch, *Fast and Feast*.

[80] See Simmons, *Global Environmental History*, 196–197.

archbishop of Ravenna, reserved for themselves the largest species (trout, pike, and eel) or just the biggest of the day's catch from waters they controlled. Fish eating was an acknowledged mark of elite lay and monastic status among late Anglo-Saxons. But Christianity was no prerequisite: notably large or rare specimens of mammals and fishes were offered for sacrifice and consumption at the tenth-century pagan Slavic ritual complex at Arkona on the Baltic island of Rügen.[81] Eleventh-century monk Ulrich of Cluny, who compiled that house's sign language dictionary, interpreted the sign for salmon or sturgeon (made with the general sign for fish, an undulating hand, followed by a fist with an erect thumb on the chin) as "signifying pride, since the very proud and rich are accustomed to have such fish."[82] Indeed, great prelates feasting on salmon and pike became a stock image for high medieval commentators.[83]

Lay magnates took no second place, with numerous references to sumptuous winter banquets where pike and sturgeon were served. Material corroboration might come from Boves castle in Picardy, where tenth- to thirteenth-century kitchen wastes contain both those varieties as well as eel, perch, and some trout or salmon.[84] In that very cultural terrain romance writer Chretien de Troyes (*c.* 1135–1183) imagined the court of legendary King Arthur, unsurpassed for its chivalry, at a Saturday dinner of pike, perch, salmon, and trout. Probably just a few years after the literary event, the young William Marshal, gaining celebrity status as a champion on the tournament tour of northern France, earned for his prowess on one field a trophy "pike of more than two and a half feet," which earlier that day great princes had exchanged as a token

81 Squatriti, *Water and Society*, 103; Fleming, "The new wealth," 4–8; Müller, "Zoological and historical interpretation of bones," 190.

82 "… *quo superbia significatur, quia superbi maxime et divites pisces solent habere*" (Jarecki, ed., *Signa loquendi*, 123). For the monastic practice see Bruce, *Silence and Sign Language*, *passim*, and the passage from Ulrich p. 177; and compare Zimmermann, *Ordensleben*, 81–87.

83 Thirteenth-century reformist cleric Jacques de Vitry (*Exempla*, ed Greven, 26) claimed that Judgement Day would find St. Benedict's 'true sons', the Cistercians, full of beans and vegetables, but the stomachs of traditional Benedictines stuffed with the "*delicata cibaria*" of pike and salmon, which two species Alan of Lille had named a century before as favourites of clerical gourmands (Alanus, *De planctu*, ed. Häring, tr. Sheridan, 173–174). Thompson, "The case of the salmon" is a satiric Latin poem anthologized before 1200 about a salmon put on trial for bursting the stomach of a gluttonous abbot. The considerable basis for these jibes is sustained by consumption records of, for instance, thirteenth-century bishops of Winchester (Roberts, "Bishop of Winchester's fish ponds," 125–130) and fourteenth-century popes resident at Avignon (Richard, "Transports par eau"; Grava, "Notes martégales"; and Weiss, *Versorgung des päpstlichen Hofes*, notably 197–202 and *passim*).

84 Clavel and Vorenger, "Quelques données sur la pêches," 61–66. Recall as well the early medieval sturgeon remains mentioned in the Introduction.

of esteem and courtesy.[85] As a later German proverb put it, "*Wildbret und Fisch gehren auf der Herren Tisch*" ("Game and fish belong on the lord's table").[86] From a trophic perspective in the words of Flemish archaeozooologist Anton Ervynck, castles held "the top-predators of the feudal system."[87]

By around 1300, however, the sturgeon and other big freshwater varieties once distinctively eaten at the count's castle in Namur had been replaced by carp, cod, haddock, whiting, plaice, and flounder, while ordinary folk in the town below were eating herring and various small freshwater fishes. This pattern predominated in late medieval northern France as well, where seigneurial establishments enjoyed fresh coastal fishes and several kinds of cods but lesser urban households just the occasional herring. In late medieval and sixteenth-century Holland rich people thought carp the best among fish from fresh water and stocked up on haddock and cod for their Lenten dining.[88] In northern Italy, however, Lenten weddings at Piacenza followed a different festive custom keyed around fish. First on the menu came drinks and sweets, then figs and almonds, followed by a 'large' fish in pepper sauce. Next a spiced rice and almond milk soup with cured eel, and then fried pike in vinegar

85 "*Ce fu un samedi a nuit qu'il mangièrent poissons et fruit, luz et perches, saumons et truites, et puis poires crües et cuites.*" *Erec et Enid*, ll. 4237–4240 (Chretien, *Les romans*, vol 1; tr. Comfort, pp. 55–56; as noticed in LeGoff, "Vestimentary and alimentary codes," 146). William's much later, though still lively, memory set his incident with "*un luz … de plus de deus piez e demi*," in 1177 in Champagne (*L'Histoire*, ll. 2875–3164, ed. Mayer, vol. I, pp. 111–115; *History*, ed. Holden, 154–155). Much-traveled thirteenth-century Italian chronicler Salimbene took pains to comment on the unusual value placed on pike in France (cited in Montanari, *L'alimentazione contadini*, 293).

Noble Alsatian Herrad of Landsberg, abbess of Hohenbourg (1130–1195) depicted as luxurious fish banquets in her heavily illustrated encyclopedia for nuns, the *Hortus deliciarum*, at least four biblical scenes: Queen Esther entertaining the Persian king, parables of the Great Supper and of Lazarus and Dives, and Solomon's Feast (ed. Green, vol. II, fols. 60v, 119v, 123r, and 204v).

86 Schreiner, "Zisterziensisches Mönchtum," 109. The sentiment is amplified in thirteenth-century German literature like *Meier Helmbrecht* (ed. Bell, ll. 464, 780, and 1606) and echoed by fifteenth-century poet Heinrich Wittenwiler in his comic epic *Ring*: "*Visch … es ist ein Herrenspeis*" (Wittenwiler, *Der Ring*, ll. 2905–2906). Wiessner, *Kommentar*, 119, gives a long list of parallel literary references, and Schulz, *Essen und Trinken*, 143–145 and 251–255, more elaborated discussion. Dyer, "Consumption of fresh-water fish," 28–31; Woolgar, "Fast and feast," 7–25, and "Feasting and fasting," 171–172, together confirm similar English and more general practice. Freedman, "Lamprey and herring," 209–211, describes upper-class gourmandizing with fish in mainly Mediterranean Europe.

87 Ervynck, "Medieval castles," 151.

88 Pigière et al., "Status as reflected in food refuse," 233–243; Clavel and Yvinec, "L'archéozoologie du Moyen Âge," 82–85; Dam, "Feestvissen en vastenvissen," 483–489, slightly revised as "Fish for feast and fast."

or mustard sauce, washed down with mulled wine. Nuts and fruits finished the repast.[89]

Gifts of fish were meant to honour the recipient. Back in the sixth century Avitus of Vienne commonly exchanged fishy presents with his fellow Gallo-Roman landowners.[90] English King Henry III (1216–72) presented to bishops, barons, and courtiers the live pike and bream from his ponds at York and other manors.[91] When Charles VI visited Amiens in March 1386, the town council offered their king thirty-eight big carp, forty pike, and forty eels.[92] Soon thereafter the city of Kraków and its suburban municipalities were habitually giving distinctive fishes, sturgeon, pike, and salmon, to win the favour of visiting dignitaries, high officials, and their king Władysław IV Jagiełło. So likewise did Tortosa, which offered its celebrated sturgeon, shad, and lamprey from the Ebro.[93] The ambitious merchant of Prato, Francesco Datini, and his astute wife Margherita sent prized sturgeon, tench, and other fishes to friends and business or political associates.[94] A highlight at the banquet the Duke of Milan gave for his daughter Bianca's wedding to Emperor Maximilian was the salad of pike – Milanese salads were famous – surrounded with effigies in marzipan of members of both courts.[95]

Several recent interpretive perspectives bear on the way medieval elites so eagerly and publicly devoured fish. One view emphasizes the consequent

[89] Montanari, *Culture of Food*, 71–72, quotes the local chronicler Giovanni de Mussis.

[90] Avitus of Vienne, Epistolae 72 and 83 (*Letters and Selected Prose*, tr. Shanzer and Wood, pp. 250 and 333). A generation earlier another aristocratic Gallo-Roman bishop did likewise: see Sidonius Apollinaris, *Sidonius Poems and Letters* (ed. and tr. Anderson), e.g. Poem XXI. And a millennium later the Farnese of Renaissance Rome used gifts of eel to sustain ties among family and friends (Luiten, "Friends and family, fruit and fish," 353–356).

[91] McDonnell, *Inland Fisheries*, 12 and 20; Steane, "Royal fishponds," 45–46 and 50; Serjeantson and Woolgar, "Fish consumption," 126. Dyer, "Consumption of freshwater fish," 33–34, has more English examples.

[92] Cloquier and Clavel, "Consommation d'animiaux aquatiques," 84.

[93] Piekosiński and Szujski ed. *Najstarsze księgi*, 2:225–345; Chmiel, ed., *Księgi radziecki Kazimierskie*, 69, 460, and 535; and generally for Poland, Dembinska, *Food and Drink in Medieval Poland*, ed. Weaver, 105–110, and Cios, *Ryby w życiu Polaków*, 32–35. Curto Homedes, "El consum de peix a la Tortosa," uses only some of the Tortosa records given in Pastor y Lluis, "La pesca de la saboga." Mid-thirteenth-century Bologna sent various fishes to the pope, the papal governor of Romagna, and other rulers (Pini, "Pesce, pescivendoli e mercanti," 335), while popes themselves exchanged fish with their guests at Avignon (Weiss, *Versorgung des päpstlichen Hofes*, 261 and 393).

[94] Datini, *Letters to Francesco*, 56, 99, 115.

[95] Toussaint-Samat, "Gastronomie et fastes culinaires," 113. In the same Italo-German borderland but at the other end of the nobility, the minor Tirolian lord Peter Schlandersberg bought fish at Christmas, 1369, "with which to do honour to all his good friends" (*damit er etlichen seinen guten friunden eren mit tet*; Ottenthal ed., "ältesten Rechnungsbücher," 587).

diversion of elite consumption away from the meat of domestic mammals and the product of arable lands, especially as supplies shrank towards the end of winter.[96] Other students stress how the variety of species and preparations encouraged display of the unusual, expensive, and, therefore, almost self-signifying. For Count Robert of Artois (1250–1302) and his daughter and successor, Mahaut, serving diverse fishes from their artificial ponds was but one way they used the park Robert had built at Hesdin to show off their wealth, rank, and power.[97] Count and countess would surely have admired the party piece chef Eberhard of Landshut later described: an entire pike, still whole, prepared in three different ways, poaching the head and middle in two different sauces while simultaneously roasting the tail. This corresponds with the mid-fourteenth-century culinary invention of *entremets,* elaborated dishes such as *gelantines de carpe* or mock sturgeon from veal, which served as diversions between regular courses.[98]

### 2.3.2 *Scales of Value*

Distinctions between 'ordinary' fishes and those for luxury consumption may be reflected in some of the price differentials which began to be recorded. Late twelfth-century authorities in Venice priced sturgeon, trout, turbot, and pike at three denarii per pound, another dozen varieties at two, and tench and "other fresh- and saltwater fish" at one. The king of Castile thought trout worth 25–50 percent more than barbel of the same size, and officials for the count of Flanders in 1187 valued a plaice the same as ten herring.[99]

Comparable ratios persist in later records. We already observed big fresh local fishes (pike, carp, bream, tench) during the Council of Constance at a premium of 25–30 percent over whitefish; imported preserved herring or cod cost double. In estate accounts from inland Staffordshire in 1461, pike, bream, and tench were all worth more than dried cod, and eel three times as much as plaice. Those prices were for

[96] Grant, "Food, status and religion," 143–145, and "Animal resources," 175; Pleij, *Dreaming of Cockaigne,* 100–101.

[97] Farmer, "Aristocratic power and the 'natural' landscape", 659–662; Abigail Dowling further develops this perspective in "Landscape of luxury," 367–387.

[98] Eberhard's recipe, Feyl ed., "Kochbuch des Eberhard," recipe #17; Plouvier, "Le gastronomia," 154. More generally, Hundsbichler, "Nahrung," 229; Jaritz, "Zur Sachkultur," 153–157; and Scully, *Art of Cookery,* 104–109.

[99] Papadopoli, *Le Moneta,* 307–311; Faugeron, *Nourrir la ville,* 31–33 and 311–317, provides valuable context for the *Lex annona* of 1173 despite dubious identification of some fish species. Powers, ed., *Code of Cuenca,* 215, translates Ureña y Smenjaud, ed., *Fuero de Cuenca,* 816–819. Delatouche, "Importance relative des divers produits," 30.

live fish; dead ones went at a heavy discount.[100] Big urban markets around the shores of the Tyrrhenian Sea rated salt tuna, eel, and tuna roe the most valued fish products, which also got top price from consumers on the Atlantic side of Spain in Cordoba. But the latter still offered a large premium for fresh inland fishes over marine varieties, while preserved forms of either were much cheaper. In contrast, to the north in Navarre salmon and dried conger both normally went for twice the price per pound of fresh conger or hake, while herring had an intermediate value.[101] Regional differences are plain, yet all local cultures placed special value on certain fishes, most often fresh specimens of large varieties. (Further in Supplement 2.3.2.)

Some preferences were plainly matters of taste and social fashion. Two deserve special mention. For one, many late medieval elites despised herring, especially in its most common preserved form, ordinarily the least expensive way to obtain culturally approved fish protein. Bishop Matthias von Rammung of Speyer left no doubt in his 1470 kitchen ordinance: "Herring shall not be considered a fish dish, [for] one can in no way make them acceptable,"[102] and his contemporaries at Westminster Abbey strongly endorsed that view, refusing all but the best fresh-smoked red herrings at the peak of their season.[103] As agreed by commentators from Alan of Lille in the 1160s to Burgundian ducal counsellor Philippe de Mezieres in 1389 and beyond, herrings were a

100 Richental, *Konzilschronik*, facsimile ed. Feger, fol. 25b (translated in Loomis, *Council of Constance*, 101, although I doubt Loomis's handling of some fish names); Stolz, *Geschichte der Gewässer*, 377; Dyer, "Consumption of fresh-water fish," 31. French princely households about 1600 took 80% off the contract price for fresh fish delivered dead (Couperie, "Les marchés de pourvoirie").

101 Bresc, "Pêche et coraillage,"108–109; Hernandez Iñigo, "La pesca fluvial," 1089–1106; Hamilton, *Money, Prices, and Wages*, 227–260. The kitchen of King Carlos III of Navarre favoured salmon but otherwise fresh marine fishes, notably hake, over preserved or freshwater varieties (Serrano Larráyoz, *La mesa del rey*, 100–106). At inland Madrid conger then went for 17–20 maravedis per pound and sardine were the cheapest fish at 6–7 mvd (Puñal-Fernandez, *El Mercado en Madrid*, 201–202). Larger market demand and the risk of spoilage may explain those instances where preserved product was valued above fresh.

Discussion of fish prices here has been confined to relative values attributed at the same place and time to different fish varieties. Comparison of fish in general to other foodstuffs appears at pp. 86–87 below, while short- and long-term fluctuations are studied further in Chapters 4, 6, and 8. For general discussion of sources and use of prices as evidence see online Supplement 0.2.2.

102 "*Heringe sollen nit fur vische geachtet werden, man mochte es dan nit mol gebessern*" (Fouquet, "Wie die kuchenspise sein solle," 27).

103 Harvey, *Living and Dying*, 48. Monks of Westminster ate less herring per capita around 1500 than they had around 1300.

dish for the poor, not for kings and nobles.[104] In this context medieval poverty could be voluntarily ascetic – as reflected in the herring served with exclusive regularity at eleventh century Benedictine Fleury – or truly material – as dispensed to poor Parisians by the king's almoner or purchased singly in 1440s Coventry for one-sixth the price of an eel.[105] French literary imagery came to portray herring in a firmly negative light, using the phrase 'not worth a pickled herring' to describe an object of no value.[106] Finds by French and English zooarchaeologists clearly reflect this disdain in long-term trends of less herring at elite sites and increased association with middling urban and, later still, poor rural households.[107] Deeper inland, however, the more exotic import retained some of its cachet.[108]

While herring fell from fashion, at least some medieval elites actively developed a taste for carp. As early as 1300 Count Robert of Artois was presenting carp as honorific gifts. A century or so later the household of William II (1365–1417), duke in Bavaria and count in Holland, Zeeland, and Hainaut, ate during one thirty-nine-week period (including Lent) 10,757 carp, 542 pike, and a few dozen each of several other species.[109] Cluniac monks at La Charité sur Loire replaced their eclectic fourteenth-century diet of native local fishes with almost exclusive reliance on carp by the late fifteenth. Since 1439 Cistercians at Altenberg in the Rhineland enjoyed a special annual memorial feast which began with

[104] Alanus, *De planctu*, ed. Häring, tr. Sheridan, 173–174; Philippe, *Songe du viel pelern*, tr. Coopland, 129–130. Also on herring as a mark of poverty see Dam, "Feestvissen en vastenvissen," 489–491, and Dam, "Fish for feast and fast," 327–328. Like social prejudice applied in medieval Byzantium against the salt fish (here mackerel and sardines) eaten by poor people (Dagron, "Poissons, pêcheurs, poissoniers," 67–68). Early fourteenth-century Humbert II of Viennoise simply refused all preserved fish (Laurioux, "Table et hierarchie sociale," 104–105).

[105] Novices at Fleury learned with their sign language that herring were the only fish they would henceforth see (Jarecki, *Signa loquendi*, 252). Bautier and Maillard, "Les aumones du roi," 41, 43, and 57; Dyer, "Consumption," 31; Freedman, "Lamprey and herring," 202–206.

[106] Rassart-Eeckhout, "Pratiques alimentaires," 157. For the negative view of poet Eustache Deschamps (1346–1406), see Uytven, *De Zinnelijke Middeleeuwen*, 187.

[107] Clavel, *L'Animal*, 176–187; Clavel and Cloquier, "Contribution des sources documentaires et archéologiques," 208; Clavel and Yvinec, "L'archéozoologie du Moyen Âge," 82; Locker, *Role of Stored Fish*, 97, 150–153, and 167; Locker, "Decline in the consumption," 100–101; Hardy et al., *Ælfric's Abbey*, 395–396.

[108] Ervynck et al., "Beyond affluence," 431–432; Rehazek and Nussbaumer, "Fish remains from a 16th century noble household." The question remains, however, what share, if any, of herring at elite sites was consumed by the head table and what a ration for servants (see below).

[109] Farmer, "Aristocratic power and the 'natural' landscape"; Uytven, *De Zinnelijke Middeleeuwen*, 186–187. See also the fish-filled meals Louis II, viscount of Thouars, gave for Cesar Borgia during the latter's visit in 1498 (Favreau, "Fêtes et jeux," 33–34).

herring and stockfish, but continued to a dish of peppered carp, and then to a main course of roast carp and salmon.[110] Town notables at Tours in the 1480s gave their policemen rations of salt herring and the occasional dried hake, but themselves dined on fresh carp, pike, and lamprey. A roast carp of three to seven pounds gave mid-sixteenth-century Italian physician and autobiographer Girolamo Cardano his favourite dish.[111] The notion that carp tasted 'mossy' or 'muddy,' though hinted by Hildegard of Bingen and grumbled by some monks at Tegernsee – whose alternative was whitefish from the Alpine lake at their doorstep – gained no widespread assent in France or Germany until quite late in the eighteenth century.[112]

In sum, then, serving and eating the freshest, largest, and most elaborately prepared fish publicly asserted high social standing. For a long time a special cachet attached to sturgeon, salmon, and pike, that is, to large species from clear and cold waters, two of them diadromous in their habits.[113] Where the seacoast was accessible, big fresh cod or conger might have comparable significance. Preserved fish, also the object of physicians' warnings, had distinct associations with poverty, whether real or culturally assumed, as by some ascetics.

## 2.4 A Stratified Structure of Demand

Plainly medieval Europeans in general were prepared to eat fish as an alternative to meat, and were encouraged by religious prescriptions of abstinence to do so. Their choices reflected medico-dietary and folkloric[114] preferences, plus the social prestige derived from consumption of fish itself, when fresh, and of certain varieties. But by the later Middle

[110] Audoin-Rouzeau, *Ossements animaux du Moyen-Age*, 146–147; Bérard, "La consommation du poisson," 174–176; Mosler, ed., *Urkundenbuch der Abtei Altenberg*, II, #126. Jaritz, "Zur Sachkultur," 152, reports like developments at Klosterneuburg.

[111] Chevalier, "Alimentation et niveau de vie," 145; Cardan, *Book of My Life*, 28–31.

[112] Bérard, "La consommation du poisson," 176–178; Cnopf, *Entwicklung der Teichwirtschaft*, 140–148; Abad, *Conjuration contre les carpes*. Compare Hildegard, *Physica*, Lib. 5:11 (ed. Hildebrandt and Gloning, vol. 1: 283; tr. Throop, p. 168); and Kisslinger, *Chronik der Pfarrei Egern*, 96.

[113] Is it also significant that all three varieties were native to all major river systems in northern Europe and only salmon was missing from large Mediterranean rivers as far south as the Tiber? I am further inclined here to group with the salmon, as medievals often did, the big lake-dwelling trout (*Salmo trutta lacustris*, etc.) and perhaps huchen, which also ascend streams to spawn.

[114] For folk beliefs that eating fish promoted human fertility and affected pregnancy see Jeay, ed., *Evangiles des Quenouilles*, 84 and 87; Jeřábek, "K studiu rybářství," 285–291; and Rösener, *Bauern im Mittelalter*, 191. Earlier, Burchard of Worms and some other writers of penitentials thought women used fish in quasi-magical ways to secure the faithfulness of their husbands (Filotas, *Pagan Survivals*, 297–298).

Ages and probably earlier fish were an expensive food. Few people in the socio-economic hierarchy of medieval Europe could afford to eat favoured fishes with regularity. Many needed lower-priced fare.[115]

### 2.4.1 *Costly Food at Any Level*

For much of the Middle Ages direct expression of the high cost of fish took the form of incidental complaints or acknowledgements. Tenth-century Anglo-Saxon monks worried over excessive luxury and their counterparts at St. Gallen griped about the high cost of fish from their own lake fishery; in the early twelfth century Abelard thought fish so dear he advised Heloise to serve them to the Paraclete convent only three times a week.[116] Recognition by the *Tractatus* cookbook of the late 1200s that fish belonged among "food for the rich" was echoed in numerous especially German literary sources.[117] But inland Poles reported in surprise from early twelfth-century Pomerania that "you can buy a whole cartload of herrings for a penny."[118] So how did that compare to pork, beans, or porridge?

It is long difficult to move beyond anecdotal records or comparisons, but eventually both price-fixing legislation and prices actually paid establish the consistent pattern that, relative to other foodstuffs, fish were a high priced commodity on the market. The best evidence allows comparing prices by weight or what a given sum could buy at a specific time and place. Venetian market regulations in 1173 set fish between 1 and 3 denarii (d) a pound while a pound of prime beef was 2d and standard beef just over 1d.[119] A few human generations later in the 1260s at Rouen, Normandy, one salmon commanded half the price of a pig and much later in the 1410s, twice the price of a pig.[120]

Late medieval examples abound. While fourteenth-century popes lived at Avignon, a pound of fresh eel at 12d cost a labourer's daily wage and three times the price of a pound of meat; the 6–8d asked for a pound of

[115] Galloway, "Fishing the Thames estuary," 265–270, remarks on the multi-layered demand for fish in late medieval London. Singer, "Use of fish remains" and "Threshold of affordability," correlates household income and the prices of fish excavated at North American historic sites.

[116] Frantzen, *Food, Eating, and Identity*, 238–245; Duft, *Bodensee in Sankt-Galler Handschriften*, 21–23 and 90–91, believes Ekkehard's tale (*Casus Sancti Galli*, '105, ed. Haefele, 212–213); Zimmermann, *Ordensleben*, 61–64.

[117] Mulon, "Deux traités inédits," 369–370.

[118] *"Carratamque pro denario recentis acciperes allecis."* Leciejewicz, "Z denara otrzymasz," 103–104, referred to Herbord's *Life* of Otto of Bamberg (*Dialogus*, lib. 2, cap. 41, ed. Liman and Wikarjak, p. 141).

[119] Faugeron, *Nourrir la ville*, 311–317 (original text in Papadopoli, *Le Moneta*, 307–311).

[120] Halard, "La pêche du saumon," 176–177.

salt eel could instead buy a dozen eggs or six loaves of bread.[121] By that same century's end in Kraków, seat of the Polish monarch, a cask (roughly 272 liters) of herring or eel at 100 local groszy (gr) was worth more than an ox (60gr) and 40 groszy could get the buyer either a salmon or a pig.[122] During the council at Constance 14 local pfennig bought a pound of preserved herring, and less of fresh carp, tench, or pike, but as much as 3.5 lb. of pork, 4.7 lb. of beef, or sixteen loaves of 'good white bread.' Meanwhile the median price in England for a fresh pickerel (small pike) or a salted cod, 8 pence, would otherwise buy thirty-two loaves of bread or 8 gallons of ale.[123] And by the early 1500s in Old Castile, fresh fish in the range of 6.9–9.1 maravedis (mvd) the pound came to half again and more the price of mutton at 4.5–5.7 mvd.[124] (See more in Supplement 2.4.1.)

Fish did not generally provide cheap calories in medieval Europe. Poor people with empty stomachs were unlikely to buy fish to fill them. When peasants, the urban poor, or even landowning families and institutions did obtain fish, they may not have used the market at all. Indeed the mechanisms which came to determine prices of fish will need study below both for their organizational forms and for their evolution from the predominantly non-market arrangements which served earlier medieval consumers of fish.

Medieval European demand for fish modulated general human metabolic needs according to characteristic medieval cultural representations and programs. Demand for a meat substitute was ubiquitous, with seasonal features arising from its ideological basis and with some selectivity regarding fish varieties. People in the earlier Middle Ages ate fishes more or less indiscriminately, though favouring the large pike, salmon, and sturgeon which lived in or migrated through fresh and estuarine waters. Some regions began around 1000 CE to alter their eating habits, so that a portion of the population eventually obtained from fish a significant share of their nutrition. By the thirteenth century effective demand, which is demand backed up by purchasing or coercive power, had clearly stratified: the wealthy wanted costly fresh fish in great variety and commonly from fresh water; the poor turned to less expensive salted or dried fishes. Fish on the market, however, were no cheap source of calories and only exceptionally of animal protein.

121 Hocquet, "Pêcheries médiévales," 53–54.

122 Data from Pelc, *Ceny w Krakowie*, tables 22 and 65–66, as summarized in Carter, *Trade and Urban Development*, 133 and 137.

123 Richental, *Konzilschronik*, facsimile ed. Feger, fol. 25b (translated in Loomis, *Council of Constance*, 101; Dyer, "Consumption of fresh-water fish," 31.

124 Hamilton, *American Treasure*, appendix III, pp. 319–334.

### 2.4.2 *Fishing for Subsistence, Sale, or Play?*

In pursuit of culturally defined purposes, medieval Europeans intentionally intervened in aquatic ecosystems, diverting through the work of fishers some portion of the natural flows of energy and materials. Fishing as an economic activity is ultimately driven by demand for fish. Efforts to meet the predominant patterns of effective demand refract through socio-economic institutions to distinguish what modern fisheries managers and historical vocabulary alike conceive as three organizational types of fisheries: subsistence, commercial, and recreational.[125] The labels identify different relations among consumer, catch, and fisher. In a subsistence fishery most of the catch is meant for consumption by the fisher's household and/or the household of the fisher's superior. It is useful in studying medieval subsistence fisheries to distinguish further between fishing for the fisher's own household consumption, here called direct subsistence, and fishing done by servants or subjects for use by a master or employer, indirect subsistence.[126] A commercial fishery involves catching fish for market sale to the eventual consumer, whether as the household enterprise of an artisan or at larger capitalist or industrial scales. Consumer and fisher may deal directly and personally or be separated by a long chain of intermediaries. A recreational fishery may also involve consuming the catch, but its purpose is more to satisfy a need for leisure pastime than for fish to eat. Recreational fishing in medieval Europe lacked evident economic or environmental significance and thus further relevance to this discussion.[127]

Real life blurs the tidy classification when servile fishers work for a lord who makes a practice of selling much of their catch or when those who control fishing rights claim for their own tables the "best" from what the fishers carry to market. Some form of subsistence fishing was, however, the overwhelmingly predominant response to most European demand for fish in the early Middle Ages and to a still considerable part of that demand for much longer. That human work to make aquatic organisms satisfy human wants is the topic of the next chapter.

125 Brandt, *Fish Catching Methods*, 2–4; McCullough, *Commercial Fishery*, 7–8; compare Everhart and Youngs, *Principles of Fisheries Science*, 21–29.

126 The distinction, but not the exact wording, is adapted from Slicher van Bath's classic *Agrarian History*, 23–25.

127 Hoffmann, "Fishing for sport," might now be augmented but it suffices to demonstrate the practice and features of medieval recreational fishing. For more refined consideration of relations between capture techniques and sport see Hoffmann, *Fishers' Craft*, 191–214 and 328–349, and Hoffmann, "Trout and fly, work and play."

# 3 Take and Eat

## Subsistence Fishing in and beyond the Early Middle Ages

As peasants once told the tale along Baltic shores and in Black Sea watersheds, God created the fishes intending to provide them the protection of many legs, defensive weapons, or a habitat high in the trees like birds. The foolish fishes refused, trusting, said some, in their slippery speed, and, said others, in the watery depths.[1] God conceded but at a price: If the men catch me alive, so also shall they scrape my scales off alive, cut me up alive, and cook me.[2] Popular wisdom running back plausibly, if unprovably, into the Middle Ages grasped an essential old relationship between humans and fish, capture of local populations for prompt consumption as food.

This chapter first establishes that the fishes early medieval Europeans ate came from their own neighbouring waters. The same would remain true for at least some of their descendants. The peasant majority themselves fished to supplement cereal diets with whatever aquatic foods they had the time, opportunity, and ability to catch. They valued their access to common resources. Sometimes ordinary peasants and, more importantly, a few select individuals almost everywhere also laboured to set fish before their lords, wealthy men and women of rank and power in the area. Regardless whether the actual fish-catchers or their social superiors topped the local food web, from the beginning medieval subsistence fisheries relied on small-scale, often passive, capture techniques well suited to limited and seasonally accessible local resources. Technical expertise derived from the long experience of ordinary people in their local environment. Simple means of preservation extended the use people made of recurring surpluses. Social and technical features of

[1] Birds and fish were created on the same day. Learned medieval authors observed that fish were quick and slippery like the water where they lived (Glass, "*In principio*"; Zahlten, *Creatio Mundi*, 187–191). Peasants concurred.

[2] Dähnhardt ed., *Natursagen*, 3:178–179, provides four variants from Romania and what was then "West Prussia." Another tale from the Upper Palatinate said fish were butchered alive because they had played merrily in the water while Christ hung on the cross (*ibid.*, 2:226).

subsistence fisheries tended to stabilize their effect on participant communities, human and natural.

## 3.1 Local Supply

Medieval Europeans exploited the fish populations of all nearby waters, no matter how small. Indeed, fish from local resources predominated through much of the Middle Ages and remained significant past the period's end. Essential to understanding of medieval fisheries is the sheer ubiquity of this activity and, especially in the early Middle Ages, of its economic role.

Surfacing all over the written record of charters, cartularies, and estate surveys, for all its fragmentary and unrepresentative quality, are words like *piscatio*, *piscatoria*, *piscatura*, or *piscaria*, terms of legal art and regional but rarely wider precision for "fishing right," "fishing place," "fisheries installation," or the financial returns from the same.[3] Other texts just mention weirs, fishers, or renders of fish. Early medieval Europeans had no speedy overland transport and few ways to preserve fast-spoiling fish or to acquire them in trade. Most people lived away from the sea and so had to rely mainly on inland, rather than even coastal marine resources. When the most elaborate preindustrial shipping facilities, commercial infrastructures, and strong government support could during the thirteenth–eighteenth centuries just barely get fresh fish in an edible state the 150 kilometers from the Norman coast to Paris, medieval Europe's richest concentration of consumers (see below, Chapter 4, pp. ***–***), that distance can be thought an effective limit for any regular movement of fresh fish (Map 3.1). For most of Europe, then, regular fish eating would mean using local freshwater or diadromous varieties.

[3] In a "technical index" Perrin, *Seigneurie rurale*, attempted precise and abstract legal definitions, though these lacked clear consistency even in his texts from ninth- to twelfth-century Lorraine and plainly failed to encompass the usages in, for instance, the English Domesday survey of 1086 (Darby, *Domesday England*, 279–286). Squatriti, *Water and Society*, 104–105, and Lampen, *Fischerei und Fischhandel*, 29–32, encountered comparable ambiguities. In common medieval use such words derived from *pisc-* (and vernacular equivalents) normally marked an activity and not the type of water in which it was practiced. On the other hand the Latin term *piscina*, nominally 'fishpond', could, as thirteenth-century scholar Bartholomeus Anglicus acutely remarked, denote at times "gathering of waters without fish, … contrary to meaning, as Isidore says" (Trevisa tr., *On the Properties of Things*, ed. Seymour, 1:661–662). Legal and technical particulars are explored below and in future chapters; the point here is the widespread presence of what were recognized as fisheries resources.

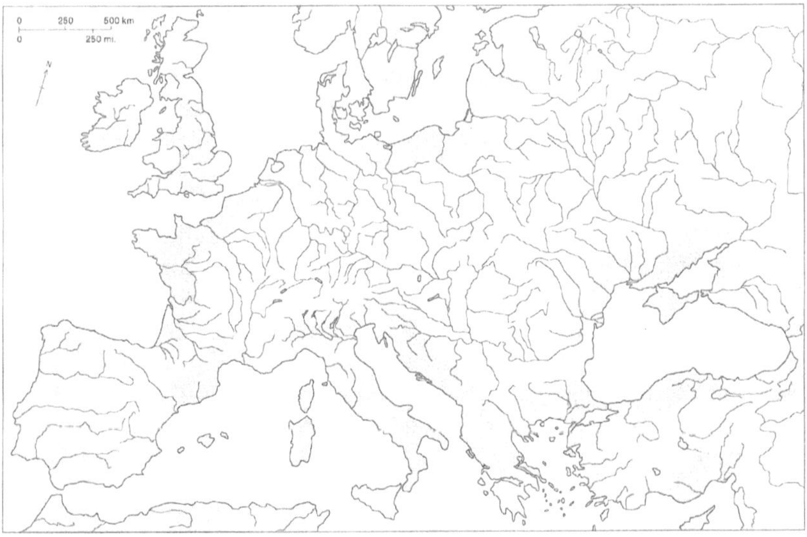


Map 3.1 A 150-kilometer range for delivery of fresh marine fish.

The result is manifest in specific remains recovered from human food waste and in verbal records of the fishes which identifiable human groups ate or meant to eat. At sixth- through eleventh-century socio-natural sites across Europe people of all ranks consumed animals naturally present in waters near them and not exotic varieties from other ecosystems.

There are collective overviews. Meta-analysis of published results from sixty-five medieval Baltic sites and forty-five on the North Sea identified local fish fauna as the principal feature of sixth- through twelfth-century assemblages.[4] Similar review of finds from a score of eighth/ninth- through thirteenth-century settlements in present-day Austria and Hungary reveals ubiquitous dominance of four taxa, sturgeons, catfish, carp, and pike, all common to still and running waters of the Danube system from the Alps and Carpathians to the Iron Gates. Among the several thousand identified medieval fish remains in this composite sample, a single herring bone was the sole trace of a fish from elsewhere.[5] An archaeological survey of early

[4] Enghoff, "Fishing in the Baltic," and "Fishing in the southern North Sea." Lõugas, "Fishing and fish trade," emphasizes entirely native fresh and brackish-water species throughout Viking and early medieval times at even coastal sites from Sweden to Estonia.

[5] Galik et al., "Austrian and Hungarian Danube." The gross pattern is as true of the mainly sieved Austrian sites as of the unsieved ones from Hungary, but closer examination notes the migratory beluga sturgeon upstream into Austrian waters, while the resident sterlet was being consumed only below the Danube's inland delta. Austrian sites, moreover,

medieval coastal communities along both sides of the southern North Sea and English Channel found them full of locally caught inshore varieties, notably small gadids, flatfishes, eel, and herring.[6]

Findings from particular sites varied with ecological circumstances, as early medieval people used their own regionally distinctive aquatic communities. Select examples show this irrespective of date, watersheds, and consumers' status (more cases are in Supplement 3.1).

Near the centre of Merovingian Frankland, the founding generation of eighth-century aristocratic nuns at Saint Irminen convent in Trier left in their garbage almost a thousand identifiable fish bones from eight taxa, just over a half of them cyprinids, one in five catfish, and one in twenty each pike or salmonids. Small numbers of perch, eel, and shad confirm the whole assemblage, freshwater and diadromous alike, came from the Moselle. Marine species were not to be found.[7]

At ninth-century San Vincenzo al Volturno, equidistant between Rome, Naples, the Adriatic and the Tyrrhenian seas, Benedictine monks did consume mullets, sea bass, and some sea bream from Adriatic lagoons, yet 78 percent of the fish bone in their trash was tench and other cyprinids from the limited fresh waters of central Italy.[8] The whole reflected a general post-Roman shift of Italian fish consumption from marine to inland and inshore varieties.[9]

Eighth- to tenth-century high-status Anglo-Saxons resident at Flixborough in the floodplain of the river Trent about eight kilometers inland from the Humber estuary may have shifted between secular and monastic foodways, but little varied the fishes they ate. The twenty-eight taxa identified from this wet-sieved site were dominated by only seven, all freshwater or diadromous: eel, smelt, flounder/plaice, pike, small

contain remains of cold- and/or faster-water fishes such as salmonids, perch, and rheophilic cyprinids which are absent from the waters of the Pannonian plain. The same exclusively local consumption pattern prevailed further down the Danube at the southern frontier of the medieval Hungarian kingdom, an area not included in the composite study (Bartosiewicz, "Pontes").

6 Loveluck, *Northwest Europe*, 186, 193, 198–200, 202, 207, 211, 250–252, and works there cited. Holmes, *Animals in Saxon and Scandinavian England*, 47–52, saw only local freshwater and inshore species at all types of early and middle Saxon sites.

7 Heinrich, "Fischresten aus St. Irminen."

8 Marazzi and Carannante, "Dal mare ai monti."

9 Montanari, *Culture of Food*, 28–29; Donati, "Dal mare al fiume," 26–27; and Squatriti, *Water and Society*, 102–105, concur that an early medieval (sixth–tenth-century) preference for freshwater fishes differed from Roman times, with the latter offering several cultural as well as security reasons for the shift. Relying more on recent archaeozoological evidence Salvadori, "La pesca nel Medioevo," 300–303, and Varano et al., "The edge of the Empire," agree.

cyprinids, and salmonids were all readily accessible in nearby waters. Even the incidental species conform to the norm of local use.[10]

Food remains from late Saxon (tenth–eleventh-century) London show an evolving but similar pattern: of twenty-eight taxa identified in more than four thousand bones from a half-dozen separate sites, nine were freshwater, five diadromous, and fourteen marine organisms, notably herring, but most of the bones came from eel and all varieties frequented the Thames river, estuary, and inshore coastal waters. The rise in English marine fish consumption around 1000, dubbed by zooarchaeologist James Barrett the "Fish Event Horizon,"[11] there meant mainly an increase in herring, followed by a slow rise in local cod from the southern North Sea.[12]

Research on the Viking Age (eighth- to mid-eleventh-century) entrepôt Haithabu/Hedeby, on the Baltic estuary of the Schlei but with easy access to the North Sea, recovered and identified 13,842 fish bones. These included twenty-six varieties, but 39 percent were herring, 25 percent perch, and 11 percent each pike and native cyprinids (roach, rudd, tench, bream). Only a few thinly represented taxa were not then available in and near the Schlei, being brought, perhaps by visitors, from the North Sea.[13]

Like dwellers beside Baltic waters those who lived scores of kilometers inland at sites now well excavated with rich early medieval fish remains consumed the occasional herring but mainly relied on stocks native to their own estuaries, rivers, and lakes. Sieved contexts from the eighth to tenth centuries at the princely stronghold of Ostrów Lednicki midway between Gniezno and Poznań contain 30–57 percent bones of sturgeon,

[10] Dobney et al., *Flixborough*, 36–58 and 199–212. Sykes, *Norman Conquest*, 56–58, finds early–mid Anglo-Saxons generally to have eaten a lot of eel, salmon, pike, and small flat fishes, with major consumption of herring postdating the mid-ninthth century.

[11] Barrett et al, "'Dark Age economics' revisited." Sykes, *Norman Conquest*, 56–58, points out that rise in English consumption of marine fish initially accompanied an increase in freshwater varieties as well, trends which neither began nor changed with arrival of the Normans.

[12] Locker, "Middle Saxon" and "Peabody Site," 150: "These results confirm a general picture for the Middle Saxon period, both at this site and others in the *Lundenwic* settlement, that the fish consumed were largely the result of local fishing in the Thames, with some exploitation of inshore marine waters, but no deep-sea fishing of any sort." Orton et al., "Fish for the city"; and Orton et al., "Fish for London," 207, further refine the mass of data from London.

[13] Lepiksaar and Heinrich, *Fischresten aus Haithabu*, 119–120. For similar proportions in later finds from the harbour, see Schmölcke and Heinrich, "Tierknochen aus dem Hafen," 220–233.

followed by tench, pike, and cyprinids from local waters.[14] While estuarine and river sites in this region favoured sturgeon and pike-perch, people on lakes from interior Wielkopolska west to the Grosse Ploner See in east Holstein ate more perch, pike, and catfish, species of lentic habitats.[15]

Returning to the post-Carolingian west, the oldest fish remains recovered from the Scheldt basin in Flanders are from emerging urban ninth/tenth-century Ghent (review Fig. 2.2). More than 90 percent are species taken in fresh water and the few marine species primarily coastal flatfishes. During the ensuing two centuries the share of flatfishes more than doubled and that of herring – also then taken along Flemish beaches – rose to 20, then 30 percent. Up to 1200, other marine varieties totalled less than 10 percent. Meanwhile fish from fresh water dwindled proportionately, though still contributing more than four remains in ten.[16]

About the year 1000 a community of what modern researchers have called 'peasant knights' settled on a shoreline shelf beside Lake Paladru in the Isère valley some way north of Grenoble. Two generations later rising waters inundated the site, driving the settlers to higher ground and leaving their farmsteads to be studied by modern submarine archaeologists. Besides fishing gear (see Figure 3.1) they recovered 421 identifiable fish bones and more than 8,000 scales. Ninety-five percent of the bones were perch and just under 4 percent still-water cyprinids, with bare traces of pike and a salmonid. Ninety percent of the scales, however, were cyprinid, suggesting different taphonomies and thus original handling of the two principal taxa. All species taken and consumed by the settlers still inhabit the lake.[17]

This brief tour might end with the written record of late eleventh-century Cluniac monks in Burgundy and the Black Forest. Ulrich of Cluny's initial 1079 dictionary of their sign language named six taxa, five of them native to the Rhône and upper Loire (some fifty kilometers away); the only exotic was cuttlefish. A few years later William of Hirsau, who carried Cluny's reforms to his community east of the Rhine, reworked Ulrich's list and raised the number of fishes to fifteen:

[14] Makowiecki, *Hodowla oraz użytkowanie*, 40–41. Similar finds at contemporary through mid-twelfth-century Szczecin are summarized in Rulewicz, *Rybołówstwo Gdańska*, 72–88.

[15] Requate, "Jagdtier in den Nahrungssytem"; Kaj, "Szczątki rybne," 74–75; Iwaszkiewicz, "Szczątki ryb," 306. General discussion in Makowiecki, "Early medieval aquatic environments in Poland."

[16] Neer and Ervynck, "Rise of sea fish," 159–161.

[17] Colardelle and Verdel, eds., *Les habitants du Lac*, 116–120, and their more popular *Chevaliers-Paysans*, 38–40. Fish remains were analyzed by Jean Desse and Natalie Desse-Berset. Note the trophic pyramid from cyprinids to carnivorous perch capped by predatory trout and pike.

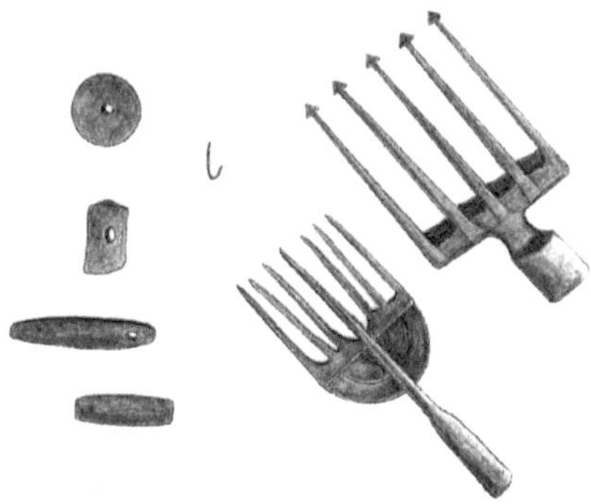

Figure 3.1 Eleventh-century fishing equipment from Lac Paladru. Gear recovered from the settlement of 'peasant knights' included (left column) floats (top) and weights (bottom) for nets; a bronze hook, *c.* 3 cm (centre), and two iron fish spears (right). Selections from a photograph in Colardelle and Verdel, *Les Habitants*, 320, of a museum display at the site. Redrawn for R. Hoffmann by D. Bilak.

now thirteen were migrants or residents in the upper Rhine watershed and six of the additions cyprinids from there; the only non-natives were Ulrich's original squid and herring, conceived as salted.[18]

Commonality of diversity has been the point of the past few pages. For the first half and more of the medieval millennium peoples of Latin Christendom ate a great variety of fishes, but in each location from the more limited biota of their own nearby natural aquatic ecosystems. Fresh, brackish, or salt, still or flowing, warm or cold habitats made no detectable difference to consumers. Rich people and poor alike ate the familiar fishes of neighbouring waters. Well into the high Middle Ages residents in even large urban centers still relied on these local sources of supply. In most localities that meant native freshwater varieties and elsewhere those available as residents or migrants from estuaries and nearshore marine waters. How were these food resources procured? For a long time mainly by efforts of consumers or their servants.

## 3.2 Direct Subsistence Fishing

Consumers can cover their own demand without a market. Economic actors assess their wants and direct the natural resources, land, and

[18] Ulrich, "Consuetudines Cluniacensis," lib. 2, c. 4, in Jarecki, *Signa loquendi*, 122–124 (translated in Bruce, *Silence and Sign*, 177–178); William's *Constitutiones Hirsaugienses*, lib. 1, c. 8, in Jarecki, *Signa loquendi*, 165–168.

labour at their command to satisfying them. In that case the catching and the eating went to the same economic account. Such self-sufficiency nevertheless still involves implicit choices allocating scarce productive factors to one priority or another. People of unequal wealth (access to productive factors) also have different patterns of effective demand, for those who more easily obtain basic food and shelter can turn surplus energies at their command to other kinds of wants. Rich or poor, fully self-sufficient consumers must be generalist producers to obtain the variety of goods and services they need.

Early medieval society may usefully be thought to comprise two socio-economic orders, the powerful and the poor. Direct subsistence is an affair of the latter, people whose economic capabilities depended on the working capacity of simple family households. Most of them we can call peasants, for they lived from their own mixed farming and in some kind of subordination to people of power who, with or without some cover of legality, regularly seized a share of the peasants' productive capacity. Direct subsistence fishing was one part of a survival strategy for people trying to meet basic needs from the relatively meager human and material resources they controlled. Later in the Middle Ages rural- or town-dwellers with other occupations but access to aquatic resources also occasionally covered their own demand for fish. Like most activities of ordinary folk in early medieval society, direct subsistence fishing rarely attracted the eyes or pens of literate groups. As a result, while early sources establish the prevalence of subsistence fishing, only later texts convey more of its technical particulars and organizational context.

### *3.2.1 Fishing "for Their Own Table"*

People plainly catching their own food and eating their own catch nonetheless appear surprisingly often in the sparse and cryptic early medieval record. Incidental references in the seventh century *Life* of sixth-century St. Columba depict a peasant farmer who used a spear and Irish ascetics who operated nets and traps to take salmon to eat. Gall's fishing in Lake Constance (see Introduction) followed that template.[19] So, independently, did his younger contemporary Sigiramnus (St. Cyran, Siran, d.655), who fished to feed his monasteries in the river-laced "wasteland of Brenne" (*in saltu Brionae*). When monks complained that Sigiramnus was too able a fisher, "for one became nauseous from eating them every day," the saint departed to live as a solitary and so feed himself and those

[19] Adomnan, *Life of Columba*, ed. Anderson, 364–366, 413–415, and 534; Krusch, ed., "Vita Galli," 289–294 and 365–367 (Joynt tr., 72–81 and 103–105).

who took him in.[20] On the other side of the Alps early tenth-century tenants on at least two estates of St. Giulia, Brescia, paid dues from both farming and fishing; the operator of a layman's property near Ferrara could keep one-fifth of his take in fish and game.[21] Perhaps those holdings resembled what excavators found at the Thames-side village of Wraysbury, Berkshire, where most farmsteads held fish bones, notably of eel but also resident freshwater species plainly taken from the river. Far to the east at Drohiczyn on the Bug, peasant houses, dated like those at Wraysbury to the ninth–twelfth centuries, also held scattered remains of locally procured fish and, one, an iron fish hook.[22] By that time fishing was also recognizably a seasonal secondary occupation of farmers and pastoralists in Bohemia, Norway, Iceland, and coastal lagoons of Languedoc and Tuscany.[23]

More forthcoming later records often treat peasant fishing for household use as a normal activity. This was certainly the case when the administratively precocious Teutonic Order of crusading knights established control in once-pagan Prussia. The charter they issued for Chełmno in 1233 became the prototype for the rights of their subjects: each household could catch its own fish by any means except the great seine net called *niewod*. After generations had followed those provisions, Grand Master Conrad von Jungingen specified in 1406 that the right meant fishing with "small gear" such as hand-held nets, baited hooks, and basket traps, and even with weirs and fish-fences which did not fully block streams, "but not otherwise than for their table."[24] Catching the very essence of direct subsistence, the phrase "for their table" in a 1342 customal had also described how established peasant tenants could

[20] "... *acsi piscator peritus de amne ad litus extraens pisces, non solum ad eorum supplendam, necessitudinem caperet, verum ad nausicam cotidie comedentibus usque deferret*" (Krusch, ed. "Vita Sigiramni," cap. 15 and 19–22). Benarrous, *Grande Brenne*, 318–329 and 338–342, deploys written, charcoal, and palynological evidence to show early medieval Brenne was covered by woods and meadows drained by many streams but lacked its later famous wetlands and ponds. Monks at eighth-century Stavelot fished the Meuse for food (*Miracula S. Remacli*, 3:23–24 in *Acta sanctorum*, Sept., I, p. 701).

[21] Castagnetti ed., *Inventari altomedievali*, 60–61 and 64–65.

[22] Coy, "Fish bones"; Iwaszkiewicz, "Szczątki ryb w Drohiczynie."

[23] Graus, *Dějiny venkovského lidu*, vol. 1:97–98; Ebel, "Fischerei und Fischereimethoden," and works there cited; Perdikaris and McGovern, "Codfish and kings," 199–203; Bourin-Derrau et al., "Littoral languedocien," 382–387; Garzella, "*In silva Tumuli* e *in Stagno*," 145–147.

[24] "*kleyne gheczow, ... yo nicht anders wen cz irme tissche*," Benecke, "Beiträge zur Geschichte," 307, with selected earlier grants on 301–305 and later at 308. Willam, "Fischerei des Deutschen Ordens," 78–85 and 95, and Kisch, *Fischereirecht*, 160–168.

fish at Alrewas, Staffordshire.[25] Back in the mid-eleventh century, a man of the abbey of Marmoutier in Berry passed the ordeal of hot iron to vindicate abbey subjects' right freely to fish for their own use on waters claimed by the seigneur of Château-Renault; only fish taken for sale need pay a seigneurial tax.[26] Holders of tenements beside the lake of Monte Sorbo, Lazio, could fish if they turned half the catch over to their lord. An Alsatian rental from about 1160 allowed only local villagers, subjects of Bouzonville abbey, to fish at Offwiler and Obermodern.[27]

Peasants fishing legitimately for domestic consumption remain visible in fifteenth- and sixteenth-century records from Iberia to the Carpathians.[28] In the Vercors, Haut Dauphiné, a 1449 charter confirmed the free fishing which inhabitants of four villages had long enjoyed: those seeking their own food were unregulated but those who sold fish had to pay for an annual licence.[29] Along the meandering rivers of the late medieval Hungarian plain, fishing was a peasant by-occupation. As water levels dropped each summer, village communities joined to block the *fok*, narrow channels cut between oxbow lakes and the main river, and while releasing small fish to grow, held the large ones for local consumption.[30] The Portuguese *Cortes* recognized river fisheries as open commons, which in practice meant nominal dues to the crown. Subsistence fishing there was associated with the cooler seasons and fish taken by angling were exempt from tithe.[31] All these common folk plainly had access to fisheries resources.

### 3.2.2 *Mutual Regulation and Local Ecological Knowledge*

To a modern viewer medieval fishing for direct subsistence is often obscure. It neither much engaged relations between lords – whose own dealings created records of property rights and political obligation – nor passed through markets where traders and tax-collectors could count it. It was rather an object of long-unwritten custom which in fact if less often in law recognized commoners' access to fishing waters for the needs of their households. In the unlettered medieval countryside the old and

[25] Dyer, *Standards of Living*, 157. Thirteenth-century tenants on Eden and Derwent estuaries had common rights to fish (Winchester, *Landscape and Society*, 111).

[26] Querrien, "Pêche et consommation … en Berry," 432–433.

[27] Vendittelli, "Diritti ed impianti di pesca," 427–428; Perrin, *Seigneurie rurale*, 728–729.

[28] LeRoy Ladurie, *Peasants of Languedoc*, 18; Ambros, "Zvieracie zvyšky"; Beck, *Eaux en Bourgogne*, 232–233.

[29] Sclafert, *Le Haut-Dauphiné*, 145–146.

[30] Mákkai, "Economic landscapes," 28–31; Zatykó, "Fishing in medieval Hungary," 402–404; Rácz, *Steppe to Europe*, 56–58 and 185.

[31] da Cruz Coelho, "A pesca fluvial," 81–84 and 89–95.

customary called for no documentation.[32] Later urban and territorial administrations were more apt to call attention to these uses if only, as in Piedmonte, by exempting subsistence fishers from the licences and regulations they imposed on commercial fishing. Fifteenth-century Venetian authorities publicly proclaimed the fishery of Lake Garda "to be common to all."[33] Statutes of the Perugia fishers' guild from 1296 had simply conceded that locals caught fish which did not enter the market.[34]

When people sensed limits to the resource and wanted to protect it, common access to direct subsistence fisheries in no way prevented mutual regulation. The council of newly formed Villa de Mar (now Saintes-Marie-de-la Mer) on the Rhone estuary's Petit Camargue in 1307 consulted the citizens before allowing conditional use of a new type of net in the salt ponds, marshes, and channels under town control. In 1341 they agreed with the archbishop of Arles to close all fishing during the spring spawning run.[35] Lake Garda's fishery for lake trout and cyprinids was closely regulated "to establish the common benefit and the multiple abundance of the aforesaid fishes."[36]

Through ubiquity and persistence the direct subsistence fishery shaped and transmitted dietary, technical, and environmental knowledge in peasant society. Irish archaeologist Aidan O'Sullivan has made a strong case that the families who built and rebuilt, sometimes since even long before the Middle Ages, the hundreds of riverine and estuarine fish weirs still detectable around the British Isles "passed down lore of place and practice through the generations."[37] Fish spears collected from country people in early twentieth-century Pomerania go back in design and markings to early medieval prototypes. In England, Spain, and the German-speaking Alps the oldest known instructions for making and using fishing baits and basket traps come directly from the medieval vernacular.[38] French rural culture drew a plain distinction between indigenous freshwater

32 Birrell, "Common rights"; Cahn, *Recht der Binnenfischerei*, 61–70; Montanari, *L'alimentazione contadini*, 280–282.

33 "*ut piscatio in dicto lacu esset comunis omnibus*," Butturini, "La pesca sur lago di Garda," 147–149; Nada Patrone, *Il cibo*, 320–324.

34 Scialoja, *Statuta et ordinamento*, 863.

35 Amargier, "Pêche en Petite Camargue," 334–336 and 338.

36 "*ad comune beneficium et copiam piscium praedictorum multiplicem conditum fuisse*," Butturini, "La pesca sur lago di Garda," 147–149. On Hungarian peasant practices see: p. 98 above.

37 O'Sullivan, "Place, memory and identity," 463. See also *ibid.*, 451 and 461, with emphasis on transmission of learning gained through work in the natural world. For labour on such weirs see pp. 104–105 below.

38 Nowakowski, "Rybackie narzędzia kolne"; Hoffmann, *Fishers' Craft*, and "Haslinger Breviary fishing tract."

organisms, known and used by generations of peasants, and the cultivated fish of intensively managed artificial ponds.[39]

Chance survivals from scraps of oral folklore reflect dense empirical familiarity with local ecosystems. Those who handled many pike knew the complex markings on the creature's head: legends from France and Brabant, Swabia and Austria, Elbe Slavs and Prussians associated the shapes with the instruments of Christ's Passion; Latvians saw in them the fisher's own tackle.[40] Coast dwellers from Pomerania to Flanders and Scotland shared tales explaining the twisted mouth of flatfish (plaice, sole, flounder) from its sneering complaint after losing a swimming race to determine the king of the fishes. In the nearly fresh eastern Baltic the pike was the envied winner and elsewhere the more marine herring.[41]

Even long before the folklore collectors, some popular knowledge of fish anatomy, behaviour, and seasonal life cycles had at the very end of the Middle Ages already precipitated from oral vernaculars into surviving written texts. While Chapter 2 observed popular views of dietary value, late medieval vernaculars also marked behavioural and physiological features of fish species. One folk compendium from the upper Rhine valley connected food preference to spawning seasons: "Bream and nose are good in February and March [just before they spawn] and at their best when the willows are dripping wet." Another likened the predatory pike to a robber, living by what he grabs; the nose is a scribe carrying his own ink [a black swim bladder]. Collectively both texts accurately catalog riverine fish communities in the upper Rhine and not elsewhere.[42] An independent English tract gives spawning times and occasional biological comments for twenty-six freshwater and inshore marine organisms then living in southern Britain.[43] None of that came from learned or scientific texts. Written records were for people without experience.

Ordinary people possessed unwritten traditional ecological knowledge shaped by generations of catching and eating their own fish from nearby

[39] Bertrand, "Pour une histoire écologique," 66–68. But why have we recovered no instructional texts from France?

[40] Dähnhardt, ed., *Natursagen*, 2:226–227 and 3:491.

[41] *Ibid.*, 4:192–196. The herring's royal rank was in places contested by the ribbonfish or giant oarfish (*Regaleus glene*), a 3–8 m monster with a red crest on its head that occasionally appears among the plankton-eating schools (Heinrich, "Information about fish," 18–19, and Jagow, "Hering im Volksglaube," 220–221).

[42] Hoffmann, *Fishers' Craft*, 32–33 and 92–97, treats and reproduces the versions printed in 1493 but connects them to older and more recent analogs handled in Schultze, "Ein mittelaterlicher Fischkenner," and "Ein Strassburger Handschrift"; Wickersheimer, "Zur spätmittelalterlichen Fischdiätetik," 412–414; and Gessner, *Historia animalium*, vol. 4, pp. 26, 594, 705–706, and 1206.

[43] BL MS Additional 25238, fol. 56r–v.

aquatic environments. For most Europeans in the early Middle Ages and later still especially for the rural poor, those were the only fish they ate. Direct subsistence fishing is the historical background and sociocultural context for all other fisheries in medieval Europe. Cryptic early medieval signs of its presence are fleshed out from later records of practice.

### 3.2.3 *Defending Fisheries Commons*

But there was a catch. In an ironic twist, the genuine "tragedy" of the medieval commons was piecemeal loss to private interests of common access to many fisheries.[44] Like other customals written down in the later Middle Ages for many south German villages, the one from Kitzbühel in Tirol confined peasant fishing to but a few days a year and to angling or a hand net. At Stams they were forbidden any use of the local brook.[45] Chapter 6 below returns to the context, motives, and environmental ramifications of such privatization. Salient now is peasant response, which should signal their concern for the fishing. In fact, European peasants long and bitterly resisted what they saw as lords' usurpation of their own proper use of local waters.

Collective and violent action against landlord control over fisheries reverberates across medieval centuries. Popular resistance to the early medieval trend of Italian inland fisheries to exclusive private rights is especially visible in the good Carolingian records from around 800. Subjects of Duke Guinichis of Spoleto assaulted the *piscaria* of Farfa abbey, "and they tore the nets of the monastery and took the fish and beat its men," while a generation later groups of fishers near Reggio Emilia and Piacenza went unsuccessfully to court against claims by Nonantula and San Fiorenzo abbeys, in both cases failing for lack of documentary proof. Royal officials could themselves suppress common rights, as men of Istria complained of their own duke's behaviour "in the truly public sea, where all the people fished commonly, [so] now we scarcely dare to fish, for they [duke's men] beat us with sticks and cut our nets."[46] The source closest to a revolt of Norman peasants *c.* 1000, William of

[44] Not, of course, a problem peculiar to fisheries commons (see Hoffmann, *Environmental History*, 247–263).

[45] Stolz, *Geschichtskunde der Gewässer*, 366; Lindner, *Deutsche Jagdtraktate*, 146. Cahn, *Recht der Binnenfischerei*, 61–70, clarifies the important distinction between access (use) and legal ownership of fishing rights.

[46] Detailed discussion and documentation in Squatriti, *Water and Society*, 111–113.

Jumièges, writing about 1070, emphasized their demands for traditional access to waters and woodland.[47]

Among the richest expressions of this sentiment came at the close of the Middle Ages in events now understood as presaging the German Peasants' War. An agitator at Niklashausen in Franconia, where the Tauber enters the Main, called in 1476 for common possession of all fish and game,[48] and restive villagers in 1502 at Untergrombach, a Neckar valley lordship of the bishop of Speyer, would revoke all private rights over woods and waters "so that each countryman could hunt and fish, where or when he wished, without interference or prohibition at any time or place whatsoever."[49] A programmatic statement the peasant army at Memmingen issued in March 1525 argued that to stop commoners taking fish from running waters violated the Creator's grant to all mankind of power over beasts on the land, fowl of the air, and fish in the water. Christian charity would let everyone fish.[50] By late April ideas had become deeds. To rebel against the lord of Schleiz villagers from Liebengrün and Lubeschitz, Thuringia, smashed the equipment of his private fishers and themselves waded into the river Saale to catch and eat a salmon. The whole community of Neustadt on the Orla, a Saale tributary, twice paraded with pipes and drums to go fishing and returned with the booty for a public feast.[51]

Local reactions best show how loss of access hit home and those could occur far from the famous general revolts. Altzelle abbey received fishing rights and lordship over the Mulde at Roßwein from the Markgraf of Meissen in 1293 and refused access to members of the community. This the people opposed for ninety years before a court decision let them fish from the town fields every Monday and Friday morning and any time the river flooded.[52] Not all were so patient. On Sunday 25 February 1408, a few days before the start of Lent, a half-dozen named men from the isle of Mersea, Essex, invaded the fishery called Seward's flats ("Sywardesflete")

[47] Van Houts, ed., *"Gesta Normannorum ducum,"* vol. II, pp. 8–9 (William lib. 5, cap. 2). Dating and authorship are in vol. I, pp. 3–7 and 82–94. Further discussed in Arnoux, "Classe agricole," 45–58, and Gowen, "996 and all that."

[48] Franz, ed., *Quellen*, 66. The authoritative study is Heimpel, "Fischerei und Bauernkrieg."

[49] "*ut cuique rustico liceret venari atque piscari, ubi et quando voluerit, sine impedimento vel prohibitione cuiuscunque omni tempore et loco*" (Franz, ed., *Quellen*, 76). Complaints by Styrian rebels in 1515 remained pragmatic: landowners' tighter control over fisheries had harmed the well-being of tenants and commoners (Benecke, *Maximilian*, 76).

[50] Franz, ed., *Quellen*, 177; Heimpel, "Fischerei und Bauernkrieg," 354. Like attitudes in Upper Austria are in Zauner, "Die Beschwerden,"114–115.

[51] Heimpel, "Fischerei und Bauernkrieg," 367–368.

[52] Beyer, *Cistercienser-Stift Alt-Zelle*, 25, 418–420, and 569.

of Bartholomew Bourgcher, knight, and there took fish and forty quarters of oysters; before the court they asserted this was in tidal waters and thus "from times beyond memory … all lieges of the lord king had common fishery there."[53] In 1490 rebellious subjects of the abbey of St. Gallen – more than a half-millennium removed from that original hermit's hut – ripped out the weirs and traps with which their lords obtained the fish prohibited to them.[54]

The long history of overt struggle against loss of access itself confirms that medieval peasants continued to value fishing as a way to supplement their usual foodstuffs. The same conclusion may be drawn from the many widely but still rarely documented incidents of illegal individual use of local fisheries resources, which is to say poaching of fish. A lawyer's model dated from the late thirteenth or early fourteenth century of pleas for an English manorial court envisioned a poor peasant swearing

> how I went the other evening along the bank of this pond and looked at the fish which were playing in the water, so beautiful and so bright, and for the great desire I had for a tench I laid me down on the bank and just with my hands quite simply, and without any other device, I caught that tench and carried it off; and now I will tell thee the cause of my covetousness and my desire. My dear wife had lain abed a right full month, as my neighbours who are here know, and she could never eat or drink anything to her liking, and for the great desire she had to eat a tench I went to the bank of the pond to take just one tench; and that never other fish from the pond did I take.[55]

Premeditation or sudden inspiration? An experienced steward might have some doubts, for tench are dark-coloured bottom dwellers with notably slimy skin, hard to spot or to grab. Thirteenth-century abbots with monopoly fishing rights on the then Gulf of Arras, in an oxbow of the Marne near Meaux, and to L. Bientina beside Lucca all struggled with local poachers.[56] Big visible fish like pike or spawning salmonids could be speared or netted: a Prussian peasant caught spearing pike in the Teutonic Knights' waters in 1453 sparked a political crisis when his

[53] London TNA Plea Rolls (KB 27/588 (Pasche 9H4) m 44 Essex (with thanks to Stuart Jenks and Suzanne Jenks).

[54] Müller, ed., *Rechtsquellen*, 272–273.

[55] Maitland and Baildon, eds., *Court Baron*, 54–55. An earlier item from the same text (pp. 37–38) suggests this defendant already had a record for poaching fish. *Ibid.*, 75, has another model precedent from a tract written at Oxford in the 1270s/80s, and actual cases from a Cambridgeshire court of the bishop of Ely dating from 1316 to 1318 (note peak famine time!) are on pp. 122–124. Bennett, *Life on the English Manor*, 270, cites more cases elsewhere.

[56] Coopland, *Abbey of St. Bertin*, 59; Endrès, "Un vivier naturel"; Onori, *L'abbazia di San Salvatore*, appendix p. 14 (a pledge exacted from three illegal fishers).

landlord broke him out of jail.[57] In some present-day situations pragmatic poaching is part of a continuum, with destructive exploitation and even attacks on the resource itself following local alienation due to loss of subsistence access.[58] Might some unusually articulate set of late medieval judicial and financial records reveal the same response from dispossessed peasants?

## 3.3 Indirect Subsistence Fishing

Medieval elites did not work. Work was for common people. The power of demand joined to the power of command over resources and men produced indirect subsistence fishing. Fish on the tables of the powerful came by effort of the poor, whether they worked as occasional forced labourers, as part-time servile specialists, or as full-time estate servants. The predominant means of satisfying elite wants in the early Middle Ages, concerted fishing by subordinates of great lay and clerical households, persisted in some circumstances into much later times.

### *3.3.1 Obligated Peasant Workers*

Even the comparatively skimpy sources of the first millennium occasionally reveal peasants obliged to regular or seasonal work in their lord's fishery. Prüm abbey in Luxembourg listed among its properties in 893 two fish weirs on the Moselle and three on the middle Rhine. Tenants on the Prüm estate had to maintain and operate those installations. Besides agricultural work and produce a fellow named Eurihc (*sic*) from Mötsch owed for this purpose a hundred poles like those for fences and a day's labour at the weir. Free tenements at Rheingönheim (now overrun by Ludwigshafen) provided some posts for a weir and had to transport two salmon to the abbey 150 kilometers away.[59] Eurihc's Anglo-Saxon contemporaries at Bewdley, Worcestershire, had likewise to "make a hedge to capture fish."[60] Tenants on several mid-tenth-century properties which the bishop of Verona had along the Po owed

[57] Burleigh, *Prussian Society*, 98–99, skips some of the report in Berlin PKB, OBA 11871, but the case has clear parallels elsewhere. See Supplement 3.2.3.

[58] Jacoby, "Class and environmental history." The mass of known medieval fishing violations better conform to this model, including its form of public destructive protest, than to the elite feuds which Manning, *Hunters and Poachers*, found in hunting crimes of Tudor–Stuart England.

[59] Schwab ed., *Prümer Urbar*, 182 and 250–255. Two other Rhine villages owed stakes but not carrying service.

[60] Salisbury, "Primitive British fishweirs," 76–77.

"week work in the fishery" (*opera ebdomada … in piscatione*), some of them alternating between that and the hay meadows and others paired off with neighbours who did the latter job.[61]

In certain settings elite exaction from ordinary farming households of prototypically servile payments in kind and in work for the lord's fish lasted into the central and even later Middle Ages, though not always without challenge. Folk living on the Ile d'Aix off the mouth of the Charente owed the castellan of Chatelaillon a one-third share of their take in fish and seabirds.[62] Forced labour in the fishery helped spark a 1440 revolt against Ermland's cathedral chapter, whose claims even noble arbitrators found too harsh.[63] But in 1453 the steward at Żarnowiec convent in Royal Prussia assigned specific peasants to fish Lake Dobre under supervision of expert fishers, and in Westfalia those waters which Marienfeld abbey had not leased out were harvested by labourers whom the fisheries manager had conscripted from seven servile farms.[64]

### *3.3.2 The Lord's Expert Servants*

So far this chapter has seen part-time, occasional, at most seasonal, fishing by people recognized as agriculturalists or with some other occupation. The last two examples, however, identified some fifteenth-century individuals as 'fishers'. In fact, specialists whose sole or principal productive work was in the fishery were present throughout the Middle Ages. The fisher Tatwine so well knew the miry Lincolnshire fens that he could lead pious and fearful Guthlac (d.714) to a secluded hermitage. Skilled household servants built the fish trap saintly Hubert, second bishop of Tongres-Maastricht, wanted in the Meuse.[65] Before the end of the eighth century, Charlemagne was warning his stewards to be sure there were able fishers and net makers on the royal estates.[66] In 832 his son Louis had thirty-two families of fishers on the Weser alone.[67] Until about 1000 these artisans were occupied primarily with indirect subsistence fishing.

61 Castagnetti, ed, *Inventari altomedievali*, 102–108.
62 Forquin, "Le temps de la croissance," 393.
63 Burleigh, *Prussian Society*, 22–24 and sources there cited.
64 Dąbrowski, *Rozwój wielkiej własności*, 87; Vahrenhold, *Kloster Marienfeld*, 129.
65 Colgrave, ed., *Felix's Life*, 87–88; *Vita Hugberti*, c. 8 (MGH, SrM, 6, 487–488).
66 "*bonos habeat artifices, id est … piscatores, … retiores, qui retia facere bene sciant, tam ad venandum quam ad piscandum sive ad aves capiendum.*" (Cap. *de villis*, §45, Boretius, ed., *Capitularia*, #32).
67 Cahn, *Recht der Binnenfischerei*, 92.

Non-agricultural specialists posed a problem for the early medieval economy. Time is needed to develop skills and knowledge, to prepare equipment, and to do the work. But preoccupied specialists had little time to get cereals, fibers, and shelter for themselves. Markets allow exchange of specialized products – metal, pottery, even fish – for basic consumables but neither surplus everyday items nor an exchange of them were then dependably available. People of power, whose wealth left something over after mere survival, offered the only significant source of effective demand for most specialized output. Medieval elites and their unusually skilled subordinates worked out in practice a mix of ways to support specialists. From the consumers' point of view they could either take certain specialized products as dues from selected otherwise generally self-supporting dependent households or they could themselves provide direct support to specialists who exercised skills for them. At one extreme bird-catchers or leather-tanners, even warriors, could live from an endowed peasant farmstead *and* work part-time on their specialty for the lord; at the other end of the spectrum a goldsmith or potter might be kept as a servant in the lord's household and/or receive a stipend in kind or cash to procure his own food. Like other rare non-agricultural specialist producers then, fishers worked for lords as obligated part-timers or as full-time professional servants or employees. Whatever mix of land, goods, or salary sustained them, specialist fishers earned their keep by supplying their catch to the lord who employed them.

Reasons why medievals organized any given indirect subsistence fishery in a particular way are almost never known to us. The outcomes of such choices and accommodations are merely uncommon. Demand structures and distribution of the resource look to have played off against regional habits of economic organization.

During and after the Carolingian age most Italian church corporations with well-documented fisheries associated their exploitation with certain dependent tenures from which the lord claimed a share or, more often, a fixed quantity of the fish taken. In 813 Nonantola abbey received half the catch of everyone who fished the river Mincio in the county of Mantua.[68] Monasteries with estates in the lake district used them for fish: ten of S. Giulia of Brescia's fifty-eight tenements at one place on Lake Iseo payed 1,200 fish a year; Bobbio's property on Lake Garda yielded trout and eel.[69] Further down the peninsula in the marshy Rieti basin servile

68 Mira, *La pesca nel medioevo*, 5.

69 Castagnetti ed., *Inventari altomedievali*, 57–65 and 138. Squatriti, *Water and Society*, 115, points out that some of these catches were surplus to the lords' needs and sold either by them or by fishers who had then to pay dues in cash. Chapter 4 below discusses further this ragged transition to commercialized arrangements.

fishers attached to domainal units of Farfa abbey paid dues in kind. The bishop of Lucca received fish weekly from at least two nearby tenants and semi-annually from a more distant one.[70] Monks and bishops thus left operation of the fishery to local experts.

Some lords and fishers in France had similar arrangements. In the early 950s Benedictine monks at Homblières in Vermandois received from a local magnate a property on the Somme at Frise some thirty-five kilometers away for the sake of the fish it could supply. At least one tenement there was later paying a hundred eel each Christmas.[71] Troarn abbey in coastal Normandy then also had one fisher whose tenure required him to provide a set quantity of fish from his use of nets in the sea.[72] Up to the 1020s a lay landowning family owned a hereditary serf who fished the Rhône not far from Lyon.[73]

Meanwhile other fishers worked directly as servants of their lords. As early as the sixth century, according to Gregory of Tours, Nicetius, bishop of Trier (*c.* 535–566), who needed a gift for his king, called upon his men to get fish. When told a flood had broken their trap he sent his reluctant servants anyway and they found a wondrously great catch.[74] Local fishers supplied ninth-century monks at San Vincenzo al Volturno with their steady diet of cyprinids, but what their chronicler called "fishers and sites for taking fish" (*piscatori, et aras ad pisces prendendos*) on Adriatic lagoons provided inshore marine varieties.[75] Around 1060/70 servants of the priory of St.-Marcel-lés-Chalon, across the Saône from the count's town, violently smashed the boats of fishers from Chalon whom they accused of fishing illegally.[76] Abbey servants (*famuli monasterii*) at Fleury were accustomed to fish in the main channel of the Loire and, unused to miracles, resisted St. Odo's advice to try elsewhere.[77]

[70] Toubert, *Structures du Latium*, 475, 607, and 672, considered fishers the epitome of such servile specialists; Sardi, *Le Contrattazione agrarie*, 96–101; Venditelli, "Diretti," 477–478.

[71] "*pro commoditate piscium eidem ecclesiae*" in Evergates, ed., *Cartulary of Homblières*, 41–43, with further references to that fishery at 43–45, 74–75, 95–104, 116–120, and 230–231.

[72] Hocquet, "La pêche," 105.

[73] Bernard, *Cartulaire de Savigny*, 2:669–670. Déléage, *La vie économique*, 164–166, found such arrangements the norm in tenth–eleventh-century Burgundy.

[74] Gregory of Tours, *Liber vitae patrum*, XVII, 4 (*Lives and Miracles*, ed. Nie, 250–251).

[75] Marazzi and Carannante, "Dal mare ai monti," 111–113.

[76] Bouchard, ed., *Cartulary of St.-Marcel*, doc. 12, pp. 36–37. In all likelihood the abbey fishers themselves operated the 36 m wooden weir found there by archaeologists and $^{14}$C-dated to the early eleventh century with repairs into the thirteenth (Bonnamour, "Pêche en Saône").

[77] John of Salerno, *Vita Odonis* 3:11 (*PL* 133, cols. 82–83). The eleventh-century *vita* of Abbot Maiolus of Cluny (*c.* 909–994) likewise credited the saint's powers with capture of a unexpected salmon on behalf of a disciple (*Vita brevior sancti Maioli*, cap. 23).

When St. Anselm needed a fish dinner, a Norman monk conveyed his instructions to a fisherman who then netted the biggest trout he had seen in twenty years' experience on that river.[78] Just as the risen Christ rightly instructed Peter and other disciples to cast their nets on the other side of the boat (John 21:3), spiritually gifted holy men had a reputation for confounding fishers whose knowledge was merely practical.

Large resources in subalpine lakes early excited special attention. Two noble brothers from the Sundgau reportedly had prior knowledge of the Tegernsee fishery before founding an abbey there in the 740s.[79] Monastic settlement on Mondsee dates to the same period. A century later Mondsee abbey and its ecclesiastical superior, the archbishop of Salzburg, heard testimony from an experienced local inhabitant, Heribald, about the fisheries of the Wolfgangsee (then called Abersee) in the Salzkammergut. He reported that abbey, archbishop, and local castellan each ran fishing boats there, but only the first two could legally take the lake trout during their fall run and the "*albuli pisces*" (probably cyprinids) which ran in the spring.[80] Such arrangements continued for more than a half-millennium. Since the 1490s – and earlier records simply do not survive – abbey and archbishop each paid and outfitted full-time fishers in ten and five boats respectively.[81]

A 1023 inventory by Tegernsee cellarer Gotahalm reveals the tackle such fishers could deploy. Alongside tools for other crafts, he counted thirty-four basket traps, seventeen nets of various kinds, winding reels, ropes, lines, and six fishing boats in use by abbey fishers.[82] For the next three centuries and more Tegernsee abbey, with exclusive rights to fish on the entire lake and its feeder streams, itself equipped fishers who worked under direct supervision but also claimed large numbers of whitefish as dues from designated subject tenures. After thorough

[78] Eadmer, *Life of Anselm*, 26–27. A later example of the generic topos had St. Richard of Chichester (1197–1253, canonized 1262) direct failed fishers for the archbishop of Canterbury to a fine catch from the river Ouse at Lewes, Sussex (Salzman, "Sussex Miracles," 71–72).

[79] *Passio Quirini* §15, Krusch ed. (MGH, SrM, 3:12); compare Weißensteiner, *Tegernsee*, 13–14. Bowlus, *Franks, Moravians, and Magyars*, 78, remarks on the attraction of mountain fisheries to eighth-century Franks and Bavarians.

[80] *Salzburger Urkundenbuch*, 1:907–908 and 914–915; Sonnlechner, "New units of production," 32–33.

[81] Linz OÖLA Stiftsarchiv Mondsee: Akten Bd. 406, Bd. 407 nr. 1, and Bd. 411, nr. 11; Handschriften Bd. 282. But a Mondsee rental from 1547 to 1560 (OÖLA Stiftsarchiv Mondsee: Handschriften Bd. 86) also still shows peasant tenures obliged to pay rents in fish.

[82] Munich BSB Clm 18181, a Latin text with German glosses, was dismembered for publication in Steinmeyer and Sievers, eds., *Die althochdeutschen Glossen*, 3:657 and 4:562–563.

religious and administrative reform in 1427 the monastery abandoned individual payments of fish and firmed up or enlarged its cadre of permanent servants. From 1443 until secularization in 1803 Tegernsee employed a team of six, seven, and eventually eight or nine full-time fishermen on annual and continuing contract with sole license to exploit the lake. For annual stipends in cash and kind, men from lakeside families worked in crews of two or three with the abbey's gear, some of which they were paid extra to construct, and turned all of their catch over to the master cook.[83]

Like monastic houses on alpine lakes, other corporate owners of highly productive fisheries long retained direct exploitation through their own servants and subjects. Even at the Venetian epicentre of the commercial economy, account books of the 1460s from San Giorgio in Alga show the canons meeting routine fish needs from their own capture and storage facilities in the lagoon and turning to the market only for Lenten treats of mackerel from the open sea.[84] At the same time institutionalized relicts of compulsory tenant fishing were supplying Scottish Cistercians at Coupar-Angus with scores, even hundreds, of fresh salmon from the river Tay.[85]

Fisheries were an important element in the organization of services required by princes in early medieval east central Europe. As patriarchal states were there created during the ninth through eleventh centuries, skilled servitors – falconers, smiths, cooks, etc. – were designated to supply the castle towns where the ruler lived. Along rivers in Great Poland, Mazovia, and Pomerania traces of settlements providing fishers' service are fairly common. They housed unfree ducal servants, tied to the nearby castle, exempt from normal public obligations, and allotted enough farmland to support their fishing for the prince. Many of those people now known entered the historical record only as the ruler transferred them to private, especially ecclesiastical, ownership during the twelfth and early thirteenth centuries.[86] By perhaps surprisingly naming those individuals many such charters fleetingly lend personal identity to

[83] Hoffmann, "Craft of fishing Alpine lakes," and Hoffmann, "Fishers in late medieval rural society," treat fishing in fifteenth–sixteenth-century Tegernsee from technical and socio-economic perspectives. Fleeting glimpses of earlier conditions occur in Munich BHSA KL Teg 1; KL Teg 3, fols. 9r–12r; KL Teg 4, pp. 36–37 and 41–50; and KL Teg 94.

[84] Faugeron, *Nourrir la ville*, 740–748. More on the heavily capitalized Venetian *valli di pesca* is in Chapter 4 below.

[85] Rogers, ed., *Rental Book of Cupar-Angus*, items 20, 42, 299, 319 *et passim* in vol. I, pp. 118–318, with detailed discussion in Hoffmann, "*Salmo salar*," 360–361; Hoffmann and Ross, "This belongs to us!" 462–463; and Hodgson, "To the abbottis profeit."

[86] Górzyski, *Zarys historii rybołówstwa*, 14–24.

otherwise anonymous fishers. In a typical instance from 1210 the duke of Kalisz endowed a new Cistercian house at Przemet: "These are the names of the peasants we gave them … In Domnik: Radoch, Swiątosz, Plewna, Nudasza, each of whom ought to give 12 fish of the length of a forearm three days a week, and further three jars of honey a year; Nowosz and Radzlaw with their sons should fish every day. … In Dłużyn: Zwan, who ought to fish every day …"[87] More abstractly, place names formed with the Slavic root *rybitw* ('fishers') also identify service groups. Old-settled Pannonian Slavs provided enough of a model for the fishing service organized by conquering Hungarians that the Magyar place name *ribar* ('fishers') occurs only at those communities. Royal dependants still fished the Danube in central Hungary in the 1320s.[88]

Institutions to supply fish to princely courts in central Europe differed from those further west chiefly in administrative particulars and in a few centuries' time lag, which may have produced sources with greater human detail. Quite analogous structures had been anticipated in Charlemagne's 794/5 instructions to estate managers. They echoed still in the eleventh century at the Italian royal palace in Pavia, where dues on commercial fishers recalled a time when as many as sixty boats brought fish from the Po to a resident king.[89] A 1220 survey of royal estates in northern Portugal identified twenty-one of thirty lordships where some or all tenants owed fishing service; the most general obligations occurred at inland riverine sites.[90] Absent dependable markets supplying fish, large secular households had to make their own arrangements.

Even in the twelfth- and thirteenth-century west the same kinds of people continued to fish local waters for elite masters or employers. Cluny, with more than 400 monks by around 1100 and always many visitors, served fish on Sundays, Thursdays, and many holidays. Ulrich's customal describes the storage tanks along the river Grosne where the cellarer went each evening to supervise selection and preparation. Some

87 "*Hec vero sunt nomina rusticorum quos eis dedimus: … in Dominiz: Radoch, Zvantos, Plefna, Nudassa, quorum quilibet tribus diebus in ebdomada XIIcim pisces ad longitudinem ulne dare debet et insuper tres urnas mellis annuatim; Novos et Radzlaws cum filiis cotitidie (sic) devent piscari; … in Dyznik: Zvan qui cotidie debet piscari; …*" (Zakrzewski and Piekosiński, eds., *CdMP*, Nr. 66). Duke Henry of Silesia's gift to Trzebnica convent of named and obligated fishers at Kotowice in 1202–1204 is recorded in Appelt and Irgang, eds., *SUB*, nos. 83 and 93.

88 Györffy, *Wirtschaft und Gesellschaft*, 36–39 and 76–92; Bartosiewicz, *Animals in the Urban Landscape*, 65.

89 Capitulary *de villis*, arts. 21, 44, 45, 62, and 65 (Boretius ed., *Capitularia*, #32); Brühl and Violante, eds., "Honorantie Civitatis Papiae," 20–23 and 60–61; Squatriti, *Water and Society*, 97.

90 da Cruz Coelho, "A pesca fluvial," 81–84.

fish came by purchase or special gift, but a fragmentary survey of Cluny's incomes from 1155 noted important resources on at least three estates in the region: at Beaumont, where the Grosne entered the Saône, seven fishers with "the office of fishing" (*officium piscandi*) each owed a "bundle" or "string" (*cordata*) of fish five times a year; some fisheries at Arpayé on the Saône paid a half or a third of the catch from mid-November to late April and another a "bundle" of fish each week; those at Montberthoud on the edge of the Dombes had been let to go derelict.[91] Across the Channel knight Robert de Ros had the right in perpetuity to send two boats of fishers from his household to Alemar, Yorkshire, whenever he was in that county. Simon 'the Fisherman', who had been serving the count of Leicester for an annual stipend since at least 1259, led the men who fished on the count's behalf during the last two weeks of Lent 1265.[92]

Full-time fishers under a lord's direct authority could meet the continual need of elite households for fish to eat. Workers in indirect subsistence fisheries shared the low social standing of their peasant neighbours or fellow servants, but gained special expertise in exploiting a non-agricultural resource in inland or coastal waters. In certain well-documented situations these arrangements persisted into early modern times. More often and by means the next chapter will trace, they likely provided the structural opportunity and initial human resources for the small-scale commercial fishing and trade in fish which had begun by the eleventh and twelfth centuries to emerge in parts of western Christendom. But for a much longer time medieval Europe's direct and indirect subsistence fishers also shared characteristic techniques and gear.

## 3.4 Compatible Technologies

Careful historians rightly resist the urge to read better-known methods of late medieval commercial fishers back into earlier centuries and concentrate instead on what verbal sources and excavation reports actually show

[91] Ulrich, "Consuetudines," lib. 3, c. 18 (Migne *PL*, vol. 149, cols. 760–762); Bernard and Bruel, *Recueil des chartes*, nr. 4143 and 4132 note (the count of Boulogne's gift of 20,000 herrings a year); Zimmermann, *Ordensleben*, 60–61 and 260–262; Bruce, *Silence and Sign*, 81–82. Evans, *Monastic Life*, 71–73, is unaware of the difference between artificial fish culture (not visible at Cluny) and the capture, storage, and use of wild fish. Organized exploitation of wild stocks also fed large monastic communities at twelfth-century Reichenau (Rösener, *Grundherrschaft im Wandel*, 226–227) and thirteenth-century Prüm (Schwab, ed., *Prümer Urbar*, 181, 194–195, 232–233, and 244).

[92] Lancaster, ed., *Chartulary of Fountains*, 2:831; Turner, ed., *Manners and Household Expenses*, 16 and 39; Woolgar, *Great Households*, 121.

of subsistence fisheries during and after the early Middle Ages.[93] That record presents techniques of catching and processing fish which were compatible with the needs of local consumers and the abilities of local producers. Both had adapted to their environments and available fish species. These methods formed the technological basis for medieval relations with aquatic ecosystems and, of course, for subsequent developments.[94]

### *3.4.1 Small Gears for Household Use*

The simplest small-scale methods required little technology but considerable environmental knowledge and skill. Capturing fish with the bare hands – like the English poacher heard pleading his case[95] – often also entailed substances meant to incapacitate fish en masse without impairing their use as food.

Poisons and explosives are traditionally associated with clandestine fishing. Classical authors had known several herbs with piscicidal properties. More recent ethnography describes whole communities – often especially women and children – using these methods in, for instance, the traditional Balkans. Regardless of the active ingredient, piscicides work best when low and warm summer streams concentrate fish in quiet pools.[96] Among medieval sources, the mid-eleventh-century fairy tale *Ruodlieb*, a work of Bavarian provenance, provides a full literary description of fishing with powered extract of *Anchusa officinalis* (Common Alkanet, called 'Ochsenzunge' and 'Buglossa'). It is the first record of a poison later widely mentioned in Latin and German manuscripts.[97] Use for this purpose of 'taxus' (yew, *Taxus* spp., or Great Mullein, *Verbascum thapsus*, called '*tassus*') was banned by Frederick II in his Sicilian

[93] Particular examples in what follows are from what look like subsistence, not artisanal, fisheries.

[94] The more extensive inventory of capture techniques found in Hoffmann, "Medieval fishing," 343–372, needs little significant revision.

[95] Instructions for manual capture are given in Salzburg Universitätsbibliothek Codex M III 3 [a collection of medical tracts and recipes dated 1439], fol. 291v, and limited use of the technique was allowed in fifteenth-century village customals (Weistümer) from Lower Austrian lordships of Anspang, Gutenstein, and Neusiedel Weidmannsfeld (references to Winter, ed., *Niederösterreiche Weistümer*, vol. I, pp. 19, 356, and 364, thanks to Jaritz and Winiwarter, eds., *Historische Umweltdatenbank*, nos. 77, 146, and 147.)

[96] Zaunick, "Fischerei-Tollköder"; Gunda, "Fish poisoning."

[97] *Ruodlieb*, ed. Haug and Vollman, Fragment II, ll. 1–26 and Fragment X, ll. 1–58; Zaunick, "Fischerei-Tollköder," 634–663; Hoffmann, *Fishers' Craft*, 44, 87, 171, and notes. In the *Ruodlieb* passage, one of the few to describe rather than just ban use of a piscicide, all of the fish are collected and consumed.

"Constitutions of Melfi" (1231), and then by a long run of Italian and Spanish laws.[98] The oldest references to poisoning fish with caustic quicklime also go back to thirteenth-century Italy, while subsequent fish-catching manuals and other evidence confirm this practice elsewhere.[99] Well before the end of the Middle Ages western Europeans further knew how to use that substance to detonate a stunning underwater blast.[100] Then about 1500 European popular cultures acquired a deadly piscicide of East Indian botanical origin, *Animirta cocculus* [*indicus*], which quickly replaced indigenous herbal agents.[101] These normally illegal techniques killed many fish indiscriminately.

As quiet as poison and in the right circumstances as lethally productive was the most primitive fishing technology, the spear. Multi-pronged, often barbed, fish spears (leisters) of varying design and type are well known archeologically and in medieval representations.[102] Two iron ones were found at the eleventh-century village site on Lac Paladru in Dauphiné (Figure 3.1). Indeed a thirteenth-century French poet thought fish spears common in peasant households, and tenants in the Aveyron did use them for lamprey each spring.[103] Elsewhere spearing eel was commonly acceptable but less so salmon or pike. The right to spear salmon was assured in thirteenth-century charters along tributaries of the upper Rhine but night spearing forbidden as a poacher's trick on the Traun in 1418.[104]

If legal doubts shadowed the spear and piscicides, a consensus acknowledged widely acceptable household tackle. About 1060 Sigebert of Gembloux hailed in verse Nature's blessings on Metz and turned to its

98 *Liber Augustalis*, lib. 3, cap. 72 (tr. Powell, 144); Balletto, *Genova nel duecento*, 188 (a 1274 contract to fish with "*erba que vocatur tassus*"); Abad Garcia and Peribáñez Otero, "Pesca fluvial," 163–164. For illicit later use of piscicides and explosives in Mediterranean coastal fisheries, see Faget, "Le poison et la poudre," and Garrido Escobar, "Anar al petardo."

99 Crescenzi, *Ruralia commoda*, 10:30 (ed. Richter, vol. 3, p. 209); Zdekauer, *Statutum Pistorii*, 131. Compare Trexler, "Measures against water pollution," and Sznura, "Veleni e '*nobilissimi pesci*'," 271–279. Florentines suspected clerics to be common offenders.

100 Quicklime sealed with primitive gunpowder into a slowly leaking jar heats in water to set off an explosion and shock wave. Hoffmann, *Fishers' Craft*, 64, 87, and 103–104.

101 *Ibid.*, 323 and 329–330. Further records of use appear in Cocula-Vaillieres, *Un fleuve et des hommes*,133–134.

102 Heinrich, "Fischerei und Fischereimethoden," 132–133; Steane and Foreman, "Medieval fishing tackle," 140; Kraskovska, "K otázke lovu," 151; Wundsch, "Aalspeere."

103 Colardelle and Verdel, *Les Habitants*, 320; Nyström, *Poèmes français*, 57; Boscus, "Le fief des Malhols," 257–258. Crescenzi, *Ruralia commoda*, 10:30 (ed. Richter, vol. 3, 208–210) also describes use of the leister.

104 Amacher, *Zürcher Fischerei*, 71; Scheiber, *Zur Geschichte der Fischerei*, 152. See also pp. 95–95, 99, and 103 above.

Moselle, where "I fish with the hook, you with a basket, he with nets."[105] Legal, not literary, aims in 1406 moved the Prussian grand master to define the small gear for household fishing "to be hand nets, scoop nets, gill nets, dip nets, cast hook, basket traps, catch nets, and the like."[106] Across medieval Europe hooks, traps, and nets in one or two fishers' hands were recognized as the norm for feeding a domestic group.

As today, fishing with hook and line came quickly to medieval minds (see Figure 3.2). About 1180 the subsequently reputed philosopher of love, Andrew the Chaplain (Andreas Capellanus), cribbed from Isidore of Seville to derive *amor* from the word for hook, [*h*]*amus*, and explained that "just as a skilful fisherman tries to attract fishes by his bait and to capture them on his crooked hook," the lover lures a person to his union of hearts.[107] The baited hook was a stock literary conceit and visual representation of fishing, but clichés lack technical or socio-economic context. Hooking techniques show many special adaptations in the Middle Ages.[108]

Fish hooks themselves came straight or curved (Figure 3.3, also Fig. 3.1 middle). The straight hook or gorge is a double-pointed cylinder bound to the line at its midpoint and concealed in a bait. After a fish swallows the bait a pull on the line jams the gorge across its throat. Predator species such as pike, pike-perch, cod, or catfish gulp their prey whole and are well suited for gorge fishing. Wood, horn, and bone gorges three to ten centimeters long have been recovered from, for instance, ninth- through twelfth-century strata at Wolin on the Polish coast.[109]

Curved hooks have a relatively straight shank or shaft, a bend, and a point, which may or may not be barbed. A curved hook is attached opposite the bend by binding the line behind an enlarged area or knotting the line through a hole or loop (eye) in the hook material. Medieval European fish hooks survive in wood, iron, and bronze; less durable materials may also have been used. The bronze hooks from Lac

105 "*Hamis piscor ego, tu vimine, retibus ille.*" Sigebert, "Vita … Deodorici," MGH, SS, 4:477–479, l. 92. A more elaborate literary inventory of small-scale equipment is in Supplement 3.4.1.

106 "*Wyr halden vor kleyne gheczow. handwate. stoknetze. klebenetze. hame. worfangil. rewse. Wenczer.* [compare Middle High German *vencvach*, 'catch net'] *und semelichen.*" (Benecke, "Beiträge zur Geschichte," 307).

107 Andreas Capellanus, *De amore*, I:3 (ed. Trojel, p. 9; *Art of Love*, tr. Parry, 31).

108 Hurum, *History of the Fish Hook*, is too untidy for incautious use on medieval topics, but see regional discussions in Steane and Foreman, "Medieval fishing tackle," 142–148; Steane and Foreman, "Archaeology," 90–91; Lampen, *Fischerei und Fischhandel*, 101–103; and Abad Garcia and Peribáñez Otero, "Pesca fluvial," 162–163.

109 Rulewicz, *Rybołówstwo Gdańska*, 98; for English examples see Steane and Foreman, "Medieval fishing tackle," 142.

Figure 3.2 St. Peter angling with rod, line, and hook from a tenth-century Anglo-Saxon manuscript.
Matthew 17: 24–27 specifies that Peter was to cast out his hook (Lat. *hamum*) to catch the fish, although other representations of the apostles fishing are with a net. Malibu, California, J. Paul Getty Museum, 85 ms 79, fol. 2r. Getty Museum open content.

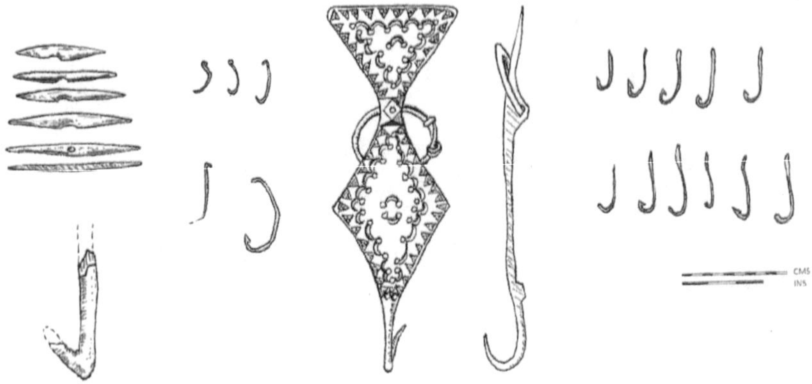


Figure 3.3 Medieval fish hooks, straight and curved (predating 1100). Representative selections of straight (left) and curved (centre) hooks in wood and bronze and an artificial lure, bronze, *c.* 8.5 cm (centre) from Wolin and (right column) of hooks from Great Yarmouth (top) and London (bottom). Selections from illustrations in Rulewicz, *Rybołówstwo Gdańska*, pp. 98, 112–113, and 129, and in Steane and Foreman, "Medieval fishing tackle," p. 147, redrawn for R. Hoffmann by D. Bilak.

Paladru (Fig. 3.1) measure only 2.2–3.5 cm shank length. Other specimens from especially coastal locations (Fig. 3.3) run larger: 5–6 cm wooden hooks from tenth-century Wolin; iron and bronze hooks of about 5–7 cm from eleventh–twelfth-century Szczecin and Great Yarmouth.[110] What served for trout or plaice would not for pike or cod. No actual fishing lines survive, though bits of hemp and other twine occur in some sites and use of braided horsehair lines by inland anglers is well documented by the fifteenth century. A hand line mounted on prototypical wooden winding frame was good enough for Norse god Thor – as pictured on a tenth-century carved stone from the Anglo-Saxon church at Gosforth, Cumbria.[111] Poor fishers from the French shore of the Channel used single-hooked hand lines in the 1100s.[112] Set lines anchored to shoreline features or a thrown stone and carrying many

[110] Colardelle and Verdel, *Les Habitants*, 208; Górzyński, *Zarys historii rybołówstwa*, 37; Heindel, "Tordierte Haken"; Schmidt, ed., *Leips*, fig. 34; Rulewicz, *Rybołówstwo Gdańska*, 99–130; and Steane and Foreman, "Medieval fishing tackle," 146–147.

[111] Wheeler and Jones, *Fishes*, 171; Kmieciński, "Spręt rybacki," had the same design from twelfth-century Gdańsk.

[112] Hocquet, "Les pêcheries médiévales," 75–76. So too, thought one twelfth-century artist, did a person fishing in the moat of Hartmannsberg Castle near Salzburg (Noichl, *Codex Falkensteinensis*, Tafel VIII, Abb. 3).

hooks could be a legitimate or poacher's method. Fishing poles shown incidentally in many early and high medieval illuminations are shorter than the people. Purposeful late medieval illustrations of fishing more often have poles equal to or much longer than the height of the people, which generally corresponds with advice in contemporary how-to manuals.[113] Reels were unknown in Europe before the seventeenth century.

Practical fishing manuals would recommend selecting baits for the fishes sought, the waters, and the season. Earlier, the seventh–tenth-century Byzantine estate manual *Geoponika* had recommended natural organisms and also baits concocted from animal and plant materials. In the thirteenth century men from Cotum, Yorkshire, gathered baits from tidal Guisborough Sands.[114] Artificial lures like the fish-shaped wobblers or jigs of lead, tin, or bronze recovered from early medieval sites in Poland attracted pike, pike-perch, or inshore cod (Figure 3.3 centre).[115] Literary allusions to feathered hooks begin with the thirteenth century and later these simulated insects were plainly part of small-scale fisheries in the Alps, England, and Spain.[116]

Anglers await a fish biting the hook, and trapping gear likewise depends on a fish voluntarily entering a device which holds it until removed by the fisher. Some important passive technologies were easily operated by individuals. As ubiquitous as hooks and lines in medieval Europe were portable traps or pots, called variously *retia*, *retz*, *Reussen*, *netz*, *weels*, *charpagne*, *verveaux*, *hoop nets*, *batrón*, *armadijo*, or the like (Figure 3.4, also Figures 3.7 and 3.8 below). Funneled openings on rounded baskets, bags, or cones of wicker, rushes, framed netting, or other materials allowed entry but no easy exit. Some types relied on a

[113] Hoffmann, *Fishers' Craft*, 125 *et passim* and contemporary illustrations in Emperor Maximilian's *Tiroler Fischereibuch*, fols. 3v, 12r, and 26r, and his *Weisskunig*, fol. 169v (Musper, ed., plate 43), as well as Jacopo da Bassano's 1538 triptych of San Zeno in the parish church of Borso del Grappa near Treviso (Masseini, "Fly fishing in early Renaissance Italy?"). Peter's long pole in Figure 3.2 is exceptionally early, while the frontispiece of Wynkyn de Worde's 1496 printing of the English *Treatyse* in the Second Boke of St Albans shows a rod no longer than the angler, which was by then surely not usual practice. Artists have different licence than fishers.

[114] *Geoponika*, 20:1–3, 10, 12, 14 (Beckh, ed., 1895, 511–522, and Dalby, tr. 339–348); Lancaster, ed., *Chartulary of Fountains*, 1:306, a charter dating about 1229. Bartholomeus Anglicus, Bk. 18, c. 115, suggested earthworms (Trevisa, tr., 2:1264–1265).

[115] Rulewicz *Rybołówstwo Gdańska*, figs. 21, 25, and 26. The lure shown in Fig. 3.3 is 8.5 cm long.

[116] Hoffmann, *Fishers' Craft*, 126. Subsequent discovery of precise recipes for artificial flies and other baits in an Austrian codex from the 1450s is reported in Hoffmann, "Haslinger Breviary." The artificial fly was not a medieval invention, Roman author Aelian having described its use in third-century Macedonia.

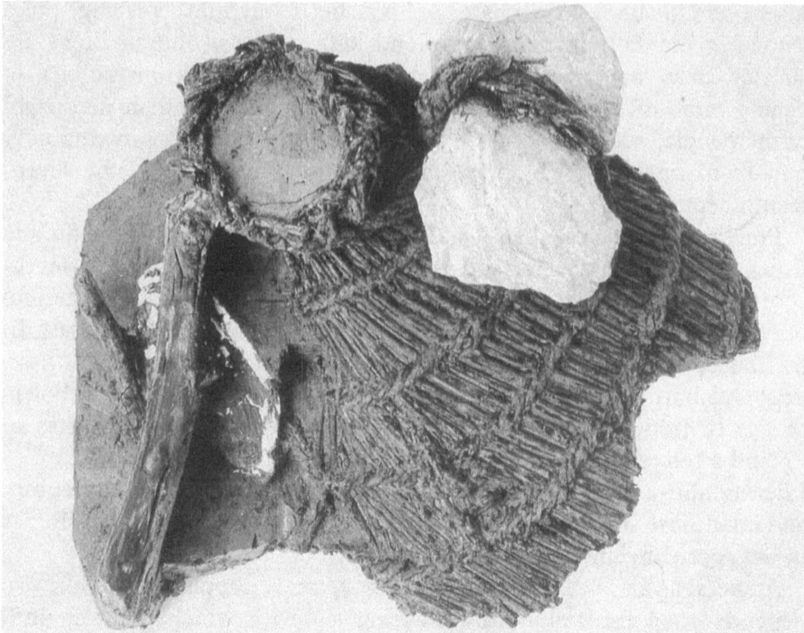

Figure 3.4 Remains of a wicker fish trap ('pot gear'). Photograph and permission provided by Dr. Anton Ervynck, Flemish Heritage Agency.

narrow neck, others a one-way entrance of sharp inward-pointing rods.[117] Just after 1300 Bolognese agricultural expert Pietro de Crescenzi distinguished a wide form for fish attracted by an enclosed bait and a narrow one for varieties trying to hide.[118] In the British Isles distinctive funneled *weels*, *kidells*, or *putts* – good archeological remains of medieval date have been recovered from the lower Trent – were set to catch salmon or eel as they tried to pass up or down through openings in weirs.[119] Choice of design and location depended on the fisher's knowledge of his quarry's habits.

117 Brinkhuizen, "Some notes on fishing gear," 38–50.

118 Crescentiis, *Ruralia commoda*, 10:28 (ed. Richter, 3:204–207): "*... sed duarum formarum fiunt, una forma est, quod sit interius multum ampla rotunda, in cuius fundo ponitur creta mollis & grana ei annexa, atque intrant quaedam genera piscium causa cibi, & exinde exire nesciunt. Alia forma est tota stricta & longa sed in introitu mediocriter aperta, & in medio ualde stricta, deinde lata, & in cauda strictissima, in quam intrant non causa cibi, sed ut ibi occulte moretur, nec de ipsa sicut de prima exire sciunt.*"

119 Steane and Foreman, "Medieval fishing tackle," 170–178; Salisbury, "Primitive British fishweirs"; Jenkins, "Trapping of salmon"; Carville, "Economic activities"; Winchester, *Landscape and Society*, 108–110; Cooper and Ripper, "Fishing and managing."

Early medieval records highlight the basket traps of lords' and landowners' indirect subsistence fisheries. Germanic law codes protected this form of property.[120] Eleventh-century Tegernsee abbey possessed thirty-four of them and twelfth-century subjects of Reichenau had to deliver cartloads of willow rods to make *reussen* and stakes to anchor them.[121] Later sources confirm peasants using pots both as a legitimate customary means of exploiting commons in, for instance, thirteenth–fourteenth-century Lorraine or early fifteenth-century Prussia and as a favoured and forbidden tool for poaching in the lord's private waters.[122]

Woven mesh was an ancient, familiar, and effective tool for capturing fish. Nets were important to medieval subsistence fishing but precise uses are hard to pin down. Besides many verbal and visual allusions, actual fragments of hempen twine, netting, and net-making tools survive from several northern continental sites. Some such scraps, descriptions, and representations reveal a distinct technique, others are no longer discernible. This is notably true of images where people in a boat haul a net over the side (e.g. Figures 4.1, 4.6, and 7.6 below). Medieval writers and artists were often unaware of the components and operation of complex fish-catching devices, even though back in the 890s Frankish administrators had acknowledged the special skills needed to make and use them. Fishers or their families commonly constructed and repaired their own nets.[123] Many net-fishing methods used a head line of floats and/or a weighted bottom line (Figure 3.5 and the left column of Fig. 3.1). The floats (wood, bark) and weights (stone, ceramic, lead) are often enough recovered from medieval shoreline sites that both English and Polish archaeologists have established regional typologies and inferred netting techniques from them.[124]

Small netting gear certainly aided active pursuit of schooling or individual fishes. What would now be labeled scoop or dip nets with a bag or basket on a handle (*truble*, *pern*), lift nets set horizontally on a frame, circular cast nets to throw over visible fish in shallows, even seines with

[120] Rothair's *Edict* §299 (Drew ed., *Lombard Laws*, 111); *Pactus legis Salica* Title 27 §28 (Drew ed., *Laws of the Salian Franks*, 91); Tischler, "Fische: Sprachliches," 138–139. Compare Cassiodorus, *Variae* 5:20 (Fridh, ed., 198–199) from Visigoth-ruled Italy.

[121] Munich BSB Clm 18181, fol. 118v; Rösener, *Grundherrschaft im Wandel*, 223–227.

[122] Collin, "Ressouces alimentaires," 43; Benecke, "Beiträge zur Geschichte," 307. Compare Blary, *Domaine de Châalis*, 95–99; Olson, *Chronicle of All That Happens*, 182–183; Coldicott, *Hampshire Nunneries*, 78.

[123] Cap. *de villis*, c. 45 (Boretius ed., no. 32); compare discussions in Lampen, *Fischerei und Fischhandel*, 103–105, and Steane and Foreman, "Medieval fishing tackle," 156–170.

[124] Steane and Foreman, "Medieval fishing tackle," 162–170 and 178–180 (although the many weights recovered from the Thames cannot be dated); Rulewicz, *Rybołówstwo Gdańska*, 240–273.

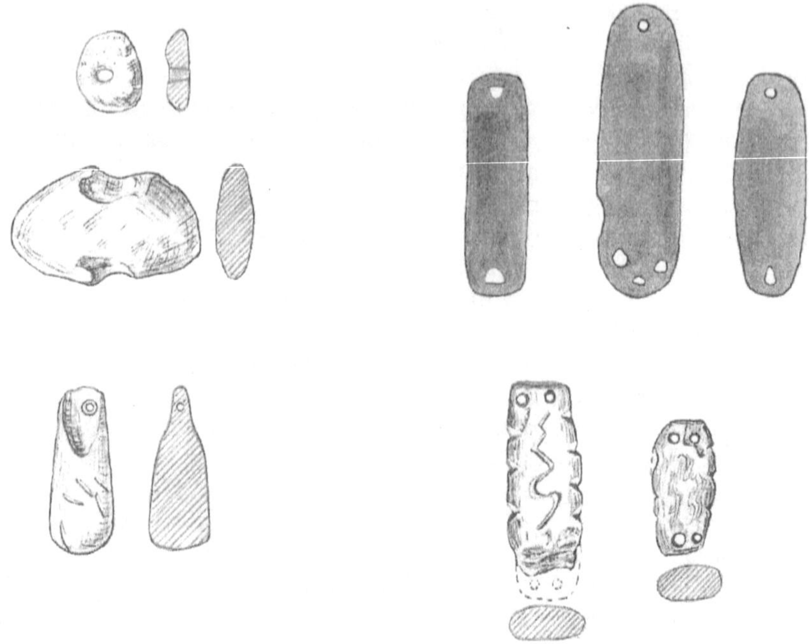

Figure 3.5 Net weights and floats from Pomeranian ports, seventh to eleventh centuries.
Selected (left) weights in stone, 5 and 10 cm, and lead, 5 cm, from Kołobrzeg and (right) floats in wood, 10–12 cm, from Gdańsk (top) and pine bark, 10–15 cm, from Szczecin (bottom). Selections from illustrations in Rulewicz, *Rybołówstwo Gdańska*, pp. 142, 161, 190, and 264, redrawn for R. Hoffmann by Donna Bilak.

elongate panels sized so one or two fishers could circle through the water to corral fish are all encountered often enough in medieval and early modern illustrations (see Fig. 7.7 below). Vernacular dialect names for these devices varied from place to place with or without – as remains true in modern ethnographies – corresponding differences in design, type, or use.[125] Ephemeral in construction and indistinguishable at historical distance, this equipment, though appropriate only for small-scale operations, is not easily associated with particular users.

Passive netting techniques also included forms small and mobile enough for one or two fishers to set and haul. The trammel net (*tremaille,*

[125] Compare illustrations and descriptions in Mane, “Images médiévales,” 244–246, with Brinkhuizen, “Some notes on fishing gear,” 9–29, Höfling, *Chiemsee-Fischerei*, 60–62, or Amacher, *Zürcher Fischerei*, 55–61.

*tramallia*, Italian *gorro*), a distinctive arrangement of three closely parallel mesh curtains, has a long medieval record since the sixth-century Frankish Salic code.[126] The gill nets essential to large-scale late medieval herring fishing and to seasonally important whitefish catches in alpine lakes were special entangling gear for schooling fishes of closely similar size. Although gill nets for herring have been suggested a medieval invention, specialized whitefish nets were already present in the Tegernsee inventory of 1023 and Sidonius Apollinaris described setting nets at night in lakes of fifth-century Gaul.[127]

### *3.4.2 Crew-Served Equipment and Installations*

Set or drift nets might still lie within the capacity of a single fisher, but contrasting active techniques of moving a large net through the water demand more hands, traction, and equipment. This technology has two basic forms: a seine surrounds fish in the water; a trawl is drawn through the water to filter fish from it (Figure 3.6 and the latter also in Figure 6.5 below). Genuine bag-shaped trawls, pulled open-mouthed on a course through the water, are nowhere visible before the later Middle Ages, when some were plainly thought innovations in commercial fishing.[128]

By whatever name, most medieval *sagenae* (*saine*, *Segen*, seine) are plainly seines. Pietro de Crescenzi described to near perfection a beach seine, which he called a "*scorticaria*": a boat circled out from shore and back again while paying a long net out behind it; the circle completed, both ends were pulled to bring the entire loop to the beach.[129] It is often difficult to determine actual examples from early centuries. Later seines are associated with fisheries in control of lords, such as the *sagena* Bury St. Edmunds had on Soham Mere in 1085 or those used in the thirteenth century to take bream for the bishop of Winchester and trout and charr from lakes owned by the abbey of Furness. The beach seine with which

126 *Pactus legis Salicae*, Title 27, §28 (Drew ed., 91); Tischler, "Fischerei und Fischereimethoden," 138–139; Crescenzi, *Ruralia commoda*, 10:28 (ed. Richter, 3: 204–206).

127 Steane and Foreman, "Medieval fishing tackle," 159; Hoffmann, "Craft of Alpine lakes," 310; Brandt, *Fish Catching Methods*, 358; Sidonius *Epistolae.*, lib. 2, no. 2 (ed. Loyen, p. 256). Compare Höfling, *Chiemsee-Fischerei*, 53–58.

128 Compare Bresc, "Pêche et coraillage," 110, or the "wondyrchoun" of which English fishers complained in 1376/77 (Given-Wilson, ed., *Parliament Rolls*, membrane 2:369, Edward III, 1377 January, 50. XXXIII; further in Jones, "'Lost' history," 204–208). For medieval resistance to environmentally destructive trawl fisheries see Chapter 6 below.

129 Crescenzi, *Ruralia commoda*, 10:28 (ed. Richter, 3: 204–206). The *boulieg* used in Languedoc lagoons was similar (Larguier, "Des lagunes à mer," 197).

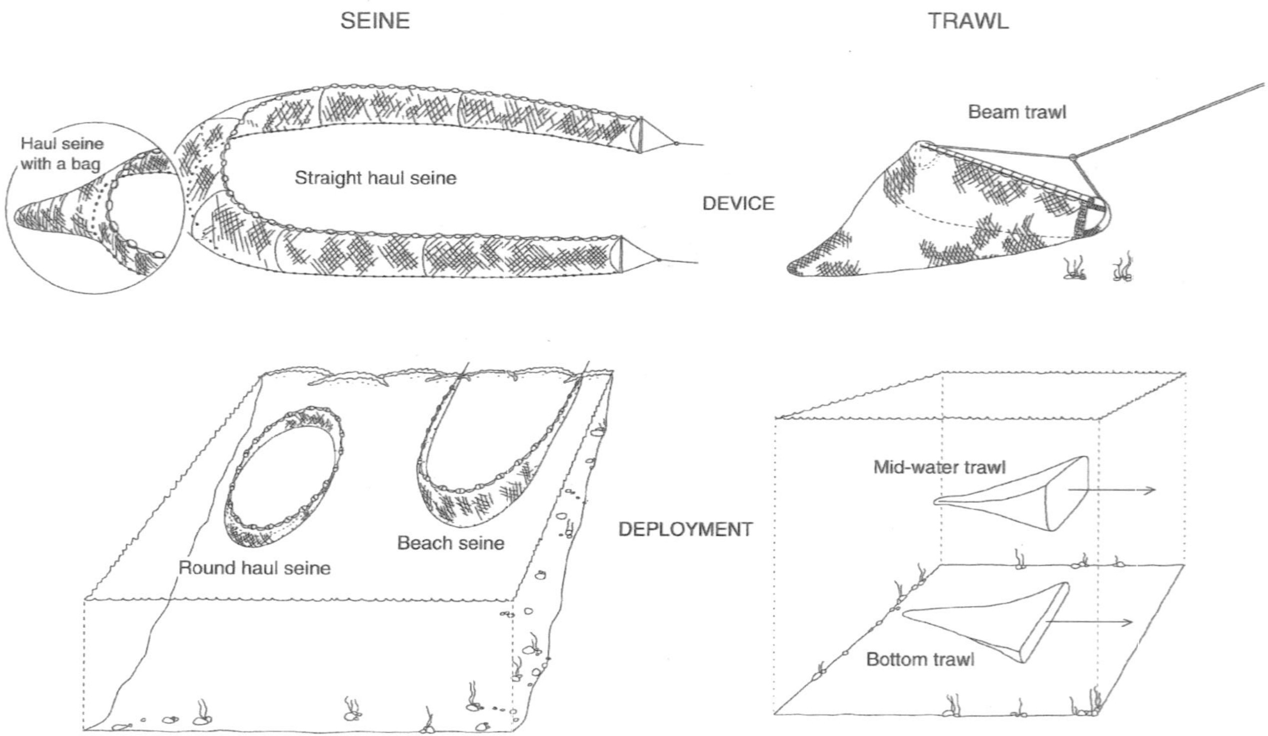


Figure 3.6 Schematic illustration of seine and trawl technologies. Drawn for R. Hoffmann by Cartographic Office, Department of Geography, York University. © R. Hoffmann.

subjects of Coupar Angus caught salmon at the Campsie site on the river Tay measured 33 fathoms (about 57–66 m) long and tapered from 4 fathoms at the centre to 3½ (from 7 or 8 to a bit more than 6 m) at the outer ends.[130] A much-recorded "*sagena magna*" of the medieval south and east Baltic was known in indigenous Balto-Slavic vernaculars as the *niewód*. This method of choice in the Teutonic Order's own late medieval fisheries had earlier been specified in twelfth–thirteenth-century Pomeranian charters and, judging from finds of locally standardized weights and floats from most urban sites along that coast, originated by at latest the eleventh century. With wings of nearly eighty meters and a depth of eighteen, the *niewód* marked an extreme scale for subsistence fishing.[131]

Major early medieval fisheries relied greatly on passive techniques using large fixed installations. More or less permanent structures which blocked and held migrating fishes should be recognized in many of the "fisheries" conveyed as appurtenant to landed property. Barrier fishing called for knowledgeable siting and timing plus sophisticated but simple engineering. Actual construction, less complex than making nets, was well suited for unskilled labour, which could also handle regular maintenance. Work and wealth spent on construction, upkeep, and annual operation produced large seasonal yields of favoured fishes, notably salmon, sturgeon, shad, lamprey, eel, mullet, and herring. Medieval records do not consistently distinguish between barrier devices (generic 'weirs') which concentrated fish and large enclosure traps which prevented their escape.

Some barriers providing important fishing opportunities were natural – falls, rivermouth bars, etc. – and others built primarily for a reason other than catching fish. The most widespread barrier fishing took place at mill dams and so multiplied during medieval centuries in tandem with the watermill. The downstream migration of adult eel gave each miller – or his master – a lucrative seasonal opportunity to put basket traps or a lift net in his sluice. Reforming abbot John of Gorze (960–973) supplied his

130 Bond, "Monastic fisheries," 81 and sources there cited; Roberts, "Bishop of Winchester's fishponds," 130–135; Winchester, *Landscape and Society*, 110; Hoffmann, "*Salmo salar*," 360 and sources there cited. Compare indirect subsistence use of seines in Kempf, "L'économie et la société," 44; Bertheau, "Wirtschaftsgeschichte Preetz," 113, or Hoffmann, "Craft of Alpine lakes."

131 Willam, "Fischerei des Deutschen Ordens," 140–142; Seligo, "Zur Geschichte des Fischerei," 17–19; Benecke, "Beiträge zur Geschichte," 307; Łęga, *Obraz gospodarczy Pomorza*, 28–33; Rulewicz, *Rybołówstwo Gdańska*, 252–277. Sixteenth-century illustrations of the *almadraba* used in the commercial fishery for bluefin tuna in the Gulf of Cádiz depict an even larger beach seine (see Chapter 8 and Figure 8.4a below).

Basket traps ('pot gear') for fish and eel placed above the mill dam in the early 14th century Luttrell psalter

British Library Luttrell psalter; add ms 42130, fol. 181r. Used with permission of the British Library

Figure 3.7 Basket traps ('pot gear') for fish and eel placed above the mill dam on the Lincolnshire estate of Sir Geoffrey Luttrell, *c.* 1330. Reproduced with permission of the British Library.

monks with fish taken at the dams of his newly built mills. Tenth- and eleventh-century Catalan charters carefully itemized the fishing rights of mills.[132] Mills and eel were closely joined in the English Domesday Book (1085) and in twelfth-century records of milling partnerships on the Garonne at Toulouse.[133] The early fourteenth-century Luttrell Psalter famously illustrates the local watermill and the basket traps set above the dam to catch adult eels migrating downstream (Figure 3.7).

Weirs built with express intent of catching fish played a large role for indirect subsistence. Fish weirs[134] across streams, rivers, or tidal channels, whether permanently built of stone and timber or of hurdles and

[132] Tischler, "Fischerei und Fischereimethoden," 139, cites MGH, SS, 4; Riera i Melis, "Sistemes alimentaris," 23–24, with cases there cited.

[133] Darby, *Domesday England*, 279–280; Sicard, *Moulins de Toulouse*, 118–128.

[134] General regional coverages appear in Lampen, "Medieval fish weirs"; Lampen, *Fischerei und Fischhandel*, 105–110 and 116–118; Salisbury, "Primitive British fishweirs"; Steane and Foreman, "Medieval fishing tackle," 170–176; Went, "Ancient Irish fishing weirs for salmon"; Jenkins, "Trapping of salmon"; Willem, "Fischerein des Deutschen Ordens," 138–140; Rippon, *Transformation of Coastal Wetlands*, 220–221; and O'Sullivan, *Foragers, Farmers and Fishers*.

wattle for a season, could concentrate and funnel migratory fishes to other trapping devices or just to a place where they might easily be scooped or speared. Under various temporally or territorially different names – *gurges*, *vertevolum*, *venna*, *banna*, *wera*, *obstacula*, *captura*, *clausura*, *exclusa*, *saepes*, *fach*, *wehr*, *cruive*, *croha*, *paxeria*, *jazy*, and more – weir fisheries had long-standing economic and hydrological importance.[135] Already in the sixth century Cassiodorus denounced fishing with *saepes* (literally 'fences') for endangering shipping in the Mincio, Ollio, Serchio, Tiber, and Arno rivers. The Frankish laws set compensation for damage to a *vertevolum*. One *venna* at a royal estate in the Ardennes is documented in a charter from 644 and another on the Weser can be traced in Corvey's possession from 832 until after 1158.[136] Recorded active weirs on the rivers of Gdańsk Pomerania number thirty-one during the twelfth and thirteenth centuries alone.[137]

Most such records denoted what early twelfth-century Czech chronicler Cosmas of Prague called an "assembly of timbers for fishing."[138] The several *vennae* of Prüm's ninth-century estate survey were made from stakes, poles, and large rods provided for the purpose (*ad vennam*, *quibus venna paratur*) by the abbey's subjects.[139] A servant of St. Hubert injured his hand driving log pilings for a *captura piscium* in the Meuse.[140] In north Yorkshire around 1200 Robert de Daiville and the Cistercians at Byland negotiated over a reach of the river Swale with a structure of two or three bays, each ten feet wide and repairable with beams and nails, available for fishing with a net.[141] Material traces corroborate the verbal. Remains of securely dated medieval fish weirs have been recovered from the rivers Trent, Severn, Witham, Charente, Dordogne, Cher, Saône, and several Swiss tributaries of the upper Rhine. Found in the Trent gravels near Colwick was a 14-meter double row of oak and holly posts set vertically into the river bed about a meter deep and a half meter apart, braced on the downstream side with oblique posts, and filled with panels of woven hazel twigs (Figure 3.8). Dendrochronology and radiocarbon put

135 Remains of weirs from the Neolithic, Bronze, and early Iron Ages have been found in Europe and Britain.

136 Cassiodorus, *Variae* 5:20; Pactus legis Salica, 27:28 (Drew ed., 91); Halkin and Roland eds., *Recueil des chartes*, I: nr. 1. In 979 Emperor Otto II settled a dispute over a *gurgustium* on the river Hörsel (Sickel ed., *Urkunden Otto*, no. 209).

137 Łęga, *Obraz gospodarczy Pomorza*, 19–24.

138 "*structuram lignorum ad piscandum*," Cosmas, *Chronica Bohemorum*, ed. Bretholz, 245.

139 Schwab, ed., *Prümer Urbar*, 176, 250–255, and 183. The same tasks were required of unfree tenants on the Tidenham estate of Anglo-Saxon Bath abbey (Bond, "Monastic fisheries," 85–86).

140 *Vita Hugberti*, ch. 8 (MGH, SrM 6, 487).

141 Burton, *Cartulary of Byland*, nos. 305 and 486–488.

Figure 3.8 Artist's reconstruction of an Anglo-Saxon fish weir from the river Trent.
Based on remains of wooden posts in river gravels near Colwick, Nottinghamshire, dated *c.* 810–880. From Steane and Foreman, "Medieval Fishing Tackle," p. 171. Drawing by C. Salisbury.
Reproduced with permission of BAR Publishing, www.barpublishing.com.

construction in one season and upkeep through a human generation between about 810 and 880. One wing of this Anglo-Saxon device – the other wing has not been found – probably ran diagonally across the river for some thirty-five or more meters. A nearby structure with radiocarbon dates of *c.* 1070–1200 had one long wing and two perpendicular short ones oriented at 45 degrees to the current, so making a funnel or 'V' shape pointed downstream.[142] (For well-dated and productive weirs elsewhere see online Supplement 3.4.2.)

As distinct from generic barriers with associated catching devices or activities, more specialized large-scale fish traps diverted fish into an enclosure whence they could not escape. Fishers at Kotowice on the Odra in Silesia in 1203 used an "enclosure to catch fish" (*clausuram pro capturam piscium*) to supply their lord several times a week. The phrase

[142] Losco-Bradley and Salisbury, "Saxon and Norman fish weir"; Salisbury, "Primitive British fishweirs."

echoes, too, in eleventh- and twelfth-century charters from Poland's other great river, the Wisła.[143] Early medieval records depict at least three specific types, viz foreshore traps, ditches, and labyrinths.

Domesday Book mentions a *heia maris* ('sea hedge') on the Suffolk coast partly owned by Bury St. Edmunds Abbey. Foreshore traps known from archaeological and ethnographic evidence are horseshoe- or crescent-shaped walls of wood and wattles or stone set in the tidal zone with the open end toward the shore; fishes moving with the ebb are held for removal at low water. Those in Ireland's Shannon estuary go back into deep prehistory and subsequently several groups are clearly dated to 400–800 and 1100–1350.[144] Similar constructions of wood and stone in the coast of lower Normandy have calibrated $^{14}$C dates between 580/630 and about 970. Under the name *gamboa* or *camboa* such structures were the oldest purpose-built fisheries installations in Galicia. Before 1200 Breton counterparts served local subsistence needs.[145]

"A structure commonly called a *verra* ... so constructed that fishes could enter that ditch but were not able to get out" vexed neighbours of the Clairmarais Cistercians in 1210.[146] Earlier tenth- and eleventh-century charters from the lower reaches of Seine and Loire refer plainly to "sluices" (*clusa*, *sclusa*) and "ditches" (*fossatis piscatoriis*) as means for catching fish.[147]

Complex enclosures of embankments, wooden sheeting, and wood or reed grillwork blocked the Tiber at Rome and the mouths of coastal ponds further south in Lazio, supplying sustenance for fishers and Romans and valued incomes to great churchmen since at least the tenth century.[148] A vast linear array of closely driven stakes still straddles the mouth of the Schlei at Kappeln, subtly guiding estuary-spawning herring

143 Appelt and Irgang, eds., *SUB*, 1: no. 83; Górzyński, *Zarys historii rybołówstwa*, 21, and sources there cited.

144 Darby, *Domesday England*, 285; Bond, "Monastic fisheries," 78; O'Sullivan, *Foragers, Farmers and Fishers*; Salisbury, "Primitive British fishweirs," 77; O'Sullivan, "Place, memory and identity," 449–450. Cohen, "Early Anglo-Saxon fish traps," found as many as thirty such fish traps in the intertidal zone of the Thames in the London area, with earlier ones mainly of stone, later more use of timber.

145 Catteddu, *Archéologie médiévale*, 86–87; Ferreira-Priegue, *Galicia en el Comercio*, 132; Mollat, *Europe and the Sea*, 143; Billard and Bernard, eds., *Pêcheries de Normandie.*

146 Coopland, *Abbey of St. Bertin*, 59 note: "*apparatus fecerunt qui vulgo verra dicuntur in terra illa ita dispositos quod pisces intrare possint fossatum illud sed exire non possint.*"

147 A 999 charter for Vierzon on the Cher, a Loire tributary, explained "*et nos teneamus exclusam totam sive decursus aquarum, in ea quam longe opus fuerit eam edificandi et prosequendi et foramina ad piscamentum nostrum in fluminis Cari*" (Querrien, "Pêche et consummation," 423). See also Brien, "Développement de l'ordre cistercien en Poitou," 4–5; Verdon, "Recherches sur la pêche," 346 note; and Fauroux, ed., *Recueil des actes des ducs*, no. 34.

148 Vendittelli, "La pesca nelle acque interne," 116–121; Vendittelli, "Diritti ed impianti," 392–399. Squatriti, *Water and Society*, 116–117, found equally complex structures even

into a chamber they cannot escape. Elements of the present labyrinth go back at least to the later Middle Ages and its forerunners to a millennium or more earlier.[149] No less elaborate was the maze of ponds, channels, and more than nine weirs connected to Glastonbury abbey's Meare Pool in the Somerset levels, largest wetland in southwestern England. Fish there taken with nets and traps could be stored in certain ponds and supplied live to the abbey's kitchen or processed in the adjoining fish house and preserved for later consumption.[150]

Technologies characteristic of early medieval and later subsistence fisheries were suited to make large catches from seasonal concentrations of diadromous and other migratory fishes and for smaller continuous exploitation of resident stocks. Unskilled, even forced, labour could build and operate fixed devices under expert supervision, but historic local power relations likely determined how much the latter derived from the communal tradition archaeologist Aidan O'Sullivan has suggested[151] and how much from the hierarchical authority implied by medieval texts. Certainly it took commensurate environmental familiarity and skills for consistent success with more mobile gear, especially the small angling, trapping and netting equipment apt to help feed a simple household. Direct subsistence fishing required little capital, and that chiefly from the time spent preparing simple equipment from locally available materials. Indirect subsistence fishing entailed more investment – the lord's allocating of his goods and his subjects' labour – in fixed capital tied to land. Across the gamut of medieval subsistence fishing, local environmental knowledge derived from the experience of a Heribald, Tatwine, or Simon the Fisherman and conveyed through oral transmission was essential to success. Otherwise it took miracles.[152]

### *3.4.3 Saving the Catch for Future Use*

In subsistence fisheries the cost of catching and the fish for eating belonged to the same party. Especially when reliant on local supplies,

in ninth-century Italy, but the evidence fails to support the inference of their use to *rear* fish.

149 Radke, "Bemerkungen zum Heringszaunen."

150 Rippon, "Making the most of a bad situation?," 119–122.

151 O'Sullivan, "Place, memory and identity," 461–463.

152 Exploits of Odo, Anselm, and other saints were mentioned above. St. Liudger (742–809) caught a sturgeon when his Frisian experts said the season had passed (Diekamp, ed., *Vita S. Liudgeri*, c. 29, p. 34). The merchant turned holy man Godric of Finchale (*c.* 1065–1170) repeatedly confounded servants by his ability to put salmon and other fish into their nets (Reginald, *Libellus S. Godrici*, ed. Stevenson, 123–125, 159–161, 206–207, 230–231, and 240–241).

people who could keep fish edible for even a few weeks or months after capture greatly extended the usefulness of seasonally large concentrations and eased provisioning for periods of seasonally high fish consumption. As just observed at Glastonbury, the catch might be held alive or the spoiling of its butchered flesh somehow delayed.

Temporary live storage was especially inviting where captured freshwater and coastal marine fishes could be retained in their own habitats. Tanks and cages made from non-durable materials are obscure in the early Middle Ages, though they may then have served peasant needs as they later did those of fishmongers and urban households (see next chapter).

On elite sites permanent fish tanks measuring up to some hundreds of square meters go back to the start of the Middle Ages. Cassiodorus retired in the mid-500s to his family estate of Vivarium on Italy's southern coast, named for its *vivaria*, natural rock coves improved for storing marine fish. The three basins still recognizable there measure roughly 10–12 by 4–5 m and just over 1 m deep.[153] Charlemagne's capitulary *de villis* urged regular restocking of *wiwariis* so a fresh supply was always available; a generation later the model survey, *Brevium exemplum*, noted stocked ponds inside the garden enclosure at three of four (unnamed) royal estates.[154] Structures of this kind and intent remained a feature at favoured residences. England's king Henry III (1216–1272) had *servatoria* for fish made or improved at York and in his park at Windsor. His contemporaries in Champagne, Burgundy, and Savoy just had different names for the similar facilities they had built for their chefs' use.[155]

The steady demand of religious communities supported fish storage facilities at many monastic sites. Tenth-century reformers Olpert of

[153] Cassiodorus, *Institutes*, I, 29:1 (Mynors ed., 73; tr. O'Donnell, *Cassiodorus*, 194) and Courcelle, "La site du monastére," 287–300. O'Donnell, *Cassiodorus*, 194–196 and 244–246, points out that this part of the *Institutes* circulated little during the Middle Ages.

[154] Cap. *de villis*, c. 65 (Boretius ed., no. 32), and *Brevium exemplum* in Boretius, ed., *Capitularia*, no. 128.

[155] McDonnell, *Inland Fisheries*, 20–23; Steane, "Royal fishponds," 39–40. For contemporary French princes see Nada Patrone, *Il cibo*, 320–321; Lambert, *Du manuscripts à la table*, 222; Bourquelot, "Fragments de Comptes," 67 and 71–73; Hoffmann "*Carpes pour le duc*"; Hoffmann, "Aquaculture in Champagne," 73. None of the *piscinae* and *vivaria* carefully inventoried from tenth–twelfth-century sources in Lampen, *Fischerei und Fischhandel*, 126–129, show the scale or purposeful management of fish varieties, fodder, or reproduction required for production rather than storage. Likewise, the famous *valle di pesce* of the Venetian lagoon neither so much reared fish as they trapped migratory schools and retained the adults nor did they serve primarily subsistence purposes (Faugeron, *Nourrir la ville*, 180–184), so they are discussed in some detail in Chapters 4 and 6 below.

Gembloux and Odo of Cluny each reportedly arranged a pond beside the house so monks' grudging acquiescence in dietary abstinence could be reinforced with a constant supply of fresh fish.[156] Store ponds at Cluny, which reared no fish during its prestigious eleventh-century apogee, were set along the Grosne river across from the abbey but easily accessible to the fish kitchen.[157] Elsewhere as in England, "a single stew stocked from a natural source ... is ... [the] type of pond to which most documents generally refer before the fourteenth century."[158] Especially in the early Middle Ages this temporary storage of captured fish came nowhere near purposeful rearing of domesticated varieties, but did promise fresh fish even in seasons when the schools were not easily available.

Early medieval Europeans also knew how to extend the time and space over which a dead fish could safely be eaten. The principal methods of preservation delayed spoiling by drying the fish physically or with the aid of smoke or salt.[159] Anaerobic fermentation, brining, or marinating then had less importance. The choice of technique played environmental resources, especially climate and fish species, against available economic capital.

Many kinds of fish will keep a long time if allowed to dry thoroughly and not re-wetted until ready to eat. No additional ingredients are needed. But the unaided drying of fish calls for a consistent low humidity available only at Europe's climatic extremes, arctic winter cold or southern summer heat. And in oil-rich fishes like herring, salmonids, and the tuna family, the fat quickly oxidizes to make the flesh inedibly rancid. Codfishes, flatfishes, and pike, however, have less oily white flesh apt for drying. As observed in an introductory fish tale, since the Bronze Age and throughout pre-Viking and Viking times coast dwellers in northern Norway and their North Atlantic island colonies caught and dried local cods for their own consumption. Mediterranean hakes were parched in the summer sun.

Dense wood smoke dries fish flesh and induces chemical changes that deter bacterial action, though hot smoke and cold give different results. Smoked fish last only days or weeks, not months, but the old peasant custom was likely known to lords' cooks and tables. Fish-smoking pits at early medieval Biskupin, a northern Polish settlement, are accompanied

[156] Sigebert and Godescalc, *Gesta abbatum*, c. 33, ed. Pertz in MGH, SS. 8; John of Salerno, *Vita Odonis* (*PL* vol. 133, cols. 80 and 83; tr. Sitwell, *St. Odo*, 78–81).

[157] Evans, *Monastic Life*, 72–73; Ulrich, *Consuetudines*, III:18 (*MPL* 149, col. 760–62). Ulrich's likely model, Bernard of Cluny, "Ordo Cluniacensis," 1:6 (ed. Hergott, 147–150), specifies aspects of fish service Ulrich omitted.

[158] Chambers and Gray, "Excavations of fishponds," 115.

[159] Cutting, *Fish Saving*, and Cutting, "Historical aspects," remain authoritative on biochemical processes but obsolete and Anglocentric regarding medieval history.

by bones of pike, perch, roach, bream, and catfish. Written sources hint at a wider regional practice.[160]

Most of Europe lacked climate or fish to dry unaided, so extended storage or transport of dead fish in large quantities called for another raw material. Salt, which both absorbs water and is itself antiseptic, served all preindustrial economies as the great preservative. Romans had enjoyed *garum*, a sauce brewed of salt fish. Access to salt and fish raised the value of lagoon and estuary fisheries. But away from southern seacoasts salt had to come in large and costly quantities from isolated natural brine springs, accessible deposits of rock salt, or the fuel and space to boil down sea water. Fisheries in the Austro-Bavarian Alps were exploited in the eighth–ninth centuries together with the salteries there, and later those of inland Pomerania. Ninth century Bobbio directed payments from salt ships to its fisheries on Lake Garda.[161]

The fish most notably tied to salt was herring: untreated, it spoils in a day.[162] Though herring are a northern, not a Mediterranean animal, their Latin name *alecium* (and so also Romance forms like *hallec*) derived from an old southern term for salt fish, *allec*, and ultimately from the Greek root for salt, *hals*. Hence early medieval herring fisheries arose near salt supplies, like those of the then Bay of the Somme, and herring enter the historical record "dusted" (*sapoudre*) with salt. Domesday Book had dues in dry-salted herrings from coastal Suffolk and so did the Norman fair at Fécamp. Further south in the Atlantic, pilchards or sardines and in the Mediterranean also anchovies were handled in the same way, landed and quickly, still with heads and entrails intact, heaped up into salt-covered piles. Dry-salted herring traveled in bundles, bags, or baskets and kept some months, though not so long that great quantities reached consumers far inland. That would await methods for anaerobic brining in sealed barrels unfamiliar to early medieval fishers and fish-eaters (see Chapter 8).[163]

><> ><> · <>< <><

Working in tandem the fish-catching and fish-preserving technologies of the early Middle Ages could take only limited quantities from even

160 Bukowski, "Uwagi o konserwacji ryb."

161 Bowlus, *Franks, Moravians, and Magyars*, 43–44; Rulewicz, *Rybołówstwo Gdańska*, 345–347; Castagnetti et al., eds., *Inventari altomedievali*, 138 and 159; and Squatriti, *Water and Society*, 114–116.

162 Hocquet, "Les pêcheries," 48–49 and 79–83; Hocquet, "Des paysans de la mer."

163 Danes on Roskildefjord ate whole fresh herring in the Viking Age and gutted and brined only from the thirteenth century (Enghoff, "Medieval herring industry"; and Enghoff, "Southern North Sea," 124–125).

abundant and accessible animals and convey those in edible condition to only a limited range of consumers. Rich and poor had to eat from nearby stocks and, to a large degree, recent catches. Coupled with relatively persistent structures of elite lordship over peasant family households, this tended to stabilize relationships between subsistence fisheries and the local aquatic ecosystems they exploited. Perhaps the visible pattern was reinforced by what present-day geologists and climatologists suggest to have been relatively quiescent planetary conditions during the eighth through early eleventh centuries.[164] Absent external pressures or shocks of human or natural origin – anything from greatly increased demand or slow habitat change to catastrophic political or flood events – most mutual local adaptations were probably more or less sustainable, if surely not equitable for all participants.

Local people actively exploiting aquatic resources to serve needs of both peasants and lords were, therefore, ubiquitous throughout the early Middle Ages. Subsistence fisheries fit well into the stable institutional structures and the relatively unimpacted environment characteristic of the second half of the first millennium CE in Europe. While some of those local relationships would continue long thereafter, most fish, fishers, and fish-eaters would eventually be caught up in the large-scale socio-economic and environmental transformation which medieval Europe experienced after about 1000. Successive chapters follow the changes, their impacts, and the consequences which rippled from them.

[164] Berger and Brochier, "Rapports de la géoarchéologie"; Dotterweich and Dreibrodt, "Past land use"; and Bradley et al., "Medieval Quiet Period."

# 4 Master Artisans and Local Markets

Little more than a thousand years ago a schoolmaster named Aelfric asked his pupils at a Dorset abbey to envisage a fisher working on a river and, more timidly, the coastal sea. The man cast nets and baited hooks from his boat and took common freshwater and inshore marine fishes of southwestern England.[1] But to the question "What do you gain from your craft?" the fisher replies "a living, clothing, and money." And when asked "Where do you sell your fish?" he answers "In the city" "Who buys them?" "Citizens. I cannot catch as many as I can sell."[2] This is not the voice of a subsistence fisher. Some two centuries later and on the other side of England, a Lincolnshire folktale about "Havelok the Dane" looked back on Aelfric's contemporary, Danish immigrant Grim, tenth-century founder of Grimsby. To feed his family the exile fished in the Humber estuary and, helped by his five children and his protégé Havelok, peddled his catch door-to-door sixty kilometers away in Lincoln. In exchange Grim received coins and necessities.[3] Aelfric and Grim put human faces on the local capture and sale of fishes then gaining importance in some English diets.

[1] Hurt, *Aelfric*, 118, and Howe "Historicist approaches," 90–93, contextualize Aelfric's *Colloquy*, ll. 85–121 (Garmonsway, ed., 26–30), now assiduously analyzed for environmental content by Preston, "Fact and fish tales," 1–25. Note that Aelfric 'of Eynsham' was still at Cerne Abbey (15–20 km inland from Weymouth) in 995 when he wrote the Latin text, moving to Eynsham in Oxfordshire some years later. His responsibility for the interlinear Anglo-Saxon gloss is dubious. For this reason I treat the Latin as more authentically Aelfric's expression.

A contemporary monk in Touraine, Letaldus of Micy, also imagined a more self-confident English coastal fisher, *Within piscator*, of Rochester, who made his way home after being swallowed by a whale (Letaldus, *Within*).

[2] Aelfric, ll. 88–89: "*Quid adquiris de tua arte?* / Hpæt beȝyst þu of þinym cræfte ? // *Uictum et uestitum et pecuniam* / Biȝleofan 7 scrud 7 feoh." Subsequently ll 96–99: "*Ubi uendis pisces tuos ?* / Hpær cypst þu fixas þine? // *In ciuitate* / On ceastre. // *Quis emit illos ?* / Hpa biȝþ hi ? // *Ciues. Non possum tot capere quot possum uendere.* / Ceasterþara. Ic ne mæg sþa fela sþa ic mæg gesyllan."

[3] *Lay of Havelok*, Skeat and Sisam, eds., ll. 732–926, as noticed in Britnell, *Commercialization*, 85 and 97. The tale goes back to local legends and twelfth-century Anglo-Norman redactions.

Both Grim and Aelfric's anonymous protagonist were artisan fishers. They worked full-time to catch fish for sale, not just for their own table or that of their lord. In the hands of real men like these fictive ones medieval fishing left its subsistence orientation and became, as Parisian social thinker and theologian Hugh of St. Victor would notice by the 1120s, a "craft" (*ars mechanica*) performed by artificers or workmen.[4] Artisan fishers gained access to aquatic resources by public or private licence obtained in return for ritual gifts, shares in the catch, or monetary payments to owners of the rights. They lived by market sales of their catch. Widespread sightings of numerous artisan fishers in medieval Europe around and after 1000 CE mark an economic and social shift, the appearance in fisheries of the exchange economy identified in its urban-centred origins as "The Commercial Revolution of the Middle Ages" and by a subsequent process of 'commercialization'. Market exchange involving specialized crafts and trades penetrated piecemeal into rural and agrarian relations.[5] Artisans are specialized producers in a commercial fishery, a different and more mediated interface between humans and the natural world. Market sales of their catches meant formation of prices which registered the relative value placed on fish by consumers and fishers alike. In medieval artisan fisheries, however, capture technology changed little from subsistence fishing; nor did commercialization itself much alter intimate ecological relations between local aquatic systems and consumers, who continued to eat fishes from their own vicinity. Small scale and close relations between artisan and resource distinguish this form of commercialization from capitalism.[6]

This chapter first illustrates the temporal, territorial, and environmental variety of artisan fishing in medieval Europe and then observes some rare but informative situations where this activity can be seen evolving from subsistence fishing. We go on to explore the characteristic

[4] Hugh's new classification (*Didascalicon* 2:20–27, ed. Buttimer, 38–45, tr. Taylor, 33 and 74–75) of fishing among the mechanical arts, done for a living as already exemplified by Aelfric, contrasted with Cassiodorus' seeing a subsistence-oriented work of charity to strangers (*Institutiones*, I:28:7, ed. Mynors, 72–73, tr. Halporn, 161–162). Fishing as *ars* resonated in such varied settings as thirteenth-century Pomeranian charters (Łęga, *Obraz gospodarczy Pomorza*, 28) and a fifteenth-century vernacular English treatise on the arts (Mooney, "Middle English text," 1034–1036).

[5] A classic formation is Lopez, *Commercial Revolution*. Britnell teased out elements of the process in his *Commercialization* and amplified some in his "Commercialization and economic development."

[6] Romano, *Markets and Marketplaces*, and Howell, *Commerce before Capitalism*, respectively explore the distinction from Italian and largely Low Countries perspectives.

integration of medieval artisan fisheries into local communities, ecologies, and economies, remarking their reliance on relatively small-scale techniques and frequent signs of mutual self-regulation which supported egalitarian and sustainable norms. The fresh and lightly preserved products of artisan fishers typically supplied nearby, mostly urban, markets, which projected onto the fishery the dynamic pressure of rising consumer demand. Artisan fishing and sales on local markets constituted the structural setting where, incrementally from the tenth century onwards, fish became a commodity, adding to its qualities as natural organism and culturally defined food an equally cultural economic aspect.

## 4.1 Artisan Fisheries and Their Formation

Continental evidence lacks the charm of English fictions but records whole groups of real working artisan fishers, people who lived by catching and selling fish. Clear signs of the purpose and practice are essential markers. We can start with some precocious regional cases,[7] few of which document any prior situation, and then mark subsequent proliferation of these arrangements across Christendom. Only occasional later instances let us trace transition of a given fishery from subsistence to artisanal quality.

### *4.1.1 "To Make Their Living by Fishing"*

Medievalists now think almost commonplace the advanced state or considerable post-Roman survival of commercial activity and written record in early medieval Italy. No surprise, then, to find early signs of market fishing there about the same time as Venetians were first winning mercantile renown. The recorded artisan fishers, however, fed not the new trading city but the old seat of long-departed Byzantine governors, Ravenna. In 943 an association of fishers, the *schola piscatorum*, reached agreement with the archbishop, to whom much public authority had fallen, for exclusive access to the fishery of the Badoreno. In return the fishers' own officers would handle fixed and hereditary payment to the archbishop of every fourteenth fish taken or an equivalent share in cash and enforce his first refusal on all sturgeon over four feet long. The money would presumably come from the fishers selling to Ravenna's other inhabitants the other thirteen fish and lesser or superfluous

7 Epstein, *Wage Labor and Guilds*, 42–44, surveys several of the following cases.

sturgeon.[8] A generation later in 997 a less fully described consortium of fishers leased from the Church of Ravenna the fishery on the lagoon of Commachio for an annual rent of twenty denars.[9]

Looking across the Po plain from then marshy and estuarine Ravenna, well-organized indirect subsistence fishers on alpine Lake Garda were already noticed. Yet not everyone beside that lake laboured in service to the big abbeys. At Lasize on the southeastern shore, in 983 eighteen named local landholders had the passing Emperor Otto II confirm their rights to fish and to fortify their shoreline, along with their exemption from royal dues. Not a century later in 1077 Henry IV reiterated this privilege for the original group's twenty successors, now called "poor fishermen" (*pauperes homines piscatores*) from the perspective of the imperial court, though their numbers included local clergy and several distinct kin groups.[10] By the thirteenth century another consortium of Garda fishers was leasing priority access to the upper Adige, where the lake's salmonids migrated to spawn. The bishop of Trent could preempt the largest fish, but the fishers, who had their own equipment, freely disposed of all the rest.[11] Eventually Venetian state regulations for the Garda fishery in 1433 were explicit: the waters of the lake were common to all and many poor subjects there lived *ex arte piscandi*, "from the craft of fishing." This was feasible because, as even thirteenth-century financial accounts confirm, people like the count of Tirol's agents from Merano were eagerly buying salt trout, eel, fish in aspic, pickled fish, and other Garda products.[12]

North of the Alps fishers often appear among the earliest free artisans to achieve official acknowledgement and, hence, record. On the Rhine in 1106 Herrich, Sethwin, Satmar, and twenty more named "fishermen of Worms" agreed to provide quarterly to the bishop two salmon and to the count one salmon in return for hereditary collective control over the wholesale market in fish. If members died without heirs, the group

[8] Mira, *Pesca nel medioevo*, 29–31, and compare Squatriti, "Marshes and mentalities," 12, and Squatriti, *Water and Society*, 119–125. The Ravenna charter is there remarked for its emphasis on private rights and collective organization, which may simply reflect institutional continuity in a town where public authority survived longer than elsewhere; here the special interest is the independent fishers working for buyers, not private masters or employers.

[9] Montanari, *L'alimentazione contadina*, 291–292. Eventual well-documented artisanal exploitation of the lagoon by Venetian fishers will be discussed below.

[10] Tabacco, *Struggle for Power*, 207 and 323; MGH Dipl. 6:375, nr. 287.

[11] Stolz, *Geschichtskunde der Gewässer*, 346–347.

[12] Butturini, "Pesca sul lago di Garda" (1879); de Rachewiltz "Versorgung von Schloß Tirol," 264; Stolz, *Geschichtskunde der Gewässer*, 265; Stolz, *Rechnungsbücher*, 52–53 and 60.

coopted replacements. Other fishers were welcome to sell their own catch, but only the privileged could buy for resale, though not outside the city nor before midday.[13] Free fishers at Tulln on the Danube are known since 1270, when the regent confirmed their access to all waters of Lower Austria not privately held and posted. The Tulln fishers' initial loose fellowship became a guild before 1469, when a municipal ordinance regulated the members and their hired help who fished from boats with seines, traps, and lines, as well as through the ice. The town complained of fishers more eager to send their catch forty kilometers downriver to Vienna than to supply their neighbours. For leases on private waters and for fish in waters open to all, the half-dozen master fishers from small-town Tulln competed against their numerous rural counterparts, called *Gäufischer*.[14]

Medieval artisan commercial fishing was by no means confined to inland waters. Around the Gulf of Lions, from Marseille to Perpignan, commercial fishing emerged within the seigneurial institutional framework from the late eleventh century. Urban markets, like that at Montpellier, where the abbot of Aniane was buying mullet and sea bass in the late twelfth century, encouraged fishers settled on barrier islands to intensify their exploitation of lagoons and the open sea, while paying a share of their catch to their lords. About the same time some hundreds of Marseille fishermen had formed a trade association with its own jurisdiction and influence on municipal government.[15]

Despite Aelfric's hesitant protagonist and the fate of that real fishing boat swept with Earl Harold Godwinson from Chichester to Normandy by a storm in 1064,[16] full-time professional fishing in the Channel grew, if slowly. Shortly before 1100 Queen Mathilda freed fishers settling at Roscelin on the Norman coast near Dieppe from dues on sale of fresh or salt fish, and in the next centuries such imposts in cash or kind were being collected from fish sales at Caen, Dieppe, Pont-Audemar, and Vernon.[17] Yet not until after 1300 did a Breton and western Norman coastal boat fishery supplant peasant part-timers fishing from shore, and nearly another century passed before fiscal records and eventually a papal bull of 1428 acknowledged that in this region whole communities

[13] Boos, *Urkundenbuch*. no. 58; translated in Herlihy, *Medieval Culture and Society*, 185.

[14] Petrin, "Tullner Fischerzeche," 30–32.

[15] Burin-Derruau et al., "Le littoral languedocien," 384–387 and 443–448; Coornaert, *Corporations*, 45; Puig, "Les ressources." Organized fishers were documented at Barcelona in 1002 (Riera Melis, "Pesca en el Mediterráneo Noroccidental," 122–124).

[16] William of Malmesbury, lib. 2, c. 228 (Stubbs, ed. 1887–1889, 1:279) called it a "navigium piscatorium."

[17] Darsel, "Servitudes de la pêche," 108–109.

and "many people make their living from fishing."[18] By then nearshore marine fishing had become a full-time and nearly year-round occupation along southwestern English coasts and was developing in Sussex as well.[19]

Small-scale commercial fishing remained in the high and later Middle Ages a characteristic activity all across interior Europe, and especially within marketing distance of large towns. Milan's markets had fresh fish of seven local freshwater species, said home-town author Bonvesin da la Riva (*c.* 1240–*c.* 1315), from fishers who exploited nearby streams, rivers, and lakes.[20] Well-organized local guilds and teams of fishers along the Rhône supplied much-appreciated bream, sturgeon, eel, lamprey, shad, burbot, and pike to consumers in Lyon, Avignon, Arles, and Tarascon.[21]

At Paris up to 200,000 inhabitants by the early fourteenth century were a huge attraction for fish supplied from a distance by lords' estates, by fishmongers specializing in either marine or freshwater varieties (see below), and by fishers exploiting the nearby aquatic habitats of the Seine. As artisans go, the last are in some ways especially well served by surviving records.[22] By the 1260s numbers of individuals worked with permission of rights holders such as the monasteries of Saint-Germain-des-Prés and Saint-Magloire, the cathedral of Notre-Dame, and the king, who together controlled fishing a dozen kilometers up- and down-river from town. The "king"s fishers" were actually then licenced to catch "all kinds of fishes" by one Guérin Dubois, whose ancestors had received the fishery as a fief from King Philip II.[23] As mid-thirteenth-century academic John of Garland saw while strolling along the river, these fishers worked on foot or from boats with seines, hoop nets, pot gear, and angling rods (Figure 4.1). Large fixed traps belonged to

18 "*plurimum ex piscatione vitam sibi procurant*," Touchard, *Commerce maritime breton*, 58–60; Hocquet, "Pêcheries médiévales," 96–98.

19 Kowaleski, "Expansion," 430–433 and 440–441; Kowaleski, "Commercialization," 202–211; Mate, *Daughters, Wives and Widows*, 44–45, and works there cited. More English coastal fishing is in Chapters 5 and 8.

20 Bonvesin de la Riva, *De magnalibus Mediolani*, lib. 3:30 (Chiesa, ed., 92–93).

21 Rossiaud, *Rhône*, 160–161, 248–249, 252–253.

22 Key normative documents are published in Lespinasse and Bonnardot, eds., *Métiers et corporations*, 212–214; Laurière, ed., *Ordonnances des roys*, II, 583–586; and Lespinasse, *Métiers et corporations*, 465–472, which include introductory essays. This and further data are discussed in Franklin, *Dictionnaire historique*, 556–557; Bossuat, "Pêche en Seine," 61–65; Egbert, *Bridges of Mediaeval Paris*, 26–27; Benoît, "La pêche de Paris"; and Cayla, "La pêche à Paris." Compare Coornaert, *Corporations*, 239.

23 By the fifteenth century the Paris Hôtel-Dieu owned fishing rights on seven leagues of the Seine, from which they received dues in cash and, twice a year, in fish (Hohl, "Alimentation et consummation," 195).

Figure 4.1 Artisan fishers in the Seine at Paris, 1317.
Miniatures of rod fishing and fishing with a seine net from small boats in the Seine below the bridges of Paris, as depicted in a marginal illumination done at St. Denis in 1317. Other pictures in the same source show angling and spearing from similar vessels.
From Paris BN, MS fr. 2090–2092 "Life of St. Denis," II, fol. 97r detail.

shoreline property owners and holders of high justice, who also leased out shoreline access.[24]

Freshwater fishers of Paris displayed their catch for sale daily at the city's "pierres à poisson," designated markets along the Seine where they and the poulterers squabbled over space to work, or peddled their goods through the streets from carts or baskets. One market was on the right bank at the Castelet, the headquarters of the king's governing provost, and the other beside the Petit Pont on the left.[25] Fishers were forbidden to buy fish for resale, which was the exclusive privilege of the freshwater fishmongers, of whom more below.

Already by the 1260s the freshwater fishers formed one and later two officially recognized corporate communities, the *pêcheurs à verge* and *pêcheurs à engins* . A privilege renewed in 1515 further describes their

[24] Laurière, ed, *Ordonnances des roys*, V, 207–208; John of Garland, *Dictionarius*, §72, in Hunt, ed., *Teaching and Learning*, vol. I, 202, and compare variants in II, 141, 148, 152, 150. Hunt's manuscript source differs from that edited by Géraud, *Paris sous Philippe le Bel*, 608.

[25] Bossuat, "Pêche en Seine," 75; Auzary-Schmaltz, "Les Contentieux," 59–60.

organization, treasury (filled by own dues), fraternal obligations, and the right of a widow to succeed her deceased husband so long as she did not marry outside the "rank." A 1292 register of taxpayers, which omitted people too poor to pay, named ten such *peschéurs*.[26] That likely comprised but a small minority of a group later commonly described as "poor fishers … who fish to earn their poor living and the maintenance of their wives and children."[27]

Towns dotting the banks and terraces of the Rhine from Basel to Mainz offered tens of thousands of potential buyers, so late medieval artisan fishers spread all along the upper river. About 1300 a friar from Colmar guessed as many as 1,500 fishers had been working along only the Alsatian shore.[28] At the largest settlement in that reach, Strasbourg, by 1315 a fishers' guild with its own statutes shared oversight with municipal authorities. On the opposite bank, three kilometers downstream others fished under the aegis of the local lord. Further below in the Palatinate free fishers were organized in another thirteen or more communities.[29]

Medieval artisanal fishing may attain some technical zenith in the *valli da pesca* of the Venetian lagoon, an elaborate adaptation to and modification of the seasonal movement of local fish stocks. Diverse species tolerant of broad variations in salinity inhabited the nutrient-rich continental shelf waters of the upper Adriatic and the rivers which feed it.[30] Early each spring various mullets, sea bass, flatfishes, and sturgeon leave the open sea for the faster-warming lagoon, where they can feed and spawn. Juveniles and adults depart the lagoon as it cools in autumn to winter over at sea. People learned to exploit these seasonal concentrations of fishes.[31]

26 Géraud, *Paris sous Philippe le Bel*, 531. Another unpublished tax return from 1297 shows fishers and boatmen as most common occupations in the poorest parish at the downstream end of Paris (Geremek, *Margins of Society*, 76).

27 "*povres pecheurs … que pescher pour gaigner leur povre vie & le gouvernement de leurs femmes et enfans*," Laurière, ed, *Ordonnances des roys*, V, 207–208; a privilege from 1358 (Lespinasse, *Métiers et corporations*, 466–467) has similar vocabulary.

28 Jaffé, ed., "Annales Colmarienses," 236.

29 Mone, "Ueber die Flussfischerei," 73 and 81–82; Dettmering, *Zunftgeschichte*, 28 *et passim*.

30 As earlier remarked, the northern Adriatic contains the largest shelf in the Mediterranean, and the Po, Adige, Brenta, Neretva, Drin, and other tributaries together discharge into the Mediterranean more fresh water than any other source.

31 For initial understanding of the *valli* and their medieval origins and use I am indebted to my former student, Dr. Cristina Arrigoni-Martelli, and her research on commercial hunting of ducks in the lagoon which became a chapter of her 2015 PhD dissertation, "Ducks and deer, profit and pleasure." General aspects discussed below are in Rosa Salva and Sartori, *Laguna e pesca*, 9–18; Zug Tucci, "Pesca in Laguna," 491–502; Crouzet-Pavan, "Mythes et réalités," 104–105; and Faugeron, *Nourrir la ville*, 180–184, 311–317, and 544.

Sparse early medieval sources, Cassiodorus among them, depict the Roman refugees who settled the barrier islands and elsewhere in the lagoon taking fish migrating through inter-island channels and into tributary rivers. The knowledge so accumulated evolved into construction of *valli da pesca*, complexes of wooden fences and reed (cane) grillworks (*grisuole* or *velledelli*) meant to guide the arriving fish into temporary enclosures of lagoon waters. Using as bases more permanent salteries erected on islets or the shallowest mudflats, fishers could thus harvest mature and young fish with relative ease for much of the year. By the eleventh and twelfth centuries, extant private charters document *piscarias* with seasonal management of channels, opening access at the turn of January and February, starting to fish a few weeks later, then shipping fresh fish to local settlements and salt mullet, for instance, to nearby Veneto and Friuli. The norm came to be keeping the *valli* enclosed from July to Christmas. The community on Malamocco required fishers in its public waters to bring their catch to its public market. The 1173 excise schedule for the Rialto market named six primarily freshwater, two diadromous, and six marine species, all at least seasonally native to the lagoon and its feeders.[32] By the 1200s owners of private salteries were employing their own fishers or leasing exclusive access in return for fixed payments of fish, daily during Lent or seasonally.

With consolidation of rule out of Rialto (modern 'Venice'), a confraternity of *Pescatori di San Nicolo e di San Rafaele* formed before or in the early thirteenth century and obtained an exclusive right to fish 'public' waters.[33] In return these *Nicolotti* annually presented the doge with 2,400 salted mullet which he distributed to officials. The republic licenced individuals or partnerships to operate *valli* – now become the predominant capture system in the northern lagoon – and to use other specific gears for particular species or in certain seasons. Governmental action thus brought into the written record what fishers themselves had learned long before.

On what remained an artisanal and at most a regional scale, *vallicoltura* comprised a cultural adaptation to the ecosystems of the lagoon. The fishers did not practice aquaculture, but cleverly manipulated wild populations. The fish stocks confined each summer to the *valle* were neither selected nor of non-native origin; they were neither artificially fed nor

32 Papadopoli, *Le Moneta*, 307–311: trout, tench, chub, riffle dace, barbel, and (dried) pike; sturgeon and eel; gilt-head sea bream, scorpion fish, gurnard, flounder, sole, and turbot. *Megle* remains obscure. See also Faugeron, *Nourrir la ville*, 31–33.

33 Zago, *I Nicolotti*.

their reproduction controlled. Even the Venetians' purposeful use of barriers and enclosures simply but knowingly extended the weir and trap technologies well recorded elsewhere. What likely began as primarily subsistence-oriented arrangements were by around 1000 gaining a strong market role and by 1200 a strong professional identity.

### *4.1.2 Transitions: From Servants to Sellers*

While the evolution of fishing in the Venetian lagoon may be reasonably inferred, only rarely do surviving records elsewhere offer a glimpse of the actual processes which at various times turned especially onetime providers of lords' indirect subsistence into market-oriented catchers of fish for sale. Historian Paolo Squatriti reasonably speculates that some ninth-century landowners both north and south of the Alps controlled more fish production than they could consume, so sold their surplus and valued their fisheries in monetary, not dietary terms.[34] Another intermediate or transitional condition is suggested in late tenth-century Pavia, where operators of sixty boats on the Ticino were no longer obliged to supply fish to the royal palace. Instead the chief fisher (*magister pischatorum*) saw to their payment all year long of cash dues to the palace steward, who was to use the accumulated fund to buy fish on Fridays whenever the royal household was in residence.[35] The market provided a means to adjust the seasonality of nature – fish to catch – and society – fish to buy – with the erratic presence of court demand. By 1179 the same group of Pavian fishers was negotiating rights of access with competitive monastic fishers.[36]

Late thirteenth-century fishers at Klosterneuburg, a priory and subject small town on the Danube above Vienna, experienced a like transition from obligations to fish to obligations to pay from the proceeds of fish sales. Their written customal recognized that those who used seines, lines and pots, or other boat-operated gear would pay the convent ten pennies annual dues and those who fished from shore or by wading only five. The priory had first refusal on all "large and strong fish" (perhaps referring to sturgeon) before a fisher could "sell to others" (*vendere aliud*). If fishers wanted to build a special trap, the prior might pay half the costs and get half the catch, but if he did not contribute, the canons received

[34] Squatriti, *Water and Society*, 115.

[35] Brühl and Violante, eds., "*Honorantie Civitatis Papiae*," 20–23; Mira, *Pesca nel medioevo*, 31–33.

[36] Occhipinti, "Fortuna ... Morimondo," 320–321. By the fourteenth and fifteenth centuries Pavian fishers worked within an elaborate regulatory regime (Pavesi, ed., *Ordini e statuti*).

only the catch of every ninth day and night. By the 1320s the monastery's account books show large and regular cash expenditures for pike, sturgeon, and other local fishes, but even at the end of the fourteenth century all sturgeon were still subject to its preemptive claims.[37]

Local emergence of artisan fishing is also visible in rural settings, but still with an orientation to urban demand. On the Odra river near Wrocław two villages named Kotowice, one belonging to the convent of Trzebnice, the other to the bishop, were both largely populated by fishers. In both places thirteenth-century documents detail fixed dues of measured quantities of fish payable on several specified days each week. By the mid-fourteenth century, however, the bishop was claiming from his subjects only an annual thirty-groschen "fisher's rent" and in the early sixteenth century the nuns were content with leasing the fishery out for seven marks a year. In both instances, the fish were being peddled in Wrocław.[38]

At Katzwang on the Rednitz just south of Nürnberg a legally hereditary tenure called "Das Fischhaus" had by the 1450s been in Fischer family hands for at least three generations. It included fields, meadows, and exclusive fishing rights in a designated reach of the Rednitz, five named old-channel lakes, and an excavated "Fischgrub" probably meant for temporary storage. Each successive tenant owed Ebrach abbey forty florins entry fine, ten florins annual rent, and "customary" services including six services of fish worth thirty pfennig each. The abbey further claimed first refusal to purchase "the fish that he would carry to Nürnberg for sale."[39]

This chapter opened with fictive but credible tales of artisanal fishers in late Anglo-Saxon England. Historians Maryanne Kowaleski and Harold Fox have drawn attention to the much later regional transition from subsistence to market-oriented fishing in coastal Devon. Successive fourteenth-century customals from the lay-owned manor of Stokenham mark the slow shift of a well-adapted local fishery. In 1309 unfree tenant farmers already possessed equipment to take from Start Bay porpoise, salmon, plaice, sea bream, skate, conger, and mullet for their own use. Their lord, however, claimed a third of their catch, his choice whether in fish or cash, and some preemptive right over the rest. Furthermore before

[37] Zeibig, ed., *Urkundenbuch … Klosterneuburg*, II, 170, 254, 262, and 265–266 (compare Perger, "Klosterneuburg," 198–199, and Cahn, *Recht der Binnenfischerei*, 103). Winter, ed., *Niederösterreiche Weistümer*, I, 972–977; Röhrig, "Die materiellen Kultur"; Holubar, "Spital des Stiftes," 36–43.

[38] Hoffmann, *Land, Liberties, and Lordship*, 50, 255–256, and 365, with sources there cited.

[39] "die Fisch so er nach Nürnberg führet zu verkaufen," Weiss, *Zisterzienserabtei Ebrach*, 71–72.

and during Lent a rota of the 160 farms had to provide three crews of nine villeins each to fish mullet for the lord, with spotters atop the cliffs to signal the seiners to surround the schools. By 1360, however, the obligation had been halved and the beaches where the fishers launched and landed become regular sites for sales of their catches.[40] An incipient similar transition is visible in the 1365 pipe roll of Irish Cloyne, where ten fishers in cottages at Ballycotton in East Cork paid cash rents to their lord but also:

> All these are fishermen [who] must serve the lord with fish, and what is worth 12d the lord can have for 8d. And in the season when they take langes (ling), the lord can have one for 2d, a mulewell (cod) for 1.5d, three haddocks for 1d. And the lord must not take more than is needed for himself and his house.[41]

The latter unintended echo of limits on direct subsistence fishing reflects the market's tilt of the balance in lord–tenant relations.

## 4.2 Household Enterprises in Local Communities

Artisanal fishing subtly altered connections among consumers, fishers, and things of nature, but medieval artisan fishers remained in many respects ordinary members of their local rural or urban societies. They spoke of fishes, boats, and gear in local dialects of regional vernaculars. Fresh local fishes in Piedmont had a bewildering mix of Latinate, Provençal, and Lombard names. Sicilian fishers used a distinct maritime dialect, calling the long lines they laid overnight on the sea floor *palmagastru* in echo of their classical and Byzantine Greek heritage. Men of Abbotsbury on Lyme Bay, Dorset, agreed in 1427 on the share their lord the abbot would take from their catch of twenty-six familiar inshore taxa, but had for nine of those fishes terms otherwise undocumented at so early a date or anywhere else. Notably the European freshwater fisheries, never having modernized, retained archaic local taxonomies into the twentieth century.[42]

40 Kowaleski, "Expansion," 429–430; Fox, *Fishing Village*, 59–60, 88–89, and 122–129. Chapter 8 below treats development of urban-centred distant-water fisheries in southwestern England.

41 Breen, "Marine fisheries," 93, drew attention to MacCotter and Nicholls, eds., *Cloyne Pipe Roll*, 93.

42 Nada Patrone, *Il cibo*, 330–333; Bresc, "Pêche et les madragues," 170–173, and "Pêche et coraillage," 110–112; Hocquet, "Pêcheries médiévales," 48; Mills, "Some late Middle English fish names." The taxonomic observation in Brinkhuizen, "Notes on fishing gear," 8, is well confirmed in the progressively expanded multilingual lexica of Conrad Gessner's 1548 "Catalogus alphabeticus," fols. 227v–232r; 1556 "Teütsche nammen der Fischen"; and 1558 *Historiae animalium liber IIII*.

### 4.2.1 *Social Positions: Residence and Status*

Residential patterns of market fishers reflected the occupational clustering normal in medieval communities and earlier noted in large-scale specialized indirect subsistence fisheries like those of lakes on both sides of the Alps. In eleventh-century Iceland, a purely rural society, specialists gathered first seasonally, then year-round, at protected shore sites where they could lease land from owners and take fish to exchange with inland farmers for food and fabric.[43] At Gdańsk excavations in tenth–thirteenth-century layers have found fishers living in the port quarter beside other artisans, sometimes even in the same houses.[44] Indeed the fishers' row, quay, or *holm* remains a relict feature in many present-day European cityscapes. At Nürnberg in 1363 twenty fishers lived along the Regnitz on the upstream side of town and at Heidelberg in 1439 fifty-four clung to the bank of the Neckar.[45] Arles had only some dozens of fishing households but smaller Martigues as many as eighteen in its shore suburb in 1308.[46]

Artisan fisheries may have embodied economic innovation but they remained socially stable elements, little disruptive of local human relationships. Even on the seaboard, inshore marine fishers returned home every night or at least every few nights.[47] The scale of enterprises and of technology differed little from norms in subsistence fisheries. This was no lucrative calling. Indeed, as already seen, from eleventh-century Lake Garda to late medieval Paris contemporaries plausibly associated fishers with poverty. Galician ports ranked the status of fishermen below that of merchant sailors.[48] Until a storm flood in 1394 destroyed the

[43] Ebel, "Fischerei und Fischereimethoden," 140–141; Hastrup, *Nature and Policy*, 66–75; Amundsen et al., "Fishing booths and fishing strategies." For Norway compare Amundsen, "Coupled human and natural systems."

[44] Rulewicz, *Rybołówstwo Gdańska*, 278–307. Annual municipal tax rolls from the Kraków suburb Kazimierz during 1369–1390 name sixteen to nineteen individuals living *inter piscatores* on the banks of the Vistula, most of whom other contemporary records also show as working fishers (Chmiel, ed., *Księgi radziecki Kazimierskie*, 8, 21, 26, 167–168, and 254).

[45] Hünemörder, "Fischerei im Mittelalter," 195–196, has more examples.

[46] Stouff, *Ravitaillement*, 201.

[47] Darsel, "Conditions du metier," 475–482; Kowaleski, "Working," 7. On local connections among fishing grounds, fishers, and markets in southeastern England of the thirteenth century and southwestern into the fourteenth, see Kowaleski, "Commercialization," 190–191, and "Expansion," 430–435.

[48] Ferreira-Priegue, *Galicia*, 345–346. In Valencia coastal and estuarine fishers alike were typically poor part-time peasants, while on Mallorca many were slaves (Aparisi, "Fishing in Valencia," 216–224, and Barceló Crespí and Mas Forners, "Fishing in Majorca," 135–138). Compare royal privileges and court testimony in *MGH Dipl.* 6:375, nr. 287; Laurière, ed., *Ordonnances*, V, 207–208; Lespinasse, ed., *Les métiers*, 466–467; and Bossuat, "La pêche en Seine," 62–63, with the remarks of Mollat du Jourdain, *Europe and the Sea*, 150–152, or Bresc, "Pêche et habitat," 530–538.

community, dwellers at "Walraf's landing" (*Walraversijde*) on the open Flemish coast fished to supply the market in Ypres but also cut peat and otherwise eked out a poor living from small sandy fields.[49] Tithe payments by participants in Scarborough's North Sea fishery differentiated the majority of small men who worked skiffs close to shore from the few who ran vessels into the North Sea or to Iceland.[50] Some fishers at contemporary Arles also did well enough to join other master artisans on the municipal council and large lessees of fixed capture devices on the Seine became wealthy from feeding a Paris reinvigorated after the end of the Hundred Years War.[51]

But even men big enough to hire helpers still themselves fished beside their crew and claimed only a share in the catch, not an owner's prior right. Equal shares for each worker and for 'the boat' were recognized norms by the 1150s in Brittany and southern England. A *famulus* working for one of Pavia's fishers received one-third of the catch, but in coastal Artois of the 1180s disputes over tithes on local herring catches revolved around definition of a crewman's share.[52] A century thereafter Klosterneuburg's customal specified equal shares for all who fished collectively through the ice or built a specially constructed trap.[53] Fishers in Genoa and nearby Ligurian ports also used familiar commercial partnerships for shared ownership, operation, and returns from fully equipped fishing vessels, while familial partnerships long remained the norm for inshore boats in southwestern England.[54] Further egalitarian social tendencies in medieval artisan fisheries are remarked below.

### 4.2.2 *Small-Scale Technologies*

Matched to household-based enterprises were the proven techniques for exploiting local resources which characterized this branch of the industry. Indeed medieval fishers who were plainly engaged in supplying local and regional markets drew on the same technical repertoire already seen

49 Tys and Pieters, "Understanding a medieval fishing settlement," 92–94.

50 Heath, "North Sea fishing."

51 Stouff, *Ravitaillement*, 201; Cayla, "La pêche à Paris."

52 Pavesi, ed., *Ordini e statuti*, 58; Darsel, "Conditions du metier," 479–480; Coopland, *Abbey of St. Bertin*, 48.

53 Zeibig, ed., *Urkundenbuch Klosterneuburg*, 169–171.

54 Balleto, *Genova nel duecento*, 175–183 and 192–195; Kowaleski, "Working," 3; Kowaleski, "Commercialization," 205–207. General remarks on the prevalence of shares in late medieval Italian, French, and Dutch fishing are available in Mira, *Pesca nel medioevo*, 43–44; Bresc, "Pêche et habitat," 529–530; Hocquet, "Pêcheries médiévales," 98–105; and de Vries and van der Woude, *First Modern Economy*, 235–246.

in the hands of subsistence fishers. A livelihood for an artisan's family required consistent exploitation from well-known local aquatic systems, calling for judicious seasonal deployment of diverse equipment.[55] Aelfric's stereotypical Anglo-Saxon market fisher put out nets and hooks from his little boat and the part-time specialists who sold salmon and inshore cod to consumers in contemporary Iceland added pot gear and weirs to that simple inventory.[56] To the east where catches from the Danube supplied a Viennese market, the lines, pots, seines, and special fixed trap handled from shore or afloat by fishers at Klosterneuburg in the late thirteenth century were closely replicated in the 1469 regulations governing their counterparts and competitors upriver at Tulln.[57]

All it takes to amplify the sophisticated variety behind generic terms are opportunely detailed sources. Thirteenth-century records from Como show several variants on three basic netting techniques – "*reti a semplice maglia, reti a mantello* [trammel], *e reti a sacco*," that is passive enclosure, entanglement, and something more active, perhaps a seine – each adapted for specific fishes.[58] Not far away at Pavia market fishers' use of small boats on the Ticino and Po rivers was remarked in privileges from the early eleventh through mid-thirteenth centuries, but the rich complexity of their activities emerges only with their fourteenth- and fifteenth-century ordinances. To take pike, tench, chub, nose, eel, sturgeon, small trout or grayling, and (unidentifiable) *balbi* for sale to the townspeople these men handled baited lines and also pot traps, *bertavella*, which they anchored in the rivers and in ditches. Some *bertavellos* were large enough for sturgeon. Of netting gear they used several types: *missia* were long nets deployed overnight in the big rivers, perhaps as drift nets since objects illegally fixed in the stream caused them damage; the *cazia* or *caccia* (literally 'hunt') had mesh attached to poles at a fixed site and was large enough that several fishers might share its construction and catch; the *tragg* or *tracta*, however, suggests some kind of seine. Fishing through the ice was also a collaborative effort

55 Medieval artisan fisheries give little trace of the most opportunistic or individual techniques, capture by hand and spearing, but a commercial contract and other records from thirteenth-century Genoa did expect commercial marine fishers to use "the herb called *tassus*" as a piscicide (Balletto, *Genova nel duecento*, 187–190 and 214–215).

56 Aelfric, *Colloquy*, ll. 90–94 (Garmondsway, ed., 26); Ebel, "Fischerei und Fischereimethoden," 140–141.

57 Zeibig, ed., *Urkundenbuch Klosterneuburg*, 169–171; Petrin, "Tullner Fischerzeche," 32.

58 Mira, *La pesca nel medioevo*, 62–63. Legal records of artisanal fishers on Tuscan lakes describe trammel nets crew-served from boats, teams of fishers building enclosures of reeds, and individuals with hand nets along the shore (Franceschini, *Lago, padule, fiume*, 17–26).

using nets.[59] As many (and, to the distant observer, as obscure) particulars can be pieced together about tackle and techniques of fishers then supplying Paris from the Seine or London from the Thames estuary.[60]

Small rowboats or two–three-man sailing skiffs like those seen in medieval Pavia or Iceland also served the inshore fishers at Scarborough, who pursued skate in summer, plaice in winter, lobster during Lent, and small cod year-round.[61] Such modest investments in water craft, seines, and other moveable gear also supported artisans fishing inshore Mediterranean waters. In 1380 Antoine Sardine [*sic*], Guillaume Mathieu, and Guillaume Esimisol entered into a three-year contract with the Bishop of Antibes, who provided them with a new boat and a grubstake of oats and beans. When the boatmen engaged in coastwise shipping, they paid the bishop three-quarters of their profits; net fishing cost them only one-third but night fishing with lights half, for the bishop then also provided the fuel. Fishing *ad lumen* was important for sardine in Sicily as well as Provence, and both areas shared use of boat seines called, respectively, *xabica* or *bourgin.*[62] Calling for greater investments of capital and labour were the fixed weirs and traps taken at lease by fishers around Paris and in estuaries along the Gulf of Lions or Provence[63] or the great *niewód* seines assembled by cooperative fishing teams (*maszoperla*, *matschoppen*, etc.) to take herring off the twelfth-century south Baltic coast.[64]

Yet even the latter groups were still engaged in marketing a recently caught product to mainly local and regional consumers (in contrast to those treated in Chapter 8 below) and employing capital equipment equally familiar in indirect subsistence fisheries of nearby lords. While increased scale and intensity of fishing effort cannot be excluded in transitions from subsistence to artisan fisheries, it seems unlikely that

[59] Pavesi, ed., *Ordini e statuti*, 41–64, puts flesh on the bare bones of earlier charters and privileges found in *ibid.*, 9–10; Occipinti, "Fortuna e crisi," 320–321; and Brühl and Violante eds., "*Honorantie Civitatis Papiae*," 20–23.

[60] For methods in the Seine fishery see Lespinasse and Bonnardot, eds., *Les métiers*, 212–222; Lespinasse, ed., *Les métiers*, 13–20 and 448–472; Duplés-Agier, "Ordonnances inédites," 48–55; Laurière et al., eds., *Ordonnances*, I, 583–600; Egbert, *On the Bridges*; Géraud, *Paris sous Philippe le Bel*, 608; Franklin, *Dictionnaire historique*, 556–557 and 580–581; Coornaert, *Les Corporations*, 239; Bossuat, "La pêche en Seine," 61–69 ; and Cayla, "La pêche à Paris." Compare for London Riley, ed., *Liber Albus*, 331–334; Wright, *Sources of London English*, 54–114; Galloway, "Storm flooding," 175–177; Galloway, "London and the Thames estuary," 130–135, and "Fishing the Thames estuary," 254–265.

[61] Heath, "North Sea fishing"; Amacher, *Zürcher Fischerei*, 72–77.

[62] Bresc, "Pêche et coraillage," 110–112.

[63] *Ibid.*, 109–110; Bourin-Derruau et al., "Le littoral languedocien," 384–385 ; and Puig, "Ressources de l'étang et de la mer," 94–104.

[64] Rulewicz, *Rybołówstwo Gdańska*, 270–276 and 324–326.

the ubiquitous artisan producers of fresh or lightly preserved fish for nearby markets ever plainly used capture methods unknown to peasant and seigneurial subsistence fishers in the same neighbourhoods. Fishing for sale itself neither necessarily nor evidently called forth technological innovation and change. On the other hand, the marketing function did often (not invariably) reshape relations within household enterprises.

### *4.2.3 A Gender Division of Labour*

In most places small-scale marketing of fresh local fish involved male fishers turning their catch over to wives to sell. The social pattern, today a cliché, thus foreshadowed emergence of long-distance marine fisheries which would remove men from their communities for extended periods of time.[65] Twelfth-century municipal codes for newly Christianized interior Castile expected that wives and children of local stream fishers would peddle the catch. At Toledo these women hawked fresh fish on the street, while imports were offered in shops.[66] The same activity is well documented at Paris, where a wide-ranging tax roll from 1292 identified fish sellers among the most numerous roles of women who worked in the victualing trades. A generation earlier poet Guillaume de la Villeneuve caught voices of those like poor Norman Luce Rossel, fish peddler spouse of a blind English immigrant recently cured in a posthumous miracle of St. Louis: 'Herring, smoked and white, freshly salted,/ I want to sell my herring./ Others you hear cry "weevers!"/ And also bleak from the sea [sardines?].'[67] But women from the families of small-scale fishers on the Seine routinely walked and stood about the city pitching the tubs of live fishes they bore in their arms or on their heads, for at only three designated sites were they permitted to set up carts or stalls. The wife of Pierre le Nourrissier was standing properly to sell at the Petit-Pont in 1422, when royal officers seized her wares for her man's illegal fishing. Margot la Goujonne (*sic*) and a half-dozen other wives of local fishers

[65] Cross-cultural remarks on women fishmongers in Jordan, *Women and Credit*, 103–106, are to be read against the broad sociological perspective of Thompson, "Women in the fishing" (despite lack of deep historical dimensions) and the elaborated example of Neis, "Familial and social patriarchy."

[66] Dillard, *Daughters of the Reconquest*, 159–160; Palacín-Gálvez and Martínez-García eds., *Documentacion*, no. 414.

[67] Herlihy, *Opera Muliebria*, 131; Delaborde, "Fragments de l'enquête," 56–70; Guillaume, *Les miracles*, 153–158; Franklin, ed., *Les rues et les cris*, 154–156, ll. 17–22: "*Sor et blanc harenc frès poudré, harenc nostre ventre voudré. Menuise vive orrez crier, et puis alètes de la mer.*"

made up the majority of pedlars arrested in 1428 for defying the rules.[68] In Scandinavia likewise, Malmø's city council set up that town's first guild of petty retailers in 1534 to regulate an exclusively female trade in fresh fish.[69] In a sense, then, some women found themselves on the sharp edge of the commercial revolution and, more so than their menfolk out on the water, experienced material changes in their daily lives.

The pattern was not, however, everywhere the same. Augsburg's 1276 *Stadtrecht* prohibited woman fishmongers and the same rule applied at Rothenburg ob der Tauber. Zürich had no legal ban and some records indicate women helping their husbands, but they were there normally neither fish sellers nor commercial fishers.[70] Why did south German women plainly play in regional artisan fisheries nowhere near the role they continued to have in subsistence fishing there? For now, the operational and interpersonal features of households based on artisan fisheries – differing just enough to matter from peasant farming or handicraft production – deserve the attention of researchers in local records across medieval Europe.

### 4.2.4 *Collective Organization*

Heads of households in community-based artisan fisheries characteristically joined their fellows in mutual pursuit of shared concerns, blending what present-day commentators might distinguish as economic, social, and environmental interests. Common rights and collective uses were familiar in both rural and urban settings. These essential features had less to do with personal legal freedom or ownership of resources than with users' liberty of economic decisions and the role of (informal) community in constraining them. Indeed simplified modern economic models of common property resources poorly fit both the relations and the outcomes of small-scale medieval commercial fishing. Like other contemporary artisans, fishers organized and regulated themselves to advance group interests, mediate internal conflicts, and promote egalitarian and sustainable forms of exploiting local resources for local market consumption.[71]

[68] Bossuat, "Pêche en Seine," 74–78. In 1496 it was the fish seller Agnès la Choarde who was jailed for her "impertinent" refusal to move along at police request (Delamare, *Traité de la Police,* 116–117 as cited in Bourlet, "L'Approvisionnement de Paris," 6–7). According to Fagniez, *Etudes sur l'industrie,* 112, all in a Parisian fisher's household shared in his right of craft mastery.

[69] Jacobsen, "Women's work and women's role," 10–11, and see Dam, "Fish for feast and fast," 313, for Dutch counterparts.

[70] Hünemörder, "Fischerei im Mittelalter," 196; Amacher, *Zürcher Fischerei,* 135–136.

[71] What follows focuses on self-governance by medieval commercial fishers. The organization and regulation of markets by other participants is treated in what follows.

As discussion of technology has already made plain, individual effort caught most fish, but collective institutions of various kinds were as familiar to country watermen as to those based in towns. Fishers from each village along the Lago di Bientina near Lucca in the thirteenth century jointly leased the fishing in a distinct area, then set rules for participation and methods for the term of their lease.[72] On both sides of the English Channel men from a fishing harbour jointly contracted with a broker (*hôte*) to market their catch.[73] Fishers' confraternities on Dalmatian Rab and in the Quartier Saint-Jean of Marseille enter the written record by about 1300 and later claimed privileges not unlike those at Ravenna back in 943 or much-recorded Paris.[74] Of course the bulky ordinances of late medieval fishers' corporations – what follows draws notably on those from the Zürichsee in 1386, Pavia in 1399, Auenheim in 1442, and others more incidentally – most fully report their activities and role, but the changing functions of literacy and public administration do not obscure the importance of collective practices among artisan fishers in most corners of medieval Latin Christendom.

One long-standing historical cliché has medieval guilds and urban craft corporations epitomize narrow artisanal self-interest. Certainly fishers' organizations reinforce much of that stereotype, though only if we acknowledge that their interests, like those of their neighbours in other trades, were more than simply material. Of course a major preoccupation aimed to secure licence to fish and sell the catch: this was the sole purpose of the documents which first report the *schola piscatorum* at Ravenna in 943 and the organized fishers at Worms in 1106.[75] Rights of the Seine fishers to sell in Paris were set out since the provost's compilation of the 1260s and reiterated in royal charters throughout the fourteenth–sixteenth centuries. At Marseille and some other French coastal towns guilds enforced a special fee on outsiders who wanted to sell fish.[76] The Venetian confraternity of *Nicolotti* had formed before or early in the 1200s, by which time their annual payment of mullet to the

[72] Onori, *L'abbazia di San Salvatore*, 55–61 and 126–128. Seamen's associations in some coastal Norman villages made comparable agreements for collective possession of exclusive rights to inshore waters (Darsel, "Servitudes de la pêche," 106–108).

[73] Hocquet, "Les pêcheries médiévales," 100–103

[74] Pederin, "Commercio, economia, pesca ... in Arbe," 237–239; Fabijanec, "Fishing and the fish trade," 375–378; Sportiello, *Les pêcheurs du vieux-port*; Stouff, *Ravitaillement*, 201; Hocquet, "Pêcheries médiévales." 116–117.

[75] Mira, *La pesca nel medioevo*, 29–31; Boos, ed., *Urkundenbuch*, no. 58. At Pavia the guild gained exclusive control of the fish market in 1386 and in 1399 claimed sole authority to negotiate leases of fishing rights in the county (Pavesi, *Ordini e statuti*, 25–40 and 54).

[76] Lespinasse and Bonnardot, *Les métiers*, 213; Lespinasse, *Les métiers*, 467–472; Hocquet, "Les pêcheries médiévales," 116–117.

doge ensured them of exclusive access to all public waters in the lagoon. Surviving statutes date only from the fifteenth through eighteenth centuries.[77]

Governance and reproduction of the organization called for rules of succession to membership, selection of officials, and procurement of material and supernatural support. The *Nicolotti* gained membership by hereditary succession and those become too old to fish were privileged to join the fishmongers, making for vertical integration of the city's fish supply. On the Rhine at Auenheim guildsmen took their place by hereditary right or payment of five schillings, which was also demanded of former members seeking reinstatement. Each year they selected two "honourable men" to inspect and enforce the rules. Similar routes to membership applied in Paris and Pavia, though the latter fishers also dedicated their entire catch the first week of September to funding their association. Parisian fishers attended and served at the morning mass on the day of their patron, St. Louis, while at Zürich the guildsmen shared regular common meals and took active part in the funerals of their deceased fellows.[78]

Social fellowship reinforced solidarity. While eating in the fishers' hall at Zürich in 1446 city fisher Heini Schwab and (unrelated) country counterpart Werli Schwab came to (temporary) agreement over use of nets in the lower Zürichsee.[79] Resolving conflicts among members was always a major concern of organized artisan groups. Users of different gear had sometimes incompatible needs and expectations. The Zürichsee agreement of 1386 banned anchored gill nets and limited use of the largest open-water seines, while setting strict dimensional rules on other tackle. Guild statutes at Pavia confronted conflict directly: though lifting of another's *bertavellum* trap was heavily fined, up to four such traps could be removed to set the long drift net called *missiam* and an even larger fine applied to any fisher whose stakes, fascines, anchors, or other barriers damaged any *missiam*; on the other hand, a fisher could put his own trap in a ditch already being used by another and on three days' notice even some other device.[80]

[77] Faugeron, *Nourrir la ville*, 42 and 544. Zago, *Il Nicolotti*, is more modern ethnography than medieval history.

[78] Mone, "Ueber die Flussfischerei," 79–81. For guild officers see Sportiello, *Les pêcheurs du vieux-port*, and Ayza Roca, "La pesca en la València," 163; Pavesi, *Ordini e statuti*, 41 and 58; Lespinasse, *Les métiers*, 469–472; Amacher, *Zürcher Fischerei*, 175–178; Faugeron, *Nourrir la ville*, 544, 559, 562, and 583.

[79] Amacher, *Zürcher Fischerei*, 177–178.

[80] *Ibid.*, 387–390; Pavesi, ed., *Ordini e statuti*, 46, 52, 57, and 62.

Collective groups also oversaw access to specific sites and tried to adjudicate or alleviate disputes which arose. The Auenheim guild required those who would lease certain locations for set nets to assemble at Kitzingen before dawn the day after the Feast of Mary's birth (8 September), then within the next eight days mark each site with two special posts. Fishers on the Seine could make exclusive claims only after Easter, when demand for fish softened.[81] This did not always work. The resource potential where the Zürichsee flowed out into the Limmat provoked long antagonism between the Bachs family, who owned a great weir, and trap fishers Jäkli Buss and Johans Vischer. The 1381 election of a guild master and efforts by the Bachs faction to amend the guild charter provoked violent confrontations in first the fishers' hall, then the city council chamber. One party grabbed the charter, the other barred the doors. Before guild leaders brought temporary peace with heavy fines for all, eight fishers led by Buss and Vischer trapped Bürgi Bachs in the city bathhouse and beat him up. Still, in 1384 Bürgi's son Bertschi caught Jäkli Buss driving piles to anchor his row of fish traps and attacked him with a stake. The antagonism rumbled on as late as 1397.[82]

Egalitarian norms commonly keynote the artisan fishers' measures of technical and social self-governance. A guiding principle among regional guilds on the Provençal coast was to enable each fisher to earn his living from his craft.[83] In this spirit at Auenheim none could begin to fish through the ice until safety had been tested with an oar. Then the initial lessee of a site was obliged to post his intention to fish there and to accept three more partners at equal shares. No one could have more than one ice fishery.[84] Subject to the proviso that each newcomer must possess the necessary equipment, nearly identical rules applied in the Pavia statutes from 1399.[85] Each fisher of the Seine above Paris could claim only so many sites as he could actually work himself. Marseille forbade using two boats to pull a seine or trawl and the 1386 agreement on the Zürichsee limited perch lines to ten per fisher, each only up to thirty fathoms long.[86]

A frequent effect and occasionally an articulated intent of the constraints artisan fishers set on their own activities tended toward

81 Mone, "Ueber die Flussfischerei," 79; Lespinasse, *Les métiers*, 472.

82 Amacher, *Zürcher Fischerei*, 192–194.

83 Frangoudes, "Il *Prud'homies*."

84 Mone, "Ueber die Flussfischerei," 79. A further provision limited the number of shelter devices and the length of time any one fisher could use them to concentrate fish for his catching.

85 Pavesi, ed., *Ordini e statuti*, 48 and 63.

86 Lespinasse, ed., *Les métiers*, 466–471; Mollat du Jourdin, *Europe and the Sea*, 141–145; Amacher, *Zürcher Fischerei*, 389. Fishers at Frankfurt am Main could employ but one helper (Cahn, *Recht der Binnenfischerei*, 111).

conservation and sustainable use of local resources. Restrictions on gear and of seasons occur in all the statutes and customaries. The agreement among Traun master fishers in 1418 prohibited equipment judged "hazardous" (*schedlich*) and summer draining of side channels for its harm to roe and small fish.[87] The Pavia guild banned fishing for two spring spawners, grayling from May to November and pike from May to 24 June.[88] Rules "that the good folks of the fishery have made as to nets" and their enforcement on the Thames around London openly reasoned that undersized mesh harmed the reproduction of fish. Those at Lake Garda in 1433 aimed explicitly at the "common benefit and abundance of the aforesaid fishes."[89] When the Venetian government substituted its own legislation in 1491, local interests resisted: four lakeside communes complained that "the said rule … will be prejudicial and detrimental to the greater part of our poor subjects, many of whom make their living from the craft of fishing."[90]

Efforts to regulate and preserve fisheries transcend artisan communities, so later in this chapter we explore another angle and then revisit these topics in Chapter 6. The point just now is simply that medieval artisan fishers did use collective measures to maintain their relatively equal access to local markets and to preserve familiar local resources supplying them.

## 4.3 Urban Fish Markets

On the dry eastern flank of the Apennines medieval Bologna's fifty to seventy thousand inhabitants, twice those in contemporary Pavia, enjoyed nearby access to only two unproductive mountain torrents, the upper Reno and the Idice. Still the city counted in the late 1200s anywhere between 185 and 267 members of a "fishers' guild" (*ars piscatorum*, *societas piscatorum*, later *Compagnia della Pescatori*). Well before its municipal recognition in the 1250s and still long after direct rule by papal governors in the 1480s, this well-organized corporate group with a full catalog of officials and fraternal rights and obligations held a monopoly over legal sales of fish on the market square at the Ravenna gate and on the Piazza del Comune. Its statutes itemize the proper buying, handling, and selling of

87 Scheiber, *Zur Geschichte der Fischerei*, 152.

88 Pavesi, *Ordini e statuti*, 45. Mira, *La pesca nel medioevo*, 44–55, lists many conservation measures by Italian fishers and fisheries authorities.

89 Riley, ed., *Liber Albus*, 333–334; Butturini, "La pesca sul lago di Garda," 47–49, "*ad comune beneficium et copiam piscium praedictorum*."

90 "*dicta capitulo … esse preiudiciales et detrimentosos majori partem pauperum fidelium nostrorum qui ut plurimum vivunt ex arte piscandi*" (Mira, *La pesca nel medioevo*, 54–55).

fish but say not a word about the catching. No surprise. Bologna's "fishers" were in fact fishmongers. They bought their wares fresh from boatmen who laboured up the Reno to the Maccagnano gate on the west or arrived at the Ravenna gate from marshes of the Po delta some 40–50 km away or from more distant coastal lagoons. They imported salt fish to help meet the demand of the Bolognese and the many students, often clerics, who swarmed to the city's famous law school.[91] Simply concentrated urban market demand could pull fish from rich but distant regional waters and make into "fishers" men who never handled a net and used only coins for bait. Commercialization made markets the defining link between producers and consumers of fish.

Buying and selling of fish, like other continual commercial exchanges in medieval Europe, is largely known as an urban activity. The setting helps make marketing, at least in some respects and in later medieval centuries, a fairly well-documented aspect of how people handled fish. This section identifies and exemplifies features, variants, and implications of the medieval trade in fresh fish. At literally hundreds of locations culturally structured consumer demand interacted with the supply of local and regional fishes

### *4.3.1 Freshly Caught from Nearby Waters*

Markets with fresh and lightly preserved products from local and regional fisheries were common in medieval towns. At Arras the toll schedule from 1024 anticipated sales of sturgeon, salmon, shad, plaice, cod, herring, and porpoise from estuarine and coastal waters. Likewise when Parisian intellectuals, Alan of Lille in the mid-1100s and John of Garland a century later, would celebrate the diversity of fishes there available, they named species native to the surrounding region. Up to the twelfth century even salt herring – called in later French *harong d'une nuit*, *blanc*, or *frais-poudré* – were prepared in limited number and for only regional distribution.[92] In the 1170s the Rialto market in Venice regularly offered

[91] Pini, "Pesce, pescivendoli e Mercanti," 333–347; Pucci Donati, "Mercato del pesce." A notarized copy of Bolognese statutes and guild membership books of the fifteenth and sixteenth centuries is ÖNB Cod. vind. 14072.

[92] van Drival, ed., *Cartulaire de ... Saint-Vaast*, 166. Alanus, *De planctu*, prose 1 (Häring, ed., 817–818; Sheridan, tr., 94–98) interestingly selected only familiar taxa from the longer, more learned, list in his probable source, Bernard Silvestris, *Cosmographia*, I, 3 (Dronke, ed., 115; Wetherbee, tr., 85). John of Garland, *Dictionarius*, §72, in Hunt, ed., *Teaching and Learning*, I, 202.

Clavel, *L'Animal*, 157–161 and 176–187, provides strong archaeological confirmation of the limited inland penetration of any marine fishes, even if preserved, before the 1200s, and for continued use of fresh product long thereafter. Scant verbal records are

fifteen varieties from the lagoon and its connected waters.[93] A generation or so later an anonymous Oxford schoolman described a London fishmonger displaying two dozen different identifiable fishes of southern English waters, eight freshwater, ten marine, and six tolerant or migratory between both.[94]

Surveys of markets which enter the record a bit later, such as that at Barcelona beginning in the 1260s, also reflect fish stocks from local waters, little more.[95] Well inland, Merano's 1317 market ordinance distinguished sales of fresh local fish from preserved imports, and late fifteenth-century stewards from the Count of Leiningen's castles in the Palatinate went to towns along the Rhine –Worms, Speyer, Oggersheim, Oppau, Mannheim, and Altrip – to shop for the river's fish.[96] About the same time, Breton fishers were selling their year-round catches fresh to consumers in Anjou and Maine, while salting down for shipment elsewhere only surpluses from their less reliable summer fishery.[97] Even in the arid plateau of southern Castile the fresh fish then on offer at inland towns (Cuenca, Guadalajara, Madrid, and Toledo) were those native to the watersheds of the Júcar, Tajo, and Guadiana and perhaps some inshore species from the Valencian coast 150–250 km away.[98] Culturally similar demand on different markets elicited responses from different local ecosystems.

The fish trade was among the first in medieval towns to specialize and to have its own designated site for sales. Hamburg's oldest marketplace, which predated Danish sack of the town in 845, was called the 'fish market' into the thirteenth century. Passing documentary references confirm special fish markets operating by 1127 at Ypres, 1169 at

summarized in Cutting, *Fish Saving*, 55; Darsel, "Servitudes de la pêche," 107; Boussard, *Nouvelle histoire*, 302–303; Hocquet, "Pêcheries médiévales," 48–49 and 82.

93 Faugeron, *Nourrir la ville*, 31–33 and 312–313, with p. 313 nn. 74–75 detailing much longer but similarly constructed lists from fifteenth-century sources. See also Papadopoli, *La Moneta*, 307–311, and discussion in Robbert, "Twelfth-century Italian prices," 393. Check taxonomy in *Vocabolario della lingua italiano* (Rome: Instituto della Enciclopedia Italiana, 1986–2011).

94 Waitz, ed., "Handschriften," 340, with thanks to Martha Carlin (pers. comm.) for calling my attention to this text and its likely reference to London under Henry III.

95 Mutgé i Vives, *Ciudad de Barcelona*, 110–112.

96 Stolz, *Geschichtskunde der Gewässer*, 375–376; Bull, "Wirtschaftliche Verflechtung der Pfalz," 73–75.

97 Touchard, *Commerce maritime breton*, 59–60.

98 Nase, barbel, trout, salmon, eel, and lamprey are freshwater or diadromous endemics; a more diverse marine list included sea bream and bass (Sánchez Quiñones, *Pesca y Comercio*, 19–54 and 274–277). Similarly on the Portuguese coast fishmongers at Porto handled lamprey, shad, sardine, sturgeon, porpoise, hake, and conger (Pereira, "Pesca maritima," 64–66).

Louvain, 1191 at Ghent, and 1286 in Brussels.[99] The riverbank location where Kölners bought and sold salmon and other fresh Rhine fishes was distinctly mentioned by 1183,[100] about the same time King Alfonso VIII of Castile, in his later much-copied compilation of municipal law for Cuenca, heavily fined anyone who sold local fish outside the public market, even from his own dwelling.[101] By the end of the twelfth century fish were being sold in Rome at the "Foro Piscium" beside the church of Sant'Angelo and since 1296 the fish sellers' statutes at Perugia confined this activity to the guild hall.[102] The marketplace with inspection facilities where local fishers at Namur displayed their catch differed from that where marine fishes were sold, as also at both Lübeck and York.[103]

Sheer size meant high medieval Paris had greatly elaborated facilities. As already seen besides ambulatory street pedlars, who could stop to rest or make a sale but nowhere set up to receive customers, the city allowed retailing of fish on the "pierres au poissons" at the Castelet, the Baudet gate, and the Petit Pont. The Castelet location provided six covered positions for wholesalers, who could also pay the king for permission to set up their own stands at Les Halles, on the rue de la Cossonnerie, or de la Cochonnerie. Local fisherfolk or so-called marchands forains, actually fishers from settlements along the Seine and Marne, competed for places on the stones and, if those were filled, could still display their wares at the same locations so long as they blocked no traffic.[104]

Characteristic institutional and economic relations on the fish market arose from the encounter of consumer and supplier interests. As for other vital consumables, medieval public authorities, morally and politically obliged to seek some balance, took large responsibility for market operations.[105] This oversight might be under the prince's aegis, as in

99 Lampen, *Fischerei und Fischhandel*, 195; Uytven, "L'approvisionnement des villes," p. XI:102.

100 Lampen, *Fischerei und Fischhandel*, 195–196, and more generally Cahn, *Recht der Binnenfischerei*, 121–122 and 127.

101 Ch. XLII:18 of the Cuenca *fuero*: Ureñja y Smenjaud, ed., *Fuero*, 810–812; Powers, tr., *Code*, 213.

102 Lanconelli, "Gli *Statuta*," 89 and 94; Sanfilippo, *Roma dei romani*, 344–347 (growing Roman demand required two more sale sites by the 1400s); Scialoja, ed., "Statuta," 839.

103 Balon, "La pêche," 28 and 35–39; Goldberg, *Women, Work, and Life Cycle*, 66 and 69; Prestwich, *York Civic Ordinances*, 5 and 13; Lampen, *Fischerei und Fischhandel*, 195–197 (with other north German examples).

104 Bossuat, "La pêche en Seine," 74–79. Non-absorbent and easily cleaned stone blocks or stone-topped tables still often serve for the butchering and sale of fish.

105 Late medieval sociopolitical consensus on managing food markets for consumer protection is amply discussed in Uytven, "L'approvisionnement des villes," 75–116; Britnell, *Commercialization*, 27, and "Price-setting," 27; and Devroey, "Food and

Cuenca's twelfth-century charter, the royal provost's jurisdiction at Paris, or the monarch's control over London fishmongers' privileges. Elsewhere communal or municipal statutes held sway. It often seemed wise to download actual supervision of fish selling to self-regulating suppliers, corporate guilds of fish sellers like those remarked above at Perugia and Bologna. Different administrative arrangements generated different kinds of records, providing historians with many prescriptive and a few descriptive accounts of how fish actually reached market consumers.

### *4.3.2 Fishmongers*

Diverse vendors, not just fishers and wives with their own catches, offered consumers fresh fish from their baskets, pushcarts, benches, stalls, or stones (Figure 4.2). Men and women who specialized in retailing fish, what the Veronese called "*pischaroli vendencii pisces recentes*," were in most places plainly distinguished from the fish catchers.[106] At Zürich they handled sales for fishers from beyond the Zürichsee, and in Basel, Würzburg, and Kraków they had formed their own guilds by the late fourteenth century.[107] Such traders had been active in most middle-sized English towns for a least a generation before 1304, when York counted as many as fifty fishmongers and nine more purchasing agents ("forestallers of fishmongers").[108] In smaller places with abundant local resources such as Berlin or Tortosa, however, fishers themselves remained the main sellers of fish.[109] Permanent shops long remained rare, probably reflecting the continued seasonality of both demand and supply, which changed only gradually with the commerce in preserved fish products to be explored in Chapter 8. As market relations intensified, especially where larger urban populations concentrated demand, specialized fishmongers organized both regional supply networks and local distribution. Two

politics," 79–84. The government of Lisbon was explicit in 1435: markets were regulated "for common benefit" ("*por prol communal*") (Catarino, "Abastecimento," 27).

106 Mira, *pesca nel medioevo*, 70–72. Perugia's statutes clearly distinguish fishers from fish sellers (*ibid.*, 67; Sciolaja, "Statuta").

107 Amacher, *Zürcher Fischerei*, 126–127; Lampen. *Fischerei und Fischhandel*, 201; Piekosiłski, ed., *Kodeks dyplomatyczne*, #367.

108 Carlin, "Provisions for the poor," 37; Masschaele, *Peasants, Merchants, and Markets*, 17, 25, 69–70, 79, 100–101, and 118; Prestwich, ed., *York Civic Ordinances*, 27. Somewhat later traders acquired fish from ships docked in and around Exeter and peddled them inland, though extant records do not always distinguish between fresh local catches and mainly durable products from far away (Kowaleski, *Local Markets*, 307–312).

109 Cahn, *Recht der Binnenfischerei*, 122–124; Curto Homedes, "El consum de peix," 158–166.

Figure 4.2 Fresh fish displayed in baskets set out on stones for market sale in Strasbourg, 1517.
Woodcut by artist Hans Frank illustrating discussion of prices and morality by the famous Alsatian preacher Johannes Geiler von Kaysersberg (1445–1510), *Die brösamlin doct. Keiserspergs vffgelesen võ Frater Johañ Paulin barfůser ordẽs: Vñ sagt võ dẽ funffzehen Hymelschen staffelen die Marie vff gestigen ist/vñ gãtz von dẽ vier Leuwengeschrei; Auch von dem Wãnenkromer/ der Kauflüt, sunderlich heupsche matery bei. lxii. predigẽ/ nutzl* ... Publication: [Strassburg]: [Johannes Grüninger], [1517], part II, fol. 47v. Reproduced with permission of Temple University Library Special Collections.

extended examples, Rome and Paris, suggest the more common and a probably extreme development.

Marketing of fish to medieval Rome's many consumers began with collaboration among local fishers and retailers. By the late eleventh century a corporate group called the *Schola piscatorum Stagni* ("Association of fishers of the pond"), many of whom sold fish in the Tiber-side quarter, Trastavere, was organizing access to waters in Tiber delta lagoons and provoking tension with church corporations which owned most of the fishing rights and ran their own subsistence fisheries there. In 1158 the Schola collectively contracted with one monastery for exclusive rights to fish the *Stagno di Maccaresse* in return for a fixed rent in kind and the landowner's employing no more than two of its own fishers. By 1296, when the original contract was renewed at a slightly lower share or annual payment in cash, 160 members were claiming right of succession in the Schola. Further renewal in 1362 left this arrangement unchanged, but people in the Schola were already reordering their relationship into a marketing guild, the *Ars piscivendulorum urbis* ("Craft of fishmongers of the city"), first so named in 1363. Notarial registers show members of the *Ars* taking *piscarie* in the Tiber and along the nearby coast on five- to ten-year leases, for which they made to the religious or lay owners significant annual cash payments plus a few symbolic fishes, often the much-prized shad. Toward that century's end some leasehold contracts varied the annual rate depending on whether the pope was resident in Rome or not. Leading fishmongers such as Paolo Rossi, Nucio Gibelli, and Cola di Nucio di Cecco had also become property holders in Trastavere and around the fish market beside Sant'Angelo in Pescheria.[110]

Early records of the *Ars* are gone, but workings of the late medieval Roman fish market come clear from the 1405 revision of its statutes. Municipal authorities allowed tax-free retail sales of fish by anyone who paid for a site, but a regular and reliable supply of sturgeon, eel, shad, drum, sardines, prawns, and dolphin – all named in the statutes – was provided through the guild's two major participant groups, the *coctiatores*, large wholesale fishmongers, and many small retailers, the *pescivendoli*. *Coctiatores* contracted for delivery from fishers on the Tiber, coastal lagoons, or local store ponds, but were obliged to deal publicly with the retailers and could sell to any individual retailer no more than one-third of any shipment. The latter in turn were to offer their goods at designated public sites or, exceptionally, in specific households which had requested service, always taking personal responsibility for the quality of their merchandise. All members agreed to boycott any "magnates and nobles of the city, monasteries, churches, hospices, religious, or their servants" who failed to pay their bills to either wholesalers or retailers and also any fishers who might then turn around and offer the fish as their own. Membership in the *Ars* was hereditary or by cooption and required regular dues to fund inspection for quality, settlement of internal disputes, and an annual procession, mass, and festival on August 15, feast of the Assumption of the Virgin, the group's patron.[111]

[110] Vendittelli, "Diritti ed impianti di pesca," 409–422; Sanfilippo. *La Roma dei romani*, 352–374. Wickham, *Medieval Rome*, 101–108, argues that specialized fishers and fishmongers by the early twelfth century signal Rome's precocious development of a stable market in ordinary consumables.

[111] Lanconelli, "Gli *Statuta pescivendulorum*." Denis-Delacour, "*Fraudis* et *Piscandis*", reports similar arrangements, issues, and conflicts on the eighteenth–nineteenth-century Roman fish market.

The *Ars* (with its free market rivals) outfitted the rich Roman market in those fresh fishes which humanist papal secretary Paolo Giovio celebrated in 1524. Much of what his *De romanis piscibus libellus* reported about roughly thirty-one native central Mediterranean fishes (and five invertebrates), seven catadromous or euryhaline, and five freshwater taxa he had lifted from classical sources, but he also knew the local dialect name for Tiber sturgeon, that trout and other freshwater fishes came from interior lakes, and that pike were a more northerly species found in Italy's Alpine waters.[112]

Counterparts elsewhere to Rome's *coctiatores* and *pescivendoli* took like initiatives to reflect and satisfy local customers' needs. The Venetian guild of *Compravendi da pesca* included wholesale and retail fishmongers from its formation in 1227; only licenced members could sell in the fish market (*Pescheria*) beside the Rialto bridge. In 1304 fifty-four licensees were named. Those neither investors in lagoon fisheries nor retired Nicolotti with sons and other kin in the fishery, could connect with suppliers through a privileged inspection and brokerage agency for all foodstuffs, the *palo*.[113] Individual enterprises can be harder to track. At Arles a local fishmonger, Rostagno Laugier alias Spanhoget, contracted in early January, 1434, with two local fishers, Jean and Bertrand de Grans alias Bernoyes, for "all the good fresh fish" they would catch up until Easter from a named salt pond thirty kilometres down the estuary near Fos-sur-Mer or elsewhere. Up to the start of Lent, Rostagno would pay seventeen groats a basket and take delivery at the pond three times a week; during Lent he would up the price to eighteen groats and get his fish every day. Baskets and terms of payment – including a guaranteed minimum of ten florins, half in mid-Lent and half at Easter – were contractually defined, as was Rostagno's responsibility for all shipping costs.[114] Thirteenth-century statutes encouraged also London fishmongers to go out and make deals at actual fishing ports: economic records from around 1300 show some of them as shipowners with regular links

[112] Giovio, "*De romanis piscibus libellus*", ed. Travio and Penio. Only Giovio's short final chapter 43 treated preserved fishes, mainly regionally well-known sturgeon, sardines, salt trout from Lake Garda, and brined tuna ("which fills all the marketplaces and pickle jars of the taverns"). For exotic northern imports see Chapter 8.

[113] Faugeron, *Nourrir la ville*, 544–564, 582–606; Romano, *Markets and Marketplaces*, 234 n, 22. Judging by their assessments for late fourteenth-century taxes (*ibid.*, 617–619), Venetian fishmongers did not attain the high local standing of their Roman counterparts, no surprise in a community dedicated to lucrative long-distance commerce and with diverse local fisheries resources.

[114] Stouff, *Ravitaillement*, 422–423. *Ibid.*, 203–204, 421–422, and 423–424, treat and exemplify contracts from other Provençal fishmongers for catches from both fresh water and the open sea ("*in maribis Arelatis et alio mari*").

around the Thames-centred region and interests as far as Yarmouth, Boston, and Lynn. But aggressive efforts to channel all London's fish through fishmongers' hands by banning local fishers or non-Londoners from selling at retail provoked such opposition from fellow citizens as to render moot even occasional royal privileges to that effect.[115]

Strength of demand on the semi-arid Spanish plateau induced inland municipalities unwilling to rely on local subsistence fishing or voluntary sales from privately owned fisheries to auction off the exclusive right to sell fish on their marketplace or a specific location. *Obligados*, the merchants or elite investors who took such contracts for the Lenten season or up to two years, were obliged to sell contracted quantities at pre-set prices but commonly exempt from sales tax (*alcabala*).[116] They secured their inventories by digging streamside holding tanks or enclosing pools with netting (facilities called in either case *Xudrias*, *judrias*, or sometimes *sudrias*).[117]

Dealers in many places transported and held live fish for future sale. Where waterways were available, boats with a perforated compartment in the hull, what the Dutch called 'waterschip', circulated new water among the living cargo. There first recorded in 1339, hulks of at least ten medieval examples have been identified in fields once beneath waters of the Ijsselmeer (Zuider Zee). Some carried eel from coastal fisheries to consumers in Flemish towns and even London.[118] On the Rhine traders hauled salmon in such craft and on the Rhône and Saône they were common enough to be assessed a special toll.[119] More direct overland routes called for pack trains, whether equipped with wooden tanks or moving just fast enough that during cool seasons fish carefully packed in straw and baskets remained edible. Sicilian mule trains thus carried fresh fish the twenty kilometers from Sferracavallo to Palermo, while 'panier ponies' likewise linked fishing ports along the Yorkshire coast to inland

115 Riley, ed., *Liber Albus*, 323–331; Unwin, *Gilds and Companies*, 37–42; Thrupp, *Merchant Class*, 95–96; Campbell et al., *A Medieval Capital*, 81–86.

116 Sánchez Quiñones, *Pesca y Comercio*, 226–240 and 259–350; Puñal Fernández, *El Mercado en Madrid*, 189–194. Municipalities in Aragon did the same (Rodrigo Estevan, "Fresco, frescal, salado, seco, remojado," 565–571). All were modeled on arrangements for cereals as detailed in de Castro, *El pan de Madrid*.

117 Sánchez Quiñones, *Pesca y Comercio*, 105–107; Puñal Fernandez, *Mercado en Madrid*, 175–180

118 Hutchinson, *Medieval Ships*, 139–141; Dam, *Vissen in veenmeren*, 169–171.

119 Mone, "Ueber die Flussfischerei," 75–77; Richard, "Commerce du poisson," 191–192; Richard, "Etangs et le commerce," 40–44; Rossiaud, *La Rhône*, 253. Eurhaline species could be transported with relative impunity, but most obligate marine and freshwater varieties will not long survive in the 'wrong' water.

Figure 4.3 A sale from live storage, Paris, 1317.
A Parisian fish seller displays to a prospective customer a lively fish from his floating storage tanks on the Seine, as depicted in a marginal illumination done at St. Denis in 1317. From Paris BN, MS fr. 2090–2092, II, fol. 129r detail.

towns.[120] Could Havelok and Grim have used such means to supply Lincoln back in the tenth century?

Like Castilians, London fishmongers kept their stock alive in the 'stews' or 'stanks' some had dug into the riverbank or built from timber and netting on the other side of the Thames at Southwark.[121] In Kraków only local merchants with a storage facility (*halder*) were allowed to carry unsold fish over until the next fast day.[122] A display manuscript given the French king in 1317 depicts in one marginal illustration (Figure 4.3) a vendor on a floating store tank moored beneath a Parisian bridge, who lifts a wriggling fish for inspection by a prospective customer.[123] Fishmongers in towns of the Hungarian kingdom from Bratislava to Obuda caged off *vivaria* or *reservacula* in protected reaches of the Danube to keep their stock alive through the winter for Lenten consumption. Such facilities in the Rhône at Lyon or Avignon were worth as much as a good city house.[124] Big fishmongers controlled significant capital investments. Their storage facilities and supply chains aimed to anticipate and satisfy customer demand with some freedom from natural seasonality or the mere luck of the day's haul.

[120] Bresc, "La pêche et les madragues," 177–178; Robinson and Starkey, "Sea fisheries," 128–130. For more English examples see Masschaele, *Peasants, Merchants, and Markets*, 202, with similar overland cartage of fresh fish in Provence (Stouff, *Ravitaillement*, 207–208) and Valencia (Aparisi, "Fishing in Valencia," 235).

[121] Kelly, "Bishop, prioress, and bawd," 350–354 *et passim*; Carlin, *Medieval Southwark*, 38 and 49–50.

[122] Piekosiński, ed., *Kodeks dyplomatyczne*, #262 and #299.

[123] Egbert, *On the Bridges*, 40–41.

[124] Szende, "Stadt und Naturlandschaft," 394–395, and "Sopron fish market," 161; Rossiaud, *La Rhône*, 258–259.

While the buyer from the store tank on the Seine was getting a freshwater fish in particular, he and many fellow Parisians generally enjoyed what may have been the widest access to fresh fishes in all medieval Europe.[125] Their regional supply network reached so far with a complex institutional arrangement called the *chasse marée* as to cap with products of the distant sea those creatures which flowed in from the fresh waters ringing western Christendom's richest concentration of market consumers.[126]

Artisan fishers of Paris and environs, already met earlier in this chapter, offered the city their own local catches of freshwater fishes native to the Seine and its nearby tributaries. Multiple seigneuries with rights and jurisdictions in and around the city complicated the fishers' access to waters and customers. This confusion of authority – and hence records – affected even more the work and our present knowledge of the quantitatively more essential fishmongers. The latter comprised at least two and sometimes three or more organizational groupings, whose diversity reflected both political and essentially functional, even ecological, differences.

Local supplies were greatly augmented from the surrounding region by 'freshwater fishmongers'. Royal privileges from 1182/3 had conceded the right of butchers established in the area of the Castelet to sell fish – perhaps especially when sales of meat were stopped during Lent? – but by the mid-thirteenth century the king's provost, Etienne Boileau, recognized the *poissonniers d'eau douce* as a distinct craft with its own organization and regulations.[127] A householder wishing eel, barbel, carp, tench, roach, lampreys or the like might turn to their members at the "pierres au poissonniers" near the Castelet or Petit Pont, or find the least of them hawking such fish in streets across town. Some of what these fishmongers had to offer came from their own fixed fish-catching installations, permanent weirs or traps taken on lease from riparian landowners along the Seine or other streams. Others owned their own fishponds, too.

[125] My grasp of the sources and literature on marketing of fish in medieval Paris was much strengthened thanks to Mme. Geneviève Séguin of Montréal, Québec, who generously made available to me her unpublished mémoire, "L'approvisionnement de Paris en poisson à la fin du Moyen Âge. Pour une histoire juridique, économique et sociale," presented for her Diplôme d'Étude Approfondie en histoire médiévale at Université de Paris I, Sorbonne-Panthéon, December 1996.

[126] Farmer, *Surviving Poverty*, 11–32, paints an especially lush picture of the wealth of late thirteenth-century Paris.

[127] Boileau, title 100 (Lespinasse and Bonnardot, eds., *Les métiers*, 214–218). For Boileau's career and relation to King Louis IX see Jordan, *Men at the Center*, ch. 2. Subsequent records of the freshwater fishmongers are in Lespinasse, ed., *Les métiers* 19–20 and 450–457. See also Boussard, *Nouvelle histoire*, 300; Coornaert, *Les Corporations*, 79; and in general Auzary-Schmaltz, "Les Contentieux," 59–60.

The largest such men were more strictly importers and buyers, exercising their exclusive corporate privilege to buy fish within two leagues (roughly 10 km) of Paris and sell them in the city. This right the *poissonniers* had to exercise where the fish were caught and not from people who were already carrying their own fish to Paris. Up to ten in the morning the freshwater fishmongers could legally sell only to consumers, but thereafter they could also market their supply wholesale to other retailers. Jehan de Lievre was jailed in 1473 for violating the first rule, but back in 1428 Jean Haucie, who brought several hundred carp from Picardy, was able to sell some to fellow dealers and others to retail pedlars.[128] Freshwater fish markets were officially supervised through the royal kitchen administration and received especially solicitous attention when Anglo-French hostilities and the English occupation during the 1420s–40s threatened the security of marine supplies.

*Poissonniers de poisson de mer*, narrowly defined, handled fresh, i.e. not preserved, fishes from the sea.[129] This supply came to Paris exclusively by commercial means – so not through estate-management channels – and under official regulation. Twelfth-century shippers on the Seine, the *mercatores per aquam*, likely handled some marine fishes, but the volume of consumption and supply soared thereafter and in inverse relation to limited, probably declining, freshwater supplies.[130] The provost's book of trades from the 1260s recognized a mature corporate grouping. Memoranda regarding internal tariffs on marine fish entering Paris in 1321 suggest 5,000 standard cartloads at 6,000 fish per load in that year.[131] Thirty million marine fish would be more than 150 per Parisian. Forty years later in 1361, when an emergency tax to ransom the king from English captivity was apportioned, the trade in seafish subject to royal authority was valued at 10 percent more than that in meat and ten times that in freshwater fish.[132] But no marine fish were to be had in Parisian waters, or even so near that fishers could, as in Rome or Bologna, themselves transport a fresh catch up the river to fishmongers and consumers.

Already by the 1260s an elaborate system, the *chasse marée*, was in place to provide and distribute to Parisian consumers fresh fish from the Norman and Picard coast some 150 and more kilometers away. There is

128 Bossuat, "La pêche en Seine," 74–75.

129 See in general Auzary-Schmaltz, "Les Contentieux," 60–67; Boussard, *Nouvelle histoire*, 302–303; Cazelles, *Nouvelle Histoire*, 384–386; and Crossley-Holland, *Living and Dining*, 82–86, although these authors too little distinguish between fresh marine fishes and preserved fish products.

130 Clavel, *L'Animal*, 164–174.

131 Maillard, "Tarifs des 'coutumes'," 247–248.

132 Bourlet, "L'Approvisionnement," 5–6.

no evident reason to think these arrangements had existed as recently as the start of the thirteenth century, when Philip II Augustus expelled a rival and disobedient vassal, John of England, from his hereditary possession of Normandy. The system then lasted up to the Revolution, all the while pushing preindustrial transport technologies to the limit.[133]

Such nearshore marine catches as rays, whiting, gurnards, weever, seasonally fresh herring, and salmon, arrived at the great market of Les Halles on pack horses (and only centuries later, wagon trains) from such fishing ports as Honfleur, Cayeux, and Boulogne. There wholesale buyers and shippers had arranged at each return of the local fishing fleet for the catch to be laid in specially designed baskets, lightly dusted (*poudré*) with salt, covered with straw, and sent off in the charge of specialized transporters. The convoys, and a legal case from 1361 thought fifty to two hundred horses a normal size range, made the trip in thirty-five to forty hours during the winter months and, with extra relays of fresh horses, twenty-four hours in summer, arriving at Les Halles with the morning sun. They followed set and privileged routes protected by royal authority against seigneurial tolls and imposts that might slow their progress and raise their costs. In 1315 royal officials took legal action against the prior and seigneur of Milly-sur-Thérain (Oise) for demanding tolls and in 1352 against the abbot of Saint-Denis and a half dozen other lay and religious lords for seizing fish in return for passage. Nor were fishmongers permitted to go out and make advance deals with the transporters after they had crossed the Oise some twenty-five kilometers from the city. Lest disputes over the transport and marketing make the fish worthless, these cases went straight to the king's high court, the Parlement de Paris, which by the mid-fourteenth century had set up a special panel to settle them promptly.[134]

Elaborate but practical regulations called for immediate inspection of the cargo arriving at Les Halles, payment of the shippers, and obligatory quick sale of a fast-spoiling commodity. The crown authorized sworn brokers, *vendeurs*, to approve each shipment for sale, set a just price on it, and mediate between the shippers and the fishmongers. As time passed the shippers gained a stronger voice in managing the system and removing some privately held charges against their business. In 1361 the

[133] See in general *ibid.*," 14–22; Auzary-Schmaltz, "Les Contentieux," 64–67; Boileau, title 101 (Lespinasse and Bonnardot, eds., *Les métiers*, 218–222); Laurière, *Ordonnances des roys*, II, 578–582; and Lespinasse, ed., *Les métiers*, 13–19 and 409–426.

[134] Bourlet, "L'Approvisionnement," 7–14, with a map of routes p. 13. Likewise in central Italy, mule trains based in Umbria carried freshwater fish from L. Trasimeno and other sources west to Lazio and Tuscany and east to towns on the Adriatic, continuing to the late sixteenth century (De Nicolò, "Production et consummation," 54–56).

collectivity of the *marée* was allowed to fund its own endowment with a charge of two denier on each livre of fish sold. To ensure wide and competitive distribution, no dealer, whether wholesale merchant or retailer, could take more than six pack-horse or three wagon loads in a day. In the warmer season from Easter to 1 October (St. Remi) the fish had to be sold the very day they arrived in the city; the cooler season allowed one day more. Remains of the fishes especially associated with marine fishmongers – the whiting, fresh herring, mackerel, cod, gurnard, dogfish, rays, sardines, sea trout, salmon, and porpoise named in their ordinances – are especially well represented in elite archaeological contexts of later medieval Paris.[135]

Personal names identified with the fish trades in the several books of returns on the *taille*, a tax on the wealth of non-noble Paris households between 1292/3 and 1313, totaled 648; after resolving duplicate and repeated entries, these labeled 263 individuals. Just more than 200 of those people dealt in fresh fish, and while not one in ten identified him- or herself by source of supply and hence 'craft', about half lived close to the Castelet, Petit Pont, or other locations for selling freshwater fish and the others around Les Halles, the site for handling fish from the sea. Members of both groups fell mainly among the higher, but not highest, taxpaying brackets. Jehan le Grand, a wholesaler in marine fish, was assessed thirty livres tax on his wealth in 1313 and his fellow fishmongers, Mauger of Cayeaux and Guillaume Babille, were each charged 22 lb. 10 s. – when three taxpayers in four owed less than one livre. Jehan and his wife Geneviève had recently acquired city property worth 294 livres, but no fishmonger achieved office as *echevin* or provost of the merchants, the leading positions in the mercantile community.[136]

Beside the 200 Parisian dealers in fresh fish around 1300, self-identified 'herringers' (*harenguers*) then numbered sixty-one. For some time after the mid-thirteenth century ambiguity or instability prevailed between their standing as part of the marine fishmongers or a separately organized craft. Eventually this became a free trade, what the English called 'salt fish merchants',[137] who handled not fresh fish but a preserved product. As compared to the fishmongers they served a different market

[135] Clavel, *L'Animal*, 187, assembles several such finds. Compare Franklin, *Dictionnaire historique*, 581.

[136] Séguin, "Approvisionnement de Paris," 87–89; Maillard, "Tarifs des 'coutumes'," 248; Bove, *Dominer la ville*, 55. The annual income of a Paris mason at the time has been estimated at about eight livres.

[137] Boileau, title 101 (in Lespinasse and Bonnardot, eds., *Les métiers*, 218–222); Laurière, *Ordonnances des roys*, II, 575–582, with corrections of the dating offered by Delamare *Traité de la Police*, 1–281; Lespinasse, ed., *Les métiers*, 13–19 and 409–426.

segment, faced different technical problems with the quality of their goods, used different modes and routes of transport, and ultimately drew on different animals captured by different means from different, no longer local, ecosystems. Salt herring in thirteenth-century Paris can stand for new trends in medieval fisheries to be examined at length in Chapter 5 below.

What especially struck observers of high medieval Parisian markets were the huge diversity and sheer quantity of fish on offer. In the mid-1200s John of Garland set his lesson on Latin and French vocabulary of crafts, tools, and occupations as a walking tour through Paris. Fish sellers, he said, sold salmon, trout, roach, two kinds of lampreys, eel, fresh cod, plaice, pike, tench, ray, herring, mullet, perch, gudgeon, and stickleback, plus the meat of porpoise.[138] Some 200 years later an anonymous writer on the rich food supplies of Paris near the close of the Middle Ages turned a half eager, half disgusted eye to the *chasse marée*:

> There comes to Paris from the sea, so much fresh as salted and stinking, and some mackerel, fresh and salted, some rays, big and small, so much fresh as stinking, and these arrive each day in such great amounts that it is impossible to know their numbers. And Paris is like an abyss![139]

Paris was a bottomless pit for the consumption of fish. Demand-driven local market mechanisms let Paris, like and beyond the lesser cities of the medieval west, pull live, fresh fish out of every water body within its technical reach.

138 "*Piscatores vendunt salmones, truttas, murenas, morium, pectines, anguillas, quibus associantur lucii, rocie, stincti, ragedie, allecia, mulli. Ipsi vero piscatores capiunt cum hamis et rethibus perchas, gobiones [MS gabiones] et gamaros; quia canes marini ab equore devehuntur.*" John of Garland, *Dictionarius*, §72 (Hunt, ed., *Teaching and Learning*, I, 202, and compare variants in II, 141, 148, 152, 150). Not all the fishes named in the text above are those suggested by Hunt, vol. II, who does not compare John's vocabulary with that of other contemporary Paris-savvy academic writers on fish, such as Albertus Magnus.

139 "*Il y a de marée a Paris, tant fresche que sallée et puante, et de macquereaulx frais et salez, de grans raies et petites, tant fresches que puantes, et en arrive par chascun jour en si grant quantité, qu'il est impossible d'en savoir le nombre. Et est un abisme que Paris.*" Le Roux de Lincy and Tisserand, eds., *Paris et ses historiens*, 494.

Having so visited Rome and Paris it would be otiose to multiply case studies of abundance, organizational features, and supply at further well-endowed and richly documented urban markets such as London or Venice. But see Galloway, "Fishing the Thames estuary," 265–270, and Faugeron, *Nourrir la ville*, 523–537 and 639–676, respectively. Both document the continued importance of local sources of supply.

### 4.3.3 For the Sake of Safe Abundance

As was by 1200 the general pattern for other dietary necessities, a ubiquitous regulatory climate promoting abundant, cheap, and safe supply on local fish markets reflected the strength of urban demand and the political importance of consumer interests. Whatever the issuing authority, the commonalities among rules on dozens of fish markets from one end of western Christendom to the other are remarkable. What follows tries only to describe the rough consensus on provisions and procedures which everywhere set the prevailing tone. Such provisions aimed first to maximize supply and retail competition,

Not only had fish to be sold in public, the designated open markets welcomed outside suppliers. Neither the privileged fishers at Worms in 1106 nor the powerful fishmongers' guild of thirteenth- through fifteenth-century London could bar entry to others with fish to sell. The London corporation fought the fishmongers to a standstill on this issue.[140] Bans on private dealings before fish reached the market, what the English called forestalling and regrating,[141] and on transactions between outsiders further encouraged direct consumer access to fish producers. Sellers on Perugia's market could not legally buy from lakeside peasants carrying their catches to the town, while Krakow applied the same rule to trade along the Wisła, requiring fishmongers who found supplies in rural marketplaces to provide sealed certificates of the fish's legal origin.[142]

Like effects were expected from the controls many places imposed on resale or re-export. Some like Cuenca's late twelfth-century charter, heavily fined anyone who removed fish from the district, while fourteenth-century markets at Tortosa and Zürich demanded a special licence to resell fish. In the 1520s the royal Scots burgh of Stirling forbade regrating of salmon in particular.[143] Restricting the time or quantities allowed for legitimate purchase and resale had similar motives. Midday was the earliest Trento, Rovereto, Pavia, Madrid, Grasse, Worms, or London allowed dealers to buy fresh fish for resale. The Grasse statutes of 1263 further

[140] Boos, ed., *Urkundenbuch*, no. 58; more than a century of London statutes appear in Riley, ed., *Liber Albus*, 328 and 401–402, and are contextualized by Thrupp, *Merchant Class*, 95–96 and references there provided.

[141] The terms refer to middlemen who buy outside the market, but not from the producers themselves, and then try to profit by selling therein at a higher price. For legal understandings see Seabourne, *Royal Regulation*.

[142] Scialoja, "Statuta et ordinamento," 831; Piekosiński, ed., *Kodeks dyplomatyczne*, #299 and 336; Riley, ed., *Liber albus*, 325 and 329.

[143] Cuenca code 49:19 (Ureña y Smenjaud, ed., *Fuero de Cuenca*, 812, Powers, tr., *Code of Cuenca*, 213); Curto-Homedes, "El consum de peix," 161–165; Amacher, *Zürcher Fischerei*, 126–127; Renwick, ed., *Extracts from the Records*, 9 and 27.

restricted subsequent sales to single-serving quantities.[144] All of these measures served to promote a retail buyer's market.

Freshness was an understandably high regulatory priority, vigilantly protected in customary 'burghal laws' of barely urbanizing twelfth–thirteenth-century Scotland and in communal market statutes of precociously commercialized Italy.[145] Practical measures could be ingenious. At Grasse the ordinance allowed no salt on fish being set out for sale for the first time, then required a layer of salt on subsequent days. Torino ordered the tails cut short on fish being offered for sale a second day and Leuven the seller to display a black flag. At Swiss Fribourg, fresh fish not sold the day they arrived on the market had to be donated to the local hospital. Berlin sought the same ends by allowing sale of live fish only.[146]

Public display and shaming sanctioned people who transgressed community rules by selling bad fish. At Rostock in 1338 *Luscus piscator* was jailed "because he sold rancid fish on the fishers' bridge."[147] The fish market statute of 1335 at Grasse forbade sale of '*pisces putrefactos*' under pain of 50 sols fine and the fish were seized and thrown away. Market overseers in Paris pitched '*mauveis poisson*' into the Seine, while Zürich made convicted sellers of spoilt fish themselves dump the offending product from the bridge over the Limmat. Bamberg ordered rotten fish publicly burnt.[148]

A generally hygienic marketplace was also thought essential for dealing with perishable foodstuffs. Shortly after 1300 fish sellers at Barcelona and at Malchin in Mecklenburg alike were held responsible for cleaning the market, with the latter small town charging a special fee to cover the expense.[149] In 1427 authorities at Winchester prosecuted a group of fishmongers for dumping rotting entrails of fish on the public roadway behind their place of business, while in 1481 Roman officials banned fish sales at Campofiore market because the heat made fish go bad.[150]

[144] Mira, *Pesca nel medioevo*, 76; Puñal Fernández, *Mercado en Madrid*, 182; Stouff, *Ravitaillement*, 424; Riley, ed., *Liber albus*, 330.

[145] Scottish Burgh Records Society, *Ancient Laws*, I, 35–36; Italian cases in Mira, *Pesca nel medioevo*, 73–76.

[146] Stouff, *Ravitaillement*, 424–425; Nada Patrone, *Il cibo*, 326–330, including parallels from elsewhere in Piedmonte; Champion, *Fullness of Time*, 33; Cahn, *Recht der Binnenfischerei*, 124–128.

[147] "*Propterea, quod vendidit ancidos pisces supra pontem piscium*," Lampen, *Fischerei und Fischhandel*, 199 n.

[148] Lespinasse and Bonnardot, eds., *Les métiers*, 217; Stouff, *Ravitaillement*, 425–426; Amacher, *Zürcher Fischerei*, 188 and fig. 75; Lampen, *Fischerei und Fischhandel*, 199 n. 1075.

[149] Mutgé i Vives, "L'abastament de peix," 17; Lampen, *Fischerei und Fischhandel*, 195.

[150] Coy, "Provision of fowls and fish," 36; Lanconelli, "Gli *Statuta pescivendulorum urbis*," 127–131.

Perugian statutes originating in the thirteenth century make a special point of forbidding fish sellers or anyone else urinating in the fish market (*mingere in domo pisciarie*), and repeat the stricture in a later article to include 'nor any other turpitude' (*nec aliquam turpitudinem*). Hygiene had a moral aspect, too: Perugia and Rome both forbade games of chance on the fish market, the latter asserting that this measure would stop blasphemy.[151] Consumer safety of all sorts was thus a second principal aim of market regulation.

Ongoing official concern for honest dealing reflected both cost and safety. Fish sellers had not only to use standard measures but to present their goods properly "according to their nature" (*secundum eorum naturam*) in the words of a Grasse municipal ordinance.[152] Cuenca's late twelfth-century *fuero* specified that river fish were to be sold according to the same weights as meat and grouped by size.[153] Perugian ordinances of 1279 required sales by *sauma* or *salma*, a unit of volume, defined in earlier communal statutes.[154] The latter resembled the standard fish baskets with which Parisians and Londoners were then already familiar. London authorities regularly inspected the baskets and burned in public those found illegally small, so catching four fish sellers in one incident at the time of Edward II (1307–1327).[155] Vendors were especially obliged not to mix fish varieties in their displays or in unit sales baskets. Echoing long-documented practice in places like London and Rome, a Paris court in 1393 found a fishmonger guilty for having sold whiting and cod in the same basket.[156] Plainly participants on local markets were thus expected to know and care about specific fish varieties.

Rules on at least some marketplaces further and explicitly aimed to protect local and regional fisheries resources by limiting the impact of consumer demand on what were thought to be important or threatened species. The earliest ordinances set down by Paris provost Etienne Boileau in the 1260s imposed minimum sizes for sale of six riverine species: pike, eel, barbel, tench, chub, and carp.[157] In 1313 London

[151] Articles 3, 49, and 54 (Scialoja, ed., "Statuta," 825, 845 and 847); Lanconelli, "Gli *Statuta pescivendulorum urbis*," 116. Romano, *Markets and Marketplaces*, 85–86, 127, 132, 144, and 148, sees close municipal attention to the honesty and deportment of fishmongers. Lleida also banned gambling in the fish market (Roca Cabau, "Provision and consumption," 300).

[152] Stouff, *Ravitaillement*, 427.

[153] Cuenca 43: 7–8 (Ureña y Smenjaud, ed., *Fuero de Cuenca*, 816–819; Powers, tr., *Code of Cuenca*, 215).

[154] Scialoja, "Statuta," 814 *et passim*.

[155] Riley, ed., *Liber albus*, 402.

[156] Auzary-Schmaltz, "Les Contentieux," 66; with the pertinent regulations in Lespinasse, ed., *Les métiers*, 13, 16, and 409.

[157] Lespinasse and Bonnardot eds., *Les métiers*, 213 and 216.

fishmongers took, condemned, and burned a net from the abbey of Lesnes, because it "was too close and insufficient for fishing, to the injury of the water (*pro destruccione riparie*) and common damage to the whole city and people resorting thither." A two-inch minimum mesh size for all waters upstream of London probably went back to the 1280s, and so, too, the fishmongers' opposition to more destructive gear. The city's 1419 legal compilation, *Liber albus*, reports fishmongers seizing eight such nets which "were false, to the destruction of the advantageousness of the waters of Thames, in regard to the fish breeding in the same waters, to the loss of all people, as well of the city as others, both living near to and at a distance from such city." An expert panel appointed by London council measured the nets, found four acceptable, and had the others burned.[158] Perugia's communal statutes of 1279 had already imposed protective size limits on pike, eel, trout and crayfish as part of regulating market sales "so that a greater abundance of fishes may be had in the city."[159] Tortosa set rules in 1394 and 1444 on market sales of shad in order to protect local stocks of this anadromous species.[160] At Zürich authorities marked the sale counters with the legal minimum length for several varieties and to ensure proper identification had images painted on the wall of the adjoining Rathaus.[161] The official actions here noted were all directed toward the market setting and sector. Concerns for food security and future resource sustainability were plainly not incompatible with commercialized regional fisheries and market consumption of local fishes in the Middle Ages.

Urban markets for fresh fish, early fruits of Europe's commercial survival and revival, connected artisan fishers to a largely town-dwelling clientele of courtiers, officials, clerics, merchants, and other artisans. Viewed from another perspective, local markets linked soon-growing numbers of people who lacked direct access to fisheries or to subordinated fishers with the resources which could, at some cost, meet their needs for fish. The resources – fish stocks and the human skills and labour to exploit them – were local and regional, keyed to nearby natural aquatic ecosystems. The connections were organized and maintained by people of the fish markets, fishers and fishmongers, by whose expertise and along whose networks the fish traveled from the waters to the

[158] Riley, ed., *Liber albus*, 251 and 334.

[159] "*ut in civitate Perusii maior habundantia piscium habeatur*" (Scialoja, "Statuta," 814 n. 2 and 817 n. 1; Biganti, "La pesca nel lago"; Vincenti, "La tutela ambientale."

[160] Pastor y Lluis, "La pesca de la saboga," 109.

[161] Fish sellers were further obliged to report to market inspectors any attempt to sell spoilt or undersized fish (Amacher, *Zürcher Fischerei*, 188–189). Chapter 6 below explores possibly mixed motives for all such restrictions.

kitchens and trenchers. Large markets, Paris being the extreme case, spread out the largest, most complex networks. Consumer demand drove expansion, and consumer interests – cheap, safe, sustainably abundant supplies – were defended against those who controlled and profited from the supply networks. Networks and protection alike remained based on general local knowledge and familiarity with the local ecosystem. But markets, even 'imperfect' markets affected by local power and cultural expectations, set and communicate value differently than by simply asserting and taking action to satisfy a want. Markets use prices.

## 4.4 Market Price

In 1289 French king Philip IV, perceiving a rise in fish prices, blamed the "malice of fishers" (*per maliciam piscatorum*) for shortages.[162] The king's assumption of inadequate supply driving up prices may hold some truth, but he failed to acknowledge that the prices paid on the market actually served to turn human wants into effective demand and so provided an impetus for maintaining or enlarging supply. Hopes for the latter explain why in 1381 the infirmary of Saint-Bénigne at Dijon advanced a local fishmonger twenty gold francs a month so that he would deliver each Friday and Saturday "two large pike and two large carp, two pike and carp neither small nor large, two small pike and carp, [and] two chub," plus, on Fridays only, a basket of *fritures* and thirty-four little fish.[163]

In a market economy price should somehow calibrate the value set on a commodity by its scarcity relative to effective demand for it. A price quotation has meaning only in comparison to something else, another price, some other goods, a wage or income. Chapter 2 (pp. 86–87 above) already established the relatively high cost of fish as a source of calories or protein. But the prices quoted for medieval fish varied greatly over time, space, and type. Their bewildering diversity and the gap-riddled historical record allow only a few, but important, generalizations and inferences about price formation and thus about the place of fish as a commodity on medieval markets.

### 4.4.1 *Price Formation*

When it came to fish, natural and cultural forces, the latter not necessarily or simply economic in quality, collectively formed medieval price

162 "*et inde accidit quod sint multo plus solito cariores*." Duplés-Agier, "Ordonnances inédites," 49. Done "the day after Easter," so after Lent.

163 Beck, "L'Approvisionnement en Bourgogne," 176.

relationships, especially on the urban markets which dominate the known historical record of price but then as well the market sector itself. Specific seasons, varieties, and group behaviours all mattered.[164]

On natural but also cultural grounds the season and success of catches strongly affected the price of fresh fish. Some long-favoured species were inherently rare in nature (unusually large specimens or predators atop the trophic pyramid) and others became so over time (see Chapter 5 below). Market scarcity also reflected accessibility, manifest in the knowledge, labour, energy, time, and capital it took to catch any given fish. Most fishes have a season of natural peak availability, commonly when their spawning concentrations are especially susceptible to passive capture techniques (weirs, traps, gill nets). Their continual consumption at other times of year called for more expensive active seine and trawl technologies or individual baits on hooks, while seasonal dispersal lowered the efficiency of fishers' efforts. In consequence, the fifteenth-century kitchen at the Duke of Guelders's Lobith toll station on the Rhine, for instance, bought different varieties at different seasons, while laws at Tortosa acknowledged the seasonal spawning migration as determining the fishery and market for shad. Rhinelanders had reason to call them *Maifisch*. The price of fresh herring in Artois and Flanders peaked in early August some 25–30 percent above annual means, just before the migratory schools reached the region for a season there traditionally dated 24 August through 11 November.[165] But extant data sets may lack the fine resolution needed to detect natural phenomena more subtle than the feast or famine of such migratory animals, here one week and gone the next.

On days and seasons when cultural ideology ordered fish to replace meat, however, everyone knew fish prices went up. Municipal ordinances in Sicilian towns Palermo and Catania set one level for fish on meat days and a higher level on Fridays and other fish days, while at Vercelli they distinguished the Lenten price from that of other seasons. Fishmongers at Tortosa got 25–30 percent more per pound of popular varieties in Lent

[164] Grafe's dissection in *Distant Tyranny* of markets and prices for salt cod in early modern Spain sets enviable standards not generally to be achieved for the Middle Ages. For some fishes we will revisit those ideas in Chapter 8. Supplement 0.2.2 treats issues with medieval price data.

[165] Bossche Erdbrink, ed., *Het 'Keuckenboeck'*, 1–38; Curto Homedes, "Consum de peix a la Tortosa," 152–153; Hocquet, "Pêcheries médiévales," 55; Uytven, *Zinnelijke Middeleeuwen*, 164–165.

than at other times.[166] Scottish royal accounts for 1343 clarify that the king's kitchen paid twelve pence per salmon before Easter and afterwards only four.[167]

Whatever the season, what people in various medieval communities were prepared to pay and take for different kinds of fish show certain broad parallels and some curious variations, the latter perhaps attributable to local differences in taste. Chapter 2 already observed prices differing by factors of two to four on markets in late twelfth-century Venice and Cuenca, the former ranking sturgeon, trout, and turbot the highest per pound and the latter also trout. Small fish of whatever origin were least valued.[168] So certain varieties commonly claimed a premium, as did fish of relatively greater size. How much of the latter might be attributed to the greater efficiency and less waste in consuming a larger fish? And how much derived from scarcity and display values? These considerations came clearly into play in the three customary value groupings used along Breton and Norman coasts: 'royal' or 'free' taxa – sturgeon, salmon, dolphin (fish), trout, turbot, weever, red mullet, and sea bass – rated the highest, followed by fish "à lard" – marine mammals and tuna –, and at bottom all kinds of "vulgar fishes."[169]

Valuations survive from several towns and territories of the late medieval Empire. The Lübeck market featured bulk quantities of the preserved stockfish, local cod, and herring for which the town gained fame, but in the early 1400s next to the stockfish salmon took the highest price; a century later this rough rank-order still prevailed.[170] Interior sites on salmon rivers such as Cheb (Eger) in the highest reaches of the Elbe system gave salmon a 20 percent premium over pike or trout; absent the salmon at Amberg in the Danube watershed, pike led. Both there and in Tirol small river species (gudgeon, dace, etc.) got half or less of the preferred varieties.[171]

Price relations around the late medieval western Mediterranean, while reflecting the great diversity of marine life there, bore interesting parallels

166 Bresc, "Pêche et les madragues," 167; Nada Patrone, *Il cibo*, 329; Curto Homedes, "Consum de peix a la Tortosa," 152–153

167 Gemmill and Mayhew, *Changing Values*, 303–304. Kowaleski, "Seasonality," clearly lays out effects on both supply and demand in English marine fisheries.

168 Papadopoli, *Le Moneta*, 307–311, and Ureña y Smenjaud, ed., *Fuero de Cuenca*, 816–819 (Powers, tr., *Code of Cuenca*, 215).

169 Darsel, "Conditions du métier," 477; Hocquet, "Les pêcheries," 46–47.

170 Bertheau, "Wirtschaftsgeschichte des Klosters Preetz," 261. But although Lübeckers and surely also their unmarried daughters dwelling honourably at Preetz ate freshwater fishes as well (e.g., Paul, "Knochenfunde," 59–60), their late medieval financial accounts support no price comparisons.

171 Hitzbleck, *Bedeutung des Fisches*, 106–107; Benecke, *Maximilian*, 116 and 148.

of scale and ecology. Early fourteenth-century sturgeon topped the charts from Valencia to Avignon, with other species ranked also by size from tuna or grouper "and such other fishes as have that size" (*et tot altre peix de tall qui haja escata*) down to the cheapest "peix menut" weighing several to the pound and costing but 10–15 percent of the top. But red mullet and some of the larger sea breams went for more than their size might predict.[172] By the 1400s, however, price listings at Grasse no longer referred to sturgeon, ranked tuna at the top, then grouped other fishes in descending value as *peys bergo*, *peys mercant*, and, priced less than a third of the top, *peys selvage*. The upper group, going for 2.5 *patac* the pound, included red mullet, sea bass, some other mullets, brill, dentex, gilthead, and two other porgies of more modest size. Cheapest were the usual small fishes as well as whiting and dogfish.[173]

During the central and later Middle Ages, then, markets across Europe had their favoured fishes, sturgeon most generally, then other locally and regionally common large varieties: along seacoasts, tuna and big inshore species; inland, salmon and pike. But certain smaller fishes – trout, red mullet, weever – further stand out as especially valued by consumers, while in some inland areas they were also willing to pay fairly well for large cyprinids. In general, however, small fishes came cheap, which meant anywhere from one-half to one-eighth the price per pound of the most expensive ones available at the same place. Surely size alone conveyed some prestige value and the possibility of higher market returns could motivate fishers to target high-priced species and specimens.

Institutions and groups who consumed much market fish could also by their coming, going, or mere deportment sharply affect local prices. English king Edward I's residence at York during 1298–1305, with his courts of justice, exchequer, and other servants on hand for a conquest of Scotland, so inflated all prices that a joint ordinance of the royal council and civic authorities undertook in 1301 to establish price controls on all victuals, including fish.[174] Well-known high prices for fish on the market of Avignon during the pontificate of the gourmet John XXII (1316–1334) inflated leases of the fishery on the coastal Étang de Berre to ten times what they had been worth in 1300. Subsequent more abstemious and fiscally weaker papal regimes brought them down by half

[172] Ayza Roca, "La pesca," 179–180; Curto Homedes, "El consum de peix," 161–162; Mutge i Vives, "L'abastament de peix," 112; Lleonart et al., "Marine species and their selling prices"; Grava, "Notes martégales," 159–161; and Stouff, *Ravitaillement*, 203.

[173] Stouff, *Ravitaillement*, 203 and 426–427, uses market regulations from 1463, and Bresc, "Pêche et coraillage," 108–109, slightly different listings from 1493.

[174] Prestwich, ed., *York Civic Ordinances*, notably pp. 1–4, 9 and 13, is contextualized by Seabourne, *Royal Regulation*, 77.

in the 1340s–60s, only to be halved again by century's end when Avignon sheltered only a contested pontiff.[175] A century and more later provisioning contracts at Madrid rose in value when the Castilian royal court arrived and sank with its departure.[176]

A congeries of causes thus made fish prices variable and, especially for fresh fish, often volatile, sometimes fluctuating abruptly from week to week or year to year. For now the tidiest early examples of interannual variations are series for fresh herring in England and for several local species in Navarre. Because those data sets run longer and say more about long-term changes, they will appear below as Figures 6.1 and 6.2.[177] Germane at this point are simply the year-to-year variations in English herring prices by factors of two or more in the early thirteenth century. In late fourteenth-century Navarre the price per hundred fresh sardines could double or fall by more than half from one year to the next. That per pound for the other staple of the region's fishery, hake, shifted only about half as abruptly, and that for the product of local inshore fishing, conger, still less again. While short-term cultural changes of the sorts just mentioned cannot be ruled out, it is worth observing that the more pelagic fisheries for herring, sardine, and hake appear to have been less stable than those caught closer to shore. Price data alone cannot discriminate between changes in supply or demand. But plainly some short-run variability in the market for at least some fresh marine varieties was fairly common.[178] Insofar as longer-term trends might be ascertained, a much larger set of price series will be explored in following chapters.

Price fluctuations of any sort sparked public worry. In anticipation or response, authorities made price fixing a common feature of supply management on medieval urban fish markets. In this fish followed precedents set even since Carolingian times for grain, wine, and other dietary essentials. While using diverse mechanisms and criteria all such regulatory efforts rested on fundamental scholastic teachings of a 'just price'.[179]

175 Grava, "Notes martégales," 155–159 (now to be read in terms of Weiss, *Versorgung des päpstlichen Hofes*, 318–322 and 392–402). Richental was surprised that the Council's increased demand had failed to exhaust supplies of fish and other foodstuffs (*Konzilschronik*, ff. 22b–26a; Loomis, tr., pp. 98–101). Rising effective demand could call forth supply.

176 Puñal-Fernández, *El Mercado*, 193–194. So, too, when the court visited Guadalajara in 1500 (Sánchez Quiñones, "Los precios," 188–191).

177 Original data appear in the Allen–Unger database, sub. England, Herring, and in Hamilton, *Money Prices, and Wages*, appendix v. Please go to Supplement 0 for more on prices.

178 Large interannual variability deserves attention because it vitiates drawing conclusions about medium- to long-term trends from mere pairs of price quotations some decades or centuries apart.

179 See note 105 above.

Later medieval English authorities watched over the price of fish in situations less extreme than that at early fourteenth-century York. Although Parliament occasionally passed statutes, more often local officials intervened to prevent individuals or sudden events from violating what a supervised market had established as 'reasonable'. The clearest known cases, like Colchester fishmongers John Chancellor and John Crene, respectively prosecuted in 1404 and 1451 for the same offence of selling four fresh herrings per penny when others were selling five, postdate the economic troubles of the mid-fourteenth century, but the institutional infrastructure was in place much earlier.[180] Edward I had confirmed a London ordinance that "neither for anger nor for spite, shall any vendor hold his fish too dear; and if the vendors do so, the Mayor and reputable men shall assign proper persons to assess the same." By the 1380s London authorities were negotiating official prices for fish in advance of sale, a practice generalized as legislation in or before 1416.[181]

On similar grounds Venetian magistrates responsible for the fish market imposed maximum prices from before 1172 until well into the eighteenth century.[182] Fifteenth-century Roman statutes set the price for white sea bream and those in Florence for sturgeon.[183] The precedent set in Cuenca's late twelfth-century royal charter was emulated in such subsequent, highly detailed municipal ordinances as from Valencia, Tortosa, and Madrid.[184] At Paris and in the Low Countries the task was more often left to market officials.[185]

With some exceptions controls on the retail price of fresh fish in the Empire may have become the norm only in the inland south and from the fifteenth century. An ordinance for Innsbruck from 1470 established price levels for freshwater fishes sold alive or more cheaply dead, and the maxima set for all Tirol in 1493 were lower in the principal Inn valley

[180] Seabourne, *Royal Regulation*, 77; Britnell, "Price-setting," 3 and 6–7.

[181] Riley, ed., *Liber albus*, 330; Seabourne, *Royal Regulation*, 87–88.

[182] Papadopoli, *La moneta*, 307–311; Robbert, "Twelfth-century Italian prices," 393; Welch, *Shopping in the Renaissance*, 74 and 87; Faugeron, *Nourrir la ville*, 32, 237–239, 632–637 and 640–647.

[183] Ciappelli, *Carnevale e Quaresima*, 56–57; Lanconelli, "Gli *Statuta pescivendulorum urbis*," 116.

[184] Cuenca *fuero* 43:9 (Smejaud, ed. 1935, 816–819; Powers, tr., *Code of Cuenca*, 215); Ayza Roca, "La pesca," 179–180; Curto Homedes, "El consum de peix," 161–166; Sánchez Quiñones, *Pesca y comercia*, 304–324; Puñal Fernández, *Mercado en Madrid*, 186–190 and 201–202, all to be understood in terms of broader provisioning policies described in de Castro, *El pan de Madrid*.

[185] Bourlet, "L'Approvisionnement," 17–22; Auzary-Schmaltz, "Les Contentieux," 59–67; Balon, "La pêche," 33.

than in communities at higher elevations.[186] Certainly some kind of price fixing was an expected feature of medieval markets in such a basic consumable as fish.

All these real, generally demonstrable, but in most particular cases ill-specified forces – natural and cultural seasonality, the intersection of scarcity and taste, heavy market demand from certain social groups, political intervention – do make systematic comparisons across space and time among a bewildering variety of fish sales in medieval Europe a hazardous venture. Nevertheless, pending further research, the limited serial data and more that is anecdotal or incidental can probably sustain and help explain two general but significant findings, namely the comparatively high and the probably rising but volatile price of fish as food.

### 4.4.2 *Buying Fish*

Viewed as a market commodity the fish artisan fishers and fishmongers had on offer were commonly too expensive to serve basic dietary subsistence needs across even medieval urban society, much less broader social strata. Long before quantitative sources let today's historian verify this, Peter Abelard, the twelfth-century professor whose fame helped build the scholarly reputation of Paris, made his views clear in advice to the abbess Heloise, his separated and cloistered wife. Meat, he wrote, had lower cost (*minores expensas*) than the many expenses for various dishes of fish (*multis expensis diversa piscium fercula*). Furthermore, said he, fish were more difficult for the poor to obtain.[187] Market records now confirm this. Early fourteenth-century Parisians paid twelve *denier* for a piece of fresh eel, same as for a dozen loaves of bread and three times that for a pound of beef.[188] Some generations later, shopkeepers in Prato were retailing eel and brined tuna at 4 *soldi* the pound, but pork meat and pork sausages at 2–3½.[189] When, as in 1449, a policeman at Namur was paid each day only twice the price of a carp and half that of a small pike, his table saw little of those animals.[190] Compelling reasons thus made fresh market

186 Stolz, *Geschichtskunde der Gewasser*, 378; Benecke, *Maximilian*, 148. Hitzbleck, *Bedeutung des Fisches*, 106, also offers a 1438 example from Amberg in the Upper Palatinate.

187 Zimmermann, *Ordensleben*, 208–209, assembles the relevant passages from Abelard, Ep. 8.

188 Houquet, "Les pêcheries," 54.

189 Marshall, *Local Merchants*, 19.

190 Lentacker et al., "Historical and archaeozoological data," 2; Thomas, "Hygiène, approvisionnement," 283–284.

fish ever more, in the words of fifteenth-century German satirist Heinrich Wittenwiler, "*ein herren speis*."[191] Persons of wealth claimed preferred access to it. The fish market provided limited sustenance for the poor.

A caveat may, however, be in order. Remains from those diverse 'little fishes' rated at the cheapest price points in records from Normandy to Valencia or the Danube commonly occur in meticulously excavated urban waste deposits. An early fourteenth-century latrine in Einbeck, a small town in southern Lower Saxony, yielded about 20 percent bones of herring and smelt but twice that share of small locally caught cyprinids. Freshwater fishes, cyprinids, perch, sculpin, and trout, all less than ten centimeters long, are numerous, sometimes almost half the total, in twenty latrines of eleventh–seventeenth-century date in towns along the Rhine from Schaffhausen to Basel.[192] How much demand for inexpensive little fishes was latent among high medieval urban workers?

High prices for fresh fish compared to meat signal strongly that cultural (ideological) motives for fish consumption much exceeded the nutritional. No one was compelled to eat fish and only a minority did eat so much as to register in their own protein. So in some sense fish were no dietary staple for medieval Europeans. Nevertheless the non-caloric drive to consume fish was strong enough that authorities regularly engaged the market to ensure supply and moderate prices. The market for fish sustained artisans, fishmongers, and the attention even of the king of France. Consumers' desire for fish – now manifest, with whatever extra-market distortions, by price –was a (the?) dynamic element projecting human numbers and wealth (not just coercive force) on to the fishery and hence on Europe's aquatic ecosystems.

><> ><> * <>< <><

The emergence and spread of commercial fishing, even at small local scale, incrementally shifted relations between consumers and the fish they ate. Urban-based consumers provided a principal impetus for growth of fish consumption and fishing efforts since about 1000 and

[191] Wittenwieler, *Ring*, Wiessner, ed., ll. 2905–2908 (Jones, tr., p. 38). Many more south German allusions of like tone are listed in Weissner, *Kommentar*, 119. Compare Hitzbleck, *Bedeutung des Fisches*, 100–109; Dirlmeier, *Untersuchungen*, 312, 376–378, and 397–406; Dyer, "Consumption," 33–35; and Serjeantson and Crabtree, "How Pious? How Wealthy?," 137.

[192] Heinrich, "Fischknochen aus Einbeck," an unpublished excavation report (personal communication); Hüster Plogmann, "Der Mensch lebt nicht from Brot allein," 192–198. Sculpin, loaches, sticklebacks, small perch, and various small cyprinids also commonly occur in urban latrines of northwestern France (Clavel, *L'animal*, 136).

Figure 4.4 "Fresh fish" in northern Italy, *c.* 1370.
Procuring 'fresh fish' in the manuscript of *Tacuinum sanitatis* illuminated for the Visconti court at Pavia, *c.* 1370. Note keep net in the water and a storage tub on the bank. ÖNB Cod. vind. ser. nova 2644, fols. 82r 'pisces recentes'. Reproduced with permission of the Austrian National Library.

more generally by around 1200. Those buyers were rarely present at the capture of the fish. The market structures just explored intervened between catch and consumption.

Fish markets are distinctively urban sites, a point of sale separate from those of capture or landing. The latter became more a space for wholesale merchants and brokers, especially in maritime locations. Even if not so separated, the very urban setting where fish passed from fisher or fishmonger to buyer can be recognized as a step toward perceiving the animal less as something of nature and more as a cultural creation. Set on the stones or in a basket rather than at the waterside the fish takes on qualities of an artifact. While the northern Italian illuminators and editors in the 1370s of a new manuscript edition of *Tacuinum sanitatis*, a widely distributed guide to health, depicted men catching 'fresh fish' from a watercourse, 'pickled' and 'salted' fish came from shops.[193] (See Figure 4.4 and compare Figure 8.5.) Even if treated with care, at such inland locations as Paris, Liège, Kraków, or Milan, the sellers handled a perishable local product. Parisian *poissoniers de la mer* pressed against the limits of that technological envelope. And even dealers in fresh fish at maritime towns like Tortosa, London, or Lübeck confronted the brief longevity of their wares.

Artisan fishers worked within limits and occasionally reveal awareness of limits. So, too, did their mainly local customers. This understanding, however amorphous, lay behind guild- and market-focused measures to promote conservation of fish stocks. But as demand increased the limits were also contracting around them. Already artisan fisheries of the medieval commercial revolution felt larger ecological consequences of an expanding European economy and, in all likelihood, of more natural forces of change.

[193] ÖNB Cod. vind. ser. nova 2644, fol. 82r and compare 83v and 83v. Source criticism in De Battisti et al., eds. *Libro di casa Cerruti*; Schlosser, "Veronesisches Bilderbuch"; and Adamson, *Medieval Dietetics*, 83–92, all as revised by Hoeniger, "Illuminated Tacuinum."

# 5 Aquatic Systems under Stress, *c.* 1000–1350

Artisan fishers played a small yet symptomatic role in medieval commercialization, itself but a part of the historic process of European development during the high Middle Ages. From something less than thirty million in Carolingian times, the human population west of Russia grew to about fifty million around 1000 CE and over seventy million by the early 1300s. Feeding that demographic surge called for intensified use of existing arable and massive clearance of woodland to produce cereal grains. An increasing share of that larger food production supported growing numbers of non-agricultural settlements, first small, then ever larger towns, dynamic nodes of the commercial revolution. What had been insignificant urban populations increased to one in twenty Europeans, perhaps more, by the early fourteenth century.

That Christendom's long-acknowledged medieval experience of growth had broad environmental impact is now generally clear. Culturally shaped wants and needs of burgeoning Europeans pressed upon interactive relations between medieval people and fish, some in quite obvious ways, others indirectly reflecting known ecological processes. What here follows can demonstrate ecosystem changes, effects on fish species, and shifts in fisheries (the engagement of humans with local/regional fish stocks) at the level of socio-natural sites. Much is the too commonplace environmental history tale of human impacts on the natural world, some of which then bounce back on human activities, whether contemporaries were aware or not.

During those same centuries autonomous European nature continued to vary independently, itself modifying conditions for aquatic life at local and larger scales. Fluctuations in temperature and hydrological parameters worked to the advantage of certain fish varieties and disadvantage of others, including species of considerable human interest. From historical distance the interplay between more familiar cultural and less easily reconstructed natural forces is not easily traced. Probable interactions do not yield a single determined outcome, but suggest multiple perspectives on diverse natural aquatic systems, some of which medieval people used.[1]

## 5.1 Environmental Consequences of Demographic and Economic Growth

What had the medieval expansion and acceleration of economic activity to do with aquatic life? Replacement of woodlands with intensified arable agriculture changed basic hydrological conditions both directly and by proliferation of water-powered grain mills. Rising human numbers and their concentration into towns added nutrients and contaminants to watercourses, while the demand for fish as food soared. As these impacts accumulated in each region, traditional freshwater and shoreline fisheries visibly came under stress.[2]

### *5.1.1 Habitat Destruction*

One long-familiar medieval economic trend needs mere re-articulation in ecological terms: growing reliance on cereal food meant permanent plowed fields replaced woodland from central Spain to Sweden and Wales to Poland.[3] Clearance of forests, which slow runoff and maintain

[1] Readers who recall Hoffmann, "Economic development and aquatic ecosystems" will at first see in this chapter extensive carry-over from that article. But decades of evolving environmental understandings and continued research into medieval conditions have shown that humans alone, even in the newest Anthropocene, are neither the sole drivers nor determinants of environmental interaction. Despite the silent opacity of most written sources, scholars must now concede that natural and cultural forces had become well entwined long before synthetic treatment of late medieval marine fisheries becomes possible. This must not deter efforts to trace environmental changes of whatever origin.

[2] In Regier et al., "Rehabilitation," 87–88 (and see works there cited), fisheries ecologists argued that "conventional exploitative development" damages aquatic ecosystems through a sequence and synergy of excessive harvests, damming, destructive cultural practices, organic and toxic pollution, and urbanization without regard for environmental effects. Their ecosystem model comes primarily from comparative study of more or less modified river systems in the late twentieth century. A longer temporal perspective on changes humans have caused in New World marine coastal environments is laid out in Jackson, "Historical Overfishing." The deeper historical record here sketched generally confirms these models, but suggests greater nuance and different specific processes. Whether in marine or freshwater systems, the fishes first or most affected will be those 'ecological guilds' with the most vulnerable habitat requirements (see Chapter 1, p. 49 above). Further discussion of habitat issues appears in Supplement 5.1.

[3] Darby, "Clearing the woodland," is a classic and so, too, Wickham, "European forests"; Hoffmann, *Environmental History*, 119–136, attempts an update.

steady streamflow, inescapably alters the pattern of stream discharge to greater seasonality and irregularity. Rain and meltwaters run more quickly off farmland. Larger and faster runoff more forcibly abrades stream beds and channels, and then falling water levels leave a contracted stream and deposits of eroded materials. One astute observer in late thirteenth-century Alsace noted how clearance of the Vosges in his own lifetime had caused much more rapid and dangerous runoff.[4] Modern scholars detect like sequences of medieval deforestation and flooding in the Po basin and in central Poland.[5] Biologists now also know that unstable flow regimes make life hard for fishes. Those living in running water must expend more energy during floods. They lose eggs and young to winter spates and suffer high mortalities when small streams dry up in summer. Species which spawn in flooded margins are adapted to consistent seasonal patterns of rising and falling water, so instability disrupts their reproductive behaviour.[6]

Soil erosion and alluvial deposition is becoming a well-known consequence of medieval agricultural expansion. First the natural vegetative cover was removed. Then characteristic medieval farming practices disrupted the soil surface and its structure. Plowmen drove long straight furrows or pulverized the soil to prepare for autumn sowing of winter grains. Bare fallow stripped all plants from a third or a half of arable almost year-round. Large open fields broke neither wind nor water nor the creep down slopes of the soil itself. All this let topsoil flow to the watercourses, especially during heavy winter rains and snowmelt.[7] In the Leine valley of Saxony, where during the 780s and 790s Frankish conqueror Charlemagne had promoted new rural settlements and clearances, large sediments from eroded topsoil overlain with former subsoil yield radiocarbon dates between the 790s and 850s. Across the entire lower Rhine valley a major pulse of sedimentation set in around

4 Writing about 1300, a Colmar Dominican compared his own times with the less-developed Alsace his fellow friars had entered a century before, including "*Torrentes et flumina non ita magna tunc sicut nunc fuerunt, quia radices arborum fluxum nivium et imbrium per tempus in montibus retinuerunt.*" (Jaffé, ed., "De rebus Alsaticis," 236).

5 Fumagalli, *Landscapes*, 110–121. Dunin-Wąsowicz, "Natural environment and human settlement," 94–95 and 102, refers to "great changes in the hydrographic balance [viz. of Central Europe], which reached proportions of a natural disaster during the course of the 13th century."

6 Compare Hynes, *Ecology*, 327–332, and the different habits of "white fish" and "grey fish" in Regier et al., "Rehabilitation," 93–96.

7 A point emphasized in both Vogt, "Aspects of historical soil erosion," 86–88, and Bell, "Archaeology under alluvium," 272. For a global perspective see Hoffmann et al., "Human impact … during the Holocene."

1000 CE.[8] In southern England, where woodland clearance got under way in the eighth century and was complete by the eleventh, erosion and deposition rates in the river Nene and the upper Thames accelerated from around 800 to reach in late Saxon and early Norman times all-time maxima tenfold their prior Holocene average.[9] Among the hills of Dauphiné about the year 1000 farmers pioneered the wooded shores of Lac Paladru; their new grain fields quickly produced higher rates of erosion now visible in well-dated lake-bottom sediments suddenly full of loam from plowed topsoil and organic waste from cattle.[10] At the crest of medieval expansion in the early fourteenth century, researchers both along the French Alps and across central Europe find unusually heavy erosion episodes when climatic change brought heavier precipitation to precisely those landscapes most recently deforested and intensively plowed for maximum output of cereal grains.[11] Incrementally, region by region, watershed by watershed, surges of hydrographic instability, soil erosion, and deposition followed pulses of local conversion of woodlands to arable.

Soils from upriver washed down to the estuaries. The Oude Rijn mouth of the Rhine in Holland silted shut by the eleventh century, and between the thirteenth and sixteenth centuries the expanding delta of the Wisła filled in the one-time bay between Gdańsk and Elbląg.[12] Long unavailing struggles of Bruges against plugging of the Zwin, and of Ravenna against filling of the Po delta are historical commonplaces.[13]

[8] Nitz, "Feudal woodland colonization," 178; Hoffmann et al., "Trends and controls of Holocene floodplain sedimentation." A more general view of the process is in Wickham, "European forests," 534, and works there cited.

[9] Lambrick, "Alluvial archaeology," 222; Williamson, *Shaping Medieval Landscapes*, 169–173; and with such regional studies synthesized in Lewin, "Medieval environmental impacts," 277–280 and 294–299. Note the temporal coincidence of these destructive impacts on English rivers with the so-called fish event horizon there (Chapters 2 and 4).

[10] Colardelle and Verdel, eds., *Chevaliers-Paysans*, 31–32, and Colardelle and Verdel, eds., *Les habitants*, 57–60. Pollen analysis finds cultivated plants and field weeds in the new sediments. Bertrand and Bertrand, "Pour une histoire écologique," 74–80, think this a common phenomenon in France during central medieval centuries.

[11] Bravard, "Des versants aux cours d'eau"; Bork, *Bodenerosion und Umwelt*; Bork et al., *Landschaftsentwicklung*, 221–226 and 237–249; and Bork and Schmidtchen, "Böden: Entwicklung, Zerstörung."

[12] TeBrake, *Medieval Frontier*, 70 and 147–148; Filuk, "Biologiczno-rybacka charakterystyka ichtiofauny," 146–148.

[13] Large areas in peninsular Italy and other parts of the classical Mediterranean had, however, already suffered enduring damage from deforestation, overgrazing, and erosion followed by coastal deposition in Roman times (Hughes, *Pan's Travail*, 90 and 190; Hughes, *Mediterranean*, 39–44 and 54–57), so further changes during the Middle Ages (as for instance at the mouth of the Tiber) were less dramatic there. Nevertheless Ortolani and Pagliuca, "Cyclical climatic-environmental changes," summarize local

For some fish species, this could mean loss of important migratory, spawning, and juvenile habitats. Heavy silt loads make water more often turbid, reduce light penetration, and can smother fish or prey species adapted to live in weed or gravel beds. Like the more dramatic alternation of floods and low water, the effects favour certain species relative to others. A tenth-century monastic chronicler at Novaliense in the Italian Piedmont was well aware that clear mountain water held lots of fish and a muddy stream few.[14] Soils from unstable cleared lands in tenth–twelfth-century Sicily went down local rivers to trigger shifts in their fish populations.[15]

To process the new grain supplies a little-used late antique invention, the watermill, surged across the medieval landscape. From perhaps a couple hundred in King Alfred's England, they multiplied to 5,624 in the Domesday Book of 1085. In Poitou, Berry, Languedoc, Burgundy, and Lorraine mills proliferated from the tenth century through the twelfth. On the Aube, where fourteen mills are recorded in the eleventh century, sixty-two may be counted in the twelfth century, and almost two hundred in the early thirteenth. Then and later their construction became a normal part of rural development in the Egerland and in central Silesia. Late thirteenth-century Milanese writer Bonvesin de la Riva estimated the territory of his city held 900 mills running about 3,000 wheels.[16] One historian of technology summed it up this way:

> By the close of the Middle Ages watermills were in use on streams of every type. They dammed up the rivers of medieval man; they were on the banks of his brooks and creeks, in the middle of his rivers, under his bridges, and along his

investigations identifying alternation between accelerated stream erosion during wet periods (the earlier and the later Middle Ages) in the Po basin and accelerated wind erosion in Sicily during warm dry periods (eleventh–thirteenth centuries).

14 Written about 926, the *Chronicon Novaliciense*, III, viii (see Montanari, *L'alimentazione contadina*, 284), remarked on the difference between two alpine streams, one flowing into Italy, "*semper turbida, paucos ferens pisces*", and its neighbour (the Durance) bound for Provence, "*valde pisciferam et claram.*"

15 Bresc, "La pêche dans l'espace économique," 274–275.

16 Reynolds, *Stronger Than a Hundred Men*, 47–69, remains a major and still not superseded overview, but its normative enthusiasms should be balanced with more recent or regional studies such as Benoit and Rouillard, "Medieval hydraulics in France," 169–180 and 203–213; Devailly, *Le Berry*, 226–227; Maas, *Moines-défricheurs*, 71–74; Durand, *Paysages médiévaux*, 253–258; Durand, ed., *Jeux d'eau*; Richard Holt, *Mills of Medieval England*, 107–144; Langdon, *Mills*, 8–64; Rynne, "Waterpower in medieval Ireland"; Muggenthaler, *Kolonisatorische und wirtschaftliche Tätigkeit*, 129–130; Cnopf, *Entwicklung der Teichwirtschaft*, 9–23; Dembińska, *Przetwórstwo zbożowe w Polsce*, 63–175; Podwińska, "Rozmieszczenie wodnych młyno," 373–402; Hoffmann, *Land, Liberties, and Lordship*, 261–262; Bonvesin de la Riva, *De magnalibus Mediolani*, lib. IV, 14 (Chiesa, ed., pp. 114–117); Squatriti, "Advent and conquests" and *Water and Society*, 126–144; Muendel, "Grain mills of Pistoia" and "Mills in the Florentine countryside."

coastlines. They impeded navigation and created streams (in the form of mill races and power canals) and lakes (in the form of storage reservoirs behind waterpower dams) where none had existed before.[17]

In fact, as remarked in Chapter 1, stream ecosystems had long existed in most of Europe, but less often ponds. Medieval watermills commonly drove their overshot or breast wheels by using a dam or weir two to five meters high to concentrate the falling water and pond a reserve supply of it. Medieval millwrights learned to do this on ever larger rivers. Once a design was in place at a location, it rarely changed.[18] Dams blocked running water and created still water: each of the nineteen mills erected during the Middle Ages on the forty-kilometer-long Skrwa, a Masovian tributary of the Wisła, had a dam about three meters high and a pond covering up to ten hectares.[19] As moving water slows, it drops the solids it has carried in suspension. On the Derwent in the English Midlands two meters of gravel and silt alluvium eventually covered a one-meter timber mill dam, gate, and race dated by dendrochronology to the mid-twelfth century.[20] The broad surfaces of standing waters absorb more solar energy. This both warms the water and further improves conditions for growth of rooted plants. In twelfth-century Picard charters and conveyances slower, deeper, and weedier waters backed up behind mill dams and weirs all along the Scarpe, the Oise, and the Somme.[21] Ubiquitous watermills formed and multiplied a new kind of aquatic habitat, one to be probed more deeply below.

On existing watercourses and their native fish populations mills had immediate effect, for they blocked movement of migrants. Like the

[17] Reynolds, *Stronger Than a Hundred Men*, 69. Compare Lewin, "Medieval environmental impacts and feedbacks," 291–293.

[18] For technical particulars and their implications, see Lucas, *Wind, Water, Work*, whose focus is industrial applications of milling power, and the essays in Walton, ed., *Wind and Water*, whose contributors have more social interests. Langdon, *Mills*, 70–107, shows breast and overshot designs dominating England by the late twelfth century and Benoit and Rouillard, "Medieval hydraulics in France," 203–208, the same for northern France, although undershot wheels were there more common in flat terrain and horizontal wheels remained a tradition in certain Mediterranean areas and also Ireland (Rynne, "Waterpower in medieval Ireland," 9–40). Nevertheless, Chiappa Mauri, *I mulini a acqua nel Milanese*, 152–175, emphasizes an early medieval replacement of horizontal with vertical wheels, and Sicard, *Moulins de Toulouse*, 38–43, documents a shift from mills on anchored rafts to land-based mills with dams or weirs.

[19] Brykala, "Watermills' functioning."

[20] Clay, "A Norman mill dam," and works there cited. Compare Bagniewski and Kubów, "Średniowieczny młyn wody," and other medieval Polish mills described in Dembińska, *Przetwórstwo zbożowe*, 90–135.

[21] Fossier, *La terre et les hommes en Picardie*, 366–368, 380–389, and 395–397. Compare findings by Walter and Merrits, "Natural streams and the legacy of water-powered mills," of extensive siltation behind preindustrial mill dams in eastern North America.

concentrations at natural barriers, those at dams and weirs offered fishers profitable access to migratory species. The contemporary biographer of St. John of Metz (Gorze, d.974) actually thought his monks' need for fish was why he built mill dams. Even deep in central Saxony, operation of a salmon trap at the mill dam at Lauenheim on the River Zschopau fueled a century of dispute (1293–1393) between Altzelle abbey and the von Steinbach family.[22] Possession of mills was associated with the right to take eel on the Duero in Castile, the Garonne near Toulouse, the Meuse around Liege, and in the early fourteenth-century psalter illuminated for Sir Geoffrey Luttrell of Lincolnshire (revisit Figure 3.7).[23]

But impassable barrier dams kept migratory species from vital spawning habitat. Blocked runs of fish – were they trout or shad? – ascending the Sarca from Lake Garda in 1210 caused the bishop of Trento, who held sovereign fishing rights in that county, to require removal of mill dams at Arco.[24] For the sake of the salmon Scottish king William (1165–1214) established judicial precedents requiring all dams and weirs be fitted with a permanent mid-stream opening and all barrier nets be lifted from each Saturday evening until Monday sunrise.[25] An English law book from the 1290s, *Fleta*, likewise acknowledged that mill dams could damage established fisheries.[26] Dutch historical ecologists have most recently argued from sparse medieval and early modern salmon price series as well as material evidence at prehistoric and medieval sites for a correlation between construction of water mills in the Rhine basin and the first declines of salmon populations there. Others have asserted that eleventh–twelfth-century dikes built to drain marshes suppressed sturgeon stocks in the Rhine delta.[27] Impassable dams are well known to break the ecological continuity of rivers and fragment even

[22] Pertz, ed., "Vita Iohannis," MGH, SS, 4: 362, "*cum molendinis fluminibus causa piscium obcludendis*"; Beyer, *Cistercienser-Stift und Kloster Alt-Zelle*, charter nos. 216, 510, and 518.

[23] Perez-Embid Wamba, *El Cister en Castilla y Leon*, 137–138 and 178–179; Mousnier, *L'Abbaye de Grandselve*, 189–190; Sicard, *Moulins*, 118–128; Derveeghe, *Domaine du Val Saint-Lambert*, 63–66; *Luttrell Psalter*, fol. 181r.

[24] Stolz, *Geschichtskunde der Gewässer*, 346.

[25] Hoffmann, "*Salmo salar* in late medieval Scotland," 362–363, and sources there cited. The so-called Assizes of King William, art. 10, described the requisite opening as "so large that a well-fed three-year-old pig could turn about without touching its snout or its tail" (*in tantum quo unus porcus trium annorum bene pastus est longus, ita quod neque grunnus porci appropinquet sepi nec cauda* . English counterparts occur in Wright, *Sources*, 91, and Winchester, *Landscape and Society*, 111.

[26] *Fleta*, III, 110–114. Those mills and mill leets on the Garonne at Toulouse also damaged the fishing (Mousnier, *La Gascogne Toulousan*, 115). On the other hand, as in the early modern Elbe, flood events could break downstream barriers and allow the return of salmon to upriver fisheries where they had long been rare or absent (Wolter, "Historic catches").

[27] Lenders et al., "Historical rise of waterpower"; Boddeke, *Vissen*, 169–176.

populations of resident fishes, while other modern research indicates a succession of even modest two- to ten-meter barriers has cumulative negative impact on upstream populations of migrants.[28] These losses mattered because, ecologists agree, the spawning environment in fresh water determines the productivity and survival of anadromous fish.[29]

Turning from rural development to other aspects of the medieval economy, human population growth and urbanization in the typically organic-based preindustrial resource system affected both water chemistry and hydrological conditions for aquatic life. Waste from more and larger human concentrations, from rural monasteries to towns of twenty or even fifty thousand, necessarily increased the nutrient load – i.e. soluble nitrates – in watercourses.[30] The several hundred monks and lay brethren at early thirteenth-century Clairvaux were served by a diversion of the Aube river, which ran through gardens, mills, brewery, fulling mill, tannery, laundry, and latrines before rejoining the main stream.[31] Such point source pollution also typically flowed from elite lay residences.[32] Local streams likewise received the human, animal, and craft waste of towns, whether by runoff from street disposal (even with intentional diversions to flush gutters as at Milan, Strasbourg, and Goslar), by purposely emptying the contents of cesspits into flowing water below town (Köln), or by direct siting of latrines over watercourses (Rouen, Nürnberg). If, as inhabitants of the Terra Firma legitimately jibed, "*I Venexian caga in aqua*," there was yet another source of nutrients for the lagoon.[33] No wonder contemporary Italian doctors and dietitians

[28] See discussion in Jungwirth et al., "Re-establishing and assessing ecological integrity" and the comparative survey of French river systems in Merg et al., "Modeling diadromous fish loss."

[29] Schalk, "Structure," 222–224. For extended discussion of anadromous issues see Supplement 5.1.1.

[30] Grewe, "Wasserversorgung," 75, cites nineteenth-century calculations of an annual preindustrial per capita output of 34 kg of feces and 428 kg of urine, totaling 462 kg of nitrogen-rich excrement. The daily adult output estimated by Leguay, *L'eau dans la ville*, 123, totals 47.5 kg of human fecal matter and 360–550 kg of urine per year.

[31] "Descriptio ... Claraevallensis." Monastic effluents are left untreated in Lillich's classic "Cleanliness with godliness" and by Magnusson, *Water Technology*, 98–101, but get more attention from Kosch, "Wasserbaueinrichtungen in hochmittelalterlichen Konventenanlagen," notably pp. 96, 110–112, and 134–135, and Benoît and Wabont, "Wasserversorgung in Frankreich," especially pp. 195, 204, and 207–216, which include orders besides Cistercians.

[32] Grewe, "Wasserversorgung," 74–75; Benoît and Rouillard, "Medieval hydraulics in France," 180–187.

[33] Grewe, "Wasserversorgung," 75–80; Dirlmeier, "Die kommunalpolitischen Zuständigkeiten und Leistungen"; Frieser, "Abwasserkanäle und heimlichen Gemächer Nürnbergs"; Lachmann, "Die Gewässer und ihre Nutzung," 311–315; La Casva, *Ingiene e Sanità di Milano*, 64–66; Guillerme, *Age of Water*, 105–107; Leguay, *L'eau*, 247–275; Leguay, *Pollution*, 16–32. Crouzet Pavan, "Les eaux noires," details the

doubted the wisdom of eating fish from waters polluted with urban effluent.[34] But Robert Guillerme has argued that organic acid- and alkaline-based processes used by early medieval textile and leather crafts caused "the precipitation of solid organic materials in water which river currents carried beyond city limits."[35]

And what was downstream? One introductory fish tale has already told of Constance's pollution of its lakeshore with urban wastes. Along what was London's little river Fleet, sediments from a human generation or two of the mid fourteenth century show loss of molluscs requiring clean water and appearance of diatoms typical in dirty water.[36] Excavations in the bed of the Pegnitz below medieval Nürnberg recovered late medieval butchery waste and household refuse, findings which corroborate the stream's foul repute when each summer's low water left it long unflushed. By the early 1400s Parisian effluent was likewise making the Seine below town "infectée et corrumpue" every summer.[37] All are symptoms of aquatic ecosystems under stress.

Nor, despite Guillerme's optimism, did medieval industry merely add nutrients. More immediate toxic effects came from crafts such as slaughtering livestock, tanning, or extracting fibers from flax and hemp by wet decomposition ('retting'). When the latter activity killed fish near Douai in 1452, holders of fishing rights sued a clothier for damages.[38] Brewers, fishers, and ordinary consumers at Colchester in 1425 complained that the tanners and tawers caused the "impayring and corrupcion" of the river Coln and "destruction of the ffysche therynne."[39]

problems Venetian authorities faced in expelling human waste from urban waters into the lagoon.

34 Nada Patrone, *Il cibo*, 339–340. Species and stable isotopes found in fish remains from sites near Basel suggest increased nutrient loads and contamination in the fourteenth-century Rhine (Häberle et al., "Carbon and nitrogen isotopic ratios").

35 Guillerme, *Age of Water*, 97–100.

36 Schofield and Vince, *Medieval Towns*, 213. Regier, "Rehabilitation," 88, specifies the latter botanical phenomenon as characteristic of an aquatic ecosystem under stress. Where current speed and substrate prevent establishment of rooted aquatic vegetation, larger nitrate loads encourage suspended algae to grow and increase turbidity. Jørgensen, "Local government responses," highlights authorities' efforts to mitigate contamination in rivers.

37 Frieser, "Abwasserkanäle," 194–195. The Seine was so pungently described in a royal ordinance against waste disposal from 1415 (Isambert et al., eds., *Recueil général des anciennes lois*, vol. 8, p. 565, cap. 683).

38 "*Et ont esté si puantes et infectés que les puissons qui estoient es yauwes de mesdis seigneurs sont aulcuns et en grant nombre mors et les autres espars au loings en estranges yauwes*," Leguay, *L'eau*, 292, citing Plouchard, "La Scarpe et les gens de rivière," 850–851. Amacher, *Zürcher Fischerei*, 96, reports a similar fish kill in 1466.

39 Gimpel, *Medieval Machine*, 86, quotes from Page and Round, eds., *Victoria History of Essex*, vol. 2, p. 459, and further mentions thirteenth-century archival regulations against letting any tanning waste flow into Marseilles harbor. More fish kills occur in Guillerme,

Toxic heavy metals emitted from medieval mining and metallurgical processes still contaminate substrates, riverbanks, and floodplains in widespread European watersheds. Even now lead concentrations in waste deposits from medieval mines in the Pennine headwaters of the Tyne, Don, and Ouse as far downstream as York exceed those of Roman or modern industrial dates.[40] Maximum contamination in lakes and shorelines of the Harz and the Staufer basin of the Schwarzwald from tenth- through thirteenth-century processing of copper, lead, zinc, and cadmium is as much as ten times the modern legal limit.[41] Alas, I know of no analyses for heavy metals in medieval human or fish remains from these and other affected regions, but studies of human skeletons elsewhere clearly indicate generally high lead exposure among especially European urban populations of high and late medieval date. Even the avid sixteenth-century mining promoter Georg Agricola had to admit to the deadly consequences of mining and refining.[42]

Activities in the urban and commercial sector further impeded the free flow of water. Since the eleventh century castles and towns had diverted rivers to fill defensive moats; this stratagem gained popularity in the later Middle Ages.[43] Accessible markets for fuel drove extensive peat-digging which created the Norfolk Broads, smaller but more numerous *plassen* in North Flanders, and South Holland's vast but equally anthropogenic Haarlemmermeer between Haarlem and Leiden.[44] And between the eleventh and fourteenth centuries still more weirs, dams, and ponds, were built to power new industrial operations like malting, fulling, metal-working hammers and bellows, sawmills, and paper-making.[45]

The type and scale of physical and chemical changes which medieval economic development brought to European inland waters most directly

*Age of Water*, 152; and Heine, "Umweltbesogenes Recht," 123. Sawmill waste harming fish was an object of 1504 legislation in Tyrol (Stolz, *Geschichtskunde der Gewässer*, 381–383).

40 Passmore and Macklin, "Geoarchaeology of the Tyne"; Hudson-Edwards et al., "Mediaeval lead pollution"; Macklin et al., "Pollution from historic metal mining."

41 Goldenberg, "Frühe Umweltbelastungen"; Diecke, "Findings." Mining and smelting waste in the Valle d'Aosta are in Di Gangi, *L'attivita mineraria*, 81–92, and Tumiati, "Ancient mines of Servette." Claustres, "Mining legacy in French Pyrenees," reported lakes with lead concentrations in medieval layers greater than those of the nineteenth century.

42 Jaworowski et al., "Heavy metals in human and animal bones"; Rasmussen et al., "Comparison of mercury and lead levels"; Agricola, *De Re Metallica*, 5 (tr. Hoover, p. 8).

43 Guillerme, *Age of Water*, 47–50 and 118–131.

44 Lambert et al., *Making of the Broads*; Borger, "Draining–digging–dredging," 153–157; Dam, "Sinking peat bogs," and Dam, *Vissen in veenmeren*, 58–81.

45 Reynolds, *Stronger*, 69–97; Benoît and Rouillard, "Medieval hydraulics in France," 208–214.

and heavily affected small- and medium-sized watercourses. Brooks, streams, and small rivers are by their very size, high ratio of surface area to volume, and abundance in the landscape more closely tied than large rivers to their immediate terrestrial environments. Removal of bankside vegetation; local ditching, diversion, or embankment; small mill ponds and dams; and effluents from concentrations of livestock or humans have profound local impact, removing the whole waterway from its natural form and sources of energy. Large rivers, in contrast, are linked to their surroundings through their multiple channels and extensive floodplains, so simple and local changes in riparian conditions have less effect.[46] Hence the impact of preindustrial economic development differed in degree and kind from that of industrial development. Yet the finding should not be oversimplified. The much-studied Rhine and its major tributaries, for instance, are said to have suffered little from human activity before channelizing and embanking in the early nineteenth century began a total degradation.[47] But most studies of major European rivers proceed from a present and retrospective standpoint with little deeper historical knowledge or awareness. The evidence of medieval landscape change is overwhelming. Watersheds are systemic continua; what enters at the top flows all the way down. Besides dams cutting off what had once been the highest spawning sites, medieval deforestation and erosion must be acknowledged a principal cause of more erratic flow regimes and contributor of material to the many mainstream sandbars and islands known from early modern records.

### *5.1.2 Perceptions of Overfishing and the Evidence of Depletion*

For many past and present observers fishing is the most obvious and understandable human impact on local and regional aquatic ecosystems. Medieval efforts to satisfy demand for fish as food rose with and beyond rising human numbers. Widened adherence to Christian food rules encouraged fish consumption. We have seen the growth and operation of urban fish markets all across medieval Europe. Such markets and the professional fishers who supplied them were by the thirteenth century held mainly responsible for rising exploitation of fisheries in northern Flanders.[48]

[46] A general comparative point made in Regier, "Rehabilitation," 88–89.

[47] Lelek, "Rhine River"; Cioc, *The Rhine*, 21–75 and 145–171. Or are modern historians as well as biologists susceptible to a form of shifting baseline syndrome?

[48] Materné, "Beroeps- en vrijetijdsvisserij," 142–144. Subsequently Van Neer and Ervynck, "New data on fish remains," 225, argued that the relatively greater reliance on marine fish in the diets of late medieval Belgian towns as compared to rural castles

Populations of fishes which Europeans liked to eat came under evident stress. One sign is contemporary awareness of damaged resources, commonly conceived as overfishing. A remarkably explicit early articulation prefaces the first full-scale fisheries ordinance issued by French King Philip IV in 1289:

> … today each and every river and waterside of our realm, large and small, yields nothing due to the evil of the fishers and the devices of [their] contriving, and because the fish are prevented by them from growing to their proper condition, nor have the fish any value when caught by them, nor are they any good for human consumption, but rather bad, and further it happens that they are much more costly than they used to be, which results in no moderate loss to the rich and poor of our realm …[49]

Philip's remedy is examined in Chapter 6 below. A more particular case of fishing pressure is reported from the Pinzgauer Zellersee, high in the Salzburg Alps. Its rich fisheries drew a mid-fourteenth-century settlement of professionals, who paid the archbishop 27,000 whitefish and 18 lake trout a year for the right to take, smoke, and sell still more. After one human generation the whitefish catch collapsed, and replacement stockings of pike ate nearly all the trout, so the fishing community determined to rest the lake for three years and then to fish only with far fewer nets in a limited season and a restricted area.[50]

By the late fourteenth century petitioners blamed weir fishing for decline of salmon and sturgeon in the Thames estuary, and English coastal fishers a generation later conceded they had depleted local stocks – to excuse, it should be noted, their illegal fishing elsewhere.[51] About the same time Siena legislated against overfishing in the lagoon of

and monasteries marks the reduction of fish populations near those urban concentrations

49 *Cum omnia et singula flumina necnon et riparie magne et parve regni nostri per maliciam piscatorum seu excogitata ingenia sint hodie absque fructu, ac per eos impediantur pisces crescere usque ad statum debitum, nec sint alicujus valloris quando ab eis capiuntur, nec etiam prosint humano usui ad vescendum, immo pocius obsint, et inde accidit quod sint multo plus solito cariores, quod cedit in dampnum non modicum tam divitum quam pauperum regni nostri …*

Duplés-Agier, "Ordonnances inédites," 49. Jeulin, "L'élaboration par la monarchie," observes how these themes became a commonplace of French legislation. Further discussion in Rouillard, "La législation royale."

50 Freudlsperger, "Kurze Fischereigeschichte," 100.

Diagnosing overfishing from evidence of fishing and fish shortages does betray aspects of circular reasoning, but the assertion in Jäger, *Einführung*, 192, of no evidence for decline in numbers of aquatic organisms before the 1500s rests on the unrealistic assumption that only governmental records of catches constitute "scientific" and hence credible data.

51 Wright, *Sources*, 91, cites an unpublished Plea and Memoranda roll of 1386; Given-Wilson et al., eds., *PROME*, Henry V, 1416 March 16, membrane V, 33, X (*Rotuli Parliamentorum II temp. Hen. V*, vol. 4, 79).

Orbetello and authorities at Santander tried to deter local depletion by taxing nearshore catches.[52] Perceptions of damaged resources were by no means confined to inland waters.

Less subjective indicators allow closer description and diagnosis of localized fish stocks under stress. Our opening tale of the sturgeon revealed a steady reduction in size of those eaten at tenth- through thirteenth-century Gdańsk. Like trends are more vaguely visible in some other large central European fishbone assemblages predating 1200.[53] At Abbeville on what was initially the Somme estuary the share of large adult flatfishes at excavated sites fell from the twelfth century to the early fourteenth, while that of small, mainly immature, specimens rose. Those reciprocal trends reversed after mid-century human population losses, but had resumed by 1500.[54] As someone in the entourage of Philip IV already seemed to know, shrinking average size indicates a stock where more fish are being extracted before reaching their full growth.

Shifts in species composition of medieval catches are still more telling. Investigators have diagnosed a decline of anadromous and cold-water varieties during the high and later Middle Ages in several regions of western Europe. Here our sad introductory tale of the sturgeon needs only brief review. Zooarchaeologists studying Baltic sites concur in the steady decline not only in size, but in numbers and relative frequency of this taxon between the seventh/ninth centuries and the twelfth/thirteenth and point to overexploitation (Figure 5.1). Records around the North Sea trace a diminishing presence, too, although some observers argue more for loss of habitat. The European sturgeon may better have survived in the long-fished Mediterranean, although market price lists less and less often included it. From a common though costly food item, sturgeon everywhere became a rare symbol of prestigious luxury.

Atlantic salmon, like sturgeon, long served to display a host's high status as ubiquitous features of conspicuous consumption at festive elite

52 Rombai, "Le acque interne … Maremma," 38–42; Echevarria Alonso, *La actividad comercial*, 40–48 and 92. Supplement 5.1.2 offers more governmental worries about overfishing.

53 Susłowska and Urbanowicz, "Szczątki kostne ryb," 53–65. Early medieval remains of fish distinctly larger than later norms for the same species are also remarked by Kozikowska, "Ryby w pokarmie średniowiecznych (X–XIV w.) mieszkańców Wrocławia," 3–14; Paul, "Knochenfunde," 59–60; and Driesch, "Fischreste aus Hitzacker," 420–421.

54 Clavel, *L'Animal*, 146–149: European flounder and plaice are not easily distinguished archaeologically, for they share many skeletal elements and the same benthic habitat. At Abbeville especially the largest specimens (over 45 cm) had vanished by 1300, became fairly common again around 1400, and were disappearing by 1500. The average size of plaice landed by Dutch coastal fishers also began to fall from the fifteenth century (Dam, "Feestvissen en vastenvissen," 491–492).

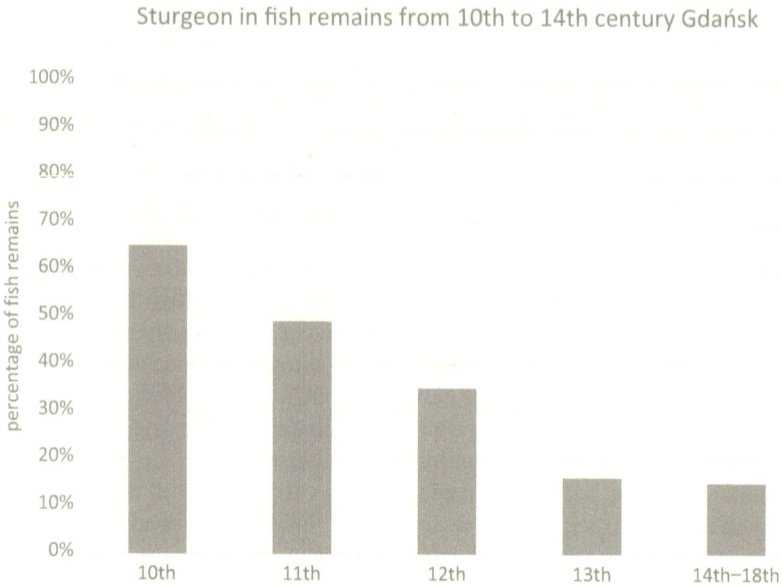


Figure 5.1 Sturgeon in fish remains from tenth–fourteenth-century Gdańsk. Data as published in Makowiecki, "Exploitation," fig. 7, and "Usefulness," fig. 3, p. 109. Graph © R. Hoffmann.

banquets, even well away from the sea.[55] As salmonid remains preserve poorly in archaeological contexts[56] it is worth observing the presence of salmon bones in excavations from Anglo-Saxon Wraysbury in Berkshire, the local Slavic prince's dwelling at high medieval Hitzacker on the Elbe, twelfth-century castles along the lower Rhine, the neighbourhood of the late medieval Louvre palace, and a contemporary house of canons at Saarbrucken.[57]

[55] Association of salmon with "the very proud and rich" in the Cluny customal was remarked in Chapter 2 above (Bruce, *Silence and Sign Language*, 177). Some other systems used the same sign for sturgeon. See further, for example: Anthimus *De observatione ciborum*, ed. Lichtenhan, pp. 18–20, or tr. Grant, §41 (pp. 66–67); Jacques de Vitry, *Exempla*, ed. Greven, p. 26; Dyer, "Household accounts," 116; Serjeantson and Woolgar, "Fish consumption," 104 and 124; Serrano Larráyoz, *La mesa del rey*, 200–203; Piekosiński and Szujski, *Najstarsze księgi*, 271–286 *passim*; Santucci, "Nourritures et symbols."

[56] Heinrich and Heidermanns, "Lachs."

[57] Coy, "Fish bones," 118; Driesch, "Fischreste," 401; Reichstein, *Untersuchungen an Tierknochen von der Isenburg*; Desse and Desse-Berset, "Pêches locales, côtières ou lointaines"; Huster-Plogmann, "Fische."

By the 1200s good anecdotal and other evidence across much of western Europe indicates decline of natural salmon stocks in especially smaller rivers and upper tributaries. Contemporaries attributed the depletion to barriers and competitive overfishing, but habitat changes resulting from agricultural clearances are also implicated. Historian Angelika Lampen traced the collapse of salmon in archives of the convent at Werden on the Ruhr from abundance in the eleventh century to absence in the fourteenth.[58] Fears of damaged salmon runs were voiced in mid-thirteenth-century Northumbria and complaints of weirs and illegal fishing killing smolts and damaging runs could soon be heard on the Thames, Severn, Wye, and Meuse.[59]. On the small coastal rivers of lower Normandy, where plowed fields and watermills multiplied through the twelfth and thirteenth centuries, the generous annual gifts of salmon offered in eleventh- and twelfth-century charters from reliable local catches were distinctly fewer by 1300. Continued 'overfishing' thereafter brought near-total destruction of the runs by the mid-1400s, with local records referring rather to gifts of individual salmon imported from Ireland and Scotland.[60] From the fifteenth-century start of suitable records along the middle Rhine the numbers and weights of salmon there taken go steadily downwards.[61] Many river basins saw by the later Middle Ages the pressure on salmon shift down to estuaries, making it out to be a marine rather than inland fish.[62] Even in wealthy Parisian households and prosperous Flemish monasteries consumption of once-favoured sturgeon, salmon, trout, and whitefish shrank to undetectable

[58] Lampen, *Fischerei und Fischhandel*, 208.

[59] Page, ed., *Three Early Assize Rolls*, 103; London TNA, Plea Rolls: KB 27/509 11R2, m1d; Wright, *Sources*, 91; Williams, *Welsh Cistercians*, 75–76; Kowaleski, "Seasonality," 132–133. The second statute of Westminster (1285) ordered protection of smolts. Mid-fourteenth-century financial accounts kept for the Counts of Namur indicate the collapse of a functional salmon fishery in middle reaches of the Meuse (Balon, "La pêche," 28–31).

[60] Halard, "Peche du saumon," 175–177. On the best-known river, catches in 1423 were less than a third those of the early 1300s, though the price per fish rose by a factor of twelve. Less grain-centred agrarian regimes and rivers unsuited to be spanned by mill dams help explain strong survival of Irish and Scottish salmon stocks (Hoffmann, "*Salmo salar* in late medieval Scotland," 64–65).

[61] Volk, *Wirtschaft und Gesellschaft*, 350–367. Lenders et al., "Historical rise of waterpower," would generalize this medieval decline across the entire lower Rhine catchment.

[62] As seen in Laurière et al., *Ordonnances des roys*, vol. 2 (1729), 578–582; Puñal-Fernandez, *Mercado en Madrid*, 169–175; Martens, *Mittelalterlichen Gartensiedlungen*, 154–156; Benecke, "Beiträge," 308–309 and 314–315; Sarnowsky, *Wirtschaftsführung des Deutschen Ordens*, 130–131; Martens, *De zalmvissers van de Biesbosch*, 41–54, 114–135, and 211–219.

by around 1500.[63] Yet despite general late medieval increase in consumption of marine fishes, salmon retained high cultural significance and commensurate prices across their entire northern and western European range.

Like the sturgeon, then, *Salmo salar* is a prime example of widespread localized medieval human impacts on European riverine, notably anadromous, fish populations. European fishers and consumers were well aware of these losses long before modern industrialization.

## 5.2 Beneficiaries?

A schedule of tolls taken around 1275 by seigneurs of Audenarde (Oudenaarde) at their bridge over the Scheldt en route to Ghent specified only four fish taxa: salmon and sturgeon paid per specimen, two and four denier respectively; eel and herring turned over a hundred fish per *los* ("*Last*"), a measure of quantity, probably 12,000, so weighing about a metric ton.[64] Different fiscal assessments reflect some of the previous two or three centuries of change in the status of fish stocks and consumption demand, contrasting the traditional elite favourites from threatened anadromous species with smaller more numerous varieties, possibly more resilient, for a broader range of consumers. Variant critical adaptations and regional ecologies across western Christendom help explain especially large increases since around 1000 CE in consumption of eel and herring, and also the territorial spread of common carp, three fishes and fisheries then on trajectories opposite to those of salmon and sturgeon. In subtle ways all had gained from socio-economic developments typical of high medieval Europe.

### *5.2.1 Eel*

Eel are as migratory as salmon but travel in reverse directions (catadromy). Unlike the sensitive eggs and young of salmon, subadult eels enjoy broad ecological tolerances and omnivorous habits during their long maturation in estuaries and waters far inland. With well-chosen techniques people could catch eels throughout their freshwater phase and when the sexually mature adults migrated downstream to spawn at sea. Despite cultural antipathies to its snake-like morphology

[63] Desse and Desse-Berset, "Pêches locales, côtières ou lointaines," 125–126; Sternberg, "L'approvisionnement de Paris en poisson"; Ervynck and Van Neer, "De voodselvoorziening," 425–426.

[64] Verriest, ed., *Le polyptyque illustré*, fol. 12r. Unless specific sources indicate otherwise, the *last* as a quantity (rather than a volume) counted ten "long thousands" of 1,200 items each.

and benthic habits,[65] this species evidently supported long-standing fisheries and human consumption in Atlantic and Mediterranean (but not Danubian or other Black Sea) drainages. Until sieving became standard archaeological practice, tiny bones and high oil content made eel remains evasive, but regular fishing of eel at weirs, mills, and in stillwater habitats everywhere supported heavy local consumption of fresh and lightly processed catches throughout the earlier Middle Ages.

Human activities described to start this chapter greatly enlarged nutrient-rich stillwater habitats, with the unplanned result of much favouring the eel and soon those who could fish for it. In precociously deforested England as early as the seventh- through mid-ninth centuries, eel was the most common taxon in all fish remains recovered from multiple contexts at York and in elite settlements at Canterbury and Lyminge, Kent. In decades around 1000 it remained as abundant at York, though yielding primacy to herring, and there continued in that second rank until the 1200s.[66] Further inland in Aelfric of Eynsham's time (*c.* 1000) eel were the single most common fish eaten in his monastery and from the twelfth century there, too, remained second only to preserved herring.[67]

Contemporary fish remains from northern France and the Low Countries point to parallel developments. Eel reached two-thirds of remains from twelfth-century Deventer, but thereafter fell back to only one-third behind undifferentiated small cyprinids.[68] In the Somme basin eel emerge as the most common fish species from the ninth century with special abundance in urban contexts at Amiens through the twelfth and thirteenth. Although herring later steadily gained pre-eminence there, even as late as 1449 Cistercians at Gard abbey on the Somme were

[65] See views transmitted by Anthimus, *De Observatione Ciborum* (Lichetenhan, ed, p. 19: Grant tr., p. 65); Hildegard of Bingen, *Physica*, lib. 5, §33 (Hildebrandt and Gloning, eds., vol. 1, pp. 283–284); Albertus, *De Animalibus*, lib. 24, §3; and legends of a saintly bishop expelling eels from Lake Lausanne (Chène, "Une sainteté exemplaire").

[66] Harland et al., "Fishing and fish trade," 174–186, and Reynolds, "Social complexities," 215–218. Dietary primacy at St. Augustine's abbey, Canterbury, shifted from eel to herring at about the same time as York (Nicholson, "Fish remains"). While eel remains at Lyminge outnumbered those of herring by ten to one, in the contemporary coastal site of Bishopstone on the east Sussex coast, eel provided only 20% of remains and herring 26%. Orton et al., "Catch per unit research effort," 15 and fig. 9, find a high medieval rise of eel consumption in London.

[67] Hardy et al., *Ælfric's Abbey*, 356–359 and 395–396. Also at Wraysbury on the Thames, eel comprised 82% of fish remains from the late ninth century through early twelfth (Coy, "Fish bones"). Holmes, *Animals*, 51–53, table 3.3, found eel at the largest share of Late Saxon and Saxo-Norman sites with fish remains.

[68] Clavel, *L'animal*, 101–102; IJzereef and Laarman, "Animal remains from Deventer," 435–436.

handling in single transactions as many as 13,000 eel. In food waste which fell through gaps in the kitchen floor in the contemporary Benedictine abbey at Ename outside Oudenaarde, eel remains at 12.6 percent of all fish (and 20 percent of those locally acquired) were also outnumbered only by herring.[69] Much further south in kitchen waste from the priory at Charité-sur-Loire eel ranked second only to small native cyprinids from the eleventh century to the sixteenth, when those monks turned to carp as the main fish on their menus.[70]

Across the whole Rhine delta region of Holland and Flanders, a great and long-term expansion of shallow estuarine and freshwater lakes since at latest the 1200s resulted from subsidence of drained peat lands, peat mining, diversion of rivers, rising water levels, and both local- and large-scale wave action. These turbid, fertile, and accessible lentic habitats soon supported large and lucrative commercial fisheries for eel. Sluices of the Spaarendam beside Haarlem in the late 1400s annually yielded 150,000 eel (ten to twenty tonne). Fishers supplied towns in the region and even exported to England.[71] Where anthropogenic environmental change came later than it had further west, as in Poland, eel populations and human use of this fish may only more slowly have expanded inland.[72]

Western Mediterranean watersheds show similar trends. In the Rhône delta eel replaced sturgeon as the principal fishery by the fourteenth century, but already in the twelfth residents were alert to the lucrative returns possible from active fishing and storage of live eel in enclosed ponds or channels.[73] Energetic Italian pursuit of this species got under way then, too. Local eel fisheries spread along Tyrrhenian shorelines from Tuscany to Sicily, into the Adriatic around the mouths of the Po, and in natural lakes and marshy river valleys of the interior, all locations where other kinds of evidence indicate increased lagoon formation,

69 Clavel and Cloquier, "Pratiques halieutiques fluviales," 207–208; Cloquier, "Pêches et pêcheries"; Ervynck and van Neer, "De voodselvoorziening," table 2 and p. 430 (although in the latter case remains of indeterminate cyprinids and flatfishes were both still more numerous).

70 Audoin-Rouzeau, *Ossements animaux*, 147. For more French eel consumption see Supplement 5.2.1.

71 Dam, *Vissen in veenmeren*, 103–121, and "Eel fishing in Holland"; Van Neer and Ervynck, "Apport de la archéologie."

72 Makowiecki, *Historia*, 145. Dembińska, *Konsumpcja żywnościowa*, 52, observes rising verbal references to eel from twelfth–fourteenth-century northern Poland (Pomerania), but the archaeological record is sparse.

73 Stouff, *Ravitaillement*, 201–203; Amargier, "La pêche en Petite Camargue," 331–336; Weiss, *Versorgung des päpstlichen Hofes*, 394–397; Berman, "Reeling in the eels."

siltation, and eutrophication.[74] So great was demand that from at latest 1275 each autumn the commune of Perugia sponsored transfers of juvenile eel (elvers) from the Chiana river into Lake Trasimeno and continued to do so even after Pope Martin IV died there in 1285 after eating [too many?] eel. During the 1350s and 60s the papal-owned fishery at the outlet of Lake Bolsena continued to ship thousands of eel a year to the curia then in Avignon.[75] Eel matched sea breams as the most common fish taxa eaten at a late fourteenth-century palace in Tarquinia, just up the coast from Rome.[76] We can but speculate on the point of balance for medieval eel stocks between anthropogenic enlarged habitats and intensified human predation.

### 5.2.2 Herring Fisheries on the Rise

People fishing for and eating herring enter the historical record of Atlantic (including North Sea and Baltic) coastal communities during the early Middle Ages.[77] Then with simultaneous rising human numbers and environmental pressures of the tenth through thirteenth centuries these activities greatly increased in scale and distribution into nearby inland districts. Herring bones are as fragile and elusive in unsieved archaeological contexts as are those of eel. (Figure 5.2) Though pelagic in habits, schools of these plankton eaters once frequented close inshore waters off northwestern European beaches and penetrated deeply into now long-obliterated estuaries there.

From the south coast of the Baltic all the way around to the English Channel seasonal spawning concentrations offered local opportunities for rich catches within sight of land.[78] Reproductively isolated estuarine stocks could be taken in fixed traps and weirs and those along open shorelines from small boats with light seines and drift nets, especially at

[74] Compare, for instance, Grove and Rackham, *Nature of Mediterranean Europe*, 290–311; Bresc, "La pêche … dans la Sicile," 167–169, *Un monde méditerranéen*, 261; and "La pêche dans l'espace normand," 280–282; Vendittelli, "La pesca nelle acque interne," 116–123, and "Diritti ed impianti," 409–422; Lanconelli, "Gli *Statuta pescivendulorum urbis*," 94–102; Vincenti, "La tutela ambientale"; Rombai, "Le acque interne in Toscana," 30–42; Spicciani, "Il Padule di Fucecchio," 64; Balletto, *Genova nel duecento*, 189–192; Onori, *L'abbazia di San Salvatore*, 55–56; and Martin, "Citta' e campagna," 333–334.

[75] Lanconelli, "I lavori alla peschiera"; Biganti, "La pesca nel lago Trasimeno," 795–797.

[76] Clark et al., "Food refuse … Tarquinia," 240–242.

[77] For more context on herring history and historiography see Supplement 5.2.2.

[78] For relevant biology see Fish Base, sub *Clupea harengus*; Hodgson, *The Herring*, 15–24; Klinkhardt, *Der Hering*; Bailey and Steele, "North Sea herring"; Krovnin and Rodionov, "Atlanto-Scandian herring," with a useful summary from a historical Anglocentric perspective in Locker, *Role of Stored Fish*, 31–44.

Figure 5.2 Archaeological herring bones, unsorted. Herring and other bones from the Blue Bridge Lane site, York, fourteenth century. Photograph © James Barrett. Used with permission.

night. Writing about 1200, Danish chronicler Saxo bragged that the fish along the Scanian coast were so abundant that they blocked shipping and could be caught by hand.[79] But these oily creatures spoil in a day unless promptly dusted with salt, smoked, or packed in simple salt brine. Such light cures make herring – 'powdered,' 'red,' or 'pickled' – an inexpensive portable food, palatable enough for several months, especially during the cold season after the autumn spawning. By the mid-twelfth century coastal artisans from Picardy to Pomerania were tapping the silvery billions to feed themselves, their neighbours, and nearby inland populations, especially in fast-growing towns.

The booming twelfth-century herring industry had emerged from a historically obscure century and more of parallel development by fishers with access to local consumers and local fish. Early medieval coast-dwellers from Sussex to Sweden ate herring they, their neighbours, or

[79] Saxo, *Gesta danorum*, preface 2:4 (Olrik and Ræder, ed., vol. 1, p. 6):

*Ab huius ortivo latere occasivum Scaniae media pelagi dissicit interruptio, opimam praedae magnitudinem quotannis piscantium retibus adigere soliti. Tanta siquidem sinus omnis piscium frequentia repleri consuevit, ut interdum impacta navigia vix remigii cònamen eripiat nec iam praeda artis instrumento, sed simplici manus officio capiatur.*

Also tr. Fisher, *History of the Danes*, vol. 1, p. 7. Albert, *De animalibus*, 24:2, observed the abundance of these tasty little fish in waters off France, Britain, Germany, and Denmark ("*allec piscis est maximae multitudinis in Occeano quod partes Galliae et Angliae et Teutoniae et Daciae attingit*").

their subjects caught from local shoreline and estuarine stocks.[80] By about 1000 these catches were also supplying nearby inland consumers.[81] At Haithabu on the Schleswig isthmus, where Viking traders gathered from about 800 until 1066, this species contributed 38 percent of the 13,842 identified fish bones. The oldest medieval herring remains in inland Flanders appear only in the late tenth/early eleventh century and those only in the mercantile settlements at Ghent and Ename.[82] Until this time herring everywhere looked like a traditional subsistence or artisanal fishery being practiced seasonally along seashores close to consumers.

The eleventh century was then plainly transitional at many places around the herring coasts, with remains of that species at York, London, and elsewhere in eastern England a principal marker for archaeozoologist James Barrett's 'Fish Event Horizon,' the significant appearance of marine fishes in local diets.[83] Some of the first strong evidence of heavy commercial use comes from the Pomeranian coast of the Baltic and southern shores of the North Sea. Before 1100 inland Poles were well aware that the beaches between the Odra and Wisła estuaries were full of fresh herrings and a source for salted ones. The rich representation of herring in food remains from coastal sites thins out toward the interior. Still, from Pomerania likely came the herring bones found in some strongholds of the emergent Polish state along the Warta *c.* 1000 and more commonly in eleventh–twelfth-century layers there and even so far inland as Wrocław. By about 1200 Silesian nuns were sending small boats down the Odra to pick up "salt fish."[84] An intensified fishery off

80 Kowaleski, "Early documentary evidence," 23–24; Loveluck, *Northwest Europe*, 198–200 and 211; Reynolds, "Social complexities"; Enghoff, "Herring and cod," 137; Barré, "Droit maritime médiéval," 524–525; Makowiecki et al., "Cod and herring," 118–119; and in many of the references to follow.

81 But not meaningfully earlier. For credible indicators of the absence of herring-eating inland, see Supplement 5.2.2.

82 Van Neer and Ervynck, "Rise of sea-fish consumption," 159–164; Enghoff, "Baltic region," 48, 56–58, *et passim*; Enghoff, "Denmark," 177 and 142–143; Lepiksaar and Heinrich, *Fischresten aus Haithabu*, 119; Heinrich, "Temporal changes,"151–156. As Locker (*Role of Stored Fish*, 114–168) argues, when archaeologists have carefully sieved, individual herring are so small that their share of fish flesh consumed was much less than their proportion of all bones recovered.

83 Barrett *et al.*, "Dark Age economics" and Barrett *et al.*, "Origins of intensive marine fishing," more fully shown in Orton et al., "Fish for the city," 517, and Harland et al., "Medieval York," 175–193. For herring at least this was a more widespread, if plainly incremental, phenomenon.

84 Leciejewicz, "Z denara" and "Zum frühmittelalterlichen Heringshandel"; Rulewicz, *Rybołówstwo Gdańska*, 68–88 and 342–347; Makowiecki, "About the history of fishing," 120; Makowiecki, "Catalogue"; Makowiecki et al., "Cod and herring," 118–123; Chełkowski and Filipiak, "Cognitive potential," 45; Kozikowska, "Ryby";

the island of Rügen, however, seems a twelfth-century development, soon contested between Slavic and Danish lords.[85]

Parallel developments along continental coasts of the North Sea and Channel from Flanders to Normandy are signaled in cartloads of herring on the market at Arras in 1024, an annual herring fair at Fecamp by 1088, and herring remains at eleventh-century sites as far inland as Compiègne, Paris, and Namur, though in the latter only at the castle, not yet the less affluent town. By the second half of the twelfth century communities along the Flemish coast had gained a papal dispensation to fish on Sundays while the run was on, and were struggling bitterly with old neighbouring monasteries over payment of tithes on their catch, which assumed a large presence on their newly established fish markets. The material results of these efforts are plain in abundant herring bones found in twelfth- and thirteenth-century urban contexts (see Figure 2.2 above).[86]

Across the narrow seas historian Maryanne Kowaleski estimates a four-fold increase in English herring catches from *c.* 1000 to *c.* 1200. Domesday Book (1085–1087) counted well over a score of large site-specific renders in herrings owed the king and East Anglian lords. This argues for a widespread fishery, if seasonal and by part-timers, and accounts for the newly high proportion of herring remains in food waste from twelfth-century Norwich, London, and other locations in eastern England. Transport some distance inland is revealed by mentions of herring in toll schedules.[87] Exploitation also of local herring stocks in

Lampen, *Fischerei und Fischhandel*, 168–171; Gallus, "Chronicon Poloniae," lib. 2, cap. 28; Herbord, *Dialogus de vita S. Ottonis*, lib. 2, c. 41 (Liman and Wikarjak, ed., p. 141); Appelt and Irgang, *SUB*, #123 and 140. Writing around 1120, chronicler Cosmas of Prague described Bishop Gebhard of Prague in 1090 distributing Lenten herring to the poor (*Chronica bohemorum*, 2:42, ed. Bretholz, p. 147).

85 Jahnke, *Silber des Meeres*, 15–38; Lampen, *Fischerei und Fischhandel*, 163–168.

86 Van Drival, ed., *Cartulaire de Saint-Vaast*, 166; Hocquet, "Pêcheries médiévales," 39–49 and 79–86; Darsel, "Le servitudes," 107 (eleventh-century Dieppe paying dues in herring to St. Catherine's of Rouen); Clavel, *L'Animal*, 161; Pigière *et al.*, "Status as reflected in food refuse," 238–241; Kapferer, *Fracas et murmures*, 93–96; Coopland, *Abbey of St. Bertin*, 47–48; Derville and Vion, *Histoire de Calais*, 14–15; Uytven, "L'approvisionnement des villes," 102; Van Neer and Ervynck, "Inland Flanders," p. 164, fig. 14.5. Fiscal evidence of the twelfth-century Flemish fishery is in Delatouche, "'Gros Brief' de Flandre," 30, but the count of Boulogne had already before mid-century donated to Cluny from his own royalties 20,000 herrings a year (Bernard and Bruel, eds., *Recueil des chartes*, #4132, p. 481).

87 Kowaleski, "Early documentary evidence," 23–31, and "Commercialization," 178–180; Locker, "Peabody Site"; Locker, *Fishergate, Norwich*, 42–44; Locker, *Role of Stored Fish*, 170–191 and 277; Riddler and Trzaska-Nartowski, "From Dover to New Romney"; Lovelock, *Northwest Europe*, 251–252 and 353. Campbell, "Domesday herring," thinks Domesday accounts for only 5% of the English catch, which would then come to more than three million fish or three hundred tonnes.

southwestern England sustained payments of 30,000 herring a year from Tidenham (where the Wye enters the Severn estuary) to the minster of Bath. When herring came to 60 percent of the fish bones gnawed by the monks of Norman Eynsham, did those rations come from the Severn or the North Sea?[88]

Most everywhere at this time the greater share of this catch was piled up whole right on the foreshore and covered with salt, making loose dry 'powdered' (*sapoudre*) herring with a few months' storage life, then sold in bundles or baskets of a thousand to consumers as much as a hundred kilometers away.[89] Already in the 1160s Alan of Lille called herring "the most common of fish, [which] by his wide availability relieved the hunger of the poor."[90] Plainly the surging exploitation of abundant herring stocks along Christendom's northwestern littoral during the eleventh and twelfth centuries contributed heavily to increased quantities of marine fishes serving to meet the growing demand for fish as food in maritime Europe.

At least into the 1200s, however, written and archaeozoological records known to date indicate that consumption of herring remained an essentially regional phenomenon which rapidly attenuated towards the continental interior. Even in the twelfth century herring bones in Parisian trash still number but two-thirds those of eel. Further inland eleventh- and twelfth-century written evidence is confined to ecclesiastical settings while material finds are absent even from well-sieved sites.[91] In the eleventh-century Rhine basin monastic writers at Lorsch and St. Gallen describe the fish in learned terms referring to Roman fish pickle, not dry salted objects, and the only physical traces come from a castle near Basel. By the mid-1100s verbal acquaintance among monks at Hirsau and of Hildegard of Bingen with the preserved product gains

[88] Lovelock, *Northwest Europe*, 207; Hardy et al., *Aelfric's Abbey*, 379–381. Fishing for nearshore migratory herring schools in the Firth of Forth is documented in the twelfth century, supported by salteries in the immediate area (Oram, "Estuarine environments," 366–367).

[89] Harland et al., "Fishing and fish trade," 189–190, makes clear that the herring consumed at eleventh–early fourteenth-century York had not been processed for packaging in barrels, so had to be consumed fresh or shipped dry-salted. Those reaching Bourges about 1100 by way of boats on the Loire paid toll by count (Querrien, "Pêche et consummation," 428–429) and so did the herring entering Flanders as late as 1252 (Lampen, *Fischerei und Fischhandel*, 186, citing *Hansische Urkundenbuch*, vol. 1, no. 432).

[90] Alanus, *De planctu*, Prose 1, tr. Sheridan, 94–98.

[91] Clavel, *L'Animal*, 154–160; Jarecki, *Signa Loquendi*, 122–124 (Ulrich of Cluny) and 252–253 (Fleury); Bernard and Bruel, eds., *Recueil des chartes*, #4132, p. 481; Sternberg, "Une spécificité de la cuisine monastique," 93–94, points out absence of all marine fish from abundant sieved kitchen remains at eleventh–thirteenth-century Tournus on the Saône and at La Charité-sur-Loire.

but tepid confirmation from rare finds in slightly later urban latrines at Basel.[92] In Bavaria no herring bones are reported in a survey of fish remains from six castles and four urban contexts predating 1200[93] nor do they leave any trace in either the Iberian or Italian peninsulas. Further east in the Baltic than Viking-Age Birka in Sweden, herring likewise occur only after 1200.[94]

Expanding eel fisheries at (western) European scale met growing medieval demand for fish by exploiting a species whose habitat itself was then growing in unintended consequence of human activities. Herring, however, epitomize intensified human use in northwestern coastal areas of an existing fish stock under conditions where larger cultural and demographic demand pressures confronted limited and probably dwindling supply from hitherto preferred varieties taken in fresh water. In a third emerging case during the same tenth through thirteenth centuries, the westward spread and human use of common carp replicated some features of the two previous, but reached across the extensive upper Danube basin into neighbouring waters then little affected by intensified consumption of eel or herring.

### 5.2.3 *Exotic Carp Invade the West*

The tale of carp in central medieval times relates the progress of what would now be called an exotic and invasive species. Throughout Greco-Roman antiquity common carp resided in European waters only in the Black Sea drainage, where remains from some Balkan sites suggest it was abundant. The northwesternmost traces of this fish up to and throughout the Roman Empire placed it in the Vienna basin, the most westerly part of Pannonia.[95] Not coincidentally, this natural post-Pleistocene range of

92 Ekkehard of St. Gallen, *Benedictiones* (ed. Egli, 285–289); Heinricus Laureshamensis, *Summarium Heinrici*, lib III, c. xvi (pp. 156–160); Jarecki, ed., *Signa loquendi*, 165–168 (William of Hirsau); Hildegard, *Physica*, lib. 5, cap. 22 (Hildebrandt and Gloning, eds., vol. 1, pp. 278–279; tr. Throop, p. 71); Hüster-Plogmann, "... der Mensch lebt nicht von Brot allein," 193–197; Deschler-Erb et al., "Tierknochen aus St. Arnoul," 529–532.

93 Pasda, *Tierknochen als Spiegel*, 106–110. Remains of other small-boned fishes were recovered.

94 Lõugas, "Fishing and fish trade," 111–112.

95 A systematic survey and interpretation of the material and verbal evidence summarized in Map 5.1 is Hoffmann, "Remains and verbal evidence," later slightly revised and augmented in Hoffmann, "Environmental change and the culture of common carp." Subsequently Enghoff, "Fishing in the Baltic Region" and Enghoff, "Fishing in the southern North Sea," confirm these findings, both generally and in the discussion pp. 100–107, as does Makowiecki, "Catalogue." I remain unconvinced by single isolated finds of alleged early carp bones found at waterside sites many hundreds of years and kilometers away from other contemporary evidence and in regions where the

a slow-water, heat-loving but otherwise broadly tolerant, species ended precisely where it encountered the fast-moving, high-gradient waters of the upper Danube above that river's 'inland delta' in the vicinity of Bratislava. Contrary to assertions by some biologists, no written or material evidence suggests that Romans dealt with carp outside that native range or there handled carp any differently than they did other freshwater fishes.[96]

In what remains the oldest known verbal European reference to this organism, in the mid-530s Cassiodorus, the learned Roman minister for the Gothic kings who ruled early sixth-century Italy, listed *carpam destinet Danubius* ("the Danube sends carp") among the several exotic fishes the king would serve to impress visiting ambassadors.[97] From Cassiodorus' northern Italian standpoint, the pertinent Danube lay to the east in Pannonia (modern Hungary or Serbia), whence the Goths had entered Italy a generation before, where they then still also ruled, and where the carp were not only native but long consumed by humans. Tellingly, the only known material evidence of this fish from medieval Italy is isolated fragments from sixth–seventh-century Comacchio and Padua,[98] precisely the time and route for such a luxury import to be transported to the king's palace.

During the half-millennium from Cassiodorus into the eleventh century, finds of carp remains continue to confirm its natural distribution, and for the first time indicate its spread (1) further up the Danube and (2) north into middle reaches of the Odra and Elbe systems (Map 5.1).

fish is a well-known later introduction – as, for example, Dobney et al., *Of Butchers and Breeds*, with a single purportedly third-century carp bone from Lincoln. More confidence-inspiring recent evidence is mentioned below.

96 Balon, "The common carp," 1–55. Indeed in this same region, the last remnant population of wild, not feral, common carp, survived into the 1960s, when they were so identified by the late biologist Eugene Balon. For the hydrological barrier see Balon and Holčik, "Gabčíkovo river barrage system," 2–4. As opposed to carp consumption in the Roman Balkans, Balon's assertions of Roman domestication of carp, as reiterated in his *Domestication*; "Origin and domestication," 21–32, and "About the oldest domesticates," 6–10, and set in the inland delta area or elsewhere, are in continued absence of supportive material or written evidence to be construed as wishful thinking. Carp remains at Roman and early medieval Austrian sites show a size distribution typical of wild populations; only from the fifteenth century do carp of uniform size indicate a farmed population (Galik et al., "Fish remains," 349–350). For much of what follows, Leonhardt, *Der Karpfen*, 12–15 and 49–54, embedded seminal ideas in much historical error.

97 Cassiodorus, *Variarum libri dvodecim*, 12:4 (Fridh, ed., p. 467). Other fishes there meant to impress visitors were to come from the Rhine, Sicily, and southernmost Italy. *Carpa* is reputedly one of only two Gothic words Cassiodorus ever wrote (O'Donnell, *Cassiodorus*, 94).

98 De Grossi Mazzarin, "I resti archeozoologici," 56–57; Gabriel, "Fish assemblages," 126–128.

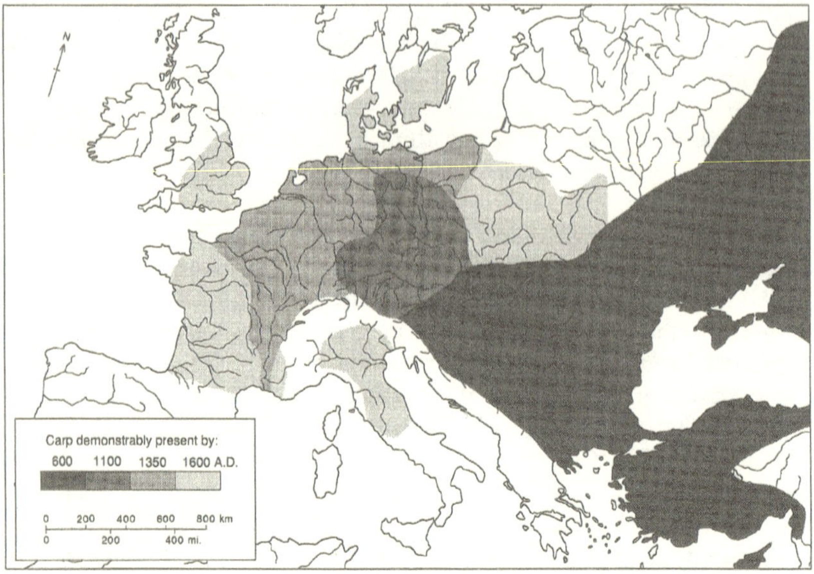


Map 5.1 The expanding range of common carp in Europe, 600–1600.

No verbal sources corroborate the northward expansion (presumably via low divides at the headwaters of the Morava River), perhaps because literate Latin Christian culture penetrated this zone only around 1000 CE and initially produced none of the requisite records of economic activity. Carp's early medieval westward push, however, did leave traces in the eleventh-century written record. About 1060 the anonymous author at Tegernsee abbey of a secular fairy tale (courtly novel), composed in Latin, listed *charpho* together with other fishes, some named in Latin, others in the vernacular but all otherwise familiar in upper Bavaria, which his hero, a knight named Ruodlieb, captures by quasi-magical means.[99] Barely a generation later, Abbot William of Hirsau (d.1091) compiled a vocabulary of signs for his monks to use during compulsory silence. To his prototype from Burgundian Cluny, William added several central European taxa, among them "the fish which is popularly called carp" (*piscis qui vulgari nomine carpho dicitur*). Did William then expect carp at Hirsau's Black Forest site in the Rhine

[99] *Ruodlieb*, Fragment X, ll. 36–48 (Haug and Vollmann eds., 2:1, pp. 136–137; Vollmann, ed. and tr., pp. 494–495; Ford tr., p. 74). This identification is undisputed, as are most of the other distinctively middle European/Danubian fishes in the list, but some of the English names used in Ford, tr., p. 74, are entirely implausible.

basin, or did he recall eating this fish during his youth at Regensburg on the Danube? Might William signify carp's movement across watersheds?[100] If so, this is corroborated by recent finds of carp remains from tenth-/early eleventh-century layers at Sulzbach castle, seat of the Count of Nordgau, in a zone of interlaced Danube and Rhine tributaries and then, barely a decade after William's death, by a bilingual lexicographer at Lorsch abbey in the northern Black Forest who glossed a fish called in Latin *carabus* with the German *charpho.*[101] Carp remains of late eleventh- and early twelfth-century date at Nürnberg castle and a house of canons at Saarbrücken further fit this scenario.[102]

Five centuries from Bratislava to the Rhine; less than two from there to Paris? Mention by Hildegard of Bingen in her *Physica* (*c.* 1160) confirms the establishment of carp (*carpo*) in the middle Rhine. The abbess likely learned from abbey fishers how the species fed on bottom organisms and vegetation in swampy and clear water and something of its spawning habits, while herself assessing its value as food and, suitably prepared, a cure for fever.[103] No remains or verbal mentions from before 1200 suggest carp culture or artificial fishponds. These were wild fish.

Further west the written and archaeozoological records seem to lag,[104] and then suddenly blossom in the mid-thirteenth century. Does this

[100] Jarecki, *Signa loquendi*, 165–168. For William's Bavarian background and subsequent experience at Cluny, whence his model for the sign language, see Jakobs, *Die Hirsauer*, 8–30.

[101] Pasda, "Tierhaltung als Spiegel," 106–109, and "Tierknochen auf Sulzbach," 254; Heinricus, *Summarium Heinrici*, lib. III: cap. xvi (ed. Hildebrandt, 159–160). Latin *carabus* was used by Pliny (*Historia naturalis* IX:LI) for a kind of crab, but in present-day scientific nomenclature refers to a family of beetles.

[102] Boessneck and von den Driesch-Karpf, "Tierknochenfunde Nürnberg," 70–72; Huster-Plogmann, "Fische," 529–532, and Deschler-Erb et al., "Tierknochen aus St. Arnoul," 529–532, and personal correspondence with Heide Huster-Plogmann, 15 April 2014. Outer limits to carp's range before the twelfth century are well established. See Supplement 5.2.3.

[103] Hildegard, *Physica*, lib. 5, cap. 5, cap. 11 (Hildebrandt and Gloning, eds., vol. 1, pp. 273–274; Throop, tr., p. 168). More recent finds by Dutch and Swiss experts in ichthyoarchaeology have better confirmed the presence of carp in the Rhine around Hildegard's time: twelfth- and early thirteenth–century remains from Utrecht are independently attested in Buitenhuis and Brinkhuizen, "Faunaresten," and Beerenhout, "Het *Huis te Vleuten*," while van Dijk and Beerenhout, "Het botmateriaal," 40–41, encountered carp bone at thirteenth-century Hoorn, on what was then a still freshwater Zuider Zee. Closer to the top of the watershed the carp remains from latrines in Schaffhausen suggested in Hüster-Plogmann and Rehazek, "Historical record versus archaeological data," to come from the twelfth century have now been redated by the same scholars to the later Middle Ages (see Hoffmann "Der Karpfen").

[104] Carp are conspicuously absent from late twelfth- and early thirteenth-century works by writers associated with Parisian schools and evidently interested in fishes: Gui of Bazoches, *Epistolae* 23 (Adolfsson, ed., pp. 89–99); Alexander Neckham, *De naturis*

mark arrival or literate recognition? A Parisian connection seems important. All three preeminent mid-century scholastic natural philosophers took cognizance of carp. Thomas of Cantimpre, who composed his *Liber de natura rerum* during the early 1240s, associated the fish with ponds and slow rivers. He had some idea of its morphology, fecundity, and ability to evade the fishers' nets, but was ill-informed of reproductive behaviour. A decade or so later Vincent of Beauvais mostly replicated Thomas. Thus when writing *De animalibus* in 1258–1262, Albertus Magnus could select from and correct his predecessors. Perhaps drawing on his own experiences in Regensburg and Köln, Albert revised errors about spawning behaviour and commented on both the carp's suitability for rearing in artificial ponds and its culinary qualities (which Albert doubted).[105] Also in 1258 managers of the estate at Igny-le-Jard belonging to Thibaut VI, Count of Champagne, "stocked 3520 *carpis* and six big pike" costing more than eighty-three livres into one of their ponds. Later that same year they spent still more to put into two ponds some unidentified 'fish' (*piscibus*), 10,000 bream and roach (*bremarum et gardonum*), and 400 *carpis*, the latter alone costing nineteen livres.[106] Igny-le-Jard, still well endowed with ponds, sits midway between Epernay and Chateau-Thierry, about three kilometers from the Marne and less than a hundred upstream of Paris. Within the next decade royal provost Etienne Boileau included *carpes* and *cuerpiaus* among the fishes of the Seine and other fresh waters which artisan fishers and fishmongers sold to Parisians.[107] And from a mid-thirteenth-century rubbish pit beside the castle of the Louvre come the oldest known carp remains in

*rerum*, lib. 2, c. 22–47 (Wright, ed., pp. 142–158); Bartolomeus Anglicus, *De proprietatibus rerum*, lib. 13, cap. 26 (Frankfurt 1601 edition, pp. 578–587). Nor do carp appear among the thirty-three fishes mustered for allegorical war in the French "Battle of Lent and Carnival," assembled in Picardy or Normandy at the start of the thirteenth century (Lozinski, ed., *La Bataille*, 121, with dating 38–45).

105 Thomas of Cantimpre, *Liber de natura rerum*, lib. 7:22 (ed. Boese, pp. 258–259); Vincent of Beauvais, *Speculum naturale*, lib. 17:40 (1624 edition, col. 1274); Albertus, *De animalibus*, lib. 24:26 (ed. Stadler, pp. 1525–1526). Intellectual context is provided by Hünemörder, "Die Geschichte der Fischbücher," 188–193.

106 Longnon, ed., *Documents*, vol. III, 17–18. Another fragmentary account from 1285 had thousands of carp going into ponds at the duke's estates in the baillage of Chaumont (*ibid.*, p. 32).

107 Lespinasse and Bonnardot, eds., *Métiers et corporations*, titles 99–100, pp. 213 and 216. The same rules also appear in a royal ordinance for marketing fish in Paris, which Laurière et al., eds., *Ordonnances*, vol. 2: 583–586, give from a confirmation of 1320, after (p. 575) expressing doubt about the 1254/8 date given that text by Delamare, *Traite de la Police*, vol. 3: 298–302. Delamare had, however, access to manuscript sources later lost (Bourlet, "L'Approvisionnement," 6–7, and Auzary-Schmaltz, "Les Contentieux," 60). Philip IV's 1289 fisheries ordinance set the value for carp at two per denier (Duplés-Agier, "Ordonnances inédites," 51).

France, followed in a human generation by the same from castles at Laarne and Londerzeel in Flanders.[108] The carp had made itself home in Europe's west-flowing watersheds.

How had this occurred? Not by entirely natural means, more as an unintended consequence of human activities. Although scholastic natural historians and Parisian regulations of the mid-thirteenth century treated carp as a wild fish, human agency with varying purposes in mind had for the previous half-millennium meshed tidily with the tolerances and tenacity of an aggressive organism to encourage its spread. As earlier here discussed regarding eel, during the sixth–eleventh centuries and later Europeans caught, prepared, and ate fish from their natural local waters. This often entailed live storage of seasonal catches in tanks, cages, or ponds close to such centers of consumption as elite residences, castles, or monasteries. Charlemagne had mandated this practice for royal estates in his capitulary *De villis* (*c.* 795), and other records suggest such *vivaria* were reasonably common.[109]

Ensuing centuries witnessed increasingly purposeful construction of individual ponds throughout western Christendom, some meant to keep fish (*vivaria, servatoriae, piscinae*), others functioning as mill ponds or moats but capable of holding fish, too. Lay seigneurs – as at eleventh-century Lanzenkirchen castle in Lower Austria – may have led this activity but churchmen like John of Metz, abbot of Gorze in Lorraine, accepted lay gifts of ponds and built more themselves. Radiocarbon dating to the period around 1000 of the oldest surviving artificial ponds in Berri matches their entry into the written record and local acceleration of woodland clearances.[110] By around 1200 – so well before carp are detected west of the Rhine – some ponds in Poitou, Maine, southern

108 Clavel, *L'Animal*, 133; Ervynck and Van Damme, "Archeozoölogisch onderzoek"; Van Neer and Ervynck, *Archeologie en Vis*. Benoît, "La carpe," provides an overview of French records.

109 Boretius and Krause, eds., *Capitularia Regum Francorum*, #32, c. 21 and 65 (pp. 85 and 89). Exemplary such *vivaria* in lay and religious possession appear in the Carolingian *Brevium exempla* (*ibid.*, #128, pp. 250–256) and a mid-ninth-century survey of imperial properties in what is now easternmost Switzerland and western Austria (Häberle and Marti-Grädel, "Teichwirtschaft," 150–151). Earlier structures were built at St. Denis ("Chronique" 1987, 179–181) and documented in Burgundy (Bouchard, *Flavigny*, #3).

110 Galik, Private correspondence, "Fischresten aus … Lanzenkirchen"; Pertz, ed., "Vita Iohannis," 362; Pichot and Marguerie, "Sur l'aménagement," 119–124. Elsewhere across the territory where carp moved westwards see mills and ponds on estates in lower Bavaria as described in eleventh–twelfth-century charters (Krausen, ed., *Urkunden … Raitenhaslach*, nos. 1, 2, 3, and 29), in Franconia (Cnopf, *Entwicklung der Teichwirtschaft*, 10–12, and Mück, "Beginn der Teichwirtschaft"), and in Burgundy (Bouchard, *St-Marcel-lès-Chalon*, #12).

England, the Ile-de-France, Champagne, and Lorraine were equipped with adjustable sluices and bypass channels to manage their drainage and simplify mass harvest of fish.[111] Even beyond the general effects of woodland clearance, watermills, and localized eutrophication, long before the carp arrived in the west, more or less unawares Europeans were preparing just the kind of habitat this fish would enjoy.

Captured wild fish stocked early medieval store ponds, whether those were isolated *vivaria*, run-of-the-river impoundments behind mill dams, or had other primary purposes. Nothing indicates any purposeful choice of variety or special care in storage; no document inventories these fishes by name. The result was inadvertent selection for tolerance of captivity, transport, and possible variations in water temperature, oxygen supply, and food. Only tough and resilient species survived, even when moved from one watershed to another, where a few escapees might colonize new territory on their own. This practice long continued for single or even multiple ponds where small wild fish were stocked for future growth, as is first explicitly described on the Count of Champagne's estate at Provins in 1217–1219 (without naming the varieties) and in the 1230s in England on the bishop of Winchester's estates with bream, perch, roach, eel, and pike.[112] But just about that time French managers were beginning to put juveniles of single named species into chosen ponds, as seen in the carp at Igny-le-Jard.

Chapter 7 will explore implications of those ponds and carp domestication. For now the point is to acknowledge how synergies among rising medieval demand for fresh fish, multi-purpose human management of watercourses, and the resilient adaptability of carp jointly enabled that species to colonize continental Europe's western watersheds by the mid-1200s.

## 5.3 Regional Manifestations of Changing Fisheries

Data presented in Table 5.1 from seven well-investigated local sets of archaeological fish remains from the high Middle Ages encapsulate the

[111] Generally for French ponds see Benoit and Wabont, "Mittelalterliche Wasserversorgung," 189–196; Gislain, "Le role des étangs," 89; and Benoît and Rouillard, "Medieval hydraulics," 177–180. Local cases appear in Delatouche, "Le poisson d'eau douce," 174–175; Sanfaçon, *Défrichements*, 26 and 85–90; Blary, *Le domaine de Châalis*, 31–40; Richard, "Le commerce du poisson," 181–197; and Grand and Delatouche, *L'Agriculture*$_2$ 544. See also Holt, "Medieval England's water-related technologies," 65–66 and 83–88, and Aston, ed., *Medieval Fish, passim.*

[112] Longnon, *Documents*, 3:1–7; Roberts, "Bishop of Winchester's fishponds," 130–135. Belliard et al., "Increasing establishment of non-native fish species," rightly consider carp the first known invasive fish species in France.

Table 5.1 *Predominant fish taxa in large bone assemblages from selected high medieval sites*

| | dated to | number of identified fish remains | Rank % Clupeids | Rank % Gadids | Rank % Flat-fishes | Rank % Smelt | Rank % Eel | Rank % Salmonids | Rank % Pike | Rank % Perch | Rank % Cyprinids | Rank % Carp | All other taxa |
|---|---|---|---|---|---|---|---|---|---|---|---|---|---|
| **York Coppergate** | 13th cent | 1262 | ❶ 54% | ❷ 24% (includes 16% haddock + 6% cod) | ○ | ○ | ③ 15% | ○ | ○ | | ○ | | 7% |
| **Paris Louvre Cour Carrée** | 13th cent | 2889 | ❶ 38% | ○ | 3% | | ③ 5% | | ○ | ○ | ❷ 33%* | ○ | 11% |
| **Charité-sur-Loire kitchen/refectory** | 11th-14th century | 1703 | | ○ (freshwater burbot only) | | | ❷ | ③ | ○ | ? | ❶ | | |
| **Mechelen/Malines *Het Steen*** | early 14th century | 161,464 extrapolated from samples | ❶ 49% (35% herring + 14% indet. herring/sprat) | ③ 12%† | 8% | 6% | 9% | ○ | ○ | ○ | ❷ 13%* | ○ | 3% |
| **Bremen Altmarkt** | late 13th century | 1255 | ❶ 41% | ○ | 4% | ③ 6% | ○ | ○ | ○ | ○ | ❷ 39% | | 10% |
| **Lekno kitchen** | 1153-14th century | 830 | ③ 14% | | | | | ○ | ❷ 21% | 10% | ❶ 53% | | 2% |
| **5 lowland Austrian sites** | mainly 13th century | 1230 | ○ | | ○ | | | 10% (trout) | ❷ 28% | ○ | ❶ 28% | ③ 27% | 7% |

blank cell indicates taxon not found ○ taxon present * includes a few carp † includes a few cod

*Sources:* York Coppergate: Harland et al., "Fishing and fish trade," table 15.4, pp. 182–183, namely the column labeled 1200–late 1200s, with proportions calculated by R. Hoffmann. When all Gadidae (haddock, cod, and other species) are totaled, they come to 24%, so still not half of the herring. Flatfishes and pike (the leading resident freshwater taxon) trail at 3% and 2%.

Table 5.1 (*cont.*)

Paris Louvre, Cour Carrée: Clavel, *L'Animal*, table IV, p. 13. The context was a trash-filled ditch beside the palace. While the author calculated percentages on all bones, I have recalculated them based on identified remains, lumping the six carp bones with the other cyprinids. In the remaining 11% of bones from Cour Carrée, no taxon or grouping exceeded 3%. No salmon or sturgeon; gadids came to 2%, mostly whiting.

Charité-sur-Loire, monastic (Cluniac) priory kitchen and refectory: Audoin, *Ossements animaux*, 146–147, enumerates no individual elements or taxa, though mentioning abundant bream among the cyprinids in early phases and barbel and ide in the fourteenth century. Carp appear only from the fifteenth century and marine species are absent.

Mechelen/Malines, *Het Steen*: Troubleyn et al., "Consumption patterns … inside *Het Steen*," tables 7–9, pp. 32–36. Context was two very large cesspits in a structure then the municipal prison. Archaeologists agree that inmates at *Het Steen* represented a cross-section of town society and ate mainly food supplied by their own households or purchased from outside.

Bremen Altmarkt: Galik and Küchelmann, "Fischreste," 215–218, notably table 2, with proportions recalculated on base of the identified bones. No carp are reported and only 1% each of eel and gadids. A smaller (104 identified bones) trash pit of similar date nearby has a similar pattern, although the particulars had not yet been analyzed for *ibid.*, 219.

Lekno monastic (Cistercian) kitchen: Makowiecki, *Historia*, Aneks 2, p. 188, item 203; Wywra and Makowiecki, "Fish in the menu of Cistercians," 65. Lekno is about 30 km south of Poznań. There are traces of sturgeon and catfish but few diadromous species and no marine other than herring. Here listed are only the well-dated and sieved finds from the 2001 excavations, not those earlier (item 202) at this site, which provide no additional taxa.

Five lowland Austrian sites, all located between Vienna and Linz: Galik et al., "Fish remains as a source," table 1, pp. 344–345, with composite calculations by R. Hoffmann. Sturgeon remains, mainly of beluga, came to 3%. Present were one bone each of herring and a flatfish, with no sign of cods, mackerel, or any other marine organisms.

conditions wrought after some centuries of human impacts and rising social demand for fish pressing against traditionally desired natural local and regional stocks. The focus here is now on the most numerous taxa: regardless whether from fresh or salt water, in each case the four most common varieties total more than 84 percent of identified remains. While lamenting the absence of comparably rich reports from Mediterranean Europe – would Spanish, Provençal, or Italian studies inform other conclusions? – inferences both ecological and economic must be drawn. Behind quite significant local variations lay several common features.

By the thirteenth century people in all sites were eating fishes from 'ecological guilds' different than had their predecessors (see Chapter 3: Section 3.1 and Supplement). The salmonids and sturgeons which lost quantitative dietary importance needed two habitats, freshwater and marine, and unimpeded access between them to sustain their anadromous life cycles; their successful reproduction depended on relatively cool, well-oxygenated, running water. In contrast the well-evidenced (and easily recognized) remains of eel, small cyprinids, and carp lead and stand for general increases in relatively more lentic and heat-tolerant freshwater varieties. Cyprinids and pike, too, spawn by choice in weedy shallows, and eel leave fresh water when ready to breed. Only at York, on an oceanic island with relatively low diversity of cyprinids compared to the continent, did that group not play a large role. Instead, any emerging gap between traditional fish stocks and rising social demand was there met from a more radical and nearby alternative, the inshore marine environment producing pelagic herring. Large quantities of the silvery plankton eaters also served consumers at Paris, Malines, Bremen, and Lekno, while at greater distance from the sea herring lacked quantitative significance at this time. Marine demersal codfishes (haddock in particular in the case of York, that and whiting at Malines) were, with the benefit of historical hindsight, just arriving as a western European dietary option, as likewise were carp at Paris and Malines – although long a staple in their older Austrian range. Essentially, Europeans were eating more fishes with broader environmental tolerances in place of traditional varieties with narrower requirements.[114]

[114] In Regier's evocative terms ("Rehabilitation," 93–96), sensitive 'white fish' were losing habitat and yielding their prior importance in human catches and consumption to more broadly tolerant guilds of 'grey fish' and 'black fish', whose favoured conditions were less damaged and which, in certain regions and localities, were thus becoming more common. Although in contrast to post-industrial impacts, preindustrial development more affected lotic components of aquatic systems, the general ecological outcome was closely similar as "through their greater flexibility [grey fish] come to dominate within modified ecosystems."

Simultaneously much consumption had shifted down the trophic pyramid. Most still-water cyprinids and carp in particular consume much plant material, the small herbivorous invertebrates living on aquatic plants, and various bottom-dwelling creatures (benthos); herring consume mainly zooplankton, so eat only one step higher. Eel are more or less omnivorous. The dietary importance of carnivorous pike and salmonids had largely vanished from England and France but remained significant in less densely developed Poland and Austria.[115]

Characteristic differences between the fishes which rose in importance and those which fell thus argue for human impacts on medieval aquatic ecosystems more complex than can be ascribed to overfishing alone. Precisely the aquatic habitats needed by species under threat were the ones being blocked or degraded by medieval agricultural, urban, and industrial developments, which were, quite without human forethought, raising the amounts of silt and nutrients in Europe's watercourses and the proportion of standing waters. But also, insofar as the fishes at risk had long been favoured for human consumption, their largest and most productive spawners, those best able to replenish the local stock, were also the least likely to survive intensified fishing pressure.

Fishing is simultaneously an ecological and an economic activity. From the latter perspective high medieval fishers responded to rising human demand against limited traditional fish stocks by blends of intensified and innovative work to exploit regional alternatives. Eel and to a lesser extent herring were old local resources now put to ever greater use. The former fishery took unconscious advantage of a likely anthropogenic increase in stocks; the latter dipped more deeply and broadly into what appeared to be unlimited marine abundance. In the Danube basin and

[115] While early medieval Europeans likely consumed relatively more predatory freshwater fishes (pike, pike-perch, trout) than did their heirs in 1200 or 1300, and eel, carp, and small cyprinids are closer to the base of the freshwater food webs, evidence now available shows no clear sign of a trophic cascade, where smaller, short-lived organisms explode in numbers and biomass as a result of fishers selectively removing large predators. Eel may have occupied and expanded a niche in part left available by diminished sturgeon, whose possible earlier keystone role in large river and estuarine ecosystems simply vanished. Was this a regime shift from one relatively stable ecosystem to another? In contrast, the salmon actually participant in freshwater ecosystems are and were not the adults, which do not feed in fresh water, but the young, which there interact with other small salmonids and fishes of comparable size, habits, and habitats. Unlike the Pacific salmons (Genus *Oncorhyncus*), Atlantic salmon adults do not in temperate Europe transport in their dying bodies ocean-gathered nutrients essential to life in infertile waters where their young must survive. European fresh waters are just more nutrient-rich and diverse, even when the salmon are removed.

other inland areas neither eel nor herring could provide actual catches nor then more than rare and occasional exotic food, so heavier use was made of native cyprinids, perch, and some whitefishes, while some people intervened more actively in the distribution and life cycle of carp. Herring and to a lesser degree eel provided protein to a larger, less wealthy, consumer base than had the traditional fisheries or the emerging ones for carp or codfishes.

But for all the reasonable likelihood that forces and activities of medieval Europeans both purposely and inadvertently drove significant alterations to the subcontinents's aquatic ecosystems and fish stocks, humans were not the only probable post-millennial engines of change. The sparse and lacunae-ridden record of the ninth through early fourteenth centuries contains signs of naturally driven environmental fluctuations affecting biodiversity and interactions among regional fish communities of interest to human fishers and consumers.

## 5.4 Natural Dynamics

Although western Christendom as a whole enjoyed remarkably stable climatic and seismic regimes during the central and high Middle Ages, regional atmospheric and hydrographic conditions necessarily interacted significantly with freshwater and marine aquatic ecosystems. In what follows, a basic understanding of climate and hydrology will frame how shifts in these natural parameters changed local land- and waterscapes with likely consequences for fish populations and their use by medieval societies. Again, shards of information long assumed irrelevant and disconnected form a mosaic when joined by known ecological relationships.

### *5.4.1 Climatic and Hydrographic Fluctuations at Multiple Scales*

Historical climatologists commonly describe the times treated in this chapter in terms of the 'Medieval Climate Anomaly' (henceforth MCA), a period in global climate history with patterns distinct from a most recent reference period (roughly 1880s–1980s/2000 or shorter) and from those of the intervening 'Little Ice Age' (henceforth LIA) (Figure 5.3). At global scale the MCA entailed a slightly warmed planet during the ninth through eleventh/twelfth centuries with peak temperatures *c*. 800–1000 and gradual cooling thereafter, "albeit with important differences regarding the timing and spatial extent" of these phenomena. Twenty-first-century climate historians stress that global or hemispheric

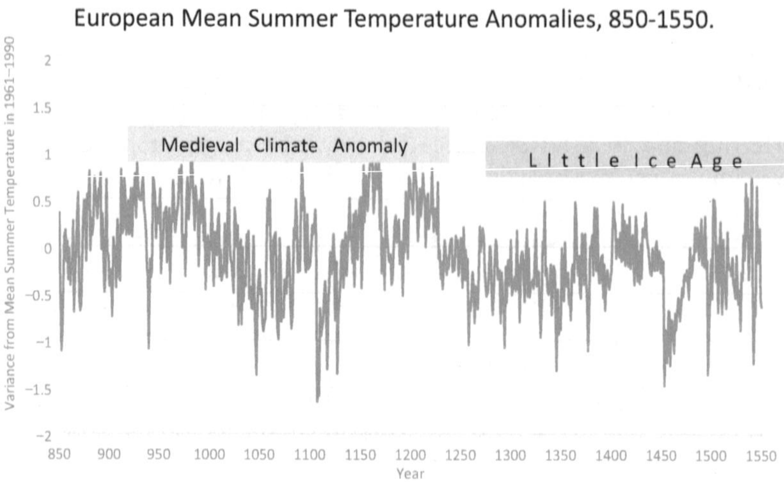


Area-weighted June-July-August mean temperature anomalies relative to Europe means, 1961-1990, as reconstructed from nine annually resolved tree-ring width (TRW), tree-ring maximum latewood density (MXD) and documentary records in Luterbacher et al., "European Summer Temperatures." Data set and permission from J. Luterbacher. Graphed for R. Hoffmann by K. Hoffmann

Figure 5.3 European mean summer temperature anomalies, 850–1550.

averages are less relevant than the regional conditions wherein humans and other organisms experience the impact of climatic anomalies.[116]

Energy flowing from the sun (solar irradiance) and terrestrial volcanism principally drove the earth's premodern climate, with regional features a result of global oceanic and atmospheric circulation.[117] The MCA coincided with high solar activity, especially during 1080–1280, and flow of energy to the earth. Only one solar minimum during 1040–1080 (the Oort) took place during the MCA, which came to an end with the Wolf minimum, 1280–1350. The ensuing LIA, a period of cooler global climates, coincided with three more minima in rapid succession.[118] Volcanic eruptions introduce aerosols and dust into the atmosphere. These reduce arrival of solar energy to the Earth's surface

[116] Quotation from Diaz et al., "Medieval warm period redux," 32. Christiansen and Ljungqvist. "Northern Hemisphere temperature," 277, confirm a greater geographic variability during the MCA than the LIA. See also Glaser and Riemann, "Thousand-year record of temperature," 446; Luterbacher et al., "European summer temperatures." Discovery of this variability led climatologists to replace the term 'medieval warm period' with MCA and calls into question scientific or scholarly explanations which assume unbroken or year-round heat at this time.

[117] Steinhilber and Beer, "Solar activity."

[118] The Spörer minimum dated to 1460–1550, Maunder to 1645–1715, and Dalton to 1790–1820. For reconstruction across the MCA and LIA see Luterbacher et al, "European summer temperatures," as here replicated in Figure 5.3.

and often have a global or hemispheric cooling effect. After an intense cluster of large eruptions and cooling during the sixth century, few volcanoes affected the northern hemisphere until a series of eruptions during the 1150s through 1260s, followed by another in the 1340s. During much of the MCA global circulation patterns produced a positive phase in the North Atlantic Oscillation (henceforth NAO), meaning westerly flows from the Atlantic protected most of western Europe from colder air out of Siberia. From 1100 to 1260 Europe's average annual temperatures surpassed twentieth-century norms. Growing seasons remained at or above twentieth-century means from the mid-tenth century through the 1250s, followed by a severe drop in the 1260s. That cool spell dissipated after 1270 when temperatures rebounded for about another human generation, followed, however, by great instability during middle decades of the fourteenth century.[119]

Planetary or even large-scale regional average temperatures are only a part, and not necessarily all that important a part, of how climate and weather patterns affect living things.[120] Fluctuations of hydroclimate (precipitation and evaporation) do not necessarily coincide with temperature in their timing nor the scale of variation. Specific regional manifestations of both temperature and precipitation differ from the large scale. Seasonal variations have the greatest impact at critical life-cycle stages for both humans and fishes. Extreme events and other environmental perturbations stress natural systems and societies dependent upon them.[121] So possible connections between climate and past fisheries need to work at smaller scale.

Neither the writer and readers of this book nor their medieval predecessors have direct physical perceptions of the abstraction 'climate' nor of its changes. Without serial records and the patterns those may reveal, humans experience meteorological events (weather), sometimes later recall extremes (cold, storms, heat, drought), and may over time adjust

[119] Goosse et al., "Origin of the European 'Medieval Warm Period'," 105–110; Trouet et al. "Persistent positive North Atlantic Oscillation"; Seager and Burgman, "Medieval hydroclimate revisited," 11–12; Ortega et al., "Model-tested North Atlantic Oscillation reconstruction"; and Franke et al., "North Atlantic circulation." Christiansen and Ljungqvist, "Northern Hemisphere temperature," 765, conclude "The two-millennia long reconstruction shows a well defined Medieval Warm Period, with a peak warming *c.* 950–1050 AD reaching 0.6°C relative to the reference period 1880–1960 AD." European averages were probably higher, but still below those of the late twentieth century.

[120] Ljungqvist et al., "Northern Hemisphere hydroclimate variability."

[121] Campbell and Ludlow, "Climate, disease and society," figure 1 and appendix 5, construct for western Europe an "Index of Environmental Instability" to help contextualize late medieval crises.

their behaviour to patterned experience. When does the snow come? How big a bridge will survive floods? When to expect the wheat harvest or the bream to assemble in shallows to spawn? Catastrophic events can transform habitats; gradual shifts in water temperature or chemistry can have similar effect, though perceived by humans only as one kind of fish replacing another. And over time human practices may change – though not due to recognition of 'climate change'.

If the intensely studied Tiber watershed in central Italy can stand for conditions in western Mediterranean Christendom, after cooler than average and increasingly wet weather from the fifth through the ninth centuries, a warming trend set in and rose to a peak between the mid eleventh century and mid twelfth only to cool again slowly into the fourteenth. The region was distinctly drier between about 1050 and 1350 than before or after; the Tiber rarely reached flood stage. Obstructed drainage, however, turned the valley floor around Rieti into a permanent wetland of slow-moving waters.[122] Surely the environment for Carolingian-age fisheries of Farfa abbey (Chapter 3, pp. 101 and 107 above) had evolved. More general surveys of Italian conditions see changing phases, with *c.* 1100–1270 typically warm and arid, while lake core samples from eastern and coastal Spain give a similar impression.[123] Subfossil remains of small vertebrate animals from sites along streams in Corsica show different patterns of diversity during droughty high medieval times than in subsequent wetter periods.[124]

North of the Alps the central European landscapes whence large rivers flow into the North Sea, Baltic, and Danube, experienced a different MCA. A period of peak temperatures was well defined in the mid-tenth to mid-eleventh centuries, followed by decadal or longer cold spells after 1050, in the early 1100s, and again in the early 1200s. Increased seasonality meant the thirteenth and early fourteenth centuries knew distinctly warm summers but chilly winters.[125] The positive NAO also produced wetter summers, although tree ring data from both Scandinavia and the

122 Mensing et al., "2700 years of Mediterranean environmental change," and Mensing et al., "Human and climatically induced environmental change," 54–57.

123 Seager and Burgman, "Medieval hydroclimate revisited," 12; Ortolani and Pagliuca, "Cyclical climatic–environmental changes"; Moreno et al., "Hydrological pattern." Written proxy sources (Camuffo and Bertolin, "Climate in the Mediterranean," figure 1, p. 127) indicate colder Italian winters set in about 1270 to 1360.

124 Vigne et al., "Sensibilité des microvertébrés."

125 Glaser and Riemann, "Thousand-year record of temperature," 444–447; Christiansen and Ljungqvist, "Extra-tropical Northern Hemisphere temperature."

Swiss Alps signals somewhat drier conditions in those boundary areas.[126] Overall warmer summer waters might inhibit successful reproduction by fishes intolerant of high temperatures while enabling successful spawning by species preferring those conditions.

On maritime fringes off northwestern Europe the warming trend of the MCA came early and may have reached its maximum sooner, too. Greenland was at its warmest *c*. 800–1000, reaching 1.8°C above the reference period. Norse settlers arrived at Iceland in 870 during a hundred-year warm spell, suffered a chilly eleventh century, but then between 1100 and the mid-1200s enjoyed the warmest summers in three centuries. After 1280 cooling set in across all seasons. In northwestern Scotland only a mid-eleventh-century drought interrupted otherwise wet conditions from around 900 to 1340. Sea surface temperatures in the North Atlantic – assessed from the barely visible fossil shells of temperature-specific diatoms in bottom cores – fluctuated at the decade scale through most of the twelfth century but then maintained high values from 1170 to 1260. A hundred years of cooling followed.[127]

Put simply, around 1100, when Mediterranean Europe was close to its warmest and aridity threatened some aquatic systems there, central Europe was already cooling down, especially in winter time, but quite well watered, and the North Atlantic coasts and islands, though very wet, retained a moderated climate, though less warm than it had been some centuries before. Organisms, ecosystems, and cultural adaptations at the edge of their critical tolerances had the most to lose or to gain from marginal changes.

Regional climate itself is but one component in the dynamic processes visible at medieval Europe's land–water interfaces, its marine estuaries, coastal marshes, and open shorelines as well as the courses and banks of its rivers. To begin at the headwaters, many preindustrial rivers took an unstable local course. Seasonal patterns of flood, drought, and (in the north) ice cover kept channels mobile and banks impermanent. During high water rivers revisited their floodplains. Above the estuaries low water commonly revealed braided structures – bars, islands, shallows, multiple channels – as well as dominant constrictions, rapids, reefs, and

[126] Seager and Burgmann, "Medieval hydroclimate revisited"; Amann et al., "Warm season precipitation"; Büntgen et al., "2500 years of European climatic variability"; Büntgen and Tegel, "European tree-ring data and the Medieval Climate Anomaly"; McCarroll et al., "A 1200-year multiproxy record of tree growth and summer temperature."

[127] Vinther, "Medieval climate anomaly in Greenland"; Campbell, *Great Transition*, 45–47 and 200–201; Cunningham et al., "Reconstructions of surface ocean conditions," 929 and appendix 1.

waterfalls.[128] To the historic interplay of climate, topography, soils, and human constructions rivers naturally responded with changes in morphology and their associated aquatic habitats.

Distinct periods of frequent and large regional inundations associated with short-term climatic instability occurred both during the MCA and as it drew to an end. Rivers in English lowlands showed rapid seasonal fluctuations and flood events during 1085–1117 and then again in the thirteenth century, which triggered structural changes into multi-channel forms.[129] On the lower Rhine in the 1150s Emperor Frederick Barbarossa instructed the bishop of Utrecht and counts of Gelders, Kleve, and Holland to take measures to protect their subjects against flooding. Two centuries later and further up the Rhine two religious institutions went to canon law courts over changes to an island.[130] Along both the upper Rhône and the rivers of the Pyrenees, after long docile centuries, frequent and disastrous flooding set in from the end of the twelfth century, rapidly increased through the thirteenth, and reached a peak during the 1350s–60s.[131] During that last wave of floods the upper Rhône transformed from a meandering to a braided morphology,[132] characterized by a higher gradient, coarser bed materials, large interannual and seasonal variability, and thus newly unstable habitats for aquatic life. Fluvial instability with human consequences even drew the mid-fourteenth-century attention of famed Italian jurist Bartolus of Sassoferrato and generations of late medieval Hungarian legists.[133]

[128] See for example Menant, *Campagnes lombardes*, 59 and 176; Fumagalli, *Landscapes*, 104–105; Pinto, "Incolti, fiumi, paludi," 1–7; Bravard, "Des versants aux cours d'eau"; Levy, "Tant va la cruche à l'eau"; Rossiaud, *La Rhône*, 93–155; Molkenthin, "Der Rhein," 49–53.

[129] Lewin, "Medieval environmental impacts and feedbacks," 277–278. Lewin (p. 297) stresses that natural physical forces drove changes in channel morphology, while human activity more resulted in floodplain wetness and sedimentation, although the scale and form of the latter also depended on local soil types.

[130] Molkenthin, "Der Rhein," 50; Trusen, "Insula in flumine nata."

[131] Berger and Brochier, "Rapports de la géoarchéologie"; Calvet et al., "Les cours d'eau des Pyrénées orientales," 286–287. Flood records from other French rivers, the Arno, Po, and those of Germany share this general chronology (Grove and Rackham, *Nature of Mediterranean Europe*, 133–136; Leguay, *L'eau dans la ville*, 399–406; Fumagalli, *Landscapes*, 88–89, 110–112, and 117–121; Camuffo and Enzi, "Two bi-millenary series"; and Bork *et al.*, *Landschaftsentwicklung*, 237–249). But a very long-term study of sediments dated by $^{14}$C found riverine flooding in coastal Iberia, southern France, and Italy most frequent during the sixth/seventh, tenth, and late fifteenth centuries, while the thirteenth and sixteenth were drier (Benito et al., "Holocene flooding," 21–23).

[132] Bravard et al., "La diversité spatiale des enregistrements morphosédimentaires"; Rossiaud, *Le Rhône*, 128–137.

[133] Bartolus wrote in his 1355 *Tractatus de fluminibus* "Travelling towards a certain villa situated near Perugia above the Tiber, I began to contemplate … the changes of the

Perturbations in riverine hydrologies flow down to estuaries and the sea. At the onset of the MCA in the tenth–eleventh centuries, upper and middle reaches of the Wisła became unusually active, with clusters of frequent flood events. In some areas a meandering morphology began to braid.[134] At the time this large basin drained into a delta and estuary extending some sixty kilometers from Gdańsk east to Elbląg with bays, wetlands, and channels reaching equally far inland and all loosely delineated from the Baltic by a string of offshore islands. Archaeozoology reveals a brackish environment with inshore schooling fishes. From the thirteenth century and into the later Middle Ages this landscape evolved into an increasingly drained delta on the west, where the river entered the sea, and to the east a separate lagoon (Zalew Wislany, Frisches Haff) fully enclosed by a continuous sand spit.[135] Fourteenth-century villagers along the lagoon shoreline there now took diadromous and freshwater fishes (salmon, pike), while by the mid-1400s sea-caught cod had become prominent in demesne fisheries of the ruling Teutonic Order.[136]

Coastal environments elsewhere experienced diverse variabilities. Average global sea level remained relatively stable throughout the mid–late Holocene, fluctuating only a meter or two. But local seismic activity, post-glacial uplift, land subsidence, shoreline erosion, terrestrial drainages, siltation, and shifts in storminess and normal wind direction variously affected medieval coastlines in several regions. Sweden's Lake Mälaren was a bay of the Baltic open to seagoing vessels in the Viking Age and until about 1200, when post-glacial rebound raised it above mean sea level so it became entirely fresh water and Stockholm replaced Birka as the principal port. Salinity in the long, narrow estuary of the Schlei fell significantly in the course of the twelfth century, so what trace element analysis indicates were fishes taken locally ceased to be the

river-bed as well as a host of unanswered questions which I had come across in practice ..." Text of Bartolus's prologue is in Cavallar, "River of law"; a complete reprint of the 1576 printed edition is Bartolus, *Tractatus de fluminibus*, ed. Astuti. I am grateful for the discussion of wandering rivers in the Pannonian plain by András Varas in his (so far) unpublished dissertation, "Who stole the water? The control and appropriation of water resources in medieval Hungary," 153–172.

134 Starkel et al., "Past hydrological events," 24, and Lewin, "Medieval environmental impacts and feedbacks," 270 and 301 (the latter citing Starkel, ed., *Evolution of the Vistula River Valley*, vol. 1 (Warsaw: Polish Academy of Sciences, 1982), a text I have not been able to consult).

135 Filuk, "Biologiczno-rybacka charakterystyka ichtiofauny," 130–159.

136 Willam, "Fischerei des Deutschen Ordens," 99–137 and 147–149; Martens, *Gartensiedlungen*, 154–156, 202–205, 276–278, and 358; Sarnowsky, *Wirtschaftsführung*, 283–284.

marine species, herrings and others, eaten at Haithabu before 1085. Instead late medieval people barely five kilometers away at Schleswig ate perch, bream, and pike.[137]

Along the southern shore of the North Sea, storms could turn land into new arms of the sea. What had into the 1100s been a complex of freshwater lakes, wetlands, rivers, and drained farmland had become by the end of the 1200s a marine embayment, locally called the 'South Sea' (Zuiderzee). Land subsidence and storm floods broke the Frisian coastal barrier at Texel in 1282, and five years later the Saint Lucia flood of 14 December washed deep inland to drown tens of thousands of people and untold livestock, replacing thousands of hectares of pasture and arable with salty waters. While subject to intermittent storm surges, the new sea's boundaries stabilized in the fifteenth century and lasted into the twentieth.[138] Storm waves from a different direction washed deep into drained wetlands of the Thames estuary in the 1230s, 1280s, 1320s,1334, and 1370s; once people stopped trying restore the damage, the newly enlarged salt marsh increased local fish habitats and fisheries there expanded, especially after 1351.[139] On the other hand, the Bay of the Somme which had reached to Abbeville twenty kilometers inland, was by 1300 receding together with its fisheries for eel and flatfishes, and so, too, further south the once even larger 'Gulf of Pictons' in Poitou. Yet on the other French coast, unstable passes and fluctuating salinity levels of Languedoc lagoons motivated fishers and rights holders to adjust their institutional relations and keep on fishing.[140]

[137] Grupe et al., "A brackish water aquatic foodweb"; Becker and Grupe, "Archaeometry meets archaeozoology." Likewise at the northern tip of Jutland deposition of eroded sand closed the Limfjord in the twelfth century, shifting its waters from a marine to a freshwater or brackish habitat (Hybel and Poulsen, *Danish Resources*, 48–49).

[138] Besides formation of the Zuiderzee, other well-known counter-examples to what happened at the mouth of the Wisła occurred during the transition from the MCA to the LIA. Large losses of human lives, arable, and villages along the North Frisian coast from 'de grote mandrenke' (the great human drowning) of 17 January 1362, and similar marine incursions recreated extensive areas of tidal flats and marshlands, productive aquatic environments of the Wadden Sea (Meier, "From nature to culture," 95–102, attends to the losses). More cases are in Supplement 5.4.1.

[139] James Galloway provides much local detail in his "Storm flooding"; "London and the Thames estuary," 130–135; "Storms, economic and environmental change," 388–391; and "Expansion or eclipse?". Bailey, "*Per impetum maris,*" describes coastal inundations elsewhere in eastern England but not their effect on local fisheries.

[140] Clavel, "Restes osseux animaux," 200–202; Clavel and Cloquier, "Sources documentaires et archéologiques," 207–208; Abel, "Defining a new coast"; Bourin-Derruau et al., "Le littoral languedocien au Moyen Âge," 349–357; Carozza et al., "Lower Mediterranean plain accelerated evolution." Provensal et al, "Geomorphic changes," explores the evolving Rhône delta.

Little recent or extant research on the history of climate or hydrology has paid direct attention to aquatic habitats. The norm in water history treats a physical fluid, not a biological substance. And indeed most any efforts to relate these variables had until recently to contend with climate data so broadly drawn (in both spatial and temporal terms) that conclusions would be crude at best and more often misguided. Even with increased palaeoscientific precision in assessing climatic variables, only very occasionally during the Middle Ages are coincident biological and/or economic archives of specific local fisheries sufficiently detailed to allow even probable hypotheses. But to assume the aquatic realm of medieval Europe was unchanging or driven only by human actions would fly in the face of basic ecological knowledge. While leaving problems of large commercial marine fisheries at the very end of the Middle Ages for treatment in Chapter 8 below, shifting ranges and stocks of several species on the continent and in the Baltic do appear consilient with climatic and hydrographic fluctuations.

### 5.4.2 *Traces of Impacts, Resilience, and Adaptation*

In the context of varying medieval climates and weather with evident or likely effects on regional and local aquatic systems, we can trace certain fairly well documented changes in specific fish stocks to prior or simultaneous natural phenomena. At least two freshwater fishes may have achieved their sometimes tenuous natural establishment in western Europe under favourable conditions of the MCA and then, lacking purposeful human intervention, barely survived the LIA there. Water temperature, salinity, and oxygen content influenced interacting species of considerable human interest in the medieval Baltic.

The rapid eleventh- through thirteenth-century spread already observed of common carp across central and western continental Europe was less evidently associated with people actually rearing fish – nearly all records before the 1250s treat carp as wild – than with their placing captured wild fish in ponds for live storage. This occurred during what is now understood as a warm phase of the MCA, when mean European annual and summer temperatures both peaked. Carp are thermophilic organisms, meaning they do like heat. In the wild this species begins to spawn in weedy shallows during May–June as water temperatures move above 18°C. Reproductive success is restricted to years when the water level starts rising in May and when high temperatures and flooding of terrestrial vegetation last for a long period during May and June. This is because carp larvae survive only in what is by modern north European standards very warm water (at or above 20°C) among shallow submerged vegetation.[141] Chapter 7 will show that later European fish farmers learned to design special spawning ponds to warm quickly and to handle the larvae with special care, but that came as the climate cooled into the LIA. Carp's prior high medieval expansion likely benefitted as much from natural heat as it did from humans moving a prospective fish dinner from one watershed to another.

Adult individuals of a small cyprinid, the bitterling, look very like juvenile carp.[142] Until about 1100 the known range of the bitterling was restricted to southeastern Europe, more or less similar to the pre-medieval range of the carp, though no one ever thought bitterling would make a palatable meal. Bitterling are perhaps more thermophilic than carp, preferring waters above 16°C by June and needing 23°C for successful reproduction. The earliest records of bitterling in central and western Europe occur in regions where carp were becoming known and carp culture would later become significant (see Chapter 7). In the mid-twelfth century Hildegard of Bingen, where the Nahe joins the middle Rhine, even knew the bitterling's love for warm water. Thereafter written, visual, and archaeozoological traces of bitterling multiply in those regions, only nearly to vanish after about 1550, while Europe endured the coldest two centuries of the LIA. Bitterling reappear and spread only from the late 1700s. While quite widely abundant in recent times, local populations decline markedly following years and decades with cold spring temperatures and revive with warmer ones as at the end of the twentieth century. Like carp, then, bitterling spread westwards during warm medieval summers with entirely inadvertent human assistance; come the cooler state of the LIA the species contracted most of its western range.[143]

141 Adult carp, however, tolerate even very cold conditions. Kottelat and Freyhof, *Handbook*, 147–148; www.fishbase.org/summary/Cyprinus-carpio.html (consulted 20 December 2016).

142 What follows summarizes Van Damme et al., "Introduction of the bitterling," with added biological information from fishbase.sub Rhodeus-amarus (consulted 20 December 2016) and Kottelat and Freyhof, *Handbook*, 82–84. Some authorities consider European *Rhodeus amarus* synonymous with *R. seriecus*, which has a widely separate Asian distribution. The little fish drew attention for the 'farting' sound it makes when handled and for the long tubular ovipositor the female deploys to insert her eggs into freshwater mussels, where the larvae live as parasites. Recent debates are over the bitterling's indigenous status in western Europe and thus its qualification for special protection under EU water regulations.

143 Dirk Heinrich explored in several articles probable connections between climatic conditions and the discontinuous range in northwestern European watersheds of another thermophilic species, catfish: "Fischreste als Quellengattung," 176–178; "Bemerkungen zur nordwestlichen Verbreitung des Welses"; "Information … from tales," 19–20; and "Methodological considerations," 163–165.

Medieval evolution of herring and cod fisheries inside the Baltic may owe more to natural variability than to human enterprise or impacts. To recapitulate and anticipate: herring were abundant in the central and southwestern Baltic (Bornholm, Pomerania, the Schlei) from the fifth/sixth centuries into the thirteenth and by the latter date also in the Danish straits (of which much more in Chapter 8). Cod were certainly present in the Baltic during the Neolithic and again from the fourteenth century, while during early and high medieval times this species is virtually absent from the written and archaeozoological record there.[144] The traditional explanation is that Slavic peoples had no taste for cod but immigrants from Germany did. Stable isotope and other studies of the cod bones themselves, however, identify the oldest medieval cod remains found at Haithabu and in the eastern Baltic as imports from the North Sea or Norway, that is, as imported stockfish. Yet by the end of the Middle Ages people in this region did plainly fish locally for cod.

More recent suggestions observe that known low salinity and hypoxic (low oxygen) conditions in deep water basins of the medieval Baltic could have suppressed cod populations. In the Baltic today strong hypoxic conditions thought to arise from a westerly flow of wind and waters from the Atlantic, a warming climate, and eutrophication from nutrient-rich runoff place a cap of warmed fresh water on top of colder saltier waters where no mixing occurs. This situation drives many free-floating eggs of cod into the deepest basins where lack of oxygen prevents larval development and so threatens cod recruitment.[145] Baltic herring, however, like their North Sea kin, breed in relatively warm, biologically productive, and often brackish surface layers. Sediment cores demonstrate that hypoxia in the Baltic is not just a modern but rather a recurring phenomenon, present

[144] Heinrich, "Fang und Konsum" and "Fishing and consumption of cod"; Makowiecki, "Studies on the evolution," 176–179, "Catalog," table 2, and "Usefulness of archaeozoological research," 109–110; Lõugas, "Fishing during the Viking Age" and "Fishing and fish trade," 111–112 and 114–115; Holm, "Commercial sea fisheries," 18–19. Absence of cod remains from the early Viking Age site of Truso and from pre-1200 Gdańsk, as well as the Atlantic origin of cod bones at Haithabu suggest that even the Norse found few of the species in the early medieval Baltic (Makowiecki, "Janów Pomorski"; Rulewicz, *Rybołówstwo Gdańska*, 61; Lepiksaar and Heinrich, *Fischresten aus Haithabu*, 119; Schmölcke and Heinrich, "Tierknochen aus dem Hafen," 220–233).

[145] The analysis here proposed applies to evident medieval fluctuations in Baltic hydrology and fish ecologies the observed late twentieth-century environmental and regime shifts there as set out in detail by Hammer et al., "Fish stock development under hydrographic and hydrochemical aspects," 557–564, and briefly by Alheit and Pörtner, "Sensitivity of marine ecosystems to climate and regime shifts," 168. These refer to work by Mackenzie et al., "Quantifying environmental heterogeneity";MacKenzie et al., "Ecological hypotheses," 177–190; Köster et al., "Baltic cod recruitment"; Zillen et al., "Past occurrences of hypoxia in the Baltic"; van der Lingen et al., "Trophic dynamics," 135 and 145; and Brander, "Impacts of climate change on fisheries," 393.

both during the early Holocene (*c.* 9000–*c.* 5000 ybp) and roughly between 550 CE and 1200 ± 50 years. Thereafter bottom waters became better oxygenated and remained so into the nineteenth century.[146]

Long-term changes in the state of the Baltic, termed a 'regime shift' by recent ecologists, have been linked to climate variations. The earlier medieval and the most recent condition of deep water hypoxia beneath a warm surface layer of low salinity result from a positive NAO characteristic of both the MCA and modern warming. Predominant westerlies then push salt water in from the North Sea and raise precipitation levels across the Baltic watersheds. Temperature and salinity differences encourage stratification. Near-surface temperatures and biological productivity are high. Fading of the MCA from around 1200 tended to bring more negative NAO, with lower salinity but also lower temperatures and nutrient levels. More balanced salinity and temperature allowed greater mixing, so conveying more oxygen to the deep basins. Distinctive onset of the LIA by around 1500 strengthened a negative NAO, produced drier and colder conditions in central European lands, and prolonged the no longer new marine trophic regime for another three centuries.

In the present-day Baltic abundant cod are the principal predator on herring, which controls the herring population. Under modern conditions of intensive fisheries, removal of predators results in superabundant prey populations, a typical form of trophic cascade. With few cod present, early medieval herring stocks could explode until limited by some other ecological factor.[147] Very recent ecological field work and theories even suggest that, as herring will themselves predate on planktonic cod eggs and larvae, the large medieval schools could (further?) have suppressed a Baltic cod stock already under stress from low oxygen in the habitats critical for its reproductive life stage. Both the abundance of adult cod and cod recruitment show negative correlation with herring biomass.[148]

146 Zillén et al., "Past occurrences of hypoxia in the Baltic," 87; Kuijpers et al., "Baltic Sea inflow regime"; Weckström et al., "Palaeoenvironmental history of the Baltic"; and Franke et al., "North Atlantic circulation." During the LIA, in contrast, east-central Europe experienced dry conditions and an unstable hydroclimate (Tylmann and Grosjean, "Climate variability in Central and Eastern Europe"). For the conceptual framework of 'regime shifts' used here return to Chapter 1, note 34.

147 Sparholt, "Fish species interactions"; Köster et al., "Baltic cod recruitment"; and Heikinheimo, "Interactions between cod, herring and sprat." Corten, *Herring and Climate*, demonstrates the positive response of herring to warmer waters.

148 Fauchald, "Predator–prey reversal"; Van Denderen and Van Kooten, "Size-based species interactions," 3; and Sánchez-Garduño et al., "Role reversal." On the negative relationship in the twentieth-century North Sea see Engelhard et al., "ICES meets marine historical ecology," 1391–1394," and works there cited. Auber et al., "Regime shift in an exploited fish community," suggest synergy between exploitation and climate change in driving regime shifts involving small pelagic species.

Based on organic and chemical markers in layered bottom sediments, the accepted chronology for shifts in oxygen and salinity levels in the medieval and early modern Baltic matches poorly with growth of human populations and the intensity of land use in that watershed. The initial conditions which would impact both cod and herring stocks were established well before high medieval clearances in central Europe and Scandinavia could much have affected the runoff regime. Indeed trends shift in the opposite direction just as the wave of settlement and clearances began to crest in the north German, Polish, and eastern river basins which provide the bulk of the sea's fresh water. Although properties of large enclosed water bodies change more slowly than do atmospheric drivers, the southern Baltic herring fishery spiraled into insignificance in the course of the 1200s, supplanted by surging growth in the Danish straits. As Chapter 8 will trace closely, the latter fishery endured some two centuries during the very time fishing for cod came forward in many Baltic coastal areas. Those local cod fisheries would then persist even as Øresund herrings faded before Dutch exploitation of North Sea stocks under now stabilized LIA conditions. In sum, given what is now known about habitat requirements and predator–prey relationships across life cycles of cod and of herring, climatic and hydrological conditions in the Baltic Sea during the central and high Middle Ages (the MCA) were propitious for the latter species and stressful for the former. Climatic forces thus appear to have been important drivers in the medieval Baltic oscillation to and then from what one ecologist calls 'a herring dominated state.'

Coastal societies exploited abundantly accessible stocks until natural conditions slowly changed. As foreshadowed earlier in this chapter and spotlighted in Chapter 8, the Baltic herring fishery of the tenth through fifteenth centuries was quite probably medieval Europe's largest single source of marine protein.

And were Mediterranean waters and fishes unaffected? Have archaeologists and historians failed to ask the right questions in the most record-rich region of medieval Europe? Or did the natural world simply interest literate medieval Italian, Provençal, and Catalan elites even less than it did those of the north? Nevertheless, good proxies indicate colder Italian winters during the early 1300s and again in certain fifteenth-century decades. We also do know that eighteenth-century fisheries for Adriatic sardines fluctuated in tandem with weather patterns.[149]

~~ <>< ~~~ ><> ~~

[149] None of the essays in Buti et al., eds., *Moissonner la mer* (2018), or Tønnes Bekker-Nielsen and Gertwagen, eds., *The Inland Seas: Towards an Ecohistory of the Mediterranean and the Black Sea* (2016), or Mylona and Nicholson, eds., *The Bountiful Sea* (2016) show any interest in Camuffo, "Freezing of Venetian lagoon," 54; Camuffo

At local and regional scale aquatic systems of high medieval Europe were subject to multiple pressures and constraints across the interplay of natural and cultural forces. Relevant drivers and effects varied from one socio-natural site to another. A widely evident rise in human environmental impact had negative consequences for fishes with strict habitat requirements, while favouring varieties more tolerant of heat, low oxygen, and high nutrient levels. Changes of natural origin, a warmer climate and more, also created, destroyed, or shifted equilibria among fish species, with at least some effect on human use. Contemporary Europeans may have been oblivious to some such variations or lacked the perspective to see them in the longer term, but people plainly did become aware of certain changes. Some traditional fisheries seemed less productive (scarce) and other local stocks to offer fishers opportunities to respond to greater demand for fish. Neither resource destruction and depletion nor the dilemmas of allocation and conservation are peculiar to present-day fisheries crises. The next chapter turns to responses in medieval communities to perceptions of limits, declines, and shortages in fishes Europeans had long liked to eat, so exploring Europeans' own cultural resilience and adaptability. Reciprocal and reiterative interactions among medieval European natures and cultures drove the larger narrative of fisheries to be apprehended as collectivities of myriad socio-natural sites.

and Bertolin, "Climate in the Mediterranean," 127; Benito et al., "Holocene flooding"; Luterbacher et al., "Review of palaeoclimatic evidence"; or even the tight correlation of Mediterranean clupeids (sprat, sardine, anchovy) with eastern Atlantic species driven by the Atlantic Multidecadal Oscillation as shown in Alheit et al., "Atlantic Multidecadal Oscillation (AMO) modulates dynamics." But see Županović, *Ribarstvo Dalmacije*, 37–131, notably 58–64, 90–98, and 145–158, despite its necessary dependence on what are now obsolete weather reconstructions.

# 6 Cultural Responses to Scarcities of Fish

Medieval Europeans were well aware of undesirable changes in their fisheries. They commonly articulated those perceptions only when they wanted to explain how they would deal with a situation, but medievalists are well acquainted with many actions taken without recorded rationale. Under pressure from rising consumer demand the dwindling yields of traditional inland and coastal fisheries called up market responses and motivated both proprietary and regulatory measures meant to control human use of aquatic resources. Medieval people thus adjusted the social face with which they confronted nature. (A different kind of response, the adjustment of nature to fit human wishes, will occupy our next two chapters.)

With this chapter the focus thus shifts from the relatively fish- and eco-centric perspectives of Chapter 5 to examine how medieval societies reacted to perceptions of insecurity and shortfall of supply. Rising demand against a limited, even diminishing, resource is a potential trigger for competition. In a market economy price mechanisms allocate scarce goods to demand backed up by purchasing power. At the same time other kinds of power came to bear on traditional fisheries as land-owners, producers, consumers, and territorial authorities vied to direct exploitation in their own interests.[1] Attempts to privatize and to regulate fisheries and fish products helped shape and constrain human environmental impact. The regional quality of habitat changes and fishing pressure suggests social reactions would vary in timing and choices.

## 6.1 Allocating Fish and Fisheries Resources

Market price and ownership rights provide two cultural mechanisms to allocate use of a good in limited supply. I pay the price and an object is

[1] The medieval situation has parallels in New Englanders' responses to destruction of anadromous Atlantic salmon populations by nineteenth-century mill dams as described in Steinberg, *Industrialization and the Waters*, ch. 6.

mine to consume. I establish possession of a productive resource and its fruits are mine. Those who cannot pay or maintain access lose out. The former has evanescent temporal scale and possible collective significance over longer time, but particulars of actual transactions and serial records materialize only well after initial phases of the medieval Commercial Revolution. The latter issue can be traced in practice across European landscapes throughout the Middle Ages but local and regional implications of owning fishing rights varied widely.

### *6.1.1 Trends of Rising Fish Prices*

Earlier discussions of medieval fish prices have set contexts for now exploring how they evolved over time. Chapter 2 used anecdotal references from specific locations and dates to show fish were a comparatively costly food throughout medieval times, not a cheap source of calories or protein. Chapter 4 reinforced this feature in the market setting of the high and later Middle Ages, identifying fresh fish of favoured large kinds as notably expensive. Most varieties offered on markets of the twelfth through fifteenth centuries so exceeded the subsistence budgets of poor people that elite demand drove their commodity value. Herring, sardine, anchovy, and small cyprinids may have been prominent exceptions in certain places and times. Beyond that consistent fact and laid over the natural variety of European fishes, seasonality (of supply and demand), local differences in taste, local shifts in purchasing groups, and official price fixing jointly lend medieval fish prices a baffling diversity and volatility. Identifying long-term changes in price levels calls for continually commensurable information scarcely heard of before the Middle Ages end. Simply few records were kept of selling and buying fish in a bewildering diversity of names, sizes, measures, and preparations; even fewer have hitherto been studied in a systematic way. Still the evidence of some temporal evolutions is compatible with market demand exceeding supply from the 1200s to around 1400.

In all likelihood the amount of coin European consumers paid for fish rose from the eleventh century until about 1400, so arguing for steady pressure of demand against supplies on the market. But given literate scribes' long disinterest in humdrum mercantile matters, prior to about 1200 the assertion of price increases rests on complaints about costly fish, not sequentially comparable data points.[2] Lack of anything

[2] Rippon's belief (*Transformation of Coastal Wetlands*, 224–225) that English fish prices rose through the twelfth and thirteenth centuries may exceed what sources can support. On price series see Supplement 0.2.2.

resembling genuine price series before the mid-thirteenth century and limited availability for the next 150 years prevents thorough testing of the proposition that human population growth, urbanization, natural limits, habitat destruction, and depletion preceded and accompanied the incremental replacement of freshwater with marine fishes in northwestern Europe. Consilience and compatibility among data sets is the better hope.

Historians in northern Europe have, however, recovered or pieced together serial records of fish prices which begin in the mid-thirteenth century and become more grounded from a century later. Inland sources lag and potential records from Mediterranean Europe remain essentially unexplored. Nevertheless the cumulative evidence prior to 1400 from several places identifies continual increase and some informative anomalies. (Chapter 8 will follow what look like different trajectories thereafter.)

So far the earliest extant fish price series are for lightly preserved salt herring, the "fish of the poor" which kept for some months during the cool season when shipped from the Channel and North Sea beaches to urban Flanders and Brabant, northwestern France, and England, a market then still plausibly regional in quality. The early-standardized herring (all of a size and sold by the thousand) provides the oldest quasi-anecdotal series of medieval fish prices modern scholars have assembled. In Figure 6.1, compiled from financial accounts of English institutional consumers, mainly hospitals and colleges,[3] the mean annual price per piece for "herring" climbs throughout the thirteenth and fourteenth centuries, so the nominal value by 1400 reached four times its starting point. Calculated in silver the price tripled, reflecting debasement of the coinage. Anomalies in the rising trend are temporary price peaks during the notoriously hungry 1310s–20s and in the 1350s, which began in plague and a serious crop failure. Could herring have substituted for the failed grain harvests and meat from diseased livestock?

Herring prices in Scotland, although but sparse from the 1260s to 1360s and then complicated by shift from dry salted to barreled product, exhibit a similar long-term increase into the early fifteenth century.[4] So, too, do those anecdotally reported from bulk sales at a production centre,

[3] Source for Figure 6.1: price data from the Allen-Unger Global Commodity Price data base: England: herring, www.gcpdb.info, last viewed 18 January 2017. Price per piece in pence and grams silver indexed to the highest prices before 1540. Despite flaws in data used by Rogers, *History of Agriculture and Prices*, vol. I, p. 641, and vol. IV, p. 545, a like pattern appears in his decadal averages for "herring" priced in shillings per [long] "thousand" up to 1400. The steady rise up to 1400 contrasts with a flatter trend but greater volatility thereafter to be examined in Chapter 8.

[4] Gemmill and Mayhew, *Changing Values in Scotland*, pp. 317–319 and table 57.

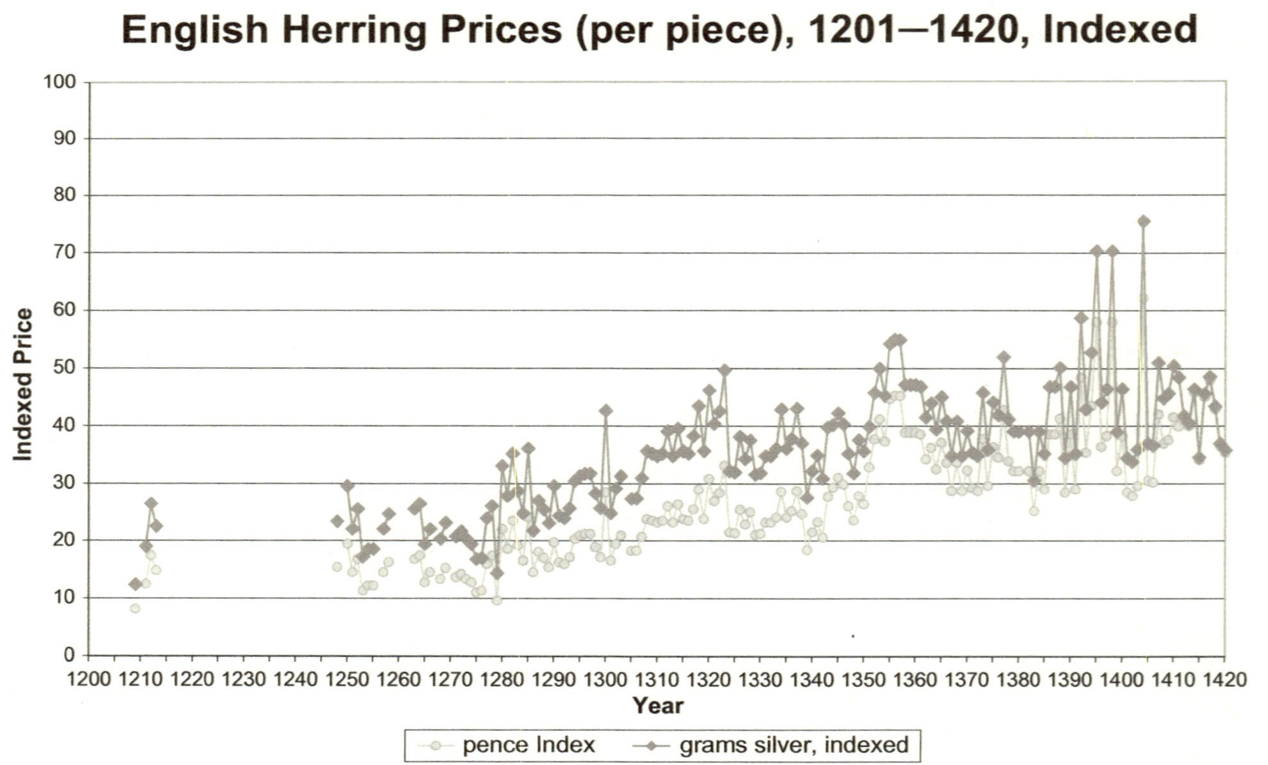


Data from Allen - Unger database. Indexed by R and K. Hoffmann 100 = 1499 pence and 1460 grams, the highest values to 1530.

Figure 6.1 English herring prices (per piece), 1201–1420, indexed.

Calais, where herring reportedly went for 15 sols the thousand in 1268, doubled to 30 sols by 1300, and then more than doubled again to 75 sols by 1342. An average daily wage in northern France bought twenty-three herrings around 1300 and nineteen a generation later.[5]

More tenuous evidence lies behind published assertions that domestic prices for dried cod in Norway and Iceland went up by two-thirds from the late twelfth century to the mid-fourteenth[6] and claims of a twelve-fold increase between the 1260s and the 1410s in what Norman salmon got at Rouen. In the latter, a fish not worth one pig had 150 years later become worth two or three.[7] Likewise, impressions of ever higher prices for trout, pike, and carp in thirteenth-century France are compatible with but not fully grounded in the most convincing sources.[8]

Further south but still on Europe's Atlantic face, genuine price series compiled from institutional sources in Navarre begin shortly after 1350 (Figure 6.2).[9] Such locally procured staple varieties as brined sardines and both fresh and dried hake exhibit the same rising late fourteenth-century trends as did herring further north, doubling and tripling by the century's end. Prices of fresh sardines, however, follow no such pattern and were much more volatile year to year.[10] If the steadier trend was driven by demand pressing against supply, rapid fluctuations suggest forces other than demand could also be at play. In any case, unless incomes rose proportionately, some purchasers may have found fish financially less attainable, while others retained access to the desired but scarce commodity.

[5] Hocquet, "Pêcheries médiévales," 53–56; Derville and Vion, *Histoire de Calais*, 45; D'Avenel, *Histoire économique*, 3:277.

[6] Urbańczyk, *Medieval Arctic Norway*, 252–253; Gelsinger, *Icelandic Enterprise*, 185–190. Nedkvitne, *Hansa and Bergen*, 501–504, and "Development of the Norwegian Stockfish Trade," 55, reiterates his 1988 finding based on English customs accounts that during 1270–1350 a hundred kilograms of Norwegian stockfish sold in Bergen for 102 grams of silver, 2.1 times that for the same weight of rye flour. Series which begin only just before or after 1400 for preserved herring and dried cod on late medieval international markets are discussed in Chapter 8 below.

[7] Halard, "Pêche du saumon," 176–178.

[8] D'Avenel, *Histoire économique*, 3:273, though evidence is only anecdotal. Prices fixed by statute at Bologna may be wishfully normative but also show an upward trend from the 1280s to 1330, stability in the 1330s to 1348, and then renewed rise to the end of the fourteenth century. A pound of sturgeon was set at 15 *denaro* in 1288 and a century later at 2 *soldi* (?24 *d*?) (Pucci-Donati, "Il mercato del pesce," tables 1–2).

[9] Hamilton, *Money, Prices, and Wages*, appendix V, pp. 227–260. Indexed by R. Hoffmann; graphics by K. Hoffmann. Good archaeological evidence establishes earlier trade in hake at several northern and even inland Spanish sites (Morales et al. "Hindcasting to Forecast," 24).

[10] Different patterns of price behaviour suggest something other than currency fluctuations drove fish prices in Navarre, although the series are too fragmentary to suggest what that may have been. Hamilton, *Money, Prices, and Wages*, found monetary inflation there set in only after 1400.

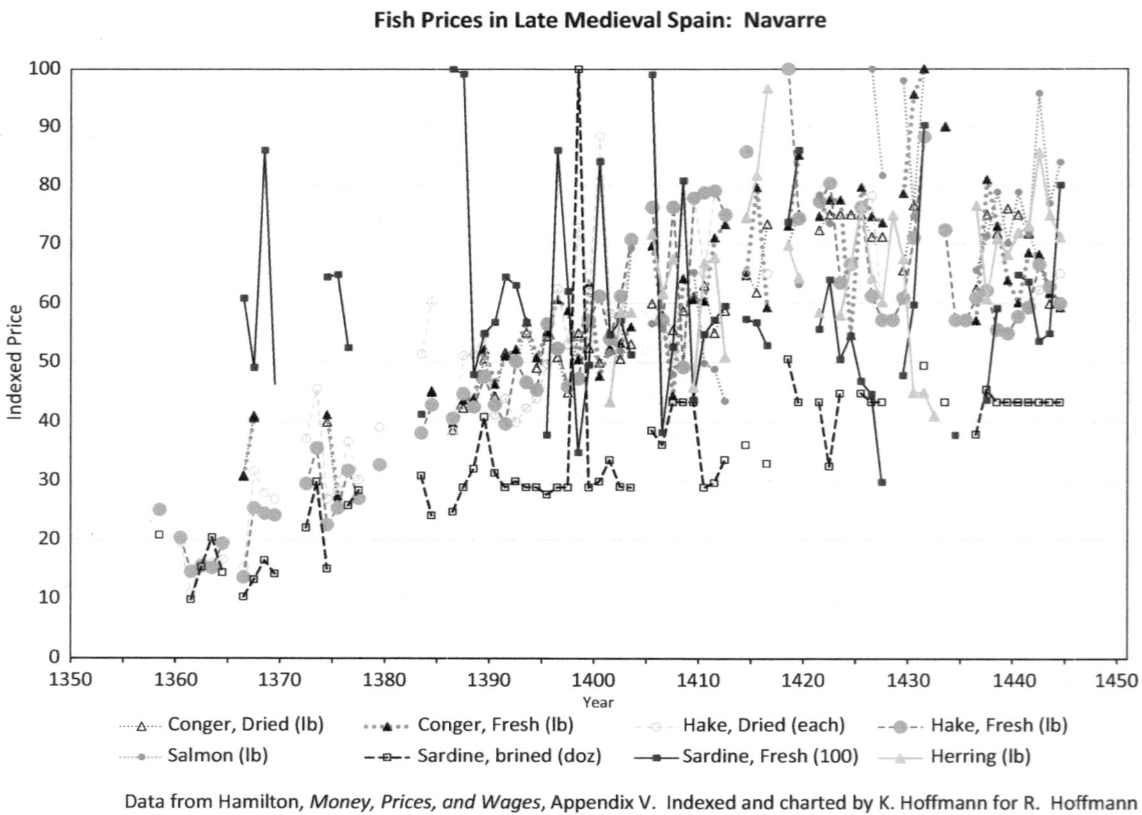


Figure 6.2 Fish prices in late medieval Spain: Navarre, 1358–1450.

Price crystallized the value medieval markets placed on fish, the product of fisheries where human action was applied to a natural ecosystem and its inhabitants. The scarcity and value of the product were thus inherently linked to the producing natural resource, lucrative but, as seen in Chapter 5, also potentially endangered. Although high and rising prices should have encouraged fishing pressure on the most-desired fish varieties,[11] barriers to market entry and expansion in turn curtailed it. Those same rising prices also reflected the increasing scarcity and income value of control over the resource itself, a possibly limited source of supply.

### 6.1.2 *Privatization of Fishing Rights*

The long-term tendency in medieval Europe to transform the right to fish from a common or public good to private property is an axiom of the legal literature.[12] Fishing rights derive economic value from demand for fish and limited opportunities to obtain them (only certain waters, sites, and seasons are productive) and not always from an absolute shortage of fish or decline in their numbers. Even in a subsistence economy fish flesh on the bank has scarcity value, use value, and value derived from at least the opportunity costs of human effort expended in its capture. So too, in prospect, have fish in the water. Privatization of traditional fisheries allocated not merely an output, fish, but control over the productive process, infrastructure, and things of Nature itself. Fishing rights thus conveyed prestige value beyond any direct correlation to the condition of fish stocks.

At the start of the Middle Ages jurisdiction over water and what was in it or done on it pertained to public or royal authority, while uncaught fish belonged to no one (*res nullius*). Sixth-century Emperor Justinian's

[11] As asserted, perhaps too categorically, by Hitzbleck, *Bedeutung des Fisches*, 83–87. I find at present no persuasive evidence that fishing for local markets by late medieval artisans as such (apart, that is, from the sheer concentration of aggregate demand in towns) damaged freshwater fish populations more than did non-market subsistence fishing. Both sorts of fishers knew well their local resource and, within the compass of that knowledge, shared comparably ambivalent incentives to maximize their catch and to preserve the resource for their future needs. As shown in Chapter 4, guild statutes simply did not countenance all-out exploitation and destruction.

[12] Cahn, *Recht der Binnenfischerei*, 13–90; Jeulin, "L'élaboration," 127–128; Mira, *Pesca nel medioevo*, 1–28. Besides cases explored below, nuanced regional examples include Łęga, *Obraz gospodarczy Pomorza*, 35–49; Freudlsperger, "Fischereigeschichte," 87–122; Grand and Delatouche, *L'Agriculture*, 538–539; Fossier, *La terre et les hommes*, 395–397; McDonnell, *Inland Fisheries*, 12–18; Materné, "De verwerving, betwisting en instandhouding van de visserijrechten"; Squatriti, *Water and Society*, 109–113; Lampen, *Fischerei und Fischhandel*, 81–97.

long-standing codification of Roman law established both points in its scenario of public rivers where all could fish and take ownership of what they caught.[13] The Germanic political order then spreading over most of western Europe preferred more personalized authority; it understood kings to be the legal possessors of waters and undeveloped or unowned land. Likewise later-emerging central European polities (Bohemia, Poland, Hungary, etc.) held the prince to own all unallocated resources. The same prevailed by right of conquest in Norman England, southern Italy, and much of Spain.[14] People who used, for instance, the fish did so more by sufferance and custom than by explicit right. In that sense, were there ever open access to medieval fisheries, it was a soon-disappearing relic of "unlorded" land or it emerged much later on offshore frontiers (see Chapter 8). The medieval regime was one of property, permissions, privileges, and liberties, not generic "liberty."

This situation came about because two intersecting trends in property rights and jurisdiction characterized the greater part of the Middle Ages: one was the establishment by the powerful of personal lordship over land and the people on it, which eventually crystallized in the seigneurie; the other was the devolution and fragmentation of public right into private hands, a condition now often called "feudalism."[15] As applied to the right to fish, the first meant that real estate, including the waters in, on, and bordering it, became an object of property rights which included authority over people living there, a power the French sources would call *ban*.

Especially in early medieval times the absorption of fishing rights into landowning had as much to do with formation of lordship as with controlling a productive resource.[16] Lords with long-standing or newly conceded property rights over land were reluctant to allow trespass by means of water and, by grant or appropriation, incorporated waters into

[13] *Corpus Iuris Civilis: Institutiones*: lib. 2, tit. I, '2: "*Flumina autem omnia et portus publica sunt, ideoque jus piscandi omnibus comune est in portu fluminibusque*"; and '12: "*Ferae… bestiae et volucres et pisces, idest, omnia animalia que terra, mari, coelo nascuntur, simulatque ab aliquo capta fuerint, jure gentium statim illius esse incipiunt, quod enim ante nullius est, id naturali ratione occupanti conceditur*" (ed. Krüger, p. 10; tr. Moyle). See Mira, *Pesca nel medioevo*, 1, and Squatriti, *Water and Society*, 100–101.

[14] Górzyński, *Zarys historii rybołówstwa*, 18–30; Györffy, *Wirtschaft und Gesellschaft*, 36–39; Graus, *Dějiny venkovského lidu*, 1:97–98. From around 1200 Danish kings asserted sovereignty over rivers, wrecks, and coastal seas (Hybel and Poulsen, *Danish Resources*, 52–53). Jaime I of Aragon, having captured Muslim Mallorca, established free marine fishing but royal and magnate lordship over lagoons and rivers in his charter *de Franquera* of 1230 (Barceló Crespí and Mas Forners, "Fishing in Majorca, 1230–1521," 123–124).

[15] Hoffmann, *Environmental History*, 241–262, offers an environmental overview of medieval property rights.

[16] An important point articulated by Squatriti, *Water and Society*, 108–110.

the property itself. For a church in Ponticello, for instance, Duke Romuald II of Benevento declared in 711 that none might enter the church's woodland in order to fish the river Calore without permission of the incumbent priest.[17] Further south in Italy Byzantines considered coastal and lagoon fisheries private, so alienable as rights or incomes: in 1038 the shoreline at Termoli was given "with its fishery" (*cum piscatione sua*). Some generations later, Norman conquerors put all aquatic space in the hands of seigneurs, so fishing, like pasturage, went for a share of the take. Fishing rights were part of donations to twelfth-century churches, although an 1122 gift to La Cava abbey was limited to shallows.[18] Early medieval Germanic law codes and dozens of ninth-century charters, from, for instance, the region of Frankfurt am Main, concur in cataloging fishing and fisheries among the normal rights of a landowner.[19] In Berry inclusion went from tacit to explicit: early tenth-century gifts of watermills to Deols abbey mention no fish, but after Pope John XI specified "you have right of exclusive fishing" (*licentiam piscandi soli habeatis*) in a protective bull of 937, all subsequent donations used the same phrase.[20] Eventually late medieval Austrian customals went so far as to specify "the fish in the water" as part of what was owned. In Iceland riparian landowners then possessed even the marine fishery as far out as the depth of a twenty-mesh seal net at low tide, but their counterparts in the older societies of mainland Scandinavia controlled coastal fishing rather through ownership of essential shore-based facilities.[21]

In the second trend early medieval rulers give favoured supporters public jurisdiction over un-owned real estate and navigable waters, with or without the land itself. When few written records were being kept in France and Germany the process is especially visible in some Italian sources. For example, users of public ponds in the Riete valley acknowledged royal authority with regular payments of "pisces publici" until 816, when Emperor Louis the Pious gave those to the abbey of Farfa and thus provided grounds for its subsequent recognition as lord over the fishers and owner of the fishing rights. So, too, the monopoly over the

[17] "… *nullus sine permissione sacerdotis nominatae ecclesiae in ipsum valdum habeat licentiam intrare ad piscandum*" (Mira, *Pesca nel medioevo*, 3). Later records of fishery ownership from the headwaters to the estuary of the Tiber are in Vendittelli, "Diritti ed impianti," 392–422.

[18] Ditchfield, *La culture matérielle*, 332–334.

[19] Cases in Cahn, *Recht des Binnenfischerei*, 14–26, echo in tenth-twelfth-century charters and inventories from France (Evergates, ed., *Homblières*, nos. 4, 6, and 27); Lorraine (Perrin, *Recherches*, 725–730); Bohemia (Graus, *Dějiny venkovského lidu*, 298–309); and Lombardy (Castagnetti et al., eds., *Inventari altomedievali*, 34, 39, and 57–65).

[20] Querrien, "Pêche et consummation" (2003), 420.

[21] Ebel, "Fischerei," 143.

Ticino which fishers at Pavia held about 1000 derived from regalian rights of former Lombard kings.[22] In a later famous Dutch counterpart, the church of St. Martin at Utrecht claimed one-sixth of the fishery at the mouth of the Rhine as a grant from one Gerulf, probably a late ninth-century Carolingian count. Extending amorphous claims of Anglo-Saxon landowners, Norman kings of England enfeoffed their tenants in chief with rights over fresh- and saltwater fishing and allowed subinfeudation.[23] The abbey of Lerins rested its claim to the marine fishery between the mainland and its isle on an explicit 1298 concession by the count of Provence.[24] Private control over access to fisheries thus became a strand in the mesh of remunerative rights being assembled into lordship over land and men.

Assimilating fishing rights to private landownership and privatizing public jurisdictions over fisheries followed neither identical nor simultaneous trajectories across western Christendom. Lac Leman (Geneva) could be thought one regional paradigm: rights long remained regalia of Burgundian kings and emperors until they were gradually parceled out to riparian landowners and public officials. Bishops of Lausanne claimed one zone on grounds of King Rudolf's granting the County of Waadt in 1011, and the archbishops of Besançon received the fishery to mid-lake from Emperor Henry III in 1043. By the thirteenth century comparable offshore rights were the acknowledged possessions of lay lords at Rolle and Prangins on the northwest shore and of the bishop of Sitten's estate at Montreux on the eastern end. Across the Alps fishing in the Bavarian and Swiss lakes, including Lake Constance, early appeared in the hands of shoreline owners, often quite without special royal concession. Equally undocumented are the legal origins of collective rights there, whether exclusive – as exercised by fishers of Constance off city beaches – or shared – as among residents of the Lindau shore.[25] In the

[22] Toubert, *Structures du Latium*, 672 and 850; Brühl and Violante, eds., *"Honoratie Civitatis Papiae"*, 60–61; Squatriti, *Water and Society*, 97. Squatriti, "Riverains et rivaux," 141–144, argues that in Italy the shift from 'public' meaning 'general access' to 'public' meaning 'royal' is visible in the 804/5 judgement by Charlemagne of the Istrian complaints against Duke John (Chapter 3, pp. 101 above).

[23] Boer, "Roerend van der visscheryen," 123–126. Anthony Scott summarized development of fishing rights in English common law in his 1986 essay "Catch quotas and shares in the fishstock as property rights," and in 1989 "Conceptual origins of rights based fishing," 15–20.

[24] Aubenas, "Droit du pêche," 3–7.

[25] For context see the Introduction, with more detail in Cahn, *Recht des Binnenfischerei*, 82–89, and the last narrated in detail by Zeheter, *Ordnung der Fischer*, 39–49. Parallel processes in Italy culminated during the eleventh century, with examples in Supplement 6.1.2.

Basque region, informal recognition of communal property rights in marine fisheries has also been traced back to the thirteenth century.[26]

In Atlantic coastal France lords' right of "wreck" (*varech*) gave ownership over whatever came on to the beach, so also all shore-based fishing activity. Every shoreline belonged to someone's lordship and that authority projected offshore as *dominium aquae*. During the eleventh and twelfth centuries the fishing along much of the Norman coast was accepted as the possession of Fécamp, Conches, and other large monastic estates. At the same time, private holders of public office alienated royal jurisdiction over coastal waters, as Count Guillaume of Poitou's 1017 donation to Cluny of half his dues from fishers on the Île de Ré. As late as 1500 the right of wreck was acknowledged to underlie the dues artisan fishers paid shoreline seigneurs in the Seine estuary.[27] Clear territorial definition of rights over the fishing in estuaries along the Gulf of Lions emerged with eleventh-century intensification of a legal regime dominated by local castellans. Already tenth-century Catalan charters for coastal seigneuries included "with the sea itself, with its ports and its fisheries" (*cum ipso mare, cum suos portos et suas piscatorias*), but otherwise fishing along that coast required licence from the prince.[28]

Contrasting with fisheries based on a lord's private *ban* were those of landowners in crusader Prussia, whose rights rested wholly on regalian grants. The Teutonic Order there built an entire social order on imperial and papal concession of full sovereignty over whatever it conquered from pagan natives. Lay and clerical landowners established by the Order could fish on their own properties, but only by sufferance of the Order, which retained superior jurisdiction plus full ownership of all other fisheries.[29]

Despite important regional nuances, about the year 1200 the legal status of fishing in western Christendom may be thought effectively encapsulated in *Sachsenspiegel*, an influential private compilation of north German legal practice. In principle, especially navigable waters were free: "every stream of flowing water is common for travel and for fishing; the fishers may also use the land up to a full stride away from a boat at the

[26] López Losa, "Derechos de propriedad informales," and as an English precis, "Informal property rights."

[27] Hocquet, "Pêcheries médiévales," 56–65 and 105–106; Darsel, "Servitudes de la pêche," 107; Lardin and Jegou, "La pêche en basse Seine." For seigneurial control over fishing rights and fishing for conger around the Channel Islands see Barré, "Droit maritime."

[28] Bourrin-Derruau et al., "Littoral languedocien," 383–387; Escobar and Alegret, "Evoluzione," 118; Riera Melis, "Pesca en el Mediterránea Noroccidental," 140–143.

[29] Kisch, *Fischereirecht im Deutschordensgebiete*; Willam, "Fischerei des Deutschen Ordens," 74–137.

legal shoreline." But the theory must be understood in terms of the preceding clause that "whosoever ... fishes in another man's natural waters owes three shillings damages under law."[30] The general norm everywhere was private possession – whether conceded or usurped – of the right to fish in inland and inshore waters. Fishing rights (habitually but not invariably called *piscatio*) were so commonly associated with riparian landownership that both an unofficial compilation of legal customs along the lower Loire, the so-called Établissements de Saint Louis, and the thoroughly royalist *Siete Partidas* assembled for King Alfonso X of Castile, thought worth remarking only the requirement that the landowner suitably publicize his forbidding others to take fish.[31]

On occasion some individuals did hold legal title to fish waters in another's acknowledged lordship, so the right could be separate from land ownership, usually through an owner's or sovereign's special grant. Sometime around 1070, after servants of St. Marcel priory "illegally took prisoner and broke the boats of fishers from Chalon" who were using a reach of the Saône, Count Hugh made a pious gift of those rights to St. Marcel. Cistercians at Merci-Dieu in Poitou had exclusive fishing rights on their lands along the river Gartempe given them by the founder, Eschivard of Preuilly, in the early 1150s. Then in 1235 Eschivard's heir and grandson, Geoffroi, donated the further right to use waters on his own properties.[32] In the last two medieval centuries the Guzmán family of magnates gained private monopoly over tuna fishing in the Gulf of Cadiz by incremental grants from Castilian kings.[33] Likewise the more rare instances of collective rights also derived from concession by a prior holder and evidently not simple survival of a primeval condition. Both sorts of multiple claims raised questions of exclusive fishing or mere

[30] Eike von Repgow, *Sachsenspiegel Landrecht* II, 28 §4: "*Swelk water strames vlut, dat is gemene to varene unde to vischene dar inne. De vischere mut ok wol dat ertrike nutten alse verne, alse he enes striden mach ut deme scepe van deme rechten stade*" is subordinated to II, 28 §1 "*Swe holt howet oder gras snidet, oder vischet in enes anderen mannes watere an wilder wage, sin wandel dat is dre scillinge, den scaden gild he oppe recht.*" Intervening clauses cover artificial waters, planted trees, etc., and such acts perpetrated at night (Eckhardt ed. 1955, 157–158). See discussion in Cahn, *Recht des Binnenfischereis*, 35–37. The variant manuscript tradition presented in Dobozy, ed., *Saxon Mirror*, 102, does not differ.

[31] *Établissements* §31 in Akehurst, ed. and tr., p. 87; Alfonso X, *Siete Partidas*, Pt. III, t. xxviii, 17, tr. Scott, ed. Burns, vol. 3, p. 824.

[32] "*ubi quadam uice seruientes Beati Marcellie naues piscatorum Cabilonensium* [i.e. Chalon] *fregerunt, ipsosque piscatores propter iniustam captionem piscium quam ibidem faciebant ceperunt*" (Bouchard, ed., *Cartulary of St.-Marcel*, nr. 12, pp. 36–37); Brien, "Développement de l'ordre cistercien en Poitou," p. 73 and appendices XII and XVI. More French conflicts over divided rights are in the Supplement.

[33] Philipps, "Who owns the fish."

shared access to it: in 1179 Morimondo abbey asserted its sole right to fish the Ticino at Pavia against claims of privileged town fishers, but a compromise recognized the shared rights of both parties; arbitration in 1247 also settled disputes over the Moere between the abbey and town of St. Omer.[34]

From the thirteenth century two distinct and divergent issues vexed the proprietary regime and dominated its further evolution. One concerned the fishing permitted in practice and the other the (re)assertion of regalian or (quasi-)public jurisdiction over it. Assessing actual medieval practice calls for distinction, if possible, among the 'public' authority to regulate a fishery (see this chapter below); the ownership of *piscatio*, the legal right to take fish and authorize or forbid others to do the same; and actual usufruct (exploitation) of the resource (fishing). Unified exercise of all levels on the model of modern 'private property rights' is all but unknown in medieval Europe.

Seigneurial ownership and control of the fishing was but the legal precondition for the actual capture of fish. Very few late medieval owners of fishing rights actually fished. Seen from below, the process pitted seigneur against "commons", whose loss of subsistence rights Chapter 3 has explored. On the other hand, owners of the water often allowed local subjects, inhabitants, or citizens access to resources providing fish for their own use. Whether tacit or contractual, the practice legitimized much of the subsistence fishing seen to continue through the later Middle Ages.[35] This was not always easy. As earlier remarked, a mid-eleventh-century subject of Marmoutier abbey had to pass an ordeal by hot iron to defeat claims of a local seigneur and vindicate the right of local people to take fish for family consumption, but not sale.[36] Where positively documented – as, for instance, in a court decision at Fontaines in Valois in 1270 or in fifteenth-century customals from Hainaut and *Weistümer* from southern Germany – the usufruct rights of ordinary villagers were ordinarily confined to small gear used personally to supply a resident household. Restraints on locations and times (seasons, days)

[34] Epstein, *Wage Labor and Guilds*, 56–57; Laplane, *Abbes de Saint-Bertin*, vol. II: 553–554. Stolz, *Geschichtskunde der Gewässer*, 345–356, emphasizes mutual rights of access in Tirol.

[35] A most active recent theoretician of private property in fisheries, Anthony Scott, simply throws up his hands at actual evidence for operative regimes under medieval English common law, which did emphasize private rights inland and, after 1216, their absence at sea (Scott, "Catch quotas and shares," 72–80; Scott, "Conceptual origins," 15–20). Yet the real medieval situation, complex and multi-layered as here shown, calls sharply into question the historical credibility of private ownership as a foolproof remedy against destructive use of fisheries.

[36] Querrien, "Pêche et consummation" (2003), 432–433.

to fish were also the norm. The "common fen" where peasants at Downham-in-the-Isle (of Ely) could fish excluded certain designated "fisheries," which were subinfeudated to individuals.[37] Similar conditions applied where municipal governments possessed the fishing rights. Burghers of Frankfurt am Oder could fish a half mile of that river for their own tables and those of Hamburg wade – but not use boats – in the Elster. At Toul all citizens could fish in the Moselle and, for a small licence fee, place one trap in a local tributary. In 1307 one of the first ordinances of the newly established town on the Rhône estuary now called Saintes-Marie-de-la-Mer allowed citizens to set a new type of net in all its waters, but only for one night at a time.[38] In the long run, then, direct subsistence use of local aquatic resources – which means the fishes naturally present in western Europe's ubiquitous streams and small rivers – was surely curbed under the increasingly enforced proprietary regime.

Even limited common access contrasted with the more or less exclusive exploitation of a fishery by the owner's employees or a contractual tenant, which turned the catch to indirect subsistence, artisanal, or maybe even larger commercial purpose. Chapter 3 established the prominence of indirect subsistence fishing during the early Middle Ages and followed some interesting cases into early modern times. But during later medieval centuries private owners of fishing rights notably conceded leases or licences for others to exploit these remunerative resources, so moving them, too, toward the status of market commodity.[39] The transition could, as earlier observed, result for a time in arrangements like those at Klosterneuburg about 1300: full-time fishers remained abbey dependants but owed the prior for their craft only first refusal rights on

[37] Blary, *Domaine de Châalis*, 95–99; Verriest, *Régime seigneurial*, 320–323; Cahn, *Recht der Binnenfischerei*, 51–56 and 67–68; Heimpel, "Federschnur," 456–464; Jaritz and Winiwarter, eds., *Historische Umweltdatenbank*, nos. 162–163 *et passim*; Coleman, *Downham-in-the-Isle*, 20, 24, and 93. Abad Garcia and Otero, "Pesca fluvial," 165–168, describe similar rules on Castilian monastic estates. Compare the disputes Bennett, *Life on the English Manor*, 270, reports from thirteenth-century English manorial courts and the peasant fishing described in Mákkai, "Economic landscapes: historical Hungary," 28–31.

[38] Cahn, *Rechte der Binnenfischerei*, 54–55; Cabourdin, *Terre et hommes*, 676–77; Amargier, "Pêche en Petite Camargue," 334–336. Supplement 6.1.2 contains more arrangements for citizen fishing of municipal waters.

[39] Cahn, *Recht der Binnenfischerei*, 91–103. As Squatriti, *Water and Society*, 113–115, illustrates, a fuzzy ambiguity distinguished independent leaseholders or licencees from subject fishers allowed access in return for customary dues. This change in operation of fisheries paralleled the well-known abandonment by lords of their direct agricultural exploitations (demesnes, granges, etc.) to take rents from tenants or lessees and similar contracts for commercial harvest of coppice wood or timber.

large fish, a share of the catch made with abbey equipment, and fixed money dues.[40] But some generations earlier, landowners all around western Europe were already letting out fisheries to specialists on fixed terms for annual money dues or, more rarely, a share of the catch. The practice could be lucrative: during 1282–1335 the prior of Ely received more from leased fisheries at his fenland manor of Lakenheath than he did from tenant rents; in Picardy revenues from fisheries exceeded those from equal areas of arable. Especially after the Black Death of 1347–1351 leases everywhere became the norm and continued steadily to rise in value.[41]Various sorts of fishers thus gained operational control over fisheries through local variants of leasehold tenure. In 1266 parcels of the River Avon at Stoneleigh, Warwickshire, were taken for life by *Gilbertus piscator* and *Normannus piscator*, whose annual dues of twenty-five and eight shillings respectively echoed the rents of neighbouring farmers. On the Lago di Bientina outside Lucca, however, the abbey of Sesto charged individual fishers fixed cash fees for delimited area licences good for three months to four years. Contracts for exclusive use were provided only to corporate groups, an arrangement also encountered at Namur.[42] Along the Atlantic beaches of Aunis and Saintonge seigneurs leased fishing access (called *courtinage*) to specialist entrepreneurs. In the 1480s one Jehan Souchet paid 100 livres for a league of shoreline close to Aytré. His exclusive access conflicted with the local custom of fishing on foot, so after finding thirty to forty people from Aytré taking fish from his nets in 1491 Jehan demanded legal redress. Not far away the offshore fishers from Les Sables-d'Olonne had since the 1100s to pay for access to the harbour a fifth of their catch to the castellan and more to local religious houses.[43]

As late as the fifteenth century an ongoing process of privatization emerges from judicial records in the Crown of Castile, often revealing complex interplay among municipalities (*concejos*), corporate religious landowners, and assertive territorial aristocrats (*grandes*). One such tale may be told of the upper Luna valley in León, where Cistercian nuns of Santa Maria de Otero, a powerful local branch of the Quinones clan

[40] Zeibig, ed., *Urkundenbuch ... Klosterneuburg*, 2: 169–171.

[41] Generally on leases and their rising value, Kilby, *Peasant Perspectives*, 146–148; Fossier, *La terre et les hommes*, 395–397, and Hitzbleck, *Bedeutung*, 83–87. Further examples: Richard, "Le commerce du poisson," 185–186; Perez-Embid Wamba, *Cister*, 178–179 and 524–525; Derveeghe, *Domaine du Val Saint-Lambert*, 110–111 and 118; Vahrenhold, *Kloster Marienfeld.*, 128–129.

[42] Hilton, ed., *Stoneleigh*, 39 and 163; Onori, *L'abbazia di San Salvatori*, 55–60 and 126–128; Brouwers, ed., *Cens et rentes*, vol. I, 105, 114, 156, and 202, and vol. II:1, lxix–lii; Balon, "La pêche," 25–27.

[43] Tranchant, "Pêches et pêcheurs des villes-ports," 83–88.

(named counts de Luna in 1462), and two rural *concejos*, Luna de Suso and Luna de Yuso, tangled repeatedly over access to rich stocks of trout, barbel, and eel. After a long slog through the courts the valley municipalities had in 1438 successfully vindicated their right to fish in their own territories for subsistence or sale free of any dues to the Quinones, long-time high royal officeholders. Then shortly before 1493 the convent sued four residents of Canales, a hamlet in the valley, for 50,000 maravedis damages for having used nets in Río Luna without permission. The lawyer for the *concejos* argued the Luna was *río publico* and free for inhabitants of the municipalities to fish. After appeal, in 1495 the royal chancellery sustained and declared specific boundaries within which the abbey's servants had exclusive fishing access from 1 April to 30 September and peasants were otherwise restricted to fishing for subsistence only with rods or small traps, not nets. The next year the countess de Luna leased her rights to one Juan Alvarez and some partners, who then sued communal residents for unlicenced fishing, fishing at night, and other violations.[44] Historian Maria del Val Valdivieso and her students have made a strong collective case that such conflicts constituted a clash between peasant economic interest in catching fish and great lords who saw fishing rights as just one component in consolidation of their private territorial authority.[45]

Essential aspects of private control are restricted access, profitability, and the conservation – or destruction – of the resource. What record had medieval private ownership? As regards the risk of depletion, many private wild fisheries were evidently exploited at sustainable levels for centuries. As regards the risk of environmental destruction, however, landowners handled resources in terms of their relative awareness of value. The fishery of Lago di Bientina, indeed the whole one-time aquatic ecosystem there, is no more because from at latest the twelfth century the abbey of Sesto and other riparian landowners actively

[44] García Cañón, "La pesca en los concejos," 191–200. Likewise from 1441 men of Olmedo clashed with Carthusians of Santa Maria de Aniago over a traditional fishing site where the Río Adaja entered the Duero, eventuating in armed occupation by town citizens in April and May 1514 and confirmation by a regional court of their right to fish from shore only (Bonachía Hernando and Val Valdivieso, "Monasterios y pesca fluvial," 26–30).

[45] Val Valdievieso, *Agua y poder*, *passim*, with many cases in Hernández Iñigo, "La pesca fluvial," 1078–1079; Abad García and Otero, "La pesca fluvial," 169–178; Bonachía Hernando and Val Valdivieso, "Monasterios y pesca fluvial," 23–58; and Sánchez Quinones, *Pesca y Comercio*, 121–206; as well as García Cañon, "La pesca en los concejos." Soens, "Capitalisme, institutiones et conflits," makes a similar point for the late medieval Low Countries, although both he and Val Valdivieso downplay the real need of workers for access to environmental resources.

promoted drainage, reclamation, and conversion of their lake into rentable farmland. The same reasoning caused landowners with private fishing rights to sponsor medieval drainage projects in the Dutch and Tiber marshes.[46] In what might be called a 'Tragedy of the Private', survival of the resource and the natural environment was always at the mercy of the owner's larger interests. When the new abbot of San Pedro de Gumiel in 1509 needed to settle a dispute with the count of Miranda, he traded away his grange at the hamlet of Milagros with water rights on the Río Rianza, an upper tributary of the Duero, depriving local people of their traditional subsistence use of its fishes.[47]

Few of the deprived could expect the marvel that antiquarian William Worcester (d.1482) related of the river Axe flowing from the Wooky Hole cave on the outskirts of Wells in Somerset. Townsfolk had long enjoyed its diverse and abundant fishes until Bishop Thomas Beckington (1443–1465) appropriated it for exclusive use by his own kitchen. Immediately the fish vanished and for two years neither bishop nor anyone else had any. But when the bishop opened fishing to the common and country folk, the fish multiplied and were again abundant.[48]

Private ownership and more tightly limited access to medieval fish more often threatened to call up a broader social animus against the aquatic resource itself. Chapter 3 showed how peasant groups closed off from fisheries made them targets of social protest. Such acute aggression likely posed less environmental danger than did what might be called 'hostile neglect'. Late medieval privatization by appropriation or exclusive leases extinguished the concept of 'res publica' on Castilian rivers and alienated rural and urban communities from their watercourses and the fish for which they had once cared.[49] Certainly the constriction or loss of common access to local fishes left members of peasant communities other than the 'lord's fisher' or leaseholding 'master fisher' less generally solicitous of their aquatic environment and more prepared to countenance loss of habitat as well as illegal and destructive fishing.

[46] Onori, *L'abbazia di San Salvatori*, 4–50; Vendittelli, "Diritti ed impianti di pesca," 409–422.

[47] Abad García and Otero, "La pesca fluvial," 170–172.

[48] Worcester, *Itineraries*, 292. Likewise in "Sabothus," a long poetic humanist Latin chorography of his native province, Silesian town secretary Franz Faber (d.1565), complained of lords' newly imposed bans on subsistence angling, but then, says he, the fish depart the closed waters: "*Durius occlusos, vetitosque severius amnes / Destituunt velut execrata piacula pisces*" (Kytzler, "*Laudes Silesiae*, II," 59–60).

[49] Abad García and Otero. "La pesca fluvial," 173.

Consider the endemic (as opposed to openly rebellious) use of poisons, night lines, and other unlawful methods.[50] Medieval poachers with herbal and chemical piscicides or underwater explosives to stun the fish had little incentive to spare what they had no right to take.

Had privatization itself demonstrable long-term consequences for the maintenance of viable and productive aquatic ecosystems? Or did it simply reallocate resources among human groups? The latter it surely achieved and, by moving effective demand for fish toward the market and toward the wealthy, lent in many places an increasingly luxury cast to especially the inland fishery.[51] Fish as *herren speis* probably lowered total fishing pressure against freshwater biomass while raising that on favoured, increasingly rare, varieties. But the bewildering diversity of tenurial regimes lends no historical support to the superior sustainability or economic viability alleged for particular forms of private or common property in fisheries. Plainly called for is precise comparative study of local legal regimes and their long-term relation to actual fishing in particular ecosystems. In all circumstances, however, shores, waters, and the fish varieties they sheltered became during the western Middle Ages ever less things of an autonomous Nature, valued and open to all, and ever more clearly, even in coastal waters, defined as possessions of identifiable human cultural agents.

Private ownership of fisheries evolved before, during, and after the period of steady medieval demographic and economic expansion, but the aligning of fish and traditional fisheries resources with market values certainly coincided with the growing impact that commercial development had on European environments. Rising prices for fish and fishing directed scarce goods in certain social directions. But markets were not the only means through which medieval societies responded to change. The reassertion of princely and public jurisdiction over fisheries was the basis for regulatory regimes as typical of the later Middle Ages as was private ownership

50 See the discussion from peasant perspectives in Chapter 3 above and incidents reported in, for instance, Bennett, *Life on the English Manor*, 94–95; McDonnell, *Inland Fisheries*, 17–18; Coldicott, *Hampshire Nunneries*, 78; Coopland, *Abbey of St. Bertin*, 59; Bidon and Bossard-Beck, "La préparation des repas," 70; Fournial, *Les villes et l'économie*, 195; Jeřábek, "K studiu rybářství"; Cocula-Vaillières, *Un fleuve et des hommes*, 132–134; Gunda, "Fish poisoning"; Hoffmann, *Fishers' Craft*, 330–332 *et passim*. Jacoby, "Class and environmental history," documented similar reactions to privatized natural resources in the nineteenth century Adirondacks of New York State.

51 As Maccarinelli, "Was pike on the menu?," sees in the freshwater varieties consumed at elite English sites.

## 6.2 Public Regulation of Fisheries

Beside exchanges at market value and private rights in fisheries, public regulation constituted medieval Europeans' third response to rising pressure against fish supplies. Previous chapters have already made familiar communal self-regulation on the part of direct subsistence fishers and through corporate actions of organized artisans. Fisheries protection also occasionally came into measures meant primarily to secure market supplies of fish. With the high and later Middle Ages, however, such acts of participants in the fishery itself were steadily absorbed, extended, and superseded by those of rulers and other public authorities. Proclaiming a more general interest in what were said to be resources under threat, the incipient state asserted control over the operation of European fisheries

The complex medieval mix of jurisdictions generated seigneurial, communal, territorial, and royal measures which from time to time reflected restricted or more general 'public' interest in fisheries. A grey area between property management and resource regulation was coloured by characteristic interpenetration of private right and public authority. On the salt ponds of coastal Languedoc in 1173 the lord of Villeneuve-les-Maguelonne forbade a new fishing technique called "*batendo et bolegando*", beating the water and scooping stunned fish with a net. Such by then frequent seigneurial enactments as the count of St. Pol's 1225 ban on trammel nets reveal ambiguous concern for the lord's private supply or profit as well as for conservation and common use.[52] Narrowly bounded but distinctly 'public' agents proclaimed comparable rules. For example, the Cumberland county assizes of 1278 imposed a closed season and oversight committee to protect salmon stocks, and statutes codified in 1322 for the Florentine commune forbade fish poisons and taking young fish. Madrid's thirteenth-century charter already made similar provision for municipal watercourses.[53]

### 6.2.1 *Authority*

'Public' fisheries regulation asserted an authority over waters and fishes not deemed to be owned or even overriding private ownership rights. (Both

[52] Bourin-Derruau et al., "Le littoral languedocien," 384–385; Fossier, *La terre et les hommes*, 397. Like examples include Verriest, ed., *Corpus des records de coutumes*, 4–5, 11, 42–43, 46, 56, 130, and 176–177; Casado Alonso, *Señores, mercaderes y campesinos*, 207–210; Gresser, "Délits dans les rivières" and *Pêche et Pisciculture*, 136.

[53] Winchester, *Landscape and Society*, 110–111; Trexler, "Measures against water pollution," 460–461; Puñal Fernádez, *Mercado en Madrid*, 175–176. Rombai, "Acque interne," 38–42, discusses efforts by the Sienese commune to protect fish stocks in the Ortobello lagoon.

traced in law or by imitation back to Roman imperial authority.) Perugia rested its jurisdiction over L. Trasimeno on an 817 imperial grant to Pope Paschal I and subsequently that of Innocent III to the free commune in 1212.[54] Alfonso VIII of Castile set rules for fishing municipal waters of Cuenca in the code he issued shortly after capturing the city in 1177.[55] Kings of France claimed imperial authority within their own domains, so Philip IV in 1289 and all his successors asserted the right to legislate for public (i.e. large navigable) waters and gradually extended this over private uses of other natural water bodies (artificial ponds were not included).[56] The oldest known Scottish regulations came as a precedent-setting judgement of King William I (1165–1214) sitting as the country's top feudal lord, but subsequent rules took the form of parliamentary statutes.[57]

Institutional articulation of fisheries regulations depended to some extent on regional governmental/administrative practice, including incremental development of literate record keeping. This evidently occurred earlier in French, British, and some Hispanic monarchies, as well as Italian city states, than in the principalities of central Europe or Scandinavia. But as already here observed in one introductory fish tale, on Lake Constance in 1481 it was two cities, two church corporations, and a secular count which together established the first interstate fisheries protection treaty, an arrangement which would evolve into the nineteenth century.[58] The long-term trend for fisheries was part of a larger expansion of public, princely, or state intervention in all sorts of environmental and resource issues.[59]

[54] Vincenti, "La tutela ambientale," 133; Biganti, "Pesca nel lago Trasimeno." For the Kingdom of Sicily Emperor Frederick II's *Liber Augustalis*, lib. 3, tit. 72 (tr. Powell, p.144), barred fishing with herbal poisons.

[55] *Fuero* chs. XXXIV: §§13–15 and XLIII: §§7–9 and 13–14 (Ureña y Smenjaud, ed., *Fuero*, pp. 728–731 and 816–819; Powers, ed., *Code of Cuenca*, pp. 191 and 215–216).

[56] Rouillard, "Législation royale" summarizes approaches and measures from texts published by Duplès-Agier, "Ordonnances inédites"; Lespinasse, *Les métiers et corporations*, I: 13–20 and 448–457; Guilhiermoz, ed., "Ordonnance inédite de Philippe le Bel"; and Jeulin, "L'ancienne reglementation de la pêche." Querrien, "Pêche et consommation," 410–412, traces the overlay of royal authority upon private rights in Berry.

[57] On the Scottish 'Assizes of King William' see Hoffmann, "*Salmo salar*," 258 and 362, with references to the legal literature there provided.

[58] Zeheter, *Ordnung der Fischer*, 44–65 and 82–100, assembled this history for L. Constance and Amacher, *Zürcher Fischerei*, 163–169, tells a like tale of collaboration around the medieval Zürichsee.

[59] Hoffmann, *Environmental History*, 263–278. For other exemplary or survey treatments of medieval regional fisheries legislation see Supplement 6.2.1. Plainly regulatory activity with protective intent was going on long before the nineteenth-century British parliamentary legislation which Bartrip, "Food for the body and food for the mind," thought unprecedented.

Authorities asserted that legislative intervention would serve their communities by resolving evident practical problems in their fisheries. In managing fisheries, as in woodlands, flood control, or questions of pollution, regulation was increasingly justified by reference to collective benefit, the 'common good' / *bien publique*, as greater than mere private profits. Philip IV in 1289 and generations of his successors asserted that excessive and improper fishing caused "no small loss to the rich and poor of our realm."[60] After consulting with the fishers' guild, in 1365 Wrocław's town council set rules for fishing the Oder "for sake of common utility" shortly before a much larger city, Venice, did likewise for fishing in Lake Garda "for common benefit and abundance of fishes."[61] King and Grand Duke Ferdinand also grounded his 1537 *Fischfangordnung* for Upper Austria in "common usefulness" (*gemainen nutz*).[62]

And what problems called for intervention? Concern for public safety led London and Austrian authorities alike to ban fish weirs as hazards to navigation in the Thames and the Danube, while Florentine councilors did likewise because of such structures' role in the Arno flood disaster of 1333.[63] Continually from 1287 Venice tried to limit the seasons and locations of *valli de pesce* on grounds they blocked tidal flow and contributed to siltation in the lagoon.[64] Guild statutes and municipal ordinances in particular – at Zürich, Madrid, and along the upper Rhine, for instance – directed attention to conflict and disorder in their fisheries and provided rules and institutional mechanisms to avoid or settle them.[65]

The most generally articulated reason for regulation was fear and perception of fisheries depleted from overfishing. Philip IV blamed fishers taking fish too small for French rivers "yield[ing] nothing."[66]

60 Rouillard, "Législation royale," with texts in Duplès-Agier, "Ordonnances inédites," 49. Wood, *Medieval Economic Thought*, 21–40, and Kempshall, *Common Good*, examine the concept.

61 Korn, *Breslauer Urkundenbuch*, Nr. 241, p. 208, '*durch gemeynis nuczes wille*'; Butturini, "Pesca sul lago," 147–149 ('*ad comune beneficium et copiam piscium*'); and for Venice in general, Faugeron, *Nourrir la ville*, 281–292.

62 Hoffmann and Sonnlechner, "Vom Archivobjekt zum Umweltschutz," 119–120.

63 Keene, "Issues of water"; Galloway, "London and the Thames estuary," 130–135; Hoffmann and Sonnlechner, "Vom Archivobjekt zum Umweltschutz," 131; Schenk, "Prima ci fu la cagione de la mala provedenza de'Fiorentini." English royal legislation against weirs blocking navigation went back to Edward the Confessor in 1065 (Cooper and Ripper, *Fishing and Managing*, 61).

64 Faugeron, *Nourrir la ville*, 181–184.

65 Amacher, *Zürcher Fischerei*, 187–194; Puñal Fernández, *El Mercado en Madrid*, 175–176; Mone, "Ueber die Flussfischerei," 79–81, 82–84, 89–94. Clashes among rights holders over access to salmon runs triggered the Cumberland county assizes to strike a commission responsible for identifying legal and illegal weirs (Winchester, *Landscape and Society*, 108–110).

66 Duplès-Agier, "Ordonnances inédites," 49.

Albrecht V of Austria in 1412 found "the waters everywhere in this land have become almost desolate and fishless" because no fishes therein could ever reach their proper growth. In the middle of the fifteenth century governments of Strasbourg and Florence shared that opinion, blaming on overfishing the remarkable declines of fish in the Rhine and in all Florentine territories. Patterned on earlier Castilian codes, the 1526 "New Charter" (*Fuero Nuevo*) for Vizcaya regulated fishing to prevent "all the rivers in Vizcaya from being destroyed and depopulated of fish."[67]

### 6.2.2 *Measures*

Medieval regulations typically established protected and open seasons for fishing, set minimum legal sizes for specified fish varieties, and restricted capture techniques. Some even worried about habitat. All measures converged in major enactments like those perhaps pioneered by thirteenth-century French authorities. Already in the 1260s it was the king's *prevôt* who compiled rules for fishing the Seine and Marne at Paris: fishers had to be licenced and observe 'customs'; barbel, eel, carp, pike, and tench had to be larger than would sell at four to the denier; fishing was closed for spawning roach between mid-April and mid-May; use of *saimes* and hoop nets was controlled along with the size of mesh in the nets.[68] King Philip's ordinance of April 1289 then applied to all waters under royal jurisdiction. To combat depletion from improper fishing royal officers were to seize and publicly burn illegal gear, impose fines, and turn contraband fish over to the poor. The king outlawed a dozen specific nets and barrier traps, and use of two more during April and May, when many river species spawn. Legitimate netting gear had to have mesh larger than a *gros tournois*, a coin of about 2.5 cm (1 inch) diameter. The closed season for roach now covered two whole months, and size limits were tailored to the varieties affected – eel at a quarter-denier, barbel and carp a half-denier, and pike at two denier. One central

[67] Hoffmann and Sonnlechner, "Vom Archivobjekt zum Umweltschutz," 117; Trexler, "Measures against water pollution," 465–466; Sznura, "Veleni e '*nobilissimi pesci*'"; Mone, "Ueber die Flussfischerei," 82; Abad García and Peribánez Otero, "La pesca fluvial," 179: "*Vizcaya se destruyan y despoblen todos los rios de pescado.*" More general discussion of late medieval regulations as responses to regional overfishing appear in Materné, "Exploitatiemetoden," and Plouchard, "La Scarpe," 849–853, while Rombai, "Acque interne," 21–22 and 39, and Sánchez-Quiñones, *Pesca y Comercio*, 110–120, draw attention to fifteenth-century fear of fish shortages in Tuscany and Castile respectively.

[68] Lespinasse and Bonnardot, eds., *Les métiers et corporations*, 212–218; Bossuat, "La pêche en Seine," 62–63.

official was named to oversee local enforcement. Later redactions transferred oversight to the royal Master of Waters and Forests, occasionally modified the vocabulary of prohibited gear, and extended the system of size limits by setting 5 inches for dace, chub, and roach.[69]

As precocious and energetic as French monarchs was the communal government at Perugia in taking charge of the fishery on L. Trasimeno. Already from 1260 cryptic references suggest fish were being stocked in the lake, and with a series of ordinances and statutes from 1276, 1279, and thereafter, the commune set up a regulatory regime that lasted until a new papal governor took over in 1568. Laws set minimum mesh sizes, restricted use of enclosure traps made from reeds, and banned netting of tench in September. The most important local fishes, pike, tench, eel, and southern roach, were protected from May 1 to August 31, though not the less-valued chub. Tench below four 'ounces', roach below one, and small eel had to be returned to the water unharmed. First mentioned in 1275, the "seeding of pike, eel, and crayfish in the lake of Perugia" became an annual enterprise, carried out in February, March, September, and October, the latter two months especially with small eel procured from the river Chiana. Since 1279 lakeside settlers were obliged every third October to *reinfrascare* the lake with bundles of reeds and branches to shelter spawning and overwintering fish. A government-appointed cleric supervised the stocking and habitat improvement work; a special communal fisheries overseer, the *tencarame*, saw to enforcement. Violators could be fined as much as 500 *lire*.[70]

Venice, too, wished to defend its fisheries, but had to juggle worries over supply with the peculiar social and biological features of the *valle di pesca* (see Chapter 4 above) and existential fears over siltation of the lagoon. A regulatory regime established towards the end of the thirteenth century was later periodically reinvigorated, notably after 1400.[71] The reed fences used to enclose the *valle* slowed currents in the lagoon and encouraged deposition, so in 1284 the Grand Council banned their use

69 Duplés-Agier, "Ordonnances inédites," 49–53; Jeulin, "Ancienne reglementation de pêche." In 1313–1314 judgements of a dispute between the abbot of Notre-Dame du Gard and Reynaud, seigneur of Picquigny, over a fishery in the Somme obliged both parties to obey the royal ban on the *harnois de seaule* and other specified weirs (Cloquier, "Pêches et pêcheries," n.p.). Rouillard, "Législation royale," observes strong protective measures under population and demand pressure in the years around 1300 and again after 1380 but less intervention when mid-fourteenth-century France experienced demographic, economic, and political troubles.

70 Biganti, "Pesca nel lago Trasimeno," 792–797; Vincenti, "Tutela ambientale del Lago Trasimeno," 134–135.

71 What follows is based on Faugeron, *Nourrir la ville*, 181–184 and 311–317.

in public waters. Soon resistance from fishing and consumer interests led to incremental easing of the ban, mainly licencing senior fishing masters to enclose specific areas during Lent so long as they left spaces between *valle* and used no site more than two years running. Decades of vacillation between enforcement and licencing ensued.

New fears of overexploitation and fish shortages emerged in a fifteenth-century setting of growing urban populations and state authority. A 1424 Senate enquiry heard even fishers blame the fences and use of certain nets for declining lagoon stocks. Worried over loss of supply, the Senate did prohibit the suspect techniques but only close to shore, which still reduced catches and supply to the Rialto market. Further fears of siltation as the fences again multiplied provoked a general ban in 1474 on fences and *valle* alike in public waters across northern reaches of the lagoon and in 1485 extension of the ban to other structures across a larger area. Violation drew higher fines and leaders of the Nicoletti were made responsible for settling conflicts. The Venetian government thus faced not only tension between regulators and traditional fishers, but also between its own aims to sustain supply and to stabilize the lagoon environment. Constraints on the fishery might prevent overfishing and deposition of silt, but exacerbate concern over insufficient fish on the market. Distinctive as well is Venetian focus on fishing structures, with little if any attention to particular fish varieties.

Relative latecomers to the story, Habsburg princes of Austria who undertook to govern fishing on the Danube and its tributaries also met opposition despite their efforts to bring fishing interests into the process. In 1418 the governor of Upper Austria confirmed in his prince's name regulations for the Traun which had been set by consensus of its thirty-two *vischmaystern*, privileged experts on a river with a complex hierarchy of operative rights. They declared three traditional types of seines and traps hazardous (*schendlich*) to the resource and banned their expansion or further use in the main river channel. None were to fish with baited set lines or at night with torches and spears. Only persons with full fishing rights could use artificial flies (*vedersnuer*), baited traps, set nets, or weirs which extended across the river. The latter were further limited to one in each designated reach, operated without hindrance to navigation. Only between St. Koloman (13 October) and St. George (24 April) might nets have small mesh; at other times they had to meet a larger standard. Summer regulations closed all fishing for small grayling (a regional delicacy) and required the release of those trout, pike, and huchen which might sell for less than a pfennig. Nor were any river side channels to be blocked and drained before 8 September, "so as not to damage the seed of the fish in the water." Violators suffered a heavy fine of sixty pounds pfennig to the duke and thirty-two pounds to the governor.[72] What began as collective self-regulation in the Traun fishery was being assumed by the prince's officer.

After some holders of fishing rights resisted regulations for two ensuing generations, in 1499 the duke of Upper Austria, Emperor Maximilian, had his governor intervene directly, "so that the fish resources in the Traun do not become barren."[73] The official now sought advice of the master fishers but issued a revised and strengthened ordinance on his own authority. Meanwhile Maximilian's father and predecessor, Frederick III (1452–1493), had himself set out a general ordinance (which does not survive) for fisheries across Upper and Lower Austria. In ensuing decades monarchs themselves declared revised fisheries regulations for the Traun and for both provinces and appointed a royal *Fischmeister* to enforce them.[74]

While Venice tried to manage fisheries in the framework of its lagoon policies, French, Perugian, and Traun river regulations establish general patterns and point at the importance of local particulars in each pronouncement.[75] In a broader sampling of the medieval record, seasonal restrictions visibly worked the interlaced calendars of fishes' natural reproduction and vulnerability and of human fish consumption. French kings closed fishing for roach during their April and May spawning time. At Ivrea and on Lake Garda fishing for (fall-spawning) trout was closed during October and November, while a 1472 law on the Swiss Bielersee protected sculpin (*bundollen*) after Easter, when demand for fish fell off and this species was spawning.[76] Fishing for conger off the Cotentin peninsula was closed from Easter to John the Baptist (24 June) and already in 1307 Piran closed fishing in its part of the Gulf of Trieste from Michaelmas to the start of Lent. The Teutonic Order in Prussia established seasonal fish sanctuaries off the mouths of rivers to coincide with important runs of migratory fishes and, since 1426, applied a general closure during spawning.[77] During the summer dry season up to 29 September no one was to fish inland waters of Sardinia.[78] In the

[72] "*damit das der sam der visch in dem wasser nicht zu laid werde.*" Text and discussion in Scheiber, *Zur Geschichte*, 30–35 and 152.

[73] "*als daß die Fischwaiden in der Traun nicht in Ödung kommen*," *ibid.*, 35–39.

[74] On the Traun, see *ibid.*, 39–49 and 153–154, and more generally Hoffmann and Sonnlechner, "Vom Archivobjekt zum Umweltschutz." Zeheter, *Ordnung der Fischer*, 82–100, traces comparably evolving rules for Lake Constance.

[75] In addition to what follows in text and notes to this section see Supplement 6.2.2.

[76] Nada Patrone, *Il cibo*, 326; Butturini, "Pesca sul lago di Garda," 158 and compare 173 note; Cahn, *Recht der Binnenfischerei*, 137.

[77] Hocquet, "Pêcheries médiévales," 75; Iona, "Istituti e alimenti," 628; Benecke, "Beiträge," 312; Seligo, "Zur Geschichte," 25–26; Kisch, *Fischereirecht im Deutschordensgebiete*, 191.

[78] Fois, "Annotazioni sull'alimentazione," 189.

1250s Alfonso X of Castile closed fishing during spawning and growing seasons especially for the sake of small salmon, and centuries later his successor Carlos I (Emperor Charles V) banned taking *madrillas* (Iberian species of nose) in March and April, barbel during April–May, and salmonids during November–December.[79]

Size limits might protect juveniles – "inrudines" in the Florentine statutes, "alevins" in Lorraine's *Loi de Beaumont*[80] – or in other instances, as already seen, measure more precisely. French ordinances first set a minimum size by market price and later measured in 'inches' (*pouces*). Brescia eventually required a minimum weight for pike, tench, and trout. At Nice mullet and sea bream were legal at a 'half foot' (*semissa*) and trout in Pedraza, Castile, at a quarter of the municipal 'yard' (*la quarta de la vara de concejo*).[81] A Douai ordinance of 1391 set pike at ten 'inches'. To avoid ambiguity Bavarian and Austrian rulers since the mid fifteenth century had used special illustrated documents, followed by printed posters to show the varieties in question and the smallest legal size for each from the Danube and its navigable tributaries (Figures 6.3 and 6.4).[82] On bag limits or quotas directly limiting kill, however, medieval sources seem mute

Capture techniques could be regulated in several ways. As earlier noticed, a minimum legal size of mesh gave an easily verified index of selectivity for small fish. London's 1279/80 legislation demanded two inches for the 'peternet' used in the Thames estuary.[83] A Scottish statute

79 Sánchez Quiñones, *Pesca y Comercio*, 118. Compare Peribañez Otero and Abad Alvarez, "La pesca fluvial," 179.

80 Trexler, "Measures against water pollution," 461; Collin, "Ressources alimentaires," 43.

81 Butturini, "Pesca sul lago di Garda," 175; Nada-Patrone, *Il cibo*, 326; Peribañez Otero and Abad Alvarez, "La pesca fluvial," 179, but not far away in Burgos trout were legal at a "palm" (Casado Alonso, *Señores, mercaderes y campesinos*, 210).

82 The unique patent issued for Upper and Lower Austria by Emperor Maximilian in 1506 (Figure 6.3) has precise coloured drawings of eight species and of the standard for measuring mesh, all meant to facilitate planned consultations between the prince's *Fischmeister* and local stakeholders about regional regulations. See Hoffmann and Sonnlechner, "Vom Archivobjekt zum Umweltschutz," for discussion and older literature there cited. Earlier and later Bavarian examples put the names of the fish on oblong bars of appropriate length. Ordinances from 1489 and 1490 survive in München, Bayerische Hauptstaatsarchiv, Ältere Bestände, Staatsverwaltung, Bd. I, Nrs. rot 1608 and 1612b, and Generalregister Fasc. 444 Fischwesen Nr. 1, Nrs. 16–17, 18–19, and 389–398 and there is the 42 × 60 cm poster printed in 1528. Figure 6.4 is reproduced too small for legibility but is meant to show the format presented to the public by the new technology. An image capable of enlargement to full size is available in Supplement 6.2.2. It is worth observing that all the medieval minimum limits so far known fall below the modern norms for those species and offered little protection for fish not yet up to their spawning size.

83 London Letterbook A, fols. 89v–91v, as cited in Wright, *Sources*, 68–69, with also later rules on mesh size. More elaborate regulations of net size, form, mesh, and other features are published in Riley, ed., *Liber Albus*, 331–334.

Figure 6.3 Emperor Maximilian's *Patent* instructing his *Vischmeister* to negotiate regulations for the Danube fishery, 1506.
Illustrated with distinctive coloured images of eight fish varieties (pike, carp, barbel, huchen, burbot, catfish, trout, and, in the left margin, a species of *Zingel*), each about 20 cm long, plus a gauge to measure mesh of a net. All fishes are native to the middle Danube and, except for the *Zingel*, named in the *Patent* as objects of official concern. Wiener Stadt- und Landesarchiv, Hauptarchiv Urkunde Nr. 5825 (A). Reproduced with permission of WSLA.

of 1318 had the same means to protect "little salmon or smolts or fry of other kinds of fishes of the sea or fresh water" and such provisions became the norm in fourteenth- and fifteenth-century regulations from German-speaking lands. Officials along the north slope of the Alps carried a locally standardized gauge called *prittlmas* to measure the openings (bottom right in Figure 6.3).[84] The duke of Brittany, who claimed

[84] Brown et al., eds., *Records of the Parliaments of Scotland*, 1318/13 [19] www.rps.ac.uk (consulted September 2011): "*salmunculi vel smolti seu fria alterius generis piscium maris vel aque dulcis*"; Cahn, *Recht der Binnenfischerei*, 137. Mone, "Ueber die Flussfischerei," 78, has the 1434 rules from Strasbourg.

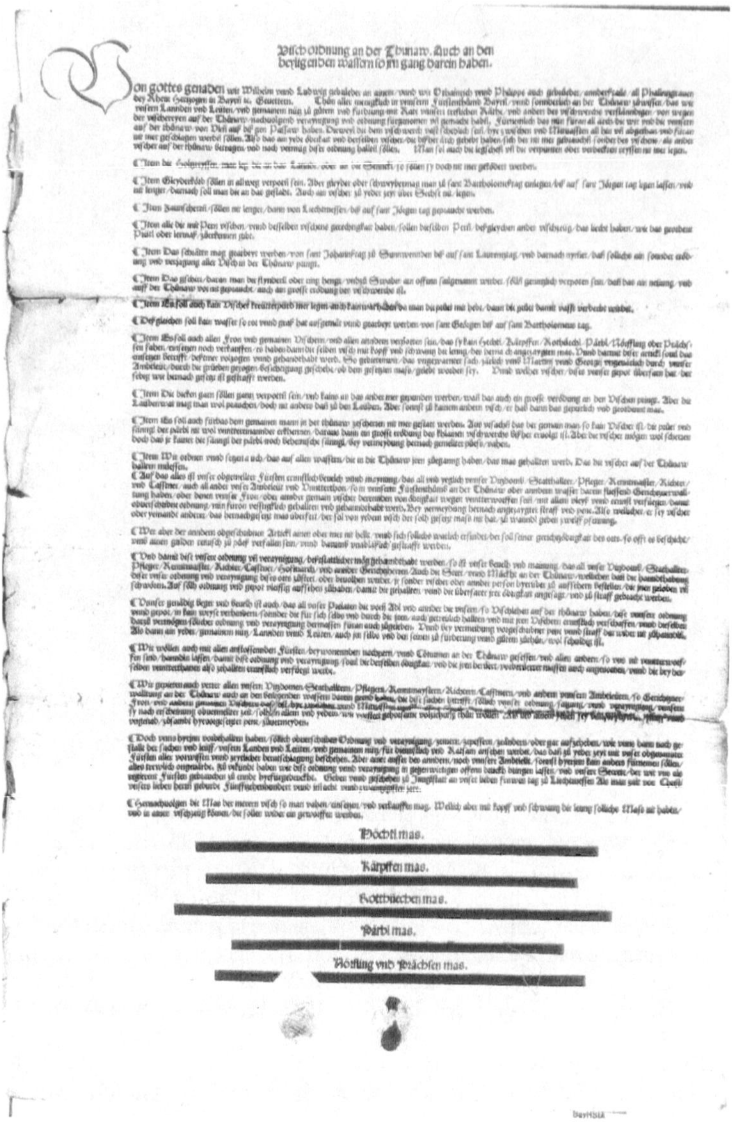

Vischordnung an der Thonaw. Auch an den beyligenden wassern so jm gang darein haben.

Huchen mas.

Karpffen mas.

Rottbuechen mas.

Parbl mas.

Nöfling und Prachsen mas.

Figure 6.4 Fisheries ordinance for Bavarian waters, 1528 (see also Supplement 6).
Printed *Vischordnung* for Danube and tributaries issued jointly by the four Wittelsbach co-Counts Palatine and Dukes of Bavaria. The last article shows the minimum legal sizes in printed bars measuring from 42 cm for huchen down to 18 cm for ide and bream. Surviving exemplars include München, Bayerische Hauptstaatsarchiv, Ältere Bestände, Staatsverwaltung, Bd. I:2, rot 1612d and Generalregister Fasz 444:1, 18v–19ar and Fasz 444:2, 374–380. Reproduced with permission of BHSA.

royal jurisdiction up to thirty leagues offshore made his tax collector in Nantes responsible for inspecting mesh.[85]

Other measures targeted named forms of weirs, traps, or nets, forbidding them entirely or sharply restricting them to certain seasons, waters, or privileged users. The *fuero* for Cuenca allowed no seines or trammel nets within nine paces of a mill or in a channel, and in 1444 the queen regent of Aragon ordered Tortosa's municipal council to control certain nets which were harming the run of shad in the Ebro.[86] The town of Cannes refused to licence more than four boats to use the *lamparo* technique of netting by torchlight in coastal waters and Villa de Mar ordered removal of mid-channel traps in the Rhône's estuarine marshes in the spring when fishes entered the salt ponds to spawn.[87] The large *niewod* seine of the Baltic littoral was often banned outright – as to town-based fishers in fourteenth-century Poland – or used only by special licence of the sovereign.[88] Governments across Mediterranean Europe outlawed piscicides, led by Frederick II's Sicilian legislation of 1231, Madrid's municipal charter a generation thereafter, and continually reiterated enactments at Florence and other places.[89] A variant strategy specified what was permitted, as the 'pridenet' on London waters in 1279/80 during eight days only of the autumn season between Michaelmas and Martinmas, the angling or simple basket traps conceded to townsfolk at Toul, or the baited hook, rod, and horsehair line used at Vercelli.[90]

In some instances protective measures extended to fish habitat. As earlier remarked, Perugia obliged shoreline residents every third October to 'refresh' shallow-water fish cover in L. Trasimeno. Madrid banned private diversion channels and the assembled Traun fishers required summer maintenance of flow in them, in both cases to ensure survival of juveniles. As count of Holland, Philip of Burgundy ordered full protection of riparian vegetation in the Haarlemmermeer and other

85 Hocquet, "Pêcheries médiévales," 73–75 and 105–106.

86 Cuenca *fuero*, xxxv: 15 (ed Ureña i Smenjaud, 728–731; tr. Powers, 191); Pastor i Lluis, "Pesca de la saboga," 108–109.

87 Aubenas, *Droit de pêche*, 10–12; Amargier, "Pêche en Petite Camargue," 338, and "Notes sur l'ichthyophagie," 318–319.

88 Górzyński, *Zarys historii*, 37–38; Kisch, *Fischereirecht im Deutschordensgebiete*, 173–182; Seligo, "Zur Geschichte," 17–19; Sarnowsky, *Wirtschaftsführung*, 208–210; Benecke, "Beiträge zur Geschichte," 302–305.

89 *Liber Augustalis*, tr. Powell, 144; Puñal Fernández, *Mercado en Madrid*, 175–176; Trexler, "Measures against water pollution," 460–461: "*Nullus audeat vel presumat pisces in flumine, vivariis, puteis, vel aliis quibuscunque aquis tossicare …*" from 1322 and later laws ever more explicitly directed against the use of lime as a fish poison.

90 London Letterbook A, fols. 89v–91v, as quoted in Wright, *Sources*, 71; Caboudin, *Terre et hommes*, 657; Nada Patrone, *Il cibo*, 325.

waters north of Leiden, hoping this cover would harbour enough forage fish to keep pike in those waters and sustain the value of Leiden's fishery.[91] Habitat and forage mattered to the city of Narbonne, too, when in 1227 it barred from its lagoon a bottom trawl or seine, the *bourgin*:

> because such a net takes away from the whole pond [and] lays waste all types of nourishment for fishes both large and small, and thus the pond would be emptied out and desolate of all fishes, and not any fishes would be found there and it would be sterile and yield nothing.[92]

The damage such trawls caused in coastal waters provoked concern in the north, too. Fishers from Essex petitioned the English parliament in 1377 against a device called the 'wondyrchoun', said to

> strike so evenly and hard against the bottom being fished that it destroys the living slime and the flowers of the ground under the waters there and also the young of the oysters, mussels, and other fishes on which the large fish are there accustomed to live and be nourished.[93]

It is hard not to see clear reference to ideas now thought as ecological. The same fears engaged some Dutch fishers in 1540, who appended to

[91] Biganti, "Pesca nel lago Trasimeno," 792–795; Vincenti, "La tutela ambientale," 134-135; Puñal Fernández, *Mercado en Madrid*, 180; Scheiber, *Zur Geschichte*, 152; Dam, "Feestvissen en vastenvissen," 489, and "Fish for feast and fast," 326.

[92] "*Nam tali rete brugina tali modo piscari ut multorum didicimus veridico testimonio totum exhaurat stagnum eremat omni genere piscium nutrimento tam magnorum quam parvorum et ita esset stagnum exhaustum et heremum omnibus piscibus, non invenirentur ibi aliqui pisces et esset sterile et infructuosum.* " (Bourin-Derruau et al., "Littoral languedocien," 384). The bishop of Agde in 1321 ruled against the same gear in the Étang de Thau: "since the aforesaid devices are of such arrangement that they would scrape even a denier up from the bottom of the pond and thus totally destroy the offspring and seed of fishes, which would be too much loss to the whole public interest" (*cum predicte artes sint tales conditionis quod levarent de solo dicti stagni unum denarium si esset et sic procreatio sive semina piscium totaliter destrueretur quod esset toti rei publice non nimium dampuosum*) (Puig, "ressources de l'étang et de la mer," 118–119, and sources there cited).

[93] "*Et outre ce, le feer grant et long du dit wondyrchoun voet si owelment et durement desur la terre ent peschant, q'il destruit la slym crascete et flurs de la terre desouz la eawe illeoqes, et auxint l'espat des oistres, musklys et d'autres pessons, parount les grantz pessons soleient vivere et illeoqes estre nurriz.*" (Given-Wilson et al., eds. *PROME*, membrane 2:369, Edward III, 1377 January, 50. XXXIII). The translation published with the original edition contains inaccuracies. The complaint further avers that the wondyrchoun took so many little fish that people used them to fatten pigs, all "*a grant damage de tout la commune de roialme et en destruccion des pescheries en lieux semblables.*" Outcome of the investigative commission struck some months later is unknown (*ibid.*, appendix January–March 1377 No. 15, from Calendar of Patent Rolls, 1374–7, pp. 489–490). See also Jones, "'Lost' history," 206–298. Bailey, "Coastal fishing," 110, reports damage from trawls on the Suffolk shore in the 1380s and 1389 parliamentary legislation against gear "which might destroy the fry of fish."

Figure 6.5 Illustrated complaint to Emperor Charles V about damage caused by bottom trawl in Zuider Zee, 1540.
Coloured picture of a Dutch fishing boat and sample of the trawl net sent to the sovereign in a petition against overfishing of the Zuider Zee by fishers from Amsterdam. Österreichische Staatsarchiv/Haus-, Hof,- und Staatsarchiv MEA RTA 59-1, fols. 309–312. Reichstagsakten, Supplikationen 2. Teil. Reproduced with permission of the director of HHFSA.

their complaint to Emperor Charles V a clear watercolour and actual sample of the device they would ban from the Zuider Zee (Figure 6.5).[94]

Not so far seen, however, are restrictions on the quantity of fish taken and only most rarely the designation of protected areas where all fishing

[94] The original petition now in the Austrian National Archive system, Haus-, Hof-, und Staatsarchiv (AT-OeStA/HHStA, MEA, RTA 59-1, fols. 308–312),and reproduced with permission, was identified thanks to the generosity of Christoph Sonnlechner.

An unusual protective measure, though one resonating with present-day fishers in central Europe and elsewhere, was directed against natural predators by Emperor Charles IV, who in 1377 directed the Wrocław city council to kill *wasserraben* (cormorants) and destroy their nesting colonies "for the great damage they do to the fish in the water" (Korn, ed., *Breslauer Urkundenbuch*, nr. 305, p. 250).

was forbidden.[95] It is also the case that some gear restrictions applied only outside the Lenten season. When religious practice forbade most other protein-rich foods, people who could afford fish ate them eagerly, and fishers and governments alike were pressed to maintain supply.

Nevertheless, medieval fisheries regulations were no mere expressions of pious intent. Madrid named inspectors to check for legal tackle and Namur made it the job of a municipal official.[96] Charged to enforce French ordinances were royal officers or, as on the Seine near Paris, subordinates of those who held royal rights in fief. Princes along the upper Rhine appointed "Counts of the Rhine" (*Rheingrafen*) to investigate and judge violations and disputes in that heavily exploited resource. In 1490 Perugia placed an armed galley on L. Trasimeno to patrol the fishery.[97] Surviving judicial records show some success. London authorities in 1320 seized and burned sixteen illegal traps from locations along the tidal Thames and in 1444 jailed a local brewer for using illegal gear.[98] In autumn 1460 their counterparts in Florence caught Buono di Francesco Beroni poisoning fish.[99] At Paris in 1422 royal officials arrested Pierre le Nourrissier and his three crewmen for fishing in the city moat with illegally small mesh and catching the fry of protected varieties, which the officers then confiscated, precipitating a judicial wrangle over boundaries of public waters.[100]

### 6.2.3 *To What End?*

Like modern regulatory regimes, medieval ones projected tension and incompatibility between professed and tacit purposes of each measure and their evident effects. Laws hoped variously to secure consumer

[95] A possibly mythic exception is Salzburg's Pinzgauer Zellersee, where fifteenthcentury sources report that fourteenth-century depletion of the whitefish from overfishing caused the fishing community to rest the lake for three years and then ban fishing in known spawning areas (Freudlsperger, "Kurze Fischereigeschichte," 100–102).

[96] Puñal Fernández, *El Mercado en Madrid,* 176–178; Thomas, "Hygiene, approvisionnement," 265–268 ; Lentacker et al., "Historical and archaeozoological data."

[97] Puñal Fernández, *El Mercado en Madrid,* 176; Mone, "Ueber der Flussfischerei," 70; Vincenti, "La tutela ambientale," 135. The duke of Bavaria's 1517 ordinance for the Ammersee assigned supervisory authority to a *Seerichter* (Cahn, *Recht der Binnenfischerei,* 145).

[98] Wright, *Sources,* 58. Persons using illegal fine mesh "to the great destruction of the fish" in common fisheries were fined at Carshalton, Surrey (cited in Bennett, *Life on the English Manor,* 95).

[99] Trexler, "Measures against water pollution," 465. Schwarz, "Umweltstrafen," 92–96, describes handling of fisheries offences in late medieval Austrian lordships.

[100] Bossuat, "La pêche en Seine," 69–74.

supply, conserve sustainable populations of desirable fishes, or allocate a productive resource among interest groups, but had trouble balancing those priorities.

Concern for cheap abundance of fish to eat justified Philip IV's size regulations and rules for capture methods in 1289 and the outlawing of piscicides at Pistoia in 1378 and Florence in the 1450s,[101] just as the same dietary rationale elsewhere motivated rules to open market access and force quick sale of well-conditioned fish (see Chapter 4). The Castilian *Cortes* meeting in Madrid in 1435 even reiterated laws against use of quicklime and other toxicants on grounds they killed trout and could poison consumers.[102] Florentines occasionally may have aimed further to benefit specifically "those fish which are called trout, and noble fish they are indeed, ... that therefore the said kind of fish might be better preserved."[103] With such words legislators transcended simple provision of a dietary commodity to voice a conservation ethos allowing desired species to reproduce and grow. The city on the Arno could claim no priority in this perception. Edward I closed fishing off the Cotentin "for the salvation of the congers," and his subjects in Cumberland controlled exploitation to protect salmon stocks.[104] Indeed the sole thrust of Scottish fisheries legislation from the late twelfth into the early sixteenth century was to protect the salmon.[105]

Broad conservation intent was most often concretized to issues of reproduction and recruitment. Thus since at latest 1414 the seigneurs of Void set mesh regulations on the Meuse "for the sake of the little trout," while city authorities at Strasbourg in 1434 justified new rules on grounds that it was "harmful to catch all those young fish called *salm* [i.e. juvenile salmon] and unspawned fish, so that the fishes and the fishery

[101] The Florentine statement "*et cumsequenter reddatur civitas nostra cetereque partes predicte copiose et habundantes talibus piscibus*" (Trexler, "Measures against water pollution," 463–464) uses the same terms as a Pistoia statute from 1387 (Archivio di Stato di Pistoia, Provvisioni et statuti, vol. 27, 211v–212r) which John Muendel generously drew to my attention.

[102] *Cortes* also then protected spawning fish (Sánchez Quiñones, *Pesca y comercio*, 110–116; Izquierdo Benito, "Precios y salarios," 124 n. 289.

[103] "*pisces qui dicuntur trote, et sunt valde nobiles pisces ... Ut igitur dictum genus piscium magis conservetur*" (Trexler, "Measures against water pollution," 464). Rulers of fifteenth-century Florence worried continually about depletion and legislated to protect not only trout but also tench (Sznura, "Veleni e 'nobilissimi pesci'," 272–282).

[104] "*Pro salvacione congrorum*" (Hocquet, "Pêcheries médiévales," 75; Winchester, *Landscape and Society*, 100.

[105] Hoffmann, "*Salmo salar*," 362–364.

have almost vanished."[106] Castilian municipalities protected spawning channels and forbade consumption of salmon eggs.[107] English parliamentary statutes in 1489 outlawed certain nets and 'engines' for similar reasons and so did Emperor Maximilan, who explained his intervention in 1506 because of population losses due to the destruction of young fish.[108]

Regulation of catch techniques – responding, for instance, to complaints about weirs on private riparian sites or relatively well-capitalized seines or trawls – could, however, slide ostensible conservation into actual diversion of resources from one user group to another. Only occasionally is the context still plain enough to see. Although in 1321 on grounds of harm to public interests officials of the bishop of Agde could imprison local fishers for using the long-prohibited heavy trawls to damage benthic habitat in the Étang de Thau, this effort to protect the resource fell on deaf local ears; by 1328 the citizens had gained from regent Philip of Valois a renewed right to use the equipment.[109] Likewise men of Woolwich, Erith, and Barking, vills along the tidal Thames below London, violently resisted London's efforts to enforce the ban on weirs.[110] On the other hand, the abbey of Lagrasse did manage in 1376–1377 to win a hard-fought victory against townspeople who had begun to fish the river Orbe with trapping gear allegedly so non-selective that the river "was at all points depopulated of its fish."[111] Avowed concern to protect the resource justified exclusive licence regimes on inshore marine waters under municipal authorities

[106] Cabourdin, *Terre et hommes*, 676–677: *truitelles*. At Strasbourg: "*Wir habent … vernomen, daz zemale schedlichen sye, daz man allerleye junge vische, die man nennet sel[m]en und ungeminte vische vahen soll, dodurch die vische und daz weidenwerck vast abegat …*" (Mone, "Ueber die Flussfischerei," 78–79).

[107] Abad García and Peribañez Otero, "La pesca fluvial," 179. Wishing to protect the trout, the town council of Teruel closed fishing during the spawning season (Rodrigo Estevan, "Fresco, frescal, salado, seco, remojado," 552).

[108] Galloway, "London and the Thames estuary," 130–135; Hoffmann and Sonnlechner, "Vom Archivobjekt zum Umweltschutz," 111–112, 122–123, and 131 ("*damit dye unntzeitigen unnd das pruet der visch also nicht verderbt und gefanngen unnd der gros schaden …*"). Our introductory tale of the Lake Constance fishery already reported the head of the Constance fishers' guild enthusiastically endorsing the 1531 fisheries ordinance because its protection of young fish would soon restore stocks in the whole lake (Zeheter, *Ordnung der Fischer*, 7: "*volab der jugent vischenten dieser [lake] werd vischrich so er sunst vast erlärt ist*").

[109] Puig, "Ressources de l'étang et de la mer," 118–119, and sources there cited.

[110] Galloway, "Expansion or eclipse?".

[111] "*estre de tous points deppepulée de poisson*" (Mahul, ed., *Cartulaire et Archives*, II, 507–515). Likewise in 1507 the municipal council of Madrid obtained a royal order against certain *xudrias* (a type of diversion trap or holding tank) installed by landowners for causing the fishery to decline (Puñal Fernández, *Mercado en Madrid*, 179).

at Marseilles, Calais, and Tréport in Normandy,[112] and that also in fact enjoyed by the handful of 'full fishers' under princely jurisdiction on the Traun. A 1533 agreement between the archbishop of Salzburg and the duke of Tirol curbed the gear and opportunities for peasant fishing "so that the Ziller is not depleted like before."[113] From historical distance it is next to impossible to test the differential impact of such restrictions on the aquatic resource, but perhaps easier to see how regulations favoured some users over others.

In the long term from about 1200 rulers increased intervention in fisheries 'for the common good', often but not exclusively in response to perceived depletion. Whether driven by ecological, sociopolitical, or other variables, the particulars of regulations reveal great local and regional diversity, but share close familiarity with each immediate situation and, as study of their content reveals, parallel problems and repertoires of potential remedies.

Laws arguably indicate the probable (or at least then traditional) presence of named taxa in local waters. With exception of the invasive carp all the fishes mentioned were indigenous to the watersheds and habitats involved.[114] Seasonal and technical particulars suggest drafters of the regulations did not rely on learned texts but had access to considerable expertise about local fishes and methods of capture. Huchen but not eel are present in the Austrian Danube and the reverse is true for the Rhine, Seine, central Spain, and central Italy. Coregonid whitefishes occur in Alpine lakes. Of course the converse is not true: important local species such as sturgeon in the middle Danube and salmonids in the Seine system simply do not appear in the laws.

So the fisheries regulations do not reflect local aquatic ecosystems alone, but also some kind of 'political' significance. The named species and methods had to matter to some influential group or individual. Each had acquired some sort of cultural, commercial, or economic value in a regional medieval society. Legislative concern for Scottish salmon has directed research attention to their export value and surprisingly extensive fiscal records of their delivery to the late medieval Low Countries (see Chapter 8 below). Perugia's diligence in preserving and enhancing diverse species in L. Trasimeno for the sake of sales to larger towns has not. The ubiquity of pike in ordinances across its range surely bears some relation to this fish's potentially impressive size, aspect, and mythic

112 Hocquet, "Pêcheries médiévales," 116–117.

113 "*das der Ziller nicht wie bisher ausgeödet werde*" (Stolz, *Geschichtskunde der Gewässer*, 373).

114 As tabulated in Hoffmann, "Fisheries regulations," table 5.1, and now reproduced in more understandable form in Supplement 6.2.3.

reputation – though that might shrink in view of minimum legal sizes set as low as about thirty centimeters. Put simply, if medieval authorities thought certain varieties worth special treatment in their laws, lack of commensurate presence in other written and archaeological sources likely deserves closer investigation (taphonomy? ambiguous nomenclature? more prestige than economic value?). Fisheries protection was one of several priorities for legislative action. As epitomized in Venice, medieval governing elites were not ignorant of the need for preservation and wise use, but other considerations might have greater weight.

Many medieval lawmakers and their advisors were thus well aware of threats to their fish stocks and of measures to reduce or mitigate those threats, whether affecting particular species or stocks in general. In that sense they could be said to seek what is now called sustainability. But taken as a whole, the 'common good' sought by fisheries regulation was primarily secure and affordable local supply of a culturally desired commodity and secondarily maintenance of a certain social order among fishers and between the fishing industry and a larger society.[115] Fish stocks came further down the list of priorities. The latter were, however, a genuine concern for would-be rulers over inland and inshore fisheries, though not on the open sea.

Hence medieval authorities purposely exercised a wide range of oversight and regulatory control over fisheries, employing both means and reasoning still familiar today. They directed and compensated for the social need for fish exceeding traditional supplies. Public proclamation and distribution of the ordinances reveal deliberate intervention of the late medieval state in the aquatic realm. The understanding of the regulators grasped, however, only the impact of overfishing: Philip IV knew "the evil of the fishers" was to blame in 1289. Whatever the actual intent, medieval fisheries legislation shaped patterns of human relations around the fishery as much as did the handling of fish as a market commodity or the assertion of exclusive fishing rights. The law's necessarily selective protections, moreover, also blended into semi-conscious human encroachment into natural relationships between favoured and disfavoured organisms. Yet neither environmental changes induced by human activity outside the fishery nor those of natural origin then entered into lawmakers' awareness. Nor could they see behind the greed

[115] The impression is that significant access to distant marine resources, even in preserved form, suppressed concern of public authorities to regulate inland or inshore fisheries other than salmon. Extensive codes as here explored are little evident from northern European maritime governments. Further research can test this proposition.

of the fishers to the demand which drove the fishing economy. The catchers, not the eaters, were held responsible for past damage and future hopes in the fishery.

><> ><> ><> ><> ><>

As means of handling the impact of population growth and economic development on a renewable resource, market adjustments, property rights, and public regulation all essentially take Nature as given and limited; humans adjust or adapt their own behaviour to fit the perceived limits. Another approach was to engage Nature differently, whether by modifying known Nature better to meet human needs or by satisfying wants from new, hitherto unexploited, parts of Nature. In both alternative strategies, under conditions of demand exceeding traditional and damaged sources of supply, high and later medieval Europeans began to extend their takings of fish beyond their natural local ecosystems. The final two chapters examine those innovative developmental responses and their consequences within western Christendom and on its resource frontiers.

# 7 Going beyond Natural Local Ecosystems, I

## Carp Aquaculture As Ecological Revolution[1]

### 7.1 Eating from outside Natural Local Ecosystems

On rare occasions fortuitous conjunction of a detailed written record with well-preserved and assiduously excavated archaeology can personalize a far-reaching general trend or historic transition. Such is the case of Christoffels Jans, a crossbow maker in the modest Flemish town of Aalst, midway between Ghent and Brussels and some fifty kilometers from the Scheldt estuary.[2] The Jans family diet in the 1490s highlights the material transformation of European relations with aquatic life. Late medieval consumers of all social ranks increasingly ate fish taken from other than the natural local ecosystems on which their forebears had long relied.

Jans was a skilled artisan retained on salary by his city and its shooting guild. From 1489 to 1498 he and his household lived and worked in a house with a thatched roof, brick floor, and small basement opposite the cattle market at the edge of town, close to the guild's armory. In 1497 the local Carmelite friars acquired the armory and then the house, so Christoffels moved elsewhere, though remaining until 1526 with the guild. To construct a new abbey the convent razed buildings around the market, so sealing three cesspits where the bowyer's household had discarded waste from both their craft and daily lives. Artifacts from the site plus modest written records indicate a fairly well-off urban craft establishment, described by the archaeologists as belonging to the town's 'upper middle class' meaning a settled household economy but well below the status of wealthy merchants, much less regional lower nobility. Think of this domestic group as middling urban consumers.

Meticulous recovery of remains revealed the Jans family enjoying a diverse fish diet, consuming at least twenty-five taxa, dominated by

[1] Seen in retrospect, earlier and provisional versions of this chapter and the next appeared as Hoffmann, "Carp, cods, connections" in Henninger-Voss, ed., *Animals in Human Histories*.

[2] Lentacker et al., "Dierlijke resten," 304–322, with context from De Groote and Moens, eds., *Archeologie en geschiedenis*, 45–80 and 373–430.

herring, several species of flatfishes, eel, haddock, cod, and various cyprinids, most notably common carp. In nearly every feature remains of marine fish outnumbered freshwater varieties by two to one. Other than eel, which were taken in fresh water, diadromous species were vanishingly rare. Costly fishes such as sturgeon, salmon, or large freshwater pike or catfish, left barely a trace. Little changed over time except in one pit where earliest deposits likely predate the Jans household: older cod there ran around 65 cm in length but the newer ones (?late? 1490s) around one meter. In the mix of species and in the fishes' origins, Christoffel's family shopper was finding on the local market a quite different menu than had been the case in the Low Countries two or three centuries before. The household ate an exotic species bred and reared in artificial ponds and several others brought in from the economy's distant marine frontiers. Were the Jans somehow eccentric? Not at all. This new relation to aquatic ecosystems they shared with contemporaries across western Christendom.

From North Sea shores to the Danube and south beyond the Tiber, by the close of the Middle Ages many Europeans were eating different fishes from different waters than had their high medieval forebears (compare Table 5.1). Specific novelties varied from place to place but the most noteworthy originated in local fish farms or distant marine waters, sometimes both.

Despite different socio-economic standings Flemish neighbours to the Jans household shared a quite similar fish diet. Monks at late fifteenth-century Ename priory consumed mostly herrings which had been gutted before delivery, various codfishes, and carp.[3] Not far away, the household of Countess van Buren at Eindhoven castle ate 60 percent marine species, half of them cods, and one-third fishes from fresh water, of which one in ten were carp.[4] At contemporary Flemish urban sites (review Figure 2.2) flatfishes and gadids dominated the fish waste, although cod trailed whiting and/or haddock. Herring frequencies varied, while freshwater and diadromous taxa together ran below 20 percent.[5]

[3] Van Neer and Ervynck, "Food rules and status," 155–164.

[4] Jong, "Huisdieren, jachtwild, vissen en weekdieren," 214–231, and "Fish consumption at Eindhoven," 129–138. Comparable dietary habits are visible in remains of the fifteenth–sixteenth centuries from castles at Bodenteich and Plesse in inland lower Saxony, where no consumption of marine fishes had been evident before the 1200s (Heinrich, "Fish remains of two medieval castles," 211–216, and Heinrich, "Fischreste von Bodenteich," 187).

[5] Van Neer and Ervynck, "Rise of sea fish consumption," 164–167. At Mechelen, where eel had long ruled, herring, whiting, plaice, and carp gained significance in the thirteenth century but cod only after 1450, when 80% of remains were of marine origin (Wouters, "1,000 years of fishing").

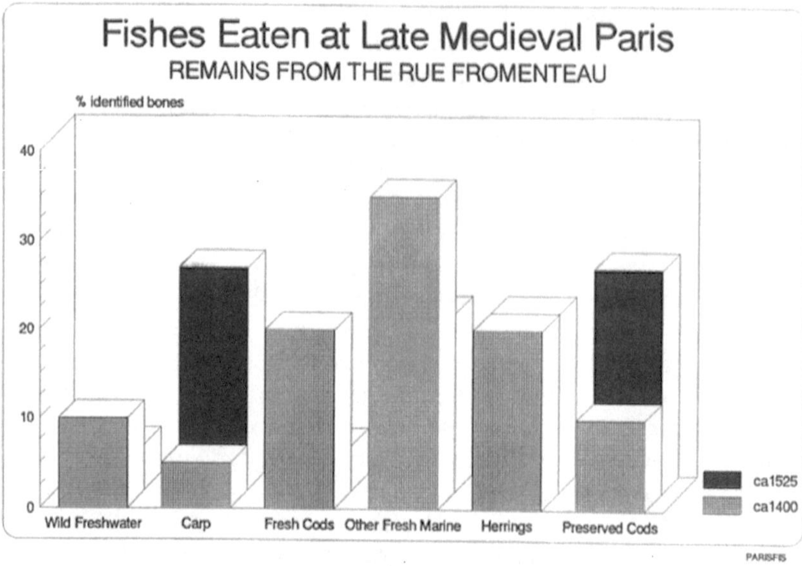


Figure 7.1 Fishes consumed at Rue Fromenteau, Paris, *c.* 1400 and *c.* 1525. Data from Desse and Desse-Berset. "Pêches locales, côtières ou lointaines," graphed by R. Hoffmann. © R. Hoffmann.

In an Amiens neighbourhood of shopkeepers and cloth workers, where eel had ruled twelfth-century tables, from the late fourteenth century through the sixteenth the most common fish had become herring.[6] Reasonably well-off late medieval Parisians who lived on Rue Fromenteau near the king's Louvre palace brushed into their garbage heaps and cesspits (Figure 7.1) 62 percent remains from fishes from the sea – after about 1450 often headless cod – and 30 percent from fresh water – after 1450 nearly all carp of uniform size.[7] Further south at Orleans, 15,004 fish bones from a latrine dated 1490–1568 at a collegiate church came from thirty species, but 52 percent were carp and 23 percent cod.[8] Up the Loire, monks at contemporary La Charité also ate more carp than all other fishes combined, but no marine organisms.[9]

6 Cloquier and Clavel, "Consommation d'animaux," fig. 2 and pp. 93–94.

7 Desse and Desse-Berset, "Pêches locales, côtières ou lointaines," 119–126. Preliminary counts from another Parisian site of the early sixteenth century show comparably numerous herrings, headless cods, carp, and pike (Sternberg, "L'approvisionnement de Paris," 127–130). Clavel, *L'animal*, 143, has the overall picture.

8 Marinval-Vigne, "Consommation d'animaux sauvages," 474–483.

9 Audoin-Rouzeau, *Ossements*, 146–147. As at Ename, these monks were Benedictines.

English experience provided a precocious precedent. In the massive assemblage of fish remains from York, freshwater and diadromous varieties, even eel, are by the fourteenth and fifteenth centuries vanishingly rare; locals threw out more herring bones than any others and many from gadids of North Sea and more distant origin. Yet unlike the continental sites just described, this fish menu differed only subtly from that in the thirteenth-century city.[10] Stable isotope studies of human remains (see Chapter 2) at York confirm one in three in the eleventh–twelfth centuries had consumed significant marine protein and by 1300 that pattern looks fairly general across northern England. At London composite analysis of carefully chosen samples from 7,270 contexts on 142 sites with 35,187 fish bones dating to medieval times identifies a dramatic eleventh–twelfth-century increase in herring remains which changed little thereafter. Cod, however, though present from the tenth century, surged in the thirteenth and, after an early fifteenth-century lull, rose again from the mid-1400s and still higher after 1500, when their remains finally surpassed those of herring in the leavings from Londoners' tables. Use of freshwater fishes had by then dwindled to quantitative insignificance.[11] England's insular situation allowed provision of marine fishes by intensified exploitation of surrounding local aquatic systems or from distant waters. What share of England's increased consumption of marine fishes, the so-called Fish Event Horizon, ought then to be traced to that distinctive ecological and economic context?

No such massive fishbone assemblages of late medieval or early modern date are available from interior continental Europe, but there are regionally interesting cases. Some 600 kilometers east from the Jens family in Aalst, young Martin Luther was then growing up in Mansfeld, a mining town on the eastern edge of the Harz. Later celebrity justified thorough local study of his roots, including meticulous excavation of waste pits at the Luther family home and at an inn just across what is now Lutherstrasse. Among both the 1304 identified fish bones from the house and the 744 from the inn, herring were most numerous, comprising a third to a quarter of remains. The moderately well-off Luther household ate 57 percent fish from the 300-kilometer distant sea, notably also flatfishes but also almost 3 percent cod, all received in preserved state. Their 40 percent from freshwater fish, plainly fresh specimens,

[10] Harland et al., "Fishing and fish trade," 178–198. Compare Table 5.1 above.

[11] Orton et al., "Catch per unit research effort," 8–15, use sophisticated sampling techniques to control for varying intensity of archaeozoological research across London's 2,000-year history. This confirmed and refined conclusions presented in Orton et al., "Fish for the city," 516–530, and "Fish for London," 207–213.

were almost equal parts pike and various cyprinids, including a few carp. Across the way the herring had less company, with few additional marine fishes and only a token trace of cod beside 53 percent freshwater varieties. Among the latter, cyprinids made up a third of the total, with one bone in five from carp, and then again a strong representation of pike but also more perch than the carp.[12]

Further southeast still, only a composite study of a few thousand remains from more than a dozen religious and secular sites in the Vienna basin and along nearby Danube tributaries hints at general developments. Carp and other cyprinids there comprise the most common single taxon, about one in every four identified remains. In diminishing order follow those of perch, sculpin, trout, and pike. Of marine species only herring approach the top five from fresh water and none but flatfishes accompany them. No bones of gadids or any Mediterranean fishes have been found in the Austrian or Hungarian sites.[13]

Across the mountains a bare whisper of archaeological record leaves mainly written sources to document arrival of Atlantic fishes on inland and Mediterranean tables. Among the more robust evidence are accounts of toll collectors at Sevilla – where the annual value of hake, conger, and sardines arriving from Galicia doubled after 1450 to reach more than nine million maravedis in the 1490s – and at Valencia – where clerks noted the actual volume of goods. Each year during the 1490s the latter port landed between seven and eleven million Galician sardines and some 200,000 dried hake (a cod relative ten times the size of a sardine), twice the number of a half-century before; conger deliveries stabilized at about twenty-two metric tonnes.[14] Pack trains conveyed these sturdy goods to fishmongers and consumers across inland Spain.[15]

[12] Heinrich, "Fischknochen aus des Gasthof" and "Fischkonsum in Luthers Eldernhaus." Martin came to Mansfeld as a one-year old in 1484 and left for school in 1497, then university in 1501.

[13] Galik et al., "Fish remains," 342–345. Oh for a large and sieved assemblage from Nürnberg, Prague, Zürich, or Lyon! Apart from Danubian Pannonia, the closest to such from late medieval interior Europe are finds from Poznań dated thirteenth-fifteenth century but excavated before 1960 (so not sieved) in one instance and published without quantitative record in the other. Neither mentions any but freshwater and diadromous varieties familiar to the central Warta drainage (summary data in Makowiecki, *Historia ryb*, Aneks items 207–208, p. 188).

[14] Ferreira Priegue, *Galicia*, 667–672 and 728–739. A team led by Arturo Morales has presented the first long-term overview of hake consumption in Spain. This includes several medieval and sixteenth-century sites, primarily along the Galician coast but some in the deep interior of the peninsula. (Morales et al., "Looking for needles in haystacks" and "Hindcasting to forecast.")

[15] Martínez García, "La asistencia material en los hospitales," 352–355; Menjot, "Marché de l'alimentation," 202–203; Rucquoi, "Alimentation des riches," 302; Vincent, "Consommation alimentaire," 449–453.

Published studies of fish remains from late medieval Italian sites present a consistent pattern of consuming local, commonly freshwater or estuarine fishes. In Rome and its environs people at distinctly littoral sites ate a lot of nearshore Mediterranean marine taxa such as mullet, sea bass, and sea breams, while only a score of kilometers inland freshwater pike and tench provided the most bones. Especially in the south, however, the share of sites with some marine fishes doubled from two in ten to four in ten by the end of the Middle Ages.[16]

Written records do indicate some changes, including the first known shipment of dried hake and brined sardines from the Atlantic to feed an Aragonese garrison at Naples in 1426.[17] A century later when papal courtier Paolo Giovio inventoried the diverse taxa on offer in Rome, he highlighted the Mediterranean marine and indigenous Italian freshwater fishes, but also marked the salted and smoked *aringhae* which he traced to Danish coasts and from Sweden and Norway the *merlucciae* (literally hake but here likely cod), dried hard and favoured by German visitors. Yet Roman households especially enjoyed salted sardines from Liguria, and the city's taverns and food stalls were replete with brined tuna from Sicily and Provence.[18] Already in the 1380s–1390s wealthy Francesco Datini, merchant of Prato and Florence, though preferring tench, pike, and eel from Tuscan waters, was also eating barreled tuna as well as dry-salted herrings. He did good Lenten business with the latter, too, importing from Southampton and especially Bruges hundreds of bundles (*balle*) each of 1,010 herrings.[19] A generation or two before Datini, however,

[16] Salvadori, "La pesca nel Medioevo," 302 and table 2; De Grossi Mazzarin, "I resti archeozoologici." De Nicolò, "Production et consommation du poisson de mer," 54–55, finds even fourteenth–early sixteenth-century Adriatic coastal towns importing freshwater fish from the interior and themselves exploiting only lagoons and other inshore waters.

[17] Henri Bresc, "La pêche et les madragues," 168–169.

[18] Giovio, "*De romanis piscibus*," cap. xlii (p. 60). As observed in Chapter 2 (p. 67 and note 39) only seven of thirty-six human skeletons from a Roman neighbourhood cemetery about 1480 show the high $\partial^{15}N$ indicative of extensive fish consumption, but accompanying $\partial^{13}C$ values leave dubious the origin of those fish in Atlantic rather than Italian fresh or Mediterranean waters.

[19] On fish consumption in the Datini household see Origo, *Merchant of Prato*, 285–286, and Margherita Datini, *Letters to Francesco*, notably nos. 15, 48, and 49 (pp. 56, 113, and 115). Nigro, "Mangiare," 119–121 and 133–135, details Datini's own large-scale trade in herrings during 1384–1410 and contrasts this with small expectations of the early 1300s. Saminato de' Ricci's 1396 *Il manuale di mercatura* likewise describes a bulk trade in herring from Bruges to Tuscany (Borlandi, ed., 129.) Both brined tuna (*tonnina*) and 'herring' (*aringa*), absent from 1383 market regulations at consumption centre Bologna, are there in similar texts from 1436 and 1482 (Pucci Donati, "Mercato del pesce," 52–55). Caution is required with Italian 'herring', for some sold in Tuscany around 1400 were said to be from Ragusa (Dubrovnik) (Marshall, *Local Merchants*, 34); as no *Clupea harengus* live in the Adriatic, were these brined anchovy or sardines?

merchant venturers like Francesco Pegolotti (active *c.* 1310–1340) had considered *allecis* or *aringhe* entirely a commodity in north European trade, so even northern Italian consumers tasted only occasional small amounts.[20] Indeed Italian cookbooks and recipe collections of the fourteenth and fifteenth centuries mention no Atlantic fishes at all.[21]

Cod, carp, tuna, even hake, sardine, and in many places herring had been absent, rare, or strictly seasonal on twelfth-century western and southern tables.[22] To serve rising demand unmet by failing favourites, high and late medieval provisioners began to substitute the flesh of creatures foreign to the consumers' homelands. This they supplied from two fundamental medieval innovations in the European fishing sector, namely the invention and inland spread of aquaculture and the large-scale commercial exploitation of offshore and distant-water resources on Europe's (mainly marine Atlantic) frontiers. Understood in context, what occurred in interior and in maritime Europe responded to different and distinct segments of medieval consumption demand. Yet, as this and the following chapter will argue, both formed new and seminal relations between people and environments, projecting human use of aquatic life beyond the historic bounds of natural local ecosystems.

Prior to this chapter we observed medieval Europeans responding to fears of a shortfall in natural local supplies of fish by trying to adjust interactions among humans, changing culturally learned behaviours through enhanced ownership rights, market mechanisms, and regulatory constraints on fishing methods. All those measures essentially took local nature as given. By some time in the twelfth and thirteenth centuries, this approach was beginning to leave some demand unsatisfied. In further response, Europeans began to engage nature differently, looking to modify it or to exploit new parts of it to meet human wants. This more aggressive alternative should be familiar, for it is the dominant modern one. People who want more than nature offers try to 'improve' nature better to serve human wishes or to discover 'untapped' resources they

[20] Pegolotti, *La Pratica*, pp. 253 and 380.

[21] Boström, ed., *Anonimo Meridionale*; Faccioli, ed., *Arte della cucina*, I: 19–204; and Platina, *De honeste voluptate* (ed. Faccioli). Johannes von Bockenheim, the German chef whom Martin V (1417–1431) acquired at the Council of Constance along with the papal title, described dried cod (stockfish) as something prepared expressly for Thuringians, Hessians, and Swabians (Laurioux, "Le 'Registre de cuisine'," 740–742). All thirty-one fish taxa named in a late fifteenth-century Neapolitan cookbook are native to the western Mediterranean (Scully, ed., *Cuoco Napoletano*, 43–105).

[22] Thirty-five archaeological settings from northern France, with 80% of fish remains of marine origin during the fourteenth – seventeenth centuries, had contained from the twelfth century only one-third marine fishes – and those almost exclusively herring – and among that earlier freshwater majority (60%) no carp at all. Clavel, *L'Animal*, 143.

can make 'useful'.[23] A different program calls for different work in the natural sphere. More intense colonization of natural ecosystems aims to channel a greater share of natural productivity to what humans value. Frontier expansion extends human efforts to spaces hitherto ignored.

Always there is a catch. What one observer might think economic development, another could call human manipulation or environmental destruction. With respect to aquatic resources, artificial fish culture constructs synthetic ecosystems to concentrate biological production on organisms humans want. Distant water fisheries export the pressure of demand to foreign ecosystems. Both strategies brought to specific socio-natural sites the major transformations in human relations with non-human nature that are ecological revolutions.[24] Driven by consumers' desire for fish to eat, each medieval strategy triggered complex cultural and environmental consequences, some of which resonate still in present-day crises of regional and global fisheries. This chapter looks at fish farming, the next goes to sea.

## 7.2 From Wildlife Management to Aquaculture

Twelfth-century western Europeans, almost certainly French, worked out ways to control water and the fish they put into it in order to procure consistent and seasonally useful food. When the technology was joined, probably not much before the mid-1200s, to that naturally well-adapted fast-growing fish of exotic eastern European origin, the carp, they together formed the dominant and most advanced artificial fish production system used from the Atlantic to the Urals well into the nineteenth century and, in some areas, still today. Carp aquaculture revolutionized local ecologies and human relations with them, forming and controlling synthetic habitats for the sake of a non-native animal and to the probable harm of some native varieties. Whether aimed at indirect subsistence or market sales, medieval fish farms made ordinary people and nature alike submit in new ways to elite cultural preferences and powers.

Pisciculture may be thought one extreme of a continuum running from wild capture fisheries to fish farms. English landscape historian Stephen Rippon refers to a progression from exploiting a natural resource to modifying and ultimately transforming it into something else.[25] That

[23] On the modernist refusal to acknowledge nature's limits see Worster, *Shrinking the Earth*, notably 3–138.

[24] Merchant, "Theoretical structure," 265–274.

[25] Rippon, *Transformation of Coastal Wetlands*, 1–53, provides a useful framework for following the coadaptations of socio-natural sites.

something else is a fully colonized anthropogenic ecosystem. Genuine fish culture entails human manipulation of all the same variables as agriculture. Aquatic systems fully colonized by humans for aquaculture share key anthropogenic features with terrestrial farms, mobilizing land, labour, and capital to produce biomass for cultural purposes, normally human food. Managers select specific varieties for production and individuals for reproduction (breeding stock). Reproductive isolation drives evolution of domesticated varieties, genetically distinct from wild populations. Habitat is closely controlled to maintain supplies and quality of water and of energy (nutrients). Such environmental variables as predators, pests, and pathogens are monitored and combated. The aquaculturist's work to manage inputs and outputs more resembles that of a shepherd or grain farmer than of a fisher trying to find and gain possession of an animal in the wild. For fisheries in particular, earlier chapters have already noted many steps along the way, perhaps better conceived as the hazy zone between historic capture techniques and habitat modifications still well short of a fully anthropogenic colonized aquatic ecosystem.

Simple measures to get more use from fisheries were always commonplace. Venetian *valle de pesce* seasonally confined migratory schools of wild fish to be harvested at need and convenience. Across Europe landowners, artisan fishers, and fishmongers used tanks, cages, millponds, moats, or even specially designed structures to keep captured fish alive and convenient for kitchen or sale. Such live storage is a functional equivalent to preservation of processed fish flesh for future consumption. Even the boundary between storage and artificial transfer of wild fish into new waters is blurry. While Charlemagne ordered estate managers to keep replenishing stocks in his fishponds,[26] the actual activity is long hard to trace. Later, about the time Pietro de Crescenzi would advise choosing local fishes for ponds,[27] we witness eel, pike, and crayfish being set to grow in L. Trasimeno[28] and pike-perch in the lakes and millponds which Preetz convent owned in Holstein.[29] Such small-scale undertakings were ubiquitous at the margin of indirect seigneurial subsistence and of

[26] *Cap. de villis*, c. 21 and 65, ed. Boretius, *MGH Legum*, II, #32.

[27] Crescenzi, *Ruralia commoda*, 9:81; like advice was compiled from late Roman sources in mid-ninth-century Byzantium as bk. 20, c. 1, of the *Geoponika* attributed to Cassianus Bassus, ed. Beckh, p. 511 (tr. Dalby, p. 339).

[28] Mira, *Pesca nel medioevo*, 48; Biganti, "Pesca nel Lago Trasimeno," 797; Vincenti, "Tutela ambientale," 124–135. In coastal Languedoc eel were transferred from one salt pond to another (Puig, "Ressources de l'étang et de la mer," 112–119).

[29] Buchwald, "Anna von Buchwald," 27–28. Thames-side drainage ditches on the bishop of Winchester's manor of Southwark were stocked with boatloads of wild-caught burbot (Carlin, *Medieval Southwark*, 38). Additional examples of stocking local wild fish are provided in the Supplement.

artisanal fisheries. Live storage, use of handy artificial watercourses, and haphazard transfers of wild juveniles belonged to the historical setting but ought not be confused with domesticated fish farming.[30]

To make clear the distinction and hence the developmental process, this chapter now identifies features of carp aquaculture as practiced in early modern East Central Europe. These are well documented in managerial records of productive enterprises and in instructional manuals prepared by knowledgeable contemporaries. It then teases out traces of how these distinctive practices had earlier evolved on high medieval estates in what is now northern France and subsequently spread eastwards. A concluding analysis of enterprises across the aquaculture regions of late medieval Europe identifies consequences for aquatic ecosystems and human societies which arose from introduction and operation of fish farming.

### *7.2.1 A Benchmark: Advanced Traditional European Fish Farming*

The history of aquaculture in medieval Europe may best be approached by establishing as a benchmark the profusely documented methods used at the close of the Middle Ages in East Central Europe, then acknowledged as leading this field.[31] Unprecedented contemporary didactic manuals describe the 'state of the art' management and techniques used by Czech, Polish, and Austrian fish farmers. First published was *De piscinis* ("On Fishponds"), which future Moravian bishop Jan Dubravius composed in the 1530s for mining capitalist and landowner Jakob Fugger. The work achieved wide circulation, with the first printing in Wrocław in 1547 (Figure 7.2), another in Zürich in 1559, and by

[30] A point well made in Lampen, *Fischerei und Fischhandel*, 125.

[31] The subject, still often poorly known elsewhere, has received much attention, sometimes repetitive, from regional writers on aquaculture and history: Teplý, "O rybnikářstvi," "Význam Vítkoviců a Viléma z Pernštejna," "Obraz rybnikářství," "Stavba rybníků," and *Příspěvky k dějinám českého rybnikářství*; Ofczarek, "Die Teichwirtschaft in Südböhmen"; Andreska, *Vývoj rybářstv*; "Development of fish-pond culture," 77–89; *Rybářství a jeho tradice*, 32–51; and *Lesk a Sláva českého Rybářství*, 57–106; Boháč, "Historical-ecological aspects," 42–47; Búžek, "Goldene Zeitalter"; Čechura, "Adliger Grundherrn," 44–77 and 99–103; Knittler, *Nutzen – Renten – Erträge*, 146–181, and "Teiche als Konjunkturbarometer?"; Brzozowski and Tobiasz, "Z dziejów rybactwa małopolskiego," 11–37; Górzyński, *Zarys historii rybołówstwa*, 41–50 and 59–60; Hurt, *Déjiny rybnikařství*, vol. I, pp. 53–219; Nyrek, *Gospodarka rybna na Górnym Śląsku*, 89–154; Szczygielski, *Gospodarka stawowa* and *Z dziejów gospodarki rybnej*. Compare Roberts, "Fish culture in sixteenth-century Poland." Sowina, *Water, Towns and People*, 121–135 (a revised translation of Sowina, *Woda i Ludzie*, 115–132) discusses rearing fish in small garden ponds around late medieval Kraków.

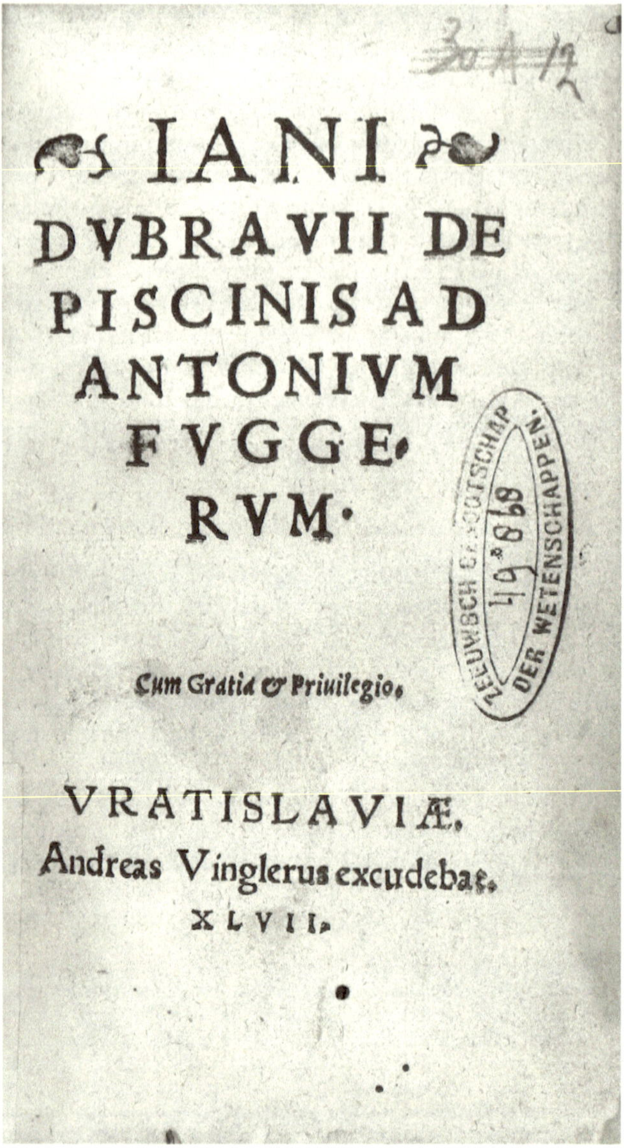

IANI DVBRAVII DE PISCINIS AD ANTONIVM FVGGERVM.

Cum Gratia & Priuilegio.

VRATISLAVIÆ.
Andreas Vinglerus excudebat.
XLVII.

Figure 7.2 Title page for the first published instructions on carp culture: Jan Dubravius, *De piscinis*. In Wratsliavia [Wrocław]: Andreas Vinglerus, 1547.
Jan Dubravius, *De piscinis*. In Wratislaviæ [Wrocław]: Andreas Vinglerus, 1547. Public domain. Image and reproduction rights courtesy of the Zeeland Library, Middleburg, Netherlands.

1600 also translations into Polish and English.[32] Olbrycht Strumieński's 1573 Polish book on ponds gained authority from the author's experience as a professional manager.[33] Recently identified manuscript instructions of slightly earlier date in Czech and German[34] augment the printed sources. Older handbooks are nowhere recorded.

Operational records from contemporary central European fishpond enterprises corroborate what the instructional literature advanced as normal and best practice. There are surveys of Polish royal estates near Kraków in 1564,[35] complete operating accounts from pond enterprises in the southwestern Polish duchy of Oswięcim-Zator during the 1510s–30s,[36] rich working archives from holdings of Bohemian magnate families Rožmberk and Pernšteyn,[37] and widespread if piecemeal documents of practice from lesser Polish, Czech, Austrian, and Franconian noble and monastic producers.[38] All these show the same procedures for an integrated production system.

At the end of the Middle Ages the best-reputed fish farmers in Europe produced fresh fish primarily for well-to-do regional consumers, both landowners and patrons of urban markets. They reared mainly common carp, with a side-crop of pike and sometimes a few bream or crucian carp. Cold-water fishes such as trout and grayling lacked importance.

[32] Dubravius, *De piscinis* (1547) and subsequent editions and translations. Compare Figures 6.3 and 6.4 as other links of fisheries to new media and management culture in late medieval decades.

[33] Strumieński, *O sprawie, sypaniu, wymierzaniu i rybnieniu stawów.* For both printed handbooks the 1960 source analysis and review by Inglot and Nyrek, "Jana Dubraviusa i Olbrychta Strumieńskiego," remains authoritative (Sawicki, "Jana Dubrawiusza," provides more biographical background but focuses on techniques for surveying, not fisheries management). Also available in print are instructions on rearing fish from the manuscript *Haushaltung in Vorwerken* (Ermische and Wuttke, eds., 195–206 and 254–257), associated with the court of Elector August of Saxony (r.1553–1586).

[34] Andreska, *Rybářství a jeho tradice*, 49–51, mentions economic instructions prepared in 1525 by Vojtěch z Pernštejna, the "Instrukce rybní pro panství Potštýnské a Litické," and a 1540 agricultural manual by Jan Brtvína z Ploskovic with chapters on pond culture. On the former see also Teplý, "Význam Vítkoviců a Viléma z Pernštejna." I hope in future to edit three texts with advice on fish culture hitherto unknown to scholars: Trier Stadtbibliothek, Hs 608 (1954) [1958/1422 4°], fols. 1–38, "Friedrich von Flersheim, Fischbuch 1530"; and two items in Vienna ÖNB, Codex vindobonensis 13103, "Anzaigung unnd underweysung der nottorfft so zu teichten und weyer steten gehoren im 1542 Jar" (fols. 7r–20r) and "Vischpuechl 1545" (fols. 1r–6r).

[35] Małecki, ed., *Lustracja województwa krakowskiego 1564*, vol. 2, 225–226 *et passim.*

[36] Rybarski, *Gospodarstwo Księstwa Oświęcimskiego*, 56–84.

[37] Šusta, *Fünf Jahrhunderte*, 1–29; Bůžek, "Goldene Zeitalter"; and Čechura, *Adelige Grundherrn*, 44–47, 70–77, 99–103, and 121–125.

[38] Besides items in note 31 above see Topolski, "Rybołówstwo i gospodarstwo rybne," and Cnopf, *Entwicklung.*

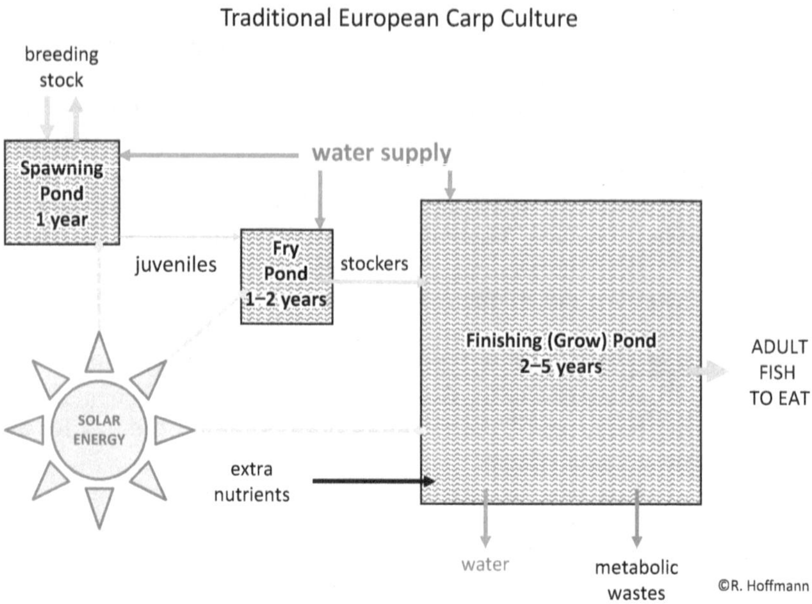


Figure 7.3 Traditional European carp culture.

Cultural methods emphasized full control over water and fish through each stage of production (Figure 7.3).

Reliably continual output required coordinated management of multiple ponds, each equipped to adjust the inflow, outflow, and level of water, from fully filled to partly empty or entirely dry, without undesired entry or exit of fish. Overland drainage (surface runoff and streams) from arable and pastures supplied the preferred nutrient-rich water. Ecological integration occurred at the landscape level by throughput of used water and sometimes removal of decomposing bottom sediments for use as fertilizer for the fields. Experts laid out ponds according to the terrain, preferring broad to deep ones even if the former required longer dams. Engineering skill predetermined the area to be flooded. In valleys a sequence of ponds, ideally with bypass channels, could feed one another downslope. Dams and dikes of beaten earth, sometimes with internal timber frames, were armoured against erosion by turf, wicker, wood, or stone sheathing. Wooden or metal grates and gates controlled passage through sluiceways made from the same selected materials.[39] Large

[39] Dams, sluices, and pipes dating from 1468/9 at Blankenheim in the Eifel are well reported and displayed in Keller, "Beobachtungen."

Figure 7.4 Bezdrev Fishpond, Czechia.
A view across Bezdrev Fishpond near Hluboká nad Vltavou, constructed 1490–1492 on the Hluboká estate of Vilém of Pernštejn. Second largest fishpond in Czechia at 520 hectares (2.008 square miles). Few medievalists have any idea of the scale of such enterprises. This artificial lake is half again larger than New York's Central Park, 20 percent greater than London's Hyde Park plus Kensington Gardens, and 2.5 times the Principality of Monaco. Some fishtank!
© R. Hoffmann.

aquaculture enterprises had use for tanks only a few dozen meters square and for regulated lakes up to 500 hectares (just over two square miles), such as Bezdrev pond built at Hluboka in southern Bohemia in 1492. The view in Figure 7.4 is well beyond what most medievalists or modern observers imagine as a 'fish pond'.

Managers bred the fish from selected stock, though not by artificial means,[40] and moved them as a uniform age group (year class, cohort) through a sequence of special-purpose ponds. As the water in a shallow and vegetated spawning or fry pond (Figure 7.5) warmed enough in late spring the pond master introduced chosen ripe adult carp to emit and

[40] Unlike trout or salmon, for instance, cyprinids do not naturally pair up and make nests for their eggs, but groups of ripe females and males gather in vegetated shallows and emit eggs and milt simultaneously, with much splashing and mixing, after which the adults leave the shallow nursery habitat.

Figure 7.5 Fry ponds at Rožmberk Fishpond, Czechia.
Fry ponds built at sixteenth-century Rožmberk pond, near Třeboň, Czechia (seen to the rear of a twentieth-century generator station).


fertilize eggs in their natural way, then removed the parents. A summer in those protected and warm conditions grew the carp larvae into tiny fish by the time they were moved en masse to deeper waters for the winter.[41] The cohort of young carp were left another year or two in a fry pond where they would reach 'stocker' (*setzling*) size of a hand's span, before being set into a much larger, recently refilled, and otherwise uninhabited finishing pond to grow big enough to harvest. Their numbers were calibrated to pond size and experience of its productivity.

Carp eat mainly small invertebrates they find on water plants or grub out of the bottom sediment and crush with the teeth in their throats. Nutrient-rich waters offer more natural food, so if necessary managers promoted growth by adding manure or channeling runoff from animal or human wastes into their ponds.[42] They also learned to concentrate the carp biomass into harvestable adults by introducing piscivorous pike to eat

[41] As remarked in Chapter 5, heat capture in spawning and fry ponds may have been crucial to consistent success rearing carp in colder areas of Europe under Little Ice Age conditions. Wintering fry in deep water further protected them from the worst cold.
[42] Handbook authors praise carp for needing no artificial food.

any fry bred by precociously mature members of the class. Depending on local temperature and climatic conditions, the carp reached a standard table size (commonly 40–60 cm (15–20 inches) long, weighing about a kilogram (2.2 lb.)) by year four, five, or six. By 1550 Polish records suggest two distinctive quick-growing domestic races of carp had evolved, 'mirror carp' with a few large scales and entirely scaleless 'leather carp'.

Demand-focused harvest took place in weekly stages during Lent when the market reached its annual peak, sometimes even where this meant breaking ice to get nets into the pond, or in a single push in the fall, when the crop not needed for Advent consumption could be stored alive in tanks for the Lenten season. Aquaculture made fresh fish available to fit a human, not nature's, calendar. In cool weather carp survive some days wrapped in damp cloths or straw for overland transport; portable tanks and boats with live wells also sometimes served. Water in the big finishing ponds, which had been designed with large warm, shallow, biologically productive areas at the periphery and a much smaller deep central channel, was lowered to concentrate the fish (Figure 7.6). When all were netted out (Figure 7.7) the flow of water was, if possible, entirely diverted and the pond site left to dry. Best practice called for the bottom to be plowed up, seeded with oats or barley, and then the grain used to pasture livestock through the summer. This fallow season or year readied the finishing pond for another cohort of young carp.

Productive operation of a late medieval fishpond enterprise thus called for integrated management of some tens or dozens of ponds and, exclusive of breeding stock, some half-dozen year classes of carp. In the first half of the sixteenth century these methods sustained, for instance, annual shipments of some 1,500 tonnes of live fish from the ponds of Upper Silesia (chiefly Oswięcim-Zator) to Kraków.[43] Enterprises in Czechia may have doubled that. But the integrated system was unknown in central Europe before the mid-fourteenth century and remained rare there until enterprises using it proliferated in southern Bohemia, Moravia, and Poland during the late fifteenth century. Having learned the characteristic multiple ponds, water control, selected breeding, and management by year classes of the mature system, we can detect their development some centuries earlier and further west. As Chapter 5 described, introduction of the exotic carp there capped cultural evolution of practices already being worked out with pond fishes native to Atlantic drainages.

[43] Nyrek, *Gospodarka*, 29–38, thinks the Polish capital thus received about 50–90% of Upper Silesian output.

Figure 7.6 Munický Fishpond, Czechia, lowered for harvest. Munický Fishpond (117 ha) lowered for harvest. Constructed for Vojtěch z Pernštejn at Hluboka nad Vltavou (then called Podhrad or Frauenberg), completed 1514. © R. Hoffmann.

### 7.2.2 *Emerging Technologies: Engineering, Practices, Fish*

Now recognizable even at historical distance, the process which resulted in traditional European aquaculture involved transition from live storage to long-term rearing of fish plus design and construction of infrastructure capable of manipulating the water and the fish for consistent growth and production. The new kinds of hydraulic engineering being installed in central France during the eleventh and twelfth centuries now deserve closer attention than Chapter 5 provided.[44]

[44] Note the timing reasonably coincident with England's Fish Remains Horizon (Barrett et al., "'Dark Age economics' revisited"). Explicit identification of eleventh-century ponds as innovations in Benoît and Mattéoni, "Conclusion" to the 2004 collective volume on freshwater fisheries in medieval France. Earlier such statements in Benoit and Wabont, "Mittelalterliche Wasserversorgung," 189–196; Gislain, "Rôle des étangs," 89; Guillerme, *Age of Water*, 54; and Benoît, *Étangs de la Dombes*, 24. Writers on ponds and fish culture in medieval France are commonly unaware of their counterparts in late medieval and early modern east central Europe (and vice versa).

Fichtenau, *Living in the Tenth Century*, 281, confirms the lack of solid evidence for early medieval fish culture. Likewise the rich records of Cluny at its height of wealth and prestige up to the 1150s refer to catching and storing fish from the Saône and its tributaries, but have no trace of their being reared (Ulrich of Cluny, "Consuetudines," II.4 (*PL*, 149, col. 703; administrative instructions, surveys, and accounts in Bernard and Bruel, eds., *Recueil des chartes*, nos. 3789, 3790, 4132, and 4143; Duby, "Le budget," 155–171, and "Un inventaire des profits," 129–140).

Figure 7.7 Harvesting a fishpond.
Harvesting a fishpond with a seine net, a view from Swabia. As illustrated in Bidpai, *Buch der Weisheit der alten Weisen*, fol. 44v. Ulm: Lienhardt Holle, 1483. Woodcut image and use with permission of Newberry Library, Chicago.

Certain disputes over water rights and flooding indicate the construction and rhythmic operation of pond systems without earlier counterparts elsewhere. Recent radiocarbon research dates the oldest such structures in Berry to the late tenth and especially eleventh centuries.[45] Even property inventories from that region mention no ponds until the eleventh century, when charters dated to the 1050s, 1081, and 1092 treat them as novelties.[46] In written records from around 1080 St. Vincent of Mans and other lay and religious lordships in Maine and Poitou, such as the lords of Laval at Marennes, undertake to excavate and build dams for multi-pond projects.[47] Monks at Ronceray complained in 1141 to Count Geoffrey IV of Anjou and canons of Bourges in 1160 to Count Étienne of Sancerre because the counts' new ponds had flooded properties of the religious.[48] While the original twelfth-century ponds at Citeaux provided water supply and drainage, with fish an afterthought,[49] Cistercians at Châalis in the Valois initiated complex pond systems shortly after 1160 and Burgundian nobles and monasteries began to do likewise between 1184 and 1206.[50]

Some charters hint how these facilities were run. A pond built by St.-Père of Chartres flooded land of a neighbour named Aucher: sometime between 1101 and 1129 the parties agreed that when the pond was filled, water and fish belonged to the abbey, but when it was dry, land and crop were Aucher's. Because some ponds on a Norman watercourse belonged to a knight and others to the monks of La Trappe, in 1215 the parties arranged to coordinate their draining and later refilling (Figure 7.8).[51] By 1263 the knight Pierre de Palluau had his ponds at Ograis in Berry emptied and dried every three years, while new constructions along the eastern edge of the Paris basin in Champagne came equipped with a full set of sluice-gates and drains.[52]

45 Pichot and Marguerie, "Approche pluridisciplinaire," 119–124; Stauner, "Les étangs de l'est du Berry."

46 Querrien, "Pêche et consummation" 2003, 412–418, with emphasis that in Berry a *lacus* was a natural water body, a *stagnum* created by an artificial barrier, and a *vivarium* a small basin for keeping live fish. The watershed of the small river Céphons, awash with more than a score of ponds in the later Middle Ages, had none before the eleventh century. Lay seigneurs took the lead here (Querrien, "Pêche et consummation" 2004).

47 Delatouche, "Poisson d'eau douce," 174–175, and works there cited; Sanfaçon, *Défrichements*, 26 and 85–89; Beech, *A Rural Society*, 38 and 106. Benarrous, *Grand Brenne*, 97–104, 130–144, and 342–344, provides detailed textual and sedimentary evidence of similar date.

48 Gislain, "Rôle des étangs," 95 n. 5; Devailly, *Le Berry*, 298, and compare 295–296.

49 Recall the reluctance of Cistercians to add fish to their diets (Chapter 2, notes 51–52).

50 Berthier, "Gestion des étangs" 2004; Blary, *Domaine de Châalis*, 31–40; Richard, "Le commerce du poisson," 181–197; Richard, "Les Etangs et le commerce," 99.

51 Grand and Delatouche, *L'Agriculture*, 541.

52 Devailly, *Le Berry*, 556–557; Maas, *Les moines-défricheurs*, 74–75. Richard, "Commerce du poisson," 187–188, details rapid construction of new ponds in the mid–late 1200s by Burgundian seigneurs around Autun.

Figure 7.8 Étang de Chaumont, La Trappe, Normandy. The Étang de Chaumont, probably of twelfth-century origin, at La Trappe abbey is and was one among the dozens of medieval ponds constructed along the little river Iton, a headwaters stream of the Eure, a tributary of the Seine. Photograph by Terryl Kinder. Used with permission.

Some of the same sorts of landowners were then also handling fish with forethought. Duke William of Normandy and monks of St.-Benoit-sur-Loire agreed in 1067 to share the fish taken regularly from one of the ponds at St.-Jacques de Beuvron.[53] Surviving scraps of financial records kept for the Count of Champagne in 1217 indicate systematic stocking of his ponds,[54] and the 1235 ordinance on fiefs of King Louis IX assumed that fishponds had a five-year production cycle.[55] These anonymous fish were probably natives, most often bream, the variety most consumed at La Charité-sur-Loire at this time.[56] In England, where after 1066 an ethnically and culturally French elite possessed pond systems wholly like those of their continental cousins, Henry III's own fishponds commonly supplied "fat mother bream" (*bremas matrices et grassas*) and pike for

[53] Gislain, "Rôle des étangs," 95 n. 6.

[54] Longnon, ed., *Documents*, III: 3–6. The scribe did not name the variety stocked.

[55] Laurière et al., eds., *Ordonnances des roys*, vol. I: 55–56.

[56] Audoin, *Ossements animaux*, 147, discusses fish remains from the floor of an eleventh–twelfth-century refectory.

stocking his other ponds and those of his favourites.[57] A late thirteenth-century guide to estate management, *Fleta*, explicitly recommended bream and perch for ponds.[58] And even in 1299 French royal officials were putting hundreds of *bramis* into ponds in Auvergne.[59]

Shortly before 1240 the English-born, Paris-educated Franciscan encyclopedist Bartholomaeus Anglicus described the ponds to which his contemporaries were accustomed:

A *piscina* is water gathered to nourish fish; although through ironic contradiction, as Isidore says, a gathering of waters without fishes is often called *piscina*. However for good quality a pond needs firm ground, pure inflowing water, and continuous flows. For where the ground is slimy and marshy, tasty fishes are no way to be nourished. Also where fresh water does not flow, the standing and stagnant water is easily corrupted. And therefore for renewal of the *piscina* sweet and fresh water is brought in through channels and pipes, with banks and walls protecting the boundaries of the pond lest the incoming water flow on out. From the ponds also are brought rivulets to irrigate gardens.[60]

'Hard' engineering would manage water for fish or other purposes, but the handling of the 'wet ware', living fish, remained somewhat haphazard.

Chapter 5's discussion of carp as an invasive species in medieval western Europe sets the stage for its purposeful management and domestication (recall Map 5.1). To reiterate, from an ancient native range confined to the Balkans, between the sixth and eleventh centuries carp spread west and northwards to the middle Odra and Elbe and into right-bank tributaries of the Rhine. By the mid-twelfth century the newcomer had reached the main stem of the middle Rhine, but not likely further. Only in the mid-thirteenth century does evidence of carp's presence explode across northern France and the Low Countries. Paris-based

[57] Steane, "Royal fishponds," 45 *et passim*; McDonnell, *Inland Fisheries*, 9–12; Roberts, "Bishop of Winchester's fishponds."

[58] *Fleta*, ed. Richardson and Sayles, bk. II, c. 73 (p. 247). Contra Grand and Delatouche, *L'Agriculture*, 544, this text has no carp.

[59] Fawtier and Maillard, eds., *Comptes royaux*, vol. 3, Tome 1, nos. 10713, 10717, and 10718.

[60] *De rerum proprietatibus*, ed. Pontanus, lib. 13, cap. xiiii, p. 566. *De piscina* : *Piscina est aqua ad nutriendum pisces collecta, quamuis per antiphrasim aquarum collectio non habens pisces, piscinis saepius nominetur, ut dicit Isidorus. Ad bonitatem autem piscine exigitur fundi soliditas, aquae influentis puritas, & influentiae continuitas. Vbi enim fundus est limosus & paludosus, pisces saporosi nullatenus nutriuntur. Vbi etiam recens aqua non fluit, aquae stantes & non motae de facili corrumpuntur. Et ideo ad piscinae renouationem per canales & fistulas aquae dulces & recentes inducuntur, vallibus & aggeribus ne influentes aquae effluant, piscinarum termini minuuntur. De piscinis etiam riuuli ad hortorum irrigationem deducuntur.*

English translator John Trevisa (*On the Properties of Things*, ed. Seymour *et al.*, I: 661–662), in 1398/9 gave *nutriendum* as "fedynge" and *nutriuntur* as "ynorisshed." The passage ought not be construed as referring to purposely breeding fish, selective or not.

scholastic encyclopedists in the generation after Bartholomaeus associated the wild fish with slow rivers and still waters. Contemporaries around 1260 themselves named carp and other species as being stocked in ponds in Brie Champagnoise and sold on the Paris market. Consumers discarded the bones into waste heaps near the Louvre where modern archaeozoologists identify the same taxa. Until that time the carp's westwards expansion had looked like an unintended consequence of human activity under favourably warm climatic conditions: across the heartlands of Latin Christendom whatever bodies of water people used or built to store live fish or for other purposes proved well suited as habitats for a hardy, omnivorous, and adaptable warm-water fish. Uneaten survivors could be transferred or simply escape and start new wild populations.

Hence with the 1200s seigneurs and their estate managers with experience handling stocks of native fishes in controllable multi-pond systems could identify and turn to the more fecund newcomer. Especially those mid-century encyclopedists confirm a new contemporary awareness of carp in the French heartland. Thomas of Cantimpré, Vincent of Beauvais, and Albert the Great, respectively sons of the Low Countries, northern France, and near Augsburg, spent much of their careers in Paris, where between the mid-1240s and early 1260s each compiled Aristotelian observations on natural history which then circulated widely. All three describe carp as a fish of still waters adept at evading nets. Thomas and after him Vincent appealed to popular knowledge ("*Opinio vulgi est…*") in relating how this fish spawned in water weeds, its young liked warm shallows, and the sickness they suffered in stagnant August conditions was cured by fresh river water. Albert added that carp spawn in certain waters and grow large in others, the best having clay bottoms seeded to wheat and plowed over before flooding.[61] All this highlighted features of carp aquaculture. With carp's entry into these widely distributed reference books, after the 1260s mere learned mention is no longer evidence for local presence of the species, but other signs of purposeful management proliferate.

From the Paris basin to Burgundy and from Hainaut to the middle Loire, by the turn of the thirteenth century carp seem almost ubiquitous – carp growing in ponds, little carp for stocking, big carp for breeding, carp

[61] Thomas of Cantimpré, *Liber de natura rerum*, 7:23, ed. Boese, 258–259; Vincent of Beauvais, *Speculum naturale*, 17:40 (1624 ed., col. 1274); Albert the Great, *De animalibus*, 24:26, ed. Stadler, 1525–1526.

in elite kitchens and on elite tables.[62] On most elite estates – royal, aristocratic, monastic – those fish lived in habitats equipped for full control of water supply and levels.[63] Agents for the count of Namur inventoried his dozen ponds in 1289 and specified each was stocked with carp for harvest on a four-year cycle.[64] In 1319 the pond at Millançay in Sologne was fished every two or three years followed by one dry year before restocking.[65] Already in 1285 the Count of Champagne's pond at Chaumont was producing carp by the thousands, and in both 1339 and 1345 the duke of Burgundy's Saint-Seine pond at Laperrière-sur Saône yielded about 16.5 metric tonnes of adult carp and of other varieties only a fraction of that.[66] While most of those big carp from Saint-Seine were destined for live storage and the duke's palace table at Argilly, output

[62] Find carp recipes and menus in the oldest late thirteenth-century form of what would come to be called the "Viandier" of Taillevent (Scully, ed., *Viandier.* 1–31 and notably recipe 102), in the "Enseignements" of *c.* 1300 (in Lozinski, ed., *La Bataille*, 185–186), and in the literary satire on court customs written in Paris in 1316, the "Roman du Comte d'Anjou," l. 1132 (*ibid.*, 192–193. Commentary by Planche, "La Table comme signe)," 253–254).

[63] Examples of dam construction, adjustable sluices, bypass channels, and other engineering particulars appear in Rouillard and Maupoume, "Les étangs royaux"; Hoffmann, "*Carpes pour le duc*," 38–39; Deligne, "Carp in the city," 284–286; Berthier, "La gestion des étangs" 2006, 281–286; Hoffmann, "Aquaculture in Champagne," 76; Querrien, "Pêche et consummation" 2003, 418. But not all ponds were so equipped: dams on those built about 1290 for the display gardens at Hesdin, though harbouring carp and other managed fishes, had to be 'ruptured' to drain for harvest or maintenance (Farmer, "Power and the 'natural' landscape," 659–662). Some Burgundian ponds in the 1340s also had to be 'broken' for harvest (Richard, "Commerce du poisson," 188–189).

Further human colonization of the natural sphere included the predator control exercised by otter catchers deployed on estates of the count of Artois (Farmer, "Power and the 'natural' landscape," *ibid.*) and the dowager queen Jeanne d'Évreux (Longnon, ed., *Documents*, III: 439–440) and the heron hunters hired by the count of Bar in 1334 (Collin, "Le train de vie," 804).

As the above references imply, present knowledge of medieval French fish culture, especially that before *c.* 1350, rests almost exclusively on extant financial accounts from the highest elite estates. While this mostly reflects the precocious development and fortuitous survival of administrative records from that sociopolitical level, it also indicates the reluctance of historians to pursue what may survive from smaller properties and to study pre-plague conditions as thoroughly as they have records from *c.* 1350–1500. Some resulting lacunae are also mentioned later in this chapter.

[64] Brouwers, ed., *Cens et rentes*, vol. 2, pt. 2, pp. 433–434; compare discussion in Balon, "La peche et le commerce," 27–28.

[65] Guérin, *La vie rurale*, 140. Besides what appears in this paragraph, Supplement 7.2.2 has further evidence of rotations, output, and disposition of fish from French seigneurial ponds before 1350.

[66] Longnon, ed., *Documents*, III: 17–20. Accounts for Saint-Seine unusually describe the carp in terms of measured size, notably mature fish "*d'un piez et pleine palme de long*", so about 40 cm, a size at which modern feral carp weigh about 1.5 kg. In each harvest year the pond yielded more than 11,000 such fish (Hoffmann, "*Carpes pour le duc*," 35–38).

from fish farms belonging to the king, to the queen dowager, and, it appears, to the abbey of Citeaux was also sold in varying proportions to local merchants and some with Parisian connections. Other large numbers of small fish (*norriens*, *alevins*) went to restock the owner's own or neighbours' recently refilled ponds.[67]

Transfers of juvenile carp obscure the extent of selected breeding then carried out by these enterprises. In 1347 managers of the queen dowager's ponds in Brie returned 100 of the 900 large carp harvested from the Parc pond near Coulommiers in order to repopulate it.[68] Sixty years earlier assessors on the count of Namur's estate found the pond Huve near Tavier exclusively dedicated to "large carps chosen to restock the other ponds."[69] Did selective breeding also imply specialized spawning/fry ponds? Such facilities may have existed before 1350 on the Citeaux estate, where they are recorded by the end of that century, while on properties of the duke of Burgundy and neighbouring princes they occur, along with specified 'mother carps' (*carpes meres*) only shortly after 1370.[70] Despite the sparse early evidence of reproduction control, well before the Black Death of 1348–1351 marked a secular turn in Europe's economic climate, clerks for great and lesser lords in northern Francophone Europe were surveying and keeping account books for carp-rearing enterprises closely resembling those in Bohemia or Poland two centuries later.[71]

Not so across the Channel. English pond managers also kept fine accounts in the thirteenth century, but they never once mentioned a

[67] Hoffmann, "*Carpes pour le duc*"; Hoffmann, "Aquaculture in Champagne," 75; Rouillard and Maupoume, "Etangs royaux." Berthier, "Gestion des étangs" 2006, 286–289, documents such sales by Citeaux immediately after the plague epidemic and implies they occurred earlier as well. This abbey's aquaculture enterprise faced the unusual demand each autumn to feed abbots from all the Order's houses attending the compulsory general chapter.

[68] Longnon, ed., *Documents*, III: 448–449.

[69] "*queil vivier on a mis d'an en an [carpes] de chief et kewe por rapissonner les autres viviers*" (Brouwers, ed., *Cens et rentes*, fol. 2, pt. 2, pp. 433–434 (with brackets *sic* in the published text). Even in the absence of carp the selection principle was visible in late twelfth- and early thirteenth-century English royal gifts of 'mother breams' (*bremias matrices*) (McDonnell, *Inland Fisheries*, 11–12).

[70] Berthier, "Gestion des étangs" 2006, 287–289. For clear late fourteenth-/early fifteenth-century documentation of ponds dedicated to spawning and larval carp see Beck, *Eaux et forêts*, 301–311, and Gresser, *Pêche et pisciculture*, 193–206, while discussions in Benoît, *Étangs de la Dombes*, 50–55, and Deligne, "Carp in the city," 289–291, are more ambiguous. Chapter 5, p. 226 above pointed out the importance of this technique in reducing risk of reproductive failure during a climatic transition to the Little Ice Age.

[71] Hence Amacher, "Teichwirtschaft," 71–73, and other writers who rightly observe expansion of carp culture after 1350 in their own regions, should not deny its significant earlier presence in France, the adjoining Low Countries, and southern Germany (see also pp. 293–95 below).

carp. When King Edward II and his wife Isabelle of France visited Paris for the knighting of her brothers in 1313, her father, Philip IV, provided the couple with the food for a banquet to thank their hosts. The gift included 200 large and 40 small pike and 160 carp.[72] Yet the species leaves no trace in the British Isles before the very end of the fourteenth century, and well into the sixteenth was there acknowledged a rare and recent introduction (also the date of the earliest credible bone finds).[73] Delayed entry of carp to England contrasts with Anglo-Norman or Angevin introduction of exotic fallow deer, rabbits, and probably peafowl, all for the sake of elite luxury food and entertainment.[74]

### *7.2.3 Diffusion of Innovations*

Well-evidenced signposts on the continent mark the subsequent spread of the entire integrated aquaculture system (multiple ponds used in rotation to rear chiefly carp in segregated size groupings) from northern France across other regions of transalpine Europe. One vector went east from ducal Burgundy into Franche-Comté – where production accounts dating 1338–1383 report all the now-familiar features[75] – and down the Saône/Rhône (or up the Loire) into Dombes, Forez, and west into Brenne. The latter regions experienced two waves of pond construction. Early initiatives in the 1230s were emulated on a large scale only fifty and more years later, when lords invested in pond systems as one form of market-linked agricultural improvement. Further purposeful flooding after the mid-fourteenth century commonly tried to get some use out of abandoned farmlands by serving regional demand for fish. Large-scale market-oriented carp production by a wide range of social groups became central to regional economies – though the best operational records are, of course, for the nineteen large and more small ponds of the count of Forez and the equivalent estates of the lords of Thoire-Villars in Dombes and the duke of Savoy in Bresse.[76]

[72] Fawtier and Maillard, eds., *Comptes royaux*, nos. 27701 and 27760–27764. Edward III also consumed carp on the continent (Serjeantson and Woolgar, "Fish consumption," 126).

[73] Further detail and evidence regarding the belated entry and environmental insignificance of medieval carp in England and Mediterranean Europe appear in the Supplement for this chapter.

[74] Sykes, *Norman Conquest*, 63 and 76–84; Grant, "Food, status and religion," 141–143; Rackham, *History of the Countryside*, 47–50 and 123–125.

[75] Gresser and Hintzy, "Les étangs du domaine comtal"; Gresser, *Pêche et pisciculture*, 151–225.

[76] Benoît, *Les étangs*, 23–31, documents the appearance of ponds in the Dombes, but see also the more interpretive ideas of Perceveaux, "Essai sur l'origine," 81–90, and "Structures et relations économiques." Egloff, *Paysan Dombiste*, 89–94, makes the region's long-lasting pond-centred society peculiarly accessible. Conditions in Forez are treated in Durand, "De l'établissement des étangs," 101–108; Fréminville,

While aquaculture in Hainaut and Brabant seems participant in the French experience, its more tenuous spread further north in the Low Countries is so far better traced through the proliferation of carp in food remains and customs schedules.[77]

The trail of aquaculture spreading eastward into central Europe might see one signpost at Heilsbronn, a Cistercian monastery located about twenty-five kilometers southwest of Nürnberg on the Schwabach, a headwaters stream in the Main–Rhine drainage. This lordship roughly midway between France and Bohemia can stand for a whole belt of ponds in Upper Franconia.[78] During the 1260s Heilsbronn, already more than a century old, began acquiring first natural, then artificial ponds in its general environs. By the 1340s the monks were buying and rearing stocker carp for a multipond enterprise which extended to more than thirty kilometers from the convent and still mainly served their own consumption needs. In the late fifteenth and sixteenth centuries, however, and eventually as a state-managed foundation, Heilsbronn's large output joined that of many neighbouring landholders who invested in aquaculture to supply urban consumers, especially in prosperous Nürnberg.[79]

Adjacent to Upper Franconia are the Czech lands, where pond construction for aquaculture first surged under the auspices of Emperor

"Comptes du maître des étangs"; Fournial, *Les villes et l'économie*, 687–690; and Mattéoni, "La pêche des étangs." Fernand Braudel (*Identity of France*, 204–206) turned this data into an intriguing case study. Benarrous, *Grande Brenne*, 224–227, can document coincident increases in pond construction but only lament the lack of surviving medieval evidence of practice (pp. 227–256).

77 Deligne, *Bruxelles et sa rivière*, 131–178, and "Carp in the city," 289–291, dates expansion of carp culture around Brussels *c.* 1250–1380, with peak prosperity from supplying the fourteenth-century city. Then see van Mieris, ed., *Groot charterboek*, vol. II, p. 656. To early finds in Chapter 5, note 108, add those in Brinkhuizen, "Preliminary notes on fish remains," 83–90, and "Visresten uit twee middeleeuwse vindplaatsen," 19–20; Jong, *Tolbrugstraat, Breda*; Ervynck and Van Neer, "A preliminary survey," 304; and Seeman, "Monnickendam," 125–134

78 Patrick Götz compiled in his unpublished *Zulassungsarbeit*, "Karpfen in Franken im Mittelalter: Eine wirtschaftsgeschichtliche Untersuchung," Friedrich-Alexander Universität Erlangen-Nürnberg, Institut für Geschichte, Sommersemester 2009, a thorough overview of the regional aquaculture which emerged in medieval upper Franconia. I am grateful to the author for sending me a copy.

79 Heilsbronn's interest in fish culture coincided with that of Salem mentioned in the Introduction, p. 13. Findings of Heidacher, *Entstehungs- und Wirtschaftsgeschichte*, 118–119, need adjustment in light of the chronology of land acquisition revealed by charters published in Schuhmann and Hirschmann, eds., *Urkundenregesten Heilsbronn*, and of the management information from 1338–1374 in Staatsarchiv Nürnberg, Klosterverwalteramt Heilsbronn, Rechnungen, Bd. 1. Cnopf, *Entwicklung*, 26–82, examines operations of the seventy-one ponds Heilsbronn had in the 1500s. To compare the Swiss situation see the Supplement.

Charles IV (r.1347–1378), who also had family and personal connections in France. Charles is even reputed to have urged pond building not only "to hold fish for people's food and to drain water from swampy areas," but also to retain rain and snow melt for flood protection.[80] Dwarfing all known earlier hydraulic works in the region, at least eighty-seven new projects are known before the Hussite revolt in 1418 seized Czech attention.[81] Though late fourteenth-century pond masters there favoured hill country, which made for cheaper short dams but less productive deep ponds, some built the oldest large ponds (100 ha plus) which still survive, installed elaborate sluices and valves, populated ponds with age-class stockers, and selected fish for breeding. Their output went to towns in Bohemia, Germany, and Austria.[82] Czech politics calmed again after 1450 – when the Rožmberk estate at Třeboň already had three large ponds covering 700 hectares and seventeen smaller ones – and gave occasion for more large engineering works in both southern Bohemia and the flat lands of the upper Labe (Elbe) basin. Then the Pernštejns, for instance, improved their Hluboka estate with the huge Bezdrev pond (see Figure 7.4) and master Štěpánek Netolický secured stable water supplies for Třeboň, which he managed from 1516 to 1530, by diverting a river through a forty-five-kilometer artificial canal, the Zlata Stoka, 'Golden Drain'.[83] Such were the enterprises familiar to Jan Dubravius.

Fish culture facilities were being built in central Silesia, then a dependency of the Czech crown, even during the difficult 1430s and 40s,[84] and by the 1450s comparable ones had come into the principalities of Oświęcim-Zator, at the low divide between the upper Odra and Wisła basins where Bohemian and Polish influence intersected. As earlier observed, carp from here supplied Kraków in the early sixteenth century.

[80] Boháč, "Historical-ecological aspects," 25–26, and comments by contemporary chronicler Beneš of Veitmíle, "Chronica ecclesiae Pragensis," 516. But several alleged particulars of Charles's involvement, though repeated by solid Czech historians since the nineteenth century, cannot now be traced to any known medieval source (see Bauch and Labbe, "Karpfen mit Spätburgunder," 8–10).

[81] Graus, *Dějiny venkovského lidu*, vol. 2: pp. 32–35, 345–346, and 483–486; Šusta, *Fünf Jahrhunderte*, 2; Boháč, "Historical-ecological aspects," 42–47.

[82] The general narrative found in Kalný, "Stará práva rybolovu," 67–74; Andreska, "Development of fish-pond culture," 77–80; Andreska, *Lesk a Sláva Českého Rybářství*, 57–106; and Búžek, "Goldene Zeitalter," needs the balance of the well-grounded historical evidence in Charvátova, "Manorial farms," 133; Neumann, *Prameny*, nr. 20, pp. 130–133; Šusta, *Purkrabské*, 34; and Hemmerle, ed., *Deutschordens-Ballei Böhmen*, 139 and 154.

[83] Šusta, *Fünf Jahrhunderte*, 4. The chronology in Lower Austria matched that in Bohemia (Knittler, *Nutzen – Renten – Erträge*, 152–158). Purkarthofer, "Teichwirtschaft Herberstein," 97–99, reports sixteenth-century pond creation further south in Styria.

[84] Hoffmann, *Land, Liberties, and Lordship*, 365–367.

That territory also probably transmitted state of the art techniques still further farther east and north in Poland to the zone between Kraków and Sandomierz and the region of Kalisz-Sieradz, where production grew after 1550.[85]

So what socio-natural implications had this four century wave of local aquacultural innovation?

## 7.3 Aquaculture As Ecological Revolution

Fishponds fascinated foreign visitors. A follower of exiled Ottoman prince Jem (Çem Sultan), held in the mid-1480s as an honoured hostage or political pawn at Bourganeuf in Limousin amidst pond-filled Brenne, Berri, and Forez, told Turkish readers the French reared fish in lakes they created by damming watercourses and bringing in stock on pack horses. After four or five years the harvest was proclaimed, the pond drained, and people travelled some days to buy "pieces of gold of fish" by the thousands. Owners kept the best to repopulate the lake and sold the rest to town dwellers who had storage tanks to fatten the fish for later consumption.[86] Not quite a century later Venetian ambassador Giovanni Michiel reported the ponds of Bohemia so teemed with fish that they contributed a large share of the country's wealth.[87] The demand structure for this high-status food combined with natural and social production requirements to transform regional environmental relations in distinctive ways.

### *7.3.1 Demand: Live Fresh Fish for Inland Elites*

The segment of consumption demand which medieval aquaculture served shaped its character and structured its impacts. Fish farms provided live fresh fish for inland Europe's well-to-do. When the duke of Burgundy was in his titular principality, his pond masters delivered thousands of carp to his castles. When the ducal family went to Bruges in coastal Flanders, they ate fresh seafood, but a two-day journey inland put the pond fish back on their menu again.[88] Thirteenth-century English ponds delivered fresh bream and pike to the king and other

[85] Szczygielski, *Gospodarka stawowa*, 21–23.

[86] Vatin, "A propos de l'exotisme," 241–242; Vatin, "Pratiques agricoles en Limousin," 265, has the Turkish text and French translation. For context see Freely, *Jem Sultan*, 118–162.

[87] As cited in Montanari, *Culture of Food*, 81.

[88] Beck, "L'Approvisionnement en bourgogne ducale," 175–176; Hoffmann, "*Carpes*"; Sommé, "L'Alimentation quotidienne."

magnates as did those of the Polish king in 1420.[89] At Horn castle in Lower Austria, kitchen accounts from 1444 to 1446 show the resident von Puchaim family buying live carp during winter and Lent for up to twice what they paid for the ordinary 'river fish' eaten in other seasons.[90] When the ponds' product went on the market it was not cheap, even for fish. At Namur in 1356 a hundred carp for a wedding feast came at twice the price of a cow.[91] During the Council of Constance (1414–1418) a pound of carp cost the same as four pounds of beef or twenty loaves of bread.[92] Throughout this period fish from the ducal ponds in Burgundy sold as luxury items on regional markets. Late fourteenth-century carp averaged 1.5 *gros* and pike two to three, which matched the daily wages of pond workers (fishers, carpenters, earth movers). On the fifteenth-century market in Dijon, 100 carp went for four to twenty francs (48–240 *gros*) and single pike for 1.5–10.5 *gros* (30–210 *denier*), when a fat capon cost 3 *gros* (60 *d*) and a skinned rabbit 30*d*.[93]

Medieval fish farming acquired strong regional features. It belonged especially to interior Europe. From upper Poitou and the great arc of the middle Loire all the way to central Poland, zones famous for fish culture (Map 7.1) appeared inland of the roughly 150 kilometers across which medieval transport technology could safely haul fresh marine fish (compare Map 3.1). Resource endowments and consumption centers played further locational roles. Where large natural lakes offered ample and tasty competition, few fishpond enterprises endured.[94] Nor was all inland Christendom equally inundated. Areas distinguished for pond culture share impermeable soils and abundant water supplies. In Franconia, Burgundy, and Lorraine springs seep from discontinuities between porous and solid strata on valley slopes, while southern Bohemia, Dombes, Brenne, and Sologne rely on surface runoff and locally

[89] Steane, "Royal fishponds," 49–50; Roberts, "Bishop of Winchester's fishponds"; McDonnell, *Inland Fisheries*; Bond, "Monastic fisheries," 92–95; Currie, "Role of fishponds," 153–160. Compare Piekosiński, ed., *Rachunki dworu króla*, 545 *et passim*.

[90] Archiv Horn-Rosenburg, Hs. Horn 44 (used with thanks to Herbert Knittler). In contrast, barely fifteen years earlier high officials at the duke of Gelders' Rhine border station at Lobith ate no carp but during Lent enjoyed fresh sea and river fish as well as stockfish (Bosscha Erdbrink, ed., *Het 'Keuckenboeck' van Lobith*).

[91] Balon, "La pêche," 31. [92] Hitzbleck, *Bedeutung des Fisches*, 105–108.

[93] Beck, *Eaux et forêts*, 323–324. Likewise in late sixteenth-century Paris, noble households contracting in advance for bulk food deliveries paid thirty-two times more for foot-long carp than for a pound of beef. Couperie, "marchés de pourvoierie," 245–246. More examples of relatively expensive carp appeared in Chapter 2, p. 82 above.

[94] Artificial ponds built near Tegernsee in 1455 were soon abandoned because the brethren found the fish "too mossy" beside their familiar native whitefish and trout (Kisslinger, *Chronik*, 96).

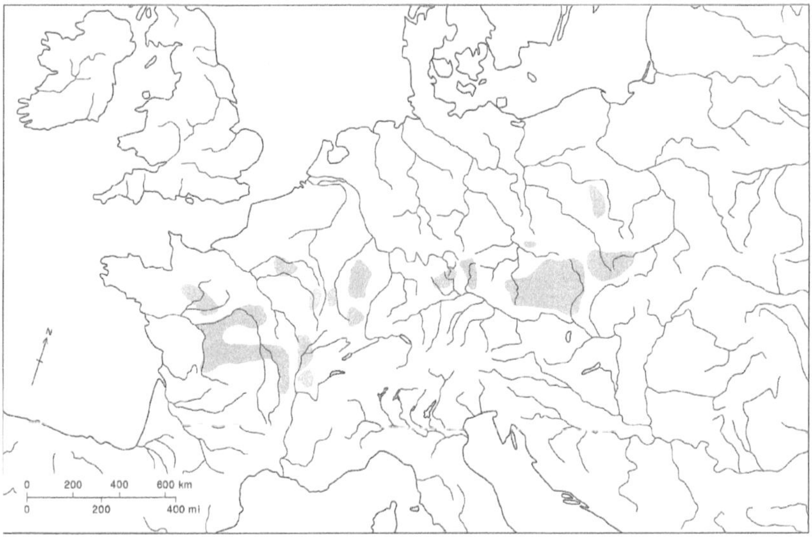


Map 7.1 Major regions of carp culture in late medieval Europe.

disordered drainage.[95] Natural potholes and intermittent wetlands made good sites for ponds. But for fish as for medieval commercial viticulture,[96] concentrated demand affected those suitable landscapes most accessible to it. Princes in Burgundy, Forez, and southern Bohemia and such monastic establishments as Citeaux, Maulbronn (Württemberg), and Halesowen (Worcestershire) created pond systems to supply their large domestic needs.[97] The papal household at Avignon sent special agents up the Rhône and Saône to buy and ship the output of Burgundian ponds.[98] Good-sized towns concentrated a well-off clientele: Lyon bought the fish of Dombes, Poitiers those of Brenne, and Brussels the carp grown in

[95] Devailly, *Le Berry*, 62–65; Guérin, *La vie rurale*, 131–135; Benoît, *Les étangs*, 22–24; Cnopf, *Entwicklung*, 23–26. The Lužnice basin around Třeboň is peat over impermeable clay.

[96] The great wine producers of medieval Europe were located, not in the best areas for the vine, but as close as possible (in terms of bulk transport) to thirsty northern consumers.

[97] Beck, "Pêche et étangs" and *Eaux et forêts*, 271–295; Hoffmann, "*Carpes pour le duc*"; Mattéoni, "La pêche des étangs"; Berthier, "Gestion des étangs" 2004 and "Gestion des étangs" 2006, 8–9; Grewe, "Wasserversorgung," 45–48; Bond, "Monastic fisheries," 98.

[98] Richard, "Commerce du poisson," 191; Richard, "Etangs et le commerce," and Richard, "Transports par eau," 40–44. Weiss, *Versorgung des päpstlichen Hofes*, 397–402, has the consumer's perspective.

periurban waters of the Senne.[99] Nürnberg, Bamberg, and other south German towns were supplied by Franconian ponds but also by those of southern Bohemia – insofar as the latter had surplus from the demand of local elites, Prague, and Vienna. Greatest consumer of all, Paris drew on the Seine basin (Valois, Champagne) and the Loire (Berry, Sologne). Medieval fish culture is associated with special resource endowments accessible to inland concentrations of wealth.[100]

### *7.3.2 Exercise of Elite Power*

Aquaculture was shaped and spread in medieval Europe by exercise of elite power, the same agent then contesting customary access to wild fisheries. Regional elites satisfied demand for fish by imposing a new aquatic regime on reluctant neighbours and subjects.

Fishponds were private property. Most belonged to lords of the land. They were expensive investments. Assembling the acreage for a finishing pond at Massileux in Forez cost 1,000 livres and as much more to build it. Work for the pond made at Chomutov, Bohemia, in 1408/9 cost two or three times the owner's annual receipts from an entire agricultural village.[101] The standard production system promised no returns for two to five years, and thereafter also continuing costs of operation and maintenance.[102] Nevertheless under the right conditions the returns could be lucrative. In 1520 the seigneur of Mèziéres, leading landholder in Brenne, sold to a local merchant the entire output from ponds Piégu and Picadon for 1,600 livres, more than twice what building the latter pond had cost in 1494/5.[103] In those same decades around the turn of the fifteenth/sixteenth centuries lords in southern Bohemia and Lower

[99] Benoît, *Etangs de Dombes*, 30; Benarrous, *Grande Brenne*, 244–250; Deligne, *Bruxelles et sa rivière*, 177–178, and "Carp in the city," 289–293.

[100] English experience validates by contrast the importance here ascribed to inland wealth. During the twelfth and thirteenth centuries state-of-the-art fish culture enterprises were there installed on manors supplying royal, magnate, and monastic household needs (Serjeantson and Woolgar, "Fish consumption," 104) but most fell into disuse and were abandoned during the fourteenth century, when consistent commercial supplies of fresh marine fish became competitive throughout the country (see Locker, *Role of Stored Fish*). Certainly no known late medieval or early modern English fish farm compared in scale with major continental enterprises. Why should they?

[101] Fournial, *Les villes et l'économie*, 691; Hemmerle, ed., *Deutschordens-Ballei Böhmen*, 93.

[102] Mercier, *Seigneurie de Jaucourt*, 77–80, closely follows expenses for upkeep of ponds in the late fourteenth century. Beck, *Eaux et forêts*, 329–340, thinks production costs a major factor in the mid-fifteenth-century closure of aquaculture on Burgundian ducal domains.

[103] Benarrous, *Grande Brenne*, 247.

Austria who had invested heavily in pond construction also enjoyed very high returns.[104] Dubravius attributed to Vilém of Pernštejn (1449–1518), who had created the large aquaculture enterprises at Hluboka and Pardubice, a boast that all his great wealth flowed from his ponds.[105]

Yet beyond the prospect of domestic fish supplies and returns on sales, men of power built fishponds to show off their social superiority. Ponds were an essential feature of the estate park Robert II, Count of Artois, created at Hesdin in the 1290s to display by artifice and exoticism the gulf setting him above all lesser folk. Here the regular production cycle was rescheduled to entertain a royal visit and drainage postponed lest a bare and muddy expanse offend the eye of the countess. Robert's daughter and heir, Mahaut (1302–29) reveled in the luxury of eating fish from her own estate and watching her people work on the pond.[106] The symbolic function of ponds in Brenne confined their construction to seigneurs with high justice and the right of warren, a privilege still being enforced in the fifteenth century. Size mattered, too: in 1455 the Seigneur de Mèziéres began constructing what he called 'la Grant Mer de Brenne'.[107] His contemporary, John Howard, first Duke of Norfolk, paid close personal attention to the management of his ponds, even though he sold none of their output.[108]

Wherever the local innovators and owners of aquaculture enterprises are collectively identifiable, they prove to be a socially representative sample of regional landholders and not especially the monks of modern myth. In Dombes, for instance, lay owners of fishponds always outnumbered clerics by more than two to one and were in all centuries after the

[104] Knittler, *Nutzen – Renten – Erträge*, 156–165, and "Teiche als Konjunkturbarometer?" 209–213; Čechura, *Adelige Grundherrn*, 44–47 and 99–103; Búžek, "Goldene Zeitalter," 91–92.

[105] Dubravius, *De piscinis*, lib. 1: cap. 3 (Schmidtová, ed., p. 24).

[106] Farmer, "Power and the 'natural' landscape," 659–662; Dowling, "Landscape of luxury," 375–379. Further on parks and ponds as expressions of aristocratic power see Creighton, *Designs upon the Land*, notably 122–166.

[107] Benarrous, *Grande Brenne*, 129 and 152–153. The duc de Berry had fishers netting the pond at his Château of Dourdan depicted in the April page of his *Trés riches Heures* (1411/16?).

[108] Turner, ed., *Manners and Household Expenses*, 560–564. Ponds were clearly a status symbol in late medieval England (Woolgar, *Great Household*, 68–70), but these records smack of self-indulgence, as also those of early Tudor churchmen (Hickling, "Prior More's fishponds," criticized in Currie, "Role of fishponds," 158–159), and eventually of well-known Elizabethan enthusiasm for imitation and experiment (as shown in the 1599 translation of Dubravius and in John Taverner's *Certain Experiments Concerning Fish and Frvite* from 1600).

thirteenth a majority among known builders of new ponds.[109] Surviving German documents likewise first associate ponds with religious houses and prelates, but from the 1200s noble landowners and both town citizens and corporations came to the fore.[110] Fourteenth-century Czech chronicler Beneš of Veitmíle was fully aware that the monarch, Charles IV, personally promoted pond construction and was then emulated by high aristocrats, lesser nobles, churchmen, and commoners alike.[111] A century later the post-Hussite burst of activity began with middling barons like Vilém of Pernštejn at Pardubice and Hluboka nad Vltavam and quasi-royal potentates like the Rožmberks around Český Krumlóv and Třeboň, but also attracted municipalities such as Česky Budejowice and Vodňany. Charles, at least, reputedly made plain his intent to transform landscapes, ordering construction of fishponds "to drain water from swampy ground and for the sun and warm winds to evaporate water from stagnant areas, which will benefit plants thereabouts."[112]

Waters manipulated for lords and their fish threatened the interests of others. In a litany of power, tropes of conflict and compensation run everywhere through detailed records. In mid-twelfth-century Berry, local seigneur Hervé Guitier took offence when Knights Templar of Villefranche made their dam so high they flooded the houses and roadway of Hervé's men, while the cathedral canons of St. Étienne at Bourges complained to Count William of Sancerre that his dam at Beaulieu had drowned their land.[113] Across the Channel a generation later Abbot Samson of Bury St. Edmunds stirred up resentment in and outside the

[109] Benoît, *Étangs de la Dombes*, 22–24; the same ratio of two lay to one clerical describes the thirty known builders of twenty-six new ponds in fourteenth–fifteenth-century Sologne (Guérin, *La vie rurale*, 157–160). While into the 1200s written records cover churches more fully than they do lay estates, from the start in the eleventh–twelfth centuries lay seigneurs in Berry led innovations in pond construction and management (Querrien, "Pêche et consummation" 2003, 412–418 and 434–435). Judicial and other records from Brenne likewise show religious and lay landowners engaged with ponds each in proportion to their holdings in the area (Benarrous, *Grande Brenne*, 153–202).

[110] Lampen, *Fischerei und Fischhandel*, 130–138.

[111] Beneš of Veitmíle, "Chronica ecclesia Pragensis," 516. Šmelhaus, "Vývoj nížinného rybničního hospodářství" tabulated lay and ecclesiastical pond owners in fourteenth-century eastern Bohemia.

[112] Instructions to the Czech estates in 1356, cited in Boháč, "Historical ecological aspects," 25. Compare Šusta, *Fünf Jahrhunderte*, 2, on Charles building ponds "*ut regnum nostrum piscibus et vaporibus abundaret.*"

[113] Devailly, *Le Berry*, 296 and 361. Blary, *domaine de Châalis*, 56, and Maas, *Moines-défricheurs*, 75, have comparable tales from Valois and Champagne. Such conflicts continued: the collegial church of Levroux and the lords of Moulins reached an agreement in 1369 but were at odds again in 1515–1518 over damages the latter's Marmagne pond on the river Céphons did to land and a mill belonging to the chapter (Querrien, "Pêche et consummation" 2003, 416); in Brenne the pond called Cinq Bondes of the Hospitaller Commandery of Blizon stood dry for more than a century

abbey when he raised the dam of his fishpond at Babwell and inundated meadows, pastures, orchards, and arable of neighbours and other monks alike.[114] In upper Saxony the abbey of Altzelle was continually at odds with its small-town neighbours in Roßwein, whose lands Altzelle flooded; with a convent at Nimbschen, whose woodland Altzelle drowned raising a pond in 1495; and with another nearby town, Grimma, whose own pond washed out the abbey's mill.[115] When Ješek of Kosovy Hory, lord of Lomnice, built Dvořiště pond in 1367, the waters drove peasants off more than 300 ha of arable. By the early 1370s more places in southern Bohemia were described as "drowned by the pond". New constructions around Krumlóv in 1479–1484 inundated one-fourth of the estate's own village Radošovic – where the tenants received an equivalent rent reduction – and all of neighbouring Hummo – for which the Cistercians of Vyšší Brod got two villages in return.[116]

Where aquaculture became a major interest, normal practice encouraged entrepreneurs to flood first and deal with local objections later. Unauthorized constructions were confirmed retroactively, not removed. Seigneurs in much of central and southeastern France enjoyed a recognized customary *droit d'inonder* (*d'inondation*): they could build dams on their own land at will and then compensate neighbours for what the pond flooded, "for the construction of such a pond provides more benefit to the builder and the community than the inconvenience of neighbours whose properties are drowned in the flooding of the water."[117] The legal

because it had inundated the place where the abbey of Fontgombault meant to install their own pond (Benarrous, *Grande Brenne*, 199–200).

114 Jocelin of Brakelond, *Chronicle*, tr. Butler, 130–131. Carpenter, *Minority of Henry III*, 357, relates a similar story from chronicler Matthew Paris.

115 Beyer, *Cistercienser-Stift Alt-Zelle*, 176, 183–184, 420, 422, and 703–704.

116 Ofczarek, *Teichwirtschaft*, 35–37 and 44–45. Boháč, "Historical-ecological aspects," 44–45, has more such Czech cases. Dubravius, *De piscinis*, advised tenants be compensated rather than deprived of resources or simply expelled.

117 "*dummodo etiam ex constructione talis stagni afferatur majus commodum construi facienti et reipublicae, quam sit incommoditas vicinorum quorum proprietates ex inundatione aquae submerguntur*" (Durand, "De l'établissement des étangs," 105–107, quotes the "Decisiones Gratianopolitanae", questio 91, from the Dauphiné). Perceveaux, "Structures et relations économiques," 348, describes the "Coutume de Villars" of Dombes as parallel to those in Forez and Nivernais; Benoît, *Les étangs*, 64, adds Berry, Orleans, Lorris, and Montargis. The area of Brenne spread across three customary codes, with Touraine most favouring the arbitrary power of a seigneur and Berry and Poitou more open to side deals or licences (Benarrous, *Grande Brenne*, 147–153). French precedents thus suggest Jan z Rožmberk did peasants at Bugóv no special favour by letting them pasture livestock on the bottom of his pond when it was dry (Ofczarek, *Teichwirtschaft*, 29). No such right existed in England: by the late 1100s law, legal commentary, and evident practice there concurred that flooding without prior consent was a *novel disseisin* nuisance and required the perpetrator to restore prior conditions (Langdon, *Mills*, 259–268).

regime divided ownership of the waters and fish from that of the flooded land. Barring other compensation involuntary victims thus retained only the crop from years the pond was dry and the right to collect shoreline vegetation. But cases from thirteenth- and fourteenth-century Dombes, for example, show pond builders offering neighbours cash, land, or a share in the yield of fish.[118]

Some flooded Czech peasants were compensated by their lords, but practicing aquaculture at peasant expense was also one way lords wielded authority in east central Europe around 1500. To take but one example from a court book of the Kraków cathedral chapter, villagers at Choczianowice who complained in 1502 that the lord's fishpond "frequently damages their fields and floods their meadows," blamed the local administrator for having raised the dam and, in preparation for another pond, cut down the local woodlot.[119] On another lordship near Sandomierz already in 1405 the lord's steward was conceded "the right to make fishponds wherever he wishes".[120] Protest was vain. Peasants undoubtedly suffered when their superiors turned land into water. As one disgruntled Czech put it in 1508, "what was left after wars, fires, and plague is now mostly inundated by ponds".[121]

### 7.3.3 *Adaptive Economic Structures*

Depending on surrounding economic institutions, medieval fishpond enterprises operated in the customary economy, the market economy, or in east-central Europe, the newly coercive economy of neoserfdom. These cultural contexts shaped the supply of labour, but not the need for expertise, and constrained a lord's choice among indirect subsistence, market sales, or ending direct management in favour of leasing out individual ponds or even entire enterprises.

Whatever the setting, fish culture called for both heavy labour and, on the part of managers, high technical skill and environmental knowledge. Around 1300, tenants at Gressenhall manor in Norfolk were obliged to perform labour services '*facient stagnum*' or, in the 1315 English-language survey, 'damyng'. At the same time their counterparts on the count of Bar's Lorraine estates did compulsory service carting fry to stock

118 Benoît, *Les étangs*, 59–66. Compare Helmlinger, *La Dombes*, 25–26, and Benarrous, *Grande Brenne*, 151–152.

119 Ulanowski, ed., *Księgi sądowe wiejskie*, vol. II, pp. 430–432. Other instances of lords' fishponds harming peasant resources are *ibid.*, 418, 421, 425, and 469–472; Górzynski, *Zarys historii rybołówstwa*, 48–49; and Hurt, *Dějiny rybnikařství*, 104–108.

120 "*ius faciendi piscinas ubi placet*" (Ludat, *Lubuser Stiftsregister*, 71).

121 Boháč, "Historical-ecological aspects," 45.

ponds.[122] Workers were later turned out to harvest fish from ponds for Ebrach, a monastery in Upper Franconia, as a thoroughly old-fashioned obligation on certain peasant tenements, limited to the specific task and with equally customary remuneration of meals and two fish.[123]

While expert monastic managers are curiously hard to identify,[124] a long catalog could be assembled of lay specialists, some employed on limited-term contract, others as long-time, high-ranking salaried servants with titles like 'pond master' or 'fish master'. We have already met Štępánek Netolický building the Zlata Stoka at Třeboň for the Rožmberks in 1516. Having learnt the craft in years assisting his predecessor, during his own fifteen-year tenure he designed and supervised construction of nine large and thirty-seven small ponds, traveled to advise the archbishop of Salzburg and the count of Salm, and for one tough engineering problem called in as consultant his counterpart from the Poděbrady estate. At a time when contract pond masters in Bohemia were getting 3 grosz a day the Rožmberks were paying Netolický 2,400 grosz a year. His retirement house still stands in a suburban part of Třěboň.[125] But such experts can be tracked just about as far back as detailed records of fish culture go. A team run by the clerk Colard de Coilly and bailiff Hugues Bernard stocked, harvested, and maintained ponds for the count of Champagne in 1258–1259 and 'Master Nicholas the Fisherman' toured southern England for at least eighteen years (1244–1262) for sake of the bishop of Winchester's ponds. During the 1330s–60s Reynaud le Tarroillon had charge of work on many ducal ponds in Burgundy.[126]

Wage labourers evidently did much of the physical work to multiply and maintain fishponds in key areas of late medieval France. Reynaud and his successors in ducal Burgundy recruited locally for construction and for harvesting the fish, but eventually came to rely on regional pools of workers. The duke's officers paid by the task or a daily rate, food and

[122] Turville-Petre, "Earliest English manorial survey," 75; Collin, "Les ressources alimentaires," 64–65.

[123] Weiss, *Zisterzienserabtei Ebrach*, 77–78. Customary tenants on a Polish estate of the bishop of Lebus in 1405 likewise owed two days a year of work on the fishpond (Ludat, *Lubuser Stiftsregister*, 46).

[124] Berthier, "Gestion des étangs" 2006, 286–287, found a layman managing fish culture at late fourteenth-century Citeaux.

[125] Netolický was one in a succession of trained pond masters who worked for the Rožmberks from the third quarter of the fifteenth century through the early seventeenth: Šusta, *Fünf Jahrhunderte*, 3–15; Ofczarek, *Teichwirtschaft*, 52–87; Bůžek, "Goldene Zeitalter," 82–85; Čechura, *Adelige Grundherrn*, 99–103.

[126] Longnon, ed., *Documents*, III: 17–21; Roberts, "Bishop of Winchester's fishponds," 130–135; Hoffmann, "*Carpes pour le duc*," 42–43; Beck, *Eaux et forêt*, 281–285. Supplement 7.3.3 notes more expert builders and managers elsewhere.

lodging plus 12 denier to 1¼ gros in the fourteenth century and 2 to 3 gros in 1450.[127] The same daily rate then went also to the local peasants whom Citeaux abbey hired for fishing its ponds, while using permanent servants and independent craftsmen for more skilled tasks.[128] Czech pond-builders worked for a wage, too: one grosz a day in the late 1300s and as much as two grosz a century later. But by that time forced labour service was becoming a norm in east-central Europe, also for heavy fishpond work. Account books for estates of Teutonic Knights and Hospitallers at Jindřichův Hradec often record forced labour (*robot*) on the ponds.[129] A survey of the Kraków chapter estates in 1531 made this obligation plain for villagers of Dłotów: "furthermore, the aforesaid men are bound to work at the fishing in the fishpond, however often it is demanded of them, and likewise on the repair of the dam of the fishpond and hauling the fish to the manor house at Pabianice, however often it is demanded of them."[130]

Cultured carp and other fish from the ponds might likewise be delivered in kind to the lord's household – a more dependable form of the indirect subsistence fisheries familiar in the earlier middle ages – or sold by auction or prior contract to merchants from town. Like elite claims on wild catches (see Chapter 3) primary household consumption often coincided with reliance on servant and obligatory peasant labour during early regional development of aquaculture. But relict practices persisted. Recorded shipments from thirteenth-century English ponds went almost exclusively to the king's, bishop's, or lord's residence, while during 1345, with the duchess of Burgundy dwelling in Argilly castle, more than 11,000 large carp arrived there from the St. Seine pond. Much later in 1522 the Polish king's ponds at Oświęcim put 32,640 carp and 6,045 pike into their master's kitchen.[131] As with wild fisheries, monastic establishments with large and localized demand for fish long retained this

[127] Beck, *Eaux et forêt*, 283–285. For like hiring in Sologne see Guerin, *La vie rurale*, 134–135.

[128] Berthier, "Gestion des étangs" 2006, 283–287.

[129] Ofczarek, "Teichwirtschaft," 18–22.

[130] Ulanowski, ed., *Księgi sądowe wiejskie*, vol. II, p. 472: "*praeterea fassi sunt homines praedicti teneri ad piscandum in piscina, quoties illis mandatur, similiter ad reformandum piscinae aggerem et ad deferendum pisces ad curiam Pabyanicze, quoties illis mandatur* ..." For Czech cases see Čechura, "Lohn oder Fron?" and Búžek, "Goldene Zeitalter," 87–89. Górzyński, *Zarys historii rybołówstwa*, 50; Hurt, *Dějiny rybnikařství*, 124–128; Szczygielski, *Gospodarka stawowa*, 150–164; and Míka, "České rybnikářství" all think this normal practice. The same applied at Altzelle (Beyer, *Cistercienser-Stift Alt-Zelle*, 428–429).

[131] Hoffmann, "Carpes pour le duc," 39–40; Rybarski, *Gospodarstwo Księstwa Oświęcimskiego*, 75. For distribution of many carp from enterprises elsewhere or later see Beck, "L'Approvisionement" and Supplement 7.3.3.

autarchic strategy: fifteenth-century monks at Citeaux ate nearly all of their own production and so did those at Heilsbronn and Chaalis. But lay lords of ponds in late medieval Brie Champenoise also favoured direct household consumption.[132]

Commercial sales of cultured fish are apparent in western Europe before 1300 and thereafter gained importance and spread eastwards. Coincident with the first artificial ponds in Berry, bream appear in 1100 as the only obligate freshwater fish in market regulations for Bourges.[133] Later ponds sent mainly carp and pike. Financial accounts kept under Philip IV for the nine royal pond complexes within 100 kilometers of Paris show varying divisions between supplying the court and markets: over a three-year period the western ponds run by Jean Harenger [!] gained 61 percent of their income from Parisian merchants. Managerial expectations are suggested by accounting for fish transferred to the king's *hotel* or other royal households in the same way as for sales elsewhere.[134] Every three to five years between 1386 and 1422 one pond at Saix in Dombes yielded daily sales in quantities like 200 carp, 515 carp, 975 carp and 12 pike, 650 carp, etc.[135] Carp from Burgundy supplied the papal court in Avignon and towns along the way. One sale at l'Eperviére-sur-Saône handled 3,400 carp and 190 pike, representing about 60 percent of that pond's harvest for 1353.[136] Historian Corinne Beck found in the annual accounts of Burgundian castellans a change of priority in the 1390s: henceforth most ducal ponds were managed for commercial production and sale, even of those fish designated for the court but left unused. This coincided with the duke's diminished presence in his principality.[137]

A mixed model of domestic consumption and market sales became the norm for central European pond enterprises from perhaps the fourteenth and indubitably the fifteenth century. Quantitative evidence lags. Dubravius had Vilém of Pernštejn brag that he made more from sale of fish (alone) than the entire wealth of ancient Roman senators; his Hluboka estate shipped to Prague and over the border hills to Bavaria

[132] Berthier, "Gestion des étangs," 288–289; Blary, *Domaine de Châalis*, 31–40, 46–81, 87–99; Cnopf, *Entwicklung der Teichwirtschaft*, 29–32; Heidacher, *Entstehungs- und Wirtschaftsgeschichte*, 118–119 and 131; Bauchet-Cubbada, "Patrimoine piscicole des seigneurs."

[133] Querrien, "Pêche et consummation" (2003), 427.

[134] Rouillard and Maupoune, "Etangs royaux." [135] Benoît, *Les Étangs*, 67–68.

[136] Richard, "Le commerce du poisson"; Richard, "Les Étangs"; and Richard "Transports par eau." At the receiving end see Weiss, *Versorgung des päpstlichen Hofes*, 319–320 and 397–402.

[137] Beck, *Eaux et forêts*, 291–295.

and Austria.[138] As early as 1522 Silesian noble landowners around Wrocław were complaining to their king Louis of Bohemia that city officials interfered with "our fish on the fish market... which we ship there" (*unseren Fyschen vff dem Fyschmargkt ... dy wir hynein furen...*). In 1534 the Polish king's Oświęcim ponds sold 26,100 carp, only 20 percent less than went to the royal household.[139] Land tax assessments from Lower Austria a generation later do indicate some lordships getting up to a quarter of their annual cash returns from sales of cultured carp.[140] Historians of early modern east-central Europe equate fish culture with more widespread local brewing monopolies and commercialized production of cereals for export sales as examples of 'landowner capitalism' or 'market feudalism'.[141]

Market conditions writ large thus affected the economic viability and even survival of regional aquacultures. Carp farming appeared and expanded where inland elites with access to land, water, and cheap labour demanded fresh fish for their own consumption and became aware of urban markets willing to pay for any surplus. When these conditions failed, princes and other large estate holders phased out their own participation and administrative documentation. Where the ponds then survived, their operation and management become more difficult to track.

The inland principalities of Brabant and Hainaut, centred around Brussels, provide a case in point. Aquaculture on ducal and other estates there expanded dynamically from the late thirteenth into the mid-fourteenth century. Elite investment met landowner needs and local market opportunities for fresh fish. Strong commercial involvement – investment by urban merchants, multi-purpose use of moats, specialized fry and storage ponds, markets in carp fry, etc. – continued for another generation or two. But about 1400 big pond owners were experiencing

138 Dubravius, *De piscinis*, lib. 1, c. 3 (ed. Schmidtova, 23–24); Čechura, *Adlige Grundherrn*, 44–47 and 125. Annual sales from the Třeboň estate during the last decade of Netolický's management averaged 118,520 carp (Čechura, *Adlige Grundherrn*, 44–46 and 99–103; Bůžek, "Goldene Zeitalter," 121–124). Kunst and Galik, "Essen und Fasten," 252–253, find carp from Bohemia on Vienna's fish market.

139 Klose, *Inneren Verhältnisse*, 32; Rybarski, *Gospodarstwo Księstwa Oświęcimskiego*, 76. Compare the estimates in Nyrek, *Gospodarka rybna*, 40–43, or Szczygielski, *Gospodarka stawowa*, 41–47 and 220–247, but note that in their regions of Poland counts of fish sold (in contrast to those stocked) occur only in later records.

140 Knittler, *Nutzen – Renten – Erträge*, 162–166.

141 An unresolved interpretive debate in Czech historiography can be traced through Míka, "České rybnikářství"; Válka, *Le grand domaine feodal*; Boháč, "Historical-ecological aspects," 45–48; Čechura, "Lohn oder Fron?"; and Čechura, *Adelige Grundherrn*. Elsewhere see Szczygielski, *Z dziejów gospodarki rybnej*, 52–63, and Knittler, "Teiche als Konjunkturbarometer?".

labour shortages, rising labour costs, and a greater supply of marine fishes competing for consumer choice. Burgundian succession in 1430/32 deprived most towns of resident courts and their business. By about 1500 large lay landowners, especially around Brussels, could supply their tables from the market and had shut down or leased out their ponds. Marine fish supplanted freshwater varieties in regional diets. Many ponds reverted to marsh or meadow. Fish culture became an affair of religious houses and poorly documented professional urban fishmongers.[142]

While leasing out arable demesne lands is a familiar trend across most of high and late medieval western Europe, short- or long-term leases of aquaculture enterprises appear related to establishment of stable market supplies of fish accessible to consuming households. Costs and reliability of labour for stocking, harvest, and maintenance also played a role. Leasing cut the landowner's administrative expenses, lowered risk, and promised a reasonably steady return on investments. The latter might or might not include continued supply of fish for household needs. Yet how much, if at all, did leasing out of big pond systems actually change the production system? Comparatively small private ponds had likely existed all along and eighteenth-century records reflect practices very like those of the fourteenth.

In certain interior parts of France late medieval economic and political changes eroded the importance of large princely fish farms, leaving the sector a thinly recorded commercially oriented activity of lesser seigneurs and urban-based dealers in fish. In Forez the comital enterprise lost its raison d'etre with the 1430 move of the family to Moulins, 150 kilometers and a different drainage basin from the ponds. The management office atrophied and in a generation all the count's ponds were out on lease. Urban interest in aquaculture already dated back a generation or more. In the 1480s an influential merchant of Montbrison, Pierre Cochard, held a nine-year lease on the main water bodies.[143] In ducal Burgundy by the late 1300s the resource office (*gruerie*) faced rising logistical problems – labour costs, shortage of *nourriens*, unstable weather, erratic output – and ever more frequent absence of the princely household. Some individual ponds were out on lease before 1400. In 1442/3 Duke Philip let the entire resource base go for twelve years to a private

[142] Deligne, *Brussels et sa rivière*, 136–174 and 177–178, and "Carp in the city," 289–293 and 301–302. A similar mix of large ecclesiastical foundations consuming the production of their own domains and small commercial enterprises also developed around fifteenth-century Paris (Benoît, "La pêche dans le domaine de la ville").

[143] Fournial, *Les villes et l'économie*, 194–195 and 691–692; Durand, "De l'établissement des étangs," 104–105.

consortium of ducal officials. When the term expired the lessees had lost money and failed to maintain fish stocks. From 1454 then, ponds went out in smaller units for longer periods to groups of local urban merchants and lesser seigneurs with earlier ill-documented experience in local fisheries and markets.[144] Brenne, in contrast, was a land of smaller seigneuries (and hence no surviving financial accounts) where a slow fourteenth-century increase in pond leases became more general after 1400. Most were taken up by town-based partnerships which by that century's end even invested in new pond construction. The student of Brenne's pond-filled landscape, Renaud Benarrous, even refers to a 'democratization' of pond ownership.[145]

Institutional variability at the ownership level and even in the disposition of the fish produced likely more affected contemporary and surviving documentation than it did material interaction between a human technology and the things and forces of nature it sought to manage. Once technologies for fish culture and markets for its product had been established, changes at the managerial level little affected the broader consequences. Despite differences in social stratification the landscape of Třeboň replicated the *paysage piscicole* of Brenne and human relations with those landscapes had close similarities.

### *7.3.4 Colonized Ecosystems*

Hence technically advanced aquaculture could meet elite demand for fresh fish in late medieval interior Europe along a spectrum from old-fashioned seigneurial self-sufficiency to up-to-date commercialization. At all points, however, this response to wants backed up by power and wealth set off localized and small-scale but numerous and widespread 'ecological revolutions'.[146] Medieval programs and work to colonize nature for fishponds drove environmental change at numerous inland socio-natural sites. By transforming ecosystems aquaculture produced winners, losers, and changes in human experience.

144 Beck, *Eaux et forêts*, 308, 315–317, and 340–344. *Gruerie* officials in adjacent Franche-Comté saw their lord even less and had many comital ponds out at lease by 1450 or simply gone derelict even before the destructive French conquest of 1478/80 (Gresser, *Pêche et pisciculture*, 151–167 and 323–335).

145 Benarrous, *Grande Brenne*, 224–226 and 344–358. Aquaculture in Dombes and Sologne likewise retained into early modern times more diverse ownership structures and strong links to urban markets (Benoît, *Étangs de la Dombes*, 30–33; Guérin, *La vie rurale*, 148–150).

146 The term is used advisedly to stress the scope and ramifications of "transformations in relations of humans with non-human nature" commensurate with the now-classic model advanced by Merchant, "The theoretical structure."

Medieval fish culture purposely furthered and locally intensified the changes medieval economic development in general was perpetrating on European watercourses (recall Chapter 5). Investors aimed to multiply formerly uncommon eutrophic stillwater habitats. Ponds sustained ecosystems unlike those of earlier running streams or seasonal wetlands. Stillwater organisms – rooted water plants, midges, mosquito larvae, perch, pike, roach, tench, bream – gained living space and notably those of moving water – gravel-loving invertebrates, gudgeon, grayling, trout, barbel, various migratory fishes – lost. The scale of transformation is impressive, for the fish culture regions acquired vast numbers of ponds. Careful local experts concur on the order of magnitude: 25,000 ponds in Bohemia and 22,000 in upper Franconia; 25,000 hectares of ponded surface in upper Silesia; 40,000 hectares in central France.[147] With ample evidence of modern drainage of medieval ponds in areas as far apart as France and Poland, it is hard to dispute Robert Delatouche's guess that the historic province of Maine, for instance, had contained at the end of the Middle Ages twenty-five times more ponds than 500 years later.[148]

Principal intended immediate beneficiary of this transformation of aquatic habitat was the carp. Fish culture meant the large-scale human introduction of an exotic animal across large areas of western Europe where it had not previously lived. Just the surviving written records from the thirteenth and fourteenth centuries count hundreds of thousands, perhaps millions, of small carp going into ponds. Pond masters saw to their food supply, protected them from predators, and arranged their reproduction.[149] As objects of controlled monoculture these fish were becoming domesticates. Select domesticated races of carp certainly had emerged in Europe by the sixteenth century.[150] Yet more telling, physiological features symptomatic of domestication remain visible in all wild-living European carp populations west of the middle Danube.[151] In other

[147] Andreska, "Development of fish-pond culture"; Nyrek, *Gospodarka rybna*, 42–43; Bautier, *Economic Development*, 198–199; Cnopf, *Entwicklung*, 22. In the early twenty-first century water still covers more than 8,000 ha in each of the southern Czech Republic, Aischgrund, Upper Lusatia, and Brenne (Benarrous, *Grande Brenne*, 61–62).

[148] Delatouche, "Le poisson d'eau douce," 173–175; Szczygielski, *Gospodarka stawowa*, 19. Compare Benoît, *Étangs de Dombes*, 12–13.

[149] See, for examples, Benarrous, *Grande Brenne*, 227–230, and Beck, *Eaux et forêts*, 266–269.

[150] Galik, "Historical and ichthyological evidence," found scales of mirror carp in fifteenth-century layers at Lanzenkirchen castle, Lower Austria, a site destroyed in the sixteenth century.

[151] Balon, "The common carp"; Ervynck and Van Neer, "Preliminary survey"; Boddeke, *Vissen en vissen*, 124–130.

words, even the 'wild' carp of central and western Europe are feral descendants of domesticates. Given a start, carp themselves became a dynamic natural element, colonizing wild habitats so promptly that astute medieval observers soon recognized them in nature. Both late thirteenth-century French fisheries regulations and the first Polish writer on fishes, Stefan Falimirz in 1534, grouped carp among wild fishes of the river as well as cultivated ponds.[152]

Together and separately the ponds and the carp worked wider change on the larger environment. Some of this followed an intentional cultural program: like other agroecosystems aquaculture is meant to control and simplify environmental variables so biomass concentrated in one desired species. But then this exotic animal 'went native' with what are now predictable – if historically undocumentable – consequences for the aquatic ecosystems receiving it.[153] Biologists watching the carp spread into new continents during the late nineteenth century found it unusually quick to colonize temperate still-water environments – which medieval economic development was proliferating – and to disrupt relatively simple fish populations which were already under stress from overfishing and environmental change – like the native cold-water fishes of medieval Europe. In wild ecosystems carp behaviour harmed native species by accelerating changes to habitat and biomass.[154] Had French king Philip IV, for instance, a freshwater research station, historians could now hope for more than just proliferating documentary references to carp.

More visible is the environmental impact of the ponds themselves. Medieval records let modern scholars see the eastern edge of the Paris basin (Champagne–Lorraine) and the valley lands of Sologne and Brenne transformed from seasonally flooded woods to arable and ponds, permanently open water where there had been little or none. Shaded springs and trickling streams became sunny or mist-shrouded expanses of water and reeds. Czech archaeologists provide copious evidence of

[152] Lespinasse and Bonnardot, eds., *Les métiers et corporations*, 212–218; Rostafiński, *Średniowieczna historia naturalna*, vol. 1, pp. 70–71.

[153] "Successful establishment of an exotic species must necessarily precipitate changes in the physical and biological characteristics of the aquatic ecosystem receiving the introduction ... That no effects should result from such perturbations strains one's confidence in ecological principles" (Taylor et al., "Known impacts of exotic fishes," 323 and 352).

[154] Wheeler, *Fishes of the British Isles*, 178–179, summarizes carp biology, and more is in Heuschmann, *Die Weißfische*, 53–63, and the discussion and references provided by Kottelat and Freyhof, *Handbook*, 146–148. Welcomme, "International transfers of inland fish species," 33–36, reviews exotic fish introductions; Taylor et al., "Known impacts of exotic fishes," classify ecological effects (324–326) and catalog those of carp (335, 336, 342, 345, and 349).

how late fifteenth century pond construction changed drainage and vegetation patterns in southern Moravia. Around Třeboň thousands of hectares of heath drowned beneath the Rožmberk ponds alone.[155] Like other aspects of medieval economic development this multiplied eutrophic still-water habitats in regions where they had been rare. Pond ecosystems differed from those of streams or seasonal wetlands. Inhabitants of still water gained living space and those of moving water lost. Organisms adapted to slow annual oscillation of water levels now encountered flooding year-round, occasionally punctuated by the artificial drought of a pond drained for harvest. Silt accumulated behind dams on the river Allier.[156] In Brenne the ponds themselves raised the water table and fed newly emergent wetlands.[157]

Charles IV himself reportedly testified (p. 300 above) to the ponds' humidifying effect on microclimates, but this minimizes the change. Landscapes had been simplified and reordered at large scale. Ten percent of the surface in five rural communes of central Brenne is still submerged and a late seventeenth century map shows water covering more than 25 percent of Třeboň lordship.[158] Folk culture in Brenne comprehended the change by populating the ponds with beings inimical to humankind: the *Grande Bissexte* reared its upper body out of the water to seize unwary humans, drag them down, and devour them; a white doe chased and drowned those who attempted a nocturnal crossing along the dam of Mer Rouge, the region's largest impoundment.[159] Later more enlightened observers would deplore the insalubrious damp of the pond-filled regions. Dombes, Sologne, and Forez in particular had by the nineteenth century well-deserved malarial reputations, a disease later learned to spread not by 'bad air' but by mosquitos bred in standing water.[160]

Lakes and ponds are, moreover, ephemeral landscape features always fated to fill with sediment and revert to dry land. Like all artificial

155 Maas, *Moines-défricheurs*, 78, sees a 'symbiosis'; Devailly, *Le Berry*, 567, eschews progressive claims. Petřík et al., "Rybník Jako Součást Hospodářství Vrchnostenského … a Indikátor Podoby Krajiny."

156 Defosse, "Pêche et pêcheries." In the 1440s workmen had to remove more than a meter of silt from three small ponds on the manor of Baddesley Clinton, Warwickshire (Aston and Bond, "Warwickshire fishponds," 429).

157 Benarrous, *Grande Brenne*, 318–326, emphasizes the absence of *marais* before the fifteenth century.

158 Coulon, "Étangs de la Brenne," 15; Andreska, *Lesk a Sláva Českého Rybářství*, 74.

159 de La Véronne, *La Brenne*, 61–65.

160 Grand and Delatouche, *L'Agriculture*, 540; Perceveaux, "Essai sur l'origine des étangs" and "Structures et relations économiques"; Durand, "De l'établissement des étangs"; Braudel, *Identity of France*, 204–206; Guérin, *La vie rurale*, 131.

impoundments, each pond had to be maneuvered between that risk of deposition and normal biological succession on the one hand and the opposite hazard of erosion downcutting the barrier dam to restore running water. Without continual upkeep dams crumbled, sluices decayed, vegetation encroached, and many artificial fishponds ceased to be functional still-water ecosystems. Annual maintenance and decade-scale rejuvenation projects permeate Dubravius's list of necessary tasks and likewise records of practice since those of Philip the Fair's Norman ponds.[161] Managers also knew that organically enriched water from densely stocked ponds could inhibit growth of fishes downstream, and set up their water systems to reduce that threat to their crop. Even more than agriculture in general, lacking regular inputs of labour and materials, aquaculture was dubiously sustainable.

Aquaculture transformed human experience of the natural world. After 1292 residents of Hesdin, Artois, no longer lived among suburban fields and meadows but saw or had to work in ponds and woodlands as key elements of a landscape created for luxury display. Across Brenne the river Claise and other streams with vegetated banks where Merovingian St. Cyran (Sigiramnus) had sheltered to fish and high medieval countrymen pastured their livestock,[162] sank beneath permanent water bodies in the rising waves of the late fourteenth century and then the 1450s–1550s. The seasonal wetlands along the Lužnice where in 1379 men of Třeboň still paid for licence to fish,[163] began to vanish. A century later it took only one or two generations to turn the valley into a network of more than fifty interconnected artificial ponds, some small, some immense, and a share of which were at any given time barren dry beds. We have already seen the new pattern of ecological knowledge and physical work required to operate the ponds.

[161] Dubravius, *De piscinis*, lib. 4, chs. 1–2, 5, and 7, and lib. 5, cap. 9. Work on royal ponds at Breteuil, Verneuil, and Glapon in 1313/14 employed the king's *pionnier*, Jean de Moustier, the royal carpenter, and uncounted earth movers at a cost of more than 2,000 livres (Rouillard and Maupoume, "Étangs royaux"); Benoît, *Étangs de Dombes*, 48–49 and 56–57, details maintenance operations on ponds of the lords of Thoire-Villars during decades around 1400. More ongoing maintenance work and expenses appear in Steane, "Royal fishponds," 50–52; Beck, *Eaux et forêts*, 288–291; Benarrous, *Grande Brenne*, 218–224; and Deligne, "Carp in the city," 291–293.

[162] Benarrous, *Grande Brenne*, 262–268 and 315–330, uses pollen profiles, charcoal analysis, and other evidence to correct the myth of a waterlogged early medieval Brenne, so it was rather the region's natural network of streams which had supported the saint's well-remembered fishing (Krusch, ed., "Vita Sigiramni abbatis," caps. 19–22 (pp. 617–619); Laugardière, *L'Église de Bourges*, 179–188; Coulon, "Les étangs," 10–11)

[163] Henningsen, *Besitz und Einkünfte*, 25–26 and 55–59; Búžek, "Goldene Zeitalter," 82–83.

Long-term social effects were destabilizing, too. Besides the professional pond masters, some ordinary rural people did find opportunity in the fishpond sector servicing large producers or small local markets. Fishers and fishmongers who had worked for or with princes in Burgundy eventually took leases on fragments of the former estates.[164] Wider social circles acquired ponds just in time to drive the post-1450 expansion in Brenne. In Silesia, Poland, and south Germany village notables prominently appear in possession of one or two small ponds. On big estates this could add up. When Himmelskron abbey was secularized in 1547, eighty-two ponds were inventoried on its peasant tenements. Twenty villages in the Pszczyń district of Silesia had in 1536 seventy-one peasant ponds with capacity to produce almost 100,000 juvenile carp.[165]

But for understandable reasons peasants' resistance to aquaculture did not abate. It rather intensified their familiar opposition to privatized natural fisheries. The twelfth-century writer Wace was sure that Norman peasant rebels back in 997 wanted not the generic water rights reported in his source but in particular "to seize the fish from the fishponds."[166] Against theft of fish from an artificial pond laws customary and written made special provisions – *Sachsenspiegel*, for instance, set fines ten times higher[167] – and lords posted special guards, especially when water was lowered for harvest. Who watched the watchers? In Franche-Comté two guards, Fourquon Boison and Jehan Beire of Dampierre, hired in 1371/2 to protect the count's pond, instead themselves broke the dam and caught the fish. A few years later tenants were refusing to work harvesting carp and others to cart *norriens* between ponds.[168] Complaints about illicit and clandestine removal of fish from ponds are a commonplace of court records. Some labourers beat up Pierre de l'Aleu in April 1387 when he tried to stop them fishing in his pond at Neuvy-en-Sologne.[169] In Dombes, Jean-Claude Schmitt has

164 Beck, *Eaux et forêts*, 304–305 and 329–345.

165 Schmidt, "Himmelkron," 51; Nyrek, *Gospodarka rybna*, 55. Fournial, *Les villes et l'économie*, 689–690, found some peasant pond owners marketing fish in late fourteenth-century Forez.

166 "*Es viviers prendre les peissuns.*" Wace, *Roman de rou*, bk. 2, line 891 (ed. Andresen, vol. II, p. 64).

167 Eike von Repgow, *Sachsenspiegel Landrecht*, II: 28, 1–2 (ed. Eckhardt, 157). Like provisions in the Charter of Beaumont, a model of customary law in Lorraine, are remarked in Collin, "Les ressources alimentaires," 43.

168 Gresser, *Pêche et pisciculture*, 300–308, with several more cases. See also Gresser, "Les délits." Similar events in the duchy of Burgundy are reported in Beck, *Eaux et forêts*, 275–281.

169 Guérin, *La vie rurale*, 150. More cases are in Gislain, "Rôle des étangs," 92; Querrien, "Pêche et consommation" 2003, 433; McDonnell, *Inland Fisheries*, 18; Aston and

argued, thirteenth-century peasant anger against widespread flooding sublimated into legends where evil emanated from the lord's pond. In 1388 and 1440, however, rebellious peasants there just publicly broke the dams.[170] Collective fishing and destruction of ponds followed by common public meals of the lord's fish are well known from the German Peasants' War – where the social content of such acts is blatant – but also from feuds between Staffordshire factions in the 1530s.[171]

Finally thousands of localized ecological revolutions coloured general cultural style and expectation. The magnate offering 'fish from our own ponds' asserted social power with at least tacit environmental implications: people and nature were both subject to human control and turned to private purpose. Sellers and buyers of pond-reared fish dealt in a standardized commercial product. Dubravius discusses regional customs of selling carp by measure (volume) or by count, favouring the latter to avoid dispute "where carp of equal size are sold."[172] As the Moravian described, so the Polish and the Austrian pond masters behaved, selling their carp and pike in units of sixty (*kop*, *schock*, *sexagena*).[173] Earlier French fish farmers counted by hundreds. Nature had become a unit of account.

><> ><> ><> ><> ><> ><> ><> ><>

Fish farms served an elite segment of medieval demand for fish. Medieval aquaculture intentionally transformed fish and landscape to fit human ends. Colonizing interventions in the environment achieved intended results and others not intended as well. What nature failed to provide was remedied by altering nature, in this instance by establishing artificial ecosystems to domesticate a useful animal and introduce an exotic life form into still-natural surroundings. Neither environmentally nor socially neutral, the aquaculture response – intensified development – to rising European demand for fish modeled an expectation that fisheries

Bond, "Worcestershire fishponds," 440 and 442; Hartley, "Leicestershire," 294; and Górzyński, *Zarys historii*, 49–50.

170 Schmitt, *Holy Greyhound*, 164–165. The 1370s–80s were an especially restless time in both England and France, but poaching and attacks on seigneurial property in the latter need study.

171 Heimpel, "Fischerei und Bauernkrieg"; Currie, "Early history of the carp," 102–103. In the tension-ridden England of 1376 sixty people attacked a park of Evesham Abbey near Ombersley in Worcestershire and took 100 shillings' worth of fish from the pond (Dyer, "Consumption," 35).

172 "*Statim emptor et sine lite cum uenditore transigit, ubi pares statura Cyprini uenundatur*," Dubravius, *De piscinis*, lib. 5, cap. 7 (ed. Schmidtova, 65).

173 Rybarski, *Gospodarstwo Księstwa Oświęcimskiego*, 75; Małecki, ed., *Lustracja województwa krakowskiego*, 241.

in particular and nature in general were subject to human authority and invention. The ponds created in Brenne, around Třeboň, and elsewhere in late medieval *paysages piscicole* remain strong human imprints on regional landscapes, now so distinctive as to be protected natural parks and UNESCO biosphere reserves. Again, an east-central European source articulates what cryptic western charters and account books can only suggest had been going on since the high Middle Ages: an inspection team touring lordships of the Kraków cathedral chapter in 1533 was instructed "if there are no fish in a pond, ask who took them out."[174] Fish are present or absent by human agency. It encapsulates a fundamental assumption of the modernist managerial mind.[175]

[174] Ulanowski, ed., *Księgi sdowe wiejskie*, vol. II, p. 481: "*quodsi in eis pisce modo non sunt, inquiratur diligenter, quis eos expiscatus fuerit.*"
[175] Compare Towle, "Authored ecosystems" or McNeill, *Something New*, 325–336.

# 8 Going beyond Natural Local Ecosystems, II

## Over the Horizon toward Abundance and 'Tragedy'[1]

While medieval Europeans inland constructed new habitats and fisheries at home, coast dwellers were reaching out into new marine ecosystems and pulling new populations of fishes into the task of satisfying western Christendom's hunger for fish. The internal frontier of intensification was more than matched by external frontiers of untapped resources found in distant waters. Like aquaculture, the new marine fisheries were powered by demand – but theirs was a less elite and potentially much larger demand. Like inland ponds, coastal fishing evolved from older traditions to innovations – but the coasts offered more than one promising opportunity and the changes were as often institutional as technological. Like interior farmers and consumers, fishers and eaters of sea fishes learned different expectations – but 'over the horizon' fewer lessons were in manipulating a familiar nature and more in beating out other humans to take a piece from a nature ever less connected to what fishers and consumers called home.

Once the elements were assembled carp culture made a single narrative. Marine ecosystems presented several distinct and unequal opportunities to increase output (Map 8.1). Passing significant representative regional cases in review will establish particulars for interpretation in ecological and cultural context.[2] Supported by greater investment in

[1] A premature discussion of this material appeared in Hoffmann, "Carp, cods, connections," 20–55.

[2] Because economically significant fisheries are often assumed to be marine and national in character, aspects of their past have long drawn attention from historians and others. Some of this is careful scholarship well versed in medieval Europe; some is not. Much of the best marine environmental history falls into the latter group. Ojaveer and MacKenzie, "Historical development of fisheries in northern Europe," introduced a special 2007 issue of *Fisheries Research* filled with excellent papers. Yet except for a paper treating 7000–3900 BCE all begin well into the LIA. The intervening 5,000 years, including the medieval millennium, remain unexplored. So, too, the 2016 collection by Engelhard et al. in *ICES Journal of Marine Science*, billed as "ICES meets marine historical ecology," lacks genuine premodern studies, providing an almost stereotypical case of shifting baseline syndrome. Even more than elsewhere in this book this section aims to ground knowledge and

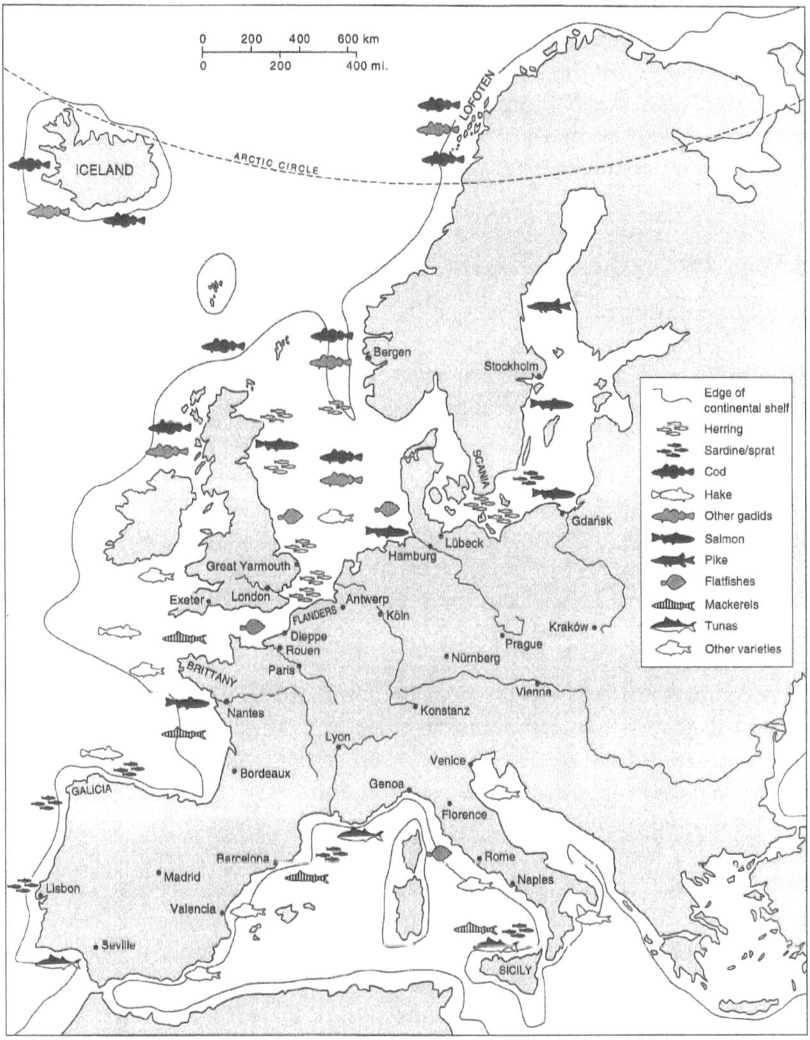


Map 8.1 Major European marine fisheries at the end of the Middle Ages.

interpretation, historical and ecological, on critically credible scholarship and selected primary evidence, not to cover every known marine fishery. Unlike fisheries histories that get underway at the end of the Middle Ages, an intent here is to recognize the maritime frontier as pushing out from long-accumulated medieval experience.

technology and marketing, coastal fishers extended their efforts offshore into untouched stocks, species, and ecosystems. They extracted products to preserve for distant consumer markets. Market-oriented expansion generated complex organizations, conflicts, and a view of limitless further frontiers over the horizon – even as the first such frontiers foreshadowed a narrowing future.

## 8.1 Innovation on Marine Fisheries Frontiers

Of course Europeans were taking fish from salt waters before and since the early Middle Ages, even if fears for physical security limited coastal settlement and thus opportunities to use local marine habitats. Sixth-through eighth-century fishers and their salt-water catches have appeared in earlier chapters off Carthage; in the Venetian lagoon and salt ponds at the mouth of the Tiber; on Öland and Bornholm; and at the mouths of the Eider and the Charente. We have seen an Anglo-Saxon fishing boat blown across the Channel in the early 1060s and people eating herrings beside the same narrow seas then, too. The twelfth-century Venetian market was full of fishes from the sea, as were trash heaps at contemporary Ghent.

So a discovery of sea fishing is not at issue, but rather late medieval development of large-scale, heavily commercialized arrangements to catch fish from an area at some distance or formerly little exploited and to transport and sell those fish to consumers elsewhere. Some economic historians usefully distinguish this 'second stage commercialization'[3] from what we have called artisanal fishing. Simultaneously fueling and driven by a stepwise spread of marine consumption from the coasts inland, certain regional fisheries achieved wide and distant markets. A prevalent historiographic focus on local and regional stories, however, can obscure interactions and competition among them. What follows looks not only at the narratives but would gauge the extraction of biomass from the oceans and its distribution over broader areas for socially shaped consumption. Transformation of herring fisheries and markets from regional to international scale may be the earliest and largest clear example. Others would follow on all Europe's coasts.

[3] Gardiner, "Character of commercial fishing," 80–81 and 88–89, credits the core concept to Kowaleski, "Commercialization," 220–227, and Unger, "Netherlands herring," 345–349. Note that the three authors respectively treat Iceland, England, and Dutch cases to be examined below.

### 8.1.1 *Networks for Silver*

As laid out in Chapter 5, by the twelfth century people all round the shores of northwestern Europe were exploiting herring schools to provide nearby townsfolk, notably 'the poor', with whole fish 'powdered' with salt or, less often, smoked or brined. Writing in the 1250s, well-traveled Dominican friar and scholastic natural philosopher, Albertus Magnus, voiced contemporary learned understandings:

The *allech* [Albert uses the classical Latin term for fish preserves] is extremely abundant in the ocean that touches parts of France, Britain, Germany, and Denmark. It is a fish about one palm long which, so long as it swims in a whole school, cannot be caught due to its great numbers. It is caught after the autumn equinox when the ranks split up. And even then, when the fish are enclosed in many large seines tied together, the lines of the nets must sometimes be cut because it is not possible to pull the nets in.

This is a scaly and tasty fish, having no intestines except the jujenum. Thus nothing is found in its stomach ...[4]

Small but remarkably abundant, the toothsome herring were for Albert especially associated with the North Sea. But why the limited and empty entrails? Had he seen not intact but gutted herring? If so, Albert becomes an unwitting witness for the future (pp. 324–326 below).

*8.1.1.1 Early Export Centres* Expansion of the herring industry under way from around 1200 centred on large easily accessible nearshore spawning agglomerations, encouraging regional concentration of fishing, processing, and marketing activities. Two areas stood out, the southern North Sea region associated with Great Yarmouth in Norfolk, and the Danish straits, notably the Øresund along southwestern Scania.[5] In the North Sea large autumn schools of herring along the coast of East Anglia attracted fishers and buyers from England, Flanders, Germany, and France to an international fishery. Using mainly drift (gill) nets, the fishers took herring in hours of darkness as the fish followed their plankton prey toward the surface. After a century's growth, peak years *c.* 1340 of the herring fair at Yarmouth drew some 500 ships, and fragmentary English customs accounts record annual exports of 700–1,800 *lasts* (at 12,000 herrings the *last*,[6] 8.4–21.6 million fish), so perhaps 5,000

[4] *De animalibus*, lib 24, §2. For greater precision I have mildly emended the English of Kitchell & Resnick, tr. p. 1660, from the Stadler ed., p. 1518, which appears in the Supplement.

[5] What is now southernmost Sweden was part of Denmark until the seventeenth century.

[6] Defined in *Fleta*, 2:12, ed., Richardson and Sayles, 119. Childs, "Eastern fisheries," 243 n. 11, indicates local and temporal deviations from this norm.

metric tonnes (mT). A large but uncounted share of the catch went to Flanders and surely still more to English domestic consumption. Written records suggest most of Yarmouth's herring were dry salted ('powdered') but from around 1300 some proportion were smoked to become 'red herrings' and a growing number packed in barrels.[7] From schools in the same waters fishers of Calais smoked ten million herrings in 1321 and exported twenty million. More than thirty million a year traveled by barge up the Seine to Paris, feeding the massive increase in herring remains seen there and at all other well-excavated northern French sites.[8] Taking English and French exploitation together, it is not unreasonable to think that each year around 1340 they may have removed from the southern stock at least 15,000 mT of fish. Quantities taken and shipped by Yarmouth and other southern centers shrank after the 1360s.[9]

Herring schools off Scania had much earlier attained mythic scale.[10] About 1215 Danish historiographer Saxo (*c.* 1160–*c.* 1220) prefaced his *Gesta danorum*:

> An arm of the sea pushes through to part its [Sjaelland] eastern side from the west coast of Scania; this is accustomed to drive the largest amount of booty to the fishermen's nets every year; the whole sound [Øresund] is so frequently packed full with fish that sometimes boats striking them have difficulty in rowing clear and no fishing gear but the hands is needed to take them.[11]

Even as Saxo wrote, the historic local subsistence use of Scanian herring was undergoing quick commercialization under the growing influence of merchants from north Germany who, contemporary chronicler Arnold of

[7] Saul, "Herring industry," where the Yarmouth product is described as dry salted; Saul, "Yarmouth and the Hundred Years War"; Locker, *Role of Stored Fish*, 55–56 and 77–78; Kowaleski, "Commercialization," 181–189; Hybel, "Sildehandel og sildefisker." Locker, *Role of Stored Fish*, 167, could not investigate evidence of gutted herring in the English remains, but brining in barrels is said to extend to ten months the shelf life of even ungutted herrings, so markedly longer than the dry-salted form.

[8] Hocquet, "Pêcheries médiévales," 48, 53–56, and 79–86; Bourlet, "L'Approvisionnement de Paris" ; Maillard, "Tarifs"; Clavel, *L'Animal*, 159–162. Boussard, *Nouvelle Histoire*, 302–303, finds shipments up the Seine by the late twelfth century.

[9] Kowaleski, "Commercialization," 191–192; Childs, "Eastern fisheries," 21. If actual English exports of some 20 million herrings weighed some 5,000 mT, Calais exports were similar and surely contributed to supplying Paris with about 7,500 mT. Now add counted English domestic consumption. For arguable causes of this and other declines in catches, see p. 392 below.

[10] Unless otherwise noted, what follows is based on Jahnke, *Silber*, 39–226; Hybel and Poulsen, *Danish Resources*, 374–376; Jahnke, "Medieval herring," 161–167; and Holm, "Commercial sea fisheries," 15–17.

[11] Saxo, *Gesta danorum*, preface, 2:4 (ed. Friis-Jensen, vol. 1, pp. 10–11). For a more literal rendition I have mildly emended the English of Fisher's translation provided there. Supplement 8.1.1.1 has the Latin original.

Lübeck acknowledged, brought great wealth to Denmark in return for the god-given fish.[12] Amply provided with salt from Lüneburg and hungry buyers in central Europe, what would become the Hanse towns gained territorial concessions on the Skanör peninsula and elsewhere nearby and built up exclusive access to the fresh catch. During the run of August to October merchants on the beach bought herrings caught just offshore by what became five to seven thousand small open boats each initially crewed by five to eight Danish peasants. Their self-organized temporary boat associations made good use of a seasonal gap in the agricultural routine. Fishers set drift nets overnight and fixed nets on posts by day. A fee to the king, lord of the beach, licenced them to make temporary shelters, land their catch, and freely sell to the highest bidder. Each could salt but six barrels for family consumption. With the industry's growth, by the fourteenth century the labour supply increasingly drew upon migrant workers from interior Denmark, and even Germans, Dutch, and English, some of them brought in by the Hansard merchants, who also hired Danish and German women to salt and pack the catch.[13] Each barrel was inspected for quality and branded with the merchant's mark. After sales at the September Skanör herring fair, the product was shipped south to towns along the Baltic or around Jutland to those on the North Sea. Wider distribution by, for instance, 1252, had Øresund herrings entering Flanders assessed per thousand for tolls.[14]

Tolls paid at Lübeck, chief of the Baltic Hansa towns, in later decades of the fourteenth century provide the only substantive grounds for estimates of this fishery's scale: in 1368 the town received 76,000 standard Rostock barrels (117 kg legal weight, 100 kg of that wet fish, the rest salt; 900 herrings then, but up to 1,200 something more than a century later). Annual receipts during 1398–1400 ran from 81,172½ to 69,975½ barrels. At twelve barrels the *last* and the *last* calculated at 12,000 fish, a modest 70,000 barrels held 84,000,000 fish … or something over 7,000 metric tons of fish flesh. Yet, as economic and environmental historian Poul Holm calculated in 2016, Lübeck itself handled only about

[12] Arnold, *Cronica*, lib. 3, cap. 5 (Lappenberg and Pertz ed, p. 77; tr. Loud, *Chronicle*, p. 99). Note that, contrary to some references, Arnold here neither refers to the fish as 'silver of the sea' nor to catching them by hand.

[13] Jahnke, *Silber*, 189–190, and on this point expanded in Jahnke, "Medieval herring," 164–165. Jahnke's discussions of processing (*Silber*, 218–222; "Medieval herring," 165) depend primarily on the Danish code of 1389, without exploring earlier development of key practices.

[14] Lampen, *Fischerei*, 186 (citing *Hanschisches Urkundenbuch*, vol. 1, nr. 432). Merchants from the Hanse towns were peddling herring in eastern English ports before the end of the thirteenth century (Jahnke, *Silber*, 251–253).

one-third of the deliveries to Hansa towns in the Baltic, implying at least 225,000 barrels sent in that direction. Holm then posits another 50,000 barrels sent westwards to the Hanse's members in the North Sea and some further number directly to Flanders, to bring exports up to 300,000 barrels. If Danish domestic consumption came to another 100,000 barrels, the total processed each year was 400,000 barrels from an annual late fourteenth-century Scanian herring catch exceeding 40,000 mT live weight (or, considered another way, 400,000,000 fish).[15]

Noble Burgundian councillor Philippe de Mézières toured the area in 1389 and marveled:

> ... the herring makes its passage through the strait from one sea to the other in marvellously great number... so great one could cut through them with a sword. ... the bounty of God ... furnishing an abundance of herrings by which all Germany, France, England and many other countries are fed in Lent, for poor Christians can have a herring who cannot afford a big fish.[16]

Indeed in this very decade Dordrecht imported nearly 1.5 million fish a year and the Flemish staple at Damme taxed more than sixteen million in only the fall quarter of 1387, in both cases primarily carried by Hansards.[17]

None would doubt that the Øresund yielded Denmark's greatest medieval export and in all likelihood this fishery was the largest anywhere in medieval Europe. But the late 1300s were its apogee: by early decades of the following century the Scanian schools were in trouble, noted as failing in 1402, 1425, and years after 1436. In the 1490s Lübeck received from all Denmark barely 20 percent of the herring it had a century earlier from the Øresund alone. Despite a debased coinage Danish crown incomes from beach licences were down 30 percent. To replace lost royal incomes already in 1429 King Eric instituted tolls on shipping through the straits. Despite a brief revival around 1500 neither the fishery nor the trade ever fully recovered.[18]

[15] Holm, "Commercial sea fishing," 15–17, reworks calculations from his "Catches and manpower," 177–180. But compare the perspectives of Nedkvitne, "Fishing, whaling, and seal hunting"; and MacKenzie *et al.*, "Ecological hypotheses," 175–176; and even Jahnke, *Silber*, pp. 421 (table VII), and 431 (appendix XIV). Additional early production centres are described in the Supplement. As already hinted in Chapter 6 and as will further emerge below, the ecology and economy of medieval European herring fisheries were more complex than can so far easily be untangled or summarized.

[16] Philippe de Mézières, *Songe du viel pèlerin*, ed.Blanchard, vol. 1, pp. 228–229; tr. Coopland, vol. 1, pp. 129–130.

[17] Uytven, "L'Approvisionnement," p. XI:103 [*sic*].

[18] Holm, "Catches and manpower," 179–180; Jahnke, *Silber*, 90–119; Jahnke, "Medieval herring," 177–178; Holm, "Commercial sea fisheries," 16–17. Sound Toll registers survive only since 1497.

In retrospect by about 1300 effective mercantile organization had plausibly brought the herring industry of northwestern Europe near limits of its traditional techniques. Coastal concentrations of fish could be exploited effectively with a modicum of operating capital, much of it going for salt, and thus supply inexpensive Lenten protein to large regional concentrations of Europeans. Dependence on a few regional spawning stocks left the industry susceptible to ordinary environmental fluctuations affecting the year-to-year presence of the schools and possibly vulnerable to larger changes. As already mentioned, from the mid-thirteenth century continued pressures of demand drove herring prices rapidly upward, doubling at Calais, for instance, between 1268 and 1300 and doubling again by 1341.[19] Short-term changes in supply superimposed year-to-year price variations exceeding 25 percent (Figures 6.1 and 6.2). The critical bottleneck was time, the need for quick processing which anchored this fishery within sight of the beach, and the limited durability of its product, which hindered access to consumers well inland. Writing in the 1330s Francesco Pegolotti warned his fellow Italian merchant travelers to be sure they got only good-smelling herring from the most recent pack.[20] How did this change?

*8.1.1.2 The Interplay of Technologies and Regional Success* Successive waves of economic and technical innovation lapped across the herring industry during late medieval centuries. A historically critical antecedent, of course, was thirteenth-century involvement of long-distance merchants, English and Flemish traders shipping out of Yarmouth to Bruges and to Bordeaux, the importers to Paris, and most famously those proto-Hansards in the Baltic. To different degrees each commercial interest could reach more customers with a more durable and portable product, but for much of the thirteenth century and beyond most herring continued to be salted whole and handled dry in bundles. The fish whose remains are recovered from sites on the twelfth–thirteenth-century Øresund are as whole as those from Haithabu 200 years earlier.[21] Traders from Köln, who distributed Scanian herrings into the Rhine basin and areas to its south and east, then dealt in *korbherring*,

19 Hocquet, "Pêcheries médiévales," 53–56. For movement of herring prices in changed post–Black Death conditions see pp. 365–367 below.
20 Pegolotti, *Pratica della Mercatura*, 380.
21 Enghoff, "Fishing in the Baltic region," 60, refers to finds at Simrishamn on the Baltic coast of Scania and at Helsingborg on Øresund shore.

herring in baskets.[22] Consumers of the 'powdered' herring of Paris and its region left only skeletal remains of entire fish throughout the thirteenth and into the fourteenth century.[23] Managerial accounts for 1289–1291 from a herringry Durham priory possessed on the coast southeast of Newcastle detail a product salted, dried, and packed into baskets with straw. Archaeozoologist Alison Locker's most widespread survey of medieval herring remains in England finds entire, not gutted, fish prior to *c.* 1300. The most recent study of massive numbers of bones dating 1360–early sixteenth century at York reports no evidence of gutting or use of barrels and infers fish of English origin, even if 1370 legislation called for Hanseatic methods. Historian Maryanne Kowaleski has wisely judged that 'white' herring were certainly salted, "but there is considerable difference of opinion about whether they were fresh, lightly salted, heavily salted or packed in brine …"[24]

A first series of waves spread gutting, barrels, and packing in brine fitfully across regions of production. Deftly carving out the gill area of each fish (*kehlen* in Danish, *kaaken* in Dutch) extracted most of the entrails, directly exposed more flesh to the salt, and left for archaeozoologists distinct evidence of the practice on individual skeletons and the collective mass of herring remains at a site. (Particulars of processing fish little interested medieval writers, but not some sharp-eyed graphic artists as in Figure 8.1.)[25] Packing in watertight barrels kept air from dry- or wet-salted fish and eased their handling in bulk. Salt brine required a sealed container but ensured full and continual contact of the fish with the preserving agent.[26] The three techniques need not be applied together but came to be so in the late medieval industry.

Indubitably some people in the Baltic had known for a long time how prompt butchering of herring extended the usefulness of their product. Herring remains with many vertebrae but no diagnostic skeletal elements have been found at Truso, the Viking Age trading post on the Wisła

[22] Kuske, "Kölner Fischhandel," 230–235. [23] Clavel, *L'animal*, 159–161.

[24] Kowaleski, "Early documentary evidence," 35. Locker, *Role of Preserved Fish*, 56; Harland et al., "Fishing and fish trade," 187–192; Cutting, *Fish Saving*, 73; Saul, "Herring industry," 35.

[25] Notably the gutting process removed a bone called the cleithra from what zoologists call the shoulder area and laypeople might think rather the throat or neck of the fish, leaving this element massively present in processing sites and absent from places where the processed fish were consumed. In visual representations such a fish has a notch removed from the ventral area immediately behind the head, clearly visible in Figure 8.1.

[26] Absorption of brine by the fish and leakage and evaporation from barrels made customary their repacking at intermediate stages of distribution such as Flanders and Köln.

Figure 8.1 Brined herring, kaaken and barreled.
"Salsus autem in vsum hominum vitra quoquis aliis pisces sanus durare potest" [When salted the herring remains suited for human use longer than any other fish]. *Allec* woodcut from *Hortus sanitatis*, cap. III, fol. 274r [p. 547]. Strassburg: Johann Prüss [not after 21 October 1497]. First used in a printing by Jacob Meydenbach at Mainz, 1491, this image appeared widely in multiple subsequent redactions of the *Hortus* by various Rhineland printers dating to and after the 1490s. Licence cc0 / Public domain. Image reproduced with permission of Technische Universität Darmstadt, Universitäts- und Landesbibliothek Signature Inc IV 203.

delta, and at a Slavic stronghold of ninth–thirteenth-century date located beside present-day Kołobrzeg on the Pomeranian coast.[27] Adoption of the practice further west is evident at Selsø-Vestby, a production site

[27] Makowiecki, "Cod and herring," 121–122, and "Badania archeoichtiologiczne."

radiocarbon-dated 1290/1380 on Roskildefjord, close to the Øresund,[28] so reinforcing the assumption that much of the Danish catch was so processed by the early fourteenth century. Köln merchants were handling barreled herring by the same time. Herring barrels of Baltic oak with tree rings from fourteenth-century timber harvests are well known in Flanders (recycled to line wells and latrines). The English industry adopted barrels – at least for the ungutted smoked Yarmouth 'red herrings' – in the 1300s but tried the full 'Scanian cure' only at their end. In Scotland herring barrels appear in records from 1360 and become normative in the 1400s.[29] Scholarly consensus now probably acknowledges that Hansards handled mostly gutted, brined, and barreled herring by the peak period of Scanian production in the late fourteenth century but adoption further west was delayed. Coupled with Hanseatic inspections and regulations to ensure a product of durable quality this difference helps explain the dominance of Øresund herring on contemporary international markets.

A further pulse of technical and economic innovation then transformed the herring fishery.[30] Flemish fishers, once leading foreigners off the English coast, followed by their northern country cousins from Zeeland and Holland, all squeezed out of the Øresund by the Wendish towns, most eagerly adopted the *kaaken* process to North Sea stocks,.[31]

Flemish fishers further led the Netherlanders in a gradual transition from coastal to offshore ('deep sea') fishing for herring. This entailed two further interlocking technical changes. First, probably before the end of the fourteenth century they discovered that the fish kept and tasted better if they were gutted and salted down while still at sea, then reprocessed and packed on shore. Hauling salt required a larger ship, soon built for the purpose as a "herring *buss*," but then a larger crew could stay at sea catching and packing herring for a week or more. The tether to shore

[28] Enghoff, "Medieval herring industry," described the site in detail, noting that Viking Age herring were whole, but later ones gutted. Enghoff, "Fishing in the Baltic region," 50–51, subsequently redated the later context as thirteenth century, and then Enghoff, "Herring and cod," 142–143, provides the radiocarbon dates here cited.

[29] Saul, "Herring industry," 35; Gemmill and Mayhew, *Changing Values*, 317–323.

[30] What follows uses essentials from Unger's important economic analysis in "Netherlands herring" and "Dutch herring," plus later refinements by Munro, "Patterns of trade," 161–163, and in de Vries and van der Woude, *First Modern Economy*, 243–254. It takes further into account more recent understandings of technical issues related above and southern antecedents to early modern preeminence of the Hollanders in particular. Schubert, *Essen und Trinken*, 145–146, would more emphasize changing German tastes, for which see below. Laget, "Géographie du hareng," provides a more Francocentric interpretation of late medieval and early modern herring.

[31] Sicking, *Neptune*, 132–142; Sicking and van Vliet, "Our triumph of Holland," 337–351.

undone, Netherlanders could depart contentious and perhaps failing English shores for unexploited spawning stocks in the central and northern North Sea waters they called *Noordover*, and eventually off Shetland and Iceland. Higher salinity fed these fish to larger size than in the Baltic. Ostend alone landed 2,400 tonnes in 1467.[32]

Unlike the rowing skiffs or flat-bottomed vessels launched over the beach for day voyages after herring – still the norm in early fifteenth-century Holland and elsewhere – the *buss* was a keeled vessel of thirty to sixty tonnes capacity, requiring crew of a dozen and harbour facilities (Figure 8.2). The specialized ship design and infrastructure to support it were developed at the start of the fifteenth century in fishing ports along the coasts and Maas estuary of Flanders and Zeeland. By 1476 Flanders had 125 busses and Zeeland 150, but Holland only 100. Subsequently endangered by the Habsburg–Valois wars, the centre of the industry slid northwards, first to places around Brielle (Brill) on the Maas in south Holland and eventually to the more secure towns of the Zuider Zee. About 1520 the Dutch herring ports along the Maas were landing about 12,000 *last* (144,000 barrels) each year and those around Enkhuizen on the Zuider Zee still only about 1400 *last* (16,000 barrels).[33] The proverbial Dutch herring kings of the sixteenth and seventeenth centuries were but the last step in a century-long process.

At the start of the sixteenth century, then, just one of the Low Countries' several herring-fishing regions was landing twice the fish of Lübeck at its peak (see p. 321 above). For some generations the Netherlanders had backed up their technical superiority with investments – salt, nets, ships, barrels – and with government policy – safe conducts, convoy defence.[34] But in the words of historian Richard Unger, "the greatest impetus to the use of all the superior methods was the presence of a market for the preserved herring and a market that had potential for growth."[35] When the herring were unloaded, some went straight to nearby urban consumers. At Deventer, where next to no eating of herring is visible in food waste from the the ninth through fourteenth centuries, in the 1400s and

[32] Ervynck, Van Neer, and Pieters, "How the North was won," 234–236; Van Neer and Ervynck, "Zooarchaeological reconstruction," 98–99; Mollat, *Commerce maritime Normand*, 314n.

[33] Bruijn, "Dutch fisheries," 105–120; de Boer, "Roerend van de vischeryen," 115–118 and 139–140; van Bochove, "Hollandse Haringvisserij."

[34] Sicking, *Neptune*, 142–201; Sicking and van Vliet, "Our triumph of Holland," 339–351.

[35] Unger, "Dutch herring," 256; de Vries and van der Woude, *First Modern Economy*, 235–237, continue to stress urban demand as the engine driving development of the Dutch fishery.

Figure 8.2 A Flemish herring *buss*, *c.* 1480.
Thought to be the earliest image of this special-purpose ship design. Print of engraving by so-called Master W with the Key. Bruges, *c.* 1480. Now in Rijksmuseum, Amsterdam, RP-P-2014-30. Reproduced with permission of Dr. Jeroen ter Brugge, curator, Maritime collections, Rijksmuseum.

1500s herring contributed 15 percent of the recovered fish bones.[36] More of the catch was washed and layered in barrels with salt for an extremely durable product, edible for a year and more. Price advantages on the Antwerp market offered consumers good reason to substitute domestic for Hanseatic herrings, so by the mid-fifteenth century Flemings, Zeelanders, and Hollanders were supplying the shores of the North Sea, the Rhineland, and, after failure of the Scanian spawning stock, even the Baltic. By century's end they provided most of the herring imported by the English.[37]

*8.1.1.3 An Evolving Consumer Base* So where were those late medieval consumers whose demand drove the fishing and the trade in herring? While thirteenth- and fourteenth-century evidence of eating herring in England, the Low Countries, and northern Germany scarcely needs further mention,[38] the course of distribution and demand in Europe's interior and more southerly regions should be established by more than occasional trade agreements. Many diverse source materials trace the penetration and limits of eating this preserved fish far from its native seas.

Herring remains increase greatly in thirteenth-century contexts at all sites in northwestern France.[39] To the south both archaeological and written records thin out. Herring were on markets in Berry at Déols in 1235 and Vierzon in 1430.[40] In most years between 1306 and 1362 the papal household in Avignon spent heavily to transport large quantities by way of La Rochelle, Toulouse, and Bordeaux for Lent, but these fish were also ordinary fare for the chapter of Saint Trophime at Arles in 1352.[41]

36 IJzereef and Laarman, "Animal remains from Deventer," 435–436, analyze 10,028 fish vertebrae recovered by sieving. In the inland southern Low Countries, bone samples from a poor neighbourhood in Namur contained but 1% herring (the only marine fish!) in the twelfth century and 25 percent herring around 1500 (Lentacker et al., "Historical and archaeozoological data," 90–94).

37 Herring barrels of Baltic oak stop being available for recycling in Flanders by the mid-fifteenth century. Some others than Netherlanders who learned the technology and exploited northern schools are noted in Supplement 8.1.1.2.

38 Locker, *Role of Stored Fish*, 166–191 and 223–247; Van Neer and Ervynck, "Rise of sea-fish," 162–167 (and see Figure 2.2 in Chapter 2 above); Ayers, *German Ocean*, 85–89; Schubert, *Essen und Trinken*, 131–136. Riddler and Trzaska-Nartowski, "From Dover to New Romney," report fish remains from Dover, 1150–1300, as 80% herring; Lampen, *Fischerei*, 32–80, identifies (p. 50) herring in alms given to the poor by the abbey in Halle in 1310 and 1314. Uytven, *Zinnelijke Middeleeuwen*, 186–187, has anecdotal evidence of socially stratified herring consumption in the Low Countries. Jahnke, *Silber*, 227–262, traces Danish exports through trade treaties, not consumption.

39 Clavel, *L'animal*, 163; Clavel and Cloquier, "Contribution," 208.

40 Querrien, "Pêche," 429.

41 Weiss, *Versorgung des päpstlichen Hofes*, 392–402 and 531–532; Stouff, *Ravitaillement*, 231.

A century later the comital household of Angoulême ate herrings when in residence at Cognac during the cooler months. At Tours in 1480 municipal guards got these fish as a regular ration; notables, however, took only a symbolic 'pittance' to open a banquet.[42] Fourteenth-century references in Burgundy are merely occasional, but in 1406/7 the toll station at Chalons passed "306 milles &/26 *tonetz*(?) [*sic*] de harenz tant blans que sors."[43]

Hanse towns from Lübeck east to Reval (Tallin) handled distribution around the Baltic and its affluents, as far upstream as, for instance, Kraków and Novgorod.[44] In the Empire's interior, however, Germany's then largest city and an early but distant affiliate of the Hanse, Köln, claimed staple rights on goods moving along the Rhine. This enabled it to control trade in herring throughout that river basin and further to the east. Köln merchants handled Scanian herrings from the 1200s, but from about 1370 also a competitive product from the North Sea. At first the fish were designated as 'basket herring'; barrels appeared in the trade shortly before 1300 and became the norm a generation or so later. From this node on the lower Rhine the herring, sometimes inspected, repacked, and resealed, went by water and overland to Frankfurt am Main, Strasbourg, Nürnberg, and beyond, even to the skirts of the Alps.[45] In the 1320s herring likely from this commercial network fed mercenary soldiers at Metz and inmates in the hospital at Klosterneuburg.[46] They were the only imported fish bought by members of the Tirolian noble Schlandersberg family between 1394 and 1401 and by their von Peucheim counterparts at Horn castle on the Danube for Advent and Lent in 1444–1446. At the very time herring appeared on dining tables and price-fixing lists for the Council at Constance (1414–1418) – where contemporary illustrations show open barrels of unmistakably gutted herring – a retail price series (herring sold by the piece) becomes available for transactions at Augsburg.[47] It is said that south German consumers preferred the unripe

42 Maillard, "Les Despenses," 122–123; Chevalier, "Alimentation," 145.

43 Beck, "L'Approvisionnement," 175–176.

44 Herring arrived in thirteenth-century Kraków via Szczecin and subsequently through Gdańsk (Carter, *Trade and Urban Development*, 134–139); trade in herring was common enough to start a price series in 1389 (Pelc, *Ceny w Krakowie*, table 22, pp. 38–39). More generally see Lampen, *Fischerei und Fischhandel*, 180–188, and Jahnke, *Silber*, 230–243.

45 Kuske, "Kölner Fischhandel," 230–260.

46 Collin, "Ressources alimentaires," 63–64; Holubar, "Spital," 42–43; Zeibig, ed., *Urkundenbuch*, 282–292. Around Basel Heide Hüster Plogmann and her team found traces of herring remains in late eleventh–thirteenth-century latrines at elite castle and lay urban sites (Hüster Plogmann, "Der Mensch lebt nicht von Brot allein," 193–197; Hüster Plogmann and Rehazek, "Historical record versus archaeological data"; Ervynck et al., "Beyond affluence," 432).

47 Ottenthal, "Ältesten Rechnungsbücher," 603; Archiv Horn-Rosenburg, Hs. Horn 44; Richental. *Konstanzer Konzilschronik*, fol. 25b (Loomis, tr., p. 101); Elsas, *Umriss*, I:

herrings Netherlanders caught in June and July, earlier than Hanse rules allowed. Smoked as *bocking* or dried as *matzginshering* (modern *Matjes*) they could arrive by way of Köln well before any fish from the Baltic.[48]

Crossing the Alps or sailing into the Mediterranean with loads of northern fish was another story, so fourteenth-century penetration of those markets long remained modest. Stuck in Avignon a homesick Pavian cleric, Opicino de Canistris (1296–1336) missed the *allecia desiccata* and other fishes he'd known back in Ticino but contemporary Florentine market customs expected them only in small quantities.[49] As Pegolotti completed his mercantile handbook about 1340, he advised buying only those freshly processed salt herring taken in the North Sea between England and Flanders. His benchmark price at Antwerp was two denarii per small barrel (*tinello*).[50]

Conditions changed after the Black Death. In 1384 Francesco Datini, merchant of Prato and trader in Florence and elsewhere, could import 643,000 herring from Southampton to Genoa and Pisa. The venture went so well that over the next twenty-five years he brought in thirty-nine shipments of 200–600 *balle* ('bundles', each of a thousand fish) from England or Flanders, always timed to hit the Lenten market.[51] A century after Datini, papal courtier Paolo Giovio was thoroughly familiar with their regular availability in Rome: "from Jutland's shores are brought to us herrings, foot-long fishes in baskets [crates?] preserved with salt and smoke."[52] Certainly by the fifteenth century Italian consumers and likely those elsewhere were aware that their salty little fishes could originate from various exotic northern waters.

So unlike Thomas Aquinas, who had needed a miracle to get a herring in late thirteenth-century Lazio, over the following 200 years consumers on markets across Latin Christendom could likely obtain a herring if they wanted it enough. People at greater distance from the North Sea or Baltic found their opportunities later, more dispersed, and more expensive.

394–395 and 615–616 (as shown in Figure 8.7). Kunst, "Medieval urban animal bone," 11–12, reports the so-far earliest herring bones at Vienna from late fifteenth-century latrines.

48 Schubert, *Essen und Trinken*, 145–146, refers to local studies published around 1900.

49 *Anonymi Ticinensis Liber*, 24; Nigro, "Mangiare," 120–121.

50 "*Aringhe insalate che si pigliano nel Mare Miano intra Inghilterra e Fiandra* ..." Pegolotti, *Pratica della mercatura*, 253 and 380.

51 Datini's records of the herring business as reported in Origo, *Merchant of Prato*, 285, and Nigro, "Mangiare," 121 and 133–135, are amplified by references in his wife's letters (Datini, *Letters to Francesco*, 56, 113, and 115). For Italian contemporaries of Datini also trading in herring, see Supplement 8.1.1.3.

52 Giovio, *De romanis piscibus*, cap. 42 (p. 60): "*Ex Cymbricis quoque litoribus aringhae, pedales pisces, in cratibus sale et fumo inveterati nobis afferuntur* ..."

They may also have appreciated them more. Popes paid heavily to get herring to Avignon. In the1390s Margherita Datini in Prato and the brothers Kaspar and Sigmund von Schlandersberg in the Vischgau (Val Venosta, Alto Adige) waited impatiently for the herrings' arrival.[53] However French courtier, diplomat, and poet Eustache Deschamps (1346–1406/7), envisaged "those spoilt herrings, pickled and smoked, yellow, black, and stinking," and monks at Eynsham in Oxfordshire consumed in the fifteenth century only a fraction of the herrings that they had in the thirteenth.[54] Was the reluctance in later medieval England, France, and the Rhineland a matter of taste and fashion (see Chapter 2, pp. 83–84 above) or of having more options? Did greater abundance drive the wealthy away from the 'fish of the poor'? Did more widespread distribution and deeper markets balance less demand from elite consumers? Alison Locker has shown that, taking into account the relative size and weight of food per individual fish and thus numbers of fish remains, the trend since the fourteenth century was for herring to yield to cod as the largest single source of fish flesh in English diets.[55]

To recapitulate before turning to those codfishes, European herring fisheries went from broad use of local coastal stocks in the twelfth century through dependence on southern North Sea and Øresund schools to exploitation of distant-water populations by the end of the Middle Ages. Over several hundred years, at each successive stage of expansion of the industry merchant entrepreneurs supplied ever more distant eager consumers with larger production from stocks further offshore. The evolution was closely linked to a market which called forth technical improvements, greater commercial complexity, and higher investment.

### 8.1.2 *The Stockfishsaga and other Tales of Codfishes*

The tale of the Atlantic cod, third of this book's introductory fish tales (pp. 14–17), bound an expanding web of human relationships around a central strand, the story of the dried cod of the north, stockfish, called by its Norse producers *stokfisk* or *skreið*. But like medieval Norse literary narratives, the expanded tale now needs also consider subsidiary encounters

[53] Weiss, *Versorgung des päpstlichen Hofes*, 392–393; Datini, *Letters to Francesco*, 56 and 113; and Ottenthal, "Äältesten Rechnungsbücher," 587 and 603. Ervynck et al., "Beyond affluence," 432, and Van Neer and Ervynck, "Remains of traded fish," 210–211, point out that herring on the table in Switzerland meant something other than it did in Flanders.

[54] "*ces mauvais harens, caqués et sors, jaunes, noirs et puens*" (Uytven, *De Zinnelijke Middeleeuwen*, 187); Hardy et al., *Ælfric's abbey*, 396.

[55] Locker, *Role of Stored Fish*, 277–282.

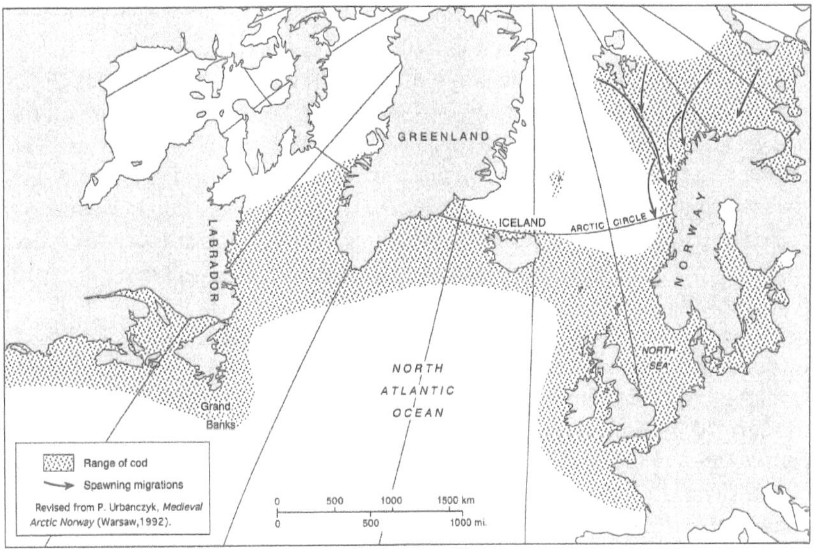


Map 8.2 The range of cod.

between coastal communities and regionally important fish varieties. Most such interactions between people and the cod family lay somewhere along a spectrum between local use of local marine ecosystems and more specialized and commercialized handling of certain fishes for consumers at greater distance. All entail incidents or phases of people intensifying exploitation and transforming fishes into objects of cultural desire.

The collective weight of medieval codfishes and their biological or culinary associates possibly approached that of herring and certainly covered comparably broad spaces. Atlantic (northern) cod is a common demersal predator in cold waters of the continental shelf (Map 8.2). The species has many breeding populations (races) with behavioural adaptations to particular environments. Local stocks are ubiquitous in near-shore habitats, but the largest fish – to 1.5 meters and 40 kg – remain in deeper waters of the shelf down to 600 meters.

Two salient features of cod in the European far north are less immediately apparent:[56] (1) adult members of the large arctic population migrate each winter in a southwesterly direction along northern Norway into

[56] Much of what follows was first summarized by Urbańczyk, *Medieval Arctic Norway*, 29–34, and Christensen and Nielssen, "Norwegian fisheries," 146–149, who synthesized material and debates from scattered Norse-language books and papers, many of them unpublished, and more recently deepened and nuanced in studies assembled by Barrett and Orton, eds., *Cod and Herring*, as cited below. To be read

spawning areas south of the Lofoten archipelago;[57] and (2) no salt is there needed to preserve cod. Decapitated, gutted, split, and hung on racks in the arctic wind and sun, the cod's oil-free white flesh "becomes as dry as wood", as a Venetian castaway on Lofoten in 1432 described the all but imperishable "stocfisi."[58] The process discarded heads and forward vertebrae and shipped off to the consumer a bundle of dried slabs still holding the rear part of the backbone (Figure 8.3a) Hence, as with herring, different patterns of cod remains distinguish production and consumption sites and only places where fish were consumed fresh will have full representation of skeletal elements.

In conditions more humid than in the far north but still cold, cod dried under cover with some salt ('dry-salting') becomes a product resembling stockfish. While this 'salt cod' became the standard shipped from early modern Newfoundland, medieval Norwegians did not produce what they call *klippfisk*. Ling, torsk (cusk), haddock, and saithe are often taken with cod and handled the same way. Other important gadids such as whiting or hake prefer different, often shallower or more southerly, habitats.[59]

*8.1.2.1 Norse Fisheries and Trades* Long familiar in the north, stockfish came south when Viking Age northerners trading to Danish Haithabu brought some as ship's rations.[60] However nothing before the eleventh century indicates northern Norwegians placed any special

with caution for the medieval period are the credulous popularization in Kurlansky, *Cod*, 17–45, and the more careful Fagan, *Fish on Friday*, 59–90 and 175–192.

[57] Like movements occur along Iceland, Greenland, and Newfoundland.

[58] Bullo, *Il viaggio Querini*, 68–69: "*Stocfisi seccano al vento et al sole senza sale, et, perche sono pesci di poca humidità grassa, diventano duri come legno.*" Norse prepared stockfish two ways: *rundfisk* were gutted and the heads removed but retained most vertebrae and two fish were tied together by the tails and hung to dry; *råskjær* (Hanseatic *rotscher*) were split more deeply and most of the vertebrae removed, then hung individually by the tail so they dried more quickly and thoroughly. The generic name refers to the racks and posts on which the fish were dried. Lampen, *Fischerei und Fischhandel*, 142–144, provides more detail on the Hansards' specialized trading and marketing vocabulary as does a mid-fifteenth-century English merchant's handbook, "The Noumbre of Weyghtys" (Jenks, "Werkzeug," 304) .

[59] Hufthammer, "Fish trade," 223–224; Wubs-Mrozewicz, "Fish, stock, and barrel," 191–192; Cutting, *Fish Keeping*, 118–122.

[60] Perdikaris, "From chiefly provisioning"; Nielssen, "Early commercial fisheries," 43–44 and 47–48; Hufthammer, "Fish trade," 221–222 and 225–229. While the DNA of five cod bones from the harbour at Haithabu (datable only to *c*. 800–1066) with a genome of Lofoten origin does confirm contact between the far north and the Danish realm (Star et al, "Ancient DNA"), the small representation of all gadids and cod in particular among the fish remains from that site (7.5% of 16,749 identified remains from the harbour – Schmölke and Heinrich, "Tierknochen aus dem Hafen," 220–233 – and 109 bones among the 13,842 from terrestrial contexts – Lepiksaar and Heinrich, *Untersuchungen aus Haithabu*, pp. 17 and 119) makes any serious trade improbable.

Figure 8.3 Drying fish in medieval Scandinavia, images from Olaus Magnus:
a. Norwegian stockfish: codfishes decapitated and bundled whole. Detail from *Carta marina* 1539. Reproduced with licence for public use/publication courtesy of Per Cullhed, Uppsala University Library, Uppsala, Sweden.
b. Swedish Bothnia: fish air-dried on the rocks. Illustration from *Historia*, 1555, lib. 2, cap. 6, p. 65. Reproduced with licence for public use/publication courtesy of Per Cullhed, Uppsala University Library, Uppsala, Sweden.

priority on cod fishing or carried on any trade in this product. When available, local cod, fresh, dry-salted, or salted, was also eaten along European coasts further south.[61] So not all cod eating is stockfish nor marks long-distance trade.

Development of Norse society and the cultural integration of elites into Europe during the eleventh and twelfth centuries – a process symbolized after the 1030s by official Christianity – modified demand for northern goods in, at first, the Norwegian west (Bergen) and south (Oslofjord).[62] To the furs and hides of marine mammals which the Vikings had sought in the far north, their descendants added fish, of which traces occur around Trondheim and Bergen before 1100. In 1103/7 King Eystein, brother of a crusader to Jerusalem, imposed a tax in kind on northern fishing and soon thereafter signs appear of English and north German demand for Norwegian cod.[63] A settlement at Vågan on the main Lofoten land mass became the only quasi-urban centre in the north. It served for tax collection, fish processing, and trade with merchants from the south.[64] Norwegians themselves carried stockfish to England well into the thirteenth century, but Hansards came to Bergen for it. Irrelevant here is the process whereby the Hanse established economic hegemony in later medieval Norway, save to observe that the Germans more than once violently drove English rivals from Bergen and had their return outlawed. But Norwegian monarchs did normally sustain sullen German and English acquiescence in the Bergen staple, the rule barring foreigners from trading beyond that city to Lofoten a

[61] Stable isotope identification of regions of origin for cod remains from England and Flanders dating before 1000 indicates only local catches and up to 1200 still only traces of arctic cod (Barrett et al., "Medieval isotopes," 1521–1522). For differential use of local cod populations see Harland et al., "Fishing and fish trade," 196–198; Orton, "Fish for London," 20–211; Heinrich, "Kabeljau," abridged as "Fishing and consumption of cod," and Heinrich, "Fischreste als archäologische Quellengattung," 174–175.

[62] Much here highlights key features of an interpretation developed in Bertelsen, "Northeast Atlantic perspective," 22–26; Urbanczyk, *Medieval Arctic Norway*, 42–78 *et passim*; Nedkvitne, "Fishing," "Trade," and "Development"; Ebel, "Fischerei und Fischereimethoden"; Perdikaris, "Scaly heads and tales" and "Chiefly provisioning"; Christensen and Nielssen, "Norwegian fisheries," 147–149; Perdikaris and McGovern, "Codfish and kings"; Nielssen "Early commercial fishing," 45–46; and summarized in Barrett, "Medieval sea fishing," 257–262.

[63] Still unsettled is what share of the cod remains found in late eleventh-century England, for example, are already Norse exports or if overseas trade developed only in the twelfth. See Barrett *et al.*, "Dark Age economics" and "Origins of intensive marine fishing"; Nedkvitne, "Early commercial fishing," 50–53; and Harland et al., "Fishing and fish trade," 191–196 with tables 15.2–15.3.

[64] Urbańczyk, *Medieval Arctic Norway*, 121–163; Nielssen, "Early commercial fisheries," 46–48.

thousand kilometers north or, into the fifteenth century, to Iceland a thousand kilometers west. Norse stockfish went to European tables through Bergen.[65]

Stockfish delivered to Bergen in summertime had dried in the sub-freezing overnight temperatures of an arctic spring. Men who were otherwise peasants fished through the late arctic night and brief dawn of February and March.[66] The inshore stock of smaller cods which had served the original subsistence fishery could not sustain commercial demand. Fishing zones shifted from coasts near the farms to offshore of southeastern Lofoten, where the big migratory cod arrived to spawn.

Structural constants and shifts in the fishery fill in the broader picture.[67] Early trips by northerners to Bergen gave way in the 1200s to Bergen merchants dealing in north-country harbours. Only after Norway's mid-fourteenth-century loss of economic and political strength could Hansard merchants successfully evade Norwegian protectionism and deal directly with northern processors of the fish. Most business still occurred in Bergen, whence men of Lofoten sailed each May, but late fourteenth- and fifteenth-century sources document numbers of illicit ventures to the far north, too. In both venues merchants advanced to northerners goods on credit for future delivery of fish, a good indicator of the Hansards' drive to lock down future supply.[68]

When it comes to counting stockfish production and thus gauging the scale of the Norwegian fishery, medieval records are obscure. Knowing that the value of stockfish entering late fourteenth-century Lübeck each year was one-quarter that of the herring[69] simply omits too much. Extrapolations by Norwegian scholars make something in the order of 3,000–4,000 tonnes of *skreið* out to provide half the value of Norway's exports about 1300, with half or more of that weight going to England. Output likely peaked then: a century later Europe's demographic crash and competition from other fisheries are thought to have cut the volume by a third to a half. The next grounded estimates for 1518–1521 come to annual exports of about 1,500 mT and about 2,800 mT in the 1560s–70s. Nevertheless economic historian Arnved Nedkvitne describes 1350–1550

[65] Perdikaris and McGovern, "Viking Age economy"; Wubs-Mrozewicz, "Fish, stock, and barrel"; Øye, ed., *Bergen and the German Hansa*; Nedkvitne, *Hansa and Bergen*, 277–496.

[66] Venetian castaways on Lofoten in 1431–1432 provide the earliest known description of the actual fishery (Bullo, *Il viaggio Querini*, 68–69).

[67] Notably developed in Urbańczyk, *Medieval Arctic Norway*, 132–147, and summarized in Christensen and Nielssen, "Norwegian fisheries," 149–150.

[68] Nedkvitne, *Hansa and Bergen*, 535–568.

[69] Dollinger, *German Hansa*, 242.

as the 'Golden Age' of Norwegian stockfish.[70] The assertion rests not on the size of the catch but on changing values in late medieval European markets.

Northern regions better survived Norway's fierce late medieval economic and demographic crisis than did the country's south. Deteriorating climatic and economic conditions for northern agriculture and for a mixed subsistence strategy highlighted the continued relative strength of demand for fish. The grain which a fixed weight of stockfish could obtain in Bergen rose rapidly after 1350. Relying on what are sparse price references from Bergen, Nedkvitne found a kilogram of stockfish worth only 2.1 kg of rye flour before the Black Death but averaged 7.3 kg in the following century (1351–1440) and slid only slightly to 6.3 kg of rye by 1500.[71] Even exports only half those of around 1300 thus earned around 1400 half again more of the essential cereal little grown in the north. Settlements in the far north moved toward the coast. Some places and men came to specialize in fishing. Land rents, hitherto calculated in weights of butter, became – the first evidence is from 1432 – expressed in weights of *skreið*.[72]

Icelanders, mid-Atlantic cousins to Norwegians, increased their fishing for subsistence but not for export up to 1264, when their independent Commonwealth acquiesced to Norwegian suzerainty. Local chiefs sponsored seasonal fishing for this purpose. Overseas markets were attracting effort by 1300, when first isolated cryptic references then direct evidence establish Icelandic *skreið* as strong competition to the Norwegian product.[73] With dried fish supplanting coarse woollen textiles as the island's

[70] Nedkvitne, "Development," 53–57, and *Hansa and Bergen*, 91–95, 240–249, and 265. While some Norwegian researchers debate the precise numbers (see Christensen and Nielssen, "Norwegian fisheries," 145–146), their order of magnitude would make the weight of Norwegian cod to England about the same as Silesian carp to Kraków (see p. 306 above), though dried fish offer more concentrated food than do live ones. On the other hand, the proposed peak medieval output is but 10% of Holm's estimate (p. 332 above) for the Scania catch of herring.

[71] Nedkvitne, "Development," 53–56, repeats ratios he published in 1983 (in Norwegian) and barely revised in *Hansa and Bergen*, 496–511. The initial version was reported by Urbańczyk in *Medieval Arctic Norway*, 253–260, and further discussed by Christensen and Nielssen, "Norwegian fisheries," 150–151. Falling post-plague cereal prices drove the change, so by the mid-sixteenth century cod–grain price ratios again resembled those of the thirteenth century.

[72] Urbańczyk, *Medieval Arctic Norway*, 240–261; Christensen and Nielssen, "Norwegian fisheries," 150–152. For a precise local exemplar of sociocultural effects from changing opportunities in the Norwegian stockfish industry, see Sørheim, "Birth of commercial fisheries," which treats the Borgundfjord, the southernmost area suitable for stockfish production, closer to Bergen than to Lofoten.

[73] Perdikaris and McGovern, "Codfish and kings," 204–207, and "Viking Age economics"; Vésteinsson, "Commercial fishing." Ebel, "Fischerei und Fischereimethoden," 140–143, provides an overview. Dufeu, *Fish Trade*, would argue that some signs of

principal export commodity, a trend considered 'recent' in a Bergen charter from 1340, the fishery challenged pastoralism as an axis of Icelandic culture. At least for trading purposes Icelanders also abandoned their traditional standard of value, a unit of native cloth, in favour of a quantity of fish. The price of *skreið* relative to cloth in Iceland rose by 34 percent during the first half of the fourteenth century and, after a plateau, by another 70 percent during the fifteenth. Lucrative prospects help explain why four of five animal remains in Icelandic archaeological sites from the fifteenth and sixteenth centuries are of fish, mostly heads and anterior skeletal elements of cods, the waste from stockfish production.[74]

Distant resources were pulled into the service of European demand for fish and so distant societies were pulled in as well. Norwegian and Icelandic scholars advance stockfish exports as the chief economic element integrating these peripheries into the developed and expanded Europe of the later Middle Ages.

Recognition that Europe was reaching out has equal validity and wider resonance. King Håkon V acknowledged the driving force of European demand in the famine year 1316, when he required any merchant seeking dried fish or other goods in Norway to arrive with grain.[75] Just after 1400 English traders, their search for stockfish at Bergen thwarted by German competition, turned to Iceland. So did English fishers, who in 1416 explained that failure of fisheries in home waters had sent them (illegally) to the abundant resources off Iceland.[76] In tacit, sometimes open, defiance of the Bergen staple and the express commands of Danish and English monarchs alike, both trade and fishery grew. English buyers bid twice what *skreið* directed to Bergen might get and eventually offered

distinct processing and consumption sites and movement of preserved fish within Iceland show 'commerce' in dried cod well before 1250. Indirect subsistence seems more likely. Harrison et al., "Gásir in Eyjafjörður," do, however, provide fourteenth-century evidence of extensive inland trade based at a seasonal market in northern Iceland.

74 Gelsinger, *Icelandic Enterprise*, especially 181–194, and Hastrup, *Nature and Policy in Iceland*, especially 66–75 and 235–275, respectively provide classic economic and cultural approaches, while Amorosi, "Icelandic Archaeofauna," 276–281, details the dramatic shift in animal remains. Note that Gelsinger's table of prices uses the year 1200 as the base index. Ogilvie and McGovern, "Sagas and science," suggest climate change as another agent in the shifting balance between pastoralism and fishery, and Ólafsdóttir et al., "Historical DNA," offers genetic evidence supportive of such changes to the cod stock.

75 Urbanczyk, *Medieval Arctic Norway*, 141.

76 *Rotuli Parliamentorum.*, vol. 4, pp. 79–80, now Given-Wilson et al., *PROME*, Henry V, 1416 March, Membrane 5, 33; Marcus, *Conquest of the North Atlantic*, 131. The classic Carus-Wilson, "Iceland trade," is updated in Childs, "England's Icelandic trade," and "Fishing and fisheries," 22–23; and Gardiner, "Character of commercial fishing."

Icelandic fishers goods in advance for future delivery. English fishers worked deeper water with longer lines and more hooks than did the natives. They arrived in late spring and stayed to midsummer, often using salt to cure their catch on board and ashore. After 1420 some wintered over. By the 1430s an anonymous English political poet could assert "Of Yseland to wryte is lytill nede / Save of stokfische."[77] Risky business tantalized with rich rewards: individual ships returned to Hull in 1460/1 laden with 60,000–90,000 stockfish.

Not all welcomed the English. Icelandic annals and Danish diplomatic correspondence indict English mariners for stealing fish, looting, violence, and even murder of the Danish governor in 1467. Landowners fearing loss of labour banned full-time fishing as well as the overwintering foreigners who might offer such work. Meanwhile Hanseatic merchants and seamen also wanted what Icelandic waters offered and they boasted better official relations. A naval war ensued in the 1470s and 1480s, with armed fleets from Hamburg facing English escort vessels. Icelandic sources record eight armed clashes among Germans and English between 1486 and 1532. English traders, many from Bristol, reduced operations. English fishers, many from the east coast, hung on through the 1520s, even though in 1500 Iceland declared English long-liners outlaw, freely open to attack.[78] As observed in Chapters 4 and 6, medieval fishing communities in England and Germany had customs protecting and allocating the resources of home waters. Wealth over the horizon belonged to the one who could take it.

*8.1.2.2 Who Ate Which Medieval Codfishes?* For all that stockfish production and trade contributed to medieval development of northern resource-based economies and inspired tenacious ambition among Hansards and English, stockfish consumption lagged well behind that of herring. Smaller volumes reached more limited markets. Widespread local stocks of gadids and possibilities for salt cures elsewhere competed with the northern product. Ambiguous, even erroneous, nomenclature and incautious use of zooarchaeology blur distinct use of different ecosystems.

Stockfish abundantly filled a major dietary role in the north: Scandinavia, the Baltic, north Germany, and England. Even there, however, local gadid fisheries could often provide rival fresh and salted

77 Warner, ed., *Libelle of Englyshe Polycye*, ll. 798–799.

78 Magnússon, *Northern Sphinx*, 106–107; Marcus, *Conquest of the North Atlantic*, 125–154; Hastrup, *Nature and Policy in Iceland*, 73; Kowaleski, "Commercialization," 196–201; Jones, "England's Icelandic fishery"; Gardiner, "Character of commercial fishing." Further in Supplement 8.1.2.1.

alternatives. In southern Norway, Sweden, and the east Baltic the initial large headless imports of the eleventh–thirteenth centuries were later replaced by smaller locally caught cods.[79] In England the first surge of gadid remains during the eleventh and twelfth centuries identifiably came from the southern North Sea. Overall the big specimens of Norwegian origin became numerous only during the 1200s, earlier in London than in York. While this dominance may have faded some at York, headless long-distance imports remained strong in London contexts until the early 1400s. Recovery a human generation thereafter was likely associated with Icelandic supplies and then, after 1500, with a rising supply of salt cod (not stockfish) from both Iceland and Newfoundland. These were the fish challenging herring for primacy on English tables.[80]

Even before 1200, kitchen middens in northern Germany and the Rhineland were filling with bones from big headless cods,[81] but southwards of roughly the river Main physical remains are lacking and the written record (cookbooks, kitchen accounts, etc.) is thin or negative past 1400.[82] As earlier observed, the chef for the pope chosen by the Council of Constance found even his recipe for stockfish appealed mainly to people from western and central Germany.[83]

The Flemish menu of seafoods lacked gadids before the eleventh century. Throughout the ensuing 400 years recovered remains of the

79 Enghoff, "Southern North Sea," 55–56, and "Baltic region," 58–61; Makowiecki, "Cod and herring," 125–129; Lõugas, "Fishing and fish trade," 115. Orton et al., "Stable isotope evidence," reports a Norwegian origin up to the fourteenth century for cod remains in Eastern Baltic sites, and local catches thereafter.

80 Harland et al., "Fish and fish trade," 191–196 and tables 15.2–15.3; Orton et al. "Fish for the city"; Orton et al., "Fish for London"; Locker, *Role of Stored Fish*, 282.

81 Heinrich, "Kabeljau" and "Fishing and consumption of cod"; Lampen, *Fischerei und Fischhandel*, 51, 65–67; Enghoff, "Southern North Sea," 65–79; Paul, "Knochenfunde"; Pudek, "Untersuchungen"; Galik and Küchelmann, "Fischreste." But contexts dated *c.* 1180–1210 at both Bremen and Lübeck yield no cod (Galik and Küchelmann, "Fischreste"; Lynch and Paap, "Untersuchungen"). Later inland finds are reported in, for example, Lampen, *Fischerei und Fischhandel*, 174–175; Heinrich, "Fischreste … Bodenteich" and "Fish remains … castles."

82 This lacuna might partly stem from the sparse zooarchaeological research carried out in central and southern regions of the medieval German empire, but the absence of cod and paucity of other marine fish remains at well-managed excavations in Switzerland, Bavaria, and Austria is surely telling. Likewise the discussion of 'interior trade' in Lampen, *Fischerei und Fischhandel*, 88–202, though replete with records of herring, mentions no stockfish anywhere south of Köln or Hildesheim. Unlike herring, no form of cod appears in kitchen accounts of the 1320s from Klosterneuburg (Fritsch, *Refektorium*, 46–47). Galik et al., "Fish remains," 343, 348, and table 1, does record post-medieval gadid and other marine remains from latrines in Salzburg, St. Pölten, and the Vienna Stallburg.

83 Laurioux, "Le 'Registre de cuisine'," pp. 741–742 (recipe nr. 69). On the market for stockfish at Constance see Richental, *Konzilschronik*, Feger ed., fol. 25b (tr. Loomis, p. 101), where Loomis inattentively translates *stockvisch* as 'salt cod'.

most common representatives of the family, haddock and whiting from the southern North Sea, still rank well behind those from herring and flatfishes. Cod numbers grew but slowly, with isotope and skeletal evidence indicating none but local cods until the very end of the Middle Ages and only thereafter the appearance of fishes originating in the North Atlantic. Yet kitchen accounts for 1428/9 at the duke of Guelders' border post on the Rhine at Lobith show consistent purchases and inventory of *stockvisch* .[84]

Writing in the 1390s the Parisian *Menagier* articulated the potential ambiguities of combining medieval names and bones of codfishes. *Morue*, says he,

> If it is not salted, it is never called 'morue' in Tournai, since the fresh item is called 'cabeleaux', and is eaten and is cooked as will be said here below about 'morue'. Item, when that 'morue' is taken from the frontiers of the sea, and meant to be kept ten or twelve years, it is gutted, and the head removed, and it is dried in the air and sun, and not put in the fire or smoked; and, that done, it is called stofix.[85]

The householder was well informed about both preparing stockfish and their origin. His knowledge anticipates a shift in the archaeological evidence across northwestern France, where up to his time cod in any form is conspicuously ill-represented at all kinds of sites while other gadids, notably whiting, are often quite common. Only in the course of the fifteenth century do some contexts in Paris acquire large numbers of cod remains in the headless state indicating a preserved product.[86] Cod were by no means absent from Norman and northern Breton coastal fisheries and the fish markets and tables of that region, but further south all along the French Atlantic front hake and whiting were the gadids which mattered.[87]

[84] Van Neer and Ervynck, "Rise of sea fish," 162–167; Bosscha Erdbrink, ed., *Het 'Keuckenboeck'*, 1–38; Van Winter, "Nahrung," 341–343.

[85] Author's translation from *Menagier*, II:v, §194 (Brereton and Ferrier, eds., p. 237); the translation by Greco and Rose, p. 305, omits the original writer's knowledge of where stockfish came from. But why Tournai? For further records and thoughts on medieval taxonomy of cod, see Supplement 8.1.2.2.

[86] Clavel, *L'Animal*, 162–163, based on site reports pp. 11–56. A Paris fishmonger was prosecuted in 1393 for mixing cod and whiting in the baskets he set out for sale (Auzary-Schmaltz, "Les Contentieux," 64–67).

[87] Hocquet, "Pêcheries médiévales," 46–47, 54–56, 86–88, 109. Fishers from Dieppe pursued cod in the Channel for much of the fifteenth century but then the stock reportedly disappeared (Mollat, *Pêche a Dieppe*, 12–13). Touchard, *Commerce maritime Breton*, 58–61, emphasizes the absence of cod from Breton fishing and from the markets up the Loire which Brittany supplied throughout the fourteenth and fifteenth centuries. Only in generations after the 1490s do cod remains appear in a latrine deposit at Orleans (Marinval-Vigne, "Consommation d'animaux sauvages," 478–482). But see further on Breton markets below.

The same was true of medieval Spain and Italy. Though stockfish were likely known to some, not until the sixteenth century did preserved cods from distant waters there join hake (*merluza*, *pixota*) and other regional catches, whether fresh or preserved, as significant dietary elements.[88] Castaway Querini in 1432 assumed his fellow Venetians were unfamiliar with the "stockfish dried in the wind" (*stocfisi seccano al vente*) he encountered on Lofoten and as late as 1516 exploration chronicler Peter Martyr d'Anghiera thought it necessary to tell his initially Spanish and Italian readers that cod resembled tuna (!).[89] Subsequent early modern references to *bacalao* are vague as to whether these fish were air-dried and so 'genuine' stockfish from arctic waters or the then proliferating salt fish of English, Dutch, Icelandic, or New World provenance.[90] As earlier quoted, Paolo Giovio conflated hake and stockfish, but did associate the board-like objects with Scandinavian marine frontiers and northern European tastes. Southern European demand for *bacalá* drove no medieval program. That is an early modern fish tale, the sequel to the medieval one of commercializing cods for northern European consumers.

### 8.1.3 *Diverse Opportunities for Innovative Competitors*

Neither stockfish and other preserved codfishes nor herring swam as late medieval commercial fishers' sole quarry nor lay alone on fishmongers' or householders' tables. Under appropriate natural and market conditions innovators along several coasts found other fishes to take in quantity and process into forms which consumers far away found worth eating.

*8.1.3.1 In Eastern Atlantic Waters* Enterprising fourteenth-century fishers found offshore opportunities south and west of the British Isles. They were less tied than the northern commercial fisheries to a single species[91] – though the close cod relative, hake, and the sardine

[88] No preserved gadids are mentioned in monastic kitchen accounts from Benevento or Naples (Fritsch, *Refektorium*, 46–47). For Spanish Atlantic fisheries see pp. 345–346 below, including Ferreira Priegue, *Galicia*, 142–144 and 667–672; Childs, *Anglo-Castilian Trade*, 99; Serrano Larráyoz, *La mesa del rey*, 203–206; Morales et al., "Sobre la presencia." Grafe, *Distant tyranny*, 52–61, establishes the mid-1500s as the earliest significant appearance of (salt) cod on Iberian markets.

[89] Bullo, *Il viaggio Querini*, 68–69; Peter Martyr of Anghiera, *De Orbe Novo Decades* (Alacalá, 1516), Dec. III, lib. vi, f. 52, as cited in Williamson, *Cabot Voyages*, 266–268. Cod like tuna? Did Martyr mean to convey cod were not small fish, even though half or less the size of a bluefin?

[90] Lange, "Når kom norsk tørrfisk til Italia?" covers the debate.

[91] More species, each with less biomass, distinguish warm southern from cold nutrient-rich northern waters, and transitional regions like the outer Channel and Bay of Biscay are commonly even more diverse (see Chapter 1).

or pilchard, kin and ecological counterpart to the northern herring, played major roles.[92] These and other regional fishes engaged the attentions of coastal communities around the Channel approaches and Bay of Biscay. All participants need not be surveyed here, nor are all so far thoroughly studied by historians. The new fisheries built on prior experience with inshore and estuarine resources.

Along the English south coast increased landings of hake and cod appear in fourteenth-century written and archaeological records from Exeter to Southampton. Hake especially was here an offshore species caught during March–June in the western approaches to the Channel (the Celtic Sea) south and west of Cornwall. Already by the 1350s hake of Cornish origin were reaching Avignon by way of Bordeaux.[93] This was but one part of a new marine fishery in southwestern England which exploited diversified ecosystems along the continental shelf. Each spring and summer sharp-eyed spotters posted on shoreline cliffs in Devon and Cornwall signaled boats with seines to surround schools of migratory pelagic pilchards (adult sardines). Workers on shore, many of them women, dry-salted the fish for a month, then washed and pressed them under weights for ten days before barreling for export, as well documented by 1450. A byproduct, 'train oil', was valued as fuel for lamps and a lubricant. Other crews went off with salt in their holds to spend long summer weeks on the west coast of Ireland fishing herring, flatfishes, and more hake and using landing sites like those of English in Iceland to salt them down. Conger offered another option. Such multi-species fisheries in seasonal succession could support full-time professional fishers and more capital-intensive infrastructure.[94] Fishing

92 Modern fisheries scientists rate cod at a trophic level of 4.1 ± 0.2 and hake at 4.4 ± 0.0, but cod prefer waters in the range of 0.5–10.3°C and hake enjoy 6.9–15.4°C. Sardine like about the same temperature (7.1–17.4°C) as hake but eat much lower on the trophic scale, 3.1 ± 0.1, while herring at 3.4 ± 1.0 prefer water of 0.5–11.2°C, just about the same temperature as the cod. See www.fishbase.org/search.php (visited 12 February 2018). Learned Nürnberger humanist traveler Hieronymus Münzer, who encountered sardines in Portugal in the 1490s, thought them simply herring under a different name (Tavares, "Estancia e imagen de Portugal," 478–479).

93 Weiss, *Versorgung des päpstlichen Hofes*, 399–401. As early as 1269/70 the Hampshire abbey of Beaulieu was receiving conger, hake, and "little fish from a seine" (*minutis piscibus … provenientibus de sagena*, so pilchards or herring?) as tithe payments from Cornwall (Hockey, *Account-Book of Beaulieu*, 103).

94 Publications by Maryanne Kowaleski are authoritative: *Local Markets and Regional Trade*, 307–312; "Expansion," 429–436 and 440–448; "Commercialization," 204–227; "Fishing and fisheries," 24–28; "Peasants and the sea," 365–369; "Seasonality," 128, all usefully correct Robinson and Starkey, "Sea fisheries," 128–130. See also Gray, "Inshore and local fisheries," 82–83. For material evidence of this fishing see Wilkinson, "Fish remains"; Coy, "Medieval records versus excavation results"; and Serjeantson and Woolgar, "Fish consumption," 115–116. From Irish perspectives see

families began to cluster into new harbour- and seaside communities. The fisheries participated in the synergy among multiple marine and terrestrial sectors which pushed the southwest from a backwater to a leading region in the sixteenth-century English economy.

Breton fishers, who began only in the 1300s to leave sight of land, targeted much the same waters and species.[95] From March to June they sailed for a day or more, fished night and day for hake, then returned. Papal purchasing agents shipped thousands of Breton hake to Avignon in the 1370s–80s, and a generation later the royal household of Navarre was buying "*merluz seco de Bretagne*."[96] Those fish came from coastal villages on both sides of the peninsula where men were said to have no employment other than at sea and the fishery was helping to drive the maritime economy. In 1427 papal permission for Sunday departures legitimized boats making two voyages a week. Using the product of local brine springs and of more southerly French coasts, the salting season for hake and conger lasted from Easter to Michaelmas. All around the coast dozens of sites with sheds and racks for drying hake and conger were a lucrative seigneurial monopoly. Heavy fines coerced fishers to deliver those species for a set price. The duke leased his drying facilities to consortia of merchants in 1279 from Bayonne and later from Nantes, where his inspector also checked the mesh in fishing nets. The fishers themselves salted or brined mackerel and sardines on board ship, so evading this seigneurial impost. Only toward the end of the fifteenth century did Bretons along the northern coast start to sail farther toward Cornwall or elsewhere to catch cod, a variety absent until then from the markets up the Loire which they supplied. By 1515, however, English authorities in Ireland were blaming Breton fishermen for depleting stocks of salmon, herring, ling, and hake to the disadvantage of English ships.[97]

On another western extremity of the European mainland, Spanish Galicia, an abrupt coastal topography and narrow (20–30 km) continental shelf put deep water unusually near shore. What had earlier been inshore subsistence fisheries grew rapidly after Spanish Christians gained

Breen, "Marine fisheries and society," 94–97, and McAlister, "Castles and connectivity," 639–40 and 653–7.

95 Touchard, *Commerce maritime breton*, 58–61; Hocquet, "Pêcheries médiévales," 86–87, 96–98, and 11–112; Kerhervé, *L'État breton*, 686–692.

96 Serrano Larróyoz, *La mesa del rey*, 203. The importance of hake, commonly salted and/or dried, and absence of cod from medieval southwestern French coasts convinces me that the large purchases of *merlucii* in La Rochelle, Toulouse, and Bordeaux made during 1316–1362 by agents for the papal household at Avignon were not stockfish, as assumed by Weiss, *Versorgung des päpstlichen Hofes*, 399–401, also referred to as 'salt fish' (531) and 'dried fish' (538), but the dominant regional catch, hake.

97 Breen, "Marine fisheries and society," 95.

control of the south and opened regular sea routes to the Mediterranean. Large-scale fishing in the late thirteenth century established a base for new or re-established coastal towns – La Coruña, Pontevedra, etc. – to integrate economic activities in each Galician estuary.[98] Many aspects – not least the target species and ways of preservation – are reminiscent of stockfish and herring industries in the north. Hake (*pixota*) were taken offshore on baited lines from small boats. One royal charter made very plain that these were to be dried, not salted, for commerce. During summer and autumn, sardines (juvenile 'pilchards') frequent plankton-rich western Iberian estuaries and inshore waters. Skilled crews hired by town-based companies caught them in costly encirclement nets and delivered the catch to processing plants on land, where women removed the heads and viscera before stacking the fish for dry salting (*sardina de pila*) or racking them to smoke before brining (*sardina arencada*). Full-time sardine workers for half the year might in the other season fish hake or conger but also tend fields and vineyards. Exports to other Iberian regions began to grow in the late fourteenth century, achieved market dominance after 1400, and expanded still more after 1450. In the 1490s annual shipments of Galician sardines unloaded at Valencia alone held twice the fish in Dieppe's best fifteenth-century herring catch; at least comparable quantities were then also going to Portugal (itself a sardine producer and exporter), to Sevilla, and to Barcelona. Quantities of larger fishes, hake and conger, were commensurate.[99]

[98] Though appearing fjord-like on a map, the *rias* are drowned rivers, not glacial formations.

[99] Ferreira Priegue, *Galicia*, 130–156, 345–346, 661–664, 667–672, and 728–739; Gautier Dalché, "La Galice"; Ferreira Priegue, "Pesca y economia." Already in the mid-thirteenth century dried Galician hake were appearing on markets in northern Castile (Martinez and Carbajo, "L'alimentation des paysans castillans," 345). Fifteenth-century shipmasters from Galicia and Portugal delivering sardines by the millions and hake by the thousands to Barcelona and other Catalan ports are documented by Salicru i Llunch, "En torna al comercia," 167–170, while Rodrigo, "Fresco, frescal, salado, seco, remojado," 558–563, has commensurate customs accounts for their import in Aragon. For the archaeozoological record of hake see Morales-Muñiz et al., "Hindcasting to forecast." Capture of cod by Galician fishers is most unlikely and no shipments of that species to the Mediterranean are recorded before 1500 (Morales et al., "Sobre la presencia de bacalao," 21–23). Recent archaeozoological research reported in Rosello Izquierdo et al., "Iberian medieval fisheries," suggests coastal settlements consumed a great variety of species while those inland ate mainly sardines. High seas cod fishing and bases in Newfoundland would therefore be a new departure for Gallegos, who arrived in the New World in 1518, a decade after more experienced Portuguese, Normans, and Bretons.

While it is discomfiting to omit Portugal from this tour of commercialized distant water fishing off late medieval southwestern Europe, a decision to do so rested on more than a need to limit this chapter's length. Portuguese whalers and fishers did take early advantage of New World opportunities but the prior process remains clouded. While

*8.1.3.2 From Local Abundance to Distant Tables* Subsistence or artisanal fisheries with potential for commercial expansion in the high and later Middle Ages were not necessarily all that far 'offshore', at least from the fishers' point of view. Besides herring off Danish beaches and pilchards in Cornish coves, consider briefly two more products then reaching far-off consumers from what Parisians or Londoners surely would have thought northern 'frontiers of the sea'.

What clerks and officials from Stockholm to Kraków called *strekfusz* (alias *stracfuss*, *strakus*, etc.) and clearly distinguished from stockfish draws attention to the importance of fish-drying in the northern maritime interior as well as ocean shores.[100] Prussian chronicler Simon Grünau, who treated seventy-two fish varieties in a 1526 description of his province, identified *strekfusz* as dried eel, sturgeon, salmon, herring, whitefish, and bream, saying the name came "from the places where one first dries them in the air." Grünau mentioned Prussian exports to German, Polish, and Czech lands from the Elbe to the Carpathians.[101] Sources going back to the 1320s from Prussia and Poland reveal "dried fish called strekfuss" (*sicci pisces strekfussy dicti* ) everywhere from royal households to local marketplaces and more often identified as pike than any other named fish.[102] Oil-free as cod, pike were common in the many lakes of Prussia and Finland as well as the Gulf of Bothnia, the nearly fresh northern arm of the Baltic. In excavations at medieval Uppsala the big Norwegian stockfish bones from thirteenth-century layers give way after about 1325 to fewer small Baltic cods and many large pike, typically cut for hanging to dry.[103]

I may sadly have missed some historical research with economic and ecological perspectives, works that have come to my attention describe marine resources and activities much like those in Galicia, if less oriented to exports. Most recently (2021) de Costa Dominguez, "Harvesting in holy waters," 161–163, while lamenting recent decades with little new research, concludes that Portuguese fisheries began major development only after the oceanic discoveries. For more detail please see Supplement 8.1.3.1.

Absent from the late medieval record of long-distance fishing are the Basques, and so too consumption of distant-water fishes in their market hinterland of Navarre. While Basque whaling and local fisheries are well in evidence, including rare references to voyages as far as the North Sea for hake and herring (see, for example, Hocquet, "Pêcheries médiévales," 114), trips far offshore or to the New World lie in the realm of myth. For references and discussion see Supplement 8.1.3.1.

100 What follows is a precis from Hoffmann, "*Strekfusz.*"

101 Grünau, *Preussische Chronik*, ed. Perlbach, 46–47.

102 Seligo, "Zur Geschichte." 27; Sarnowsky, *Wirtschaftsführung des Deutschen Ordens*, 130–131, 625–626, and 814–825; Piekosiński and Szujski, eds., *Najstarsze księgi Krakowa*, 235–241; Piekosiński, ed., *Rachunki dworu*, 22, 310 *et passim.*

103 Enghoff, "Fishing in the Baltic," 58–61; Jonsson, "Finska gäddor och Bergenfisk." Sten, "Trading with fish," 65–66, reports a similar pattern of finds from a hospice in Stockholm.

When Swedish internal customs records begin in the mid-sixteenth century, salmon and dried pike make up 80–90 percent of shipments to Stockholm from Bothnia. Contemporary northern historian Olaus Magnus observed the lucrative profits gained in this export trade and how from the acres of fish spread out to dry at one Bothnian processing site, Bjuröklubb, "there rises such a stench of fish that far out to sea sailors as they approach are aware of it flying out to meet them" (see Figure 8.3b).[104] So also brackish or freshwater frontiers could support fisheries of near mythic scale.

Salmon from Scotland, another resource as locally traditional as pike in eastern Baltic lands, emerged as an important export industry during the fifteenth century, in part by replacing sales by count with packaging in salt and standard barrels. While runs elsewhere dwindled, Scotland's less intensive agrarian regime and regulatory protection sustained local stocks. Royal grants had given lay and ecclesiastical magnates rights over specific estuarine and riverine locations to deploy nets and traps and take adult fish returning from the sea to spawn. Rights holders employed obligated peasants or let the fishing out to consortia of free commoners for shares of the catch, but churchmen also claimed one fish in ten as tithe. Export agents at Aberdeen and Leith consolidated local surpluses.[105]

Principal export markets for Scotland's salmon developed where local stocks had been depleted all around the southern shores of the North Sea. Scots first shipped salmon to the Low Countries in the fourteenth century, then expanded sales to eastern England and, especially after 1450 to France. Fragmentary surviving customs accounts show Aberdeen alone exporting 200–500 barrels a year in the late 1420s–early 30s and the national total going over 2,000 barrels by the 1470s and 3,000 in the 1530s–40s. Receipts for import tolls on the principal shipping channel to Antwerp corroborate such numbers. Prices rose almost continually from the 1420s into 1580s and profits from this trade helped

[104] As translated in Magnus, *Historia … Description*, vol. 1, pp. 99–100, from original Latin in Magnus, *Historia*, bk. 2, ch. 6, pp. 65–66. Bjuröklubb is about 700 km north of Stockholm. In a separate systematic discussion of northern fishes, Magnus stressed the importance of the pike fishery and told how these fish were dried for storage and shipped in bundles or "stacked like great piles of logs" (*ibid.*, bk. 20, chs. 1–2 and 8–9, pp. 697–699 and 704–705). Friberg, *Stockhom i bottniska farvatten*, 196–199, has the official customs returns.

[105] See in general Hoffmann, "*Salmo salar.*" Hoffmann and Ross, "This belongs to us!," relates conflicts between the royal borough of Stirling and neighbouring Cambuskenneth abbey over salmon in the river Forth, while Hodgson, "To the abbottis profeit," delineates the rising role of salmon from the Tay system on the estate of Cistercians at Coupar Angus. Since the 1420s cleaned and salted salmon in 'Hamburg barrels' (calculated at twelve to the *last*, so fourteen gallons or fifty-four liters each) were a standard commodity.

underwrite Scots' overseas credit operations.[106] But the gustatory appeal of those salmon which each year passed through the northern entrepôt faded on the long further journey to Rome: Paolo Giovio consigned this import to consumption by commoners, "for the salted ones lose their original nobility."[107]

*8.1.3.3 On the Southern Frontier* Distinctive environmental structures of Mediterranean waters held maritime opportunities and challenges different from those of northern seas. Diverse local fisheries have already been examined, but the region offered few of the large-scale single-species commercial possibilities which so excited the north. A major exception was the marine migration of pelagic bluefin tuna, a top oceanic predator. Mobile schools of 50- to 200-kilogram parcels of rich and oily flesh gathered annually off southwestern Spain before entering the western Mediterranean to spawn and to forage on mackerel and sardine in a counterclockwise gyre past the northern coast of Sicily, along Sardinia, and then westwards off Provençal and Catalan shores. Other populations occur in the Adriatic and eastern reaches of the sea.[108] Owing in part to their being an unusual homeothermic ('warm-blooded') fish, bluefin move too fast for effective trawling with wind-powered vessels or more than incidental capture by hook and line, but if the schools can be stopped or contained, hard and bloody work with harpoons and gaffs yields a huge return.

Principal late medieval and early modern fisheries for bluefin arose along the north coast of Sicily and the southwestern corner of Castile (coastal Andalusia). Both production centres emerged during the thirteenth century, with the Sicilian industry likely leading growth in the 1300s and Andalusian output playing a larger role by the late 1400s. Capture techniques revived or improved upon ancient forms of barrier fishing. Spanish *almadrabas* were permanent locations where, as well known in the classical Mediterranean, upon sight of the tuna, boats directed by signals from shore-based observation towers moved out to

106 Rorke, "Scottish overseas trade, 1275/86–1597," vol. II, pp. 564–571 and 662–669; and compare Unger, *De tol van Iersekeroord*, 294–299 and 504–516. If the numbers even approximate a lower limit, 3,000 barrels came to exports in the range of 150 mT. Innes, ed., *Ledger of Andrew Halyburton*, illustrates the dealings of just one Scottish merchant in the Low Countries. Gemmill and Mayhew, *Changing Values*, 303–317. If, as I suspect, salmon of Irish origin also entered interregional trade, I have failed to find evidence for it.

107 Giovio, *De romanis piscibus*, cap. xliii (p. 60): "*Salmones etiam Gallia belgica quotannis mittit, sed in plebis usum, quum saliti pristinam nobilitatem amittant.*"

108 Zambernardi, "Dernier 'paysans de la mer'," 86–93, and Felici, *Thynnos*, 39–62, provide biological and pre-medieval details.

deploy a massive vertical net in front of the school. Stopped and then surrounded by a wall of mesh the fish were hauled to shore behind teams of oxen and there butchered. Early modern descriptions and images depict immense beach seines (Figure 8.4a).[109] Sicilian *tonnara* were semi-permanent installations of a type first recorded in tenth-century Byzantium but also known to contemporary Arabs. At chosen coastal sites a long leader fence was anchored perpendicular to the shoreline to divert migrating schools into successive impounding chambers, ending in the 'death chamber' (*camera della morte*) where men raised the net floor and took the fish with barbed harpoons (Figure 8.4b). Documented enterprises multiplied from perhaps a dozen in the thirteenth century to about thirty in the fifteenth, when the fishery expanded westwards from Palermo to Trapani.[110]

Whether taking the fish by seine or trap, each of these enterprises employed 200 or more workers, seamen, butchers, cookers, coopers, packers, spotters on cliffs or towers, and armed guards, for neither coast was without danger from North African corsairs or Christian rivals, local or dynastic.[111] Both generated significant demand for salt and fuel supplies and otherwise stimulated local economic activity, but did so from distinctive positions in political economy. The Sicilian crown retained tuna trapping rights but leased individual sites to entrepreneurs who selected a knowledgeable, quasi-hereditary labour chief, the *rais*, to assemble and operate the catching crew. This team formed a share association for the duration of the tuna season, mid-April to late June. Shoreward of each *tonnara* salaried and wage labourers butchered the tunas, then cooked, brined, and barreled the flesh. Investors from Palermo's civic nobility, from Genoa, and from other towns in Sicily itself, financed the enterprise by advance purchase of its catch or its product, *tonnina*, the processed and packaged tuna meat of various grades. In post-Reconquista Andalusia great landed families (grandees)

[109] Ladero Quesada, "Las almadrabas"; Phillips, "Who owns the fish," 80–87; García Vargas and Florido del Corral, "Origin and development."

[110] Zambernardi, "Derniers 'paysans de la mer'," 93–96, uses the technically ambiguous French term *madrague*. Bresc, "Pêche et les madragues" and *Un monde méditerranéen*, 264–272, thoroughly cover the Sicilian *tonnara*. García Vargas and Florido del Corral, "Origin and development," 212–213, find no classical Greco-Roman reference to such fixed structures but nevertheless persist in applying the Spanish term *almadraba* to what elsewhere were distinctly traps not seines. Arabic terms applied in *tonnara* fishing indicate presence of this technique before the eleventh-century Norman conquest of the island (see also Felici, *Thynnos*, 103–112).

[111] Secure Castilian control of Gibraltar came only in 1340, which may help explain earlier relative obscurity of the Andalusian fishery. Dynastic conflict between Aragonese Sicily and Angevin-ruled Provence and Naples was a constant of late medieval Mediterranean politics.

Figure 8.4 To catch the bluefin tuna.
a. Sixteenth-century representation of the *almadraba* in Bay of Cádiz.
*Almadraba* depicted in Georg Braun and Franz Hogenberg, *Civitates orbis terrarum* 1572–1618, vol. 5, image 19. Original engraving attributed to Georgius Houfnaglius. Original is Public Domain. Reproduction thanks to the Geography and Map Division, Library of Congress.

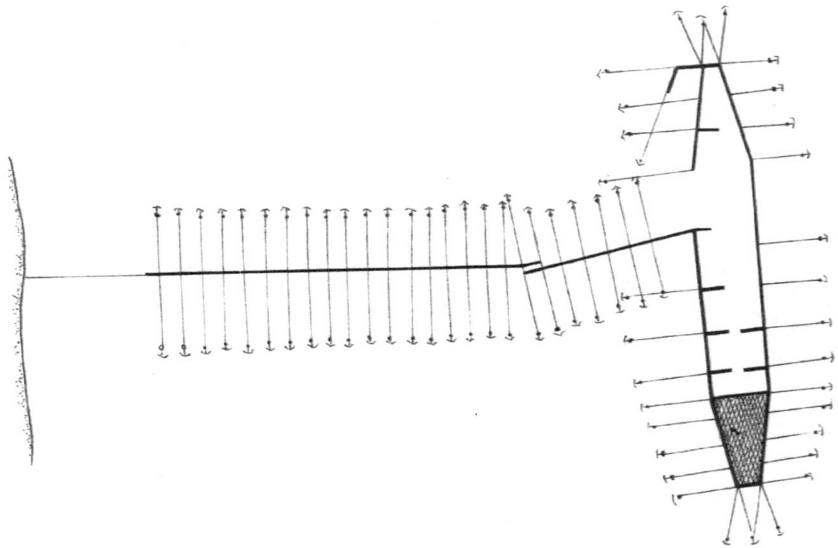

Figure 8.4 (*cont.*)
b. *Tonnara* trap (modern schematic).
Schematic of *tonnara* after Brandt, *Fish Catching Methods*, fig. 274, and similar published images elsewhere, redrawn by D. Bilak for R. Hoffmann.

early gained control of *almadrabas*, along the Bay of Cádiz. But between 1299 and 1444 successive Castilian monarchs rewarded the loyalty of the Guzmán family, counts of Niebla and since 1445 dukes of Medina Sidonia, with what became a claim to exclusive private rights over all tuna fishing along the entire shore from the Portuguese border to Gibraltar. Struggles for control – even to private naval bombardment of a rival's working *almadraba* – suggest the value these magnates placed on the fishery, whether for its material product or as a sign of power and prestige. From the 1520s or so and into the eighteenth century the house of Medina Sidonia and its servants actively and effectively enforced their monopoly, assembling the capital, hiring the specialized workers, employing numbers of slaves, and marketing the product.

Managerial accounts for medieval *tonnara* are not known to survive. Fragmentary fifteenth-century financial returns let historian Henri Bresc estimate each killed one or two thousand fish a year and packed roughly the same number of standardized barrels, so forty to eighty tonnes of marketable *tonnina*. If so, the thirty Sicilian traps operating in the fifteenth century annually turned thirty to sixty thousand bluefin into 1,200 to 2,400 tonnes of product.[112] Individual Spanish *almadrabas* seem to have taken more and larger fish than did their Sicilian counterparts. When the Cádiz enterprise was temporarily in royal hands during the 1510s, the five years of full accounts averaged annual capture of 6,000 tunas yielding seventy-eight tonnes of processed meat and marketable byproducts. That may have been an unusually low return. Subsequently the two *almadrabas* run by the ducal house –which kept precise records – together averaged 29,487 fish a year during the 1520s–30s; if those enterprises handled their catch as had their rivals in Cádiz, this meant an annual output of about 400 tonnes. But when the same Medina Sidonia tuna fishery peaked out in the 1540s–60s it caught more than 100,000 fish and preserved for sale some 10,000 tonnes of mostly salted and brined product;[113] at that rate the smaller catches of the 1520s–30s would have come to almost 3,000 tonnes per annum. Processing and marketing destroyed or dispersed the skeletal material, so tuna bones are rare in archaeological contexts and can neither directly verify nor quantify the written references to these fisheries.

[112] Bresc, *Un monde méditerranéen*, 271–272, also suggests fifteenth-century Sicilian production was (like that of Norwegian cod) smaller but more valuable than in pre-plague times; Ladero Quesada, "Las almadrabas," 353–354.

[113] Phillips, "Long-term profitability," table 1 and figure 1; and Phillips, "Who owns the fish," 84–87. The marked discrepancy between Sicilian and Andalusian yields of fish and of product remains puzzling. Is something omitted from Bresc's Sicilian records? Of course neither set of modern scholars has been attentive to the other.

Besides extensive regional distribution to coastal cities and their hinterlands, preserved tuna from Sicily and Andalusia was marketed and eaten around the western Mediterranean. In peacetime from the early fourteenth century on, Palermo annually shipped thousands of barrels to Naples and Tuscany – where Pegolotti knew them in the 1330s – to Liguria and Rome – where Paolo Giovio found brined *tonnina* in all the taverns in 1524 – and in the fifteenth century, even 'Spanish style' to Sevilla.[114] Andalusian exports to Catalonia, insignificant until a 1374 break in dynastic wars, grew strongly from the first half of the fifteenth century, when the traps also proliferated along the Catalan coast.[115] Though the capture sites were more removed than Sicily from Italian urban markets, by the early sixteenth century Castilian, Genoese, Florentine, and Catalan merchants were handling large volumes of Andalusian tuna to Cagliari, Naples, Livorno, and Genoa. No quantitative assessments of this trade have been proposed.

><> ><> ><>

If modern estimates are to be believed, even just as approximate orders of magnitude, a minimal late fifteenth-century year's catch of Atlanto-Mediterranean tunas, now also including those from Provence, Catalonia, and the Adriatic,[116] may have yielded in the range of 3,000 tonnes of processed fish. Higher Spanish yields would have put the total over 5,000 tonnes. This was more than three times the contemporary Norwegian output of stockfish and about equal to the latter's peak two centuries before. But brined tuna in barrels begins to approach the

[114] Bresc, "La pêche et les madragues"; Bresc, *Un monde méditerranéen*, 264–272. Epstein, *Island for Itself*, 294–295, doubts that the tuna industry relied principally on extra-Sicilian markets, but does concede (pp. 284 and 346) its rapid growth, capital intensity, and profitability. Pegolotti, *Practica della mercatura*, 181, 205, and 380; Giovio, *De romanis piscibus*, cap. xlii (on pickled fishes), p. 60. Salt and fresh tuna appeared seasonally on monastic Lenten menus in fourteenth-century Naples (Fritsch, *Refektorium*, 48). Supplement 8.1.3.3 has more on Italian consumption.

[115] Diago Hernando, "Relaciones comerciales," 31–34, 44–45, and 50–51; Garrido i Escobar and Pujol i Hamelink, "Changements techniques," 26–27. Sánchez Quiñones, "Los precios," 182–183, reports a retailer of tuna in Guadalajara in 1485. Further on the tuna fishery in Provence and Catalonia, which took less desirable spawned out fish, is in Supplement 8.1.3.3.

[116] Also from the fifteenth century Dalmatian fishers exploited a distinctive stock of bluefin in the Adriatic, replacing seines with traps only in the late sixteenth century. Venice, followed by other eastern Italian ports, provided their principal market. One estimate suggests an annual late fifteenth-century catch in the range of 30,000 Venetian pounds (15 tonnes?). The tuna probably lagged behind total Dalmatian output of sardines, mackerel, and diverse inshore species, the latter mainly serving local consumers. Fabijanec, "Fishing and the fish trade," 369–375.

weight of the live animal, while a given kilogram of dried and mostly deboned stockfish had been at least 4.5 kilograms of live cod. In the latter terms, then, the wild biomass of cod and tuna taken around 1500 were roughly equivalent (5,000–7,000 mT) while about 1300 the Norwegian take of cod had been two or three times larger (13,000–18,000 mT). Yet both paled by comparison to herring where processed and live weights more closely aligned. As earlier remarked, North Sea herring catches of the early fourteenth century reached or exceeded 15,000 mT and at the end of that century the Øresund take at 40,000 mT was three times greater still, so two to five times larger than probable maxima taken in the great medieval single species fisheries elsewhere.[117] Sadly, though wisely, historians have attempted no similar calculations for the herring catch in the fifteenth-century North Sea nor that of sardines and pilchards across the waters between Cornwall and Portugal. Surely, however, from before 1300 to beyond 1500 aggressive expansion of European fishing into hitherto little-used stocks and waters on all of Christendom's maritime frontiers brought large new marine biomass into European diets and enlivened many regional economies.

## 8.2 Markets and Ecosystems, Expectations and Experiences

New frontier fisheries of the later Middle Ages established distinctive relations between markets and ecosystems. Unlike traditional local catches, fish from 'over the horizon' seldom reached the consumer fresh.[118] Most output from the new marine fisheries was preserved for consumers distant in time and place from the animals and ecosystems on which those people now fed. Separation of fish eaters from the environments which produced fish coincided with evolving rearrangements in

[117] Recall the calculations of Holm, "Commercial sea fisheries," 15–20, as discussed pp. 321–322 above. Peak Dutch catches of North Sea herring in the early seventeenth century went to or slightly above 75,000 mT (Holm et al., "Marine animal populations," 9; Poulsen, *Dutch Herring*, 43–45), which Nedkvitne, "Development," 55–57, approximates as between 800,000 and 1,200,000 barrels. By comparison, might we guesstimate annual carp production in early sixteenth-century Bohemia and Poland (see Chapter 7) in the range of 3,000 mT?

[118] One possible exception was England, small enough for pack horse trains from the coast to reach the interior with still-edible seafood, especially in cooler seasons (Dyer, "Régimes alimentaires"; Dyer, "Consumption," 30; Serjeantson and Woolgar, "Fish consumption," 105, have adequate examples). But Locker, *Role of Stored Fish*, 93–96, rightly points out how the cost and unreliability of fresh supplies encouraged consumption of the preserved. The remarkable effort it took to supply rich Parisians with fresh marine fish (Chapter 4) sets the 150 km distance overland as a practical maximum for any fresh fish, whatever the source. Ordinary consumers either ate from a much narrower radius or they ate preserved fishes.

the fisheries themselves. In a natural world subject to both autonomous and anthropogenic stresses, the consequences could multiply, some of them neither wholly intended nor easily anticipated.

### 8.2.1 *Distinctive Market Features*

Overview of major and representative new fisheries repeatedly observed three characteristic features: preserved fish, mass markets, and consumers elsewhere. Each had implications for fisheries, merchants, and ultimate purchasers. More can also be learned from a (modest) proliferation of price series for late medieval fish.

*8.2.1.1 Preserved Fish* The big commercial fisheries increased the durability of their products with two major technologies, conceivable as 'wet' and 'dry'.[119] The former method mostly applied to fishes with oily flesh – herring, sardines, tuna, salmon – and the latter to gadids, flatfishes, and pike, species which store energy-rich fats and oils elsewhere. Wet brining processes sealed the oils away from the air or induced chemical changes to more stable compounds. Most effective on large scale, these methods required economical access to copious supplies of salt. About 1400 it took a tonne of salt to pack every five tonnes of Scanian herring and more for the Dutch method; 500 barrels of *tonnina* required 15.5 cubic meters of salt.[120] In some southern fisheries olive oil could replace salt brine. Drying fish had climatic preconditions, freezing cold or blazing hot, so long as the humidity was low, but the process was cheap and the product kept well in dry conditions. A combined procedure, dry salting, put the processor back into dependence on the salt trade and, to judge from the history of herring and cod, gave a less durable but often cheaper product.

In medieval towns of appreciable size, salt, dried, and pickled fishes were the stock-in-trade of 'salt-fishmongers' or 'herringers' (*harengier, allecistae, śledzowniki*), retail merchants distinct from purveyors of fresh, preferably live, local fish.[121] Manuscripts of the health manual *Tacuinum*

[119] Cutting, *Fish Saving*, 2–3; Cutting, "Historical aspects of fish," 8–16; Mollat, *Europe and the Sea*, 63–65; Hundsbichler, "Nahrung," 203; Locker, *Role of Stored Fish*, 51–67.

[120] Bresc, *Un monde méditerranéen*, 269. To assess commercial quality Pegolotti, *La practica della mercatura*, 380, applied similar criteria to the pickled herring of the north as he did to Mediterranean *tonnina*.

[121] Géraud, *Paris sous Philippe le Bel*, 516; Lespinasse and Bonnardot, eds., *Les métiers et corporations*, 218–222 ; Laurière et al., eds., *Ordonnances*, vol. II, 575–582 (cf. Delamare, *Traité de la police*, as discussed in Chapter 4, p. 167, note 137, above); Lespinasse, ed., *Les métiers et corporations*, Tom. 1, 13–19 and 409–426; Lentacker *et al.*, "Historical and archaeozoological data"; Piekosiński, ed., *Kodeks dyplomatyczne miasta Krakowa*, #262 and 299; Ptaśnik, *Cracovia artificum*, #201–2, 218, and 230.

*sanitatis* illuminated in Lombardy in the late fourteenth century depict shops retailing different fishes as *pisces saliti* (Figure 8.5).[122] While herring or sardine might be eaten straight from the container or simply warmed, other preserved varieties demanded special culinary preparation as well. Eberhard, chef to Duke Henry 'the Rich' of Bavaria–Landshut (1404–1450) asked "Would you like to make a good stockfish?" Soak it for half a week and then cook in lard with ginger and saffron. The Parisian compiler of the *Menagier* alternated soaking the stockfish and beating it with a mallet. *Tonnina* from the cask had to be elaborately washed in fresh water and so, too, the most strongly brined or dry-salted herring.[123] Then recall (Chapter 2) that a chorus of physicians and dietitians – Eberhard among them – condemned all preserved fish on medical grounds. No wonder elites uniformly sneered and thought consumption an act of penance. Bishop Matthias von Rammung of Speyer left no doubt in his 1470 kitchen ordinance: "Herring shall not be considered a fish dish, [for] one can in no way make them acceptable."[124] To varying degrees and as compared to freshwater fishes if available, this was no high-status food. Its purveyors rather sought much larger markets.

*8.2.1.2 At Unprecedented Scale* Mass markets for preserved fish were served by shipments and sales in bulk, "by the last and by the hundred" as some English accounts put it. Indeed the standard measure for Galician sardines was the 'thousand' by count and actual shipments of sardines and herrings alike were, as previously observed, in many,

122 Hoeniger, "Illuminated *Tacuinum sanitatis*," p. 61; Schlosser, "Ein veronesisches Bilderbuch"; or De Battisti, ed., *Il libro di casa Cerruti*. Compare Kightly et al., *1465 Walraversijde*, p. 78 (from the Liege ms), with a different iconography done in Milan, not, sadly, the Low Countries. Shops and barrels contrast with the open market and baskets for fresh fish in Figure 4.3.

123 Feyl, "Kochbuch des Eberhard," recipe #23; *Ménagier*, ed. Brereton and Ferrier, 237 (§194). The *ménagier*'s technique may descend from that Pliny reported (*Hist. nat.*, lib. IX, xxxii) from Ibiza for handling dried gadids, in all likelihood hake, "which cannot be cooked unless beaten with a stick" (*qui nusquam percoqui possit nisi ferula verberatus*). Johannes von Bockenheim just soaked and boiled his stockfish (Laurioux, "Registre de cuisine," 741–742), but contemporary Flemish cooks also beat it first (Braekman, ed., *Een nieuw zuidnederlands Kookboek*, recipes 206–208). On the elaborate handling of *tonnina*, see Platina, *Il piacere*, cap. 353 (ed. Faccioli, 216–217). Lampen, *Fischerei und Fischhandel*, 191–193, discusses the craft of *lotor allecium* or *herinchweschere*.

124 "*Heringe sollen nit fur vische geachtet werden, man mochte es dan nit mol gebessern*" (Fouquet, "Wie die kuchenspise sin solle," 27). A satirical Nürnberg poem from the 1490s, "Kunz Haß," railed against the hazard of consuming any preserved fish (Schmauderer, *Lebensmittelwissenschaft*, 260–261). Cutting, *Fish Saving*, 4, 27, and 30–32, reiterates the general point. Locker, *Role of Stored Fish*, 277–282, argues that English elites sharply cut back their consumption of herring in the last medieval centuries, preferring ever more strongly stockfish or other less durable but more palatable preparations and varieties.

Figure 8.5 The saltfish monger.
As represented in a manuscript of *Tacuinum sanitatis* illuminated in Lombardy *c.* 1370. ONB cod vind. ser. nova 2644, fol 82v 'pisces saliti'. Reproduced with permission of the Austrian National Library.

many thousands. As early as 1321 an estimated thirty million herrings entered Paris; with time and freshness no constraint, most were sailed or hauled slowly but cheaply up the Seine. A century later Portuguese and Galician shipmasters each year delivered sardines in the millions and hake in the tens of thousands to Barcelona and other Catalan ports. Merchants come to Kraków from elsewhere could offer salt fish at retail for just three days – though they could await a fish day – and thereafter sell only by the cask or keg.[125]

Bulk packaging dominated. At customs frontiers of landlocked Aragon in 1447–1448 sardines arrived in baskets, dried and smoked herring in crates (*pirotas*), and brined tuna in barrels. While stockfish and some other dried products traveled in bundles or bales, barrels of diverse but often closely regulated sizes became the standard unit for preserved fish. In the north Hansards likely set a pattern which Netherlanders, English, and Scots emulated. Hanseatic legislation set the 'Rostock barrel' at 117 kilograms. Scottish parliamentarians repeatedly legislated use of the 'Hamburg barrel' of 14 gallons (*c.* 60 liters?) for salmon and a smaller size for herring. On the outer estuary of Rhine and Scheldt late fifteenth-century officials tolled herring, salmon, and other fish arriving in barrels on most vessels making for Antwerp. Standard Sicilian barrels for *tonnina* held 45 and 60 kilograms and in Andalusia *barriletes* measured one quintal (*c.* 46 kg) and *barriles* four (184 kg).[126] By any measure that was a lot of food.

The mass market had social substance, too, with herrings acknowledged a fish of 'the poor' by Alan of Lille in the mid-twelfth century and Philippe de Mèzieres at the end of the fourteenth. Portuguese King João II had the same view of sardines.[127] Actual rations suggest these observers were correct, or at least came to be. Up to the thirteenth century in northern France and Flanders alike herring remains are found among all

[125] Given-Wilson, "Purveyance for the royal household," 147–150, and more generally Locker, *Role of Stored Fish*, 74–82; Ferreira Priegue, *Galicia*, 147; Bourlet, "L'Approvisionnement de Paris," 6; Salicru i Llunch, "En torna al comercia," 167–170; Piekosiński, ed., *Kodeks dyplomatyczne miasta Krakowa*, #336, p. 459. Herring priced by the barrel were the norm in Kraków from the late fourteenth through mid-sixteenth century (Pelc, *Ceny*, 38–39).

[126] Rodrigo Estevan, "Fresco, frescal, salado, seco," 560–563; Wubs-Mrozewicz, "Fish, stock, and barrel," 190–192; Holm, "Commercial sea fishing," 15–17; Brown et al., eds., *RPS*, for instance, 1478/6/87, 1487/10/20, or 1493/5/24; Gemmill and Mayhew, *Changing Values*, 303–323; Rorke, "Scottish overseas trade," 192–198; Unger, *De tol*, 294–316 and 504–516 *passim*; Bresc, *Un monde méditerranéen*, 269; Ladero Quesada, "Las almadrabas," 351–356. Interdisciplinary and comparative study of medieval coopers and their wares would benefit many kinds of researchers.

[127] Alan, "De planctu naturae" tr. Sheridan, 95; Philippe, *Le Songe*, 129–130; da Cruz Coelho, "Apontamentos sobre a comida e a bebida," 95.

social ranks near the coast but inland only at high-status sites (castles, monasteries); thereafter from around 1300 herring become a general feature even of poor urban households throughout the area. In 1390 the almoner to the king of France bought 78,000 herrings on the Paris market and distributed them to hospices and poor households.[128] Further south sardines were the one fish commonly eaten by late medieval Portuguese workers and peasants. Recipients of charity dinners in Valladolid chewed hake and sardines.[129] The spread of preserved marine fishes to more modest consumers ever further inland was associated with improvements in durability and with general advances in transport infrastructure, both of which would cut the cost of fish actually eaten (not spoiled in transit).

Distribution costs along the supply chain certainly affected monetary values of preserved marine fish compared to other foods. At Constance during the Council herring sold for 14 pfennig the quarter-pound when in good supply and 36–48 pfennig more normally, while similar-sized local *ganckvisch* went for 4 pfennig the quarter or 12 pfennig the pound.[130] There close to the head of Rhine navigation and some 500–600 kilometers from the nearest herring fishery, herring cost four to ten times more than its local freshwater counterpart. Stockfish, however, at 2 schilling for a small and 3 schilling for a large (1.0–1.5 kg?) may have more closely matched the local pike, carp, or tench at 18–20 pfennig the pound – still way over the price of meat. Nevertheless, by calculation of historian Arnved Nedkvitne, about the time of the Council the markup inland consumers paid over Bergen prices for stockfish had dropped by half from that of a century before (down to 250 percent from almost 500).[131]

Few medieval consumers could live beside a large, fish-rich lake. For the majority north of the Alps preserved herring offered more for the money than did most other fish. Fiscal assessments and tolls bore this

[128] Clavel, *L'Animal*, 176–187; Van Neer and Ervynck, "Remains of traded fish," 209–211; Douët-d'Arcq, *Comptes de l'Hotel des rois*, 266 (as cited by Coopland in Philippe, *La Songe*, 130n.). Other respectable poor in France eat herring in Endrès, "Alimentation d'assistance à l'Hôtel-Dieu de Meaux" and Hohl, "Alimentation et consommation à l'Hôtel-Dieu de Paris." English villeins doing compulsory harvest work were fed herrings in the thirteenth century, as were later recipients of charity (Serjeantson and Woolgar, "Fish consumption," 122; Dyer, "Régimes alimentaires," 209; Locker, *Role of Stored Fish*, 95–96).

[129] Oliveira Marques and Ferro, "L'alimentation au Portugal," 286; Rucquoi, "Alimentation," 302; Catarino, "Abastecimento," 19–27.

[130] Ulrich, *Konzilschronik*, fols. 23b–25b. Here and below I follow reasoning of Van Neer and Ervynck, "Remains of traded fish," 210.

[131] Nedkvitne, *Hansa and Bergen*, table VIII: 14. Further at Figure 8.10 below.

out, asking less from herring shipments.[132] But widespread consumption and a clear price advantage for herring followed noticeably upon better preserving techniques. In seasons of abstinence during 1444–1446, the kitchen at castle Horn on the Lower Austrian Danube repeatedly purchased dozens of herring at 1.5 or 2 pennies per fish, a price one-third or less than paid for any of the local fishes that also came to the table.[133] Other northern marine taxa sold at different price points than herring yet still gained against competing fish or other protein sources. When the bishop of Coventry bought food in 1461, herrings at four to the penny compared favourably to 12d for an eel; a stockfish at 3d gave more for the money than a fresh tench at 6d or a pike at 8–12d. Frankfurt's market regulations in 1487 priced a pound of herrings the same as ⅔ pound of salt cod, ⅓ pound of (local?) salmon, and a pound of beef.[134] Sardines were the cheapest and most common on markets in Piedmont and Madrid, at the latter priced some 10–20 percent less than equal weight of other preserved fishes.[135] To reiterate, no fish on medieval markets provided cheap calories, but much anecdotal and incidental evidence confirms that some preserved varieties did offer the least costly meat alternative people could buy.

*8.2.1.3 To Be Eaten Far Away* Although marine fishers could work near or far from home and preserved fish were relatively bulky low-cost commodities, all the frontier fisheries supplied consumers at great distance from sites of their catches. Cartloads of Baltic herring paid tolls 200 kilometers inland at Poznań and Gniezno by the 1240s and so likewise a generation later did river boats with them at stations on the middle Elbe. On the market in Kraków herring from at nearest Pomerania (600 km) and by 1300 more normally Scania or the North Sea were always the most common, even stereotypical, fish. As already seen further west, traders from Köln passed herrings on to centres up the

132 Delatouche, "Importance relative," 30; Hocquet, "Pêcheries médiévales," 53–56; Maillard, "'Coutumes' pour la vente du poisson."

133 Archiv Horn-Rosenberg, Hs. Horn 44.

134 Dyer, "Consumption," 31; Hitzbleck, *Bedeutung des Fisches*, 106. At Cheb (Eger) on the border between upper Franconia and Bohemia in 1465 only dried bleak came cheaper than herring and only small cyprinids could be bought fresh for less than the cost of a stockfish (*ibid.*, 107).

135 Nada Patrone, *Il cibo*, 333–339; Puñal Fernández, *Mercado en Madrid*, 195–196 and 201–202. Sardines also fell into the cheapest price class at Tortosa in 1342 (Curto Homedes, "El consum de peix," 153), but, mostly sold by count elsewhere, their values are incommensurate with fishes sold by weight (see Hamilton, *American Treasure*, 319–334). Grafe, *Distant Tyranny*, 69–79, found similar advantages for salt cod in early modern Spain.

Rhine and its tributaries. Merchants from Nürnberg had other options, too: in 1429 the Volkmeier brothers imported 36½ barrels directly from Szczecin. Some of that herring Nürnbergers re-exported to Salzburg, but men from Ulm and Linz also themselves obtained it in the Low Countries.[136] The one marine fish bought by Austrian gentry at the start of the fifteenth century, preserved herring first occur in deposits of Viennese latrines only at the century's end.[137]

Recipe collections rarely mention herring but stockfish were the one marine product named in the oldest known German-language cookbook, compiled in Würzburg before 1350.[138] Those slabs of dried Lofoten cod had been carried more than 2,000 kilometers by sea via Bergen to Lübeck, and likely by circuitous water routes again well more than the 600 straight-line kilometers from the Hanse town to their destination. The Parisian *menagier* knew well that "stofix" had come from "frontiers of the sea" and been dried to keep a very long time.[139]

Generations earlier, Gascon traders had ranked among the best customers at Yarmouth's herring fair.[140] Thanks in part to their successors, buyers for fourteenth-century popes at Avignon could find herrings and hake at La Rochelle, Bordeaux, and Toulouse. Merchants in Burgundy supplied more herrings, too. More customers than just the papal curia kept the latter trade going, witness the quantities of those fish tolled at Chalons in 1406.[141]

Traders in the English West Country contracted to ship many tons of their salted hake and other fishes into Castile.[142] Smoked and dried herrings arrived at Basque ports from northern Europe, and so did 2½ tonnes of them on a Flemish ship at Lisbon in 1402. Among imports to Aragon in 1413–1414, marine fish comprised 10 percent by value: conger, hake, herring, sardine, eel, and miscellaneous dried varieties, some having been carried about 200 kilometers over rugged Pyrenean or Cantabrian mountain passes and the rest a greater distance up the Ebro valley from the Mediterranean. People nearer Iberia's Mediterranean coast in Murcia and

[136] Jahnke, *Silber*, 227–262 (focuses on the export trade, not actual consumers); Carter, *Trade and Urban Development*, 134–139. Hemmerle, *Deutschordens-Ballei Böhmen*, 110, 114–116, 155 *et passim* has many herrings in Bohemia. As also noted below in interior France, the heartlands of aquaculture never lacked for preserved fish, but different people ate fresh carp than ate herring.

[137] Ottenthal, ed., "Die ältesten Rechnungsbücher," 603; Kunst, "Medieval urban animal bones," 11–12.

[138] Adamson, ed., *Daz buoch von guoter spise*, #20, p. 96.

[139] *Ménagier*, ed. Brereton and Ferrier, 327 (§194).

[140] Saul, "Herring industry," 36.

[141] Stouff, *Ravitaillement*, 212, and details in Weiss, *Versorgung des päpstlichen Hofes*, 392–394 and 401–402; Beck, "L'Approvisionnement," 175.

[142] Childs, *Anglo-Castilian Trade*, 99.

Valencia ate brined herrings from the North Atlantic, too, and also dried hake and salt sardines from Galicia and the Gulf of Cádiz.[143] *Aringhe* imported from Bruges were the first salt fish mentioned for Barcelona in the merchant manual a resident Florentine trader compiled in 1396.[144] But the earliest known consignment of Atlantic fishes to southern Italy in 1426 – sixty dozen dried hake and 3,550 sardines[145] – pointed straight back to Galicia. The very term southern Europeans applied since the fifteenth century to dried or salt-dried gadids, *bacalao/bacalà*, derived from a lowland Germanic (Flemish/Dutch, Low German) name for a cod. Humanist Paolo Giovio avoided the vulgar vernacular, but well knew the distant origins of both the herrings and the dried cods he could buy at Rome's market.[146] Though become elements of local cuisines from Poland to Provence to Portugal, these fishes never swam in local waters nor bore indigenous names.[147]

As emblematic of the new frontier fisheries as the stench of a Bothnian fish camp or the fish and salteries in Dieppe which assaulted noses of a Czech embassy, then, must be those headless cod skeletons in Paris or London, while the heaps of their heads moldered beside the Norwegian Sea (Figure 8.6). Quite without modern technologies the medieval fish trades separated the lives and death of these animals from their ultimate consumption and disposal.[148] How alien to a European landsman or -woman were the board-like stockfish and *strekfusz*, named not as familiar beasts but for the poles and racks on which they had dried, or the slippery *kaakharing* and *tonnina*, swimming in barrels of brine? None in Kraków,

[143] Arizaga Bolumburu, "Alimentacion en el pais vasco," 204; Oliveira Marques, *Hanse e Portugal*, 75; Rodrigo, "Fresco, frescal, salado, seco," 558; Menjot, "Marché de l'alimentation," 202–203; Aparisi, "Fishing in Valencia," 235–240. Barceló Crespí and Mas Forners, "Fishing in Majorca," 145–147, describe imported tuna and sardine as replacing local catches from the mid-fourteenth and especially fifteenth centuries. Nada Patrone, *Il cibo*, 335, found imported herrings and sardines on dining tables in Piedmonte, too.

[144] Borlandi, ed., *Il manuale di mercatura*, 129. For the same northern imports at Tortosa see Curto Homedes, "El consum de peix," 152. These extra-Mediterranean animals were less likely some generations earlier at Barcelona, when market regulations and royal household accounts mention only native fishes (Mutgé i Vives, *La ciudad de Barcelona* 19–20 and Mutgé i Vives, "L'abastament de peix," 112) and the cathedral's almoner was buying only fresh fish to feed the deserving poor (Echeniz Sans, "Alimentación de los pobres," 181–182).

[145] Bresc, "Pêche et les madragues," 169 n. 7: *merluza*, *sardina roia* [smoked sardines], and *sardina arrencada*.

[146] *De romanis piscibus*, cap. xlii, p. 60. For medieval names for cod see Supplement 8.1.2.2.

[147] Even within the Romance-speaking western Mediterranean, *tonnina* from Palermo was shipped 400 km across the Tyrrhenian to Rome and that from Cádiz well more than twice that distance.

[148] Kowaleski, "Seasonality," 139–143, demonstrates separation in time as well as space.

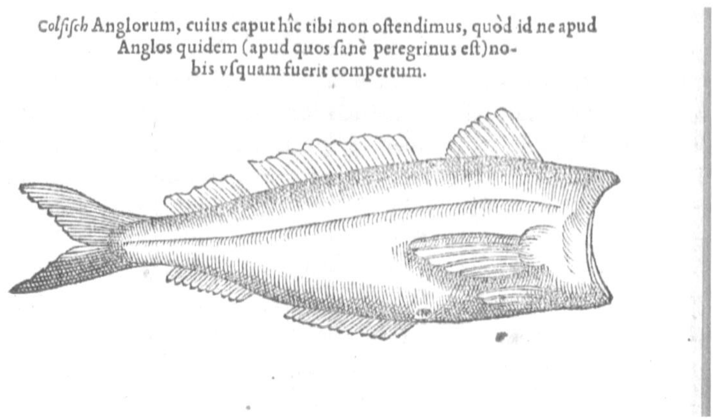


Figure 8.6 Headless codfish as commodity.
"'Colfish' of the English, whose head we do not show you here because not even among the English (among whom it is clearly foreign) has it anywhere [i.e., in any scientific work] been ascertained for us." Pierre Belon (Petrus Bellonius), *De aquatilibus, libri duo. Cum eiconibus ad vivam ipsorum effigem, quoad eius fiere potuit, expressis* (Parisis: apud Carolum Stephanum, 1553), p. 134. Note precise representation of fins and body structure of *Gadus morhua.* But *Pollachius virens,* called in England 'saithe' or, from its darkish flesh, 'coalfish', is a gadid of very similar outward appearance often taken together with cod.

Nürnberg, Lyon, Madrid, or Florence had otherwise laid eyes on their like. How did a late medieval Parisian imagine those distant seas whence he drew his food? Long before the industrial age, medieval frontier fisheries, pressed by mass demand for the least expensive fish, distanced consumers from the animal and the producing ecosystem. At the same time, requirements for bulk preservation and marketing pushed the largest offshore fisheries to rely on abundant stocks of single species. This standardization enables more extensive series of fish prices, while, as seen below, exposing certain targeted fish stocks to risks from both depletion and environmental fluctuations.

*8.2.1.4 Fluctuating Prices* Previous chapters looked at prices to show fish as generally expensive sources of calories and of protein, with herring (and sardine) 'for the poor' a partial exception. Relative values of individual species differed some from market to market but even the earliest records show high short-run variability. Anecdotal impressions of longer-term rising price levels appear by around 1100. Serial or quasi-serial records from the mid-thirteenth century up to and beyond 1400 confirm this trend for herring, cod, salmon, and Navarre sardines (Section 6.1.1 in Chapter 6 above). What now follows attempts a closer look at prices for preserved products of the growing frontier fisheries, especially in the almost two centuries between the Black Death and the so-called Price Revolution of the 1500s. As was seen for farmed carp, species with individuals of fairly uniform size and standardized packaging make for more consistent series. While improved scholarly access to late medieval prices remains limited to products as common as salted or brined herring, stockfish, and rarely other preserved fishes, evident long- and short-term variations had implications for fishers, consumers, and regional stocks of some fishes.

Late medieval fish prices must be understood against a general post-plague pattern of shrunken European populations and reduced demand for basic subsistence goods, notably grain, accompanied by the survivors' rising per capita wealth and effective demand for 'better' goods, notably animal protein, on the part of a larger proportion of those fewer consumers. Prices for grain fell everywhere and those for meat, dairy, and fish rose. The resultant 'price scissors' meant gains for protein producers and losses for those dependent on cereal sales. Hence consuming and producing perspectives differ.

Figure 8.7A–D indexes annual values from those published data sets worth calling herring price series which begin appreciably prior to 1500.[149]

[149] Figure 8.7 indexes annual values from those thirteen published data sets at ten locations which are worth calling herring price series and begin appreciably prior to 1500. Areas of both consumption and production are represented. The purpose is to compare patterns of change over time in each series, not relative prices at different places. The following are included (in chronological order):

England, 1260–1914 – annual series in the Allen–Unger Database, European Commodity Prices, from data provided by Gregory Clark (http://www.gcpdb.info/index.html, last visited 30 November 2020) supersedes decadal averages, 1250/9–1391/1400 and 1401–1582 given in Rogers, *History of Agriculture and Prices*, vol. 1, p. 641, and vol. 4, p. 545.

Scotland, 1263–1541 – Gemmill and Mayhew, *Changing Values*, 317–373 with tables 56 and 57. Different measures and packaging break even the regular record after 1364 into two series.

Rostock, 1353–1530 – Hauschild, *Studien zu Löhnen und Preisen*, 51–59, 99–102, and 160–163, with tables 67, 68, and 97, and Anlage [graph] 12. Here shown is Hauschild's composite of accounts from three municipal accounting offices.

Brussels, 1386–after 1550 – published in the Allen–Unger Database, European Commodity Prices 1260–1914, (www.gcpdb.info/index.html last visited 30 November 2020), from Van der Wee, *Growth of the Antwerp Market*, vol. I, appendix 22, pp. 279–286).

Antwerp, 1386–1540s – Verlinden, *Dokumenten voor de Geschiedenis von Prijzen*, vol. 2, pp. 731–737, graph 19; and Van der Wee *Growth of the Antwerp Market*, vol. 1, pp. 277–286 (table 22), and vol. 3, 42–43 (graph 15 and published in the Allen–Unger Database, European Commodity Prices 1260–1914, www.gcpdb.info/index.html, last visited 1 December 2020).

Separate charts for two series, England and Utrecht (Figures 8.8 and 8.9), have been extracted better to illustrate general trends. This quantified material can be augmented with shorter or less serial but seemingly well-grounded observations elsewhere. These data sets collectively tell similar stories about late medieval herring. Nominal and silver prices in England (Figure 6.1) had slowly doubled between the 1210s and 1340s, then doubled again briefly in the economic chaos of the 1350s, which provoked short-lived 1357 legislation to regulate prices at the Yarmouth herring fair.[150] After falling back they rose again from the 1360s to attain early fifteenth-century levels twice those of a century before (Figure 8.8). In the following two or three generations English consumers found herring prices little changed, but from the 1470s these fell by as much as 20 percent. New increases after 1500 were balanced by loss of silver value and the start of the general inflation of the sixteenth century.

None but anecdotal pre-1350 herring prices elsewhere leave only the observation of general parallels between England and the continent from the post-plague decades to the mid-1500s (Figures 8.7 and 8.9). Most curves peak during the first quarter or third of the fifteenth century and flatten or decline during its second half. The telling exceptions, given what is now known about shifting production zones, are from the Low Countries (Ghent, Utrecht, Leiden, Antwerp), where a positive price trend lasted into the late 1400s (Figure 8.9 provides a clear example). Although late medieval elites turned away from herring, that market deepened as improved packaging for greater durability maintained the geographic and social breadth of this still relatively cheap protein.

Kraków, 1389–1585 – Pelc, *Ceny w Krakowie*, table 22, pp. 127–131, and diagram IV, p. 162.

Leiden, 1414–1540 – Posthumus, *Inquiry into the History of Prices*, vol. 2, pp. 493–494, 498, 592–596, and table 208.

Utrecht, 1419–1530 – Posthumus, *Inquiry into the History of Prices*, vol. 2, pp. 127–131, 272–274, 368–370, and LXXIV table A.

Augsburg, 1418–1538 – Elsas, *Umriss einer Geschichte der Preise*, vol. 1, pp. 384–385 and 615–616.

Hamburg, 1443–1475 and 1500–1545 – Gerhard and Engel, eds., *Preisgeschichte der vorindustriellen Zeit*, table 03.01 (pp. 134–135).

Prices from Ghent, 1361–1398 (Verlinden, *Dokumenten voor de Geschiedenis von Prijzen*, vol. 4, p. 296) and from Klosterneuburg, 1440/50–1490/1500 (Hitzbleck, *Bedeutung des Fisches*, table 23, p. 23, as from Pribram, *Materialien zur Geschichte der Preise*, vol. 1, 627), while confirming some patterns, are too short and/or broken to graph in Figure 8.7. Sadly, herring in Jahnke, *Silber*, are priceless.

Fish price series from Navarre (Figure 6.2) as compiled by Hamilton, *Money, Prices, and Wages*, appendix V, run only 1358–1450 and contain several lacunae. The price for imported herring seems, however, to reach its maximum in the late 1420s–early 1440s, when prices of domestic species were relatively low.

150 Seabourne, *Royal Regulation*, 77 and 88.

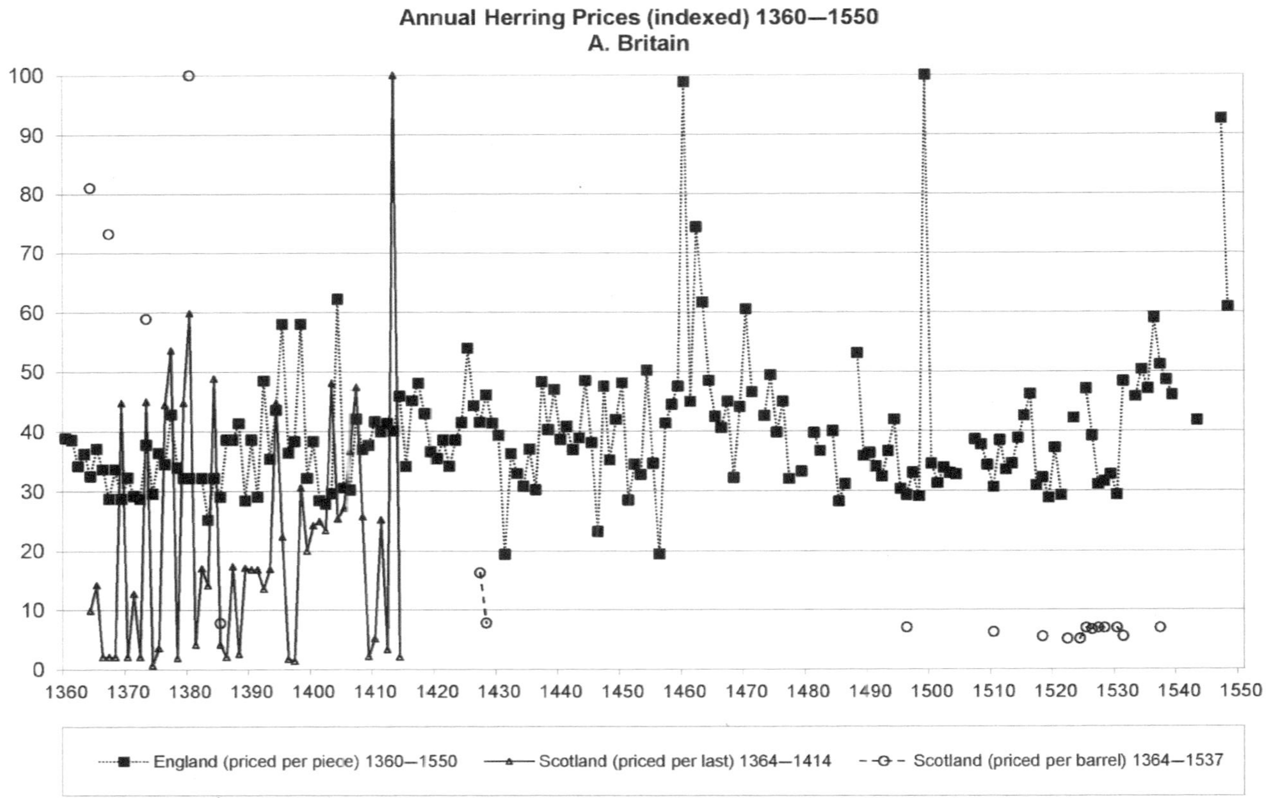


Figure 8.7 Annual herring prices (indexed), 1360–1550.
Figure 8.7A Britain.

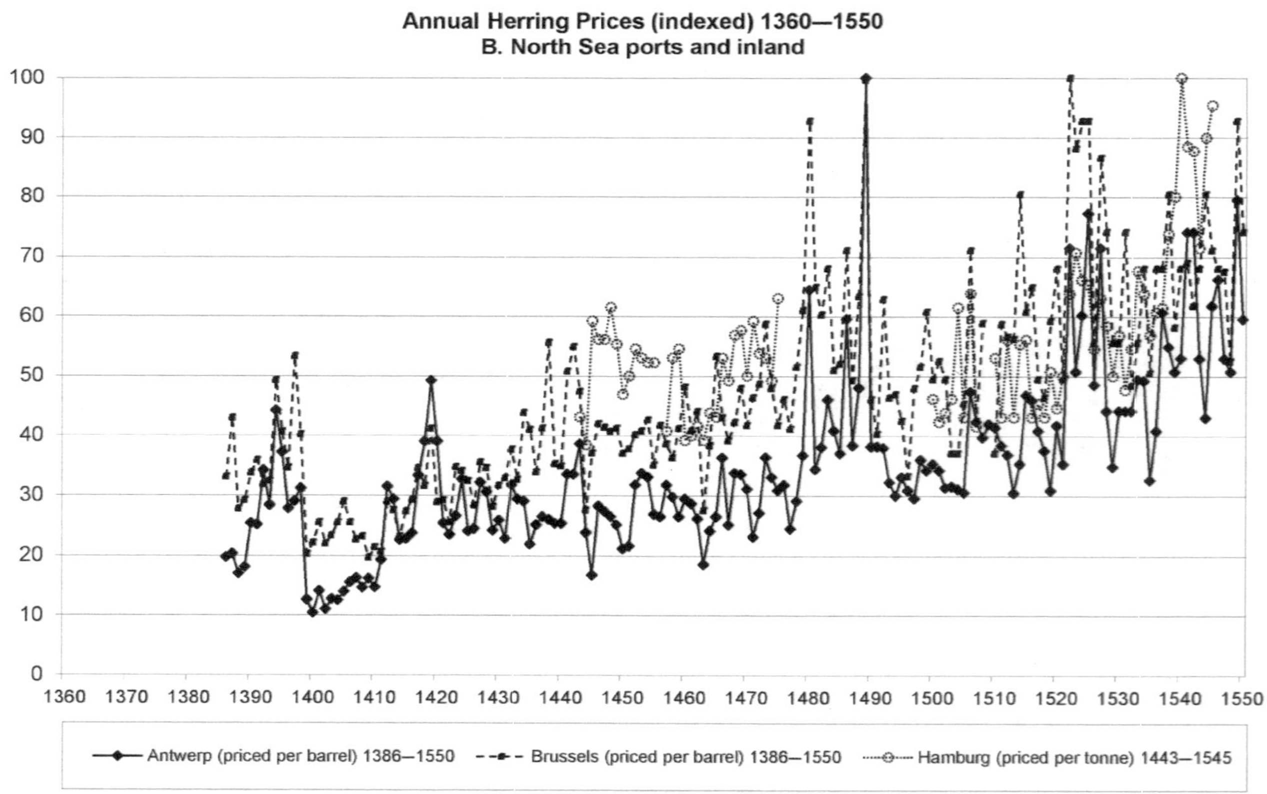


Figure 8.7B North Sea ports and inland.

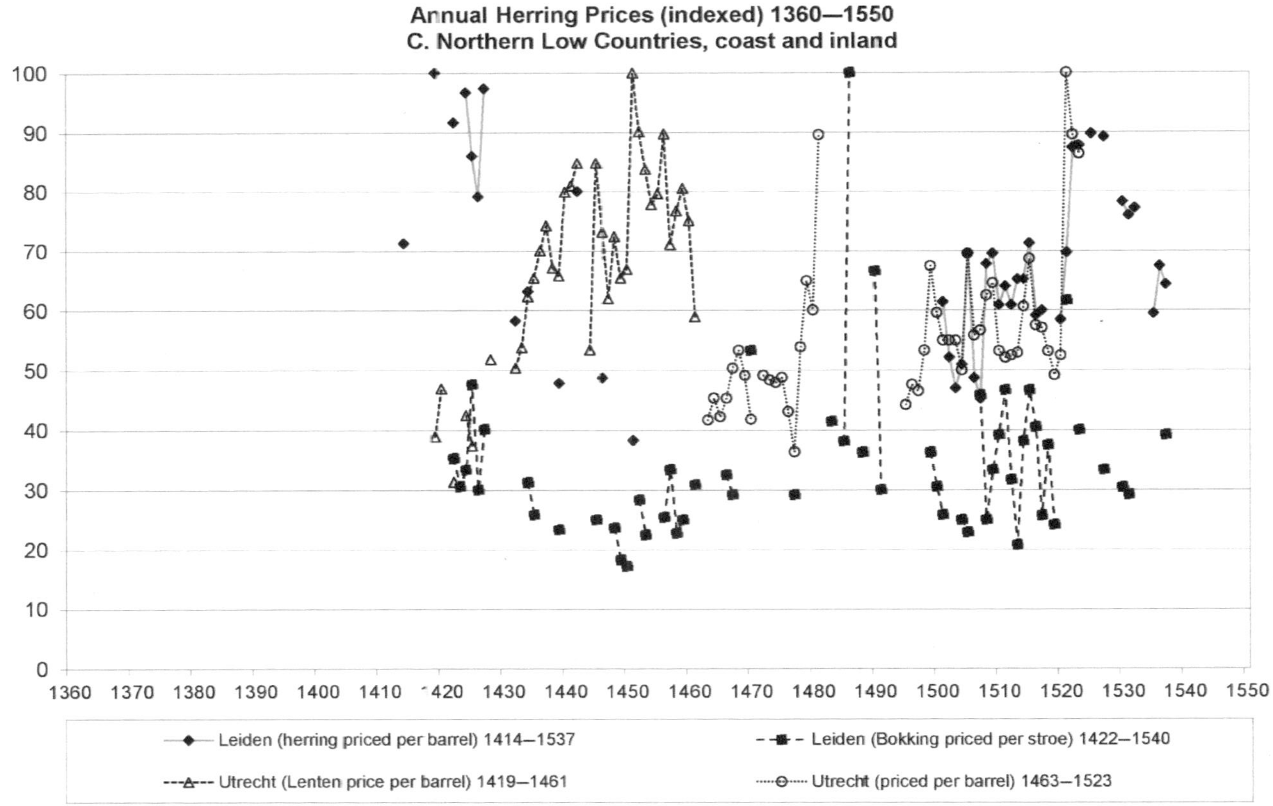


Figure 8.7C Northern Low Countries, coast and inland.

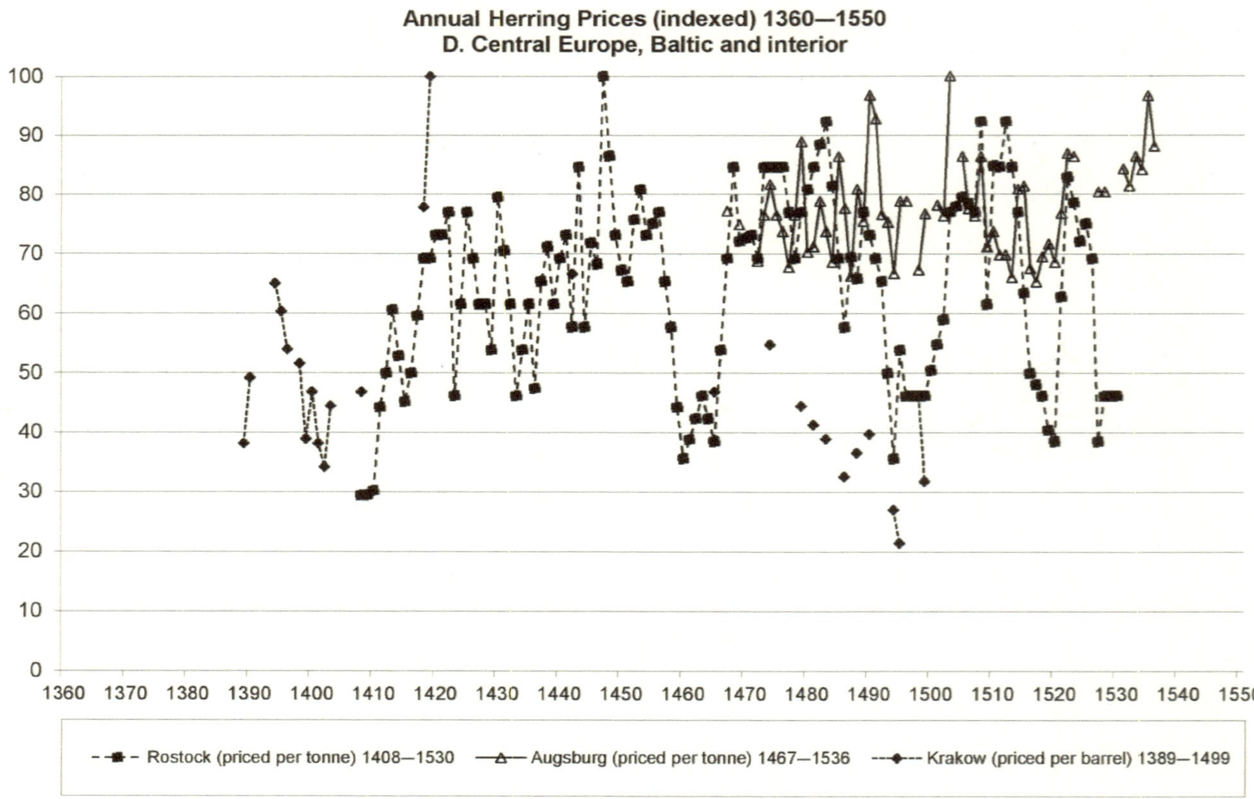


Figure 8.7D Central Europe, Baltic, and interior.

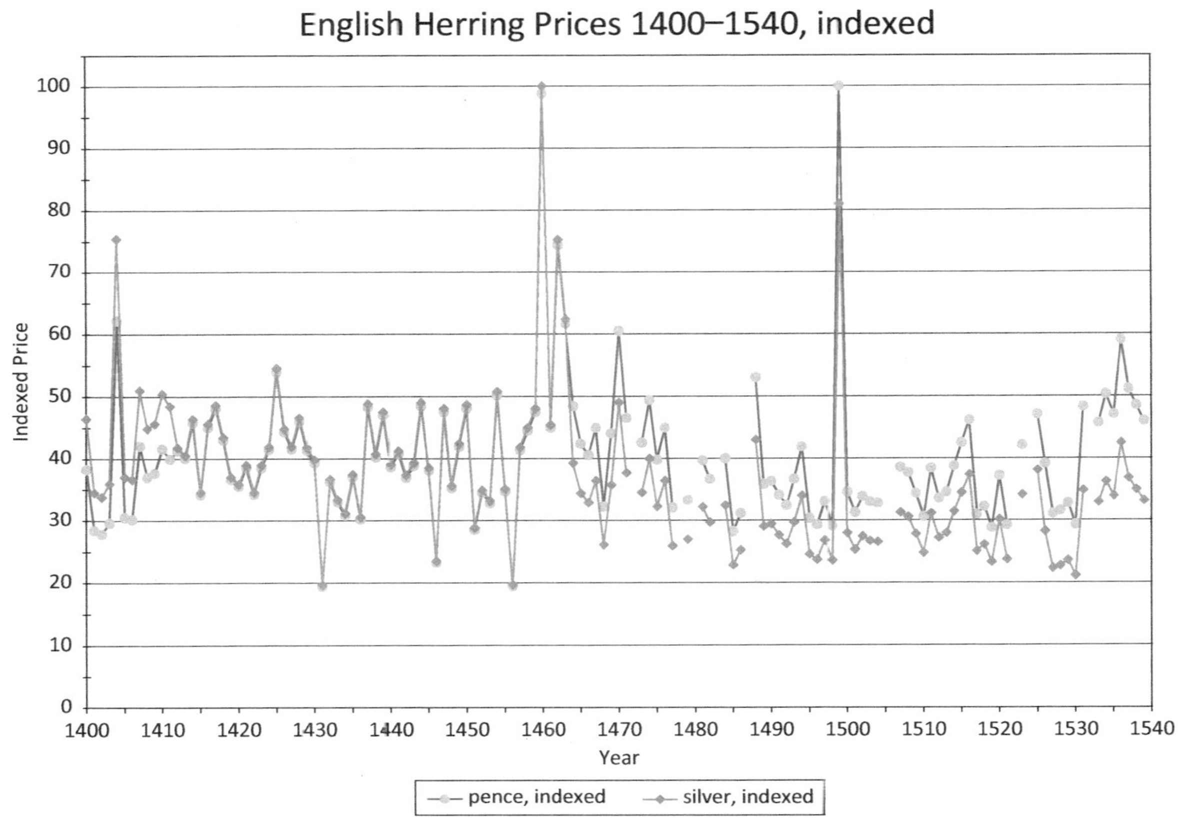


Figure 8.8 English herring prices, 1400–1540, indexed.

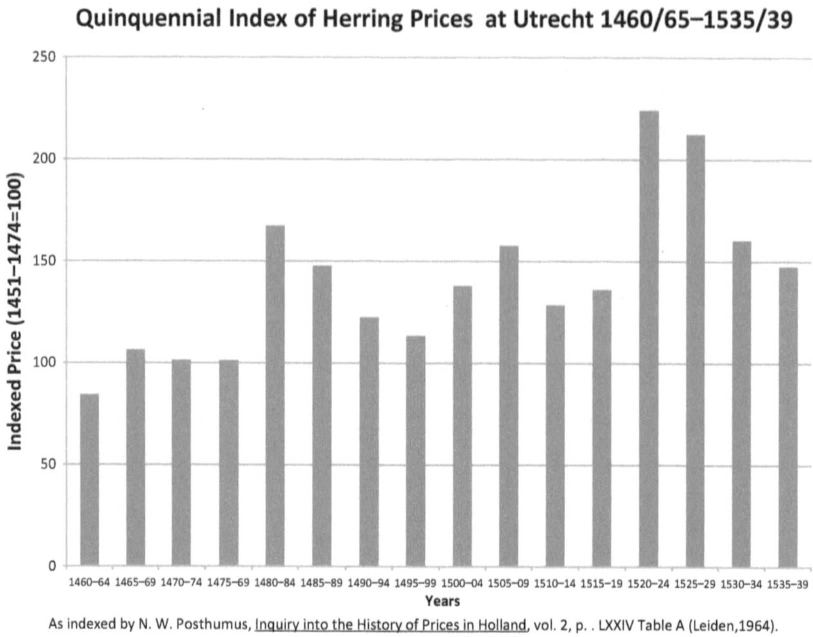


Figure 8.9 Quinquennial index of herring price at Utrecht, 1460/5–1535/9.

Herring which had been one-fourth the price of an equal weight of stockfish around 1300 came to half that level in 1550.[151]

That same 'Price Revolution' of the sixteenth century marked nominal prices for preserved cod as well, but exiguous published data obscures earlier developments in both producing and consuming regions. The likely best data set, that for Bergen stockfish published by Arnved Nedkvitne, aggregates values along the supply chain at more than generational scale. Figure 8.10 graphs these as silver per hundred fish. Consumers in inland England thus paid in coined silver an average of 30 percent more for stockfish in 1400 than they had before 1350, and sellers in Bergen received 2½ times what they had in 1300. Reflecting the reciprocal movement of two commodities, if the fishers selling in Bergen converted that silver to rye flour, they got 3½ times what their predecessors had a century before.[152] This more than compensated for lower export volumes and enabled purchases of textiles, German beer, and

[151] Nedkvitne, *Hansa and Bergen*, 508–511.

[152] Christensen and Nielssen, "Norwegian fisheries," 150–151; Nedkvitne, "Development," 53–56; Nedkvitne, *Hansa and Bergen*, table VIII, pp. 714–715.

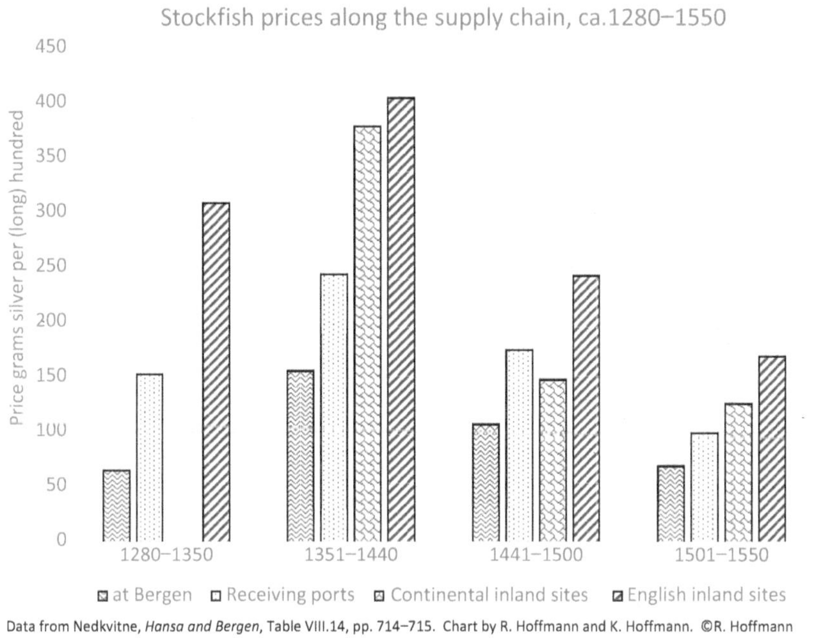


Figure 8.10 Stockfish prices along the supply chain, *c.* 1280–1550.

other enhancements to living standards in Lofoten. It was a bonanza for Norwegians in the north. Although the premium shrank in the course of Europe's fifteenth-century demographic recovery, as late as 1500 sellers of stockfish still got twice the return of their forebears two centuries earlier. As so described the late medieval situation more reflected cereal prices responding to decline in human numbers than it did the price of the fish, which moved in an opposite direction, to the benefit of late medieval Norwegian producers. When new demographic and market patterns reversed those relationships and pushed Norse product to ever lower exchange values, Nedkvitne's 'Golden Age' of Norwegian stockfish came to an end.[153]

Published stockfish prices for Iceland reveal similar clear trends in thin but consistent data.[154] During early fourteenth-century decades when Icelandic exports were taking off, the price of *skreið* rose by a third, but then held steady through the century's second half. Strong European

153 Nedkvitne, *Hansa and Bergen*, 92–95 and 496–504. Record high exports after 1550 to low-price Mediterranean and African markets could not stop regional economic depression.
154 Gelsinger, *Icelandic Enterprise*, 185–190.

demand in the 1400s almost doubled the price again. Medieval Icelandic culture and political leadership privileged a farm-centred economy where fishing provided only extra subsistence or income. Elites discouraged independence and growth of the fishing sector, so preferred stable prices.[155] Consequences for organization and labour in fisheries are observed below (pp. 377–379).

From continental consumption markets for stockfish and other preserved gadids, good price series become available only with the fifteenth century, so cast no light on pre- and post-plague market fluctuations. In England after the 1420s a downward trend of stockfish prices is counterbalanced by a rise for 'cod'. This fits anecdotal evidence that mid- to upmarket English consumers were then abandoning preserved fish for fresher forms and for meat.[156] At Rostock, however, prices for several cod products moved in reasonable unison upwards during the first half of the century and then flattened until after 1500.[157]

Prices for salted and barreled salmon rose slowly in late fourteenth-century Scotland and, from 1411, England as well. A sudden trough around 1450 is attributed in both countries to currency manipulation, followed by further rise to and beyond the 1520s.[158]

Fully conceding the fragmentary, questionable, and incommensurate quality of much available data, in the long term late medieval prices for preserved herring, cod, and salmon look driven by interplay of fluctuating human numbers with the greater per capita wealth and discretionary incomes of the post-plague period. The cheaper product, herring, responded more positively early on, while larger and more costly cods performed better during the fifteenth century. For many late medieval Europeans even the new mass-marketed fish from marine frontiers remained a commodity at the margin between subsistence necessity and an affordable luxury … or else the cheapest way to gain animal protein without flouting religious constraints. When the cost of cereals fell, as after 1350, larger discretionary incomes allowed a greater share of

[155] Hastrup, *Nature and Policy*, 66–75; Gardiner, "Character of commercial fishing," 87–89.

[156] Rogers, *History*, vol. 4, p. 545; price data from Clark is tabulated in the Allen–Unger Database, England, salt cod, 1371+ (www.gcpdb.info/index.html, last accessed 1 December 2020); Locker, *Role of Stored Fish*, 273; Serjeantson and Woolgar, "Fish consumption," 126–130.

[157] Hauschild, *Löhnen und Preisen in Rostock*, 149, re "Dorsch," "Kabeljau," and "Stockfisch," with discussion *passim*.

[158] Gemmill and Mayhew, *Changing Values*, 303–317; Allen–Unger Database, England, salt salmon (www.gcpdb.info/index.html, last accessed 1 December 2020) provides data from Clark. Given the small samples and different formats, it remains unclear whether brief episodes of variability differ in the two series.

Europe's population to eat more fish, even at greater cost, but still, for those consumers, at the lowest price points. The Christoffels Jans household whom we visited in the 1490s (pp. ***–*** above), contemporaries of Zuan Caboto, personified this situation.

Stronger prices in the Low Countries could help sustain local fishers and merchants investing in more productive new technologies and market strategies which let them undercut competitors and capture more stagnant markets elsewhere. Overall, and acknowledging such important regional anomalies as the slipping English market for preserved herrings, late medieval price trends suggest a strong large-scale demand for preserved fish products both widened and deepened in most regions north of the Alps. But as to be observed more closely below, at shorter temporal scale natural fluctuations as well as political conflicts could also affect herring prices.

Did similar conditions prevail in southern Europe before the sixteenth-century surge of preserved Atlantic and New World fish into the Mediterranean sustained a new long-term rise in effective demand and prices?[159] Useable serial data from southern Europe remains confined to two Iberian regions, nearly landlocked Navarre in the northwest (Figure 6.2) and Mediterranean coastal Valencia (Figure 8.11), where published prices begin only after 1410 and end by 1490. Both data sets, however, include key regional marine products (sardine, hake, conger) and, as already observed in fourteenth-century Navarre, moved roughly parallel to those for herring in the north. After the 1410s what had been rising prices for brined sardine and dried hake in Navarre went flat. A coincident early fifteenth-century plateau initiated the series from Valencia, followed, amidst much interannual variation, by slightly lower post-1460 values resembling those then seen for much northern herring.[160] In both instances, sardines, close ecological and dietary counterparts to northern herring, also showed the most extreme year-to-year fluctuations.

Dare we infer some medium-run late fifteenth-century balance between effective demand and normal supply up and down coastal western Europe, while only inland markets for stockfish and the North

[159] Vickers, "The price of fish," and for wider context, Richards, *Unending Frontier*, 547–573, and Grafe, *Distant Tyranny*.

[160] A dozen or so price quotations for each of sardines and sea bream provided in Toledo's cathedral accounts show higher nominal values in the 1460s than in the 1410s, but intervening debasement meant a loss of silver value. All fish prices in Toledo went up after 1470 (Izqierdo Benito, *Precios y salarios*, 127–130).

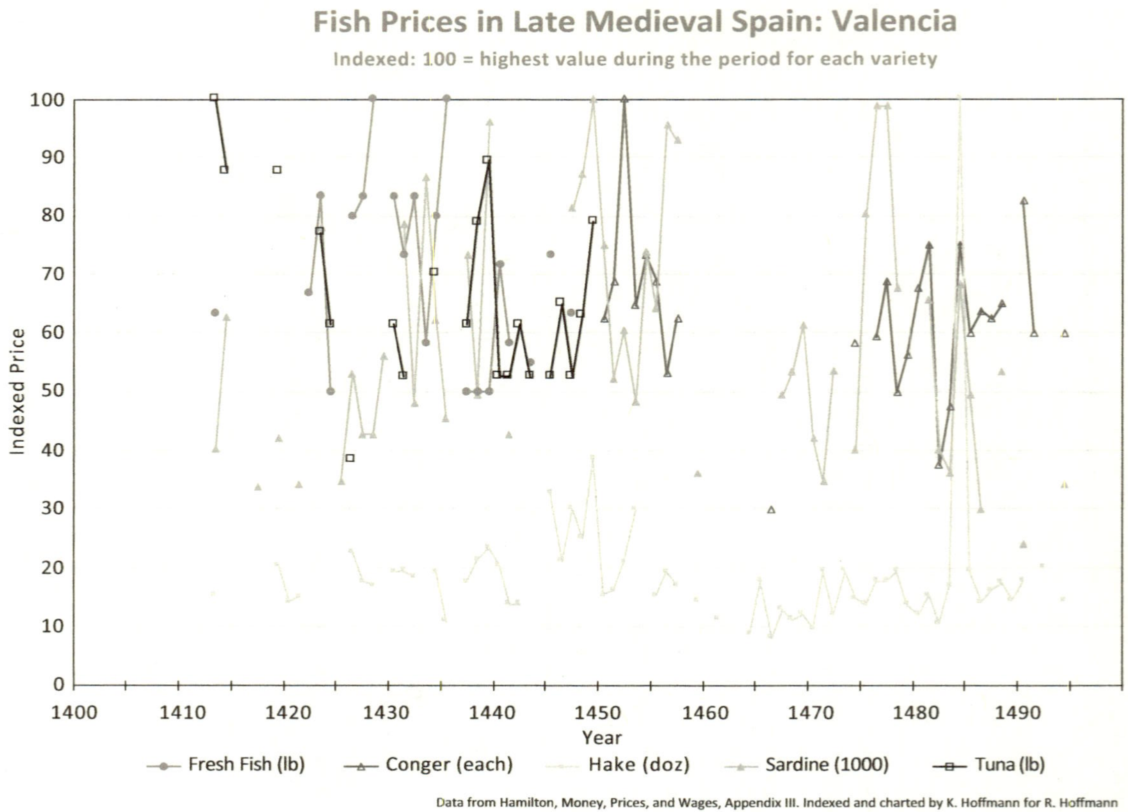


Figure 8.11 Fish prices in late medieval Spain: Valencia.
Data from E. Hamilton, *Money, Prices, and Wages in Valencia, Aragon, and Navarre, 1351–1500*, appendix III, pp. 215–221. Indexed and graphed by R. Hoffmann and K. Hoffmann. © R. Hoffmann.

Sea herring remained buoyant? The call for Scottish salmon in domestic, English, and Low Countries markets seems to have held strong. What lack of clues leaves a mystery are price relations for tuna and for those diverse products from dynamic new regional fisheries exemplified by Bretons and men of the English West Country.[161] But as indicative as long-term trends are for understanding ecological and social developments in the marine fisheries, also notable are the great volatility of prices in all the annual series and the changing fates of some regional herring fisheries still to be investigated below. Long and short run price fluctuations reflect different aspects of interactions between medieval people and the fish of the sea.

### 8.2.2 *New Structures in the Fisheries*

Not just consumers and fishes, but fishers, too, could be transformed by constraints and possibilities on the medieval maritime frontier. Market-driven expansion propelled offshore fisheries into new structures, some material, others more institutional or mental. Considered as economic activities, the rising marine fisheries shared elements grouped collectively under 'second stage commercialization'.[162] So long as cultural norms endorsed eating fish, improved methods for preservation and widening markets drove increased scale of production, longer voyages, and larger vessels. Both new techniques and further diffusion of older ones required new capabilities and an increased level of investment (capitalization) throughout the fishing sector. It took the sturdier but more expensive 'dogger' design, for instance, to project English fishing activity out of eastern coastal waters first to Dogger Bank and then as as far as Iceland[163] and the Dutch *buss* to exploit herring out in the North Sea. Investors and entrepreneurs included town-based merchants from Lübeck, Bruges, Exeter, and Palermo but also regional seigneurial elites ranging from dukes of Brittany and Medina Sidonia to manorial lords in Devon[164] and eastern Denmark. Large-scale commercial fishing was an economic dynamo with strong backward linkages. Achieving prerequisite regional mobilization of labour, skill, and capital had further strong positive feedbacks for economic growth and development. Processing herring, salmon, pilchards, and tuna called up greater production of raw

[161] Kowaleski, "Expansion" and "Commercialization." Bresc, *Un monde méditerranéen*, 271–272, speculates that Sicilian tuna followed the pattern seen for Norwegian stockfish, with lower output more than compensated by greater exchange value.

[162] See note 3 above. Bresc, "Pêche et les madragues," 174–175, writes of Sicilian tuna traps as 'precocious capitalism' comparable to that in the island's sugar cane industry.

[163] Heath, "North Sea fishing," 59–60. [164] Kowaleski, "Expansion," 445.

materials (salt), of equipment (ships, nets), and of packaging (barrels).[165] Czech envoys nauseated by herring-rich Dieppe in 1460 simply failed to recognize the scent of money.[166]

Capital requirements changed the balance of power in many fisheries. While traditional labour relations long survived in some areas, both capitalist organization and seigneurial initiative supplanted artisan guilds and household enterprises with other arrangements. By providing resources for greater output, mercantile investment tied many producers more closely to fishing as an occupation, whether year-round or seasonal. In northern Norway and in Denmark specialized full-time fishers first appeared under Hanseatic influence in the fourteenth century.[167] Diverse fisheries on the English southwest, probably also Brittany, could offer year-round maritime work, alternating fishing seasons with voyages in the carrying trade. Ventures to Ireland or Iceland meant steady work but long and hazardous absences. Ordinary hands were often village youth, taking a fishing job while awaiting an agricultural tenure.[168] To reduce precisely that competition for farm workers Icelandic authorities prohibited the landless from owning a boat, employing others at sea, or taking jobs with foreign visitors.[169] Elsewhere professionals operated the *tonnara*, while *almadrabas* drew more on unskilled labourers to haul the big nets. The latter's seasonal contracts provided food and lodging, all framed, of course, by the dukes' local power.[170]

Contract work for wages or salary began to replace hitherto customary allocation of shares in the catch or profit of a voyage or a year. Tradition prevailed at the *tonnara* and into the fifteenth century in the Flemish–Dutch herring fishery, but eventually first crewmen, then even captains became waged and salaried employees of land-based ship owners.[171] Money wages occurred occasionally in the older Yarmouth fishery and later became a norm in English voyages to Iceland. Otherwise most fishers out of Devon and Cornwall received them only as supplement

[165] Unger, "Netherlands herring" and "Dutch herring"; Mollat, *Commerce maritime*, 313–318; Bresc, *Un monde méditerranéen*; and García Vargas and Florido del Corral, "Origin and development," 217–218 and 223–224, all consider these backward linkages of their fisheries. So do Ferreira Priegue, *Galicia*, 132–139; Kowaleski, "Expansion," 445–447, and "Commercialization," 204–206 and 229–230; and Zambernardi, "Deniers 'paysans de la mer'," 95–96.

[166] Jaroslav, *Diary of an Embassy*, 45.

[167] Ebel, "Fischerei und Fischermethoden," 143–145; Christensen and Nielssen, "Norwegian fisheries," 150.

[168] Kowaleski, "Working at sea"; Fox, *Evolution of the Fishing Village*, 165–168.

[169] Hastrup, *Nature and Policy*, 66–75 and 137–139.

[170] Phillips, "Long-term profitability."

[171] de Vries and van der Woude, *First Modern Economy*, 244–245.

to their shares. Yet a slick shipowner could exploit this arrangement, too: Devon merchant John Gundy hired men for his ships on shares, but then claimed a preemptive right to purchase the men's take of fish at a price favourable to himself.[172]

An older and more common mechanism for changed relationships was advance purchase of the catch, which capitalized a fisher's seasonal venture at some risk of debt bondage. Hansards used this arrangement in both the Øresund herring and northern Norwegian cod fisheries. To obtain more herring than Danish labour could supply, early fourteenth-century German merchants outfitted migrant German fishers, lending to them a certain value of goods to be repaid from a specific and larger value of fish. To ensure returns, the lender had exclusive claim on the fisher's entire catch, then was to return to him the value in excess of the repayment due.[173] Likewise Sicilian and foreign merchants working out of Palermo or Trapani bought in March the tunas to be caught in June.[174]

The situation in northern Norway looks more ambiguous. While Hansards confined to Bergen might offer payment for the coming year's catch, they lacked enforcement power in Lofoten. Once the Germans gained direct access to the Norwegian north, they advanced grain and other goods to the fishers on credit and took their catches in payment. Nevertheless Arnved Netkvitne makes a convincing case that fishers held the stronger hand because their family subsistence rested on economic activities other than fishing for export, which yielded a discretionary income. He thus notes the considerable financial and material success of northern Norwegians during late medieval times.[175] In any case fishers were increasingly tied to distant markets and market values. At the same time advance purchase again confirms the eagerness of frontier traders – at Yarmouth, on Lofoten, in Aberdeen or Palermo – to obtain fish for sale to consumers across Europe.

While most marine expansion was urban-based, arising within the occupational structures of coastal towns like Exeter, Brill, or La Coruña, under certain conditions increased commercial opportunities encouraged new settlement patterns for full-time fishers. Whole communities centred on fishing differed from a handful of families in a village of peasant farmers and from one of several occupational groups in a larger town. Specific well-documented cases vary.

172 Kowaleski, "Working at sea" and "Expansion," 446. 173 Jahnke, *Silber*, 183–189.
174 Bresc, "La pêche et les madragues," 174–178.
175 Christensen and Nielssen, "Norwegian fisheries," 149–150; Nedkvitne, *Hansa and Bergen*, 535–558 and 572–575.

Storm floods and sand drift in 1394 destroyed a hamlet of poor local multi-tasking fishers, farmers, peat-cutters, and salt-burners who had long eked out a living on the beach beside a Flemish inlet called Walraversijde. Within five years however the hamlet was replaced inside the protective dune line by a different kind of community, professional full-time fishers (and occasional privateers). A generation later, 100 dwellings housed up to 500 people, most of them somehow engaged in the fishery or such ancillary enterprises as making rope or processing the catch. Ship captains or owners each signed on up to twenty independent crewmen to work for shares of the local catch, herring in season, otherwise mostly flatfish and haddock. By the mid-1400s their area of operations shifted to the North Sea proper, what Flemings called *Noordover*, where the boats principally targeted herring and cod. Workers in the community processed catches into commercial commodities and shipped the best to towns; resident households ate less preferred bycatch. Longer voyages and larger vessels called for more investment by ship owners, local or urban-based, who hired crew, even perhaps captains, for wages. Local elites with good regional connections might endow and embellish the village chapel but Walraversijde still lacked an actual market and its largest homes were but twice the size of smaller ones. Yet all were built in brick and contained a wide range of material goods, suggesting even crewmen's families gained from fisheries work and maritime connections. The small harbour, however, could hold no more than a landing stage, not enough to berth a *buss* and thus fully to benefit from the industry's growth. Habsburg–Valois wars after 1480 turned local waters into a combat zone. With sand again on the move, over the next century or so the village, too, drifted into oblivion.[176] Had Walraversijde endured, its potential for archaeological reconstruction five centuries later would be much less.

On the Devon coast people at fifteenth-century Stokenham exploited their potential for year-round multi-species fishing by means not unlike Walraversijde. Having outgrown earlier subsistence or even part-time artisanal work in the fishery, large tenant farmers housed and employed subtenants as full-time fishing crews through an annual round of mullet, hake, pilchard, and herring seasons in coastal waters and voyages as far as Ireland or the North Sea. Onetime seasonal encampments and out-of-season storage facilities on southwestern English beaches gradually became permanent communities who even petitioned ecclesiastical authorities for religious services more convenient than the old manorial

[176] Tys and Pieters, "Understanding," 100–117; Pieters, "Archaeology of fishery," 44–55.

parish some distance inland. Social separation between land- and sea-based cultures slowly ensued.[177] Numbers of these late medieval creations would become centres for the modern fishing industry.

In far northern Norway peasants on marginal farms began about 1250 to relocate to coastal sites beside the best fishing, even where conditions were inimical to growing grain. Among the first evidence of this shift is an episcopal grant in 1307 of parish status for Vardø, which fronts on the Barents Sea, and erection of a royal castle for tax collection soon thereafter. Immigrants arrived from the south as other parts of the country suffered post-plague depression. On northern farms women cultivated non-cereal crops for subsistence while their men worked full-time in the lucrative cod fishery for export. To livestock they fed seasonal grass, seaweed, and the fish heads left over from preparing stockfish. Symbolic of this economic transformation was the 1432 adoption of fish as a standard of value instead of assessing rents in butter. The north and coast remained Norway's most prosperous regions into the sixteenth century.[178]

Where, however, even highly productive fisheries failed to provide year-round employment, hundreds, even thousands, of workers assembled only seasonally and, the fishing over, dispersed to their permanent homes and other activities. Danish peasants and temporary migrants from Germany and the Low Countries gathered along Scanian beaches every August. During late-spring months men from hereditary dynasties of experts ran Sicilian *tonnara*, bringing in their own teams and hiring local boatmen for work on the water.[179] In 1528 the duke of Medina Sidonia recruited 225 wage workers and assigned some number of slaves to work his *almadraba*.[180] So even where the big commercial fisheries did not induce late medieval shifts in settlement structures, they shaped work, social interactions, and spatial relations between work and home for thousands of ordinary Europeans.

[177] Fox, *Fishing Village*, 122–129; Kowaleski, "Expansion," 429–430 and 442–444, and "Commercialization," 203–204. Fox, "Fishermen and mariners," 77–78, documents responses from the bishop of Exeter and the pope in 1415 and 1430 to groups seeking new churches. While scholars of several maritime regions cogently suggest that women likely gained a greater and more independent role in communities where men were often absent or even lost at sea, all concede a lack of medieval evidence for findings like those from later periods (Kowaleski, "Seasonality," 129–130; Pieters, "Archaeology of fishery," 54–55; Franco, "Dynamiques familiales").

[178] Urbańczyk, *Medieval Arctic Norway*, 233–261; Nedkvitne, *Hansa and Bergen*, 535–538; Amundsen, "Coupled systems"; Nielssen, "Early commercial fishing," 46–48. Similar cases appear in Supplement 8.2.2.

[179] Bresc, "La pêche et les madragues," 170–178, and *Un monde méditerranéen*, 264–270.

[180] Phillips, "Long-term profitability"; Ladero Quesada, "Las almadrabas."

At a larger scale the binding to markets of the large marine fisheries which lay at the core of 'second stage commercialization' is called economic integration. This the Middle Ages experienced not as 'globalization' but as 'Europeanization'. A central role for the fishery is especially posited for North Atlantic communities – Norway, Iceland, other island societies – which did join western Christendom during this period and whose economic connection entailed exchange of staple resources for European cereals and manufactures.[181] Fishing helped propel England's southwest into a leading role in the national economy. Perhaps the clearest mark of economic integration is when it worked to disadvantage, as when better and cheaper Netherlands herring around 1400 helped push the Hanse's Scanian product off the Flemish market or when Norwegian cod prices collapsed after 1567 under competition from cheaper barreled salt fish from America.[182] Dutch herring are, of course, a tale of seizing the mercantile centre, not joining the periphery. Galician and Sicilian fisheries also deserve consideration in this larger perspective, and not merely as engines of local enterprise. If Gallegos saw their dried hake driven from Mediterranean export markets by New World salt cod, they and other Iberian or French fishers could, unlike the Norwegians, respond by themselves going to the latest frontier waters.[183]

Another side of the story is competition among user groups over rights and access to frontier resources. In struggles among market-driven fishers and merchants the 'tragedy of the commons' was scripted offshore. At the domestic scale, fisheries along the west Baltic shore, including the fixed weirs that trapped estuarine herrings for local consumption, had guilds and regulations; those fishing for distant markets did not. New offshore fisheries in southwestern England likewise knew far less regulation than had those in North Sea ports.[184] Still fewer holds were barred between rival communities. Maritime conflicts among Norman, Breton, Gascon, Galician, and Portuguese fishers formed part of the prelude to Anglo-French wars of the 1290s. Violent confrontations in the Øresund that famously broke out between Danish and German fishers in

181 Bertelsen, "North-east Atlantic perspective"; Urbanczyk, *Medieval Arctic Norway*; Barrett, "Fish trade in Norse Orkney"; Barrett et al., "Archaeo-ichthyological evidence," 370–374; Barrett et al., "What was the Viking Age"; Barrett, "Medieval sea fishing," 266.

182 Holm, "Catches and manpower," 179–191; Urbanczyk, *Medieval Arctic Norway*, 261; Nedkvitne, *Hansa and Bergen*, 529 and 558–562.

183 Richards, *Unending Frontier*, 552–559 and 564; Ferreira-Priegue, *Galicia*, 147–148. Tranchant, "Pêches et pêcheurs," 13, sees similar for small French ports along the Bay of Biscay.

184 Jahnke, *Silber*, 39–48; Kowaleski, "Commercialization," 204–206.

1463 probably went back before 1389.[185] The North Sea became an arena for struggles between Norman, Picard, Flemish, Dutch, and English herring boats.[186] Anglo-Hansard antagonisms appear especially intense, first in Norway and later in waters around Iceland, where native fishers also became engaged.[187] By modern analogy this might be called 'Cod War I'.

Nor was the Mediterranean a zone of peace, even among Christian fishers. Mainlanders and those from Lerins long contested the intervening strait, while in 1469 men of Antibes even waged a miniature naval battle with their rivals from Cannes.[188] Violent confrontations punctuated establishment of Medina Sidonia's dominion over Andalusian tuna. In contrast, for all the capitalist quality of the *tonnara*, Sicilian participants in that still land-linked fishery undertook to limit their take by using some locations only in alternate years.[189] Introduction of that technology from Languedoc to Catalonia at the end of the fourteenth century provoked a generation of conflict over its alleged damage to juvenile fish and traditional fishing methods. Then during 1410–1430 communities from Rosas to Tarragona reached mutual or juridical resolutions integrating these techniques into local practice.[190] Insofar as the latter arrangements resemble the collaborative trends seen on Lake Constance, this sharpens the contrast with the competitive courses taken beyond the marine horizon.

## 8.3 Unanticipated Concomitants, Unintended Consequences

The late medieval turn to offshore resources appeared to escape the limits of deteriorating traditional freshwater fish stocks and sociocultural constraints on exploitation so as to meet insatiable European demand for fish. Yet distant retrospect sees the lurking paradox. New marine frontiers did open new stocks to feed European demand but the paradigm of frontier expansion there set loose also contained its own vulnerabilities. Even in medieval times human hubris and natural forces alike could betray the claim to human control over 'infinite fish'. There is always a catch.

185 Jahnke, *Silber*, 190.

186 Saul, "Herring industry," 38–40; Mollat, *Europe and the Sea*, 76 and 148–142; Sicking, *Neptune*, 142–201. Fishers' truces, though well known, do not a regulated fishery make.

187 Besides works mentioned in note 78 above, see Dollinger, *German Hansa*, 243 and 429–430; Lloyd, *England and the German Hanse*, 137–140, 190, 238–239, and 260; Fudge, *Cargoes, Embargoes and Emissaries*, 25–72, 90–91, 118–119, and 164–170.

188 Aubenas, *Le droit de pêche*, especially 8–9, with more in Stouff, *Ravitaillement*, 201–202.

189 Bresc, *Un monde méditerranéen*, 264–270.

190 Garrido i Escobar and Aleret, "Evoluzione," 120; Garrido i Escobar and Pujol i Hamelink. "Changements techniques e conflits," 25–26.

### 8.3.1 *Risky Business*

Marine fishing concentrated men and material where fish were seasonally abundant: herring sites in the Baltic, Øresund, and North Sea; around Iceland, the Channel approaches and certain Irish bays; and along the Gulf of Cádiz and Sicilian shores. Beyond competition among the fishers, larger political rivalries and the chance of illicit plunder and profit put fishers' installations, fleets, and lives at risk. Threats lurked over western Mediterranean horizons. Mutual hostilities of Guelfs and Ghibellines merged into dynastic conflict between the houses of Anjou (Provence, Naples) and Aragon (Catalonia, Sardinia, Sicily), keeping Provençal fishers close to home and leaving openings for North African corsairs. Naval insecurity (along with ravages of malaria near coastal lagoons) drove Sicilians from the island's southern and western shores by about 1400 and forced fortification of *tonnara* even on the safer north coast. On both sides of the Tyrrhenian, fishing villages were unknown and fishers sheltered behind town walls.[191] As French Valois and Spanish Habsburg monarchs each inherited multiple geopolitical rivalries, coastal fishers in the north at Walraversijde and elsewhere paid the price.[192] Even the Andalusian grandee's well-guarded *almadrabas* got shut down in, for instance, 1531, 1534, 1537–1538, and 1540, as the crown recruited labour for the navy.[193] Herring fishers in the Baltic and North Sea suffered through Danish–Hansard wars of 1361–1370 and 1426–1435, while during the intervening 1390s *Vitalienbrüder* pirates plundered ships and coastal warehouses alike.[194] Fisheries offered soft and potentially lucrative targets, yes, but in these situations they were mere collateral damage and fish stocks themselves little affected.

Chapter 5 has already identified high medieval changes to local and regional fish populations and some of their likely natural and anthropogenic drivers. Comparable material and written evidence also signals late medieval problems for local marine stocks. Recalling that fish grow throughout their lives, absent other influences, shrinking average size through repeated exposure and loss of older and larger specimens is a common sign of overexploitation. Among the abundant bones of plaice recovered from kitchen waste in dozens of northern French sites, those

[191] Bresc, "Pêche et coraillage," 107–109; "Pêche et les madragues," 174–175; *Un monde méditerranéen*, 271–272; and "Pêche dans l'espace économique," 532–538.

[192] Sicking and van Vliet, "Our triumph of Holland," 339–349, detail losses and costs of protection for Flemish herring fishers during Anglo-French wars in the fifteenth century and for all Low Countries fishing fleets in subsequent sixteenth-century Habsburg–Valois wars.

[193] Phillips, "Long-term profitability."

[194] Hauschild, *Studien*, 168–172.

from fish longer than 45 cm, 10 percent in the twelfth century, left but a trace in the thirteenth, while those of smaller fish rose from about 80 to over 90 percent and remained so in early fourteenth-century deposits. After 1350, however, large plaice reappeared in French catches, reaching almost 20 percent of remains, while small ones fell back to 75 percent. Archaeozoologists attribute this to reduced human predation. Late fifteenth-century demographic and economic recovery then restored pressure against the life expectancy of plaice: bigger age classes fell again by the early 1500s to the 5 percent range and almost zero by 1600.[195] Signs of overfishing and decline in haddock stocks of the southern North Sea also accompanied Flemish fishers' shift to the cod and herring of *Noordover*.[196]

Vulnerabilities in the large-scale herring industry are manifest in two salient features, the successive crises and collapses of local fisheries and the extreme short-term volatility of herring catches and prices. Well-recorded failure of Danish herring after 1420 has already been mentioned. A Lübeck chronicler in the 1430s lamented the previous decade in the Øresund:

> … there came no herring to the Sound so that the fishers could not take any all the time that they were there … and so the absence of the herring from the Sound remained long years … and they came not again. [The herring] split in the sea and a part came to Flanders and a part to Heligoland and many such places where they were caught. Nowhere, however, had they the form and goodness which they had at Scania.[197]

A few kilometers inland from the beaches, Orja farm, which had during the 1300s become a large-scale fish processing site, ceased operation in the mid-1400s and the farmstead was abandoned.[198] For perhaps two human generations Scanian catches failed repeatedly, so despite a strong post-1480 revival the take in 1494 came to less than a third that of a century before. With the 1530s the decline was permanent.

195 Clavel, *L'animal*, 146–49, 164–74, and summarized p. 189: "Le réapparition momentanée, dans la deuxième moitié du XIVe siècle, des grans poissons dans le commerce indique alors un reconstitution partielle de la population de pleuronectidés, en réponse à une baisse de la prédation." Dam, "Feestvissen en vastenvissen," 491–492, treats Dutch overfishing of plaice.

196 Ervynck et al., "How the North was won," 234–237. Northern cod (Norway, Iceland), show no signs of reduced population or size ranges, although limited nearshore stocks are said to have motivated fishers migrating to outer island settlement sites.

197 Koppmann, ed., "Rufus-Chronik," 226–227 (see original in Supplement 8.3.1). Editorial notes there provide similar statements from other local reporters. Modern research indicates that neither the Heligoland nor Flemish catches were of the Baltic herring stock taken in the Øresund.

198 Lagerås, *Environment, Society and the Black Death*, 94–95.

Before and after the breakdown in the Øresund, equally unexpected and unintended sequels to episodes in the rolling medieval herring boom include tales already told of commercial and ecological collapse of other localized stocks. Early medieval estuarine and inshore fisheries soon vanished from Picardy. During the thirteenth century once abundant Pomeranian herrings, followed a human generation later by those off Rügen, dwindled to mere local significance.[199] After the 1360s East Anglian stocks failed. The Scottish fishery in the northwestern North Sea, which developed after the 1360s, itself suffered a deep trough in the 1430s–70s, before reviving a decade or so later.[200] Each crash followed long years of peak production.

Economic historians and archaeologists faced with hoary local and scholarly myths blaming failed fisheries on departure of herring schools insulted by some human behaviour (symbolic[201] or material) or on random natural phenomena, have been quick to dismiss both human and environmental impacts on marine life during the Middle Ages. They collectively portray the frontier fisheries as essentially participants and pawns in economic and political contests among human communities. By that account the Øresund held a commercial location superior to supplanted Rügen and the Flemish and Dutch simply outcompeted English, Hanseatic, and Danish rivals. While others failed, the simultaneous growth of the Dutch herring fishery or the late fifteenth-century boom in pilchards off Cornwall are offered as evidence the natural world of the sea continued with its productive capacity unimpinged.[202]

[199] Zbierski, "Ichthyological studies"; Leciejewicz, "Zum frühmittelalterlichen Heringshandel"; and Makowiecki, *Historia ryb i rybołówstwa*, document the schools' disappearance from the area of Kołobrzeg.

[200] Gemmill and Mayhew, *Changing Values*, 319–320.

[201] Heinrich, "Information about fish from tales and myths," 18–19; Jagow, "Heringfischerei," 19–23, and "Hering im Volksglaube," 220–223, identify folk explanations known since the late Middle Ages from both the Baltic and Scotland. Further in Supplement 8.3.1.

[202] Jahnke, *Silber*, 11–13, 25–26, and 38, even attributes complaints about overfishing solely to competition, not to any possible change in the stock. Also working from a Danish perspective, MacKenzie et al., "Ecological hypotheses," 174–176, recognize the general importance of climate for Baltic ecosystems and also a periodic Bohuslan fishery for spent North Sea, not Baltic, herring. Still they ignore or politicize medieval failures in the Baltic fishery and imagine eutrophication a purely modern phenomenon. Holm, "Catches and manpower," 177–180 and 186–187, and Holm and Bager, "Danish fisheries," 107–111 and 120–122, offer more complete historical data, even for the Middle Ages, and more openness to how these might relate to climate. Locker, *Role of Stored Fish*, 35–45, acknowledges long-term late medieval cooling as possibly affecting the predator–prey relationship of cod and herring, but misconstrues how cooler water might affect herring directly and dismisses overfishing as peculiar to highly capitalized modern fisheries. Kowaleski, "Expansion," 448–449, and "Commercialization," 192–193, emphasizes Dutch and other competition, while also eliding the different

### 8.3.2 *Herring, People, Climate, and Weather,* c. *1350–1540*

Because most fifteenth-century fisheries remain poorly documented and little explored from any environmental perspective, it is the herring industry on which discussion of marine fishes entangled simultaneously with human predation and environmental fluctuation necessarily centres. Received historical opinion could not take fully into account the historical and ecological particulars of the fading medieval fisheries, the behaviour of their pelagic prey, nor what is becoming known about environmental changes during Europe's late medieval centuries. While the previous high medieval period and coincident medieval climate anomaly did not lack environmental variation affecting fisheries in several ways (recall Chapter 5), the late medieval flowering of marine fisheries coincided with sometimes dramatic instability in transition to (or the first phases of) the Little Ice Age.[203] Both medium- to long-term trends and newly typical events could affect fish stocks already stressed by human predation. Put another way, both climate and weather patterns are known to influence the availability of the animal late medieval fishers sought. Ecologists' consensus is voiced by fisheries analyst Keith Brander, "Reductions in stock productivity mean that levels of fishing to which a stock was previously resilient, become unsustainable."[204] What follows thus views late medieval herring fisheries as hybrid colonized socio-natural systems, wherein synergy between human activities and environmental variation acted upon the animal's life cycle to shift dynamics of the most heavily exploited stocks and the human arrangements exploiting them

Both the successive demise of fisheries dependent upon specific local/regional herring stocks and occasional general short-term fluctuations in catches and prices were heavily affected by herring biology and fishing practices at times of environmental change. This assertion rests first upon a closer look at herrings and the conditions under which medieval

phases and effects of climatic variability during the transition to and early centuries of the LIA. Bo Poulsen's magisterial *Dutch Herring* describes little human or natural impact on seventeenth–nineteenth-century schools, having been published just as LIA climate/weather conditions at small temporal scale were beginning to be deciphered, much less considered as triggers for biological or cultural adaptation.

203 As considered by Büntgen and Hellmann, "Little Ice Age in scientific perspective."

204 Brander, "Impacts of climate change," 398. As Alheit and Hagen, "Effect of climatic variation," 247, put it: "The dynamics of exploited fish populations are affected by both environmental variability and man-made activities (fishing, habitat alteration) …" And from an intersecting perspective ichthyologist Myron Peck finds "Small pelagics … are particularly sensitive to overfishing at low stock levels" (Peck et al., "Life cycle ecophysiology of small pelagic fish and climate-driven changes," 234). See also Auber et al., "Regime shift in exploited communities."

Europeans pursued them. Like related small pelagic (open-water) plankton-eating fishes, herrings comprise a very large biomass near the consuming base of aquatic food webs.[205] All around the world such animals and their food are subject to great and long-term cycles, as well as short-run swings, of abundance. The biomass changes relate closely to changes in oceanic temperature and nutrient levels (water chemistry). *Clupea harengus* is especially adapted to cold temperate waters. The southern boundary of its European range – and the northern limit of its warm-temperate cousin and ecological counterpart, the pilchard or sardine – lies just southerly of the British Isles. Multi-decade oscillations in the climate of the North Atlantic (the NAO) push that boundary southwards during cold periods, so herring become abundant off southern England and into the Bay of Biscay. Warm phases shift the boundary northwards and then pilchards replace herring as the dominant plankton feeder there.

Medieval Europeans fished primarily Baltic and North Sea herring populations.[206] Each large population migrates through a seasonal cycle, those of the North Sea, for instance, following a great counterclockwise gyre that takes them southwards along the British coast in summer and fall months and back north and west off the continent in winter and spring (Map 8.3). From that whole population come numerous local spawning groups. These segregate themselves into genetically distinct reproductive concentrations at particular places and seasons and, their business completed, rejoin the larger gyre.[207] Some of these groups once penetrated deeply into open bays and estuaries like the lower Somme, and in the Baltic another still enters the Schlei. Others gathered to spawn along beaches or at offshore locations. As observed above (p. 201), all the early European herring fisheries intentionally (and wisely) targeted such seasonal concentrations of inshore spawning fish. Toward the end of

205 For what follows see Checkley et al., "Habitat"; Field et al., "Variability," notably 46; Alheit, Roy, and Kifani, "Decadal scale variability"; and other papers in Checkley et al., eds. *Climate Change and Small Pelagic Fish*, as well as Bailey and Steele, "North sea herring fluctuations" (and compare other papers in Glantz, *Climate Variability, Climate Change, and FIsheries*); Alheit and Hagen, "Effects of climatic variation"; and Alheit and Hagen, "Long-term climate forcing." Corten, *Climate and Herring*, 203–206, and "Northern distribution of North Sea herring," finds movement in the twentieth century (so post-LIA) of both adult and juvenile North Sea herring at their northern boundary in response to decade-scale climatic variability.

206 Distinct territories mean the large populations interact little with one another. A third population, North Atlantic herring, was little engaged by major medieval fisheries.

207 For homing of herring populations in both the North and Baltic seas to traditional grounds where each group spawns at a different time and place see Barange et al., "Current trends," 208–210; www.fishbase.de/summary/Clupea-harengus.html (last consulted 15 September 2018); and van der Lingen et al., "Trophic dynamics," 135–137, with references there provided.

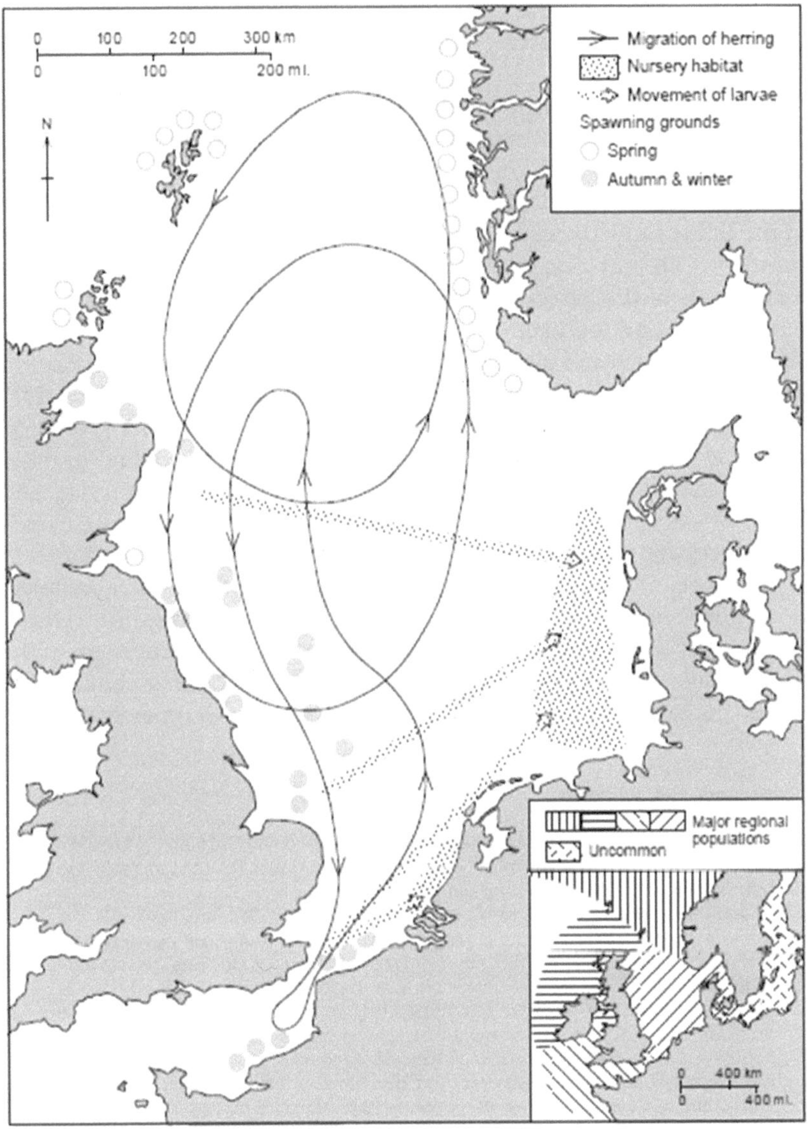


Map 8.3 Herring in the North Sea.

the Middle Ages, however, Flemish and Dutch fleets pioneered fishing offshore, where the herring schools were less closely sorted by their reproductive activity.

In late winter and spring months after the spawn, newly hatched larval and juvenile herring drift and swim into eastern parts of the North Sea, the German Bight off Dutch and German Friesland and along the Jutland coast. This is their nursery habitat. The shallow southern shore of the Baltic similarly serves that stock. Using nineteenth- and twentieth-century catch and climate data, modern research has shown that these early stages of the herring life cycle are especially sensitive to varied sea temperature and food supply. Weak year classes – which will bring low adult populations and poor fishing some two to five years later – occur in years when spring comes cold and late and the summer remains cool.[208]

Historical, biological, and climatological findings together give good reason to suspect that several spawning populations of herring (in other terms, regional fisheries) were successively under pressure (as evidenced by historically peak catches) just before their commercial collapse, which may then have been triggered by extreme environmental fluctuations of diverse sorts. When the 400- year-old Pomeranian fishery, once an object of wonder for its abundance, petered out in the late thirteenth century and that off Rügen followed only a generation later, the ecosystem of the south Baltic was no longer what it had once been. For a century and more, human agricultural development in eastern German and Polish

[208] Krovnin and Radionov, "Atlanto-Scandian herring," 243–247; Nash and Dickey-Collas, "Influence of life history dynamics and environment," and Casini et al., "Food-web and climate-related dynamics in the Baltic," 19–20. Hufnagl et al., "Unravelling the Gordian knot!" tease out mechanisms for this, notably the cold-induced failure of plankton production.

Some years of intermittent study early this century provided initial grounds to infer an association between variations in late medieval climate and those in recorded catches and prices of herring. At the time, however, both relevant data sets (herring and weather) lacked decade-scale precision and a clearly articulated link between them. During 2008–2014 I began to test some ideas on collegial audiences of medievalists, archaeozoologists, and environmental historians, before alluding to the relationship in my 2014 *Environmental History of Medieval Europe*, 332–333. Only thereafter however, did expanding awareness of an ever more detailed regional climate record; evidence of a consistent inverse correlation of preindustrial catches and prices for another well-documented north European fish, salmon; and a deep dive into new science on herring ecology together produce a plausible, I think probable, mechanism for documented historical phenomena. Meanwhile the relatively obscure essay by Philipp Gabriel, "Die Hansestadt Lübeck, der Hering und das Klima" (2016) independently proposed in particular that extensive LIA late winter–spring ice cover along south shore Baltic spawning and nursery areas generally inhibited fifteenth-century recruitment, resulting in long-term decline of those herring stocks. I had not myself grasped the role of the ice, so appreciate the opportunity to incorporate Gabriel's suggestion into the discussion which follows.

hinterlands had been stripping the forest cover and opening the soils of the Odra and Wisła watersheds which feed this part of the sea. Massive erosion and flooding followed.[209] Combined with shifts in Atlantic circulation, these events should have affected water chemistry and nutrient levels in the estuarine and nearshore areas frequented by juvenile herring. Palaeoenvironmental research has not yet fully reconstructed these conditions, but associated changes detrimental to herring have already been remarked in Chapter 5. The south Baltic shoreline was transformed as bays filled with sediment – the Wisła estuary became a great delta – and offshore barrier islands formed and consolidated into long sand spits.[210] And once the Pomeranian herrings had dwindled, despite ongoing demand even into the twentieth century none but a local fishery ever revived.

Lacking applicable sea surface temperature records, prevailing conditions in the North Sea and Baltic during the last medieval centuries must be assessed from reconstructions of preferably decade-scale seasonal fluctuations in the Low Countries, Germany, and England. This calls for climate history more fine-grained than earlier shown in Figure 5.4. Indices of winter severity for key continental regions are now available on an annual basis and ecologically critical spring conditions in multi-proxy reconstructions at close to decade scale (Figures 8.12 and 8.13).[211] English presentations remain less systematic.[212] New work on Sweden helps cover the western Baltic, where pertinent studies of the sea itself begin only in 1501.[213]

Climates of crisis stand out reasonably well. Historical climatologists now attribute ferociously unstable weather to Europe's transition into the LIA. A cooling pattern with intense storminess set in at Europe's northwestern fringe already early in the fourteenth century and by the 1390s

209 Dunin-Wąsowicz, "Natural environment and human settlement," 94–96; Bork et al., *Landschaftsentwicklung*, 221–249; Brázdil and Kotyza, *History of Weather*, 60 and 166.

210 Filuk, "Biologiczno-rybacka charakterystyka ichtiofauny zalewu wiślanego," 146–147.

211 Buisman and van Engelen, *Duizend jaar wind*, vol. 4, p. 707, and Glaser, *Klimageschichte*, 2d ed., Abb. 23.

212 Lamb, *Weather, Climate, and Human Affairs*, 50–71 and fig. 4.3; Ogilvie and Farmer, "Documenting the medieval climate," fig. 6.4; Pribyl, *Farming, Famine, and Plague*, 77–136, examines mainly eastern English growing seasons from the early thirteenth century to 1431.

213 Retsö and Söderberg, "Winter severity," 33–37 and table 2.2, cover 1305–1496; Koslowski and Glaser, "Variations in reconstructed ice winter severity," fig. 2. As displayed in Campbell, *Great Transition*, fig. 3.22, 207, the index of sea ice in the Atlantic north of Iceland shows extensive ice cover during the 1310s–40s, 1360s, and more or less continually from 1480 on. Reconstructed sea surface temperatures in the oceanic northeast Atlantic are discussed in Supplement 8.3.2. While these do accord with nearshore evidence and arguments here presented I remain reluctant to extrapolate from the outer ocean to confined Baltic and North Sea waters.

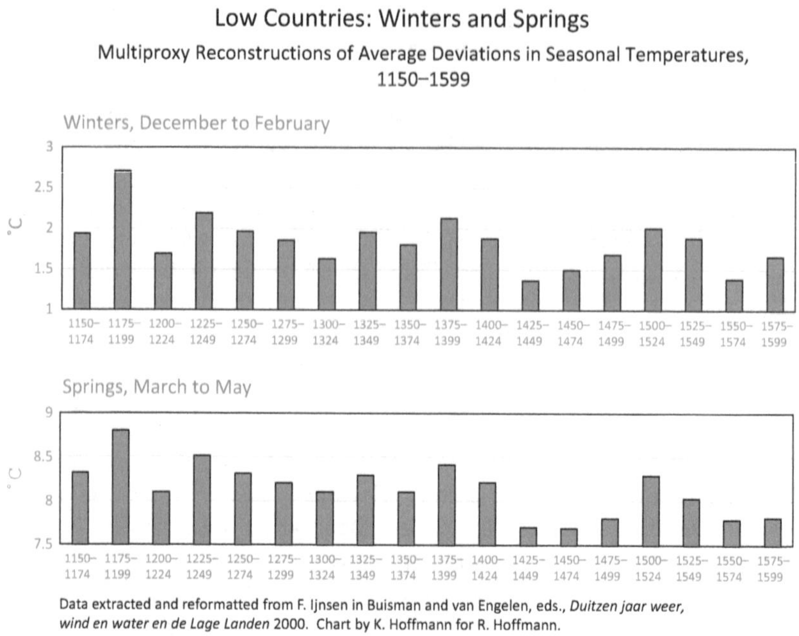


Figure 8.12 Low Countries: winters and springs. Multiproxy reconstructions of average deviations in seasonal temperatures, 1150–1599.

extended eastward into central Europe, where mean annual temperatures fell until the 1520s. Affected seasons included the springs most critical for survival of young herrings.[214]

When the heavily fished herring stocks in the southern North Sea broke down after 1360, climate scholars on all documented coasts concur that the 1360s–70s were distinctly cold, especially during the most studied winter season. German and Low Countries records indicate frequent remarkably cold springs throughout this period.[215]

214 Bailey, "*Per impetum maris*"; Grove, "Initiation of the Little Ice Age"; Pfister, Schwarz-Zanetti, and Wegmann, "Winter severity"; Glaser, *Klimageschichte Mitteleuropas*, 2d ed. rev, Abb. 23, pp. 57, 64–66, 76–78, 84–85, and 88–90; Campbell, *Great Transition*, fig. 3.20, p. 203; Luterbacher et al., "European summer," fig. 3.

215 Glaser, *Klimageschichte*, 88–90, provides a narrative. Luterbacher et al., "European summer," fig. 3; Glaser and Riemann, "Thousand year record of temperatures"; and Moreno Chamarro et al., "Winter amplification," confirm that cold decade. Swedish data shows a pronounced and long spell of cold winters from the 1340s to 1363 but then goes silent until 1390 (Retsö and Söderberg, "Winter severity," 32–37). East Anglia experienced notable cold from the mid-1360s to mid-1370s, then again during the 1380s (Pribyl, *Farming, Famine, and Plague*, 68–70, 74, 83 and 113–116).

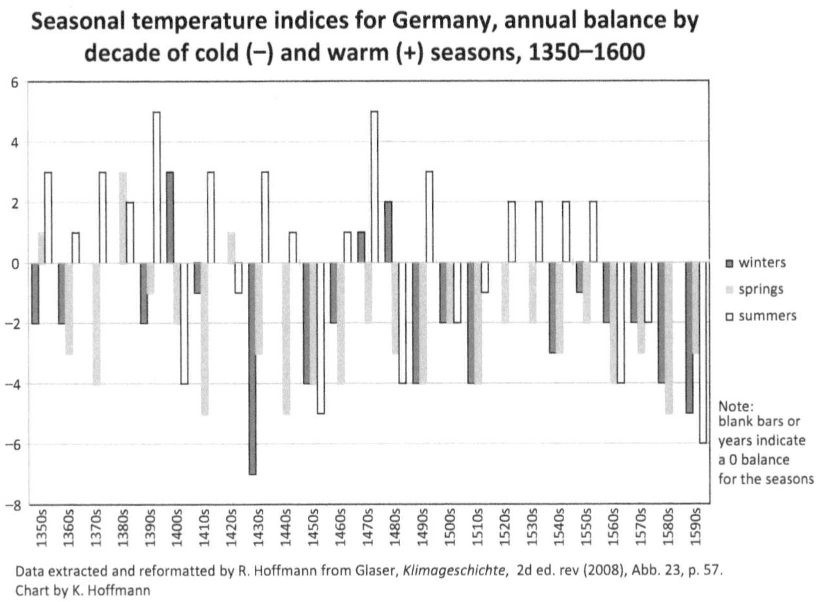


Figure 8.13 Seasonal temperature indices for Germany, annual balance by decade of cold (–) and warm (+) seasons, 1350–1600.

The half-century breakdown of catches seen in the Øresund (and nadir of those off Scotland) coincided with repeated bouts of difficult weather conditions. During the second and third decades of the fifteenth century, the Netherlands experienced many notably cold winters and a cooling trend in spring, while German accounts highlight especially cold springtimes.[216] Seasonal temperatures of early fifteenth-century England may lack scholarly consensus,[217] but English, Dutch, German, and Swedish climate historians subsequently agree on cold mid-century conditions,

[216] Glaser, *Klimageschichte*, 90–91, documents cold winters and springs past the century's midpoint, as do Buisman and Van Engelen, *Duizend jaar wind*, vol. 4, 707, and Retsö and Söderberg, "Winter severity." Camenisch et al., "1430s," 2110–2112, emphasizes a period of extraordinary variability although cold did not extend into summer. For a general and graphic overview of what has been called the definitive shift by the 1450s to a typical LIA climate with a weak and negative NAO see Campbell, *Great Transition*, 339–345 and fig. 5.2.

[217] While Ogilvie and Farmer, "Documenting the medieval climate," fig. 6.3, thought English winters of 1390–1410 relatively mild, Pribyl, *Farming, Famine, and Plague*, 115–117 and 227, finds a cold East Anglia in 1399–1411, 1421–1423, and 1428 as well as the 1430s.

including summers.[218] The latter three regions then experienced their coldest winters and springs of the entire 1400s.

Revived catches off Scotland and Scania after 1480 coincided with the start of a warmer interlude along Europe's western shores. At this time a pilchard fishery materialized along the coast of Cornwall, a good sign of warming waters there.[219]

Decline of the Øresund fishery had resumed by the 1520s. During and shortly after that decade northern European weather patterns showed some inconsistencies, suggesting a possible lag in the response of herring stocks and fisheries. While seasons during the 1520s were relatively mild, they followed repeatedly severe winters and chilly springs in the previous decade and cold conditions returned shortly after 1530. Research on the Baltic identifies remarkably severe cold and sea ice persisting deep into springtime for five winters during the 1510s and four more between 1525 and 1534.[220] Not good conditions for the planktonic prey of larval herring.

At some longer temporal scale total numbers and biomass of adult herring feeding in the North Sea may not have much changed. As observed with salmon, loss of recruits from damaged local spawning populations could be replaced by better survival of others bred elsewhere. The rising Dutch fishery at the end of the Middle Ages targeted previously unimpacted offshore populations, perhaps also less confined gene pools, and in a marine habitat less vulnerable to weather or local runoff. But, also as seen with medieval salmon, certain once heavily fished local herring stocks had become insignificant. Modern research no longer finds the schools which concentrated to spawn off East Anglian beaches. The species was not extirpated from the Baltic but commercial fishing in

[218] Osborn and Briffa, "Spatial extent"; Camenisch et al., "1430s," acknowledges the 1450s as initiating three decades of cold in interior western Europe, a diagnosis supported in Luterbacher et al., "European summer," fig. 3; Moreno-Chamarro et al., "Winter amplification," fig. 2; Retsö and Söderberg, "Winter severity," 33–34, who note the negative NAO of 1445–1460; Litzenburger, *Un ville face au climat*, 95–116; Esper et al., "Northern Hemisphere temperature anomalies"; and Cook et al., "A Euro-Mediterranean tree-ring reconstruction of the winter NAO Index." Meanwhile the North and Baltic seas were further impacted by diminished flow of tributary rivers during the megadrought of 1437–1473 visible in a wide sample of tree rings (Cook et al., "Old World megadroughts").

[219] Checkley et al., "Habitats," 30–31; Barange et al., "Current trends," 206–207. The warmer decades stand out in Retsö and Söderberg, "Winter severity," 36–37.

[220] Glaser, *Klimageschichte*, 99–109; Koslowski and Glaser, "Variations in reconstructed ice winter severity," fig. 2; Moreno-Chamarro et al., "Winter amplification," fig. 2; Camenisch et al., fig. 2. Cunningham et al., "Reconstructions of surface ocean conditions," figure 4, find the coldest North Atlantic since 1000 CE in the first half of the sixteenth century.

the main basin shifted to supply more sprats (a different small plankton-eating Clupeid, more tolerant of low salinity) and on eastern coasts to take a dwarf inshore subspecies (*C. harengus membras*, "Strömling" or "Stremling") for mainly local consumption.[221] Note, too, that adult Baltic herring in the twentieth century averaged 20 percent smaller than those consumed a millennium before.[222] Certainly something changed.

A common scenario was thus reenacted several times in the medieval herring fishery which fed European need for a relatively cheap, long-lasting, and portable fish food. A synergy of fishing and climate fluctuations acted upon genetically distinct local stocks. Under pressure from commercial export demand, long-fished local reproductive concentrations of these animals successively came under more intense exploitation, then crashed to commercial insignificance. Each collapse coincided with independently established environmental variation, whether of plausibly human cause, as the changing runoff regime from south Baltic watersheds, or of plausibly natural origin, as the climatic changes which pulsated eastward from the North Atlantic. Each new success tapped untouched stocks at greater distance from favoured markets. Medieval herring stories call to mind typical present-day fisheries crises with their market-driven technical innovations, intensification of capital, and continual move outward from commercially depleted to less accessible 'virgin' stocks. Yet even as the outward expansion of marine fisheries shaped a model for following centuries, it foreshadowed how things might go awry.

Quite apart from long-term shifts in regional sources of supply, European herring markets exhibit great short-term volatility in price (Figure 8.7A–D). Surely much local fluctuation did arise from political, monetary, or other perturbations in local supply, but a wider perspective gives reason to think that at decade scale the violent fluctuations characteristic of the herring market bore some relationship to short-term weather patterns, notably late winter and spring temperatures.

A possibly unique record of annual landings at fifteenth-century Dieppe (Figure 8.14) confirms extreme short-term fluctuations before

221 On the biology of sprat see www.fishbase.de/summary/Sprattus sprattus (last consulted 21 September 2018). Population trends of Baltic herring and sprat tend to be reciprocal, with one rising as the other declines (van der Lingen et al., "Trophic dynamics," 135–136 and 145; Barange et al., "Current trends," 212–213), while both interact with cod (Rijnsdorp et al., "Resolving the effect of climate change"; Hammer et al., "Fish stock development," 558–562; and Alheit and Pörtner, "Sensitivity of marine ecosystems," 168). For historical perspectives see Pöltsam, "Essen und Trinken," 120–121; Mand, "Festive food," 49–50 and 73; Sidrys, "Fish names in the Eastern Baltic," table 1; Hodgson, *The Herring*, 14–15.

222 Jahnke, *Silber*, 13.

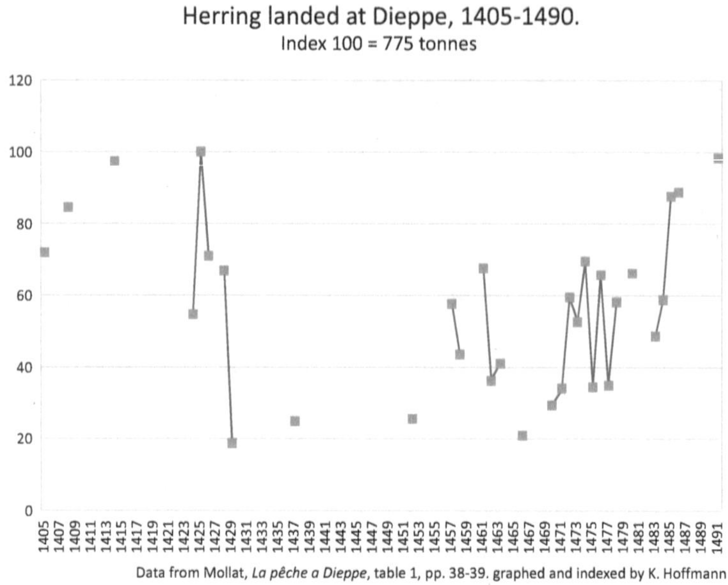


Figure 8.14 Herring landed at Dieppe, 1405–1490.

the fish reached any market.[223] Dieppe was a relatively minor player, landing well less than half the herrings of any number of contemporary Netherlandish ports, but its late medieval catches supported an active processing industry and supplied extensively to Normandy, Paris, and further inland. Annual landings averaged about 400 tonnes but peak years went five times higher than low ones and even the unbroken run of returns during the 1470s–80s often varied over 50 percent year to year. Dieppe is near the southern margin of normal herring range; its fishers worked so close to home in peak herring season they managed two trips a day. Landings there crested in the first quarter of the century and again in its last fifteen years. (Isolated mid-century numbers are hard to interpret.)

[223] Mollat, *La pêche à Dieppe*, table 1, pp. 38–39, repeated and more generally discussed in Mollat, *Commerce maritime*, 313–318; Hocquet, "Pêcheries médiévales," 102–103. In contrast, Jahnke, "Medieval herring fishery," 178–179, and "European fishmonger," 40–42, can still quantify only Danish exports of herring, so he will infer neither the total catch nor the biomass of the fourteenth-century Øresund. Although Jahnke's latest work notes "abundant" evidence for large fluctuations and the eventual decline of these stocks, he explores no biological and environmental grounds for them. Holm, "Commercial sea fisheries," 16–17, generally acknowledges climate to affect food supply and numbers of herring without relating this to any historical events or trends in Danish catches.

Elsewhere prices must stand surrogate for supply.[224] Reviewing Figures 8.7–8.9, the long-term price trends reflective of larger movements in the European economy are no longer the issue here. As total demand appeared relatively stable after the 1350s, short-run changes must arise primarily from supply. Thus bringing together price series from different markets illuminates a more fine-grained aspect of this fishery, namely its short-term volatility everywhere. Taken one by one and especially if presented only as a numerical table, individual data sets appear meaningless or at most merely reflective of local circumstances, political or otherwise.

But viewed collectively the data series in Figure 8.7 show at least four coincident price peaks, relatively short periods when four and more locations experienced prices significantly higher than in preceding and following decades. Such events may indicate more general problems with the fishery itself. They occurred in 1418–1427 (at Antwerp, Leiden, Brussels, Rostock, England, Kraków); 1444–1460 (Antwerp, Leiden, Utrecht, Brussels, Hamburg, England, Rostock, and likely also Klosterneuburg); 1479–1491 and therein especially 1487–1489 (Antwerp, Leiden, Utrecht, Brussels, England, Rostock, Augsburg); and 1520–1536 (Antwerp, Leiden, Utrecht, Brussels, Hamburg, Scotland, England, Rostock, Augsburg).[225] If the price data does so identify likely times of regional or larger shortfalls of supply between 1400 and 1540, what conditions may have preceded or accompanied these bio-social events? Always keeping in mind the potentially lagged or dampened response of ecosystems to external stimuli, how might the herring fishery fit what is now known of weather conditions?

[224] In a forthcoming contribution ("Salmon variability related to phases of the Little Ice Age: consilience from Arctic Russia to Scotland?") to a collective volume in memory of our late colleague Alasdair Ross, I demonstrate the close inverse relationship between early modern salmon catches and prices to be expected under conditions without evident changes in demand or capture techniques (as well as the relation of that species to decade-scale changes in weather/climate). Such coincident data sets for salmon are not available from the Middle Ages.

[225] The price peaks occur both in places close to production zones (Antwerp, Leiden, Rostock) and at what can only be consuming centres (Kraków, Augsburg). At decadal scale the price series for England peaks in the first third of the fifteenth century and again around mid-century; the latter maximum also occurs at Klosterneuburg (Pribram, *Materialien*, 627). I do not now know what, if anything, to make of convergent low herring prices at Rostock and Low Countries cities during the 1430s, 60s, and 90s. Hamilton's broken series of herring prices from Navarre (Figure 6.3) seems, like those further north (the probable origin of the fish themselves), to peak in the 1420s and again in the early 1440s. Note that periods of political insecurity in the Baltic during the 1390s and 1426–1435 do not coincide with high prices elsewhere.

Noteworthy first are the coincidences of three widespread price peaks (1418–1427, 1444–1460, and 1520–1536) with the narratives of regional failures in fifteenth-century Scotland and Øresund, followed by further decline of the latter after 1520. An intervening general peak in herring prices is manifest during 1474–1491, especially 1487–1489. German records show unusually frequent cold springs from the mid-1470s to the end of the 1480s and then in the 1480s cool summers as well.[226] Dutch conditions were similar, though marginally warmer than at mid-century, a trend which continued past 1500.

So some price peaks coincide with regional failures and another does not. Taken collectively, the best reconstructions of seasonal temperatures do consistently show colder, even coldest, conditions shortly before and/or during times of high prices and weak catches in the herring industries of the North Sea and western Baltic. Of the seven decades with the most cold springs in Germany (1370s, 1410s, 1440s–1460s, 1490s, and 1510s, only the 1490s failed to precede difficulties in the fishery. Every period in which records from the Low Countries and Swedish Baltic[227] report frequent severe winters – the 1350s–60s, 1430s–early 40s, 1460s–early 80s, 1510s, and after the mid-1530s – is associated with regional breakdowns and high prices for herring. These situations correspond to what occurred in the same region during the nineteenth and twentieth centuries when scientists had data to establish a close relationship between winter–spring temperatures, survival of juveniles, and subsequent catches from those year classes.[228]

Lest this appear a simple matter of climatic determinism, several caveats are in order. Documented difficulties and collapses are especially evident in traditional regional inshore herring fisheries which should be recognized as exploiting local reproductive communities of fish. Such heavily fished stocks would, as Brander and others assert, be those most sensitive to general climatic stresses. The developing Dutch fishery, however, targeted the mixed herring stocks offshore and, at least by the seventeenth century, moved along with the main feeding schools, not the local

[226] Glaser, *Klimageschichte*, 92–93. Swedish winters in the 1470s and 80s, however, were mild.

[227] Retsö and Söderberg, "Winter severity," table 2.2, with sixteenth-century ice and winter reconstructions from Koslowski and Glaser, "Variations in reconstructed ice winter," and Hansson and Omstedt, "Modeling the Baltic," figs. 6 and 9.

[228] Overland et al., "Climate controls on marine ecosystems," emphasizes the role of decadal events. Alheit et al., "Atlantic Multi-Decadal Oscillation," indicates how such short term variations buffer the effects of climate change on small pelagic fishes in the northeastern Atlantic.

spawners.[229] Human practices and decisions affected, if unknowingly, the vulnerabilities of fishes. But so, too, in several ways did the behaviour of the herring influence human situations. Larval mortalities had no immediate effect on catches of adults the following summer and fall, but rather curtailed recruitment of fish catchable some years later. Colder conditions pushed the range of herring to the south enough that catches at normally marginal Dieppe twice peaked (in 1405–1425 and 1485–1492) when northern catches were weak (this is not visible in the broken record of the mid-1400s). But under those very conditions of cold and crisis elsewhere, post-spawn adult herrings unusually concentrated in late autumn and winter in the northeasternmost North Sea. For some mid-fifteenth-century decades these fish in the Skagerrak off Bohuslan supported a large fishery.[230] Likewise the growth of a herring fishery at Heligoland is traditionally dated to the 1420s and associated by contemporaries with the collapse of stocks in the Øresund although the 'new' fish had not in fact come from the Baltic.[231] From the perspective of fisheries and wider environmental history, much still remains inexplicable.

If herring are any indicator, going over the horizon for fish did not guarantee unbroken or sustainable success, but new offshore opportunities did often enough find stocks free of previous commercial exploitation. Tales of herring highlight the broader demand-driven, incremental, and uncontrolled extension of late medieval fisheries ever farther into the sea. When any given local population went from virgin to shrinking, another was brought on line. The pattern can hardly be construed as planned; each new move by fishers or by merchants must have been more or less speculative. Venturers had learned to expect wealth, or at least the good returns seen elsewhere. Failures were unanticipated, in part because each new advance went into somehow unfamiliar situations where environmental norms had still to be learned and as yet could hardly be seen as subject to change, much less to human influence.

229 Poulsen, "Variability of fisheries," 329–331, as shown in Poulsen, *Dutch Herring*, 160–213.

230 Jahnke, *Silber*, 281–319, provides the medieval evidence and Holm, "Commercial sea fisheries," 17–18, the later story, plainly now recognized as related to cold phases of the LIA (negative NAO). Background on Bohuslan appears in the Supplement under 8.1.1.1 and 8.1.2.2.

231 Herring fishing at Heligoland may go back to the late fourteenth century but its traditional growth is dated to the 1420s. This late fall fishery for North Sea fish peaked after 1500 with annual yields barely 10% of the Øresund's late fourteenth-century maxima. Catch per unit effort there was in full and permanent decline by the 1550s. See Poulsen, "Late medieval and early modern peasants," 47–50; Poulsen, "Herring fisheries off Heligoland"; Holm, "Catches and manpower," 181–182; Jahnke, *Silber*, 319–346; Holm and Bager, "Danish fisheries," 102–105; Holm, "Human impacts," 43.

## 8.4 Infinite Fish?

Might one link some of the difference between inland and inshore attitudes and those emerging offshore to a separation of participants in the distant-water fisheries from the producing ecosystems they exploited? The expectation is born offshore: the resource is superabundant, it needs only to be taken. Historian Arnved Nedkvitne remarks on the care with which Norwegian peasants managed their farms; "Sea fish, in contrast," he writes, "would have been regarded as a resource without limits and from a medieval perspective this was correct."[232] Such was plainly the view of the English petitioners of 1416 who explained they had left depleted home waters to find "a great plenty of fish" off Iceland's shores.[233] Insofar as the long lines of baited hooks they and other newcomers set out penetrated depths beyond reach of less well-equipped Icelanders, they could have found something like pristine stocks. For a while English merchants and fishers did capture Icelandic supplies for their own market but when their position deteriorated these, too, came largely under Hanseatic influence. Where then to go? Further offshore. It has been thought that the rejected English looked further west and, with Caboto's voyage of 1497, found North America with its own potentially lucrative and "infinite" cod stocks.[234] A Milanese agent in London passed on what the returnees said they had found:

> They assert that the sea there is swarming with fish, which can be taken not only with the net, but in baskets let down with a stone, so that it sinks in the water. I have heard this Messer Zoane state so much.
>
> These same English, his companions, say that they could bring so many fish that this kingdom would have no further need of Iceland, from which place there comes a very great quantity of the fish called stockfish.[235]

There again was the unlimited abundance Philippe de Mézières and before him Saxo had seen at the peak of the Scanian herring fishery and in its earliest days. In 1502 the *Gabriel* out of Bristol brought home

[232] Nedkvitne, "Development," 57.

[233] *Rotuli Parliamentorum*, vol. 4, p. 79, now Given-Wilson et al., *PROME*, Henry V, March 1416, Membrane 5, 33.: "*ils ount trovez lieux, l'ou graunt plentie de tiel maner pesson est prise sur les costes de Island ...*"

[234] Mollat, *Europe and the Sea*, 104 and 146–147, emphasizes both the particular (compare Marcus, *Conquest of the North Atlantic*, 164–173) and a general connection between European offshore fishing and its extension to Newfoundland. For an oddly cynical yet credulous view see Kurlansky, *Cod*, 24–29, and a multi-stranded narrative, Fagan, *Fish on Friday*, 205–223.

[235] Translation as Williamson, *Cabot Voyages*, document 24, from Biggar, ed. and tr., *Precursors of Jacques Cartier*, 17–19 (see Supplement 8.4).

thirty-six tons of salt fish, the first recorded cargo of North American cod.[236] Over the horizon the frontier still abounded with what was needed to satisfy European demand.

But the tale is not quite so tidy. New perspectives on Caboto, his backers, and his voyages are emerging from not yet fully published research under way in the Cabot Project at the University of Bristol.[237] The Venetian seemingly had more experience in Mediterranean and eastern trades than formerly thought. So, too, his Bristol investors, who had by 1490 abandoned Icelandic ventures for more promising dealings in Portuguese and Biscayan ports. English trade, more than actual fishing, in Iceland remained worth the risk especially for vessels from east coast ports. Those who financed Caboto – Bristolmen, heads in England of Italian branch banks, and the English crown – placed greater hope in a potential route to Asia as promised, some thought almost achieved, by Columbus's first two voyages. Baskets full of cod were an unintended consequence, at best a consolation prize,[238] while Bristol interests continued for another decade to pursue the main chance. Still, reliance on Iceland and threat of Hansard rivals surely remained an irritant, so "no further need for Iceland" did pose a positive prospect.

Nevertheless tepid Bristol and English interest in the abundance Caboto had found sustained only a second partial load of salt fish delivered by the *Gabriel* in 1504. With accession of Henry VIII in 1509 English activity ceased for decades. Others displayed no such reluctance. By 1515 Portuguese, Biscayan, Breton, and Norman vessels had already spent more than one summer catching and drying Newfoundland cod.[239] Then a reported failure in 1517 of southern coastal cod all along the European littoral triggered a large and general turn to what were already being called *los Baccalaos*, the isles of codfish,

[236] Quinn, *North America*, 353–357.

[237] Jones and Condon, *Cabot and Bristol's Age of Discovery*, 21–48. Jones, "The *Matthew* and the financiers" provides more detail on Cabot's sponsors, superseding the now obsolete Sacks, *The Widening Gate*, 34–36. A more journalistic account is Hunter, *The Race to the New World*. The newly evolving tale of Cabot arises from long-time researcher Alwyn Ruddock ordering her unpublished findings destroyed after her 2006 death; for particulars see Jones, "Alwyn Ruddock."

[238] Did Cabot anticipate encountering cod? A crew resorting to nets and baskets seems unprepared to catch cod with the usual hooks and long lines.

[239] Pope, "Transformation of the maritime cultural landscape," 124–132, provides a chronology for European group arrivals. Leading elements came from regions long worried about inadequate local supply (see Chapter 6, pp. 265–266 above).

far away, with their "infinite fish."[240] Amidst rapidly changing economic and climatic conditions, massive growth of a fishery soon followed.

><> ><> ><> ><>

New fisheries, inland and frontier, artificial and offshore, formed a double response to a widening gap between medieval European demand for fish and the supply traditional artisans could pull from natural local fish populations and habitats. Satisfying consumers thrust fishing beyond old bounds. The response to elite inland demand for fresh fish was aquaculture, where Europeans learned to manipulate and humanize aquatic ecosystems and animals. The response to mass demand for inexpensive fish was frontier fisheries, where Europeans learned to push further, to invest more, and to fight other users for the biggest piece of a distant inexhaustible abundance. Development left natural and cultural limits behind. Or did it?

[240] Mollat, *Pêche à Dieppe*, 12–13; Ferreira Priegue, *Galicia*, 149; Williamson, *Cabot Voyages*, 266–268, from Peter Martyr d'Anghiera, *De Orbe Novo Decades*, Dec. III, lib. vi, f. 52. Note the similarity between this report of local failures in 1517 and that alleged by English cod fishers in 1416 (note 75 above). On the sudden shift in origin of cod on European tables see Supplement 8.1.3.1.

# 9 Last Casts

## Two Perspectives on Past Environmental Relations

This book ends where most histories of fisheries begin. It has sought to explore how medieval Europeans interacted with their aquatic ecosystems in a world of natural and cultural change. The conclusion returns briefly to the two historical perspectives inherent in this book. Drawing on the line of fisheries pulls up common and diverse features of medieval Christendom's evolving relations with the natural world. Needs and experiences of medieval people patterned their evolving expectations and models for dealing with creatures from the water. Sharing cultural norms of religious and social practice, medieval Europeans knew and adapted to their diverse aquatic environments. Some of those adaptations enabled long-sustained use of resources. Others damaged both natural and human conditions. Some portion of this history happened to take late medieval Europeans to an unknown New World.

As one story ends, another begins. Certain of the medieval approaches spread around the globe in the wakes of Columbus, Cabot, and their successors and emulators. Many other practices continued to shape European diets, working lives, waters, and life forms. Some of the latter eventually also had belated influences elsewhere. Working from a medieval historian's perspective has identified and narrated some elements of Christendom's millennial coevolution with its fisheries resources, whether those were grasped by contemporaries or are accessible only in retrospect. To the extent that purpose has been accomplished, historical findings can also inform a novel view of the global fisheries crisis in the incipient Anthropocene and the remedies being offered at the turn from the second to this third millennium.

### 9.1 Fishing in the Medieval Encounter with European Nature

Other than assiduously vegetarian ascetics, nearly all medieval Europeans came occasionally or regularly into contact with fish. To recapitulate and to recall some illustrative details provided earlier, many fish tales arise from the multiple forms of medieval experience and related cultural understandings of fish and fishing. There was a narrative flow and more lay below.

### 9.1.1 *Story Lines*

Historical retrospect enables this book to compile a millennial tale of interactions between aquatic nature and medieval Europeans' use of it for culturally defined purposes. The central flow of overlapping narratives here traced played out within the larger evolution of Europe's post-Roman political economy, though with strong cultural constraints and local/regional variability arising from the sub-continent's distinctive aquatic ecosystems. The hybrid story of many fishes started from the natural diversity of those ecosystems providing living space for certain fish species and interspecific dynamics (Chapter 1) and from the several cultural understandings which shaped medieval uses of fish for food (Chapter 2). Spreading conformity to Christian religious strictures encouraged consumption of fish protein in weekly and seasonal rhythms. Markers of social and political status said much about who ate how much and which fish.

At least some consumption of locally available fishes occurred everywhere across post-Roman Europe (Chapter 3). The role of direct subsistence fishing by poor people likely varied greatly from place to place, while their obligations to provide by various means fish for elite consumption is manifest wherever verbal or material records survive. Both direct and indirect subsistence fishing survived with diminishing importance throughout the Middle Ages. Seasonal abundances could even encourage preserving catches for later use. Whether inland or along the coasts, these workers' experiences in nature grounded local and traditional ecological understandings of aquatic life.

Some elements of mercantile exchange had persisted throughout early medieval centuries, but vanishingly little then involved such basic bulk consumables as fish. Where nodes of non-agricultural populations did exist, however, such as in parts of Italy and around large ecclesiastical or secular establishments elsewhere, subsistence fishers might find people willing to pay for any surplus catch (Chapter 4). The incremental emergence of artisan fishers from the tenth century onwards aligned closely with markets in small but growing towns across widening regions of western Christendom. Urban-centred consumption drove small-scale commercial fisheries in natural local ecosystems, fostering some degree of regional economic integration, and much later extension to artificial and distant waters.

Social and natural networks surrounding high and late medieval subsistence and artisan fishers become tangible in their shared toolkits and evident selection of methods apt for taking particular varieties at identifiable times and locations. While habitats and habits of local fishes shaped fishers' work experiences, these evolved within larger economic and social contexts. Relations of subordinated fishers to their lords differed only in particulars among the Chiana valley in the ninth, the Loire or Humber in the late eleventh, or the Odra in the mid-thirteenth centuries. Certain capture techniques used along the twelfth-century Rhine also took fish off fourteenth-century Liguria, and along late fifteenth-century shores of Lake Constance and the Baltic, but the artisans who then handled those gears were subject to the markets they supplied. As members of village or town communities, artisan fishers gained access by common right or private licence. Their family enterprises exploited local resources under village bylaws, guild, or municipal statutes meant to minimize conflict, reassure consumers, and perhaps maintain prospects for future catches. Lords' servants or free artisans and their respective consumers were all vulnerable to environmental variation, whether or not then recognized as such.

As demographic and economic expansion spread across high medieval Europe, so too did total fish consumption, fishing activity, and fish markets. Elaborate regional networks supplied consumption centres. Environmental effects of intensified agricultural and urban development stressed aquatic ecosystems and beyond certain thresholds threatened sensitive species (Chapter 5). Intensified fishing pressure and even modest natural environmental dynamics could also raise the vulnerability of local or regional fish stocks. Other species reciprocally gained living space. Fishers adapted their practices and fish sellers their prices. Less prestigious but more abundantly accessible fishes gained regional dietary importance. Some people along Channel, North Sea, and Baltic coasts less often ate salmon or sturgeon; they and their neighbours consumed more herring and eel. Inland areas found the latter and cyprinids more palatable when salmonids or pike perch were less to be had. Some traditional freshwater and migratory fisheries, perhaps also some local coastal stocks, approached limits or perceptibly declined.

Awareness, fear, or perhaps just an amorphous worry for shortages of culturally valued fish to eat underlay changes in European fisheries from the thirteenth century on (Chapter 6). The particular measures taken differed with regional ecosystems, economic relations, and access to resources. Some programs had the effect (and were meant) to limit or ration access to existing fish stocks; others claimed to protect and maintain the same for the common good. When well informed and enforced, both approaches had their successes with, probably, dissimilar social effects. Two distinctive late medieval innovations fractured pieces of the established pattern and, by seeking to enlarge accessible supply, triggered wider and longer consequences. Purposely building and managing artificial ponds for domesticated carp created new local colonized ecosystems (Chapter 7). Europeans' routines and expectations on such hybrid socio-natural sites differed from those on natural waters. Fish became a human artifact. The second new vector, extending human predation on wild fisheries to hitherto unexploited marine regions and species, called forth new technologies and organizational structures to preserve the catches and convey unfamiliar foods to distant consumer markets (Chapter 8). Natural boundaries seemed overcome. Yet neither enlarged form of anthropogenic influence over fishes and ecosystems could really free fisheries from natural impacts and constraints. Dutch herring, Czech carp, and salt cod from Iceland or Newfoundland remained hybrids, each a biological parcel wrapped in cultural layers to serve human wants and needs.

### 9.1.2 *Undercurrents*

Narrative lines link up much evidence and depict successive, often overlapping, waves of stability, adaptation, disruption, and change. But the narrative approach elides some important elements underlying encounters of medieval Europeans with other life forms in their aquatic environments and thus components for a present-day grasp of those encounters. These involve problems of knowledge and of identifying drivers of this history.

What are we now capable of knowing about fish and fisheries in the medieval past? Traditional written and pictorial sources had real but constrained value. For much of the Middle Ages the written record was long dominated by a small literate elite more concerned with inherited Christian and classical texts and interpretations than with contemporary material experiences and observations. Religious intellectuals legitimized fish as an acceptable substitute for animal flesh in times of abstinence and used certain cultural constructs of fish to critique or promote human behaviours. Writings on dietary rules, estate management, and legal claims over property and people sometimes yield glimpses of aquatic nature and work being done there. Literate interests slowly expanded from the twelfth century, especially in mercantile contexts, but remained fragmentary, occasional, and unsystematic. Every text or illumination calls for critical assessment of its value and deficiencies as historical evidence. So other angles of enquiry have been pursued: archaeology and notably archaeozoology unearthed material evidence of some technologies and at least samples of the fish remains which identifiable communities had discarded from their work stations and dining tables. Ichthyology reports the biology and habits of modern species, and aquatic ecology their relation to habitats, all essential to understanding the behaviour and environmental thresholds for fishers' quarry. Various palaeosciences, notably historical climatology, climate history, and historical geography but also hydrology, soil science, oceanography, and dendrochronology, help reconstruct past environmental conditions. This book has continually searched for consilience, asking whether independent sources of information yield compatible, hence reinforcing, understanding and if not, why not. Inter alia, the poor survival rates of salmonid bones in archaeological contexts help explain the small representation of what written evidence shows to have been a culturally important fish variety. Long-term shifts in the chemistry of the Baltic somehow connect with the oscillating balance between herring and cod in those waters. The result has enabled our best grasp at what was going on at medieval socio-natural sites of various sorts, including aspects outside evident medieval awareness.

But medieval awareness of the aquatic realm is itself part of this history: what did medievals know about fish and fishing? As already indicated, the written record of medieval Europe rarely and at most in fleeting glances conveys how the vast majority of people perceived their natural surroundings and their various encounters with what was extracted from their waters. This veils traditional ecological knowledge (TEK) of fish (and other material practices) present among the illiterate majority. What at least some of those people knew of what mattered locally had to underlie successful capture, preservation, and culture of the particular species they or more often their social superiors consumed. TEK must be recognized as a signal of adaptations resulting in specialized gear, seasonal practices, and evidently conservation-oriented regulations. But medieval TEK blends historically into the infamous realm of 'known unknowns', important but its substance nearly always only to be inferred from scraps of peripheral records and material remains of the work itself. At many points in the Middle Ages medieval fishers, property managers, mercantile entrepreneurs, and even consumers knew more about their own aquatic world than any present-day historian, archaeologist, or scientist can. But we now also know much the medievals could not imagine. *The Catch* has offered consilient evidence to reconstruct situations, including medieval views we now think were quite wrong ('the king of the herring'). We know what medieval people did and sometimes the outcomes, but are commonly unable to ascertain their precise motives. There are no a priori grounds to assume our thoughts were theirs.

Forces driving change or sustaining historic stability recur in our stories. The principal variables differ and take different configurations over time and space. Even persistent biological and basic cultural pressures could have varying effects. Aspects of biology/ecology always mattered, both regarding fish and humans. As noted in Chapter 1, threshold parameters such as salinity, temperature, current speed, and sometimes special requirements for spawning success and juvenile survival set limits on each species' range and stock density but can change over time, whether the animals in question are subject to human predation or not. Human organisms need calories and protein, so human numbers strongly affect total demand for food. European populations went through demographic swings from low numbers in early medieval times to a long period of increase during the tenth through thirteenth centuries. Doubling the population to perhaps seventy million by 1300 drove landscape changes affecting inland and coastal waters as well as multiplying the mouths seeking, under various circumstances, fish to eat. The abrupt generational crash of the early to mid-1300s (due to famine and plague) initiated a century and more of reduced total numbers until renewed growth after the 1450s, first local or regional, then general. If humans in some ways topped the trophic pyramid, their shifting pressures for calories affected organisms all the way down, not least the balance between cereal production and acquisition of animals to eat.

Religious and social practices exerted steady pressure on medieval relations with aquatic life. Christian dietary abstinence from meat promoted but did not require consumption of fish as a substitute. Initially a pious ascetic practice, its impact grew only generations after conversion with general acceptance as a mark of Christian identity. It is telling that both the English 'fish remains horizon' and purposeful construction by lay and religious lords of artificial fish ponds in Poitou and Berri coincide with a well-known eleventh-century surge in observant religious practice across western Christendom. Some degree of fish eating had become ubiquitous, attuned to weekly and seasonal rhythms, but long much more common among socio-economic elites, who could display their superior status with the approved costly food for themselves and the cheapest varieties for their households. Reformist twelfth-century prelates and religious orders might reluctantly add pious donations of penitent herring to their bean-rich diets but some centuries later their successors followed a token herring, if herring at all, with a fine dish of pike, salmon, halibut, or carp. Overall, however, fish on Friday and fish during Lent were the acknowledged norm among medieval Christians and this culturally shaped demand for fish solidly grounded diverse fisheries, markets and trade in fish, and eventually concern for those supply chains on the part of authorities. While the taste of inland elites for fresh fish, not least as objects of social display, drove construction of storage tanks and eventually artificial fish culture facilities, successful design and operation of the latter derived from the needs and attributes of a natural organism, the common carp.

If natural conditions, human needs and numbers, religious ideology, and social display jointly and in different local proportions moulded Europeans' engagement with fish in early medieval centuries and also thereafter, urbanization and the concomitant medieval commercial revolution became major driving forces from about 1000 CE onwards. An exchange economy based in concentrations of non-agricultural people included fishers and dealers in fish almost from the start, for their neighbours relied on the market for their food supply, fish included. Whether in English fictions or the earliest charters to sell fish in Ravenna or Worms, artisan fishers found their customers in cities. Urban consumers soon included wealthy people, who at least occasionally shared tastes and demand with other elites. To such buyers local fishers gladly offered fresh salmon or sole, turbot or pike-perch, but peddled as well smaller, cheaper varieties to ever-growing numbers of the less well-to-do. Both economic strata depended on regional supplies. Londoners and Flemings could easily get fresh eel or 'herring of one night', Lisbon, Barcelona, and Venice found mullet or sardines on their waterfronts, but inhabitants of Basel or Prague might make do with eel, perch, or small cyprinids … until over time merchants developed ways to get preserved fish to those eager mass markets. Urban governments gave much attention to supply, sanitation, and dealings on their fish markets.

The interplay of basic biology with market relations becomes clear during the demographic slump of 1350–1450, after mass mortalities left smaller numbers of Europeans in possession of the same total material wealth and newly scarce labour gained in value. Fewer mouths meant less demand for cereals and greater per capita income left survivors with more discretionary choices. Demand for fish held up better than that for bread, so fish prices continued to rise. Hence Norwegian cod fishers enjoyed a 'golden age' when in 1400 stockfish exports only half those of 1300 gained in exchange half again more in wealth or imported goods. More generally the growth and operation of late medieval markets widened distribution of fisheries products and encouraged technical innovations to raise the harvest and improve commercial durability. Cases in point famously include first the Baltic, then the Netherlanders' North Sea herring fisheries, but should also recognize Bothnian and Prussian *strekfusz* and the mackerel and pilchard of Breton and English West Country fishers. Markets for cultured carp broadened consumption while weakening the rationale for self-sufficient production on elite estates.

Medieval towns likely concentrated local pressures on fish stocks and habitats while commerce spread this more widely, but human impacts originated in rural settings too. Direct human predation (fishing pressure) likely shrank longevity and average size of Baltic sturgeons already in the eleventh–twelfth centuries, of plaice and flounder along thirteenth-century northern French coasts, and in the fourteenth century all but extirpated whitefish from a lake near Salzburg. Agricultural clearances for permanent arable, subsequent farming practices, and water-powered grain mills together reduced migratory salmon stocks in Normandy and in headwaters of the Rhine system, likewise shad in tributaries to northern Italian lakes. Waste from dense urban populations and craft production polluted fish habit nearby and downstream. By and since the mid-1200s awareness of limits and fear of shortages in this culturally essential food motivated governments at all levels to issue protective regulations of diverse sorts. French monarchs imposed size limits and English kings set seasonal restrictions on fishing conger around the Cotentin; republican Florence banned certain methods for catching trout and the count of Holland protected vegetation where pike sought their forage. But not all environmental change was anthropogenic.

While natural tolerances, habitats, habits, and life cycles of fish species provided opportunities for but could constrain success of human predation, changing climates and other environmental fluctuations forced fish and people to adapt. Riverine and coastal landscapes and habitats were subject to natural dynamics, as when siltation obliterated the once-productive bay of the Somme and others on the French Atlantic shore while marine incursions created new and well-documented fisheries in the Zuider Zee and elsewhere along North Sea coasts. Those local changes stemmed from local events but also from long-term variability, notably in climate. Europe's MCA seems fairly benign, with warmer temperatures facilitating expansion of the carp and possibly the herring-dominated regime prevalent in the ninth- to thirteenth-century Baltic. Inception of the LIA with erratic weather and a cooling climate provided the setting for breakdowns of long-exploited local herring stocks and a longer-term shift to salinity and temperature conditions favouring a cod-dominated Baltic. Allowing for lags in fish life cycles, decadal-scale weather patterns coincided with short-term fluctuations at the herring–pilchard boundary in the outer Channel and Celtic Sea and in the apparent abundance of herring stocks further north. In terms of technology and practice the chill provided impetus to improve arrangements in carp culture and several key innovations in the North Sea herring fishery.

It is, however, difficult to trace to any environmental shift the conflicting ideas that on the one hand natural limits called for protection against overfishing and habitat loss or on the other that humans held dominion over aquatic nature and its bounty. Both cultural positions evolved throughout the late MCA and early centuries of the LIA. Perceptions of domestic scarcity certainly lay behind conservation measures on Lake Constance and elsewhere. A contrasting sense of frontier abundance in present rather than retrospective terms was voiced from 1200 (Saxo on the Øresund) to and beyond 1500 (Caboto et al.) but almost exclusively 'out there', where the natural limits feared inland seemed to evaporate.

In sum, only in the rarest or most vaguely general way can any single cultural or natural condition be seen to determine the inception, operation, or fate of medieval European fisheries. Circumstances of their interactions mean medieval people and their societies set their own courses for good or for ill. Their lives were not led to prepare for modernity or the post-modern.

## 9.2 In a Longer Haul: Medieval Legacies in a Present Global Fisheries Crisis

One emerging late medieval perception, the notion that humans could determine the presence, absence, and productivity of fish and, by extension, the operation of nature, may then have been confined to tightly focused managers of fish culture enterprises, who did create local ecological revolutions. The idea of actual human dominance, however, essentially a kind of cultural determinism, would subsequently form the intellectual core for one aspect of modernity. With adequate scientific and engineering know-how humans could transform the Earth to their liking. At the same time a possibly contradictory understanding proliferated on the maritime frontiers and there long held sway: this was the conviction and operating assumption that untouched resource abundance was always to be found over the next horizon, whether geographic or technological.

Driven by cultural needs and fear of shortage, the search for abundance propelled innovation all around late medieval maritime frontiers. That Zuan Caboto and his successors had fortuitously found those 'infinite fish' in the northwest Atlantic took a human generation to grasp. Then the cod fishery exploded into massive growth amidst a rapidly changing sixteenth-century climate and a new economic setting, to which it surely also contributed. Cod became the first material object of European plundering of New World nature and this direct transfer of biomass meant for Europe the potential dietary equivalent of 'ghost acreage', a virtual expansion of Europe's resource base.[1] This acquisition coincided, moreover, with the Spanish plundering of New World cultures which directly increased European silver supplies and contributed to inflation. Both new inputs joined a demographic surge already underway in Europe to call up rising demand and prices for cereals and to reverse those late medieval price ratios which had favoured 'luxury' meat and fish over grain. If Cabot triggered an ecological revolution in the northwestern Atlantic after 1500, as seen from Europe this was but a part of larger processes of change which increasingly distanced late sixteenth- and seventeenth-century culture and environmental relations from those of 1500. While Cabot, Columbus, and their contemporaries can usefully be understood as late medieval Europeans, not so their heirs.

Left unchanged in the medium run were Europe's (and general humanity's) utter dependence on short-run flows of solar organic energy and Europe's understanding of frontier hyperabundance. The now presumed absence of natural limits opened all to a competitive scramble and the conviction that new virgin resources always lay just over the horizon. Indeed Cabot's Grand Banks confirmed this very assumption and so repeatedly would later technological and spatial expansions of the northwest Atlantic fisheries.

As much a part of modernity as the expectations of distant waters' unlimited abundance and as rooted in elements of late medieval fisheries revolutions is belief in human managerial dominance. "If there are no fish in a pond ask who took them out" voiced the converse of the engineered manipulation of nature which had put the pond and the fish there in the first place. Both cultural constructs set the active human agent against a passive nature there for the taking. This mentality may have seemed more plausible and convenient to consumers ever more distanced from aquatic nature.

A catch here is that these were not the only lessons learned by medieval fishers. Pushed aside or left behind by fish farmers and frontier fishers alike were recognition of natural limits, human environmental degradation, and the common good, which together had grounded the stewardship inherent in both regulatory measures and some evident concern for what might now be called sustainability. Diverse voices going back through the master fisher at Constance in 1531, petitioners to the English parliament against bottom trawls in 1377, to even Philip IV of France in 1289 had long articulated these precautionary positions.

[1] Holm et al. "North Atlantic Revolution." See further in the Supplement.

Thus briefly to return with historical perspective to current situations. By the end of the Middle Ages essential elements for present-day global fishery crises were in place in European waters. The intervening half-millennium has spread those elements around the world into other economies like the Japanese, East African, or South Pacific, and to the limits of one aquatic ecosystem after another. Official statistics for the world's declining fisheries of the most recent half-century again confirm the joint role of elite and subsistence demands in driving environmental change and exploitation. Expanding agriculture, forestry, cities, and industry curtail and transform aquatic habitats. Exotic species brought in to solve nature's or mankind's earlier errors have their own unforeseen effects. So, too, have climate fluctuations. As since the high Middle Ages, even groups who fiercely protect, regulate, and manage their own local ecosystems, nevertheless wage their 'fish wars' to plunder insatiably those beyond the horizon. Information about these convergent environmental and social predicaments is familiar to fisheries specialists and drawn from published sources, but the historical underpinnings have hitherto gone unrecognized.

Seen from the Middle Ages, the processes of today's fisheries destruction are not new, but in a deep historical sense derive from the circumstances and choices of European fishers who pioneered the open Atlantic. Overexploitation, habitat destruction, selective predation on large or prestigious species, and human competition without regard for the resource were all part of medieval experience. The modernist assumption of infinite fish over the horizon, open competition, commodification, and alienation of consumers from the living organism or native ecosystems might all be thought relics from what Europeans carried across marine frontiers opened up by late medieval fishers. Other medieval approaches to aquatic systems they abandoned on the foreshore. If a complex set of arrangements – heavily capitalized commercial fishing in distant waters – had its highest development potential when just beginning its growth, the Atlantic marine fisheries for herring and cod plausibly set long-lasting and much emulated precedents. Once established, a path-dependent trajectory excludes other initial alternatives, here notably those of Mediterranean, inshore, and inland waters.

Fishing pressure is demand-driven, whether for subsistence or elite fashion and power. It affects not only the target species or stock but cascades through entire ecosystems. Habitat destruction more often originates outside the fishery, whether from ignorance of consequences or from low priority granted aquatic life in particular or nature in general. Political and economic power at national and international scales structure institutional arrangements of fisheries. Awareness and priorities shape the unintended consequences.

Present-day responses to perceived shortage of fish also have medieval precedents. Rising demand brings price increases which encourage further commodification of the product and stratification of consumers. Privatization of fishing rights has like effects on the resource itself, sometimes destructive, sometimes protective, and often socially disruptive. Results of regulation depend on its purpose (allocation or protection), appropriate local knowledge, and consent of affected parties. Local knowledge and stewardship of resources can keep exploitation within sustainable levels and even repair and enhance key habitat, but greater cultural distance puts whole natural systems at risk. Sustainable management loses its raison d'etre if fish are thought infinite. The history of modern capture fisheries is iterative expansion from depleted to virgin stocks, which are then depleted in turn. Unlike medieval Europe, no virgin stocks now remain.

The legacy of medieval reengineering nature into fish farms may be more ambiguous, as aquacultures are now thought to replace diminishing natural supplies of fish protein, and not only to satisfy elite tastes. This strategy nevertheless again carries heavy intended and unintended environmental and social costs. Colonizing nature more intensely massively replaces complex natural systems with simplified and thus fragile monocultures. The physical transformations of landscapes are similar. But whereas medieval carp culture cycled local terrestrial nutrients through its ponds, present-day arrangements typically impoverish wild aquatic systems elsewhere by extracting nutrients (forage fish and invertebrates) to subsidize the artificial crop and/or damage wild stocks by emitting concentrated metabolic wastes into their surroundings. All farming of plants, livestock, or fish entails destruction of diverse ecological relationships to fabricate something of human invention. Who decides whether to pay the price?

Past experience cannot predict future outcomes but can help identify situations and forces with positive and negative consequences for natural and human communities alike. What courses are likeliest to lead toward resilience on a changing planet? Insofar as even the twenty-first-century present derives key features from European triggering of modernity, medieval experience and handling of nature in general and fisheries in particular left legacies both to overcome and to enjoy. Neither human desires nor natural forces determine environmental outcomes. Recognition and quality of interactions shape the path. That is the biggest catch even in the dawning Anthropocene.

# Appendix A Glossary of European Fishes Named in This Book

| Common name as used in this book | Scientific (Latin) name | Family | Habitat(s) | Temperature pref. |
|---|---|---|---|---|
| amberjack | *Seriola dumerili* | Carangidae | SW pelagic | warm |
| anchovy | *Engraulis encrasicolus* | Clupeidae | SW pelagic | warm |
| asp | *Aspius aspius* | Cyprinidae | FW lotic | cold to warm |
| barbel | *Barbus barbus* | Cyprinidae | FW lotic | cool to warm |
| beluga sturgeon | *Huso huso* | Acipenseridae | anadromous | cool to warm |
| bitterling | *Rhodeus amarus* | Cyprinidae | FW lentic | warm |
| bleak | *Alburnus alburnus* | Cyprinidae | FW lentic | warm |
| bogue | *Boops boops* | Sparidae | SW demersal | warm |
| bream | *Abramis brama* | Cyprinidae | FW lentic | cool to warm |
| brill | *Scophthalmus rhombus* | Scophthalmidae | SW demersal | cool to warm |
| bullhead – see sculpin | | | | |
| burbot | *Lota lota* | Lotidae (formerly Gadidae) | FW lentic | cold |
| carp (common carp) | *Cyprinus carpio* | Cyprinidae | FW lentic | warm |
| catfish (wels) | *Silurus glanis* | Siluridae | FW lentic | cool to warm |
| charr (Alpine charr also char) | *Salvelinus alpinus* | Salmonidae | FW lentic | cold |
| chub | *Leuciscus (Squalius) cephalus* | Cyprinidae | FW lotic | cold to warm |
| cod | *Gadus morhua* | Gadidae | SW benthic | cold to cool |
| conger | *Conger conger* | Apodes | SW benthic | cool to warm |
| crucian carp | *Carassius carassius* | Cyprinidae | FW lentic | cool to warm |
| dab | *Limanda limanda* | Pleuronectidae | SW demersal | cool to cold |
| dace | *Leuciscus leuciscus* | Cyprinidae | FW lentic | cool to warm |
| dentex | *Dentex dentex* | Sparidae | SW demersal | warm |
| dogfish | *Scyliorhinus* sp. | Scyliorhinidae | SW demersal | cold to warm |
| dogfish | *Squalus acanthias* | Squalidae | SW & brackish benthopelagic | cold to warm |
| dolphinfish | *Coryphaena hippurus* | Coryphaenidae | SW pelagic | warm |
| drum | *Umbrina sp.* | Sciaenidae | SW demersal | warm |

| | | | | |
|---|---|---|---|---|
| eel | *Anguilla anguilla* | Cyprinidae | catadromous | |
| flounder | *Platichthys flesus* | Pleuronectidae | SW & estuarine | cold to warm |
| grayling | *Thymallus thymallus* | Thymallidae | FW lotic | cold |
| grouper | *Epinephelus* sp. | Serranidae | SW benthic | warm |
| gudgeon | *Gobio gobio* | Cyprinidae | FW lotic | cool to warm |
| gurnard | *Trigla sp.* | Triglidae | SW sublittoral, benthic | cool to warm |
| haddock | *Melanogramus aeglefinus* | Gadidae | SW benthic | cool |
| hake | *Merluccius merluccius* | Gadidae | SW benthic | cool to warm |
| halibut | *Hippoglossus hippoglossus* | Pleuronectidae | SW benthic | cold to cool |
| herring | *Clupea harengus* | Clupeidae | SW pelagic | cold to cool |
| huchen | *Hucho hucho* | Salmonidae | FW lotic | cold to cool |
| ide | *Leuciscus idus* | Cyprinidae | FW lentic | cool to warm |
| lamprey (river lamprey) | *Lampetra fluviatilis* | Petromyzonidae | anadromous | |
| lamprey (sea lamprey) | *Petromyzon marinus* | Petromyzonidae | anadromous | |
| ling | *Molva molva* | Gadidae | SW benthic | cold |
| mackerel (Atlantic mackerel) | *Scomber scombrus* | Scombridae | SW pelagic | cool |
| mackerel (Spanish mackerel) | *Scomber colias* | Scombridae | SW pelagic | warm |
| meagre | *Argyrosomus regium* | Sciaenidae | SW sublittoral | warm |
| miller's thumb – see sculpin | | | | |
| minnow | *Phoxinus phoxinus* | Cyprinidae | FW lotic | cool |
| mullet (red mullet) | *Mullus surmuletus* | Mullidae | SW littoral | warm |
| mullet (grey or striped mullet) | *Mugil cephalus* | Mugelidae | SW littoral | warm |
| mullet (thicklip mullet) | *Chelon labrosus* | Mugelidae | SW littoral | cool to warm |
| mullet (thinlip mullet) | *Liza ramada* | Mugelidae | SW littoral | cool to warm |
| mullet (golden mullet) | *Liza aurata* | Mugelidae | SW littoral | warm |
| nase (Iberia & southern France ) | *Parachondrostoma sp* | Cyprinidae | FW lotic | warm |
| nose (*Hotu*, *Nase*) | *Chondrostoma nasus* | Cyprinidae | FW lotic | warm |
| orfe – see ide | | | | |
| perch | *Perca fluviatilis* | Percidae | FW lentic | cool to warm |
| picarel | *Spicara smaris* | Sparidae | SW pelagic | warm |
| pike | *Esox lucius* | Esocidae | FW lentic | cool |

(*cont.*)

| Common name as used in this book | Scientific (Latin) name | Family | Habitat(s) | Temperature pref. |
|---|---|---|---|---|
| pike-perch (zander) | *Sander lucioperca* | Percidae | FW | cool |
| pilchard – see sardine | | | | |
| plaice | *Pleuronectes platessa* | Pleuronectidae | SW sublittoral | cold to cool |
| pollack | *Pollachius pollachius* | Gadidae | SW benthic | cold to cool |
| rays | | Rajidae | SW benthic | cold to warm |
| riffle dace (*Strömer, saufie)* | *Telestes (Leuciscus) souffia/ muticellus* | Cyprinidae | FW lotic | cool to warm |
| roach | *Rutilus rutilus* | Cyprinidae | FW lentic | cool to warm |
| roach (southern) | *Rutilus rubilio* | Cyprinidae | FW lentic | warm |
| rudd | *Scardinius erythrophthalmus* | Cyprinidae | FW lentic | cool to warm |
| ruffe | *Gymnocephalus cernua* | Percidae | FW lentic | cool |
| saithe | *Pollachius virens* | Gadidae | SW benthic | cool |
| sand eels (4 European species) | | Ammodytidae | SW littoral & pelagic, benthic | cool |
| salmon (Atlantic salmon) | *Salmo salar* | Salmonidae | anadromous | cool |
| sardine (adult is pilchard) | *Sardina pilchardus* | Clupeidae | SW pelagic | cool to warm |
| scorpion fish | *Scorpaena scrofa* or *S. porcus* | Scorpaenidae | SW demersal | cold to warm |
| sculpin (now many localized species) | *Cottus gobio* | Cottide | FW lotic | cool to warm |
| sea bass | *Dicentrarchus labra* | Moronidae | SW littoral & estuarine | cool to warm |
| sea bream, porgy | *Sparid* sp. | Sparidae | SW littoral | cool to warm |
| shad (Allis shad) | *Alosa alosa* | Clupeidae | anadromous | cold to cool |
| shad (Twaite shad) | *Alosa fallax* | Clupeidae | anadromous, also FW lentic | cold to warm |
| skate | *Diptura (Raja) batis* | Rajidae | SW benthic | warm |
| sole | *Solea solea* | Solidae | SW benthic | cool to warm |

| | | | | |
|---|---|---|---|---|
| sprat | *Sprattus sprattus* | Clupeidae | SW pelagic | cool |
| sterlet | *Acipenser ruthenus* | Acipenseridae | FW lentic | cool |
| stickleback (10-spined stickleback) | *Pungitius pungitius* | Gasterosteidae | FW lentic | cool |
| sturgeon (European Atlantic sturgeon | *Acipenser sturic* | Acipenseridae | anadromous, also FW lentic | cool |
| sturgeon (American Atlantic sturgeon | *Acipenser oxyrinchus* | Acipenseridae | anadromous, also FW lentic | cool |
| swordfish | *Xiphias gladius* | Xiphiidae | SW pelagic | cool |
| tench | *Tinca tinca* | Cyprinidae | FW lentic | cool to warm |
| torsk (also known as cusk) | *Brosme brosme* | Lotidae (formerlyGadidae) | SW benthic | cold to cool |
| trout (brown trout) | *Salmo trutta* | Salmonidae | FW lotic, also anadromous | cold to cool |
| tuna (bluefin tuna) | *Thunnus thynnus* | Scombridae | SW pelagic | cool to warm |
| turbot | *Scophthalmus maximus* | Bothidae | SW sublittoral benthic | cool to warm |
| weever | *Trachinus sp.* | Trachinidae | SW benthic | cool to warm |
| white bream | *Blicca bjoerkna* | Cyprinidae | FW & brackish demersal | cool to warm |
| whitefishes (many localized species) | *Coregonus* sp. | Coregonidae | FW lentic or anadromous | cold |
| whiting | *Merlangius merlangus* | Gadidae | SW littoral benthic | cool |
| wreckfish | *Polyprion americanus* | Serranidae | SW pelagic | warm |

*Sources*: Froese and Pauly, *FishBase*, www.fishbase.org, version 12/20; Kottelat and Freyhof, *Handbook*; Logez et al., "Modelling the habitat requirement."

# References

## Manuscript, Archival, and Other Unpublished Materials


Additional manuscript 25238, paper miscellany, 15th century. Fols. 56r–v. Sesone and seueralte of alle ffysses.

Allen–Unger Database, European Commodity Prices 1260–1914 www.gcpdb.info/index.html.

Arrigoni Martelli, Cristina. "Ducks and deer, profit and pleasure: hunters, game, and the natural landscapes of medieval Italy." PhD dissertation, York University, Toronto, Ontario, 2015. https://yorkspace.library.yorku.ca/xmlui/handle/10315/30637

Baker, Polydora. "Subsistence in Northern Italy during the fifth and sixth centuries A.D.: analysis of Monte Barro Faunal Collection." MA thesis. Trent University, Peterborough, Ontario, 1991.

Beerenhout, B. "Het *Huis te Vleuten*; onderzoek aan vismateriaal, aangetroffen bij archeologisch onderzoek van vindplaats 6," CD-ROM 2005.

Berlin. Geheime Staatsarchiv (GSA)

Preussische Kulturbesitz (PKB), Ordensbriefarchiv (OBA) nr. 11871, 19 March 1453.

Brien, Jacques. "Le développement de l'ordre cistercien en Poitou aux XII$^{\text{ème}}$ siècle et dans la premieère moitié du XIII$^{\text{ème}}$ siècle. Essai sur l'emplacement des monastères cisterciens." Diplôme d'études supérieures, Faculté des Lettres de l'Université de Poitiers. Poitiers, 1954.

Claustres, A., "Mining legacy in French Pyrenees mountain range: new insights from environmental archives," a paper delivered at the biennial meeting of the European Society for Environmental History, Munich, 2013.

"Facing water challenges in the Po River Basin, Italy: A WWDR3 case study." waterwiki.net. 2009; Department of Marine Biology, University of Vienna. www.marinehypoxia.com/joomla (accessed 27 October 2014).

Froese, F. and. D. Pauly, eds. *Fishbase*. World Wide Web electronic publication. www.fishbase.org, version (10/2017).

Galik, Alfred. Private correspondence, 1995. Unpublished manuscript, "Die Fischreste aus der mittelalterlichen Burganlage Lanzenkirchen in Niederösterreich," used by permission of the author.

"Historical and ichthyological evidence of fish consumption in medieval Austria," a paper delivered at the 10th meeting of the Fish Remains Working Group, International Council for Archaeozoology (ICAZ), New York, 24 September—2 October 1999. Unpublished and used by permission of the author.

Galloway, James A. "Expansion or eclipse? The Thames fisheries in the later Middle Ages." Unpublished 2016, used by permission of the author.

Götz, Patrick. "Karpfen in Franken im Mittelalter: Eine wirtschaftsgeschichtliche Untersuchung." Unpublished Zulassungsarbeit, Institut für Geschichte, Friedrich-Alexander Universität Erlangen-Nürnberg, Sommersemester 2009.

Häberle, Simone, and Heide Hüster Plogmann. "Archaeological and historical evidence of fish food supply, fish farming, and fish trading in medieval and early modern Switzerland." Unpublished presentation summarized in "Fish and Fishing Communities." 19th Fish Remains Working Group Meeting. Alghero–Stintino, Italy, –7 October 2017.

Horn-Rosenburg [castle], Austria. Archiv, Hs. Horn 44. Vermerkt das ausgeben auf die kuchen, das ich Thaman Hawsschaffer zu Hornn getan hab von meiner frawn und jungenherren von Puechaim wegen … nach des heiligen krewcz tag exaltacionis … anno xliiii.

Hüster Plogmann, Heidemarie, and Andre Rehazck. "Historical record versus archaeological data. Fish remains from Roman and medieval sites in Switzerland." A paper delivered at the 10th meeting of the Fish Remains Working Group, International Council for Archaeozoology (ICAZ), New York, 24 September—2 October 1999.

Jong, Theo de. "Tolbrugstraat, Breda, de botresten. Resultaten van het archeologische botrestenonderzoek," unpublished report, Department of Archeology, Municipality of Breda, 1988.

Linz. Oberösterreichische Landesarchiv (ÖOLA).

Littler, Alison. "The fisheries industry in medieval England down to the Reformation." PhD dissertation, Swansea University, 1979.

London. British Library [BL]

London (Greenwich). The National Archive [TNA]

Plea Rolls King's Bench, KB 27/384 30E3, m13d; KB 27/588 m44; and KB 227/509 11R2, m1d.

Klosterliteralien Tegernsee (KLTeg),

1 Urbarbuech, 13th century

3 Urbarium antiquissimum, *c.* 1250

4 Urbarpuech, 11th–14th century

94 Stiftsbuch, 1357–1378

185 1/3 Richterbuch, 1483–1531

MacKinnon, Amy T. "Dietary reconstruction of medieval and early modern Spanish population using stable isotopes of carbon and nitrogen." MA thesis, Department of Anthropology, California State University, Chico, California, 2015.

Munich, Bayerische Hauptstaatsarchiv (BHSA),

Ältere Bestände, Staatsverwaltung, Bd. I, Teil 2, Jagd- und Fischerei Generalregister Fasc. 444 Fischwesen #1.

Munich, Bayerische Staatsbibliothek (BSB),

Handschriften Abteilung Clm 18181, Tegernsee 'instrumenta piscalia', 1023.

Nürnberg, Staatsarchiv,

Rep. 400 IV Klosterverwalteramt Heilsbronn. Bd. 1, Jahresrechnungen, 1338–1374.

Ofczarek, Emmerich. "Die Teichwirtschaft in Südböhmen bis zum Dreissigjährigen Krieg." [Unpublished] Hausarbeit des Instituts für Österreichische Geschichtsforschung. Vienna: typescript, 1937.

Olsen, Karyn C. "A multi-isotope investigation of two medieval German populations: insight into the relationship among diet, disease, and tissue isotopic compositions." PhD dissertation, University of Western Ontario. Electronic Thesis and Dissertation Repository. Paper 1573. University of Western Ontario (accessed November 2015).

Rehazek, André, and Marc Nussbaumer. "Fish remains from a 16th century noble household in Untersee, Bernese Oberland, Switzerland," paper delivered to the 13th International Council for Archaeozoology Fish Remains Working Group Meeting, Basle/Augst, 4–9 October 2005.

Rorke, Martin. "Scottish overseas trade, 1275/86–1597." 2 vols. PhD dissertation, University of Edinburgh, 2001. http://hdl.handle.net/1842/23178 (accessed 18 August 2021) or as www.bris.ac.uk/Depts/History/Maritime/Sources/2001phdRorke1.html (last accessed 13 November 2017). NB: This is not a consistently stable link.

Rosello Izquierdo, Eufrasia, Eduardo González Gómez de Agüero, Carlos Fernández, and Arturo Morales Muñiz. "The Iberian medieval fisheries: a search for origins." A paper presented to the 18th International Council for Archaeozoology Fish Remains Working Group, Lisbon 2015.

Salzburg. Universitätsbibliothek

Codex M III 3 Collected medical tracts and other recipes ("Wiltu vische vahen"), 1439.

Séguin, Geneviève. "L'approvisionnement de Paris en poisson à la fin du Moyen Âge. Pour une histoire juridique, économique et sociale." Unpublished mémoire for the Diplôme d'Étude Approfondie en histoire médiévale at Université de Paris I, Sorbonne-Panthéon, December 1996.

St. Gallen, Stiftsbibliothek.

Csg 26. Pgm. 8° s. XIV. Psalterium. Namen der Fische des Bodensees beigeschrieben.

Csg 919. Papier. 4° s. XIV und XV. Miscellanea monastica et historica.

Csg 1050. Papier. 4° s. XIV und XV. (Johannes Schenklin, Monastic miscellany, 1416.)

Stiftsarchiv Mondsee: Akten. Bd. 406, 407, 411, and Handschriften Bd. 86. 282.

Trier, Stadtbibliothek,

Hs 608 (1954) [1958/1422 4°, fols. 1–38: "Friedrich von Flersheim, Fischbuch 1530."

Varas, András. "Who stole the water? The control and appropriation of water resources in medieval Hungary." PhD dissertation, Central European University, Budapest, 2020.

Vienna. Österreichische Nationalbibliotek (ÖNB).

Cod. Vind. 14072. ["Statuten der Fischer von Bologna" in Italia]

Cod. Vind. 13103, fols. 1r–6r, "Vischpuechl 1545" and fols. 7r–20r, "Anzaigung… im1542 Jar."
Cod. Vind. ser. nova 2644 "Tacuinum Sanitatis."

Vienna, Österreichische Staatsarchiv/Haus-, Hof-, und Staatsarchiv
MEA RTA 59-1. Fols. 309–312. Reichstagakten, Supplikationen, 2. Teil.

Vienna, Wiener Stadt- und Landesarchiv
Hauptarchiv Urkunde Nr. 5825 Kaiserpatent Maximilians, 1506.

Weiss-Amer, Melitta. "Food and drink in medical-dietetic Fachliteratur from 1050 to 1400." PhD dissertation, University of Toronto, 1989.

Wouters, Wim. "1,000 years of fishing and fish consumption in the town of Mechelen (Belgium)." An unpublished presentation summarized in "Fish and Fishing Communities." 19th Fish Remains Working Group Meeting. Alghero-Stintino, Italy, 1–7 October 2017.

Zeheter, Michael. "Eine Umweltgeschichte der Bodenseefischerei, ca. 1350–1900." Magisterarbeit, Fachbereich Geschichte und Soziologie, Universität Konstanz, December 2005.

Zürich, Zentralbibliothek Zürich (ZBZ),
Handschriftenabteilung,
MS A 83 Gregor Mangolt 'Fischbuoch' [author holograph].
MS C 150. Kat I: Gallus Kemly, mon. Sangall.: Collectanea.
MS Simmler 425 Gregor Mangolt 'Fischbuoch' [author holograph].

## Published Sources and Studies

Aarts, Bram G. W., Fred W. B. van den Brink, and Piet H. Nienhuis, "Habitat loss as the main cause of the slow recovery of fish faunas of regulated large rivers in Europe: the transversal floodplain gradient," *River Research and Applications* 20:1 (January 2004), 3–23.

Abad, Reynald. *La conjuration contre les carpes. Enquête sur les origines du décret de dessèchement des étangs du 14 frimaire an II.* Paris: Fayard, 2006.

Abad García, Isabel, and Jesús Peribáñez Otero. "La pesca fluvial en el reino de Castilla durante la Edad Media." In M. I. del Val Valdivieso, ed., *Vivir del agua en las ciudades medievales*, 147–180. Valladolid: Universidad de Valladolid, 2006. [Authors as given in table of contents. The title page of the article (p.147) identifies the authors as Jesús Peribáñez Otero and Isabel Abad Álvarez [*sic*].]

Abel, Mickey. "Defining a new coast: G.I.S. reconstruction of Maillezais Abbey's hydraulic drainage program and the coastline it created," *Peregrinations: Journal of Medieval Art & Architecture*, 5:3 (Spring 2016), 12–53.

Adamson, Melitta A., ed. *Daz buoch von guoter spise (The Book of Good Food). A study, edition, and English translation of the oldest German cookbook.* Medium Aevum Quotidianum Sonderband IX. Krems: Medium Aevum Quotidianum, 2000.

Adamson, Melitta W., ed. *Food in the Middle Ages: A Book of Essays.* New York and London: Garland, 1995.

*Medieval Dietetics: Food and Drink in Regimen Sanitatis Literature from 800 to 1400*. Frankfurt and New York: P. Lang, 1995.

"Medieval Germany." In Melitta W. Adamson, ed., *Regional Cuisines in Medieval Europe: A Book of Essays*, 153–196. New York and London: Routledge, 2002.

ed. *Regional Cuisines in Medieval Europe: A Book of Essays*. New York and London: Routledge, 2002.

Adolfsson, Herbert, ed. *Liber epistularum Guidonis de Basochis*, Acta Universitatis Stockholmiensis: Studia Latina Stockholmiensis, 18. Stockholm: University of Stockholm, 1969.

Adomnan. *Life of Columba*, ed. and tr. Alan O. Anderson and Marjorie O. Anderson. 2d. ed. rev. by Marjorie O. Anderson. Oxford: Clarendon Press, 1990.

Aelfric of Eynsham. *Aelfric's Colloquy*, ed. G. N. Garmonsway. 3d ed. rev. Exeter: University of Exeter, 1978.

Agricola, Georgius. *De Re Metallica*. Basiliae: Froben, 1556 [facsimile reprint Brussels, 1967]. Tr. Herbert C. Hoover and Lou H. Hoover. New York: Dover, 1950.

Aichholzer, Doris. *"Wiltuy machen ayn guet essen." Drei mittelhochdeutschen Kochbücher: Erstedition, Übersetzung, Kommentar*. Bern: Peter Lang, 1999.

Airoldi, L., and W. M. Beck, "Loss, status and trends for coastal marine habitats of Europe," *Oceanography and Marine Biology: An Annual Review* 45 (2007), 345–405.

Akehurst, F. R. P. ed. and tr. *The Etablissements de St. Louis. Thirteenth-Century Law Texts from Tours, Orléans, and Paris*. Philadelphia: University of Pennsylvania Press, 1996.

Alan of Lille [Alanus de Insulis]. "De planctu naturae," ed. Nikolaus M. Häring, *Studi Medievali*, 3, series, 19:2 (1978), 797–879. Tr. James J. Sheridan, *The Plaint of Nature*. Toronto: PIMS, 1980.

Albala, Ken. *Eating Right in the Renaissance*. Berkeley: University of California Press, 2001.

*Food in Early Modern Europe*. Westport, CT: Greenwood Press, 2003.

"The ideology of fasting in the Reformation Era." In Ken Albala and Trudy Eden, eds., *Food and Faith in Christian Culture*, 41–58. New York: Columbia University Press, 2011.

Albert the Great [Albertus Magnus]. *De animalibus*. Ed. Hermann Stadler as *De Animalibus libri XXVI nach der Cölner Handschrift*. Beiträge zur Geschichte der Philosophie des Mittelalters, vols. 15–16. Münster, 1916 and 1920.

Aldebrandin of Siena [aka Aldobrandino, Alebrant]. *Le régime du corps*, ed. Louis Landouzy and Roger Pépin; preface by Antoine Thomas. Paris, 1911. Reprinted Geneva: Slatkine, 1978

Alexander, Michelle M., Christopher M. Gerrard, Alejandra Gutièrrez, and Andrew R. Millard. "Diet, society, and economy in Late Medieval Spain: stable isotope evidence from Muslims and Christians from Gandía, Valencia," *American Journal of Physical Anthropology* 156 (2015), 263–273.

Alfani, Guido, and Riccardo Rao, eds. *La gestione delle risorse collettive nell'Italia settentrionale (secoli XII–XVIII)*. Milan: Franco Angeli, 2011.

Alfonso X, King of Castile and Leon. *The Siete partidas*, tr. and notes by Samuel Parsons Scott, ed. Robert I. Burns. 5 vols. Philadelphia: University of Pennsylvania Press, 2001.

Alfonso Anton, Isabel. *La colonización cisterciense en la meseta del Duero. El dominio de Moreruela (Siglos XII–XIV)*. Zamora: Instituto de Estudios Zamoranos "Florian de Ocampo" CSIC: Diputación de Zamora, 1986.

Alheit, Jürgen, and Eberhard Hagen. "The effect of climatic variation on pelagic fish and fisheries." In P. D. Jones, A. E. J. Ogilvie, T. D. Davies, and K. R. Briffa, eds., *History and Climate: Memories of the Future?*, 246–266. New York: Kluwer, 2001.

"Long-term climate forcing of European herring and sardine populations," *Fisheries Oceanography* 6 (1997), 130–139.

Alheit, Jürgen, Claude Roy, and Souad Kifani. "Decadal-scale variability in populations." In Checkley et al., eds., *Climate Change and Small Pelagic Fish*, 64–87. 2009.

Alheit, Jürgen, and Hans-Otto Pörtner. "Sensitivity of marine ecosystems to climate and regime shifts." In P. C. Reid and L. Valdes, eds., *ICES Status Report on Climate Change in the North Atlantic*, 159–173. ICES Cooperative Research Report No. 310. Copenhagen: ICES, 2011.

Alheit, Jürgen, Priscilla Licandro, Steve Coombs, Alberto Garcia, Ana Giráldez, Maia Teresa Garcia Santamaría, Aril Slotte, Athanassios C. Tsikliras. "Atlantic Multidecadal Oscillation (AMO) modulates dynamics of small pelagic fishes and ecosystem regime shifts in the eastern North and Central Atlantic," *Journal of Marine Systems* 131 (March 2014), 21–35.

Allison, Edward H., Allison L. Perry, Marie-Caroline Badjeck, W. Neil Adger, Katrina Brown, Declan Conway, Ashley S. Halls, Graham M. Pilling, John D. Reynolds, Neil L. Andrew and Nicholas K. Dulvy, "Vulnerability of national economies to the impacts of climate change on fisheries," *Fish and Fisheries* 10:2 (2009), 173–196.

Altisent, Austí. *Les Granges de Poblet al segle XV. Assaiq d'història agrarian d'unes granges cistercenques catalanes*. Barcelona: Istitut d'Estudis Catalans, 1972.

Alves Dias, and João José. "Un Banquet royal au Portugal au XVIe siècle." In Martin Aurell, Olivier Dumoulin, and Françoise Thélamon, eds., *La sociabilité à table: Commensalité et convivialité à travers les âges*, 155–158. Actes du Colloque de Rouen, 14–17 novembre 1990. Publications de l'Université de Rouen, 178. Rouen: Université de Rouen, 1992.

Amacher, Urs. "Geschichte der Fischer und der Fischerei im Mittelalter." In Heide Hüster Plogmann, ed., *Fish und Fischer aus zwei Jahrtausenden. Eine fischereiwirtschaftliche Zeitreise durch die Nordwestschweiz*, 95–105. Forschungen in Augst, Bd. 39. Augst: Römerstadt Augusta Raurica, 2006.

"Die Teichwirtschaft im Spätmittelalter, vom Frischhaltebecken zum Fischmastweiher," *Medium Aevum Quotidianum [Krems]* 34 (1996), 68–90.

*Zürcher Fischerei im Spätmittelalter*. Mitteilungen der Antiquarischen Gesellschaft in Zürich, Bd. 63. Zürich: Verlag Hans Rohr, 1996.

Amann, Benjamin, Sönke Szidat, and Martin Grosjean. "A millennial-long record of warm season precipitation and flood frequency for the North-western Alps inferred from varved lake sediments: implications for the future," *Quaternary Science Reviews* 115 (2015) 89–100.

Amargier, Paul. "Notes sur l'ichthyophagie des serviteurs de dieu en Provence au moyen âge." In Denis Menjot, ed., *Manger et boire*, II: 313–323. Paris: Belles lettres, 1984.

"La pêche en Petite Camargue au XIVe siècle." In *Bulletin Philologique et Historique (jusqu'à 1610) du Comité des Travaux Historiques et Scientifiques*, 99ème Congrès des Sociétés savantes, Tours, 1968, vol. 1: *Les Problèmes de l'alimentation*, pp. 331–346. Paris: BNF, 1971.

Ambros, Cyril. "Zvieracie zvyšky zo stredovekého sidliska v Chlábe," *Archeologické výskumy a nálezy na Slovensku v roku 1979* (1980), 21–28.

Ammann, Hektor. "Untersuchungen zur Wirtschaftsgeschichte des Oberrheinraumes. *II. Das Kloster Salem in der Wirtschaft des ausgehenden Mittelalters," Zeitschrift für die Geschichte des Oberrheins 110*, NF 71 (1962), 371–404.

Amorim, Inês. "The evolution of Portuguese fisheries in the medieval and early modern period. A fiscal approach." In Louis Sicking and Darlene Abreu-Ferreira, eds., *Beyond the Catch: Fisheries of the North Atlantic, the North Sea and the Baltic, 900–1850*, 245–279. Leiden: Brill, 2009.

"Portuguese fisheries, c. 1100–1830." In Starkey, et al., eds., *A History of the North Atlantic Fisheries, 1. From Early Times to the Mid-Nineteenth Century*, 279–298. Bremen: Hauschild, 2009.

Amorosi, Thomas. "Icelandic Archaeofauna: a preliminary review," *Acta Archaeologica*, 61 (1990), 272–284.

Amundsen, Colin. "Coupled human and natural systems: a new perspective on early fishing and fishing cultures of Northern Norway." In Ramona Harrison and Ruth A. Maher, eds., *Human Ecodynamics in the North Atlantic: A Collaborative Model of Humans and Nature through Space and Time*, 55–78. Lanham, MD: Lexington Books, 2014.

Amundsen, Colin, Sophia Perdikaris, Thomas H. McGovern, Yekaterina Krivogorskaya, Matthew Brown, Konrad Smiarowski, Shaye Storm, Salena Modugno, Malgorzata Frik, and Monica Koczela. "Fishing booths and fishing strategies in medieval Iceland: an Archaeofauna from the [*sic.*, *recte* site?] of Akurvík, North-West Iceland," *Environmental Archaeology* 10:2 (2005), 127–142.

Andreas Capellanus. *De amore*, ed. E. Trojel. Munich: Eidos Verlag, 1964. Tr. John J. Parry, *The Art of Courtly Love*. New York: Norton, 1969.

Andreska, Jiří. "Archeologické nálezy rybářského nářadí v Mikulčicích," *Archeologické rozhledy* 27 (1975), 132–138.

"Development of fish-pond culture in Bohemia." In Bela Gunda, ed., *The Fishing Culture of the World: Studies in Ethnology, Cultural Ecology and Folklore*, vol. 1: 77–89. Budapest: 1984.

*Lesk a Sláva Českého Rybářství*. Prague: Nuga,1997.

*Rybářství a jeho tradice*. Prague: SZN,1987.

*Vývoj rybářství, Zemědělské muzeum Ohrada Hluboká n/Vlt. Pruvodce expozici*. Prague: Prague Museum, [1977].

*Anonymi Ticinensis Liber de laudibus Civitatis ticinensis*, ed. Rodolfo Maiocchi and Ferruccio Quintavalle. Rerum italicarum scriptores, vol. 11, pt. 1. Città di Castello: S. Lapi, 1903.

Anthimus. *De observatione ciborum ad Theodoricum regem Francorum epistula*, ed. and tr. Eduard Liechtenhahn. Corpus medicorum Latinorum, vol. 8:1. rev. ed. Berlin, 1963. [original ed. Berlin 1928] A colloquial modern translation is Mark Grant, ed and tr., *De obseruatione ciborum: On the Observance of Foods*. Totnes, Devon: Prospect Books, 1997.

Aparisi, Frederic. "Fishing in medieval Valencia," *Imago Temporis. Medium Aevum* 15 (2021): 213–241.

Appelt, Heinrich, and Winfried Irgang, eds., *Schlesisches Urkundenbuch* [*SUB*]. 4 vols. Vienna–Cologne–Graz, 1963–1988.

Arizaga Bolumburu, Beatriz. "La alimentacion en el pais vasco en la baja edad media: el caso de Guipuzcoa." In Denis Menjot, ed., *Manger et boire*, vol. 2, pp. 197–208. Paris: Belles lettres, 1984.

Arnold of Lübeck. *Arnoldi Cronica Slavorum*, ed. Johannes Lappenberg and Georg Heinrich Pertz. MGH SS Rerum Germanicarum, 14. Hannover: Hahn, 1868. (Tr. Graham A. Loud, *The Chronicle of Arnold of Lübeck*. London and New York: Routledge, 2019.)

Arnold, Ellen F. *Negotiating the Landscape: Environment and Monastic Identity in the Medieval Ardennes*. Philadelphia: University of Pennsylvania Press, 2013.

Arnoux, M. "Classe agricole, pouvoir seigneurial et autorité ducale. L'évolution de la Normandie féodale d'après le témoinage des chroniqueurs (X$^{e}$–XII$^{e}$ siècles)," *Le moyen âge*, 5th ser., 6 (1992), 35–60.

Aston, Michael, ed. *Medieval Fish, Fisheries and Fishponds in England*. 2 vols. paginated continuously. BAR British Series 182 (i–ii). Oxford: BAR, 1988.

Aston, Michael, and C. J. Bond. "Warwickshire fishponds." In Aston, ed., *Medieval Fish*, 417–434. BAR British Series 182. Oxford: BAR, 1988.

"Worcestershire fishponds." In Aston, ed., *Medieval Fish*, 435–455. 1988.

Aubenas, Roger. *Le droit de pêche de l'abbaye de Lérins*, fascic. no. 2. Cannes: Equipe des historiens cannois, 1953.

Auber A, M., M. C. Travers-Trolet, B. Villanueva, and B. Ernande "Regime shift in an exploited fish community related to natural climate oscillations." *PLOS ONE* 10:7 (2015), 1–18: e0129883. doi:10.1371/journal. pone.0129883.

Audoin, Frédèrique. "Les ossements animaux dans les fouilles médiévales de la Charité-sur-Loire: Conclusions relatives à l'alimentation du XI$^{e}$ au XVI$^{e}$ siècle," *Académie des inscriptions de belles-lettres. Comptes rendus des sèances (Jan–March, 1984)*, 199–217.

Audoin-Rouzeau, Frèdèrique. *Ossements animaux du Moyen-Age au monastère de la Charité-sur-Loire*. Paris: Publications de Sorbonne, 1986.

Ausonius. *Mosella* in R. P. H. Green, ed., *The Works of Ausonius*, 116–121. Oxford: Clarendon, 1991.

Austen, D. J., P. B. Bailey, and B. W. Menzel. "Importance of the guild concept to fisheries research and management," *Fisheries* 19 (1994), 12–20.

Austin, Thomas, ed. *Two Fifteenth-Century Cookery-Books*. Early English Text Society, original series, 91. London: EETS, 1888. Reprinted London:

Published for the Early English Text Society by the Oxford University Press, 1964.

Auzary-Schmaltz, Bernadette. "Les Contentieux en matière d'approvisionnement, d'après les registres du Parlement de Paris (XIV^e^–XV^e^ siècles)," *Franco-British Studies* 20 (1995), 49–67.

Avitus of Vienne. *Avitus of Vienne. Letters and Selected Prose*, tr. Danuta Shanzer and Ian Wood. Translated Texts for Historians, vol. 38. Liverpool: Liverpool University Press, 2002.

Ayers, Brian. *The German Ocean: Medieval Europe around the North Sea. Studies in the Archaeology of Medieval Europe.* Sheffield, England, and Bristol, CT: Equinox Publishing, 2016.

Ayza Roca, Alfred. "La pesca en la València de segle XIV," *L'Espill [Valencia]* 17–18 (1983), 159–180.

Bagniewski, Zbigniew, and Piotr Kubów, "Średniowieczny młyn wody z Ptakowic na Dolnynm Śląsku," *Kwartalnik Historii Kultury Materialnej* 25 (1977), 3–30.

Bailey, Geoff, James Barrett, Oliver Craig, and Nicky Miller. "Historical ecology of the North Sea Basin: an archaeological perspective and some problems of methodology." In Torben C. Rick and Jon M. Erlandson, eds., *Human Impacts on Ancient Marine Ecosystems: A Global Perspective*, 215–242. Berkeley: University of California Press, 2008.

Bailey, Mark. "Coastal fishing off south east Suffolk in the century after the Black Death," *Proceedings of the Suffolk Institute of Archaeology and History* 37:2 (1990), 102–114.

"*Per impetum maris*: natural disaster and economic decline in eastern England, 1275–1350." In B. M. S. Campbell, ed., *Before the Black Death: Studies in the "Crisis" of the Early Fourteenth Century*, 184–208. Manchester: Manchester University Press, 1991.

Bailey, R. S. and J. H. Steele. "North sea herring fluctuations." In Michael H. Glantz, ed., *Climate Variability, Climate Change, and Fisheries*, 213–230. Cambridge: Cambridge University Press, 1992.

Baker, Polydora. "Fauna." In G. P. Brogiolo and L. Castelletti, eds., *Archeologia a Monte Barro: 1. Il grande edificio e le torri*, 153–167. Lecco: Stefanoni, 1991.

"Le rôle de la chasse comme moyen de subsistence durant le haut moyen âge dans le nord de l'Italie: une comparison des sources zooarchéologiques et historiques." In Jean Desse and Frédérique Audoin-Rouzeau, eds., *Exploitation des animaux sauvages à travers le temps*, 505–516. XIII^e^ Recontres Internationales d'Archéologie de d'Histoire d'Antibes. IV Colloque International de l'Homme et l'Animal. Juan-les-Pins, 1993.

"The vertebrate remains from the longobard and 9th – 10th C. occupation at S. Giulia, Brescia." In G. P. Brogiolo, ed., *S. Giulia di Brescia, gli scavi dal 1980 al 1992. Peperti preromani, romani e altomedievale, 425–449.* Firenze: All'Insegna del Giglio, 1999.

Baker, Polydora, and Gillian Clark. "Archaeozoological evidence for medieval Italy: A critical review of the present state of research," *Archeologia medievale [Firenze]* 20 (1993), 45–77.

Balestracci, Duccio. *The Renaissance in the Fields. Family Memoirs of a Fifteenth-Century Tuscan Peasant*, tr. Paolo Squatriti and Betsy Merideth. University Park: Pennsylvania State University Press, 1999. [Original Italian edition 1984.]

Balletto, Laura. *Genova nel duecento, uomini nel porto e uomini sul mare*. Collana storica di fonti e studi 36. Genoa: D. Daste M.G., 1983.

Balon, Eugene K. "About the oldest domesticates among fishes," *Journal of Fish Biology* 65 (2004), Supplement A, 1–27.

"The common carp, *Cyprinus carpio*: Its wild origin, domestication in aquaculture, and selection as coloured nishikigoi," *Guelph Ichthyology Reviews* 3 (1995), 1–55.

*Domestication of the carp Cyprinus carpio L.* Royal Ontario Museum Life Sciences Miscellaneous Publication. Toronto: ROM, 1974.

"Origin and domestication of the wild carp, *Cyprinus carpio*: from Roman gourmets to the swimming flowers," *Aquaculture* 129 (1995), 3–48.

Balon, Eugene K., and Juraj Holčík. "Gabčíkovo river barrage system: the ecological disaster and economic calamity for the inland delta of the middle Danube." *Environmental Biology of Fishes* 54 (1999), 1–17.

Balon, Joseph. "La pêche et le commerce du poisson dans le comté de Namur au Moyen âge," *Namurcum* 19 (1942), 25–39.

Bănărescu, P. "Zoogeography and history of the freshwater fish fauna of Europe." In Jurai Holčik, ed., *The Freshwater Fishes of Europe*, vol. 1, part 2, 88–107. Wiesbaden: Aula Verlag, 1989.

Barack, K. A., ed. *Des Teufels Netz. Satirisch-didaktisches Gedicht aus der ersten Hälfte des fünfzehnten Jahrhunderts*. Stuttgart: Litterarische Verein in Stuttgart, 1863.

Barakat, Robert A. *The Cistercian Sign Language*. Cistercian Studies Series, 11. Kalamazoo: Cistercian Publications, 1975.

Barange, Manuel, and 24 co-authors. "Current trends in the assessment and management of stocks." In Checkley et al., eds., *Climate Change and Small Pelagic Fish*, 191–255. 2009.

Barceló Crespí, Maria, and Antoni Mas Forners. "Fishing in the Island of Majorca in the Middle Ages (1230–1521): state of the art and initial results of an ongoing research," *Imago Temporis. Medium Aevum* 15 (2021): 119–147.

Barinaga, Marcia. "New study provides some good news for fisheries," *Science*, n. s., 269: 5227 (25 Aug. 1995), 1043.

Barkham, Michael M. "The offshore and distant-water fisheries of the Spanish Basques, c.1500–1650." In Starkey et al., eds., *A History of the North Atlantic Fisheries*, vol. 1: *From Early Times to the Mid-Nineteenth Century*, 229–249. 2009.

Baron, Henriette. "An approach to Byzantine environmental history: human–animal interactions." In Henriette Baron and Falko Daim, eds., *A Most Pleasant Scene and an Inexhaustible Resource. Steps towards a Byzantine Environmental History*, 171–198. Byzanz zwischen Orient und Okzident, 6. Mainz: Verlag des Römisch-Germanischen Zentralmuseums, 2017.

Barré, Éric. "Le droit maritime médiéval en Normandie," *Annales de Normandie* 47:5 (1997), 511–536.

Barrett, James H., ed. *Being an Islander: Production and Identity at Quoygrew, Orkney, AD 900–1600*, Cambridge: McDonald Institute for Archaeological Research. 2012.

"Farming and fishing on medieval Westray," *Current Archaeology* 199 (September–October 2005), 336–341.

"Fish trade in Norse Orkney and Caithness: a zooarchaeological approach," *Antiquity* 71 (1997), 616–638.

"Medieval sea fishing, AD500–1550: chronology, causes and consequences." In Barrett and David, eds. *Cod and Herring*, 250–272. 2016.

"Studying medieval sea fishing and fish trade: how and why." In Barrett and Orton, eds., *Cod and Herring*, 1–12. 2016.

Barrett, James H., Alison M. Locker, and C. M. Roberts. "'Dark Age economics' revisited: the English fish bone evidence AD 600–1600," *Antiquity* 78 (2004), 618–636.

"The origins of intensive marine fishing in medieval Europe: the English evidence," *Proceedings of the Royal Society of London: Biological Sciences* 271 (2004), 2417–2421.

Barrett, James H., and David C. Orton, eds. *Cod and Herring: The Archaeology and History of Medieval Sea Fishing*. Oxford and Philadelphia: Oxbow Books, 2016.

Barrett, James H., C. Johnstone, J. Harland J., W. Van Neer, A. Ervynck, D. Makowiecki, D. Heinrich, A. K. Hufthammer., I. B. Enghoff, C. Amundsen, J. S. Christiansen, A. K. G. Jones, A. Locker, S. Hamilton-Dyer, L. Jonsson, L. Lõugas, C. Roberts, and M. Richards. "Detecting the medieval cod trade: a new method and first results," *Journal of Archaeological Science* 35(2008), 850–861.

"The medieval origins of commercial Sea Fishing Project: a preliminary synthesis." In Daniel Makowiecki et al., eds., *Fishes – Culture – Environment through Archaeoichthyology, Ethnography & History*, 29–30. 15th Meeting of the International Council for Archaeozoology (ICAZ) Fish Remains Working Group (FRWG), 3–9 September 2009, Poznań & Toruń, Poland. Poznań: Bogucki Wydawnictwo Naukowe, 2009.

Barrett, James H., D. Orton, C. Johnstone, J. Hartland, W. Van Neer, A. Ervynck, C. Roberts, A. Locker, C. Amundsen, I. B. Enghoff, S. Hamilton-Dyer, D. Heinrich, A. K. Huftmanner, A. K. G. Jones, L. Jonsson, D. Makowiecki, P. Pope, T. C. O'Connell, T. De Roo,and M. Richards. "Interpreting the expansion of sea fishing in medieval Europe using stable isotope analysis of archaeological cod bones," *Journal of Archaeological Science* 38 (2011), 1516–1524.

Barrett, James H., Rebecca A. Nicholson, and Ruby Cerón-Carrasco. "Archaeo-ichthyological evidence for long-term socioeconomic trends in northern Scotland 3500 BC to AD 1500," *Journal of Archaeological Science* 26:4 (1999), 353–388.

Barrett, James H., Roelf Beukens, Ian Simpson, Patrick Ashmore, Sandra Poaps, and Jaccqui Huntley. "What was the Viking Age and when did it happen? A view from Orkney," *Norwegian Archaeological Review* 33 (2000), 1–39.

Bartal, Ruth. "Le programme iconographique du portail occidental de Sainte-Marie d'Oloron et son contexte historique," *Cahiers de Saint-Michel de Cuxa* 18 (1987), 95–123.

Bartholomaeus Anglicus. *De proprietatibus rerum. As Bartholomaei Anglici de genuinis rerum coelestium, terrestrium et inferarum proprietatibus* ..., ed. Georg Barthold Pontanus. Francofurti, 1601. Facsimile repr. Frankfurt: Minerva, 1964.

Bartolus de Saxoferrato, *Tractatus de fluminibus seu Tyberiadis*, ed. Guido Astuti. Turin: Bottega d'Erasmo, 1964.

Bartosiewicz, László. *Animals in the Urban Landscape in the Wake of the Middle Ages: A Case Study from Vác, Hungary*. BAR International Series, 609. Oxford: Tempus Reparatum, 1995.

"Early medieval faunal remains from Pontes (Iron Gates Gorge, Eastern Serbia)," *Acta Archaeologica Academiae Scientiarum Hungaricae* 48 (1996), 281–315.

"Like a headless chicken: meaning, medium and context in medieval urban taphonomy." In A. M. Choyke and G. Jaritz, eds., *Animaltown: Beasts in Medieval Urban Space*, 19–26. BAR International Series 2858. Oxford: BAR, 2017.

Bartosiewicz, László, and Erika Gál. "The archbishop's dinner? Late medieval fish from Esztergom-Várhegy-Kőbánya, Hungary," *Archaeological and Anthropological Sciences* 13:8 (August 2021). https://doi.org/10.1007/s12520-021-01361-w.

Bartosiewicz, László, Clive Bonsall, and Vasile Şişu. "Sturgeon fishing in the middle and lower Danube region." In Clive Bonsall, Vasile Boroneanţ, and Ivana Radovanović, eds., *The Iron Gates in Prehistory: New Perspectives*, 39–54. BAR International Series 1893. Oxford: Archaeopress, 2008.

Bartrip, Michael. "Food for the body and food for the mind: the regulation of freshwater fisheries in the 1870s," *Victorian Studies* 28 (1985), 285–304.

Battafarano, Maria, and Jacopo De Grossi Mazzorin. "Analisi dei resti ittici da alcuni contesti archeologici della Puglia di età tardo-antica e medieval." In A. Tagliacozzo, I. Fiore, S. Marconi, and U. Tecchiati, eds., *Atti del V Convegno Nazionale di Archeozoologia (Rovereto, 10–12 novembre 2006)*, 289–292. Rovereto: Università del Salento, Dipartimento di Beni Culturali, 2010.

Bauch, Martin, and Thomas Labbe. "Karpfen mit Spätburgunder. Fragwurdiges zu Weinanbau und Teichwirtschaft im karolinischen *imaginaire*." *Mitropa 2019. Jahresheft des Leibniz-Instituts für Geschichte und Kultur des östlichen Europa (GWZO)*, 8–10.

Bauchet, Olivier. "Droits et structures de pêche associés aux moulins fluviaux du cours de la Marne." In Benoît et al., eds., *Pêche et pisciculture en eau douce*, 11 pages [unpaginated booklet and CD-ROM]. [Lille]: Conseil Général du Nord, 2004.

Bauchet-Cubadda, Valérie. "Le patrimoine piscicole des seigneurs laics de la Brie champenoise d'après leurs aveux et dénombrements (XIVe–XVIe siècles)." In Benoît, et al., eds., *Pêche et pisciculture en eau douce.* 2004.

Bautier, Robert-Henri. *The Economic Development of Medieval Europe.* London, Thames & Hudson, 1971.

Bautier, Robert-Henri, and François Maillard. "Les aumônes du roi aux maladeries, maisons-Dieu et pauvres établissements du royaume. Contribution à l'étude du réseau hospitalier... de Philippe Auguste à Charles VII." *Assistance et assistés jusqu'à 1610,* 37–105. Actes du 97e congrés national des Sociétés savantes. Nantes 1972. Section de philologie et histoire. Paris: Bibliothéque Nationale, 1979.

Bavinck, Maarten. "Understanding fisheries conflicts in the South – legal pluralist perspective," *Society and Natural Resources* 18 (2005), 805–820.

Bavington, Dean. "From hunting fish to managing populations: fisheries science and the destruction of Newfoundland cod fisheries," *Science as Culture* 19:4 (2010), 509–528.

*Managed Annihilation: An Unnatural History of the Newfoundland Cod Collapse.* Vancouver: University of British Columbia Press, 2010.

Beck, Corinne. *Les eaux et forêts en Bourgogne ducale (vers 1350–vers 1480). Société et biodiversité.* Paris: L'Harmattan 2008.

"Pêche et étangs ducaux en Bourgogne aux XIVe et XVe siècles: l'exemple du bailliage de la Montagne." In Benoît, et al., eds., *Pêche et pisciculture en eau douce.* 2004.

Beck, P. "L'Approvisionnement en bourgogne ducale aux XIVème et XVème siecles." In Menjot, ed. *Manger et boire,* vol. 1, 171–181. Paris: Belles lettres, 1984.

Becker, C., and G. Grupe. "Archaeometry meets archaeozoology: Viking Haithabu and medieval Schleswig reconsidered," *Anthropological and Archaeological Sciences* 4:4 (2012), 241–262. DOI: 10.1007/s12520-012-0098-z.

"Teamplayer oder Gegenspieler? Der zweifache Nutzen von Archäozoologie und Isotopenanalyse – Haithabu und Schleswig neu bewertet," *Beiträge zur Archäozoologie und Prähistorischen Anthropologie* 8 (2011): 9–27.

Bede. *Historia Ecclesiastica Gentis Anglorum, Bede's Ecclesiastical History of the English People,* ed. and tr. Bertram Colgrave and R. A. B. Mynors. Oxford Medieval Texts. Oxford: Clarendon, 1969.

Bedini, Elena. "Osservazioni preliminari sui resti faunistici." In E. Bedini and E. Micheletto, "Indagine archeologica al Castello di Manzano (Comune di Cherasco, Provincia di Cuneo). Secondo rapporto preliminare (1990–1991)," *Archeologia Medievale* 19 (1992), 223–241.

Beech, George T. *A Rural Society in Medieval France: The Gâtine of Poitou in the Eleventh and Twelfth Centuries.* Baltimore: Johns Hopkins University Press, 1964.

Bekker-Nielsen, Tonnes, and Ruthy Gertwagen, eds. *The Inland Seas: Towards an Ecohistory of the Mediterranean and the Black Sea.* Geographica Historica, 35. Stuttgart: Franz Steiner Verlag, 2016.

Bell, Martin. "Archaeology under alluvium: human agency and environmental process. Some concluding thoughts." In Stuart Needham and Mark G. Macklin, eds., *Alluvial Archaeology in Britain*, 271–276. Oxbow Monograph 18. Oxford: Oxbow Books, 1992.

Belliard, Jérôme, Sarah Beslagic, Julien Boucault, and Amandine Zahm, "Increasing establishment of non-native fish species in the Seine River basin: insights from medium and long-term approaches," *Frontiers in Ecology and Evolution* 9 (September 2021), Article 687451 (pp. 1–13).

Benarrous, Renaud. *La Grande Brenne. Histoire d'une zone humide continentale et de ses étangs: Du paysage imaginaire à l'interaction de l'homme avec son environnement, aux périodes préindustrielles*. 66e Supplément à la *Revue Archéologique du Centre de la France*. Tours: FERACF and Parc naturel régional de la Brenne, 2017.

Benecke, Berthold. "Beiträge zur Geschichte der Fischerei in Ost- und Westpreussen," *Altpreussisch Monatsschrift* 17 (1880), 300–322 and 385–416.

Benecke, Gerhard. *Maximilian I (1459–1519): An Analytical Biography*. London: Routledge & Kegan Paul, 1982.

Benecke, Norbert. "Die Fischreste aus den slawischen Inselsiedelungen in der Lieps und im Tollensesee." In Voker Schmidt, ed., *Lieps. Eine slawische Siedlungskammer am Südende des Tollensesees*, 118–121. Beiträge zur Ur- und Frühgeschichte der Bezirke Rostock, Schwerin und Neubrandenburg, Bd. 16. Berlin: VEB Deutscher Verlag der Wissenschaften, 1984.

"Die Fischreste aus der Mecklenburg, Kr. Wismar." In P. Donat, ed., *Die Mecklenburg. Eine Hauptburg der Obodriten*, 183–187. Berlin, 1984.

"Some remarks on sturgeon fishing in the Southern Baltic Region in medieval times." In D. C. Brinkhuizen and A. T. Clason, eds., *Fish and Archaeology*, 9–17. BAR International Series, 294. Oxford: BAR, 1986.

"Die sozialökonomische Interpretation von zoologischen Fundkomplexen – Ausgewählte methodologische Probleme." In *Historisch-archäologische Quellen und Geschichte bis zur Herausbilding des Feudalismus*, 78–79. Berlin: Zentralinstitut für Wissenschaften der DDR, 1983.

"Untersuchungen zum Einfluß der Bergungsmethode auf die Qualität von Tierknochen," *Ausgrabungen und Fünde* 30 (1985), 260–265.

"Zur frühmittelalterlichen Heringsfischerei im südlichen Ostseeraum – ein archäologischer Beitrag," *Zeitschrift für Archäologie* 16 (1982), 283–290.

Beneš of Veitmíle. "Chronica ecclesiae Pragensis," ed. J. Emler. In *Fontes rerum Bohemicarum*, IV: 457–548. Prague: Nákl. Nadání F. Palackého, 1884.

Benito, Gerardo, Mark G. Macklin, Christoff Zielhofer, Anna F. Jones, and Maria J. Machado. "Holocene flooding and climate change in the Mediterranean," *Catena* 130 (2015), 13–33.

Bennett, H. S. *Life on the English Manor: A Study of Peasant Conditions 1150–1400*. Cambridge: Cambridge University Press, 1937.

Benoît, Catherine. *Les étangs de la Dombes au Moyen Âge, XIII^e–XV^e siècles: Approche ethnohistorique d'un system agro-piscicole*. Mémoires et documents d'histoire médievale et de philologie, Publiés par la Section d'Histoire Médiévale et de Philologie du Comité des Travaux Historiques et Scientifiques, 5. Paris: Éditions du CTHS, 1992.

Benoît, Paul. "La carpe dans l'Occident médiévale." In Danièle James-Raoul and Claude Thomasset, eds., *Dans l'eau, sous l'eau: le monde aquatique au moyen âge*, 227–236. Presses de l'Université de Paris-Sorbonne, 2002.

"La pêche dans le domaine de la ville de Paris au XV^e siècle." In Benoît et al., eds., *Pêche et pisciculture en eau douce*. 2004.

Benoît, Paul, Frédéric Loridant, and Olivier Mattéoni, eds. *Pêche et pisciculture en eau douce: la rivière et l'étang au Moyen Âge: actes des 1^ères Rencontres internationales de Liessies: 27, 28, 29 avril 1998*. [unpaginated booklet and CD-Rom]. [Lille]: Conseil Général du Nord, 2004.

Benoît, Paul, and Joséphine Rouillard. "Medieval hydraulics in France." In Squatriti, ed. *Working with Water in Medieval Europe*, 161–216. 2000.

Benoît, Paul, and Monique Wabont. "Mittelalterliche Wasserversorgung in Frankreich. Eine Fallstudie: Die Zisterzienser." In *Die Wasserversorgung im Mittelalter* (Geschichte der Wasserversorgung, 4), ed. Klaus Grewe for the Frontinusgesellschaft, 185–226. Mainz: Phillipp von Zabern, 1991.

Benoît, Paul, and Olivier Mattéoni. "Conclusion: Pêche et pisciculture en eau douce: la rivière et l'étang." In Benoît et al., eds. *Pêche et pisciculture en eau douce*, 2004.

Bépoix, Sylvie, and Hervé Richard, eds. *La Forêt au Moyen Âge*. Paris: Belles Lettres, 2019.

Bérard, Laurence. "La consommation du poisson en France: des prescriptions alimentaires à la prépondérance de la carpe." In Liliane Bodson, ed., *L'Animal dans l'alimentation humaine: les critèeres de choix. Actes du colloque international de Liège, 26–29 novembre 1986*, 171–180. *Anthropozoologica*, num. spec. 2. Paris, 1988.

Berger, Jean-François, and Jacques-Léopold Brochier. "Rapports de la géoarchéologie à la connaissance des paysages et des climats de l'époque médiévale en moyenne vallée du Rhône." In Odile Maufras, ed., *Habitats, nécropoles et paysages dans la moyenne et basse vallée du Rhône (VII^e–XV^e siècle): contributions des travaux archéologiques sur le tracé du TGV-Méditerranée à l'étude des sociétés rurales médiévales*, 163–208. Paris: Éditions de la Maison des Sciences de l'Homme, 2006.

Berman, Constance H. "Reeling in the eels at la Trinquetaille near Arles." In Scott G. Bruce, ed., *Ecologies and Economies in Medieval and Early Modern Europe. Studies in Environmental History for Richard C. Hoffmann*, 149–164. Leiden and Boston: Brill, 2010.

Bernard of Clairvaux. *Life and Works of Saint Bernard, Abbot of Clairvaux*, tr. Samuel J. Eales, 2 vols., 2d. ed. London and New York: Burns & Oates, 1912.

Bernard of Cluny. "Ordo Cluniacensis sive Consuetudines." Ed. M. Herrgott in *Vetus Disciplina Monastica*, 136–364. Paris: Caroli Osmont, 1726.

Bernard Silvester [aka Bernardus Silvestris], *Bernardus Silvestris: Cosmographia*, ed. Peter Dronke. Leiden: Brill, 1978.

*The Cosmographia of Bernardus Silvestris*, tr. Winthrop Wetherbee. New York: Columbia University Press, 1973.

Bernard, Auguste, ed. *Cartulaire de l'Abbaye de Savigny suivi du petit cartulaire de l'Abbaye d'Ainay*. Collection de documents inédits sur l'histoire de France. 1. sér. Histoire politique. Paris: Imprimerie Impériale, 1853.

Bernard, Auguste, and Alexandre Bruel, eds., *Recueil des chartes de l'abbaye de Cluny*. 6 vols. Paris: Imprimerie nationale, 1876–1903.

Bertelsen, Reidar. "A North-East Atlantic perspective," *Acta Archaeologica* 61 (1990), 22–26.

Bertheau, Friedrich. "Wirtschaftsgeschichte des Klosters Preetz im 14. und 15. Jahrhundert," *Zeitschrift der Gesellschaft für schleswig-holsteinische Geschichte* 47 (1917), 91–266.

Berthier, Karine. "La gestion des étangs de l'abbaye de Cîteaux aux XIV$^{e}$ et XV$^{e}$ siècles." In Benoît et al, eds. *Pêche et pisciculture en eau douce*. 2004. [20 pages]

Berthier, Karine. "La gestion des étangs de l'abbaye de Cîteaux aux XIV$^{e}$ et XV$^{e}$ siècles," *Cîteaux Commentarii Cistercienses, Revue d'histoire cistercienne* 57 (2006), 281–293.

Bertrand, Georges, with collaboration of Claude Bertrand. "Pour une histoire écologique de la France rurale." In G. Duby and A. Wallon, eds., *Histoire de la France rurale, 1: La formation des compagnes françaises des origines au XIV$^{e}$ siècle*, 34–118. Paris: Seuil, 1975.

Bestehorn, Friedrich. "Die geschichtliche Entwicklung des märkischen Fischereiwesens, ein Beitrag zur Kultur- und Wirtschaftsgeschichte der Mark Brandenburg," *Archiv für Fischereigeschichte* 1 (1913), 1–199.

Beyer, Eduard. *Das Cistercienser-Stift und Kloster Alt-Zelle in dem Bisthum Meissen*. Dresden, 1855.

Bidon, Alexandre, and Corinne Bossard-Beck. "La préparation des repas et leur consommation en Forez au XV$^{e}$ siècle d'aprés les sources archéologiques." In Menjot, ed., *Manger et boire*, II, 59–71. Colloque du Nice, 1982. Paris: Belles lettres, 1984.

Biganti, Tiziana. "La pesca nel lago Trasimeno: Sfruttamento et tutela delle risorse ittiche del lago di Perugia (seccoli XIII–XV)." In Paolo Carucci and Massimo Buttazzo, eds., *Gli archivi per la storia dell'alimentazione*, 772–799. Atti del convegno, Potenza-Matera, 5–8 settembre 1988. 3 vols. paginated continuously. Publicazioni degli Archivi di Stato, Saggi, 34. Rome: Ministero per i Beni Culturali e Ambientali, Ufficio Centrale per i Beni Archivistici, 1995.

Biggar, H. P., ed. *The Precursors of Jacques Cartier 1497–1534. Publications of the Canadian Archives*, no. 5. Ottawa: Government Prnting Bureau, 1911.

Billard, Cyrille, and Vincent Bernard, eds. *Pêcheries de Normandie: archéologie et histoire des pêcheries littorales du département de la Manche*. Rennes: Presses universitaires de Rennes, 2016.

Billington, Sandra. "Butchers and fishmongers: their historical contributions to London's festivity." *Folklore* 101:1 (1990), 182–191 and 227–229.

Birrell, Jean. "Common rights in the medieval forest: disputes and conflicts in the thirteenth century," *Past & Present* 117 (Nov. 1987), 22–49.

Blanc, M., P. Banarescu, J.-L. Gaudet, and J.-C. Hureau. *European Inland Water Fish: A Multilingual Catalogue*. London: Fishing News Books, Ltd., 1971.

Blary, François. *Le domaine de Châalis, XII$^{e}$–XIV$^{e}$ siècles: Approches archéologiques des établissements agricoles et industriels d'une abbaye cistercienne*, Comité des Travaux Historiques et Scientifiques, Mémoires de la Section d'Archéologie et d'Histoire de l'Art, 3. Paris: CTHS, 1989.

Blaschitz, Gertrud. "Der Biber im Topf und der Pfau am Spieß." In C. Tuczay, U. Hirhager, and K. Lickblau, eds., *Ir sult sprechen willekommen: Grenzenlosen Mediävistik, Festschrift für Helmut Birkhan zum 60. Geburtstag*, 416–436. Bern: Peter Lang, 1998.

Blurton, Heather. "The language of the liturgy in the life and miracles of William of Norwich," *Speculum* 90:4 (2015), 1053–1075.

Bochove, Christiaan van. "De Hollandse Haringvisserij tijdens de vroegmoderne tijd," *Tijdschrift voor sociale en economische geschiedenis* 1 (2004), 3–27.

Boddeke, R. *Vissen en vissen: Ervaringen van een visserijbioloog en sportvisser.* Amsterdam: Elsevier, 1971.

Boer, Dick E. H. de. "'Roerend van der visscheryen'. Enkele aspecten van de visvangst in Holland en Zeeland tot de Sint Elisabethsvloed van 1421." In J. J. J. M. Beenakker et al., *Holland en het water in de middeleuwen. Strijd tegen het water en beheersing en bruik van het het water*, 115–140. Hilversum: Verloren, 1997.

Boessneck, J. and Angela von den Driesch-Karpf. "Tierknochenfunde von der Burg Nürnberg," *Jahrbuch für fränkische Landesforschung* 28 (1968), 73–80.

Boháč, Zdeněk. "Historical–ecological aspects of the Bohemian feudal state economy," *Historická Ekologie / Historical Ecology* [Prague] 1 (1988), 11–59.

Bökönyi, Sándor. "Die Wirbeltierfauna der Ausgrabungen in Zalavár." In Ágnes Sós and Sándor Bökönyi, eds., *Zalavár*, 313–386. Budapest: Akadémiai Kiadó, 1963.

Bolster, W. Jeffrey. *The Mortal Sea: Fishing the Atlantic in the Age of Sail.* Cambridge, MA, and London: Belknap Press of Harvard University Press, 2012.

Bonachía Hernando, Juan Antonio, and Isabel del Val Valdivieso. "Monasterios y pesca fluvial en la Castilla bajomedieval: conflictos y luchas por el poder." In Maria I. del Val Valdivieso, ed., *Monasterios y recursos hídricos en la edad media*, 11–58. Colección LAYA, nr. 40. Madrid: A.C. Almudayna, 2013.

Bond, C. J. "Monastic fisheries." In Aston, ed., *Medieval Fish*, 69–112. 1988.

Bonnamour, Louis. "La pêche en Saône à l'époque médiévale, données historiques et archéologiques." In Benoît, et al., eds., *Pêche et pisciculture en eau douce.* 2004.

Bonvesin de la Riva [della Riva]. *De magnalibus Mediolani*, ed. and tr. Paolo Chiesa. Milan: Scheiwiller, 1998.

Boos, Heinrich., ed. *Urkundenbuch der Stadt Worms.* 3 vols. Quellen zur Geschichte der Stadt Worms, vols. 1–2. Berlin: Weidmann,1886–1893.

Boretius, A., and Victor Krause, eds. *Capitularia Regum Francorum.* MGH, *Legum*, Sectio II. 2 vols. in 4. Hannover: Hahn, 1883–1897.

Borger, G. J. "Draining–digging–dredging; the creation of a new landscape in the peat areas of the Low Countries.." In J. T. A. Verhoeven, ed., *Fens and Bogs in the Netherlands: Vegetation, History, Nutrient Dynamics and Conservation*, 131–171. Dordrecht: Kluwer, 1992.

Bork, Hans-Rudolf. *Bodenerosion und Umwelt. Verlauf, Ursachen und Folgen der mittelalterlichen und neuzeitlichen Bodenerosion, Bodenerosionprozesse, Modelle und Simulationen.* Braunschweig: Technische Universität, 1988.

Bork, Hans-Rudolf, and Gabriele Schmidtchen. "Böden: Entwicklung, Zerstörung und Schutzbedarf in Deutschland," *Geographische Rundschau* 53:5 (May 2001), 5–9.

Bork, Hans-Rudolf, Helga Bork, Claus Dalchow, Berno Faust, Hans-Peter Piorr, and Thomas Schatz. *Landschaftsentwicklung in Mitteleuropa: Wirkung des Menschen auf Landschaften.* Gotha and Stuttgart: Klett-Perthes, 1998.

Borlandi, Antonia, ed. *Il manuale di mercatura di Saminato de' Ricci.* Genova: Di Stefano, 1963.

Boscus, L. "Le fief des Malhols et son église. Droits de pêche de l'abbaye de la Garde-Dieu sur la rivière d'Aveyeron," *Bulletin archéologique et historique de la société archéologique de Tarn-et-Garonne* 24 (1896), 255–259.

Bossart, Julia, and Matthias Flück. "*... dass auch die fisch feüchter und kalter natur sind.* Archäologische und historische Spurensuche durch ein Jahr im Mittelalter." In Heide Hüster Plogmann, ed., *Fisch und Fischer aus zwei Jahrtausenden. Eine fischereiwirtschaftliche Zeitreise durch die Nordwestschweiz,* 131–140. Forschungen in Augst, Bd. 39. Augst: Römerstadt Augusta Raurica, 2006.

Bosscha Erdbrink, G. R., ed. *Het 'Keuckenboeck' van de Tollenaar van Lobith 1428/29.* Fontes minores medii aevi uitgegeven vanwege het Instituut voor Geschiedenis, Rijksuniversiteit te Utrecht, vol. 18. Groningen: Wolters-Noordhoff, 1979.

Bossuat, André. "La pêche en Seine au XV^e siècle," *Bulletin de la Société de l'histoire de Paris* 90 (1963 [1965]), 60–81.

Boström, Ingemar, ed. *Due Libri di cucina. Anonimo Meridionale.* Acta universitatis Stockholmensis, Romanicia Stockholmiensia, 11. Stockholm: Almqvist & Wiksell International, 1985.

Bouchard, Constance B., ed. *The Cartulary of Flavigny, 717–1113.* Medieval Academy Books, 99. Cambridge, MA: Medieval Academy of America, 1992.

ed. *The Cartulary of St.-Marcel-lès-Chalon, 779–1126.* Medieval Academy Books, 102. Cambridge, MA: Medieval Academy of America, 1999.

Boudin, Mathieu, Pascal Boeckx, Peter Vandenabeele, and Mark Van Strydonck. "An archaeological mystery revealed by radiocarbon dating of cross-flow nanofiltrated amino acids derived from bone collagen, silk, and hair: case study of the Bishops Baldwin I and Radbot II from Noyon-Tournai," *Radiocarbon* 56:2 (2014), 603–617.

Bougard, Pierre, and Bernard Delmaire, eds. *Le cartulaire et les chartes de l'abbaye de femmes d'Avesnes-lès-Bapaume (1128–1337). Atelier de Recherche sur les Textes Médiévaux,* 19. Turnhout: Brepols, 2014.

Bourin-Derruau, Monique, Daniel Le Blévec, Claude Raynaud, and Laurent Schneider. "Le littoral languedocien au Moyen Âge." In Jean-Marie Martin, ed., *Castrum 7: Zones côtières littorales dans le monde méditerranéen au Moyen Âge: defense, peuplement, mise en valeur,* 345–423. Rome–Madrid: École française de Rome, Casa de Velázquez, 2001.

Bourlet, Caroline. "L'Approvisionnement de Paris en poisson de mer aux XIV^e et XV^e siècles, d'après les sources normatives," *Franco-British Studies* 20 (1995), 5–22.

Bourquelot, Felix. "Fragmentes de comptes du XIIIe siècle," *Bibliothèque de l'École des Chartes* 24 (1863), 51–79.

Boussard, Jacques. Nouvelle histoire de Paris, Vol. 1: *De la fin du siège de 885–886 à la mort de Philippe Auguste*. Paris: Hachette, 1976.

Bove, Boris. *Dominer la ville: Prévots des marchands et échevins parisiens de 1260 à 1350*. Histoire, 13. Paris: Comité des Travaux Historiques et Scientifiques, 2004.

Bowlus, Charles R. *Franks, Moravians, and Magyars: The Struggle for the Middle Danube, 788–907*. Philadelphia: University of Pennsylvania Press, 1995.

Bradley, Raymond S., Heinz Wanner, and Henry F. Diaz. "The Medieval Quiet Period," *The Holocene* 26:6 (2016), 990–993. DOI:10:1177/0959683615622552

Braekman, Willy L., ed. Een nieuw zuidnederlands Kookboek uit de viftiende eeu. Scripta. *Mediaeval and Renaissance Texts and Studies*, 17. Brussels: Omirel, 1986.

Brander, Keith. "Impacts of climate change on fisheries," *Journal of Marine Systems* 79 (2010) 389–402.

Brandt, Andres von. *Fish Catching Methods of the World*. 3d. ed. rev. Farnham: Surrey Fishing News Books Ltd., 1984.

Braudel, Fernand. *The Identity of France*, vol. 1: *History and Environment*, tr. Siân Reynolds. New York: Harper and Row, 1989.

Bravard, Jean-Paul. "Des versants aux cours d'eau, les implications des fluctuations paléohydrologiques à l'époque médiévale." In Michel Colardelle, ed., *L'homme et la nature au Moyen Age: Paleoenvironnement des sociétés occidentales*, 171–179. Actes du V^e Congrès Internationale d'Archéologie Médiévale, Grenoble (France), 6–9 October 1992; Société d'Archéologie Médiévale. Paris: Errance, 1996.

Bravard, Jean-Paul, and Michel Magny, eds. *Les fleuves ont une histoire. Paléo-environnement des rivières et des lacs français depuis 15 000 ans*. Paris: Errance, 2002.

Bravard, Jean-Paul, J.-L. Peiry, and P.-G. Salvador. "La diversité spatiale des enregistrements morphosédimentaires tardiglaciaires et holocènes dans quelque vallées du piémont des Alpes du Nord englacées au Würm (Arve, Haut-Rhône, Isère)." In Bravard and Magny, eds., *Les fleuves ont une histoire*, 205–213. 2002.

Brázdil, Rudolf, and Oldrich Kotyza. *History of Weather and Climate in the Czech Lands*, vol. 1: *Period 1000–1500*. Zürcher Geographische Schriften, Bd. 62. Zürich: Geographisches Institut, Eidgenössische Technische Hochschule Zürich, 1995.

Breen, Colin. "Marine fisheries and society in Medieval Ireland." In Barrett and Orton, eds., *Cod and Herring*, 91–98. 2016.

Bresc, Henri. "La pêche dans l'espace économique normand." In G. Musca, ed., *Terra e uomini nel Mezzogiorno normanno-svevo*, Atti delle settime giornate normanno-sveve, Bari 15–17 ottobre 1985, 271–291. Centro di studi normanno-svevi Università degli Studi di Bari, Atti, vol. 7. Bari: Centro di studi normanno-svevi Università degli Studi di Bari, 1987.

"Pêche et coraillage aux derniers siècles du moyen âge: Sicile et Provence orientale." In *L'Exploitation de la mer de l'Antiquité à nos jours* (Vémes

Recontres internationales d'archeologie et d'histoire d'Antibes), 107–116. Juan-les-Pins: Association pour la promotion et la diffusion des connaissances archéologiques, 1985.

"Pêche et habitat en méditerranée occidentale." In Jean-Marie Martin, ed., *Castrum 7: Zones côtières littorales dans le monde méditerranéen au Moyen Âge: defense, peuplement, mise en valeur*, 525–539. Rome–Madrid: École française de Rome, Casa de Velázquez, 2001.

"La pêche et les madragues dans la Sicile médiévale." In Micheline Galley and L. L. Sebai, eds., *L'Homme Méditerranéen et la mer*, 166–180. Tunis: Salammbo, 1985.

*Un monde méditerranéen. Économie et société en Sicilie 1300–1450*. Rome: École Française, 1986.

Brinkhuizen, D. C. "Preliminary notes on fish remains from archeological sites in the Netherlands," *Palaeohistoria* 21 (1979), 83–90.

"Some notes on recent and pre- and protohistoric fishing gear from Northwestern Europe," *Palaeohistoria* 25 (1983), 7–53.

"Visresten uit twee middeleeuwse vindplaatsen te Leeuwarden." In Planten- Vis- en Vogelresten uit vroeghistorisch Leeuwarden, 19–20. Leeuwarden: Commissie Archeologisch Stadskernonderzoek, 1983.

Britnell, Richard H. "Commercialization and economic development in England, 1000–1300." In Richard H. Britnell and Bruce M. S. Campbell, eds., *A Commercializing Economy, England 1086 to c.1300*, 7–26. Manchester: Manchester University Press, 1995.

*The Commercialization of English Society, 1000–1500*. Cambridge: Cambridge University Press, 1993.

"Price-setting in English borough markets, 1349–1500," *Canadian Journal of History* 31 (1996), 1–15.

Broadberry, Stephen, Bruce M. S. Campbell, Alexander Klein, Mark Overton, and Bas van Leuwen. *British Economic Growth 1270–1870*. Cambridge: Cambridge University Press, 2015.

Brouwers, D. D., ed. *Cens et rentes du Comté de Namur au XIIIe siècle*. 2 vols. in 3, L'Administration et les finances du Comté de Namur du XIIIe au XVe siècle, Sources, vol. 1 and vol. 2: parts 1 and 2. Documents inédits relatifs a l'histoire de la Province de Namur publiés par ordre du Conseil Provincial. Namur: A. Wesmael-Charlier, 1910–1911 and 1926.

Brown, Andrew, and Jan Dumolyn, eds. *Medieval Bruges, c.850–1550*. Cambridge: Cambridge University Press, 2018.

Brown, Keith M., Gillian H. MacIntosh, Alastair J. Mann, Pamela E. Ritchie, and Roland J. Tanner, eds. *The Records of the Parliaments of Scotland to 1707*. St Andrews: 2007–2011. Online edition www.rps.ac.uk (last accessed 2016 March).

Bruce, Scott G. *Silence and Sign Language in Medieval Monasticism: The Cluniac Tradition c. 900–1200*. Cambridge: Cambridge University Press, 2007.

Brühl, Carl-Richard, and Cinzio Violante, eds., *Die "Honorantie Civitatis Papiae": Transkription, Edition, Kommentar*. Cologne: Böhlau, 1983.

Bruijn, Jaap R. "Dutch fisheries: an historiographical and thematic overview." In Holm, Starkey, and Thor, eds., *The North Atlantic Fisheries*, 105–120. 1996.

Brykala, Dariusz. "Watermills functioning within the Skrwa river catchment area (Central Poland)." In Mauro Agnoletti et al., eds., *History and Sustainability. Proceedings of the Third International Conference of the European Society for Environmental History, Florence, Italy, February 16–19, 2005*, 176–179. Florence: Instituto di Studi sulle Società del Mediterraneo – CNR, Università di Firenze, and ESEH, 2005.

Brzozowski, Stanisław and Mieczysław Tobiasz. "Z dziejów rybactwa małopolskiego," *Studia i Materiały z Dziejów Nauki Polskiej*, seria B: *Historia nauk biologicznych*, vol. 9, 3–102. Warsaw, 1964.

Bucher, Bruno, ed. *Die alten Zunft- und Verkehrs-Ordnungen der Stadt Krakau nach Balthasar Behem's* Codex Picturatus *in der K. K. Jagellonischen Bibliothek*. Vienna: Carl Gerold's Sohn, 1889.

Buchwald, G. von. "Anna von Buchwald, Priörin des Klosters Preetz 1484–1508," *Zeitschrift der Gesellschaft für schleswig-holsteinisch Geschichte* 9 (1879), 3–93.

Buisman, J., and A. F. V. van Engelen, eds. *Duizen jaar weer, wind en water in de Lage Landen*, vol. IV, 1575–1675. Franeker: Van Wijnen, 2000.

Buitenhuis, H., and D. C. Brinkhuizen. "Faunaresten." In J. B. Hielkema et al., *Aanvullend Archeologisch Onderzoek op terrein 9 te Houten-Loerik, gemeente Houten (U.)*, 35–46. ARC-Publicaties 69. Groningen: ARC, 2003.

Bukowski, Zbigniew. "Uwagi o konserwacji ryb u słowian w świętle materiałów archeologicznych i etnograficznych," *Studia z dziejów gospodarstwa wiejskiego* 9:3 (1967), 49–81.

Bull, Karl-Otto. "Die wirtschaftliche Verflechtung der Pfalz am Ende des Mittelalters (1440–1550)." In *Beiträge zur Pfälzischen Wirtschaftsgeschichte*, 55–96. Veröffentlichungen der Pfälzischen Gesellschaft zur Förderung der Wissenschaften in Speyer, vol. 58. Speyer: die Gesellschaft, 1968.

Buller, Frederick. *Fish & Fishermen in English Medieval Church Wall Paintings*. Ellesmere: The Medlar Press, 2009.

Bullo, C. *Il viaggio di M. Piero Querini e le relazioni della repubblica Veneta colla Svezia*. Venice: Tipografia Antonelli, 1881.

Bullock, Andrea. "Evidence for fish exploitation from Tudor deposits excavated at Little Pickle, Surrey/England, in 1989." In Dirk Heinrich, ed., *Archaeo-Ichthyological Studies. Papers presented at the 6th Meeting of the I.C.A.Z. Fish Remains Working Group, Schleswig*, 267–275. Beiheft zur *Offa*, 51. Neumünster: Wacholz, 1994.

Büntgen, Ulf, et al. "2500 years of European climatic variability and human susceptibility," *Sciencexpress* 13 January 2011, 1–4. 10.1126/science.1197175.

Büntgen, Ulf, and Lena Hellmann. "The Little Ice Age in scientific perspective: cold spells and caveats," *Journal of Interdisciplinary History* 44 (2014), 353–68.

Büntgen, Ulf, and Willy Tegel. "European tree-ring data and the Medieval Climate Anomaly," *PAGES news*, 19:1 (March 2011), 14–15.

Buonincontri, Mauro P., Alessandra Pecci, Gaetano Di Pasquale, Paola Ricci, and Carmine Lubritto. "Multiproxy approach to the study of medieval food habits in Tuscany," *Archaeological and Anthropological Sciences* 9 (2017), 653–671.

Burleigh, Michael. *Prussian Society and the German Order: An Aristocratic Corporation in Crisis, c. 1410–1466.* Cambridge: Cambridge University Press, 1984.

Burnouf, Joëlle, et al. "Fluvial metamorphosis of the Loire river during the Holocene and the role of natural and anthropogenic factors: a case study from the area of Tours, France." In Andy J. Howard, Mark G. Macklin and David G. Passmore, eds., *Alluvial Archaeology in Europe: Proceedings of the Alluvial Archaeology of North-West Europe and the Mediteranian* [*sic*], 163–171. Lisse: A.A. Balkema, 2003.

Burton, Janet, ed., *The Cartulary of Byland Abbey.* Publications of the Surtees Society, 208. Woodbridge, UK, and Rochester, NY: Boydell and Brewer, for the Surtees Society, 2004.

Butcher, Andrew. *Medieval Lowestoft: The Origins and Growth of a Suffolk Coastal Community.* Woodbridge, UK: Boydell Press, 2016.

Buti, Gabriel, Daniel Faget, Olivier Raveux, and Solène Rivoal, eds., *Moissonner la mer. Économies, sociétés et pratiques halieutiques méditerranéennes (xv^e–xxi^e siècle)*, Aix-en-Provence: Aix-Marseille Université, CNRS, and Paris: Éditions Karthala, 2018.

Butturini, Mattia. "La pesca sul lago di Garda," *Archivio Storico Lombarda* 6 (1879), 147–149; 7 (1880), 73–91 ; and 8 (1881), 157–195.

Bůžek, Václav. "Das Goldene Zeitalter der südböhmischen Teichwirtschaft." In Herbert Knittler and Andrea Komlosy, eds., *Die Lainsitz. Natur und Kulturgeschichte einer Region*, 81–92. St. Pölten: Verein für Landeskunde von Niederösterreich, 1997.

Bynum, Caroline W. *Holy Feast and Holy Fast: The Religious Significance of Food to Medieval Women.* Berkeley, Los Angeles and London: University of California Press, 1987.

Cabourdin, Guy. *Terre et hommes en Lorraine (1550–1635): Toulois et Comté de Vaudémont. Annales de l'Est Mémoire* no. 55. 2 vols. paginated continuously. Nancy: Université de Nancy, 1977.

Cage, A. G., and W. E. N. Austin. "Marine climate variability during the last millennium: the Loch Sunart record, Scotland," *Quaternary Science Reviews* 29 (2010), 1633–1647.

Cahn, Ernst. *Das Recht der Binnenfischerei im deutschen Kulturgebiet von den Anfängen bis zum Ausgang des 18. Jahrhunderts*, ed. Ekkehard Kaufmann. Frankfurt a.M.: Klostermann, 1956.

Calvet, Marc, et al. "Les cours d'eau des Pyrénées orientales depuis 15 000 ans. État des connaissances et perspectives de recherche." In Bravard and Magny, eds., *Les fleuves ont une histoire*, 279–294. Paris: Errance, 2002.

Camenisch, Chantal, and 23 co-authors. "The 1430s: a cold period of extraordinary internal climate variability during the early Spörer Minimum with social and economic impacts in north-western and central Europe," *Climate of the Past* 12 (2016), 2107–2126

Campbell, Bruce M. S. *The Great Transition. Climate, Disease and Society in the Late-Medieval World.* Cambridge: Cambridge University Press, 2016.

Campbell, Bruce M. S., and Francis Ludlow. "Climate, disease and society in late-medieval Ireland," *Proceedings of the Royal Irish Academy* 120C (2020), 159–252.

Campbell, Bruce M. S., James A. Galloway, Derek Keene, and Margaret Murphy. *A Medieval Capital and Its Grain Supply: Agrarian Production and Distribution in the London Region c.1300*. Historical Geography Research Series, Nr. 30. Belfast and London: Institute of British Geographers, 1993.

Campbell, James. "Domesday herrings." In Christopher Harper-Bill, Carole Rawcliffe, and Richard G. Wilson, eds., *East Anglia's History: Studies in Honour of Norman Scarfe*, 5–17. Woodbridge, UK, and Rochester, NY: Boydell and Brewer for the Centre of East Anglian Studies, University of East Anglia, 2002.

Camuffo, Dario. "Freezing of the Venetian lagoon since the 9th century A.D. in comparison to the climate of western Europe and England," *Climatic Change* 10 (1987), 43–66.

Camuffo, Dario. and Chiara Bertolin. "Climate in the Mediterranean area during the last millennium." In Ruthie Gertwagen, T. Fortibuoni, O. Giovanardi, S. Libralato, C. Solidoro, and S. Raicevich, eds., *When Humanities Meet Ecology: Historic Changes in Mediterranean and Black Sea Marine Biodiversity and Ecosystems since the Roman Period until Nowadays. Languages, Methodologies and Perspectives*, 125–132. HMAP International Summer School, Abdus Salam International Centre for Theoretical Physics, Trieste (Italy), 3 August–4 September 2009. Rome: ISPRA Serie Atti, 2010.

Camuffo, Dario, and Silvia Enzi. "The analysis of two bi-millenary series: Tiber and Po river floods." In Philip D. Jones, Raymond S. Bradley, and Jean Jouzel, eds., *Climatic Variations and Forcing Mechanisms of the Last 2000 Years*, 433–450. NATO ASI Series. Series I, Global environmental change, vol. 41. Berlin/New York: Springer, 1996.

Canivez, Josephus-Maria, ed. *Statuta capitulorum generalium ordinis Cisterciensis ab anno 1116 ad annum 1786*. 8 vols. (*1, Ab anno 1116 ad annum 1220*). Bibliothèque de la Revue d'histoire ecclésiastique, 9–14b. Louvain: Bureaux de la revue, 1933–41.

Cardan, Jerome [Girolamo Cardano]. *The Book of My Life (De Vita Propria Liber)*, tr. Jean Stoner. New York: Dutton, 1930. Corrected reprint New York: Dover, 1962.

Carlin, Martha. *Medieval Southwark*. London and Rio Grande, OH: Hambledon Press, 1996.

"Provisions for the poor: fast food in medieval London." *Franco-British Studies*, 20: *Structures d'approvisionnement à Londres et à Paris au Moyen Age* (1995), 35–48.

Carozza, Jean-Michel, Carole Puig, Thierry Odiot, Philippe Valette, and Olivier Passarrius. "Lower Mediterranean plain accelerated evolution during the Little Ice Age: geoarchaeological insight in the Tech basin (Roussillon, Gulf of Lion, Western Mediterranean," *Quaternary International* 266 (2011), 94–104.

Carpenter, David. *The Minority of Henry III*. London: Methuen, 1990.

Carter, Francis W. *Trade and Urban Development in Poland: An Economic Geography of Cracow, from Its Origins to 1795*. Cambridge Studies in Historical Geography, 20. Cambridge: Cambridge University Press, 1994.

Carus-Wilson, Eleonora. "The Iceland trade." In Eileen Power and Michael M. Postan, eds., *Studies in English Trade in the Fifteenth Century*, 155–182 and notes 381–383. London, 1933.

Carville, G. "The economic activities of the Cistercian Order in medieval Ireland (1142–1541)," *Cîteaux: commentarii cistercienses* 22 (1971), 278–299.

Casado Alonso, Hilario. *Señores, mercaderes y campesinos: La comarca de Burgos a fines de la edad media*. Valladolid: Junta de Castilla y León, Consejería de Cultura y Bienestar Social, 1987.

Casini, Michele, Christian Möllmann, and Henrik Österblom. "Food-web and climate-related dynamics in the Baltic Sea: present and potential future applications in fish stock assessment and management." In Andrea Belgrano and Charles W. Fowler, eds., *Ecosystem-Based Management for Marine Fisheries: An Evolving Perspective*, 9–31. Cambridge: Cambridge University Press, 2011.

Cassiodorus, Flavius Magnus Aurelius Senator. *Institutiones*, ed. R. A. B. Mynors. Oxford: Clarendon, 1937, reprinted 1963. Selections tr. James W. Halporn, *Cassiodorus Institutions of Divine and Secular Learning and On the Soul*. Liverpool: Liverpool University Press, 2004.

*Variarum libri dvodecim* [letters], ed. A. J. Fridh. Corpus Christianorum, Series Latina, vol. 96. Turnhout: Brepols, 1973. Selections tr. S. J. B. Barnish, *Cassiodorus: Variae*. Liverpool and Philadelphia: Liverpool University Press, 1992.

Castagnetti, Andrea, et al., eds. *Inventari altomedievali di terre, coloni e redditi. Fonti per la storia d'Italia*, vol. 104. Rome: Nella sede dell'Istituto Palazzo Borromini, 1979.

Catarino, Maria M. "Abastecimento e consumo de pescado – alguns aspetos do quotidiano na Lisboa dos séculos XIV e XV. Supply and fish consumption – some aspects of daily life in Lisbon of the 14th and 15th centuries," *Cadernos do Arquivo Municipal* [Lisbon] 8 (2017), 17–35.

Catoni, Giulano. "'*Super facto piscium*' (Siena 1291–1486)." In Bruno Laurioux and Laurence Moulinier-Brogi, eds., *Scrivere il medioevo. Lo spazio, la santità, il cibo. Un libro dedicato a Odile Redon*, 295–303. Libri de Viella, 28. Rome: Viella, 2001.

Catteddu, Isabelle. *Archéologie médiévale en France. Le premier Moyen Âge (Ve–XIe siècle)*. Paris: La Découverte, 2009.

Cavacciocchi, Simonetta, ed. *Alimentazione e nutrizione secc. XIII–XVIII*. Istituto Internazionale di Storia "F. Datini," Prato. Firenze: Le Monnier, 1996.

Cavallar, Osvaldo. "River of law: Bartolus's Tiberiadis (De alluvione)." In John A. Marino and Thomas Kuehn, eds., *A Renaissance of Conflicts: Visions and Revisions of Law and Society in Italy and Spain*, 84–116. Toronto: CRRS, 2004.

Cayla, Olivier. "La pêche à Paris aux XV^e^ et XVI^e^ siècles à travers les comptes de l'abbaye de Saint-Germain-des-Prés." In Benoît et al., eds., *Pêche et pisciculture en eau douce*. 2004.

Cazelles, Raymond. *Nouvelle Histoire de Paris*: vol. 2: *De la fin du règne de Philippe Auguste à la morte de Charles V, 1223–1380*. Paris: Hachette, 1972.

Čechura, Jaroslav. *Adelige Grundherrn als Unternehmer: Zur Struktur südböhmischer Dominien vor 1620*. Sozial- und Wirtschaftshistorische Studien, Institut für Wirtschafts- und Sozialgeschichte, Universität Wien, Bd. 25. Munich: Oldenbourg and Vienna: Verlag für Geschichte und Politik, 2000.

"Lohn oder Fron? Grundherrschaft und bäuerliches Wirtschaftsleben im böhmisch-niederösterreichischen Grenzgebiet." In Andrea Komlosy, Václav Bůžek, and František Svátek, eds., *Kulturen an der Grenze – Kultury na hranici; Waldviertel, Weinviertel, Südböhmen, Südmähren*, 111–116. Vienna: Promedia Druck und Verlagsgesellschaft, 1995.

Cerón-Carrasco, Ruby. "Fish and marine-shell remains." In Thomas Addyman, Tanja Romankiewicz, Kenneth Macfadyen, Alasdair Ross, and Nicholas Uglow, eds., *The Medieval Kirk, Cemetery and Hospice at Kirk Ness, North Berwick*, 44–48. Oxford: Oxbow, 2013.

"The investigation of fish remains from an Orkney farm mound." In Wim Van Neer, ed., *Fish Exploitation in the Past: Proceedings of the 7th Meeting of the I.C.A.Z. Fish Remains Working Group*, 207–210. Annales du Musée Royal de l'Afrique Central, Sciences Zoologiques, vol. 274. Tervuren: Musée Royal, 1994.

Chambers, R. A., with M. Gray. "The excavation of fishponds." In Aston, ed., *Medieval Fish*, 137–186. 1988.

Champion, Matthew S. *The Fullness of Time: Temporalities of the Fifteenth-Century Low Countries*. Chicago: University of Chicago Press, 2017.

Charvátova, Kateřina. "Manorial farms of Cistercian abbeys of mediaeval Bohemia." In Jerzy Strzelczyk, ed., *Historia i kultura cystersów w dawnej polsce i ich europejskie związki*, 111–135. Poznan: Wydawn. Nauk. Uniwersytetu im. Adama Mickiewicza w Pozaniu, 1987.

Chassaing,Olivier, Nathalie Desse-Berset, Marilyne Duffraisse, Sandrine Hughes, Catherine Hänni, and Patrick Berrebi. "Palaeogenetics of western French sturgeons spotlights the relationships between *Acipenser sturio* and *Acipenser oxyrinchus*," *Journal of Biogeography* 40 (2013), 382–393.

Checkley, David M., Jürgen Alheit, Yoshioki Oozeki, and Claude Roy, eds. *Climate Change and Small Pelagic Fish*. Cambridge: Cambridge University Press, 2009.

Checkley, David M., and 14 co-authors. "Habitats." In Checkley et al., eds., *Climate Change and Small Pelagic Fish*, 12–44. 2009.

Chełkowski, Zygmunt, and Jarosław Filipiak. "Cognitive potential of bone remains of fish from archaeological excavations on the banks of the Odra River estuary." *Archeozoologia* 21 (2003), 37–48.

Chène, Catherine. "Une sainteté exemplaire: les évêques de Lausanne et l'excommunication des anguilles du lac Léman." In Bernard Andematten et al., eds., *Mémoires de cours: Études offertes à Agostino Paravicini Bagliani par ses collègues et élèves de l'Université de Lausanne*, 223–254. Cahiers Lausannois d'Histoire Médiévale, 48. Lausanne: Section d'histoire, Faculté des Lettres, Anthropole, Université de Lausanne, 2008.

Chevalier, Bernard. "Alimentation et niveau de vie à Tours à la fin du XVe siècle." In *Bulletin Philologique et Historique (jusqu'à 1610) du Comité des Travaux Historiques et Scientifiques*, 99ème Congrès des Sociétés savantes, Tours, 1968, vol. 1: *Les Problèmes de l'alimentation*, 143–158. Paris: BNF., 1971.

Cheyette, Fredric L., and Howell Chickering. "Love, anger, and peace: social practice and poetic play in the ending of *Yvain*," *Speculum* 80 (2005), 75–117.

Chiappa Mauri, Luisa. *I mulini a acqua nel Milanese (secoli X–XV)*. Biblioteca della nuova rivista storica, 36. Rome: Dante Alighieri, 1984.

Childs, Wendy. *Anglo-Castilian Trade in the Later Middle Ages*. Manchester: Manchester University Press,1978.

"England's Icelandic trade in the fifteenth century: the role of Hull," *Northern Seas Yearbook* 56 (1995), 11–31.

"Fishing and fisheries in the Middle Ages: the eastern fisheries." In David J. Starkey, Chris Reid, and Neil Ashcroft, eds., *England's Sea Fisheries: The Commercial Sea Fisheries of England and Wales since 1300*, 19–23. London: Chatham Publishing, 2000.

Chmiel, Adam, ed. *Księgi radzieckie Kazimierskie 1369–1381 i 1385–1402. Acta consularia casimiriensia 1369–1381 et 1385–1402*. Wydawnictwa Archiwum Aktów Dawnych Miasta Krakowa, vol. 2. Kraków: W druk. Uniw. Jag, 1932.

Chrétien de Troyes. *Arthurian Romances*, tr. W. W. Comfort. London: Dent and New York: Dutton, 1914 reprinted 1965.

*Les romans de Chrétien de Troyes*, ed. Félix Lecoy, Alexandre Micha, and Mario Roques. 6 vols. Paris: H. Champion, 1975–1979.

Christensen, Pål, and Alf R. Nielssen. "Norwegian fisheries 1100–1970 main developments." In Holm, Starkey, and Thor, eds., *The North Atlantic Fisheries, 1100–1976*, 145–168. Esbjerg: Fiskerei- og Søfartsmuseets Forlag, 1996.

Christiansen, B., and F. Ljungqvist. "The extra-tropical Northern Hemisphere temperature in the last two millennia: reconstructions of low-frequency variability," *Climate of the Past* 8 (2012), 765–786.

"Chronique des fouilles médiévales en France," *Archéologie Médiévale* 17 (1987), 161–289.

Ciaffi, R., R. Lelli, G. Müldner, K. Stantcheva, A. L. Fischetti, G. Ghini, O. E. Craig, F. Milano, O. Rickards, G. Arcudi, and C. Martinez-Labarga. "Paleobiology of the medieval population of Albano (Rome, Italy): a combined morphological and biomolecular approach," *International Journal of Osteoarchaeology* 25:4 (July/August 2015), 477–488.

Ciappelli, Giovanni. *Carnevale e Quaresima: Comportamenti sociali e cultura a Firenze nel Rinascimento*. Rome: Edizioni di Storia e letteratura, 1997.

Cioc, Mark. *The Rhine: An Eco-Biography, 1815–2000*. Seattle: University of Washington Press, 2002.

Cios, Stanisław. *Ryby w życiu Polaków od X do XIX w*. Olsztyn: Wydawnictwo Instytutu Pybactwa Śródiądowego im. Stanisława Sakowicza, 2007.

Clark, Gillian, Lorenzo Costantini, Angelo Finetti, John Giorgi, Andrew Jones, David Reese, Sheila Sutherland, and David Whitehouse. "The food refuse of an affluent urban household in the late fourteenth century: faunal and botanical remains from the Palazzo Vitelleschi, Tarquinia (Viterbo)," *Papers of the British School at Rome* 57 (1989), 200–321.

Clason, A. T. and W. Prummel. "Collecting, sieving and archaeozoological research," *Journal of Archaeological Science* 4 (1977), 171–175 and erratum.

Clason, A. T., W. Prummel, and D. C. Brinkhuizen. "Vogelen en vissen. Een glimp van de Nederlandse vogel- en viswereld uit het verleden," *Westerheem* 28 (1979), 9–22.

Clavel, Benoît. "L'alimentation urbaine au moyen âge: L'exemple d'un quartier rémois entre le IX$^{e}$ et le XIV$^{e}$ siècle (Reims, Marne)." In Fabienne Ravoire, and Anne Dietrich, eds., *La cuisine et la table dans la France de la fin du moyen age. Contenus et contenants du xiv$^{e}$ au xvi$^{e}$ siecle. Colloque dijon/sens, 8–10 janvier 2004.*, 107–124. Caen : Publications du CRAHM, 2009.

*L'Animal dans l'alimentation médiévale et moderne en France du nord (XII$^{e}$ – XVII$^{e}$ siècles)*. Revue Archéologique de Picardie, No. Spécial 19. [Amiens]: CRAVO, 2001.

"Les restes osseux animaux du Moyen Âge découverts Place de L'Hotel de ville à Abbeville (Somme)," *Revue archéologique de Picardie* 3:4 (1997), 193–205.

Clavel, Benoît, and Christophe Cloquier. "Contribution des sources documentaires et archéologiques à l'étude des pratiques halieutiques fluviales médiévales et modernes dans le bassin de la Somme." In Joëlle Burnouf and Phillipe Leveau, eds., *Fleuves et marais, une histoire au croisement de la nature et de la culture. Sociétés préindustrielles et milieux fluviaux, lacustres et palustres: pratiques sociales et hydrosystèmes*, 201–210. Paris: CTHS, 2004.

Clavel, Benoît, and Justine Vorenger. "Quelques données sur la pêches et la consummation du poisson en milieu castral dans le comté d'Amiens (Somme, France) du X$^{e}$ au XIV$^{e}$ siècle." In Philippe Béarez, Sandrine Grouard, and Benoît Clavel, eds., *Archéologie du poisson. 30 ans d'archéo-ichtyologogie au CNRS. Hommage aux travaus de Jean Desse et Nathalie Desse-Berseté*, 61–66. Antibes: Éditions APDCA, 2008.

Clavel, B., and J. H. Yvinec. "L'archéozoologie du Moyen Âge au début des la périod moderne dans la moitié nord de la France." In Jean Chapelot, ed., *Trente ans d'archéologie médiévale en France. Un bilan pour un avenir*, 71–87. IX$^{e}$ congrès international de la Société d'archéologie médiévale, Vincennes, 16–18 juin 2006. Caen: CHRAM, 2010.

Clay, Patrick. "A Norman mill dam at Hemington Fields, Castle Donington, Leicestershire." In Stuart Needham and Mark G. Macklin, eds., *Alluvial Archaeology in Britain*, 163–168. Oxbow Monograph 18. Oxford: Oxbow, 1992.

Cloern, James E., Paulo C. Abreu, Jacob Carstensen, Laurent Chauvaud, Ragnar Elmgren, Jacques Grall, Holly Greening, John Olov, Roger Hansson, Mati Kahru, Edward T. Sherwood, Jie Xu, and Kedong Yin. "Human activities and climate variability drive fast-paced change across the world's estuarine coastal ecosystems." *Global Change Biology* 22:2 (February 2016), 513–529. doi: 10.1111/gcb.13059

Cloquier, Christophe. "Pêches et pêcheries dans le cours de la Somme: l'approvisionnement en poisson de l'abbaye du Gard du XII$^{e}$ au XVIII$^{e}$ siècle." In Benoît et al., eds., *Pêche et pisciculture en eau douce.* 2004.

Cloquier, Christophe, and Benoît Clavel. "La consommation d'animaux aquatiques dans le bassin de la Somme durant les périodes médiévale et moderne." In Christiane Demeulenaere-Douyère, ed., *Tous à table*, 83–97. Actes des 138e Congrès national des sociétés historiques et scientifiques tenu à Rennes en 2013. [Paris]: Comité des travaux historiqus et scientifiques, 2015.

Cnopf, Willy, *Die Entwicklung und wirtschaftliche Bedeutung der Teichwirtschaft in Mittelfranken*. PhD dissertation, University of Erlangen, 1926. Also published separately Nürnberg, 1927, and as an article with same title in *Archiv für Fischereigeschichte* 11 (1927), 26–82.

Cocula-Vaillières, Anne-Marie. *Un fleuve et des hommes. Les gens de la Dordogne au xviiie siècle*. Paris: Tallandier, 1981.

Cohen, Nathalie. "Early Anglo-Saxon fish traps on the River Thames." In S. Brookes, S. Harrington, and A. Reynolds, eds., *Studies in Early Anglo-Saxon Art and Archaeology: Papers in Honour of Martin G. Welch*, 131–139. BAR British series 527. Oxford: Archaeopress, 2011.

Colardelle, Michel, and Eric Verdel, eds. *Chevaliers-Paysans de l'an mil au lac de Paladru*. Paris: Musée Dauphinois. 1993.

eds. *Les habitants du Lac de Paladru (Isère) dans leur environnement: La formation d'un terroir au XIe siècle*. Documents d'Archéologie Français, 40. Paris: Editions de la Maison des sciences de l'homme, 1993.

Coldicott, Diana. K. *Hampshire Nunneries*. Chichester: Phillimore, 1989.

Coleman, M. Clare. *Downham-in-the-Isle: A Study of an Ecclesiastical Manor in the 13th and 14th Centuries*. Woodbridge, UK: Boydell, 1984.

Colgrave, Bertram, ed. and tr. *Felix's Life of St. Guthlac*. Cambridge: Cambridge University Press, 1956.

Colley, Sarah M. "Fishing for facts. Can we reconstruct fishing methods from archaeological evidence?" *Australian Archaeology*, 24 (1987), 16–26.

Collin, Hubert. "Les ressources alimentaires en Lorraine pendant la première partie du XIVème siècle." In *Bulletin Philologique et Historique (jusqu'à 1610) du Comité des Travaux Historiques et Scientifiques*, 99ème Congrès des Sociétés savantes, Tours, 1968, vol. 1: *Les Problèmes de l'alimentation*, 37–75. Paris: BNF, 1971.

"Le train de vie d'Eduard 1er comte de Bar," *Bulletin Philologique et Historique (jusqu'à 1610) du Comité des Travaux Historiques et Scientifiques* (1969), II:793–817.

Cook, Edward R., Yochanan Kushnir, Jason E. Smerdon, A. Park Williams, Kevin J. Anchukaitis, and Eugene R. Wahl. "A Euro-Mediterranean tree-ring reconstruction of the Winter NAO Index since 910 C.E." *Climate Dynamics* 53 (2019): 1567–1580. https://doi.org/10.1007/s00382-019-04696-2.

Cook, Edward R., and 56 co-authors. "Old World megadroughts and pluvials during the Common Era," *Science Advances* 1:10 (6 November 2015). DOI: 10.1126/sciadv.1500561.

Cooper, Lynden P., and Susan Ripper. *Fishing and Managing the Trent in the Medieval Period (7th–14th Century): Excavations at Hemington Quarry (1998–2000), Castle Donington, UK*. Oxford: BAR Publishing, 2017

Coopland, G. W. *The Abbey of St. Bertin and Its Neighbourhood, 900–1350*. Oxford Studies in Social and Legal History, v. 4 [no. 8]. Oxford: Clarendon, 1914.

Coornaert, E. *Les Corporations en France avant 1789*, 2d ed. rev. and enlarged. Paris: les Editions ouvrières, 1968.

*Corpus Iuris Civilis: Institutiones*, ed. P. Krueger. Berlin: Weidmann, 1872. Tr. John B. Moyle, *The Institutes of* Justinian, 5th ed. Oxford: Clarendon, 1913.

Corten, Adrianus. *Herring and Climate: Changes in the Distribution of North Sea Herring due to Climate Fluctuations*. Groningen: Groningen University Library, 2001.

"Northern distribution of North Sea herring as a response to high water temperatures and/or low food abundance," *Fisheries Research* 50 (2001), 189–204.

Cortonesi, Alfio. "Le spese in *victualibus* della *domus helemosine Sancti Petri* di Roma," *Archeologia Medievale* 8 (1981), 193–225.

Cosmas of Prague (*c.* 1045–1125). *Cosmae pragensis chronica bohemorum*, ed. Berthold Bretholz, MGH SS rerum Germanicarum, NS 2. Berlin: Weidmann, 1923.

Cosmas of Prague.*The Chronicle of the Czechs*, tr. Lisa Wolverton. Washington, DC: Catholic University of America Press, 2009.

Coulon, Gérard. "Les étangs de la Brenne," *La Bouinotte – Revue Regionaliste Trimestrielle*, Autumn 1983, pp. 10–18.

Couperie, Pierre. "Les marchés de pourvoierie: viands et poissons chez les Grands au XVIIe siècle." In Jean-Jacques Hémardinquer, ed., *Pour une histoire de l'alimentation*, 241–249. Paris: A. Colin, 1970.

Courcelle, Pierre. "La site du monastère de Casiodore," École Française de Rome. *Mélanges d'archéologie et d'histoire* 55 (1938), 259–309.

Coy, Jennie. "Fish bones." In G. G. Astill and S. J. Lobbs, "Roman and Saxon deposits at Wraysbury," pp. 118–120 & microfiche M1/ 68–74, *Archaeological Journal* 146 (1989), 68–134. [*sic*; the Coy paper is within the paper by Astill and Lobbs.]

"Medieval records versus excavation results. Examples from Southern England," *Archaeofauna* 5 (1996) (special issue *Ichthyoarchaeology: Fish and the Archaeological Record: Proceedings of the Eighth Meeting of the ICAZ Fish Remains Working Group – Madrid, 3-11 October 1995)*, 55–63.

"The provision of fowls and fish for towns." In D. Serjeantson and T. Waldron, eds., *Diet and Crafts in Towns. The Evidence of Animal Remains from the Roman to the Post-Medieval Periods*, 25–40. BAR, British Series, 199. Oxford: BAR, 1989.

Cramer, Deborah. "Troubled waters," *Atlantic Monthly* 275:6 (June 1995), 22–26.

Crawford, Anne, ed. *The Household Books of John Howard, Duke of Norfolk, 1462–1471, 1481–1483*. Stroud: A. Sutton for Richard III & Yorkist History Trust; Wolfeboro Falls, NH: A. Sutton, 1992. Essentially a facsimile reprint of Turner, ed., *Manners and Household Expenses*, pp. 147–621, q.v., with a more complete codicological introduction but no changes to anything regarding fish and fisheries.

Creighton, Oliver H. *Designs upon the Land: Elite Landscapes in the Middle Ages.* Woodbridge, UK, and Rochester, NY: Boydell and Brewer, 2009.

Crescenzi, Pietro de [Petrus de Crescentiis]. *Ruralia commoda: Das Wissen des vollkommenen Landwirts um 1300*, ed. Will Richter, prepared for publication by Reinhilt Richter-Bergmeier. 3 vols. Editiones Heidelbergenses, 25. Heidelberg: C. Winter, 1995–1998. [Note that some earlier editions and translations number chapters differently in book 10.]

Crook, D. S., D. J. Siddle, J. A. Dearing, and R. Thompson. "Human impact on the environment in the Annecy Petit Lac Catchment, Haute Savoie: a documentary approach," *Environment and History* 10 (2004), 247–284.

Crossley-Holland, Nicole. *Living and Dining in Medieval Paris: The Household of a Fourteenth-Century Knight.* Cardiff: Wales University Press, 1996.

Crouzet-Pavan, Élisabeth. "Les eaux noires. Essai sur la miasme et la salubrité dans la ville médiévale." In Jean-François Cottier, Martin Gravel, and Sébastien Rossignol, eds., *Ad libros! Mélanges d'études médiévales offerts à Denise Angers et Joseph-Claude Poulin*, 265–280. Montréal: Les Presses de l'Université de Montréal, 2010.

"Mythes et réalités de la crise d'un écosystème littoral. Venise et sa lagune à la fin du Moyen Âge." In Jean-Marie Martin, ed., *Castrum 7: Zones côtières littorales dans le monde méditerranéen au Moyen Âge: defense, peuplement, mise en valeur*, 95–106. Rome–Madrid: École française de Rome, Casa de Velázquez, 2001.

Crumley, Carole L., Tommy Lennartsson, and Anna Westin, eds. *Issues and Concepts in Historical Ecology: The Past and Future of Landscapes and Regions.* Cambridge: Cambridge University Press, 2018.

Cunningham, Laura, William Austin, Karen Knudsen, Jón Eiríksson, James Scourse, Alan Wanamaker, Paul Butler, Alix Cage, Thomas Richter, Katrine Husum, Morten Hald, Carin Andersson, Eduardo Zorita, Hans Linderholm, Björn Gunnarson, Marie-Alexandrine Sicre, Hans Sejrup, Hui Jiang, and Rob Wilson. "Reconstructions of surface ocean conditions from the northeast Atlantic and Nordic seas during the last millennium." *The Holocene* 23 (2013), 921–934.

Currie, Christopher K. "The early history of the carp and Its economic significance in England," *Agricultural History Review* 39 (1991), 97–107.

"The role of fishponds in the monastic economy." In Roberta Gilchrist and Harold Mytum, eds., *The Archaeology of Rural Monasteries*, 147–172. BAR British Series, 203. Oxford: BAR, 1989.

Curtis, Daniel H. "Danger and displacement in the Dollard: the 1509 flooding of the Dollard Sea (Groningen) and its impact on long-term inequality in the distribution of property," *Environment and History* 22 (2016), 103–135.

Curtis-Summers, Shirley, Janet Montgomery, and Martin Carver. "Stable isotope evidence for dietary contrast between Pictish and Medieval populations at Portmahomack, Scotland," *Medieval Archaeology* 58 (2014), 21–43.

Curtis-Summers, Shirley, Jessica A. Pearson, and Angela L. Lamb. "From Picts to Parish: stable isotope evidence of dietary change at medieval Portmahomack, Scotland," *Journal of Archaeological Science* Reports 31

(June 2020). www.sciencedirect.com https://doi.org/10.1016/j.jasrep.2020.102303

Curto Homedes, Albert. "El consum de peix a la Tortosa baix-medieval." In *Actes Ir Colloqui d'Historia de l'Alimentació a la Corona d'Aragó edat mitjana*, vol. 2: *Comunicacions*, 149–166. Lleida: Institut d'Estudis Ilerdencs, 1995.

Cutting, Charles L. *Fish Saving: A History of Fish Processing from Ancient to Modern Times*. London: Hill, 1955.

"Historical aspects of fish." In Georg Borgstrom, ed., *Fish As Food*, vol. II: *Nutrition, Sanitation, and Utilization*, 1–28. New York: Academic Press, 1962.

D'Avenel, Georges. *Histoire économique de la propriété, des salaires des denrées et de tous les prix en général depuis l'an 1200 jusqu'en l'an 1800*. 4 vols. 2d ed. rev. Paris: Leroux, 1913.

da Costa Dominguez, Rodrigo. "Harvesting in holy waters: an overview of fishery in Portugal in the Later Middle Ages and Early Modern Times." *Imago Temporis. Medium Aevum* 15 (2021), 149–163

da Cruz Coelho, Maria Helena. "A pesca fluvial na economia e sociedade medieval portuguesas," *Cadernos Historicos* VI: Actas do Seminário "Pescas e navegações na História de Portugal (Séculos XII a XVIII)," Lagos, 22–24 de Maio de 1992, 81–102. Lagos (Portugal): the Seminário, 1995.

"Apontamentos sobre a comida e a bebida do campesinato Coimbrão em tempos medievos," *Revista de historia económica y social* [Lisbon] 12 (1983), 91–101.

Dąbrowski, Kazimierz. *Rozwój wielkiej własności ziemskiej klasztoru cysterek w Żarnowcu od XIII do XVI wieku*. Gdańskie Towarzystwo Naukowe, Wydział i Nauk Społecznych i Humanistycznych, Seria monografii, Nr. 40. Gdańsk: Zakład Narodowy im. Ossolińskich, 1970.

Dagron, Gilbert. "Poissons, pêcheurs et poissoniers de Constantinople." In Cyril Mango and Gilbert Dagron, eds., *Constantinople and Its Hinterland*, 57–73. Papers from the 27th Spring Symposium of Byzantine Studies, Oxford, 1993. Society for the Promotion of Byzantine Studies, vol. 3. Aldershot: Variorum, 1995.

Dähnhardt, Oskar, ed. *Natursagen; eine Sammlung naturdeutender Sagen, Märchen, Fabeln und Legenden*. 4 vols. Leipzig and Berlin: Teubner, 1907–1912.

Dam, Petra J. E. M. van. "Eel fishing in Holland: the transition to the early modern economy," *International Journal of Maritime History* 15:2 (2003), 163–175.

"Feestvissen en vastenvissen: Culturele, ecologische en economische aspecten van de visconsumptie in de Nederlanden in de Late Middeleeuwen," *Tijdschrift voor Sociale Geschiedenis* 29:4 [Themanummer Eetregimes in de Middeleeuwen] (2003), 467–496.

"Fish for feast and fast: fish consumption in the Netherlands in the Late Middle Ages." In Louis Sicking and Darlene Abreu-Ferreira, eds., *Beyond the Catch: Fisheries of the North Atlantic, the North Sea and the Baltic, 900–1850*, 309–336. Leiden: Brill, 2009.

"Sinking peat bogs: environmental change in Holland, 1350–1550," *Environmental History* 6 (2001), 32–45.

*Vissen in veenmeren. De sluisvisserij op all tussen Haarlem en Amsterdam en de ecologische transformatie in Rijnland 1440–1530.* Hilversum: Historische Vereniging Holland, Uitgeverij Verloren, 1998.

Dam, Petra J. E. M. van., and Johanna Maria van Winter. "Theorie en praktijk van eetregimes in de Middeleeuwen," *Tijdschrift voor Sociale Geschiedenis* 29:4 [Themanummer Eetregimes in de Mideeleeuwen] (2003), 385–412.

Danker-Carstensen, Peter. "Wiefiel Störfleisch verträgt die Dienstmagd in der Woche? Regional Legendenbildung in einem kurzen Kapitel Fischereigeschichte." In *Som en rejselysten flåde, tilegnet Ole Lisberg Jensen (Festkrift Ole Lisberg Jensen på 60-årsdagen 21. juni 1999*, 30–46. Orlogsmuseets Skriftenrække nr. 2. Christianshavn: Orlogsmuseets, 1999.

"Stör oder Lachs – aber auf keinen Fall mehr als zweimal in der Woche?. Legendenbildung und Erzähltradition in einem Kapitel deutscher Fischereigeschichte." In Ortwin Pelc, ed., *Mythen der Vergangenheit: Realität und Fiktion in der Geschichte. Jörgen Bracker zum 75. Geburtstag*, 265–285. Göttingen: Vandenhoeck & Ruprecht Unipress, 2012.

Darby, H. C. "The clearing of the woodland in Europe." In W. L. Thomas, ed., *Man's Role in Changing the Face of the Earth*, 183–216. Chicago: University of Chicago Press, 1956.

*Domesday England.* Cambridge: Cambridge University Press, 1977.

Darsel, Joachim. "Les conditions du métier de la mer au Moyen Age dans les ports du Ponant," *Bulletin Philologique et Historique 88^e^ Congrès du Nationale Sociétés savantes, Clermont-Ferrand 1963*, I:461–483. Paris, 1966.

"Les servitudes de la pêche en Normandie sous l'Ancien Régime," *Actes du 81^e^ Congrès national des sociétés savantes, Rouen-Caen, 1956, Section d'histoire moderne et contemporaine*, 105–109.

Datini, Margherita. *Letters to Francesco Datini*, tr. Carolyn James and Antonio Pagliaro. The Other Voices in Early Modern Europe: The Toronto Series, 16. Toronto: ITER, 2012.

Dawson, A. G., K.Hickey, P. A.Mayewski, and A. Nesje. "Greenland (GISP2) ice core and historical indicators of complex North Atlantic climate changes during the fourteenth century." *The Holocene*, 17 (2007), 427–434.

De Battisti, Mariella, et al., eds. *Il libro di casa Cerruti.* Milan: Arnoldo Mondadori, 1983. [facsimile of ONB Codex Vindobonensis Series Nova 2644, *Tacuinum in medicina*, aka *Tacuinum sanitatis*].

de Castro, Concepción. *El pan de Madrid. El abasto de las ciudades españolas del Antiguo Régimen.* Madrid: Alianza Editorial, 1987.

De Groote, K. and J. Moens, eds. *Archeologie en geschiedenis van een middeleeuwse woonwijk onder de Hopmarkt te Aalst.* Relicta Monografieën 16. Brussels: Onroerend Erfgoed: 2018. Available online as https://oar.onroerenderfgoed .be/publicaties/RELM/16/RELM016-001.pdf.

De Grossi Mazzorin, Jacopo. "I resti archeozoologici come strumento di conoscenza dell'economia alimentare nell'alto medioevo," In *L'Alimentazione nell'Alto Medioevo: Pratiche, Simboli, Ideologie, Spoleto, 9–14 aprile 2015*, vol. I, 21–81 (discussion pp. 83–85). Settimane di Studio della Fondazione Centro Italiano di Studi sull'Alto Medioevo, LXIII. Spoleto: Fondazione Centro Italiano di Studi Sull'alto Medioevo, 2016.

de La Mare, Nicholas. *Traité de la Police*. 4 vols. in 5. Paris: Chez M. Brunet, 1704–38.

De Nicolò, Maria L. "Production et consommation du poisson de mer: L'espace adriatique (xve–xviiie siècle)." In Buti et al., eds. *Moissonner la mer*, 53–68. 2018.

de Rachewiltz, Siegfried. "Die Versorgung von Schloß Tirol." In Eines Fürsten Traum Meinhard des II. – Das Werden Tirols, 258–265. Katalog Landesaustellung Stift Stams / Schloß Tyrol. Tirolo (BZ): Museo provinciale di Castel Tirolo; Innsbruck: Tiroler Landesmuseum Ferdinandeum, 1995.

de Sostoa, Adolfo, and Javier Lobon-Cervia. "Fish and fisheries of the River Ebro: actual state and recent history." In Geoffrey E. Petts, Heino Moller, and A. L. Roux, eds., *Historical Change of Large Alluvial Rivers: Western Europe*, 233–247. New York: Wiley, 1989.

Defosse, Marine. "Pêche et pêcheries médiévales et modernes sur l'Allier entre Langeac et Pont-du-Château." In Benoît et al., eds., *Pêche et pisciculture en eau douce*, 2004.

Delaborde, H.-François. "Fragments de l'enquête faite à Saint-Denis en 1282 en vue de la canonisation de Saint Louis," *Mémoires de la Société de l'Histoire de Paris et de l'Ile de France* 23 (1896), 1–70.

Delatouche, Raymond. "Importance relative des divers produits agricoles dans le 'Gros Brief' de Flandre pour 1187," *Bulletin Philologique et Historique (jusqu'à 1610) du Comité des Travaux Historiques et Scientifiques, 99ème Congrès des Sociétés savantes, Tours, 1968*, vol. 1: *Les Problèmes de l'alimentation*, 19–39. Paris: Bibliothèque National, 1971.

"Le poisson d'eau douce dans alimentation médiévale," *Comptes-rendus de l'Academie d'agriculture de France* 52 (1966), 793–798.

"Le poisson d'eau douce dans l'alimentation médiévale: L'exemple du Maine," *Bulletin Philologique et Historique (jusqu'a 1610) du Comité des Travaux Historiques et Scientifiques*, année 1967, vol. 1 (Paris, 1969), 171–182.

Déléage, A. *La vie économique et sociale de la Bourgogne dans le Haut Moyen Âge*. Mâcon: Imprimatur Protat Frère, 1941.

Deligne, Chloé. *Bruxelles et sa rivière: Genèse d'un territoire urbain (12e–18e siècle)*. Turnhout: Brepols, 2003.

"Carp in the City: fish-farming ponds and urban dynamics in Brabant and Hainault, c.1100–1500." In Louis Sicking and Darlene Abreu-Ferreira, eds., *Beyond the Catch: Fisheries of the North Atlantic, the North Sea and the Baltic, 900–1850*, 283–308. Leiden and Boston: Brill, 2009.

Dembińska, Maria. *Konsumpcja żywnościowa w Polsce średniowiecznej*. Instytut Historii Kultury Materialnej PAN. Wrocław: Ossolineum, 1963. [2d ed. rev. by William Woys Weaver, *Food and Drink in Medieval Poland. Rediscovering a Cuisine of the Past*, tr. Magdalena Thomas (Philadelphia: University of Pennsylvania Press, 1999), is a much revised version for English-language readers interested in food. It omits most of the documentary detail present in the Polish original.]

*Przetwórstwo zbożowe w Polsce średniowiecznej (X–XIV wiek)*. Wrocław: Zakład Narodowy imienia Ossolińskich PAN, 1973.

Denis-Delacour, Christopher. "*Fraudis* et *Piscandis* dans les États pontificaux. Les sanctions juridiques des marchés halieutiques et leurs objets (XVIII^e^–XIX^e^ siècles)." In Faget and Sternberg, eds., *Pêches méditerranéennes*, 143–168. 2015.

Derveeghe, Denise Van. *Le Domaine du Val Saint-Lambert de 1202 à 1387*. Paris: Belles lettres, 1955.

Derville, Alain. "Rivières et canaux du Nord/Pas-de-Calais aux époques médiévale et moderne," *Revue du Nord* 72 (1990), 5–22.

Derville, Alain, and Albert Vion. *Histoire de Calais. Collection Histoire des villes du Nord/Pas-de-Calais*, vol. 8. Westhoek: Editions des Beffrois, 1985.

Deschler-Erb, Sabine, Miki Bopp-Ito, and Heide Hüster Plogmann. "Die mittelalterlichen Tierknochen aus dem Kreuzgangbereich des Stiftes St. Arnaul." In Hans-Walter Herrmann and Jan Selmer, eds., *Leben und Sterben in einem mittelalterlichen Kollegiatstift. Archäologische und baugeschichteliche Untersuchungen im ehemaligen Stift St. Arnaul in Saarbrücken*, 523–538. Saarbrücken: [n.p.], 2007.

Desse, Jean, and Nathalie Desse-Berset., "Pêche et surpêche en Méditerranée: Le témoignage des os." In Jean Desse and Frédérique Audoin-Rouzeau, eds. *Exploitation des animaux sauvages a travers le temps*, 327–339. XIIIe Recontres Internationales d'Archéologie de l'Histoire d'Antibes. IVe Colloque International de l'Homme et l'Animal. Juan-les-Pins, 1993), 327–339.

"Pêches locales, côtières ou lointaines: Le poisson au menu des parisiens du Grand Louvre, du 14ème au 18ème siècle," *Anthropozoologica* 16 (1992), 119–126.

Desse-Berset, Nathalie. "Sturgeons of the Rhône during protohistory in Arles (6th–2d century BC)." In Wim Van Neer, ed., *Fish Exploitation in the Past. Proceedings of the 7th Meeting of the I.C.A.Z. Fish Remains Working Group*, 81–90. Annales du Musée Royal de l'Afrique Central, Sciences Zoologiques, vol. 274. Tervuren: Musée Royal, 1994.

Desse-Berset, Nathalie, and P. Williot, "Emerging questions from the discovery of the long term presence of *Acipenser oxyrinchus* in France," *Journal of Applied Ichthyology* 27 (2011), 263–268.

Dettmering, Wilhelm. *Beiträge zur älteren Zunftgeschichte der Stadt Strassburg*. Berlin: Ebering, 1903. Kraus reprint Vaduz: 1965.

Devailly, Guy. *Le Berry du Xe siècle au milieu du XIIIe. Étude politique, religieuse, sociale et économique*. Paris: Mouton, 1973.

ed. *Le cartulaire de Vierzon*. Paris: Presses universitaires de France, 1963.

Devroey, Jean-Pierre. "Food and politics." In M. Montanari, ed., *A Cultural History of Food in the Medieval Age*, 73–90. London: Berg, 2012.

Dewar, Margaret E. *Industry in Trouble: The Federal Government and the New England Fisheries*. Philadelphia: Temple University Press, 1983.

di Gangi, Giorgio. *L'attivita mineraria e metallurgica nelle Alpi occidentali italiane nel medioevo: Piemonte e Valle d'Aosta: fonti scritte e materiali*. Oxford: BAR Archaeopress, 2001.

Diago Hernando, Máximo. "Relaciones comerciales de la corona de Aragón con la Andalucía atlántica durante el siglo XIV y primera mitad del XV," Historia. Instituciones. Documentos *[Seville]* 27 (2000), 1–54.

Diaz, Henry F., R. T. Grigo, D. Marriopedro, E. Xoplaki, M. Hughes, and M. E. Mann. "The medieval warm period redux: Where and when was it warm?" *PAGES news* 19:1 (March 2011), 32.

Dickinson, William A. "Changing times: the Holocene legacy," *Environmental History* 5 (2000), 483–502.

Diecke, M. "Findings concerning the environmental history of the Harz Mountains and the utilisation of mineral resources." In C. Segers-Glocke and Harald Witthoft, eds. *Aspects of Mining and Smelting in the Upper Harz Mountains (up to the 13th/14th Century) – in the Early Times of a Developing European Culture and Economy*, pp. 66–78. Sachüberlieferung und Geschichte, Bd. 33; Arbeitshefte zur Denkmalpflege in Niedersachsen, Bd. 22. St. Katharinen: Scripta Mercaturae Verlag, 2000.

Diefenbach, Lorenz. *Glossarium Latino-Germanicum mediae et infimae aetatis.* Frankfurt a.M.: Baer, 1857; facsimile reprint Darmstadt: Wissenschaftliche Buchgesellschaft, 1997.

Diekamp, Wihelm, ed. *Die Vitae Sancti Liudgeri.* Die Geschichtsquellen des Bisthums Münster; Bd. 4. Münster: Theissing, 1881.

Dijk, J. van, and B. Beerenhout. "Het botmateriaal van de campagne 2004." In J. Van Dijk, B. Beerenhout, and A. L. De Sitter-Homans, *Het onderzoek op het terrein van de voormalige Winston bioscoop te Hoorn. Het ecologisch materiaal van de campagne 2004*, 19–75. Verslagen van de Archeologische Dienst Hoorn, 4. Hoorn: Gemeente Hoorn, Archeologische Dienst, 2006.

Dill, William A. *Inland Fisheries of Europe.* EIFAC Technical Paper, No. 51. Rome: FAO, 1990.

Dillard, Heath. *Daughters of the Reconquest: Women in Castilian Town Society, 1100–1300.* Cambridge: Cambridge University Press, 1984.

Dirlmeier, Ulf. "Die kommunalpolitischen Zuständigkeiten und Leistungen süddeutscher Städte im Spatmittlelalter (vor allem auf dem Gebiet der Ver- und Entsorgung)." In J. Sydow, ed., *Städtische Versorgung und Entsorgung im Wandel der Geschichte*, 113–150. Sigmaringen: J. Thorbecke, 1981.

*Untersuchungen zu Einkommensverhältnissen und Lebenshaltungskosten in oberdeutschen Städten des Spätmittelalters (Mitte 14.bis Anfang 16.Jahrhundert).* Abhandlungen des Heidelberger Akademie der Wissenschaften, phil.-hist. Klasse, 1978, nr. 1. Heidelberg: Winter, 1978.

Ditchburn, David. *Scotland and Europe: The Medieval Kingdom and Its Contacts with Christendom, c.1245–1545*, vol. 1: *Religion, Culture and Commerce.* East Linton: Tuckwell, 2001.

Ditchfield, Philip. *La culture matérielle médiévale. L'italie méridionale byzantine et normande.* Rome: Ecole française, 2007.

Dobney, Keith, Deborah Jaques, James Barrett, and Cluny Johnstone. *Farmers, Monks and Aristocrats: The Environmental Archaeology of Anglo-Saxon Flixborough.* Excavations at Flixborough, vol. 3. Oxford: Oxbow Books, 2007.

Dobney, K. M., S. D. Jaques, and B. G. Irving. *Of Butchers and Breeds: Report on Vertebrate Remains from Various Sites in the City of Lincoln*. Lincoln Archaeological Studies, 5: The Animal Bones from Lincoln. Lincoln: City of Lincoln Archaeology Unit, 1995.

Dobozy, Maria, ed. and tr. *The Saxon Mirror: A Sachsenspiegel of the Fourteenth Century*. Philadelphia: University of Pennsylvania Press, 1999.

Dollinger, Philippe. *The German Hansa*, rev. ed. and tr. D. S. Ault and S. H. Steinberg. London and Stanford: Stanford University Press, 1970 [original ed. *La Hanse*. Paris, 1964].

Donati, Angela. "Dal mare al fiume: la pesca sul finire dell'antichità." In Angela Donati and Paolo Pasini, eds., *La pesca realtà e simbolo: fra tardo antico e Medioevo*, pp. 7–29. Pesca e pescatori, vol. 2. Milano: Leonardo Arte, 1999.

Dotterweich, Markus, and Stefan Dreibrodt. "Past land use and soil erosion processes in central Europe," *PAGES news* 19:2 (July 2011), 49–51.

Douët-d'Arcq, L.-C. *Comptes de l'Hôtel des rois de France aux XIVe et XVe siècles*. Paris: Société de l'Histoire de France, 1865.

Dowling, Abigail. "Landscape of luxury: Mahaut d'Artois's (1302–1329) management and use of the park of Hesdin." In Albrecht Classen and Christopher Clason, eds., *Rural Space in the Middle Ages and Early Modern Age: The Spatial Turn in Premodern Studies*, 367–387. Boston: De Gruyter, 2012.

Drew, Katherine F., ed. and tr. *The Laws of the Salian Franks*. Philadelphia: University of Pennsylvania Press, 1991.

ed. and tr. *The Lombard Laws*. Philadelphia: University of Pennsylvania Press, 1973.

Driesch, Angela von den. "Fischreste aus der slawisch-deutschen Fürstenburg auf dem Weinberg in Hitzacker (Elbe)," *Neue Ausgrabungen und Forschungen in Niedersachsen* 15 (1982), 395–423.

Dubravius, Jan [Jan Skalý z Doubravky a Hradiště]. *De piscinis ad Antonium Fuggerum*. Vratislaviae: Andreas Vinglerus excudebat, 1547. Also as *De Piscinis et Piscium qui in eis aluntur naturis libri quinque … ad illustrem virum Antonium Fuggerum. Item Xenocratis de alimento ex aquatilibus Graece & Latine nun primum aeditus: cum scholijs Conradi Gesneri*. [Zürich: 1559), which further appeared in 1596, 1623, and 1671. Ed. Anežka Schmidtová in *Sborník filologický ČSAV* 1, supplement 1 (1953), 11–45. Loosely tr. as *A New Book of Good Husbandry* (London, 1599), which is available as a facsimile reprint in *Three Books on Fishing (1599–1659)*, J. Milton French, intro. (Gainesville, FL, 1962). Tr. Andrzej Progę as *Janusza Dubrawiusza o Rybnikach y Rybach* (Kraków: w drukarnia Wojciecha Siekielowica, [*c.* 1600]).

Duby, Georges. "Le budget de l'abbaye de Cluny entre 1080 et 1155. Économie domaniale et économie monétaire,' *Annales: e.s.c.* 7 (1952), 155–171 [reprinted in his *Hommes et structures du moyen âge. Recueil d'articles*, 61–82. Paris: Mouton, 1973, and later editions].

"Un inventaire des profits de la seigneurie clunisienne à la mort de Pierre le Vénérable," *Studia Anselmiana* 40 (1956), 129–140 [reprinted in his *Hommes et structures du moyen âge. Recueil d'articles*, 87–101. Paris: Mouton, 1973, and later editions].

DuCange, Charles, et al. *Glossarium mediae et infimae latinitatis conditum a Carolo DuFresne, domino Du Cange …* Paris: Fermin Didot, 1840–1850.

Dufeu, Val. *Fish Trade in Medieval North Atlantic Societies: An Interdisciplinary Approach to Human Ecodynamics*. Amsterdam: Amsterdam University Press, 2018.

Duft, Johannes. *Der Bodensee in Sankt-Galler Handschriften: Texte und Miniaturen aus St. Gallen*. Bibliotheca Sangallensis, vol. 3. 3d ed. rev. St. Gallen and Sigmaringen: Thorbecke, 1979.

Dulvy, N. K., S. I. Rogers, S. Jennings, V. Stelzenmüller, S. R. Dye, and H. R. Skjoldal. "Climate change and deepening of the North Sea fish assemblage: a biotic indicator of warming seas," *Journal of Applied Ecology* 45 (2008), 1029–1039.

Dunin-Wąsowicz, Teresa. "Natural environment and human settlement over the Central European Lowland in the 13th century." In Peter Brimblecombe and Christian Pfister, eds., *The Silent Countdown: Essays in European Environmental History*, 92–105. Berlin, Heidelberg, New York: Springer, 1990.

Duplés-Agier, H. "Ordonnances inédites de Philippe-le-Bel et de Philippe-le-Long sur la police de la pêêche fluviale." *Bibliotheque de l'École des Chartes* 14 (1852), 43–55.

Durand, Aline., ed. *Jeux d'eau. Moulins, meuniers et machines hydrauliques, XI^e–XX^e siècle. Études offertes à Georges Comet*. Aix-en-Provence: Publications de l'Université de Provence, 2008.

*Les paysages médiévaux du Languedoc (Xe–XIII^e siècles)*. Toulouse: Presses Universitaires du Mirail, 1998

Durand, Vincent. "De l'établissement des étangs dans la plaine du Forez," *Bulletin de la Diana* 5 (1889–1890), 101–108.

Dyer, Christopher C. "The consumer and the market in the later Middle Ages," Economic History Review. 2d series 42 (1989), 305–327.

"The consumption of fresh-water fish in medieval England." In Aston, ed., *Medieval Fish*, 27–38. 1988.

"Do household accounts provide an accurate picture of late medieval diet and food culture?" In Emmanuelle Rassart-Eeckhout, et al., eds., *La vie matérielle au moyen âge. L'apport des sources litteraires, normatives et de la pratique*, 109–125. Actes du Colloque international de Louvain-la-Neuve 3–5 octobre 1996. Université Catholique de Louvain, Publications de l'Institut d'Études Médiévales: Textes, Études, Congrès, vol. 18. Louvain-la-Neuve: l'Institut d'Études Médiévales, 1997.

"Les régimes alimentaires en Angleterre, XIII^ème – XV^ème siècles." In Menjot, ed., *Manger et boire*, vol. 2, 263–274. 1984.

*Standards of Living in the Later Middle Ages: Social Change in England c.1200–1520*. Cambridge: Cambridge University Press, 1989.

Eadmer of Canterbury. *The Life of St. Anselm, Archbishop of Canterbury*, Ed. and tr. Richard W. Southern. London: Nelson, 1962, and Oxford: Clarendon, 1972.

Ebel, E. "Fischerei und Fischereimethoden: Literarisches zur Fischerei im Norden." In Heinrich Beck et al., eds., *Reallexikon der Germanischen Altertumskunde*, vol. 9, pp. 140–146. Berlin and New York: de Gruyter, 1994.

Echániz Sans, María. "La alimentación de los pobres asistidos por la pia almoina de la catedral de Barcelona según el libro de cuentas de 1283–1284." In *Alimentació i societat a la Catalunya medieval*, Anuario de Estudios Medievales, Anex 20, 173–261. Barcelona: Instituto de Historia Medieval de España, 1988.

Echevarria Alonso, Maria José. *La actividad comercial del puerto de Santander en el siglo XVII*, ed. Jose Ramon Ruiz Manso. Biblioteca Navalia, vol. 4. Santander: Autoridad Portuaria de Santander, 1995.

Edelman, T. "Oude ontginningen van de veengebieden in de Nederlandse kuststrook," *Tijdschrift voor Economische en Sociale Geografie* 49 (1958), 239–245.

Egbert, Virgina W. *On the Bridges of Mediaeval Paris: A Record of Early Fourteenth-Century Life*. Princeton: Princeton University Press, 1974.

Egli, Johannes. "Ein st. gallischer Küchenzettel aus der Wende des ersten Jahrtausends n. Chr.: Kulturhistorische Studie," *Schweizer Rundschau* 9 (1908/9), 341–368.

Egloff, Wilhelm. *Le Paysan Dombiste. Étude sur la vie, les travaux des champs et le parler d'un village de la Dombes, Versailleux (Ain)*. Société de Publications Romanes et Françaises sous la direction de Mario Roques, vol. 20. Paris: Droz, 1937.

Ehlers, Anke. *Des Teufels Netz. Untersuchung zum Gattungsproblem*. Stuttgart: Kohlhammer, 1973.

Ehlert, Trudi, ed. *Küchenmeisterei. Edition, Übersetzung und Kommentar zweier Kochbuch-Handschriften des 15. Jahrhunderts. Solothurn S490 und Köln Historisches Archiv GB4°27. Mit reprographischen Nachdruck der Kölner Handschrift*. Kultur, Wissenschaft, Literatur. Beitgräge zur Mittelalterforschung, Bd. 21. Frankfurt: Peter Lang, 2010.

Eike von Repgow. *Sachsenspeigel. Landrecht, Lehnrecht*, ed. Karl A. Eckhardt, 2d ed. rev., 2 vols. MGH Fontes Iuris Germanici Antiqui, nova series, vol. 1:1–2. Göttingen: Musterschmidt-Verlag, 1955–1956.

Ekkehard IV of St. Gallen. *Benedictiones ad mensas*, ed. Johannes Egli as "Der Liber Benedictionum Ekkeharts IV. nebst den kleinern Dichtungen aus dem Codex Sangallensis 393," *Mitteilungen zur vaterländischen Geschichte*, hggb vom Historischen Verein in St. Gallen 31:4 (1909), 282–325.

*Casus Sancti Galli. St. Galler Klostergeschichten*, ed. and tr. Hans F. Haefele. Darmstadt: Wissenschaftliche Buchgesellschaft, 1980.

Elsas, Moritz J. *Umriss einer Geschichte der Preise und Löhne in Deutschland vom ausgehenden Mittelalter bis zum Beginn des neunzehnten Jahrhunderts*. 2 vols. in 3. Leiden: Sijthoff, 1936–1949.

Endrès, André. "Alimentation d'assistance à l'Hôtel-Dieu de Meaux en 1527." In *Bulletin Philologique et Historique (jusqu'à 1610) du Comité des Travaux Historiques et Scientifiques, 99ème Congrès des Sociétés savantes, Tours, 1968*, vol. 1: *Les Problèmes de l'alimentation*, 209–256. Paris: BNF, 1971.

"Un vivier naturel dans une anse de la Marne en aval de Meaux. Les marais de Lésches," *Bulletin Philologique et Historique (jusqu'a 1610) du Comité des Travaux Historiques et Scientifiques*, anneé 1967, vol. 1, 184–204. Paris: BNF, 1969.

Engelhard, Georg H., Ruth H. Thurstan, Brian R. MacKenzie, Heidi K. Alleway, R. Colin, A. Bannister, Massimiliano Cardinale, Maurice W. Clarke, Jock C. Currie, Tomaso Fortibuoni, Poul Holm, Sidney J. Holt, Carlotta Mazzoldi, John K. Pinnegar, Saša Raicevich, Filip A. M. Volckaert, Emily S. Klein, and Ann-Katrien Lescrauwaet, "ICES meets marine historical ecology: placing the history of fish and fisheries in current policy context," *ICES Journal of Marine Science* 73:5 (May/June 2016), 1386–1403.

Enghoff, Inge Bødker. "Fishing in the Baltic region from the 5th century BC to the 16th century AD: evidence from fish bones," *Archaeofauna* 8 (1999), 41–85.

"Fishing in the Southern North Sea region from the 1st to the 16th century AD: evidence from fish bones," *Archaeofauna* 9 (2000), 59–132.

"Herring and cod in Denmark." In Barrett and Orton, eds., *Cod and Herring*, 133–155. 2016.

"A medieval herring industry in Denmark: the importance of herring in eastern Denmark," *Archaeofauna* 5 (1996) (special issue *Ichthyoarchaeology: Fish and the Archaeological Record: Proceedings of the Eighth Meeting of the ICAZ Fish Remains Working Group – Madrid 3–11 October 1995)*, 43–47.

Epstein, Stephan R. *An Island for Itself: Economic Development and Social Change in Late Medieval Sicily*. Cambridge: Cambridge University Press, 1992.

Epstein, Steven A. *Wage Labor and Guilds in Medieval Europe*. Chapel Hill: University of North Carolina Press, 1991.

Erasmus, Desiderius [of Rotterdam]. *Familiarum colloquiorum formuale*, tr. Craig R. Thompson as *The Colloques of Erasmus*. Chicago: University of Chicago Press, 1965.

Ermisch, Hubert, and Robert Wuttke, eds. *Haushaltung in Vorwerken. Ein landwirtschaftliches Lehrbuch aus der Zeit des Kurfursten August von Sachsen*. Leipzig: Teubner, 1910.

Ervynck, Anton. "Medieval castles as top-predators of the feudal system: an archaeozoological approach," *Château Gaillard. Etudes de Castellologie médiévale* [Caen], 15 (1992), 151–159.

Ervynck, Anton, and D. van Damme. "Archeozoölogisch onderzoek op het Kasteel van Laarne; hoezo?" *Castellum [Laarne]* 5:1 (1988), 10–17.

Ervynck, Anton, Mathieu Boudin, Tess van den Brande, and Mark Van Strydonck. "Dating human remains from the historical period in Belgium: diet changes and the impact of marine and freshwater reservoir effects," *Radiocarbon* 56:2 (2014), 779–788.

Ervynck, Anton, Mathieu Boudin, and Wim Van Neer. "Assessing the radiocarbon freshwater reservoir effect for a Northwest-European river system (the Schelde Basin, Belgium)." *Radiocarbon* 60:2 (2018), 1–23.

Ervynck, Anton, and Wim Van Neer. "De voodselvoorziening in de Sint-Salvatorsabdij te Ename (stad Oudenaarde, prov. Oost-Vlaanderen)

I. Beenderen onder een keukenvloer (1450–1550 A.D.)," *Archeologie in Vlaanderen* 2 (1992), 419–434.

"A preliminary survey of fish remains in medieval castles, abbeys and towns of Flanders (Belgium)." In Dirk Heinrich, ed., *Archaeo-Ichthyological Studies. Papers presented at the 6th Meeting of the I.C.A.Z. Fish Remains Working Group, Schleswig*, 303–308. Beiheft zur *Offa*, 51. Neumünster: Wacholz, 1994.

Ervynck, Anton, Wim Van Neer, Heide Hüster Plogmann, and Jörg Schibler. "Beyond affluence: the zooarchaeology of luxury," *Luxury Foods, a thematic issue of World Archaeology* 34:3 (2003), 428–441.

Ervynck, Anton, Wim Van Neer, and Marnix Pieters. "How the North was won (and lost again). Historical and archaeological data on the exploitation of the North Atlantic by the Flemish fishery." In Rupert A. Housley and Geraint Coles, eds., *Atlantic Connections and Adaptations: Economies, Environments and Subsistence in Lands Bordering the North Atlantic*, 230–239. Symposia of the Association for Environmental Archaeology; no. 21. Oxford: Oxbow, 2004.

Esper, Jan, Ulf Buntgen, Claudia Hartl-Meier, Clive Oppenheimer, and Lea Schneider. "Northern Hemisphere temperature anomalies during the 1450s period of ambiguous volcanic forcing." *Bulletin of Volcanology* 79 (2017): article 41. doi:10.1007/s00445-017-1125-9.

Evans, D. "Aspects of conservation of European freshwater fishes. " In Juraj Holčik, ed., *The Freshwater Fishes of Europe*, vol. 1, part 2: pp. 107–119. Wiesbaden: Aula Verlag, 1989.

Evans, Joan. *Monastic Life at Cluny, 910–1157*. Oxford: Oxford University Press, 1931 (reprinted Archon Books 1968).

Evergates, Theodore, ed., in collaboration with Giles Constable. *The Cartulary and Charters of Notre-Dame of Homblières*. Cambridge, MA: Medieval Academy of America, 1990.

Everhart, W. Harry, and William D. Youngs. *Principles of Fisheries Science*, 2d ed. rev. Ithaca: Cornell University Press, 1981.

Fabijanec, Sabine Florence. "Fishing and the fish trade on the Dalmatian coast in the late Middle Ages." In Tønnes Bekker-Nielsen and Ruthy Gertwagen., eds. *The Inland Seas. Towards an Ecohistory of the Mediterranean and the Black Sea*, 369–386. Geographica Historica, 35. Stuttgart: Franz Steiner Verlag, 2016. [A revised version of the author's "La pesca e il commercio del pesce lungo la costa della Dalmazia allaf ine del Medioevo.." In Ruthy Gertwagen, Saša Raicevich, Tomaso Fortibuoni, Otello Giovanardi, eds. *Le interazioni tra uomo ed ambiente nel Mediterraneo dall'Epoca Romana al XIX secolo: una visione storica ed ecologica delle attivita di pesca*, 106–115. Atti del II Workshop Internazionale HMAP del Mediterraneo e Mar Nero. Chioggia, 27–29 settembre 2006. Supplemento ai Quaderni ex ICRAM. Chioggia: ISPRA – Istituto Superiore per la Protezione e la Ricerca Ambientale (Chioggia), 2008.

"Fishing, consumption, and processing of fish and shellfish in the Eastern Adriatic through the Long Middle Ages," *Imago Temporis. Medium Aevum* 15 (2021), 167–191.

Faccioli, Emilio, ed. *Arte della cucina. Libri di recetti, testi sopra lo scalco, il trinciante e i vini dal XIV al XIX secolo.* 2 vols. Milan: Il Polifilo, 1966.

Fagan, Brian. *Fish on Friday: Feasting, Fasting and the Discovery of the New World.* New York: Basic Books, 2006.

Faget, Daniel. "Le poison et la poudre. Passé du braconnage haleiutique en Méditerranée (XVIII–XXe siècles)." In D. Faget and M. Sternberg, eds., *Pêches méditerranéennes. Origines et mutations, protohistoire–XXIe siècle,* 169–188. Paris: Karthala; Aix-en-Provence: Maison méditérranéenne des sciences de l'homme, 2015

Faget, Daniel, and Miriam Sternberg, eds. *Pêches méditerranéennes. Origines et mutations, protohistoire–XXIe siècle.* Paris: Karthala; Aix-en-Provence: Maison méditérranéenne des sciences de l'homme, 2015.

Fagniez, Gustave Charles. *Etudes sur l'industrie et la classe industrielle à Paris au XIIIe et au XIVe siècle.* Paris: Bibliothéque de l'Ecole des Hautes Études, 1877. reprinted New York: B. Franklin, [1970].

FAO, *The State of World Fisheries and Aquaculture 2014.* Rome: United Nations Food and Agriculture Organization, 2014. www.fao.org/3/a-i3720e.pdf (accessed 27 July 2015).

Farmer, Sharon. "Aristocratic power and the 'natural' landscape: the garden park at Hesdin, ca.1291–1302," *Speculum* 88/3 (July 2013), 644–680.

*Surviving Poverty in Medieval Paris: Gender, Ideology, and the Daily Lives of the Poor.* Ithaca: Cornell University Press, 2002.

Fauchald, Per. "Predator–prey reversal: a possible mechanism for ecosystem hysteresis in the North Sea?" *Ecology* 91:8 (Aug. 2010), 2191–2197.

Faugeron, Fabien. *Nourrir la ville: ravitaillement, marchés et métiers de l'alimentation à Venise dans les derniers siècles du Moyen Âge.* Bibliothéque des Écoles françaises d'Athènes et de Rome, 362. Rome: École française de Rome, 2014.

Fauroux, Marie, ed. *Recueil des actes des ducs de Normandie de 911–1066.* Société des antiquaires de Normandie. Mémoires 4. sér., v. 6, t. 36. Caen: Société d'impressions Caron, 1961.

Favreau, Robert. "Fêtes et jeux à Poitou à la fin du moyen âge." In *Jeux, Sports et Divertissements au Moyen Âge et* à *l'Âge Classique,* Actes du 116e Congrès Nationale des Sociétés savantes, Chambery, 1991, 25–43. Paris: Editions de CTHS, 1993.

Fawtier, Robert, and François Maillard, eds. *Comptes royaux (1285–1314).* 3 vols. Recueil des historiens de la France, Documents financiers, no. 3. Paris: Imprimerie nationale, 1953–1956.

Felici, Enrico. *Thynnos. Archeologia della tonnara Mediterranea.* Bari: Edipuglia, 2018.

Fernández -Armesto, Felipe. *Near a Thousand Tables: A History of Food.* New York: Free Press, 2002.

Fernández-Armesto, Felipe, and Daniel L. Smail. "Shared Substance: Food." In Andrew Shyrock and Daniel Lord Smail, eds., *Deep History: The Architecture of Past and Present,* 131–159. Berkeley and Los Angeles: University of California Press, 2011.

Ferreira Priegue, Elisa. *Galicia en el Comercio Maritimo Medieval*. Santiago and La Coruña: Fondación 'Pedro Barrie de la Maza', 1988.

"Pesca y economía regional en Galicia." In *Pesca en la Edad Media*, 11–34. Madrid: Centro de Estudios Medievales de la Universidad de Murcia, 2009.

Feyl, Anita. "Das Kochbuch des Eberhard von Landshut (erste Hälfte des 15. Jhs.)," *Ostbairische Grenzmarken* 5 (1961) 352–366.

"Das Kochbuch Meister Eberhards. Ein Beitrag zur altdeutschen Fachliteratur." PhD dissertation, Freiburg im Breisgau, 1963.

Fichtenau, Heinrich. *Living in the Tenth Century: Mentalities and Social Orders*, tr. Patrick J. Geary. Chicago: University of Chicago Press, 1991.

Field, David B., Tim R. Baumgartner, Vicente Ferreira, Dimitri Gutierrez, Hector Lozano-Montes, Renato Salvatteci, and Andy Sutar. "Variability from scales in marine sediments and other historical records." In Checkley et al., eds., *Climate Change and Small Pelagic Fish*, 45–63. 2009.

Filotas, Bernadette. *Pagan Survivals, Superstitions and Popular Cultures in Early Medieval Pastoral Literature*. Toronto: PIMS, 2005.

Filuk, J. "Biologiczno-rybacka charakterystyka ichtiofauny zalewu wiślanego na tle badań paleoichtiologicznych, historycznych i wspólczesnych," *Pomorania antiqua* 2 (1968), 119–175.

Fischer-Kowalski, Marina, and Helga Weisz. "Society as hybrid between material and symbolic realms. toward a theoretical framework of society–nature interactions," *Advances in Human Ecology* 8 (1999), 215–251.

Fleming, Robin. "The new wealth, the new rich, and the new political style in late Anglo-Saxon England," *Anglo-Norman Studies* 23 (2000), 1–22.

*Fleta*, ed. H. G. Richardson and G. O. Sayles. *Selden Society*, vol. 72. London: Selden Society, 1955.

Fois, B. "Annotazioni sull'alimentazione nella Sardegna del trecento: i prodotti, le vivande, prezzi e salari,." In Menjot, ed., *Manger et boire*, I, 183–197. 1984.

Ford, Gordon B., Jr., tr. *The Ruodlieb. The First Medieval Epic of Chivalry from Eleventh-Century Germany*. Leiden: Brill, 1965.

Ford, Jennifer S., and Ransom A. Myers, "A global assessment of salmon aquaculture impacts on wild salmon," *PLoS Biology* 6:2 (February 2008). e33.doi:10.1371/journal.pbio.0060033.

Forquin, Guy. "Le temps de la croissance." In George Duby and Armand Wallon, eds., *Histoire de la France rurale*, vol. 1: *La formation des campagnes françaises des origins au XIV^e^ siècle*, 377–552. Paris: Editions du Seuil, 1975.

Fossier, Robert. "L'essor économique de Clairvaux." In Commission d'Histoire de l'Ordre de Cîteaux, ed., *Bernard de Clairvaux*, vol. 3, pp. 95–114. Paris: Editions Alsatia, 1953.

*La terre et les hommes en Picardie jusqu'à la fin du XIII^e^ siècle*. Publications de la Faculté des Lettres et Sciences Humaines de Paris – Sorbonne, Série "Recherches," vols. 48–49. Paris: Béatrice-Nauwelaerts, 1968.

Fouquet, Gerhard. "'Wie die kuchenspise sin solle' – Essen und Trinken am Hof des Speyerer Bischofs Matthias von Rammung (1464–1478)," *Pfälzer Heimat* 39 (1988), 12–27.

Fournial, Etienne. *Les villes et l'économie d'échange en Forez aux XIII<sup>e</sup> et XIV<sup>e</sup> siècles*. Paris: Les presses du palais royal, 1967.

Fox, Harold. *The Evolution of the Fishing Village: Landscape and Society along the South Devon Coast*. Leicester Explorations in Local History, 1. Oxford: Leopard's Head Press, 2001.

"Fishermen and mariners." In Miri Rubin, ed., *Medieval Christianity in Practice*, 77–80. Princeton: Princeton University Press, 2009.

Francheschini, Fabrizio. *Lago, padule, fiume: il lessico delle pesche tradizionali nella Toscana occidentale*. Perugia: RUX Edizioni, 1994.

Franco, Isabel Maria Madureira. "Les dynamiques familiales et sociales dans un village de pêcheurs des environs de Porto (1449–1497)." In C. Beattie, A. Maslakovic, and S. R. Jones, eds., *The Medieval Household in Christian Europe, c.850–c.1550: Managing Power, Wealth, and the Body*, 271–292. Turnhout: Brepols, 2003.

Frangoudes, Katia. "I *Prud'homies*: associazioni professionali di pescatori francesi nel Mediterraneo. / The *prud'homies*: fishers institutions in French Mediterranean Sea." In Ruthy Gertwagen, Saša Raicevich, Tomaso Fortibuoni, and Otello Giovanardi, eds. *Le interazioni tra uomo ed ambiente nel Mediterraneo dall'Epoca Romana al XIX secolo: una visione storica ed ecologica delle attivita di pesca*, 46–49. Atti del II Workshop Internazionale HMAP del Mediterraneo e Mar Nero. Chioggia, 27–29 settembre 2006. Supplemento ai Quaderni ex ICRAM. Chioggia: ISPRA – Istituto Superiore per la Protezione e la Ricerca Ambientale (Chioggia), 2008.

Frank, Kenneth T., Brian Petrie, Jae S. Choi, and William C. Leggett, "Trophic cascades in a formerly cod-dominated ecosystem," *Science* 308 (2005), 1621–1623.

Franke, Jasper G., Johannes P. Werner, and Reik V. Donner. "Reconstructing Late Holocene North Atlantic atmospheric circulation changes using functional paleoclimate networks" *Climate of the Past* 13 (2017), 1593–1608. https://doi.org/10.5194/cp-13-1593-2017

Franklin, Alfred. *Dictionnaire historique des arts, métiers et professions exercés dans Paris depuis le XIII<sup>e</sup> siècle*. 2 vols. paginated continuously. Paris: Welter, 1905–1906. Reprinted Marseille: Laffitte, 1987.

ed. *Les rues et les cris de Paris au XIII<sup>e</sup> siècle … d'après les documents de la Bibliothèque nationale*. Paris: L. Willem, 1874. Reprinted Paris: Editions de Paris, 1984.

Frantzen, Allen J. *Food, Eating, and Identity in Early Medieval England*. Anglo-Saxon Studies, no. 22. Rochester, NY: Boydell, 2014.

Franz, Günther, ed. *Quellen zur Geschichte des Bauernkrieges*. Darmstadt: Wissenschaftliche Buchgesellschaft, 1963.

Freedman, Paul, ed. *Food: The History of Taste*. Berkeley and Los Angeles: University of California Press, 2007.

"Lamprey and herring," *Imago Temporis. Medium Aevum* 15 (2021), 193–212.

Freely, John. *Jem Sultan: The Adventures of a Captive Turkish Prince in Renaissance Europe*. London: Harper Perennial, 2005.

Fréminville, J. de. "Les comptes du maître des étangs de Forez de 1400 à 1430," *Bulletin de la Diana* 7 (1893–1894), 5–20.

Freudlsperger, Hans. "Kurze Fischereigeschichte des Erzstiftes Salzburg," *Mitteilungen der Gesellschaft für Salzburger Landeskunde* 76 (1936), 81–128 and 77 (1937),145–175.

Friberg, Nils. *Stockhom i bottniska farvatten: Stockholms bottniska handelsfält under senmedel tiden och Gustav Vasa*. Uppsala: Almqvist und Wiksell, 1983.

Frieser, Claudia. "Die spätmittelalterlichen Abwasserkanäle und heimlichen Gemächer Nürnbergs. Ein anrüchige Geschichte." In B. Friedel and C. Frieser, eds., *Nürnberg – Archaeologie und Kulturgeschichte*, 190–198. Buchenbach: Verlag Dr. Faustus,1999.

Fritsch, Susanne. *Das Refektorium im Jahreskreis. Norm und Praxis des Essens in Klöstern des 14. Jahrhunderts*. Veröffentlichungen des Instituts für Österreichische Geschichtsforschung 50. Vienna/Munich: R. Oldenbourg, 2008.

Fudge, John D. *Cargoes, Embargoes and Emissaries. The Commercial and Political Interaction of England and the German Hanse 1450–1510*. Toronto: University of Toronto Press, 1995.

Fuller, B. T., G. Muldner, W. Van Neer, A. Ervynck, and M. P. Richards. "Carbon and nitrogen stable isotope ratio analysis of freshwater, brackish and marine fish from Belgian archaeological sites (1st and 2nd millennium AD)," *Journal of Analytical Atomic Spectrometry* 27 (2012), 807–820.

Fumagalli, Vito. *Landscapes of Fear: Perceptions of Nature and the City in the Middle Ages*. Oxford: Polity Press, 1994.

Gabriel, Philipp. "Die Hanstadt Lübeck, der Hering und das Klima." In Margit Mersch, ed., *Mensch-Natur-Wechselwirkungen in der Vormoderne. Beiträge zur mittelalterlichen und frühneuzeitlichen Umweltgeschichte*, 165–183. Göttingen: Universitätsverlag, 2016.

Gabriel, Sónia. "Fish assemblages in early medieval periods." In Alexandra Chavarría Arnau, ed., *Ricerche sul Centro Episcopale di Padova Scavi 2011–2012*, 123–129. Mantova: Società Archeologica, 2017.

Galik, Alfred, and Hans Christian Küchelmann. "Fischreste aus der mittelalterlichen Bremer Altstadt," *Bremer Archäologische Blätter*, Neue Folge 7 (2005–2008), 215–231.

Galik, Alfred, Gertrud Haidvogl, Laszlo Bartosiewicz, Gabor Guti, and Mathias Jungwirth. "Fish remains as a source to reconstruct long-term changes of fish communities in the Austrian and Hungarian Danube," *Aquatic Sciences* 77 (2015), 337–354.

Galimova, Dyliara N., Igor V. Askeyev, Oleg V. Askeyev, Danijela Popovic, and Hanna Panagiotopoulou. "A Study of the fish bones from the medieval town – Staraya Ladoga,." In *Fishing through Time: Archaeoichthology, Biodiversity, Ecology, and Human Impact on Aquatic Environments*. 63–64. Lisbon: 18th Fish Remains Working Group, International Council for Archaeozoology, 2015.

Galinié, Henri, et al. "Quelques aspects documentés des relations entretenues par les habitants de Tours avec la Loire du I[e] au XIIe siècle." In Joëlle Burnouf and Philippe Leveau, eds., *Fleuves et marais, une histoire au croisement de la nature et le la culture. Société préindustrielle et milieux fluviaux, lacustres et palustres: practiques socials et hydrosystèmes*, 127–136. Paris: CTHS, 2004.

Galloway, James A. "Fishing the Thames estuary in the later Middle Ages: environment, technology and the metropolitan market for fish c. 1250–1550," *Imago Temporis. Medium Aevum* 15 (2021), 243–271.

"London and the Thames estuary in the later Middle Ages: economic and environmental change,." In A. Wilkin, J. Naylor, D. Keene, and A.-J. Bijsterveld, eds. *Town and Country in Medieval North Western Europe: Dynamic Interactions*, 119–144. Turnhout: Brepols, 2015.

"Storm flooding, coastal defence and land use around the Thames estuary and tidal river c.1250–1450," *Journal of Medieval History* 35:2 (2009), 171–188.

"Storms, economics and environmental change in an English coastal wetland; the Thames estuary *c.*1250–1550." In Erik Thoen et al., eds., *Landscapes or Seascapes?. The History of the Coastal Environment in the North Sea Area Reconsidered*, 379–396. Turnhout: Brepols, 2013.

Gallus Anonymous, "Chronicon Poloniae," ed. Karol Maleczyński in *Monumenta Poloniae Historica*, series nova, 2. Kraków: PAN, 1952.

García Cañón, Pablo. "La pesca en los concejos de Luna de Yuso u de Suso (León) durante el siglo XV: luchas e intereses entre señores y pescadores." In M. I. del Val Valdivieso, ed., *Vivir del agua en las ciudades medievales*, 181–202. Valladolid: Universidad de Valladolid, 2006.

García Vargas, Enrique, and David Florido del Corral. "The origin and development of tuna fishing nets (*Almadrabas*)." In Tønnes Bekker-Nielsen and Darío Bernal Casasola, eds., *Ancient Nets and Fishing Gear: Proceedings of the International Workshop on "Nets and Fishing Gear in Classical Antiquity: A First Approach," Cádiz, 15–17 November 2007*, 205–227. Cádiz: Universidad de Cádiz: Servicio de Publicaciones; Aarhus: Aarhus University Press, 2010.

Gardiner, Mark. "The character of commercial fishing in Icelandic waters in the fifteenth century." In Barrett and Orton, eds., *Cod and Herring*, 80–90. 2016.

Garrido i Escobar, Alfons. "'Anar al petardo': la pêche à l'explosif sur la Costa Brava." In D. Faget and M. Sternberg, eds., *Pêches méditerranéennes. Origines et mutations, protohistoire–XXI^e^ siècle*, 189–202. Paris: Karthala; Aix-en-Provence: Maison méditérranéenne des sciences de l'homme, 2015.

Garrido i Escobar, Alfons, and Juan Luis Alegret. "Evoluzione delle tecniche di pesca e dell'uso del territorio: un panoramica a lungo termine." In Ruthy Gertwagen, Saša Raicevich, Tomaso Fortibuoni, and Otello Giovanardi, eds. *Le interazioni tra uomo ed ambiente nel Mediterraneo dall'Epoca Romana al XIX secolo: una visione storica ed ecologica delle attivita di pesca*, 116–126. Atti del II Workshop Internazionale HMAP del Mediterraneo e Mar Nero. Chioggia, 27–29 settembre 2006. Supplemento ai Quaderni ex ICRAM. Chioggia: ISPRA – Istituto Superiore per la Protezione e la Ricerca Ambientale (Chioggia), 2008.

Garrido i Escobar, Alfons, and Marcel Pujol i Hamelink. "Changements techniques et conflits de pêche dans la Catalogne de l'époque moderne." In Buti et al., eds. *Moissonner la mer*, 21–36. 2018.

Garzella, Gabriella. "*In silva Tumuli e in Stagno*: Paesaggio dell'incolto e risorse naturali lungo il litorale pisano nel Medioevo." In Alberto Malvolti and Giuliano Pinto, eds., *Incolti, fiumi, paludi. Utilizzazione delle risorse naturali nella Toscana medievale e moderna*, 143–158. Firenze: Leo S. Olschki, 2003.

Gautier Dalché, Jean. "La Galice dans le commerce maritime médiéval," *Le Moyen Âge* 104 (1998), 513–525.

Gelsinger, Bruce S. *Icelandic Enterprise: Commerce and Economy in the Middle Ages*. Columbia: University of South Carolina Press, 1981.

Gemmill, Elizabeth, and Nicholas Mayhew. *Changing Values in Scotland: A Study of Prices, Money, and Weights and Measures*. Cambridge: Cambridge University Press, 1995.

*Geoponika*, ed. Heinrich Beckh, as *Geoponika sive Cassiani Bassi Scholastici De Re Rustica Eclogae*. Leipzig: Teubner, 1895. Tr. Andrew Dalby, *Geoponika Farm Work: A Modern Translation of the Roman and Byzantine Farming Handbook*. Totnes: Prospect Books, 2011.

Géraud, Hercule. *Paris sous Philippe le Bel d'après les documents originaux, et notament d'après un manuscrit contenant le role de la taille imposée sur les habitants de Paris en 1292*. Paris: Crapelet, 1837.

*Gerefa*, intro. and ed. Thom Gobbit in www.earlyenglishlaws.ac.uk/laws/texts/rspger (accessed 18 August 2021).

Geremek, Bronisław. *The Margins of Society in Late Medieval Paris*, tr. Jean Birrell. Cambridge: Cambridge University Press, 1987.

Gerhard, Hans-Jürgen, and Alexander Engel, eds. *Preisgeschichte der vorindustriellen Zeit. Ein Kompendium auf Basis ausgewählter Hamburger Materialien*. Studien zur Gewerbe- und Handelsgeschichte, Bd. 26. Stuttgart: Franz Steiner, 2006.

Gertwagen, Ruthy, Saša Raicevich, Tomaso Fortibuoni, and Otello Giovanardi, eds. *Le interazioni tra uomo ed ambiente nel Mediterraneo dall'Epoca Romana al XIX secolo: una visione storica ed ecologica delle attivita di pesca*, 46–49. Atti del II Workshop Internazionale HMAP del Mediterraneo e Mar Nero. Chioggia, 27–29 settembre 2006. Supplemento ai Quaderni ex ICRAM. Chioggia: ISPRA – Istituto Superiore per la Protezione e la Ricerca Ambientale (Chioggia), 2008.

Gessner, Conrad [von]. "Catalogus alphabeticus animalium," fols. 221r–233v in his *Pandectarum sive Partitionum universalium … libri XXI*. Tiguri [Zürich]: Christophorus Froschouerus, 1548.

*Historiae animalium*. Liber IV. *Conradi Gesneri medici Tigurini Historiae animalium liber IIII, qui est de piscium & aquatilium animantium natura. Cvm iconibvs singvlorvm ad vivvm expressis fere omnib DCCVI. Continentur in hoc volumine Gvlielmi Rondeletii quoq[ue] et Petri Bellonii de Aquatilium singulis scripta*. Tiguri: apud Chr. Froschovervm, 1558.

"Teütsche nammen der Fischen vnd Wasserthieren" [added and running title; title page gives as "III. Eorundem Nomenclator Germanicus longe copiosissimus. Et alia quaedam ad Piscium historiam pertinenciam"], pp. 97–280 in his *De piscibus et aquatilibus omnibus libelli III. noui*. Tiguri: apud Andream, [1556]. Facsimile reprint as *Deutsche Namen der Fische und Wassertiere*, ed. Manfred Peters. Aalen: Scientia, 1974.

Gimpel, Jean. *The Medieval Machine: The Industrial Revolution of the Middle Ages*. New York: Holt, Rinehart and Winston, 1976.

Giovio, Paolo. "*De romanis piscibus libellus*. Roma: 1524." In *Pauli Iovii Opera*, vol. 9: *Dialogi et descriptiones*, ed. Ernesto Travi and Mariagrazia Penio,

11–64. Rome: Sociètà storica Comense and Instituto Poligrafico dello Stato, 1984.

Gislain, G. de. "Le rôle des étangs dans l'alimentation médiévale." In Menjot, ed., *Manger et boire*, I: 89–101. Paris: Belles lettres, 1984.

Given-Wilson, C. J. "Purveyance for the Royal Household, 1362–1413," *Bulletin of the Institute for Historical Research* 56 (1983 ), 145–163.

Given-Wilson, C., et al., eds. *The Parliament Rolls of Medieval England* [*PROME*]. Woodbridge, UK, and Rochester, NY: The Boydell Press, 2005; digital edition Scholarly Digital Editions and The National Archives, Leicester, 2005. Internet version, used at www.sd-editions.com.ezproxy.library.yorku .ca/PROME.

Glanz, Michael H., ed. *Climate Variability, Climate Change, and Fisheries*. Cambridge: Cambridge University Press, 1992.

Glaser, Rüdiger. *Klimageschichte Mitteleuropas. 1000 Jahre Wetter, Klima, Katastrophen*. Darmstadt: Wissenschaftgliche Buchgesellschaft, 2001. 2d ed rev., Darmstadt: Wissenschaftliche Buchgesellschaft, 2008.

Glaser, Rüdiger, and Dirk Riemann, "A thousand-year record of temperature variations for Germany and Central Europe based on documentary data," *Journal of Quaternary Science* 24:5 (2009), 437–449.

Glass, Dorothy. "*In Principio*: The Creation in the Middle Ages." In Lawrence D. Roberts, ed., *Approaches to Nature in the Middle Ages*, 67–104. Medieval and Renaissance Texts and Studies, vol. 16. Binghamton, NY, 1982.

Glaze, Florence E. "Medical writer: 'Behold the human creature.'" In Barbara Newman, ed., *Voice of the Living Light: Hildegard of Bingen and Her World*, 125–148. Berkeley and London: University of California Press, 1998.

Gneuss, Helmut. *Aelfric of Eynsham: His Life, Times, and Writings*. Old English Newsletter Subsidia, 324. Kalamazoo: Medieval Institute Publications, 2009.

Goldberg, P. J. P. *Women, Work, and Life Cycle in a Medieval Economy: Women in York and Yorkshire c.1300–1520*. Oxford: Clarendon, 1992.

Goldenberg, Gert. "Frühe Umweltbelastungen durch Bergbau und Hüttenwesen." In Heiko Steuer and Ulrich Zimmermann, eds., *Alter Bergbau in Deutschland*, 107–113. Archäologie in Deutschland, Sonderheft 1993. Stuttgart: Thiess, 1993.

Gomes Filho, Antônio. *Um Tratado de cozinha portuguêsa do sécolo XV*. Rio de Janeiro: Ediçöes do Departamento Nacional do Livro, 1963.

Good, G. L., R. H. Jones, and M. W. Ponsford, eds. *Waterfront Archaeology. Proceedings of the Third International Conference on Waterfront Archaeology held at Bristol 23–26 September 1988*. CBA Research Report 74. London: Council on British Archaeology, 1991.

Goosse, H., O. Arzel, J. Luterbacher, M. E. Mann, H. Renssen, N. Riedwyl, A. Timmermann, E. Xoplaki, and H. Wanner. "The origin of the European 'Medieval Warm Period'," *Climate of the Past* 2 (2006), 99–113.

Górzyński, Stefan. *Zarys historii rybołówstwa w dawnej Polsce*. Warsaw: PWRiL, 1964.

"Governing the oceans: the tragedy of the high seas," *The Economist*, 22 February 2014. www.economist.com/leaders/2014/02/22/the-tragedy-of-the-high-seas.

Gowen, Bernard. "996 and all that: the Norman peasants' revolt reconsidered," *Early Medieval Europe* 21:1 (February 2013), 71–98. DOI: 10.1111/emed.12010.

Gracia Cárcamo, Juan. "El sector pesquero en la historia del Pais Vasco: esbozo de los conocimientos actuales y problemas abiertos a la futura investigación historiográfica," *Itsas Memoria: Revista de estudios maritimos del Pais Vasco* 1 (1996), 168–214.

Grafe, Regina. *Distant Tyranny. Markets, Power, and Backwardness in Spain, 1650–1800.* Princeton and Oxford: Princeton University Press, 2012.

Grand, Roger, and Raymond Delatouche. *L'Agriculture au moyen âge de la fin de l'empire romain au XVI^e siècle*, L'Agriculture à travers les âges, vol. 3. Paris: E. de Boccard,, 1950.

Grant, Annie. "Animal resources." In Grenville Astill and Annie Grant, eds., *The Countryside of Medieval England*, 149–187. Oxford: Wiley-Blackwell, 1988.

"Food, status and religion in England in the Middle Ages: an archaeozoological perspective." In Liliane Bodson, ed. *L'animal dans l'alimentation humaine: Les critères de choix*, 139–146. Actes du colloque international de Liège, 26–29 novembre 1986 *Anthropozoologica*, num. spec. 2. Paris: CNRS, 1988.

Graus, František. *Dějiny venkovského lidu v čechách v době předhusitské.* 2 vols. Prague: Státní nakl. politiché literatury, 1953–1957.

Grava, Y. "Notes martégales sur le revitaillement et la consommation du poisson à la cour pontificale d'Avignon au cours du XIV^e siècle." In Menjot, ed., *Manger et boire*, I, 153–179. Colloque de Nice, 2. Paris: Belles lettres, 1984.

Gregory of Tours. *Lives and Miracles* [*Liber vitae patrum*], ed. and tr. Giselle de Nie. Dumbarton Oaks Medieval Library 39. Cambridge, MA: Harvard University Press, 2015.

Gresser, Pierre. "Les délits commis dans les étangs comtaux en Franche-Comté aux XIV^e et XV^e siècles." In Benoît et al., eds., *Pêche et pisciculture en eau douce.* 2004.

"Les délits commis dans les rivières comtales en Franche-Comté aux XIV^e et XV^e siècles." In R. Racine, ed., *Fleuves, rivières et canaux dans l'Europe occidentale et médiane*, 138–142. Nancy: CRDP Lorraine, 1997.

*Pêche et pisciculture dans les eaux princières en Franche-Comté aux XIV^e et XV^e siècles*, Burgundica, 14. Turnhout: Brepols, 2008.

Gresser, Pierre, and J. Hintzy, "Les étangs du domaine comtal en Franche-Comté d'après les comptes de gruerie du XIV^e siècle." In *Société d'émulation du Jura. Travaux 1975–1976*, 129–156. Dole: Société d'émulation du Jura, 1978.

Grew, Klaus. "Wasserversorgung und -entsorgung im Mittelalter – Ein technikgeschichtlicher Überblick." In Klaus Grewe, ed. for the Frontinusgesellschaft, *Die Wasserversorgung im Mittelalter*, 11–88. Geschichte der Wasserversorgung, 4. Mainz: Philip von Zabern, 1991.

Grieco, Allen J. "From the cookbook to the table: a Florentine table and Italian recipes of the fourteenth and fifteenth centuries." In Carole Lambert, ed., *Du manuscrits à la table: Essais sur la cuisine au Moyen Âge et répertoire des manuscrits médiévaux contenant des recettes culinaires*, 28 –38. Montréal: Champion-Slatkine, les Presses de l'Université de Montréal, 1992.

Grinberg, Martine and S. Kinser. "Les combats de Carnaval et de Carême. Trajets d'une métaphore," *Annales: économie, sociétés, civilizations* 38 (1984), 65–98.

Grove, A. T., and Oliver Rackham. *The Nature of Mediterranean Europe: An Ecological History*. New Haven: Yale University Press, 2001.

Grove, J. M. "The initiation of the Little Ice Age in regions round the North Atlantic," *Climatic Change* 48 (2001), 53–82.

Grumett, David and R. Muers. *Theology on the Menu: Asceticism, Meat and Christian DIet*. London: Routledge, 2010.

Grupe, G., D. Heinrich, and J. Peters. "A brackish water aquatic foodweb: trophic levels and salinity gradients in the Schlei fjord, Northern Germany, in Viking and medieval times," *Journal of Archaeological Science* 36 (2009), 2125–2144.

Guérin, Isabella. *La vie rurale en Sologne aux XIVe et XVe siècles. École pratique des hautes études – VIe Section – Centre de recherches historiques. Les hommes et la terre*, no. 5. Paris: SEVPEN, 1960.

Gui of Bazoches [Guido de Basochis] [*Epistolae*] ed. Herbert Adolfsson as *Liber epistularum Guidonis de Basochis*. Acta Universitatis Stockholmiensis. Studia Latina Stockholmiensis, 18. Stockholm: Almqvist & Wiksell, 1969.

Guilhiermoz, Paul, ed. "Ordonnance inédite de Philippe le Bel sur la police de la pêche fluviale (17 Mai 1293)," *Bibliothèque de l'Ecole des Chartes* 63 (1902), 331–337.

Guillaume de Saint-Pathus. *Les miracles de Saint Louis*, ed. Percival B. Fay. Paris: Champion, 1931.

Guillaume de Tocco, *Ystoria sancti Thome de Aquino de Guillaume de Tocco (1323)*, ed. Claire le Brun-Gouanvic. Toronto: PIMS, 1996.

Guillerme, André E. *The Age of Water. The Urban Environment in the North of France, AD 300–1800*. College Station: Texas A & M University Press, 1988.

Gunda, Bela, ed. *The Fishing Culture of the World: Studies in Ethnology, Cultural Ecology and Folklore*. 2 vols. Budapest: Akademiai Kiadó, 1984.

"Fish poisoning in the Carpathian area and in the Balkan Peninsula." In Gunda, ed., *Fishing Culture*, vol. 1, 181–188. Budapest: Akademiai Kiadó, 1984.

Györffy, György. *Wirtschaft und Gesellschaft der Ungarn um die Jahrtausendwende*. Tr. Zsusza Keszeg. Studia historica Academiae Scientiarum Hungaricae, 186. Budapest: Akademiai Kiado, 1983.

Haberl, Helmut, Marina Fischer-Kowalski, Fridolin Krausmann, Helga Weisz, and Verena Winiwarter. "Progress towards sustainability? What the conceptual framework of material and energy flow accounting (MEFA) can offer," *Land Use Policy* 21 (2004), 199–213.

Häberle, Simone, and Elisabeth Marti-Grädel. "Die Teichwirtschaft vom Mittelalter bis in die Frühe Neuzeit." In Heide Hüster Plogmann, ed., *Fisch und Fischer aus zwei Jahrtausenden. Eine fischereiwirtschaftliche Zeitreise durch die Nordwestschweiz*, 149–159. Forschungen in Augst, Bd. 39. Augst: Römerstadt Augusta Raurica, 2006.

Häberle, Simone, Benjamin T. Fuller, Olaf Nehlich, Wim Van Neer, Jörg Schibler, and Heide Hüster Plogmann. "Inter- and intraspecies variability in stable isotope ratio values of archaeological freshwater fish remains from

Switzerland (11th –19th centuries AD)." *Environmental Archaeology* 21:2 (2016), 119–132. http://dx.doi.org/10.1179/1749631414Y.0000000042.

"Carbon and nitrogen isotopic ratios in archaeological and modern Swiss fish as possible markers for diachronic anthropogenic activity in freshwater ecosystems," "*Journal of Archaeological Science: Reports* 10 (2016), 411–423.

Hagenmeyer, Christa, ed. *Das Regimen Sanitatis Konrads von Eichstätt: Quellen, Texte, Wirkungsgeschichte*. Sudhoffs Archiv Beihefte, 35. Stuttgart: Franz Steiner, 1995.

Hakenbeck, Susanne, et al. "Diet and mobility in early medieval Bavaria: a study of carbon and nitrogen stable isotopes," *American Journal of Physical Anthropology* 143 (2010), 235–249.

Halard, Xavier. "La pêche du saumon en Normandie du XIe au XVe siècle," *Journal of Medieval History* 9 (1983), 173–178.

Haldon, John, Lee Mordechai, Timothy P. Newfield, Arlen F. Chase, Adam Izdebski, Piotr Guzowski, Inga Labuhn, and Neil Roberts. "History meets palaeoscience: consilience and collaboration in studying past societal responses to environmental change." *Proceedings of the National Academy of Sciences* 115 (2018): 3210–18.

Halkin, J., and C. G. Roland, eds. *Recueil des chartes de l'Abbaye de Stavelot-Malmédy*. 2 vols. Publications de la Commission royale d'histoire, 37. Bruxelles: Kiessling et Cie, P. Imbreghts, successeur, 1909–1930.

Hallström, A. "Die Fischknochen." In J. Bossneck, A. von den Driesch and R. Steuberger, eds., *Eketorp. Befestigung und Siedlung auf Öland Schweden*, vol. 3: *Die Fauna*, pp. 422–493. Stockholm: Almqvist och Wiksell, 1979.

Hamilton, Earl J. *American Treasure and the Price Revolution in Spain 1501–1650*. Cambridge, MA: Harvard University Press, 1934, reprinted New York: Octagon, 1965.

*Money, Prices, and Wages in Valencia, Aragon, and Navarre, 1351–1500*. Harvard Economic Studies, no. 51. Cambridge, MA: Harvard University Press, 1936. Reprinted Philadelphia: Porcupine Press, 1975.

Hamilton-Dyer, Sheila. "Fish in Tudor naval diet – with reference to the *Mary Rose*," *Archaeofauna* [Madrid] 4 (1995), 7–26.

Hammer, Cornelius, Christian von Dorrien, Peter Ernst, Tomas Gröhsler, Fritz Köster, Brian Mackenzie, Christian Möllmann, Gerd Wegner, and Christopher Zimmermann. "Fish stock development under hydrographic and hydrochemical aspects: the history of Baltic Sea fisheries and its management." In Rainer Feistel, Günther Nausch, and Norbert Wasmund, eds., *State and Evolution of the Baltic Sea, 1952–2005*, 543–581. New York: John Wiley & Sons, 2008.

Hansson, Daniel, and Anders Omstedt. "Modelling the Baltic Sea ocean climate on centennial time scale: temperature and sea ice." *Climate Dynamics* 30 (2008), 763–778.

Hardy, Alan, Anne Dodd, Graham D. Keevil, et al. *Ælfric's Abbey: Excavations at Eynsham Abbey, Oxfordshire, 1989–92*. Oxford Archaeology, Thames Valley Landscapes, 16. Oxford: Oxford University School of Archaeology for Oxford Archaeology, 2003.

Harland, Jennifer. "The origins of aquaculture," *Nature, Ecology and Evolution* 3 (2019), 1378–1379.

Harland, Jennifer F., Andrew G. K. Jones, David C. Orton, and James H. Barrett. "Fishing and fish trade in medieval York: the zooarchaeological evidence." In Barrett and Orton, eds., *Cod and Herring*, 172–204. 2016.

Harland, Jennifer, C. Johnstone, and A. Jones. "A case study from the Medieval Origins of Commercial Sea Fishing project: zooarchaeological results from York (UK)." In Philippe Béarez, Sandrine Grouard, and Benoît Clavel, eds., *Archéologie du poisson: 30 ans d'archéo-ichtyologie à CNRS. Hommage aux travaux de Jean Desse at Nathalie Desse-Berset*, Actes des recontres 18–20 octobre 2007, 15–26. Antibes: Éditions APDCA, 2008.

Harrison, Ramona, Howell M. Roberts, and W. Paul Adderley. "Gásir in Eyjafjörður: international exchange and local economy in medieval Iceland," *Journal of the North Atlantic* 1 (2008),99–119.

Hart, Jeffrey A. *The Anglo-Icelandic Cod War of 1972–1973: A Case Study of a Fishery Dispute*, Research series no. 29. Berkeley: Institute of International Studies, University of California, 1976.

Hartley, Leslie P. *The Go-Between*. London: Hamish Hamilton, 1954.

Hartley, R. F. "Medieval fisheries and fishponds in Leicestershire (including Rutland)." In Aston, ed., *Medieval Fish*, 291–300. 1988.

Harvey, Barbara F. *Living and Dying in England 1100–1540: The Monastic Experience*. Oxford: Clarendon Press, 1993.

"Monastic pittances in the Middle Ages,." In Woolgar, Serjeantson and Waldron, eds., *Food in Medieval England*, 215–227. Oxford: Oxford University Press, 2006.

Hastrup, Kirsten. *Nature and Policy in Iceland 1400–1800: An Anthropological Analysis of History and Mentality*. Oxford: Clarendon, 1990.

Hauschild, Ursula. *Studien zu Löhnen und Preisen in Rostock im Spätmittelalter*. Quellen und Darstellungen zur Hansischen Geschichte, n.f., Bd. 19. Cologne: Böhlau, 1973.

Havelok the Dane. *The Lay of Havelok the Dane*, ed. Walter W. Skeat, 2d ed., rev. by Kenneth Sisam. Oxford: Clarendon, 1963.

Hayer, Gerold. *Daz buoch von guoter spîse. Abbildungen zur Überlieferung des ältesten deutschen Kochbuches. Litterae: Göppinger Beiträge zur Textgeschichte*, Nr. 45. Göppingen: Verlag Alfred Kümmerle, 1976.

Hayes, Murray L. "Active fish capture methods,." In Larry A. Nielsen and David L. Johnson, eds., *Fisheries Techniques*, 123–146. Bethesda, MD: American Fisheries Society, 1983.

Heath, P. "North Sea fishing in the fifteenth century: the Scarborough fleet," *Northern History* 3 (1968), 53–69.

Heidacher, Alfred. *Die Entstehungs- und Wirtschaftsgeschichte des Klosters Heilsbronn bis zum Ende des 15. Jahrhunderts: Gründung, Gründer, Wirtschafts- und Verfassungeschichte*. Bonn: Röhrscheid, 1955.

Heikinheimo, Outi. "Interactions between cod, herring and sprat in the changing environment of the Baltic Sea: a dynamic model analysis," *Ecological Modelling* 222:10 (24 May 2011), 1731–1742.

Heimpel, Hermann. "Fischerei und Bauernkrieg." In P. Classen, ed., *Festschrift Percy Ernst Schramm zu seinem siebzigsten Geburtstag*, vol. 1, pp. 353–372. Wiesbaden: Steiner, 1964.

Heindel, Ingo. "Tordierte Haken – Angelgeräte?" *Zeitschrift für Archäologie* 16 (1982), 185–191.

Heine, Günter "Umweltbesogenes Recht im Mittelalter." In Bernd Herrmann, ed., *Umwelt in der Geschichte: Beiträge zur Umweltsgeschichte*, 111–128. Göttingen: Vandenhoeck und Ruprecht, 1989.

Heinrich, Dirk. "Bemerkungen zur nordwestlichen Verbreitung des Welses, *Silurus glanis* L., unter Berücksichtigung subfossiler Knochenfunde," *Zoologische Jahrbücher Systematik* 121 (1994), 303–320.

"Fang und Konsum von Kabeljau im Mittelalter," *Offa* 43 (1986), 87–93.

"Fischerei und Fischereimethoden: Archäologisches." In Heinrich Beck et al., eds., *Reallexikon der Germanischen Altertumskunde*, vol. 9, pp. 130–138. Berlin and New York: de Gruyter, 1994.

"Die Fischknochen aus der Latrine des Gasthofs 'Goldener Ring' vis-à-vis von Luthers Elternhaus in Mansfeld," In Harald Meller, ed., *Mansfeld – Luther (s)stadt. Interdisziplinäre Forschungen zur Heimat Martin Luthers*, 499–518. Forschungsberichte des Landesmuseums für Vorgeschichte Halle, Bd. 8. Halle an der Saale: Landesamt für Denkmalpflage und Archäologie Sachsen-Anhalt, Landesmuseum für Vorgeschichte, 2015.

"Fischkonsum in Luthers Elternhaus als Spiegel für Fischerei und Fischhandel in der frühen Neuzeit," *Tagung des Landesmuseums für Vorgeschichte Halle* 1 (2008), 337–345.

"Fischreste als archäologische Quellengattung – Probleme und Ergebnisse," *Archäologische Informationen* 12:2 (1989), 172–179.

"Die Fischreste aus dem Hafen von Haithabu – handaufgelesene Funde." In C. Radtke, ed., *Untersuchungen an Skelettresten von Tieren aus dem Hafen von Haithabu*, 157–193. Berichte über die Ausgrabungen in Haithabu, Bericht 35. Neumünster: Wachholz, 2006.

*Die Fischreste aus der frühgeschichtlichen Marschensiedlung beim Elisenhof in Eiderstedt.* Schriften aus der Archäologisch-Zoologischen Arbeitsgruppen Schelswig-Kiel, Heft 9. Kiel: Universität Kiel, 1985.

"Die Fischreste von der Burg Bodenteich, Kr. Uelzen." In Cornelia Becker et al., eds., *Historia animalium ex ossibus: Beiträge zur Paläoanatomie, Archäologie, Ägyptologie, Ethnologie und Geschichte der Tiermedizin. Festschrift für Angela von den Driesch*, 181–187. Rahden/Westf.: Verlag Marie Leidorf, 1999.

"Fishing and consumption of Cod (Gadus morhua Linnaeus 1758) in the Middle Ages." In D. C. Brinkhuizen and A. T. Clason, eds., *Fish and Archaeology*, 42–52. BAR International Series, 294. Oxford: BAR, 1986.

"Fish remains of two medieval castles and of an urban context – a comparison." In Wim Van Neer, ed., *Fish Exploitation in the Past. Proceedings of the*

*7th Meeting of the I.C.A.Z. Fish Remains Working Group*, 211–216. Annales du Musée Royal de l'Afrique Central, Sciences Zoologiques, vol. 274. Tervuren: Musée Royal, 1994.

"Information about fish from tales and myths," *Archaeofauna* 6 (1997), 17–21.

"Some methodological considerations with regard to analyses of faunal history with special reference to fish remains." In Norbert Benecke, ed., *The Holocene History of the European Vertebrate Fauna*, 163–170. Archäologie in Eurasien, 6. Rahden/Westf.: Verlag Marie Leidorf, 1999.

"Temporal changes in fishery and fish consumption between early medieval Haithabu and its successor, Schleswig." In C. Grigson and J. Clutton-Brock, eds., *Animals and Archaeology: 2. Shell Middens, Fishes and Birds*, 151–156. Oxford: BAR, 1983.

"Untersuchungen an Fischresten des 8. und 10. Jahrhunderts aus dem Kloster St. Irminen-Oeren in Trier," *Trierer Zeitschrift* 81 (2018), 305–338.

"Untersuchungen an mittelalterlichen Fischresten aus Schleswig: Ausgrabung Schild 1971–1975," *Ausgrabungen in Schleswig. Berichte und Studien*, vol. 6. Neumünster: Wachholtz, 1987.

Heinrich, Dirk, and F. Heidermanns. "Lachs." In Heinrich Beck et al., eds., *Reallexikon der Germanischen Altertumskunde*, 2d ed. rev., vol. 17, 528–532. Berlin and New York: de Gruyter, 2001.

Heinricus Laureshamensis [*fl.* c. 1100]. *Summarium Heinrici*, ed. Reiner Hildebrandt. 2 vols. Berlin: de Gruyter, 1974–1982 [for correct dating see Wagner, "Zur Datierung" (1975).].

Helmlinger, Georges. *La Dombes. Les hommes et l'eau*. Bourge-en-Bresse: Taillanderie, 1990.

Hemmerle, Josef, ed. *Die Deutschordens-Ballei Böhmen in ihren Rechnungsbüchern 1382–1411*. Quellen und Studien zur Geschichte des Deutschen Ordens, Bd. 22. Bonn: Verlag Wissenschaftliches Archiv, 1967.

Henisch, B. A. *Fast and Feast: Food in Medieval Society*. State College and London: Pennsylvania State University Press, 1976.

Henningsen, Uta. *Besitz und Einkünfte der Herren von Rosenberg in Böhmen nach dem Urbar von 1379/84*. Historische und Landeskundliche Ostmitteleuropa-Studien, 5. Marburg: J.G.Herder Institut, 1989.

Herbord of Michelsberg. *Herbordi Dialogus de vita S. Ottonis episcopi Babenbergensis*, ed. Kazimierz Liman and Jan Wikarjak. Monumenta Poloniae Historica, series nova, VII: 3. [Pomniki dziejowe polski: Seria II: t. 7, cz. 3.] Warsaw: Państwowe wydawnictwo naukowe, 1974.

Herlihy, David, ed. *Medieval Culture and Society*. New York: Harper, 1968.

*Opera Muliebria. Women and Work in Medieval Europe*. Philadelphia: Temple University Press, 1990.

Hernández Iñigo, Pilar. "La pesca fluvial y el consumo de pescado en Córdoba (1450–1525)," *Annuario de estudios medievales* 27:2 (1997), 1045–1116.

Herrmann, Bernd, ed. *Umwelt in der Geschichte: Beiträge zur Umweltsgeschichte*. Göttingen: Vandenhoeck und Ruprecht, 1989.

Herrscher, Estelle. "Inferring diet by stable isotope analysis: a case study from the French Alps." In Martin Carver and Jan Klápšte, eds., *The Archaeology of*

*Medieval Europe*, vol. 2. *Twelfth to Sixteenth Centuries*, 139–146. Aarhus: Aarhus University Press, 2011.

Heuschmann, Otto. *Die Weißfische (Cyprinidae)*. Handbuch der Binnenfischerei Mitteleuropas, ed. R. Demoll and H. N. Maier, Bd. 3, Lfg. 8. Stuttgart: Schweizerbart, 1957.

Hickling, C. F. "Prior More's fishponds," *Medieval Archaeology* 15 (1971), 118–123.

Hieatt, Constance B., ed. *An Ordinance of Pottage. An Edition of the Fifteenth Century Culinary Recipes in Yale University's MS Beinecke 163*. London: Prospect Books, 1988.

"Sorting through the titles of medieval dishes: what is or is not a 'blanc manger'?" In Melitta Weiss Adamson, ed., *Food in the Middle Ages: A Book of Essays*, 25–44. New York and London: Garland, 1995.

Hieatt, Constance B., and Sharon Butler, eds. *Curye on Inglysch: English Culinary Manuscripts of the Fourteenth Century (including the Forme of Cury)*, Early English Texts Society, Second Series, vol. 8. London: Oxford University Press for EETS, 1985.

Hieatt, Constance B. and Robin F. Jones. "Two Anglo-Norman culinary collections edited from British Library Manuscripts Additional 32085 and Royal 12.C.xii," *Speculum* 61 (1986), 859–882.

Higginbotham, James. *Piscinae: Artificial Fishponds in Roman Italy*. Chapel Hill: North Carolina University Press, 1997.

Higounet, Charles. "Les forêts de l'Europe occidentale du V^e^ au X^ie^ siêcle." In *Agricoltura e mondo rurale in Occidente nell'alto medioevo*. XIII Settimana di studio del Centro Italiano di studi sull'alto medioevo, 343–398. Spoleto: Presso la sede del Centro, 1996.

Hildegard of Bingen. *Hildegard von Bingen's Physica: The Complete English Translation of Her Classic Work on Health and Healing*, tr. Priscilla Throop. Rochester, VT: Healing Arts Press, 1998. ( The translator lacks familiarity with European fishes and German names for them.)

*Physica. Liber subtilitatum diversarum naturarum creaturarum: Textkritische Ausgabe*, ed. Reiner Hildebrandt and Thomas Gloning. 2 vols. Berlin: De Gruyter, 2010.

Hilding-Rydevik, Tuija, Jon Moen, and Carina Green, "Baselines and the shifting baseline syndrome – exploring frames of reference in nature conservation." In Crumley, Lennartsson, and Westin, eds., *Issues and Concepts in Historical Ecology*, 112–144. 2018.

Hilton, R. H., ed. *The Stoneleigh Leger Book*. Dugdale Society. Oxford: Oxford University Press, 1960.

*History of William Marshal*, ed. A. J. Holden, with English translation by S.Gregory, and historical notes by D. Crouch. 3 vols. London: Anglo-Norman Text Society from Birkbeck College, 2002–2006.

Hitzbleck, Herbert. "*Die Bedeutung des Fisches für die Ernährungswirtschaft Mitteleuropas in vorindustrieller Zeit unter besonderer Berücksichtigung Niedersachsens*." PhD dissertation, University of Göttingen, 1971.

Hockey, Stanley F., ed. *The Account-Book of Beaulieu Abbey*. London: Royal Historical Society, 1975.

Hocquet, Jean-Claude. "Des paysans de la mer: L'exploitation du sel en Picardie au moyen âge." In Elisabeth Mornet, ed., *Campagnes mèdièvales: L'homme et son espace. Etudes offertes à Robert Fossier*, 629–638. Histoire Ancienne et Médiévale, 31. Paris: Publications de la Sorbonne, 1995.

"Les pêcheries médiévales." In Michel Mollat, ed., *Histoire des pêches maritimes en France*, 36–129. Toulouse: Privat, 1987.

Hodgson, Victoria. "'To the abbottis profeit': The Cistercian Abbey of Coupar Angus and the Scottish export economy." In Richard Oram et al., eds., *Land, Lordship and Environment in the North Atlantic: A Memorial for Alasdair Ross*. Tervuren: Brepols, 2022.

Hodgson, W. C. *The Herring and Its Fishery*. London: Routledge & Kegan Paul, 1957.

Hoeniger, Cathleen. "The illuminated *Tacuinum sanitatis* manuscripts from northern Italy ca. 1380–1400: sources, patrons, and the creation of a new pictorial genre." In J. A. Givens, K. M. Reeds, and A. Touwaide, eds., *Visualizing Medieval Medicine and Natural History, 1200–1550*, 51–81. AVISTA Studies in the History of Medieval Technology, Science and Art, 5. Aldershot and Burlington: Ashgate, 2006.

Hoffmann, Richard C. "Aquaculture in Champagne before the Black Death of 1348–1350." In P. Béarez, S. Grouard, and B. Clavel, eds., *Archéologie du poisson. 30 ans d'archéo-ichtyologie au CNRS. Hommage aux travaux de Jean Desse et Nathalie Desse-Berset*, 67–82. XXVIIe rencontres internationales d'archéologie et d'histoire d'Antibes. Antibes: Éditions APDCA, 2008.

"Carp, cods, connections: new fisheries in the medieval European economy and environment." In Mary J. Henninger-Voss, ed., *Animals in Human Histories: The Mirror of Nature and Culture*, 3–55. Rochester, NY: University of Rochester Press, 2002.

"*Carpes pour le duc* ...: the operation of fish ponds at La Perriére-sur-Saône, Burgundy, 1338–52," *Archaeofauna* [Madrid] 4 (1995), 33–45.

"The craft of fishing Alpine lakes, ca. A.D. 1500." In D. Heinrich, ed., *Archaeo-Ichthyological Studies. Papers presented at the 6th Meeting of the I.C.A.Z. Fish Remains Working Group, Schleswig*, 308–313. Beiheft zur *Offa*, 51. Neumünster: Wacholz, 1994.

"Economic development and aquatic ecosystems in medieval Europe," *American Historical Review* 101 (1996), 631–669.

"Environmental change and the culture of common carp in medieval Europe," *Guelph Ichthyology Reviews* 3 (1995), 57–85.

*An Environmental History of Medieval Europe*. Cambridge: Cambridge University Press, 2014.

"Fisheries regulations in late medieval Europe: authorities, concerns, measures." In Richard Keyser and Abigail Dowling, eds., *Conservation's Roots*, 127–153. Environments in History: International Perspectives series. New York: Berghahn, 2020.

*Fishers' Craft and Lettered Art: Tracts on Fishing from the End of the Middle Ages*. Toronto: University of Toronto Press, 1997.

"Fishers in late medieval rural society around Tegernsee, Bavaria – a preliminary sketch." In Edwin B. DeWindt, ed., *The Salt of Common Life: Individuality and Choice in the Medieval Town, Countryside and Church. Essays Presented to*

*J. Ambrose Raftis*, 371–408. Studies in Medieval Culture, 36. Kalamazoo: Medieval Institute Publications, Western Michigan University, 1995.

"Fishing for sport in medieval Europe: new evidence," *Speculum* 60 (1985), 877–902,

(partly coauthored by Peter Kidd). "The Haslinger Breviary fishing tract: an Austrian manuscript holds the oldest collection of fly-tying patterns now known," *American Fly Fisher* 42:2 (Spring 2016), 4–23.

"Der Karpfen (*Cyprinus carpio L.*): Der lange Weg eines 'Fremdlings' in die Schweiz." In Heide Hüster Plogmann, ed., *Fisch und Fischer aus zwei Jahrtausenden. Eine fischereiwirtschaftliche Zeitreise durch die Nordwestschweiz*, 161–167. Forschungen in Augst, Bd. 39. Augst: Römerstadt Augusta Raurica, 2006.

*Land, Liberties, and Lordship in a Late Medieval Countryside: Agrarian Structures and Change in the Duchy of Wrocław*. Philadelphia: University of Pennsylvania Press, 1989.

"Medieval fishing." In Squatriti, ed., *Working with Water in Medieval Europe*, 331–393. 2000.

"Remains and verbal evidence of carp (*Cyprinus carpio*) in medieval Europe." In Wim Van Neer, ed., *Fish Exploitation in the Past. Proceedings of the 7th Meeting of the I.C.A.Z. Fish Remains Working Group*, 139–150. Annales du Musée Royal de l'Afrique Central, Sciences Zoologiques, vol. 274. Tervuren: Musée Royal, 1994.

"*Salmo salar* in late medieval Scotland: competition and conservation for a riverine resource," *Aquatic Sciences* 77 (2015), 355–366. DOI: 10.1007/s00027-015-0397-4.

"*Strekfusz*: a fish dish links Jagiellonian Cracow to distant waters." In Piotr Górecki and Nancy van Dusen, eds., *Central and Eastern Europe in the Middle Ages: A Cultural History*. London: I.B. Tauris, 2008.

"Trout and fly, work and play, in medieval Europe." In Samuel Snyder, Bryon Borgolt, and Elizabeth Tobey, eds., *Backcasts: Historical and Global Perspectives in Fly Fishing and Conservation*, 27–45. Chicago: University of Chicago Press, 2016.

Hoffmann, Richard C., and Alasdair Ross. "This belongs to us! Competition between the Royal Burgh of Stirling and the Augustinian Abbey of Cambuskenneth over salmon fishing rights on the River Forth, Scotland." In Edel Bhreathnach, Malgorzata Krasnodebska D'Aughton and Keith Smith, , eds., *Monastic Europe AD1100–1700: Communities, Landscape and Settlement*, 451–475. Tervuren: Brepols, 2019. An illustrated but unannotated version for an audience of informed amateurs appeared in *History Scotland* 2018 no. 1 Jan/Feb, pp. 16–21, and 2018 no. 2 March/April, pp. 44–47.

Hoffmann, Richard C., and Christoph Sonnlechner. "Vom Archivobjekt zum Umweltschutz: Maximilians Patent von 1506," *Studien zur Wiener Geschichte. Jahrbuch des Vereins für Geschichte der Stadt Wien* 62/63 (2006/2007) [2011], 79–133.

Hoffmann, Richard C., and Verena Winiwarter. "Making land and water meet: cycling of nutrients between fields and ponds in pre-modern Europe," *Agricultural History* 84:3 (Summer 2010), 352–380.

Hoffmann, Thomas, Gilles Erkens, Renate Gerlach, Josef Klostermann, and Andreas Lang. "Trends and controls of Holocene floodplain sedimentation in the Rhine catchment," *Catena* 77 (2009), 96–106.

Hoffmann, Thomas, V. R. Thorndycraft, A. G. Brown, T. J. Coulthard, B. Damnati, V. S. Kale, H. Middelkoop, B. Notebaert, and D. E. Walling, "Human impact on fluvial regimes and sediment flux during the Holocene: review and future research agenda," *Global and Planetary Change* 72:3 (June 2010), 87–98.

Hoffmeister, Gerhard, ed. "Fischer- und Tauchertexte vom Bodensee. Teil I: Einleitung und Text." In Guldolf Keil, ed., *Fachliteratur des Mittelalters: Festschrift fur Gerhard Eis*, 261–275. Stuttgart: J. B. Metzler, 1968. E-mail correspondence with Gerhard Hoffmeister in January 2015 confirmed that the commentary and glossary ("Teil II") announced in 1968 was never published.

Höfling, Paul, with Torsten Gebhard. *Die Chiemsee-Fischerei. Beiträge zu ihrer Geschichte*. Munich: Institut für Volkskunde, 1987.

Hohensinner, Severin, and Mathias Jungwirth. "Hydromorphological characteristics of the Danube River: the historical perspective," *Zeitschrift des Österreichischer Ingenieur- und Architektur-Vereins* 154:1–6 (2009), 33–38.

Hohensinner, Severin, Christoph Sonnlechner, Martin Schmid, and Verena Winiwarter. "Two steps back, one step forward: reconstructing the dynamic Danube riverscape under human influence in Vienna." *Water History* 5:2 (July 2013), 121–144.

Hohl, Claude. "Alimentation et consommation à l'Hôtel-Dieu de Paris aux XV$^{e}$–XVI$^{e}$ siècles." In *Bulletin Philologique et Historique (jusqu'à 1610) du Comité des Travaux Historiques et Scientifiques*, 99$^{ème}$ Congrès des Sociétés savantes, Tours, 1968, vol. 1: *Les Problèmes de l'alimentation*, pp. 181–208. Paris: BNF, 1971.

Holland, Hilmar A., Bernd R. Schöne, Constanze Lipowsky, and Jan Esper. "Decadal climate variability of the North Sea during the last millennium reconstructed from bivalve shells (Arctica islandica)," *The Holocene* 24: 7 (2014), 771–786.

Holling, C. S. "Resilience and stability of ecological systems," *Annual Review of Ecology and Systematics* 4 (1973), 1–23.

Holm, Poul. "Catches and manpower in the Danish fisheries, *c*1200–1995." In Holm, Starkey, and Thor, eds., *The North Atlantic Fisheries, 1100–1976*, 177–206. 1996.

"Commercial sea fisheries in the Baltic Region, c. AD 1000–1600.." In Barrett and Orton, eds., *Cod and Herring*, 13–22. 2016.

"Human impacts on fisheries resources and abundance in the Danish Wadden Sea, c1520 to the present.." In Heike K. Lotze and Karsten Reise, eds., *Helgoland Marine Research* 59:1 (April 2005) [special issue 'Ecological history of the Wadden Sea], 39–44.

Holm, Poul, and Bager, Maibritt. "The Danish fisheries, c. 1450–1800: medieval and early modern sources and their potential for marine environmental history." In Holm, Smith, and Starkey, eds., *The Exploited Seas*, 97–122.

St. John's: International Maritime Economic History Association and Census of Marine Life, 2001.

Holm, Poul, and David J. Starkey, eds., *North Atlantic Fisheries: Markets and Modernization*. Studia Atlantica, 2. Ejsberg: Fiskerei- og Søfartsmuseets Forlag, 1998.

Holm, Poul, Francis Ludlow, Cordula Scherer, Charles Travis, Bernard Allaire, Cristina Brito, Patrick W. Hayes, et al. "The North Atlantic Fish Revolution (ca. AD 1500)," *Quaternary Research* 91 (2019), 1–15. https://doi.org/10.1017/qua.2018.153.

Holm, Poul, Anne H. Marboe, Bo Poulsen, and Brian R. MacKenzie, "Marine animal populations: a new look back in time." In Alasdair D. McIntyre, ed., *Life in the World's Oceans: Diversity, Distribution, and Abundance*, 3–23. Chichester: Blackwell, 2010.

Holm, Poul, Tim D. Smith, and David J. Starkey, eds., *The Exploited Seas: New Directions for Marine Environmental History*. Research in Maritime History, vol. 21. St. John's: International Maritime Economic History Association and Census of Marine Life, 2001.

Holm, Poul, David J. Starkey, and Jon Th. Thor, eds., *The North Atlantic Fisheries, 1100–1976: National Perspectives on a Common Resource*. Studia Atlantica, 1. Esbjerg: Fiskerei- og Søfartsmuseets Forlag, 1996.

Holm-Olsen, I. M. "Economy and settlement pattern 1350–1600 AD, based on evidence from farm mounds," *Norwegian Archaeological Review* 14 (1981), 86–100.

Holmes, Matilda. *Animals in Saxon and Scandinavian England: Backbones of Economy and Society*. Leiden: Sidestone Press, 2014.

Holt, Richard. *The Mills of Medieval England*. Oxford: Basil Blackwell, 1988.

"Medieval England's water-related technologies." In Squatriti, ed., *Working with Water in Medieval Europe*, 51–100. 2000.

Holubar, Karl. "Das Spital des Stiftes Klosterneuburg," *Jahrbuch des Stiftes Klosterneuburg* 15 (1994), 7–96 and 273–281.

Horden, Peregrine, and Nicholas Purcell. *The Corrupting Sea: A Study of Mediterranean History*. Oxford: Blackwell, 2000.

Howe, Nicholas. "Historicist approaches." In Katherine O'Brien O'Keeffe, ed., *Reading Old English Texts*, 79–100. Cambridge: Cambridge University Press, 1997.

Howell, Martha C. *Commerce before Capitalism in Europe, 1300–1600*. Cambridge and New York: Cambridge University Press, 2010.

Hubert, Wayne A. "Passive capture techniques." In Larry A. Nielsen and David L. Johnson, eds., *Fisheries Techniques*, 95–122. Bethesda, MD: American Fisheries Society, 1981.

Hudson-Edwards, Karen A., et al. "Mediaeval lead pollution in the River Ouse at York, England," *Journal of Archaeological Science*, 26:7, (1999), 809–819.

Huet, Marcel. "Profiles and biology of Western European streams as related to fish management," *Transactions of the American Fisheries Society* 88 (1959), 155–163.

Hufnagl, Marc, Myron A. Peck, Richard D. M. Nash, and Mark Dickey-Collas. "Unravelling the Gordian knot! Key processes impacting overwintering larval survival and growth: a North Sea herring case study," *Progress in Oceanography* 138 (2015), 486–503. http://dx.doi.org/10.1016/j.pocean.2014.04.029.

Hufthammer, Anne Karin. "Fish trade in Norway AD 800–1400: zooarchaeological evidence." In Barrett and Orton, eds., *Cod and Herring*, 221–230. 2016.

Hugh of St. Victor. *Hugonis de Sancto Victore Didascalicon de studio legendi: A Critical Text*, ed. Charles H. Buttimer. Studies in Medieval and Renaissance Latin, 10. Washington, DC: Catholic University of America Press, 1939. *The Didascalicon of Hugh of St. Victor: A Medieval Guide to the Arts*, tr. Jerome Taylor. New York: Columbia University Press, 1961. Reprinted Columbia University Press, 1991.

Hughes, J. Donald. *The Mediterranean: An Environmental History*. Santa Barbara: ABC-CLIO, 2005.

*Pan's Travail: Environmental Problems of the Ancient Greeks and Romans*. Baltimore: Johns Hopkins University Press, 1994.

Hull, Bradley, D., and Tomsin C. O'Connell. "Diet, recent evidence from analytical chemical techniques." In H. Hamerow, S. Crawford, and D. Hinton, eds., *The Oxford Handbook of Anglo-Saxon Archaeology*, 667–688. Oxford: Oxford University Press, 2011.

Humphries, Paul, and Kirk O. Winemiller. "Historical impacts on river fauna: shifting baselines and challenges for restoration," *Bioscience* 59:8 (2009), 673–684.

Hundsbichler, Helmut. "Nahrung." In Harry Kühnel et al., *Alltag im Spätmittelalter*, 196–231. Graz: Verlag Styria, 1984.

Hünemörder, Christian. "Fischerei im Mittelalter. Wissenschafts-, wirtschafts- und technikhistorische Aspekte," *Deutsche Schiffahrtsarchiv* 9 (1986), 189–198.

"Die Geschichte der Fischbücher von Aristoteles bis zum Ende des 17. Jahrhunderts," *Deutsche Schiffahrtsarchiv* 1 (1975), 185–200.

Hunt, Tony, ed. *Teaching and Learning Latin in Thirteenth-Century England*. 2 vols. Woodbridge, UK, and Rochester, NY: D. S. Brewer, 1991.

Hurt, James. *Aelfric*. New York: Twayne, 1972.

Hurt, Rudolf. *Dějiny rybnikařství na Moravě a ve Slezsku*. 2 vols. Ostrava: Vydalo Krajské nakladtelstvi v Ostravè, 1960.

Hurum, Hans J. *A History of the Fish Hook and the Story of Mustad, the Hook Maker*. London: A & C Black, 1977. Original Oslo, 1976.

Hüster Plogmann, Heide. "Fische." In Sabine Deschler-Erb, Miki Bopp-Ito, and Heide Hüster-Plogmann, "Die mittelalterlichen Tierknochen aus dem Kreuzgangbereich des Stiftes St. Arnaul,." In Hans-Walter Herrmann and Jan Selmer, eds., *Leben und Sterben in einem mittelalterlichen Kollegiatstift. Archäologische und baugeschichteliche Untersuchungen im ehemaligen Stift St. Arnaul in Saarbrücken*, 529–532. Saarbrücken: [n.p.], 2007.

"… der Mensch lebt nicht von Brot allein. Gesellschaftliche Normen und Fischkonsum." In Heide Hüster Plogmann, ed., *Fisch und Fischer aus zwei Jahrtausenden. Eine fischereiwirtschaftliche Zeitreise durch die Nordwestschweiz*,

187–199. Forschungen in Augst, Bd. 39. Augst: Römerstadt Augusta Raurica, 2006.

Hutchinson, Gillian. *Medieval Ships and Shipping*. London: Leicester University Press, 1994.

Hutchinson, William F., Mark Culling, David C. Orton, Bernd Hänfling, Lori Lawson Handley, Sheila Hamilton-Dyer, Tamsin C. O'Connell, Michael P. Richards, and James H. Barrett. "The globalization of naval provisioning: ancient DNA and stable isotope analyses of stored cod from the wreck of the *Mary Rose*, AD 1545." *Royal Society Open Science* 2 (2015), 150–199.

Hutton, Diane. "Women in fourteenth century Shrewsbury." In Lindsey Charles and Lorna Duffin, eds., *Women and Work in Pre-Industrial England*, 83–99. London: Routledge, 1985.

Hybel, Nils. "Sildehandel og sildefiskeri i den nordvestlige Nordsø i begyndelsen af det 14.århundrede." In Hans Jeppesen, ed., *Søfart Politik Identitet: tilegnet Ole Feldbæk*, 27–42. Handels- & Søfartsmuseet på Kronborg. Søhistoriske skrifter, vol. 19. Viborg: Falcon, 1996.

Hybel, Nils, and Bjørn Poulsen. *The Danish Resources c.1000–1550: Growth and Recession*. Leiden and Boston: Brill, 2007.

Hynes, H. B. N. *The Ecology of Running Waters*. Toronto: University of Toronto Press, 1970.

IJzereef, G. F., and F. Laarman. "The animal remains from Deventer (8th –19th centuries AD)," *ROB. Berichten van de Rijksdienst voor het Oudheidkundig Bodemonderzoek* 36 (1986), 405–443.

Immonen, Visa. "Monasticism in a border landscape: religious orders in medieval Finland." In Edel Bhreathnach, Malgorzata Krasnodebska D'Aughton, and Keith Smith, eds., *Monastic Europe AD1100–1700: Communities, Landscape and Settlement*, 303–330. Tervuren: Brepols, 2019.

Inglot, Stefan, and Aleksander Nyrek, "Jana Dubraviusa i Olbrychta Strumieńskiego działa o gospodarce rybnej. Studium porówznawcze,." In Ewa and Karol Maleczyński, eds., *Studia z dziejów polskich i czechosołowackich*, vol. 1, 249–282. Wrocław: Zaklad Narodowy im. Ossolinskic, 1960.

Innes, Cosmo, ed. *Ledger of Andrew Halyburton. Conservator of the Privileges of the Scotch Nation in the Netherlands, 1492–1503*. Edinburgh: Her Majesty's General Register House, 1867.

International Union for Conservation of Nature. "The IUCN Red List of Threatened Species. Version 2015.2." www.iucnredlist.org (accessed 11 August 2015).

Iona, Maria Laura. "Istituti e alimenti dagli statuti dell'area friulano-giulana." In Paolo Carucci and Massimo Buttazzo, eds., *Gli archivi per la storia dell'alimentazione*. Atti del convegno, Potenza-Matera, 5–8 settembre 1988, 608–636. 3 vols. paginated continuously. Pubblicazioni degli Archivi di Stato, Saggi, 34. Rome: Ministero per i Beni Culturali e Ambientali, Ufficio Centrale per i Beni Archivistici, 1995.

Irsigler, Franz. "Ein großbürgerlicher Kölner Haushalt am Ende des 14. Jahrhunderts." In Edith Ennen and Günter Wiegelmann, eds., *Studien zu*

*Volkskultur, Sprache und Landesgeschichte. Festschrift für Matthias Zender*, vol. 2, 635–668. Bonn: Ludwig Röhrscheid Verlag, 1972.

Isambert, François A., et al., eds., *Recueil général des anciennes lois françaises depuis l'an 420 jusqu'à la Révolution de 1789*, vol. 8. Paris: Belin-Le-Prieur [etc.], 1822; reprinted Ridgewood, NJ: Cregg Press, 1964.

Isidore of Seville. Etymologiae. Isidore de Seville, Etymologies, general eds. Jacques Fontaine and Yves Lefevre. *Livre XII: Des animaux*, ed. and tr. [French] Jacques André. Paris: Belles lettres, 1986.

*The Etymologies of Isidore of Seville*, tr.[English] Stephen A. Barney, Jennifer Beach, Oliver Berghof, and Wendy Lewis. Cambridge: Cambridge University Press, 2006.

Iwaszkiewicz, Michał. "Szczątki ryb w materiale wykopaliskowym grodu i osady podgrodowej w Drohiczynie pow. Siemiatycze," *Materiały Wczesnośredniowieczne* 6 (1969), 237–239.

"Szczątki ryb z podgrodzia w Bninie." In Jan Żak, ed., *Materiały do studiów na osadnictwem bnińskim (Podgrodzie)*, 305–307. Warsaw–Poznań: Wydawnictwo Naukowe UAM, 1979.

Izdebski, Adam., K. Holmgren, E. Weiberg, S. R. Stocker, U. Büntgen, A. Florenzano, and Y. E. Xoplaki. "Realising consilience: how better communication between archaeologists, historians and natural scientists can transform the study of past climate change in the Mediterranean." Quaternary Science Reviews *(Special Issue: Mediterranean Holocene Climate, Environment and Human Societies)* 136 (March 2016), 5–22. https://doi.org/10.1016/j.quascirev.2015.10.038.

Izquierdo Benito, Ricardo. *Precios y salarios en Toledo durante el siglo XV (1400–1475)*. Toledo: Caja de Ahorro de Toledo, 1983.

Jackson, Jeremy B. C., et al. "Historical overfishing and the recent collapse of coastal ecosystems," *Science*, 293:5530 (27 July 2001), 629–638.

Jacobs, A. "China's appetite pushes fisheries to the brink," *New York Times*, 30 April 2017, www.nytimes.com/2017/04/30/world/asia/chinas-appetite.

Jacobsen, Grethe. "Women's work and women's role: ideology and reality in Danish urban society, 1300–1550," *Scandinavian Economic History Review* 31 (1983), 3–20.

Jacoby, Karl. "Class and environmental history: lessons from 'The War in the Adirondacks'," *Environmental History* 2 (1997), 324–342.

Jacques de Vitry. *Die Exempla aus den Sermones feriales et communes des Jakob von Vitry*, ed. Joseph Greven. Heidelberg: C. Winter, 1914.

Jaffé, P., ed. "Annales Colmarienses majores,." In *MGH, Scriptorum*, vol. 17, pp. 207–232. Hannover: Hahn, 1861.

ed. "De rebus Alsaticis ineuntis saeculi XIII,." In *MGH, Scriptorum*, vol. 17, pp. 232–237. Hannover: Hahn, 1861.

Jäger, Helmut. *Einführung in die Umweltsgeschichte*. Die Geographie. Einführungen in Gegenstand, Methoden und Ergebnisse ihrer Teilgebiete und Nachbarwissenschaften. Darmstadt: Wissenschaftliche Buchgesellschaft, 1994.

Jagow, Kurt. "Die Hering im Volksglaube und in der älteren Forschung," *Archiv für Fischereigeschichte* 6 (1915), 213–247.

"Die Heringsfischerei an den deutschen Ostseeküsten im Mittelalter," *Archiv für Fischereigeschichte* 5 (1915), 1–45.

Jahnke, Carsten. "The European fishmonger: the great herring fishery in the Øresund c. 1200–1600." In Fredrik Björk, Per Eliasson, and Bo Poulsen, eds., *Transcending Boundaries. Environmental Histories from the Øresund Region*, 34–47. Skrifter med historiska perspektiv, 9. Malmö: Malmö University, 2009.

"The Medieval herring fishery in the Western Baltic." In Louis Sicking and Darlene Abreu-Ferreira, eds., *Beyond the Catch: Fisheries of the North Atlantic, the North Sea and the Baltic, 900–1850*, 157–186. Leiden and Boston: Brill, 2009.

"Regulating the medieval fisheries of the Baltic region: the 'urban' and 'international market' compared." In Poul Holm and David J. Starkey, eds., *North Atlantic Fisheries: Markets and Modernization*, 39–58. Ejsberg: Fiskerei- og Søfartsmuseets Forlag, 1998.

*Das Silber des Meeres. Fang und Vertrieb von Ostseehering zwischen Norwegen und Italien (12.–16. Jahrhundert)*. Quellen und Darstellungen zur hansische Geschichte, N.F. 49. Cologne: Böhlau, 2000.

Jakobs, Hermann. *Die Hirsauer, ihre Ausbreitung und Rechtsstellung im Zeitalter des Investiturstreites*. Cologne: Böhlau, 1961.

Jarecki, Walter. *Signa loquendi: Die cluniacensischen Signa-Listen eingeleitet und herausgegeben*. Saecula spiritalia, ed. Dieter Wuttke, vol. 4. Baden-Baden: Koerner, 1981.

Jaritz, Gerhard. "Die Reiner Rechnungsbücher (1395–1477) as Quelle zur klösterlichen Sachkultur des Spätmittelalters." In *Die Funktion der schriftlichen Quelle in der Sachkulturforschung*, 145–249. Veröffentlichungen des Instituts für mittelalterliche Realienkunde Österreichs, 1 'Sitzungsberichte der Österreichischen akademie der Wissenschaften', phil-histl Klasse, 304/4. Vienna: ÖAW, 1976.

"Zur Sachkultur österreichischer Klöster des Spätmittelalters." In *Klösterliche Sachkultur des Spätmittelalters*, 147–168. Veröffentlichungen des Instituts für mittelalterliche Realienkunde Österreichs, 3 'Sitzungsberichte der Österreichischen Akademie der Wissenschaften', phil.-hist. Klasse 367. Vienna: ÖAW, 1980.

Jaritz, Gerhard, and Verena Winiwarter, eds. *Historische Umweltdatenbank Österreichs*, public domain database. Krems and Vienna: Institut für mittelalterliche Realienkunde Österreichs, 1992.

Jaroslav [the Secretary]. *Diary of an Embassy from King George of Bohemia to King Louis XI of France in the Year of Grace 1464*, tr. A. H. Wratislaw. London: Bell, 1871.

Jaworowski, Zbigniew, Francis Barbalat, Claude Blain, and Evelyn Peyre. "Heavy metals in human and animal bones from ancient and contemporary France," *Science of the Total Environment* 43:1 (1985), 103–126.

Jeay, Madeleine, ed. *Les Evangiles des Quenouilles*. Paris: Librairie philosophique J. Vrin ; Montréal: Presses de l'Université de Montréal, 1985.

Jenkins, J. Geraint. "The trapping of salmon in the rivers of England and Wales." In Gunda, ed., *Fishing Culture*, vol. 1, 239–250. 1984.

Jenkins, James T. *The Herring and the Herring Fisheries*. London: King, 1927.

Jenks, Stuart. "Werkzeug des spätmittelalterlichen Kaufmanns: Hansen und Engländer im Wandel von memoria zur Akte (mit einer Edition von The Noumbre of Weyghtys)," *Jahrbuch für fränkische Landesforschung* 52 (1992), 283–319.

Jenn, Jean-Marie. "Défrichments cisterciens dans la région de Colmar au douzième siècle," *Annuaire de la Société historique et litteraire de Colmar* 19 (1969–1970), 42–48.

Jeřábek, Richard. "K studiu rybářství v oblasti moravskoslezských Beskyd a Javorníku," *Česky Lid* 50 (1963), 285–291.

Jeulin, Paul. "L'élaboration par la monarchie de l'ancienne règlementation de la pêche fluviale," *Revue historique du Droit français et étranger*, ser. 4, 40 (1962), 127–128, and 42 (1964), 728–730.

Jocelin of Brakelond. *The Chronicle of Jocelin of Brakelond concerning the acts of Samson Abbot of the Monastery of St. Edmund*, ed. and tr. H. E. Butler. London: Nelson, 1949.

John of Salerno. *Vita Odonis* in *PL*, vol. 133, cols. 43–86. Paris: Garnier, 1881.

*St. Odo of Cluny*, ed. and tr. Gerard Sitwell. London and New York: Sheed & Ward, 1958.

Jones, Andrew K. G. "Bulk-sieving and the recovery of fish remains from urban archaeological sites." In A. R. Hall and H. K. Kenward eds., *Environmental Archaeology in the Urban Context*, 79–85. Council for British Archaeology Research Reports, 43, 1982. London: Council for British Archaeology, 1982.

"The survival of fish remains at monastic sites." In Roberta Gilchrist and Harold Mytum, eds., *The Archaeology of Rural Monasteries*, 173–184. BAR British Series 203. Oxford: BAR, 1989.

Jones, Evan T. "Alwyn Ruddock: 'John Cabot and the Discovery of America'," *Historical Research* 81 (2008), 224–254.

"England's Icelandic fishery in the early modern period." In David J. Starkey, Chris Reid, and Neil Ashcroft, eds., *England's Sea Fisheries: The Commercial Sea Fisheries of England and Wales since 1300*, 105–110. London: Chatham, 2000.

"The *Matthew* of Bristol and the financiers of John Cabot's 1497 voyage to North America," *English Historical Review* 121 (2006), 778–795.

Jones, Evan T., and Margaret M. Condon. *Cabot and Bristol's Age of Discovery*. Bristol: The Cabot Project, University of Bristol, 2016.

Jones, Peter. "The long 'lost' history of bottom trawling in England, c.1350–1650," *International Journal of Maritime History* 30:2 (2018), 201–217.

Jong, Theo de. "Fish consumption at Eindhoven Castle: archaeological remains versus historical sources." In Wim Van Neer, ed., *Fish Exploitation in the Past. Proceedings of the 7th Meeting of the I.C.A.Z. Fish Remains Working Group*, 129–138. Annales du Musée Royal de l'Afrique Central, Sciences Zoologiques, vol. 274. Tervuren: Musée Royal, 1994.

"Huisdieren, jachtwild, vissen en weekdieren: een weerspiegeling van gevarieerde maaltijden." In Nico Arts, ed., *Het Kasteel van Eindhoven. Archeologie,*

*ecologie en geschiedenis van een heerlijke woning 1420–1676*, 214–231. Eindhoven: Museum Kempenland Eindhoven, 1992.
Jónsson, Hannes. *Friends in Conflict: The Anglo-Icelandic Cod Wars and the Law of the Sea*. London: Hurst, 1982.
Jonsson, Leif. "Finska gäddor och Bergenfisk ett försök att belysa Uppsalas fiskimport under medeltid och yngre Vasatid." In N. Cnattingius and T. Neréus, eds., *Från Östra Aros till Uppsala. En samling uppsatser kring det medeltida Uppsala*, 122–139. Uppsala stads historia, 7. Uppsala: Almqvist & Wiksell, 1986.
Jordan, William C. *Men at the Center: Redemptive Governance under Louis IX*. Natalie Zemon Davis Annual Lectures, 6. Budapest and New York: Central European University Press, 2012.
*Women and Credit in Pre-industrial and Developing Societies*. Philadelphia: University of Pennsylvania Press, 1993.
Jørgensen, Dolly. "Cooperative sanitation. Managing streets and gutters in late medieval England and Scandinavia," *Technology and Culture* 49 (2008), 547–567.
"Local government responses to urban river pollution in late medieval England," *Water History* 2 (2010), 35–52.
Joynt, Maud, ed. and tr. *The Life of St. Gall*. London: SPCK, 1927.
Jungwirth, M., S. Muhar, and S. Schmutz. "Re-establishing and assessing ecological integrity in riverine landscapes," *Freshwater Biology* 47: 4 (April 2002), 867–887.
Jungwirth, Regina. "Fischereigeschichte des Stiftes Wilhering," *Analecta Sacri Ordinis Cistercensis* 57 (2007), 195–370.
Kaiser, Knut, Nora Keller, Arthur Brande, and Oliver Benz. "A large-scale medieval dam–lake cascade in central Europe: water level dynamics of the Havel River, Berlin – Brandenburg region, Germany," *Geoarchaeology* 33:2 (March 2108), 237–259. DOI: 10.1002/gea.21649.
Kaj, Józef. "Szczątki rybne w materiale kostnym wykopaliskowym z grodzisk kultury 'łużyckiej' i wscesnohistorycznej w Biskupinie w powiecie żnińskim," In Józef Kostrzewski, ed., *III. Sprawozdanie z prac wykopaliskowych w grodzie kultury łużyckie w Biskupinie w powecie żninskim za lata 1938–1939 i 1946–1948*, 72–78. Poznań: Nakladem Polskiego Towarztstwa Prehistorycznego, 1950.
Kalný, Adolf. "Stará práva rybolovu na tekoucích vodách bývalého třeboňského panství." In *Československá Akademie Zemědělská, Současnost, perspektivy rozvoje a tradice československého rybářství*, pp. 67–74. Prague: n.p., 1978.
Kapferer, Anne-Dominique. *Fracas et murmures. Le bruit de l'eau dans un Moyen-Age picard et boulonnais*. Amiens: Trois Cailloux, Maison de la Culture d'Amiens, 1991.
Keene, Derek. "Issues of water in medieval London to c.1300," *Urban History* 28 (2001), 161–179.
Keller, Christoph. "Beobachtungen zum mittelalterlichen und neuzeitlichen Staudammbau. Das Beispiel Blankenheim," *Bonner Jahrbücher 214* (2014) [copyright 2015], 193–219.

Kelly, Henry A. "Bishop, prioress, and bawd in the stews of Southwark," *Speculum* 75 (2000), 342–388.

Kempf, Jean-Pierre. "L'économie et la société aux XIIe et XIIIe siècles d'après le cartulaire de l'abbaye cistercienne de Cherlieu en comté de Bourgogne," *Bulletin de la Société d'agriculture, lettres, sciences et arts du Département de la Haute-Saône* 9 (1973), 1–55, and 10 (1975), 55–121.

Kempshall, M. S. *The Common Good in Late Medieval Political Thought*. Oxford: Clarendon Press, 1999.

Kerhervé, Jean. *L'État breton aux XIVe et XVe siècles: les ducs, l'argent et les hommes*. 2 vols. paginated continuously. Paris: Maloine, 1987.

Kightly, Charles, Marnix Pieters, Dries Tys, and Anton Ervynck. *1465 Walraversijde. De bloeperiiode van een vissersdorp aan de zuidlijke Noordzeekust*. n.p.: Institut voor het archeologisch patrimonium, Provincie West-Vlaanderen, 2000.

Kilby, Susan. *Peasant Perspectives on the Medieval Landscape. A study of Three Communities*. Studies in Regional and Local History, vol. 17. Hatfield: University of Herefordshire Press, 2020.

King, Sarah J. *Fishing in Contested Waters: Place and Community in Burnt Church/ Esgenoôpetit*. Toronto: University of Toronto Press, 2013.

Kirk, R. E., ed. *Accounts of the Obedientiars of Abingdon Abbey, 1322–1479*. Camden Society Publications, n.s., vol. 51. Westminster: Camden Society, 1892. Reprinted New York, 1965.

Kisch, Guido. *Das Fischereirecht im Deutschordensgebiete. Beiträge zu seiner Geschichte*, 2d ed. rev. [substantively unchanged from 1932 original]. Forschungen und Quellen zur Rechts- und Sozialgeschichte des Deutschordenlandes, vol. 3. Sigmaringen: Thorbecke, 1978.

Kisslinger, Johann N. *Chronik der Pfarrei Egern am Tegernsee*. Oberbayerisches Archiv für vaterländische Geschichte, 52:3. Munich: G. Franz, 1907.

Klein, Emily S., and Ruth H. Thurstan. "Acknowledging long-term ecological change: the problem of shifting baselines." In Kathleen Schwerdtner Máñez and Bo Poulsen, eds. *Perspectives on Oceans Past: A Handbook of Marine Environmental History*, 11–29. Dordrecht: Springer, 2016.

Klinkhardt, Manfredt. *Der Hering. Clupea harengus*. Magdeburg: Westarp Wissenschaften; Heidelberg: Spektrum Akademischer Verlag, 1996.

Klose, Samuel B. von. *Darstellung der inneren Verhältnisse der Stadt Breslau vom Jahre 1458 bis zum Jahre 1526*, ed. Gustav A. H. Stenzel. Scriptores rerum Silesiacarum, vol. 3. Breslau: Josef Max & Komp, 1847.

Kluge, Friedrich, et al., eds. *Etymologisches Wörterbuch der deutschen Sprache*, 19th ed. rev. by Walther Mitzka. Berlin: De Gruyter, 1963.

Kmieciński, Jerzy. "Spręt rybacki i organizacje rybołwstwa w Gdańsku w XII–XIII w. w świetle prac wykopaliskowych w latach 1948–1951," *Studia wczesnośredniowieczne* 3 (1955), 212–216 and tables 19–26.

Knittler, Herbert. *Nutzen – Renten – Erträge: Struktur und Entwicklung frühneuzeitlicher Feudaleinkommmen in Neiderösterreich*. Sozial- und wirtschaftshistorische Studien, 19. Vienna: Verl. für Geschichte und Politik, 1989.

"Teiche als Konjunkturbarometer? Das Beispiel Niederösterreich.." In Jan Klápště, ed., *Water Management in Medieval Rural Economy / Les usages de l'eau en milieu rural au Moyen Âge*, 208–221. Památky archeologické, Supplementum 17, Ruralia V. Prague: Institute of Archaeology, Academy of Sciences of the Czech Republic, 2003.

Knottnerus, Otto S. "Reclamations and submerged lands in the Ems River Estuary (900–1500)." In Erik Thoen et al., eds., *Landscapes or Seascapes? The History of the Coastal Environment in the North Sea Area Reconsidered*, 241–266. Turnhout: Brepols, 2013.

Koch, Wilhelm. "Die Geschichte der Binnenfischerei von Mitteleuropa." In Reinhard Demoll, Hermann N. Maier, et al., eds., *Handbuch der Binnenfischerei Mitteleuropas*, vol. 4, pp. 1–52d. Stuttgart: Schweizerbart, 1925.

Kok, Ruurd. *Wonen op het veen. Archeologisch en ecologisch onderzoek van een twaalfde-eeuwse boerderij in de oostpolder te Gouda.* Gouda: Gemeente Gouda, Afdeling Stadsvernieuwing Volkshuisvesting en Monumentenzorg, 1999.

Koppmann, Karl, ed. "Der sog. Rufus-Chronik zweiter Theil von 1395–1430." In *Die Chroniken der niedersächsischen Städte. Lübeck*, Bd. 3, pp. 1–342. Die Chroniken der deutschen Städte vom 14. bis ins 16. Jahrhundert, ed. Königlich Bayerische Akademie der Wissenschaften, Gesamtreihe Bd. 28. Leipzig: S. Hirzel, 1902. www.geschichtsquellen.de/repOpus_04161.html (accessed 1 November 2019).

Korn, Georg, ed. *Breslauer Urkundenbuch*, Erster Theil [only one that appeared]. Breslau: Wilh. Gottl. Korn, 1870.

Kosch. Clemens. "Wasserbaueinrichtungen in hochmittelalterlichen Konventenanlagen Mitteleuropas." In Klaus Grewe, ed. for the Frontinusgesellschaft, *Die Wasserversorgung im Mittelalter*, 89–146. Geschichte der Wasserversorgung, 4. Mainz: Philip von Zabern, 1991.

Kosiba, Steven B., Robert H. Tykot, and Dan Carlsson. "Stable istotopes as indicators of change in the food procurement and food preferences of Viking Age and Early Christian populations on Godland (Sweden)," *Journal of Anthropological Archaeology* 26 (2007), 394–411.

Koslowski, Gerhard, and Rüdiger Glaser, "Variations in reconstructed ice winter severity in the Western Baltic from 1501 to 1995, and their implications for the North Atlantic Oscillation," *Climatic Change* 41 (1999), 175–191.

Köster, F. W., C. Mollmann, H.-H. Hinrichsen, K. Wieland, J. Tomkiewicz, G. Kraus, R. Voss, A. Makarchouk, B. R. MacKenzie, M. A. St. John, D. Schnack, N. Rohlf, T. Linkowski, and J. E. Beyer. "Baltic cod recruitment – the impact of climate variability on key processes," *ICES Journal of Marine Science* 62:7 (October 2005), 1408–1425.

Kottelat, Maurice, and Jörg Freyhof. *Handbook of European Freshwater Fishes.* Cornol, Switzerland: Publications Kottelat, and Berlin: Freyhof, 2007.

Kowaleski, Maryanne. "The commercialization of the sea fisheries in medieval England and Wales," *International Journal of Maritime History* 15:2 (2003), 177–231.

"The demography of maritime communities in late medieval England." In Mark Bailey and Stephen Rigby, eds., *Town and Countryside in the Age of the Black Death: Essays in Honour of John Hatcher*, 87–120. Turnhout: Brepols, 2012.

"The early documentary evidence for the commercialization of the sea fisheries in medieval Britain." In Barrett and Orton, eds., *Cod and Herring*, 23–412016.

"The expansion of the southwestern fisheries in late medieval England," Economic History Review, 2d series, 53 (2000), 429–454.

*Local Markets and Regional Trade in Medieval Exeter*. Cambridge: Cambridge University Press, 1995.

"Peasants and the sea in medieval England." In Maryanne Kowaleski, John Langdon, and Philip R. Schofield, eds., *Peasants and Lords in the Medieval English Economy: Essays in Honour of Bruce M.S. Campbell*, 353–376. Turnhout: Brepols, 2015.

"The seasonality of fishing in medieval Britain." In Scott G. Bruce, ed., *Ecologies and Economies in Medieval and Early Modern Europe. Studies in Environmental History for Richard C. Hoffmann*, 117–147. Brill's Series in the History of the Environment, vol. 1. Leiden and Boston: Brill, 2010.

"Working at sea during the Middle Ages: changes in remuneration and labor conditions." In Simonette Cavachiocci, ed., *Ricchezza del Mare, Ricchezza dal Mare, Secc. XIII–XVIII, Prato, 11–15 Aprile 2005*, 907–936. Prato: Istitute Internazionale di Storia Economica "F. Datini," 2005.

Kozikowska, Zofia. "Ryby w pokarmie średniowiecznych (X–XIV w.) mieszkańców Wrocławia na Ostrowie Tumskim jako wskaźnik gatunków łowionych w wodach danych okolic lub dobierających tam drogą handlu," *Acta Universitatis Wratislaviensis* 223 (Prace Zoologiczne 4) (1974), 3–14.

Kraskovská, Ludmila. "K otázke lovu a rybolovu u Slovanov," *Časopis Moravského Musea (Brno)* 57 (1972), 147–156.

Krausen, Edgar, ed. *Die Urkunden des Klosters Raitenhaslach 1034–1350*. 2 vols. Quellen und Erörterungen zur bayerischen Geschichte, n.s. vol. 17. Munich: Beck, 1959–1960.

Kristiansen, M. S. "Fish for peasants and kings – a Danish perspective. " In J. Klapště and P. Sommer, eds. *Processing, Storage, Distribution of Food: Food in the Medieval Rural Environment*, 213–218. Ruralia VIII, 7–12 September 2009, Lorca, Spain. Turnhout: Brepols, 2011

Krovnin, Andrei S., and Sergei N. Rodionov. "Atlanto-Scandian herring: a case study." In Michael H. Glantz, ed., *Climate Variability, Climate Change, and Fisheries*, 231–260. Cambridge: Cambridge University Press, 1992.

Krusch, Bruno, ed. "Vita Galli confessoris triplex.." In *MGH, SrM*, vol. 4: 229–337. Hannover: Hahn, 1902.

ed. "Vita Sigiramni abbatis Longoretensis." In *MGH SrM*, vol. 4: 606–625. Hannover: Hahn, 1902.

Kubasiewicz, Marian. *Badania Archeozoologiczne na terenie Gdańska IX–XIV wieku*. Gdańskie Towarzystwo Naukowe wydział i nauk społecznych i humanistycznynch, Prace komisji Archeologicznej Nr. 9. Gdańsk: Gdańskie Tow. Naukowe, 1977.

*Küchenmeisterei. Nürnberg: Peter Wagner, [ca. 1490]*. Facsimile edition with introduction by Hans Wegener, Veröffentlicheungen der Gesellschaft für Typenkunde des XV. Jhs., Bd. 3. Leipzig: Harrossowitz, 1939.

Kuijpers, Antoon, Helmar Kunzendorf, Peter Rasmussen, Marie-Alexandrine Sicre, Ullah Ezat, Assia Fernane, Kaarina Weckström. "The Baltic Sea inflow regime at the termination of the Medieval Climate Anomaly linked to North Atlantic circulation," *Baltica* 25:1 (2012), 57–64.

Kunst, Günther Karl, and Alfred Galik. "Essen und Fasten in mittelalterlichen Klöstern aus archäologischer Sicht." In *Fundort Kloster: Archäologie im Klösterreich. Katalog zur Ausstellung im Stift Altenburg vom 1. Mai bis 1. November 2000*, 249–258. Österreichische] Bundesdenkmalamt, Abteilung für Bodendenkmale. Horn: Verlag Ferdinand Berger & Söhne, 2000

Kunst, Karl. "What makes a medieval urban animal bone assemblage look urban?" In A. M. Choyke and G. Jaritz, eds., *Animaltown: Beasts in Medieval Urban Space*, 9–18. BAR International Series 2858. Oxford: BAR, 2017.

Kurlansky, Mark. *Cod: A Biography of the Fish That Changed the World*. Toronto and NewYork: Knopf, 1997.

Kuske, Bruno. "Der Kölner Fischhandel vom 14. bis 17. Jahrhundert," *Westdeutschen Zeitschrift für Geschichte und Kunst* 24:3 (1905), 227–313.

Küster, Hansjörg. "Mittelalterliche Eingriffe in Naturräume des Voralpenlandes." In Bernd Herrmann, ed., *Umwelt in der Geschichte: Beiträge zur Umweltgeschichte*, 63–76. Gottingen: Vandenhoeck und Ruprecht,1989.

Kytzler, Bernhard. "Laudes Silesiae II. Franz Fabers 'Sabothus'," *Jahrbuch der schlesische Friedrich-Wilhelms Universität zu Breslau* 26 (1985), 51–63.

La Casva, A. Francesco. *Ingiene e Sanità negli Statuti di Milano del Sec. XIV*. Milan: L. Hoepli, 1946.

La Ronciere, Charles-M. de. *Prix et salaires à Florence au XIVe siecle (1280–1380)*. Rome: École française de Rome, 1982.

La Véronne, Chantal de. *La Brenne: Histoire et traditions*, 2d ed. rev. Tours: Gibert-Clarey, 1971.

Lachmann, Till. "Die Gewässer und ihre Nutzung." In B. Friedel and C. Frieser, eds., *Nürnberg – Archaeologie und Kulturgeschichte*, 311–315. Buchenbach: Verlag Dr. Faustus,1999.

Ladero Quesada, Miguel A. "Las almadrabas de Andalucia (siglos XIII–XVI)," *Boletín de la Real Academia de la Historia* 190:3 (1993), 345–354. Repeated with the same title in *Villes et sociétés urbaines au Moyen Âge. Hommage à M. le Professeur Jacques Heers*, 299–306. Cultures et civilisations médiévales, XI. París: Université de Paris-Sorbonne, 1995.

"La caza en las ordenanzas municipales de Andulacia, siglos xv y xvi." In *La chasse au moyen âge*. Actes du Colloque de Nice (22–24 juin 1979), 237–251. Centre d'études médiévales de Nice. Publications de la Faculté des lettres et des sciences humaines de Nice, 20. Paris: Belles lettres, 1980.

Lagerås, Per, ed. *Environment, Society and the Black Death. An Interdisciplinary Approach to the Late-Medieval Crisis in Sweden*. Oxford and Philadelphia: Oxbow Books, 2016.

Laget, Frédérique. "Géographie du hareng à la fin du Moyen Âge: les mers du Nord, des lieux de production?,"*Anthropozoologica* 53 (2018), 81–86.

Lagler, Karl, John E. Bardach, Robert R. Miller, and Dora M. May Passino. *Ichthyology*, 2d ed. New York: Wiley, 1977.

Lamb, Hubert H. *Weather, Climate, and Human Affairs. A Book of Essays and Other Papers*. London: Routledge, 1988.

Lambert, Carole, ed. *Du manuscrits à la table: Essais sur la cuisine au Moyen Âge et répertoire des manuscrits médiévaux contenant des recettes culinaires*. Montéal: Champion-Slatkine, les Presses de l'Université de Montréal, 1992.

ed. *Le recueil de Riom et la Manière de henter soutillement: Un livre de cuisine et un réceptair sur les greffes du XV^e siècle. Le moyen français*, fasc. 20. Montréal: CERES, 1988.

Lambert, J. M., et al. *The Making of the Broads: A Reconsideration of Their Origin in the Light of New Evidence*. Royal Geographic Society Research Series, no. 3. London: RGS, 1961.

Lambrick, George. "Alluvial archaeology of the Holocene in the Upper Thames Basin 1971–1991: a review." In Stuart Needham and Mark G. Macklin, eds., *Alluvial Archaeology in Britain*, pp. 209–228. Oxbow Monograph 18. Oxford: Oxbow, 1992.

Lampen, Angelika. *Fischerei und Fischhandel im Mittelalter: Wirtschafts- und sozialgeschichtliche Untersuchungen nach urkundlichen und archäologischen Quellen des 6. bis 14.Jahrhunderts im Gebiet des Deutschen Reichs*. Historische Studien, Bd. 461. Husum: Matthiesen Verlag, 2000.

"Medieval fish weirs: the archaeological and historical evidence." *Archaeofauna* 5 (1996) (special issue *Ichthyoarchaeology: Fish and the Archaeological Record: Proceedings of the Eighth Meeting of the ICAZ Fish Remains Working Group – Madrid 3–11 October 1995)*, 129–134.

Lancaster, William T. ed. *Abstracts of the Charters and Other Documents Contained in the Chartulary of Fountains Abbey*. 2 vols. Leeds: Whitehead and Son, 1915.

Lanconelli, Angela. "Gli *Statuta pescivendulorum urbis* (1405). Note sul commercio di pesce a Roma tra XIV e XV secole," *Archivio della Societa romana di storia patria* 108 (1985), 83–131.

"I lavori alla peschiera del Marta. Contributo alla storia della pesca nel Lazio bassomedioevale." In *Scritti in onore di Giuseppi Marchetti Longhi*, vol. II, 233–249. Biblioteca di Latium, 11. Anagni: Istituto di storia e di arte del Lazio meridionale, Centro di Anagni, 1990.

"La pesca nelle acque interne: fiumi e laghi,." In Duccio Balestracci and Paolo Pasini, eds., *Pesca e pescatori dal tardo medioevo all prima età moderna, 7–25. Pesca e Pescatori*, vol. 3. Milan: Leonardo Arte, 2001.

Langdon, John. *Mills in the Medieval Economy: England, 1300–1540*. Oxford and New York: Oxford University Press, 2004.

Lange, Marit. "Når kom norsk tørrfisk til Italia?," *Historisk Tidskrift* 58:1 (1979), 89–97.

Laplane, Henri de. *Les abbes de Saint-Bertin d'après les anciens monuments de ce monastère*. 2 vols. Saint-Omer: Chanvin fils, 1854–1855.

Lardin, Philippe, and Laurent Jegou. "La pêche en basse Seine à la fin du Moyen Age." In Benoît et al., eds., *Pêche et pisciculture en eau douce*. 2004.

Larguier, Gilbert. "Des lagunes à la mer: Pêche et société littoral autour du golfe du Lion (xv^e^–xviii^e^ siècle)." In Buti et al., eds. *Moissonner la mer*, 195–210. Aix-en-Provence: Aix-Marseille Université, CNRS, and Paris: Éditions Karthala, 2018.

Lasalle, G. M. Béguer. L. Beaulaton, and E. Rochard, "Diadromous fish conservation plans need to consider global warming issues: an approach using biogeographical models," *Biological Conservation* 141 (2008),1105–1118.

Laugardière, Maurice de. *L'Église de Bourges avant Charlemagne*. Paris and Bourges: Tardy, 1951.

Laurière, Eusèbe J. de, et al., eds., *Ordonnances des roys de France de la troisième race recueillés par ordre chronologique*. 21 vols. Paris: Imprimerie royale, 1723–1849.

Laurioux, Bruno. *Une histoire culinaire du moyen âge*. Sciences, Techniques et Civilisations du Moyen Âge à l'Aube des Lumières, 8. Paris: Honoré Champion, 1995.

"Le 'Registre de cuisine' de Jean de Bockenheim, cuisinier du pape Martin V," *Mélanges de l'Ecole française de Rome* 100 (1988), 709–760.

"Table et hiérarchie sociale à la fin du Moyen Age." In Carole Lambert, ed., *Du manuscrits à la table: Essais sur la cuisine au Moyen Âge et répertoire des manuscrits médiévaux contenant des recettes culinaires*, 87–108. Montréal: Champion-Slatkine, les Presses de l'Université de Montréal, 1992.

Le Roux de Lincy, Antoine J. V., and L. M. Tisserand, eds. *Paris et ses historiens aux 14^e^ et 15^e^ siècles: Documents et écrits originaux recueillis et commentés* ... Paris: Impr. impériale, 1867.

Le Roy Ladurie, Emmanuel. *The Peasants of Languedoc*, tr. John Day. Urbana: University of Illinois Press, 1976

Leciejewicz, Lech. "Z denara otrzymasz wóz świeżych śledzi." In Stefan K. Kuczyński et al., eds., *Nummus et historia. Pieniądz Europy Średniowiecznej*, 103–109. Warsaw: Polskie Towarzystwo Archeologiczne i Numizmatyczne, Komisja Numizmatyczna, 1985.

"Zum frühmittelalterlichen Heringshandel im südlichen Ostseegebiet," *Zeitschrift für Archäologie*, 25 (1991), 209–214.

Łęga, Władysław. *Obraz gospodarczy Pomorza Gdańskiego w XII i XIII wieku*. Prace Instytutu Zachodniego, nr. 14. Poznań: Prace Instytutu Zachodniego, 1949.

LeGoff, Jacques. "Vestimentary and alimentary codes in Erec et Enide." In his *The Medieval Imagination*, tr. Arthur Goldhammer, 132–150. Chicago: University of Chicago Press, 1988.

Leguay, Jean-Pierre. *L'eau dans la ville au Moyen Âge*. Rennes: Presses Universitaires de Rennes, 2002.

*La pollution au moyen age: dans le royaume de France et dans les grands fiefs*. Paris: Editions Gisserot 1999.

Lelek, Antonin. "The Rhine River and some of its tributaries under human impact in the last two centuries." In D. P. Dodge, ed., *Proceedings of the International Large River Symposium*, pp. 469–487. Canadian Spec. Pub.

Fisheries and Aquatic Science, 106. Ottawa: Department of Fisheries and Oceans, 1989.

Lenders, H. R. J., T. P. M. Chamuleau, A. J. Hendriks, R. C. G. M. Lauwerier, R. S. E. W. Leuven, and W. C. E. P. Verberk. "Historical rise of waterpower initiated the collapse of salmon stocks," *Scientific Reports* 6:29269. DOI: 10.1038/srep29269. www.nature.com/scientific reports, published online 20 July 2016.

Lentacker, An. "Preliminary analysis of the fish remains from the medieval Moorish town of Saltés (Huelva, Spain)." In D. Heinrich, ed., *Archaeo-Ichthyological Studies. Papers presented at the 6th Meeting of the I.C.A.Z. Fish Remains Working Group, Schleswig*, 284–287. Beiheft zur *Offa*, 51. Neumünster: Wacholz 1994.

Lentacker, An, Anton Ervynck, and Wim Van Neer. "Dierlijke resten." In K. De Groote and J. Moens, eds., *Archeologie en geschiedenis van een middeleeuwse woonwijk onder de Hopmarkt te Aalst*, 299–348. Relicta Monografieën 16. Brussels: Onroerend Erfgoed, 2018. Online as https://oar.onroerenderfgoed.be/publicaties/RELM/16/RELM016-001.pdf

Lentacker, An, Wim Van Neer, and Jean Plumier. "Historical and archaeozoological data on water management and fishing during medieval and post-medieval times at Namur (Belgium)." In Guy De Boe and Frans Verhaeghe, eds., *Environment and Subsistence in Medieval Europe: Papers of the "Medieval Europe Brugge 1997" Conference*, 9, 83–94. I.A.P. Rapporten, 9. Zellik: Instituut voor het Archeologisch Pastrimonium, Wetenschappelijke instelling van de Vlaamse Gemeenschap, 1997.

Leonhardt, Em. Erich. *Der Karpfen. Geschichte, naturgeschichte und wirtschaftliche Bedeutung unseres wichtigsten Zuchtfisches*. Neudamm: J. Neumann, 1906.

Lepiksaar, Johannes. *Die spät- und postglaziale Faunengeschichte der Süsswasserfische im Schweden. Ubersicht der subfossilen Funde und der Versuch einer faunengeschichtlichen Analyse der rezente Artareale*, rev. ed. by Dirk Heinrich. Kiel: Oetker-Voges Verlag, 2001 [original typescript Göteborg, 1983].

Lepiksaar, Johannes, and Dirk Heinrich. *Untersuchungen an Fischresten aus der frühmittelalterlichen Siedlung Haithabu. Berichte über die Ausgrabungen in Haithabu*, vol. 10. Neumunster: K. Wachholtz, 1977.

Leppäkoski, Erkki, Stephan Gollasch, and Sergej Olenin, eds. *Invasive Aquatic Species of Europe: Distribution, Impacts, and Management*. Dordrecht and Boston: Kluwer Academic Publishers, 2002.

Lespinasse, René de, ed. *Les métiers et corporations de la ville de Paris*. Tom. 1, *XIV–XVIIIe siècle: Ordinances Générales, Métiers de l'alimentation*. Paris: Imprimerie Nationale, 1886.

Lespinasse, René de, and F. Bonnardot, eds. *Les métiers et corporations de la ville de Paris: Le Livre des métiers d'Étienne Boileau*. Paris: Imprimerie Nationale, 1879.

Letaldus of Micy [Letaldo di Micy]. *Within Piscator*, ed. and tr. [into Italian] by Ferruccio Bertini. Biblioteca del Medioevo Latino. Florence: Giunti, 1995. An English translation from this edition is in Ziolkowski, *Fairy Tales from before Fairy Tales*, pp. 243–247.

Levy, Brian J. "'Tant va la cruche à l'eau …': Parodies et périls hydriques dans les fabliaux français." In Bernard Ribémont, ed., *L'eau au moyen âge: Symboles et usages*, 41–60. Actes du colloque, Orléans – mai 1994. Orléans: Paradigme, 1996.

*The Liber Augustalis or Constitutions of Melfi Promulgated by the Emperor Frederick II for the Kingdom of Sicily in 1231.* Tr. James M. Powell. Syracuse, NY: Syracuse University Press, 1971.

Lewin, John. "Medieval environmental impacts and feedbacks: the lowland floodplains of England and Wales," *Geoarchaeology* 25:3 (May 2010), 267–311.

Lightfoot, E., M. Šlaus, and T. C. O'Connell. "Changing cultures, changing cuisines: cultural transitions and dietary change in Iron Age, Roman, and Early Medieval Croatia," *Americal Journal of Physical Anthropology* 148 (2012), 543–556.

Lillich, Meredith. "Cleanliness with godliness: a discussion of medieval monastic plumbing." In Benoît Chauvin, ed., *Mélanges à la mémoire du Père Anselme Dimier*, vol. 3 (part 5), pp. 123–149. Arbois: privately printed B. Chauvin,1982.

Limburg, K. E., and J. R. Waldman, "Dramatic declines in North Atlantic diadromous fishes," *Bioscience* 59 (2009), 955–965.

Lindner, Kurt. *Deutsche Jagdtraktate des 15. und 16. Jahrhunderts.* 2 vols. Quellen und Studien zur Geschichte der Jagd, Bd. V–VI. Berlin: de Gruyter,1959.

Linke, M. "Medieval deserted fields and deserted villages in the Lower Harz Mountains – an interim report." In B. K. Roberts and R. E. Glasscock, eds., *Villages, Fields and Frontiers. Studies in European Rural Settlement in the Medieval and Early Modern Periods*, 293–305. BAR International Series, 185. Oxford: BAR, 1983.

Litzenburger, L. *Une ville face au climat: Metz à la fin du moyen âge, 1400–1530.* Nancy: PUN–Editions Universitaires de Lorraine, 2015.

Ljungqvist, Fredrik C., Paul J. Krusic, Hanna S. Sundqvist, Eduardo Zorita, Gudrun Brattström and David Frank. "Northern Hemisphere hydroclimate variability over the past twelve centuries," *Nature* 532 (2016), 94–98.

Lleonart, Jordi, Francesc Maynou, Roser Salicrú i Lluch. "Marine species and their selling prices in the Crown of Aragon. An initial approach with some examples from the 14th to the 17th Centuries." In Buti et al., eds. *Moissonner la mer*, 159–174. 2018.

Lloyd, Terrence H. *England and the German Hanse, 1157–1611: A Study of Their Trade and Commercial Diplomacy.* Cambridge: Cambridge University Press, 1991.

Lobon-Cervia, Javier, Elvira Benigno, and Pedro A. Rincon. "Historical changes in the fish fauna of the River Duero Basin." In Geoffrey E. Petts, Heino Moller, and A. L. Roux, eds., *Historical Change of Large Alluvial Rivers: Western Europe*, 221–232. New York: Wiley, 1989.

Locker, Alison. "The decline in the consumption of stored cod and herring in post-medieval and early industrialized England: a change in food culture." In Barrett and Orton, eds., *Cod and Herring*, 99–107. Oxford and Philadelphia: Oxbow Books, 2016.

"Fish bones." In Brian S. Ayers, Excavations at Fishergate, Norwich, 1985, 42–44. East Anglian Archaeology Report No. 68. Norwich: Field Archaeology Division, Norfolk Museums Service, 1994.

"The fish bones." In Robert Cowie and Robert L. Whytehead, "Two Middle Saxon occupation sites: excavations at Jubilee Hall and 21–22 Maiden Lane," *Transactions of the London and Middlesex Archaeological Society* 39 (1988), 149–150.

"Fish bones." In R. Daniels, "The Anglo-Saxon Monastery at Church Close, Hartlepool, Cleveland," *Archaeological Journal* 145 (1988), 201.

"Fish bones." In R. Daniels, "The Anglo-Saxon Monastery at Church Close, Hartlepool, Cleveland," *The Archaeological Journal* 147 (1990), 396–398.

"Fish bones." In Robert L. Whytehead et al., "Excavations at the Peabody Site, Chandos Place, and the National Gallery," *Transactions of the London and Middlesex Archaeological Society* 40 (1989), 148–151.

"The role of stored fish in England 900–1750 AD: the evidence from historical and archaeological data." PhD dissertation, Department of Archaeology, University of Southampton, 2000. Published in facsimile, Sofia, Bulgaria: Publishing Group Limited, 2001.

Logez, M., P. Bady, and D. Pont. "Modelling the habitat requirement of riverine fish species at the European scale: sensitivity to temperature and precipitation and associated uncertainty." *Ecology of Freshwater Fish* 21 (2012), 266–282.

Lohrmann, Dietrich. "Zwei Mühlenweistümer von 1188 und 1212 aus der Picardie." In Elisabeth Mornet, ed., *Campagnes médiévales: L'homme et son espace. Etudes offertes à Robert Fossier*, 217–229. Histoire Ancienne et Médiévale, 31. Paris: Publ. de la Sorbonne, 1995.

Longnon, Auguste, ed. *Documents relatifs au Comté de Champagne et de Brie, 1172–1361.* 3 vols. Paris: Imprimerie national, 1901–1914.

Lopez, Robert S. *The Commercial Revolution of the Middle Ages 950–1350.* Englewood Cliffs: Prentice-Hall, 1971.

López Losa, Ernesto. "Derechos de Propriedad Informales y Gestion Comunal de las Pesquerias en el Pais Vasco: un Enfoque Ecologico-Institucional," *Revista de Historia Económica* 21:1 (2003), 11–48.

"Informal property rights and common management in the northern Spanish fisheries until the twentieth century." In Leoš Jeleček et al., eds. *Dealing with Diversity. 2d International Conference of the European Society for Environmental History, Prague, 2003, Proceedings*, 83 –86 Prague: Charles University Faculty of Science, 2003.

Losco-Bradley, P. A. and C. R. Salisbury. "A Saxon and a Norman fish weir at Colwick, Nottinghamshire." In Aston, ed. *Medieval Fish*, 329–352. 1988.

Lõugas, Lembi. "Fishing and fish trade during the Viking Ages and Middle Ages in the eastern and western Baltic Sea regions." In Barrett and Orton, eds., *Cod and Herring*, 111–116. 2016.

"Fishing during the Viking Age in the eastern and western Baltic Sea." In Philippe Béarez, Sandrine Grouard, and Benoît Clavel, eds., *Archéologie du poisson. 30 ans d'archéo-ichtyologogie au CNRS. Hommage aux travaus de Jean*

*Desse et Nathalie Desse-Berset*, 27–34. Actes de recontres 18–20 octobre 2007. Antibes: Éditions APDCA, 2008.

Loveluck, Christopher. *Northwest Europe in the Early Middle Ages, c. AD 600–1150: A Comparative Archaeology*. Cambridge: Cambridge University Press, 2013.

Loyn, H. R. and J. Percival, eds. and tr. *The Reign of Charlemagne: Documents on Carolingian Government and Administration*. London: Edward Arnold, 1975.

Lozinski, Grégoire, ed. *La Bataille de Caresme et de Charnage*. Paris: Champion, 1933.

Lucas, Adam. *Wind, Water, Work: Ancient and Medieval Milling Technologies*. Leiden and Boston: Brill, 2006.

Lucy, Frances E., and Vadim E. Panov, "Keep beating the drum: ICAIS confirms aquatic invasive species are of continuing concern," *Aquatic Invasions* 9:1 (2014), 239–42.

Ludat, Herbert. *Das Lubuser Stiftsregister von 1405. Studien zu den Sozial- und Wirtschaftsverhältnissen in mittleren Oderraum zu Beginn des 15. Jahrhunderts*. Giessener Abhandlungen zur Agrar- und Wirtschaftsforschung des europäischen Ostens, Bd. 9. Wiesbaden: Otto Harassowitz, 1965.

Ludwig, A., L. Debus, D. Lieckfeldt, I. Wirgin, N. Benecke, I. Jenneckens, P. Williot, J. R. Waldman, and C. Pitra. "Fish populations: when the American sea sturgeon swam east," *Nature* 419 (2002), 447–448.

Ludwig, A., D. Makowiecki, and N. Benecke. " Further evidence of trans-Atlantic colonization of Western Europe by American Atlantic sturgeons," *Archaeofauna* 18 (2009), 185–192.

Luiten, Loek. "Friends and family, fruit and fish: the gift in Quattrocentro Farnese cultural politics," *Renaissance Studies* 33 (2018), 342–357.

Luterbacher J., J. P. Werner, J. E. Smerdon, L Fernández-Donado, F. J. González-Rouco, D. Barriopedro, F. C. Ljungqvist, U. Büntgen, E. Zorita, S. Wagner, J. Esper, D. McCarroll, A. Toreti, D. Frank, J. H. Jungclaus, M. Barriendos, C. Bertolin, O. Bothe, R. Brázdil, D. Camuffo, P. Dobrovolný, M. Gagen, E. García-Bustamante, Q. Ge, J. J. Gómez-Navarro, J. Guiot, Z. Hao, G. C. Hegerl, K. Holmgren, V. V. Klimenko, J. Martín-Chivelet, C. Pfister, N. Roberts, A. Schindler, A. Schurer, O. Solomina, L. von Gunten, E. Wahl, H. Wanner, O. Wetter, E. Xoplaki, N. Yuan, D. Zanchettin, H. Zhang and C. Zerefos. "European summer temperatures since Roman times," *Environmental Research Letters* 11:2. (2016), 1–12. doi: 10.1088/1748-9326/11/2/024001 (accessed 8 December 2016). http://iopscience.iop.org/article/10.1088/1748-9326/11/2/024001

Luterbacher, J., and 44 co-authors. "A review of 2000 years of palaeoclimatic evidence in the Mediterranean." In Piero Lionello, ed., The Climate of the Mediterranean Region: From the Past to the Future, 87–185. San Diego: Elsevier, 2012.

*The Luttrell Psalter* [facsimile]. London: British Museum, 1932.

Lynch, A., and N. Paap. "Untersuchungen an botanischen Funden aus der Lübecker Innenstadt. (Ein Vorbericht)," *Lübecker Schriften zur Archäologie und Kulturgeschichte* 6 (1982), 339–360.

Maas, Walter. *Les moines-défricheurs. Études sur les transformations du paysage au Moyen Age aux confins de la Champagne et de la Lorraine*. Moulins: Impr. du Progrès de l'Allier, 1944.

Maccarinelli, Angela. "Was pike on the menu? Exploring the role of freshwater fish in medieval England," *Archaeological and Anthropological Sciences* 13:133 (August 2021) https://doi.org/10.1007/s12520-021-01373-6

MacCotter, P. and K. W. Nicholls, eds. *The Cloyne Pipe Roll*. Cork: Cloyne Literary and Historical Society, 1996.

MacKenzie, Brian R., J. Alheit, D. J. Conley, P. Holm, and C. C. Kinze. "Ecological hypotheses for a historical reconstruction of upper trophic level biomass in the Baltic Sea and Skagerrak," *Canadian Journal of Fisheries and Aquatic Sciences* 59 (2002) 173–190.

MacKenzie, Brian R., H. H. Hinrichsen, M. Plikshs, K. Wieland, and A. S. Zezera. "Quantifying environmental heterogeneity: habitat size necessary for successful development of cod *Gadus morhua* eggs in the Baltic Sea," *Marine Ecology Progress Series* 193 (2000), 143–156.

Macklin, M. G., K. A. Hudson-Edwards, and E. J. Dawson. "The significance of pollution from historic metal mining in the Pennine orefields on river sediment containment fluxes in the North Sea," *Science of the Total Environment* 194/195 (1997), 391–397.

Magnus, Olaus. *Historia de Gentibus Septentrionalibus. Romae 1555*, ed. John Granlund. Copenhagen: Rosenkilde and Bagger, 1972. [Facsimile.] Tr. Peter Fisher and Humphrey Higgins, ed. Peter Foote, *Historia de Gentibus Septentrionalibus. Romae 1555. Description of the Northern Peoples. Rome 1555*. 3 vols. The Hakluyt Society, 2d series, no. 182, 187, 188. London: Hakluyt Society, 1996–1998.

Magnusson, Roberta J. "Water and wastes in medieval London." In T. Tvedt and E. Jakobsson, eds., *A History of Water*. Vol. 1: *Water Control and River Biographies*, 299–313. London: I.B. Tauris, 2006.

*Water Technology in the Middle Ages: Cities, Monasteries, and Waterworks after the Roman Empire*. Baltimore: Johns Hopkins University Press, 2001.

Magnússon, Sigurdur A. *Northern Sphinx: Iceland and the Icelanders from the Settlement to the Present*. Montréal: McGill-Queens University Press, 1977.

Mahul, Alphonse J., ed. *Cartulaire et Archives des Communes de l'ancien diocèse et de l'arrondissement administratif de Carcassonne*. 7 vols. Paris: Didron, 1857–1882. Reprinted in 3 vols. Rennes: Le châateau, 1980.

Maillard, François. "Les Dépenses de l'Hôtel du Comte Jean d'Angoulême pour le second semestre 1462." In *Bulletin Philologique et Historique (jusqu'à 1610) du Comité des Travaux Historiques et Scientifiques*, 99ème Congrès des Sociétés savantes, Tours, 1968, vol. 1: *Les Problèmes de l'alimentation*, 119–127. Paris: BNF, 1971.

"Tarifs des 'coutumes' pour la vente du poisson du mer à Paris au début du XIVe siècle," *Bulletin Philologique et Historique (jusqu'à 1610) du Comité des Travaux Historiques et Scientifiques, 97e Congrès National des Sociétés savantes, Rennes,1966*, 247–258. Paris: BNF, 1968.

Maitland, Frederic W., and William P. Baildon, eds. *The Court Baron: Being Precedents for Use in Seignorial and Other Local Courts with Select Pleas from the*

*Bishop of Ely's Court of Littleport*. Publications of the Selden Society, vol. 4. London: Quartich, 1891.

Mákkai, Laszlo. "Economic landscapes: historical Hungary from the fourteenth to the seventeenth century." In Antoni Mączak, Henryk Samsonowicz, and Peter Burke, eds., *East-Central Europe in Transition: From the Fourteenth to the Seventeenth Century*, 24–35. Cambridge: Cambridge University Press and Paris: Éditions de la Maison des Sciences de l'Homme, 1985.

Makowiecki, Daniel. "About the history of fishing and its development in prehistoric and mediaeval times in Poland." In *Beiträge zum Oder-Projekt*, vol. 5, 113–127. Berlin: Deutsches Archäologisches Institut, Römisch-Germanische Kommission, 1998.

"Badania archeoichtiologiczne szczątków ze stanowiska Janów Pomorski 1." In Mateusz Bogucki and Beata Jurkiewicz, eds., *Janów Pomorski Stan. 1 / Site 1: Wyniki ratowniczych badań archeologicznych w latach 2007–2008*, vol 1:3: *Analizy*, 313–335. Studia nad Truso, 1. Elbląg: Muzeum Archeologiczno-Historyczne w Elblągu, 2012.

"Catalogue of subfossil fish remains from Poland," *Archaeofauna* 9 (2000), 133–149.

"Exploitation of early medieval aquatic environments in Poland and other Baltic sea countries: an archaeozoological consideration." In *L'acqua nei secoli altomedievali*, 753–778 [with discussion]. Settimane de studio della fondazione centro italiano di studi sull'alto medioevo, LV, Spoleto, 12–17 aprile 2007. Spoleto: Fondazione Centro Italiano di Studi sull'alto medioevo, 2008.

*Historia ryb i rybołówstwa w holocenie na Niżu Polskim w świetle badań archeoichtiologicznych*. Poznań: Instytut Archeologii i Etnologii Polskiej Akademii Nauk, 2003.

*Hodowla oraz użytkowanie zwierząt na Ostrowie Lednickim w średniowieczu: Studium archeozoologiczne*. Bibliotka studiów lednickich, tom vi. Poznań: Muzeum Pierwszych Piastów na Lednicy, 2001.

"Some aspects of studies on the evolution of fish faunas and fishing in prehistoric and historic times in Poland." In Norbert Benecke, ed., *The Holocene History of the European Vertebrate Fauna: Modern Aspects of Research, Workshop 6th to 9th April 1998, Berlin*, 171–184. Rahden/Westf.: Verlag Marie Leidorf GmbH, 1999.

"The usefulness of archaeozoological research in studies on the 'reconstruction' of the natural environment," *Archeozoologia* 21 (2003),103–120.

Makowiecki, Daniel, David C. Orton, and James H. Barrett. "Cod and herring in medieval Poland." In Barrett and Orton, eds., *Cod and Herring*, 117–132.

Makowiecki, Daniel, and Marzena Makowiecka. "The character of animal exploitation and the environment at the Polish/Prussian frontier in the medieval period: a case study," *Archaeologia Baltica* 20 (2013), 91–116.

Malaj, Egina, Peter C. von der Ohe, Matthias Grote, Ralph Kühne, Cédric P. Mondy, Philippe Usseglio-Polatera, Werner Brack, and Ralf B. Schäfer, "Organic chemicals jeopardize the health of freshwater ecosystems on the continental scale," *PNAS* 111: 26 (2014) 9549–9554.

Małecki, Jan, ed. *Lustracja województwa krakowskiego 1564*. 2 vols. Lustracje Dóbr Królewskich, XVI–XVIII wieku: Małopolska. Warsaw: Państwowe Wydawn. Naukowe, 1962–1964.

Malvolti, Alberto. "I proventi dell'incolto: note sull'amministrazione delle risorse naturali del commune di Fucecchio nel tardo Medioevo." In Alberto Malvolti and Giuliano Pinto, eds., *Incolti, fiumi, paludi. Utilizzazione delle risorse naturali nella Toscana medievale e moderna*, 247–272. Firenze: Leo S. Olschki, 2003.

Mänd, Anu. "Festive food in medieval Riga and Reval," *Medium Aevum Quotidianum* 41 (1999), 43–93.

*Urban Carnival: Festive Culture in the Hanseatic Cities of the Eastern Baltic, 1350–1550*. Turnhout: Brepols, 2005.

Mane, Petrine. "L'alimentation des paysans en France et en Italie au XII^e et XIII^e siècles à travers l'iconographie des calendriers, sculpture, fresque, mosaïque et vitrail." In Manjot, ed., *Manger et boire*, I. Colloque de Nice, 1982, 319–333. 1984.

Mane, Petrine. "Images médiévales de la pêche en eau douce," *Journal des savants*, July–December 1991, 227–261.

Mangolt, Gregor. *Fischbouch. Von der natur vnd eigenschaft der vischen insonderheit deren so gefangen werdend im Bodensee* ... Zürich: Andreas and Jakob Gessner, 1557. Reprinted as "Gregor Mangolts Fischbuch. Zürich gedruckt von Andreas und Jacob Gessner, 1557," ed. Johannes Meyer, *Thurgauische Beiträge zur vaterländischen Geschichte*, 45 (1905), 119–185.

Maniatis, George C. "The organizational setup and functioning of the fish market in tenth-century Constantinople," *Dumbarton Oaks Papers* 54 (2000), 13–42.

Manning, Roger B. *Hunters and Poachers. A Social and Cultural History of Unlawful Hunting in England, 1485–1640*. Oxford: Clarendon Press, 1993.

Marazzi, F., and A. Carannante. "Dal mare ai monti: l'approvvigionamento ittico nelle cucine nel monastero di San Vincenzo al Volturno nel IX secolo." In G. Volpe, A. Buglione, and G. De Venuto, eds., *Vie degli animali, vie degli uomini. Transumanza e altri spostamenti di animali nell'Europa tardoantica e medievale*. Atti del Secondo Seminario Internazionale di Studi (Foggia, 7 ottobre 2006), 107–118. Bari: Edipuglia, 2010.

Marcus, Geoffrey J. *The Conquest of the North Atlantic*. New York: Oxford University Press., 1980.

Marinval-Vigne, Marie-Christine. "Consommation d'animaux sauvages en milieu ecclésial à Orléans au XVI^e siècle: données archéozoologiques et livres de cuisine." In Jean Desse and Frédérique Audoin-Rouzeau, eds. *Exploitation des animaux sauvages à travers le temps*, 474–483. Juan-les-Pins: EPDCA, 1993.

Marshall, Richard K. *The Local Merchants of Prato: Small Entrepreneurs in the Late Medieval Economy*. The Johns Hopkins University Studies in Historical and Political Science, 117th Series (1999), 1. Baltimore and London: Johns Hopkins University Press, 1999.

Martens, Jürgen. *Die mittelalterlichen Gartensiedlungen in Preussen*. Lüneburg: Institut Nordostdeutsches Kulturwerk, 1997.

Martens, P. J. M. *De zalmvissers van de Biesbosch: Een onderzoek naar de zalmvisserij op het Bergse Veld 1421–1869. Bijdragen tot de geschiedenis van het zuiden van Nederland*, 93. Tilburg: Stichting Zuidelijk Historisch Contact, 1992.

Martin, Jean-Marie, ed. *Castrum 7: Zones côtières littorales dans le monde méditerranéen au Moyen Âge: defense, peuplement, mise en valeur.* Actes du colloque international organisé par l'École française de Rome et la Casa de Velázquez, en collaboration avec le Collège de France et le Centre inter-universitaire d'histoire et d'archéologie médiévales (UMR 5648 – Université Lyon II – C.N.R.S. – E.H.E.S.S.), Rome, 23–26 octobre 1996. Rome–Madrid: École française de Rome, Casa de Velázquez, 2001.

"Cittá e campagna: economia e societá (secc. VII–XIII)." In Giuseppi Galasso, ed., *Storia del Mezzogiorno*, vol. 3, *alto Medioevo*, 259–382. Naples: Ed. del sole, 1990.

Martin, Jean-Marie, and Ghislaine Noyé. "Les façades maritimes de l'Italie du sud: Défense et mise en valeur (IVe–XIIIe siècle)." In Jean-Marie Martin, ed., *Castrum 7: Zones côtières littorales dans le monde méditerranéen au Moyen Âge: defense, peuplement, mise en valeur*, 467–512. Rome–Madrid: École française de Rome, Casa de Velázquez, 2001.

Martínez García, Luis. "La asistencia material en los hospitales de Burgos a fines de la edad media." In Menjot, ed., *Manger et boire*, I, 349–360. 1984.

Martínez-Jarreta, B., C. Sosa, C. Laliena, B. Budowle, and R. E. M. Hedges. "Stable isotope and radiocarbon dating of the remains of the medieval royal House of Aragon (Spain) shed light on their diets, life histories and identities," *Archaeometry* 60:2 (April 2018), 366–382.

Martínez Sopena, P., and J. Carbajo Serrano. "L'alimentation des paysans castillans du XIe au XIIIe siecle d'après les 'fueros'." In Menjot, ed., *Manger et boire*, I, pp. 336–347. 1984.

Mary, Yannick, Frédérique Eynaud, Sébastien Zaragosi, Bruno Malaizé, Michel Cremer, and Sabinet Schmidt. "High frequency environmental changes and deposition processes in a 2 kyr-long sedimentological record from the Cap-Breton canyon (Bay of Biscay)." *The Holocene* 25 (2014), 348–365.

Marzano, Annalisa. *Harvesting the Sea: The Exploitation of Marine Resources in the Roman Mediterranean.* Oxford Studies in the Roman Economy. Oxford: Oxford University Press, 2013.

Masschaele, James. *Peasants, Merchants, and Markets: Inland Trade in Medieval England, 1150–1350.* New York: St. Martins, 1997.

Masseini, Alvaro. "Fly fishing in Early Renaissance Italy? A few revealing documents," *American Fly Fisher* 25:4 (Fall 1999): 10–11.

Mate, Mavis E. *Daughters, Wives and Widows after the Black Death: Women in Sussex, 1350–1535.* Woodbridge, UK, and Rochester, NY: Boydell Press, 1998.

Materné, Jan. "Beroeps- en vrijetijdsvisserij op de Vlaamse binnenwateren tussen Antwerpen en Gent vóór de industriële revolutie," *Bijdragen tot de geschiedenis* 71 (1988), 141–172.

"Exploitatiemetoden, uitrustung en reglementering in de vlaamse binnenvisserij. Een regionale analyze: De Wase durmestreek van de late middeleeuwen tot op het einde van het ancien régime," *Annalen van de koninklijke oudheidkundige kring van het land van waas* 91 (1988), 205–266.

"De verwerving, betwisting en instandhouding van de visserijrechten in de Wase Durmestreek, 13de–18de eeuw," *De Soevereinen* 22:3 (1991), 1–14.

Mattéoni, Olivier. "La pêche des étangs du domaine comtal en Forez à la fin du Moyen Age,." In Benoît et al., eds., *Pêche et pisciculture en eau douce*. 2004.

Matthews, David R. *Controlling Common Property: Regulating Canada's East Coast Fishery*. Toronto: University of Toronto Press, 1993.

Maunier, Francis. "Évolution de l'économie de pêche de l'embouchure du Rhône au golfe de Marseille." In *Actes du 83^e^ congrés des Sociétés savantes*, Aix-Marseille, 1958; Section de Géographie, 183–194. Paris, 1959.

Maximilian I. *Das Tiroler Fischereibuch Maximilians I*, ed. Franz Unterkircher, 2 vols. Vienna and Graz: Verlag Styria, 1967.

*Weisskunig*. In *Lichtdruck-Faksimiles nach Fruhdrücken mit Hilfe der Max-Kode-Foundation, Inc. New York*, ed. H. T. Musper et al. 2 vols. Stuttgart: W. Kohlhammer,1956.

Mayer, Theodor. "Konstanz und St. Gallen in der Frühzeit," *Schweizerisches Zeitschrift für Geschichte* 2 (1952), 473–524.

Mays, S., and N. Beavan. "An investigation of diet in early Anglo-Saxon England using carbon and nitrogen stable isotope analysis of human bone collagen," *Journal of Archaeological Science* 39 (2012), 867–874.

McAlister, Victoria. "Castles and connectivity: exploring the economic networks between tower houses, settlement, and trade in late-medieval Ireland," *Speculum* 91:3 (2016), 631–659.

McCarroll, Danny, Neil Loader, Risto Jalkanen, Mary Gagen, Håkan Grudd, Björn Gunnarson, Andreas Kirchhefer, Michael Friedrich, Hans Linderholm, Markus Lindholm, Tatjana Boettger, Sietse Los, Sabine Remmele, Yuri Kononov, Yasuhiro Yamazaki, Giles Young, and Eduardo Zorita, "A 1200-year multiproxy record of tree growth and summer temperature at the northern pine forest limit of Europe," *The Holocene* 23:4 (April 2013), 471–484.

McCullough, A. B. *The Commercial Fishery of the Canadian Great Lakes*. Studies in Archaeology, Architecture and History, National Historic Parks and Sites, Canadian Parks Service, Environment Canada. Ottawa: Environment Canada, Canadian Parks Service, 1989.

McDonald, John D. *The Origins of Angling*. Garden City, NY: Doubleday, 1963.

McDonnell, J. *Inland Fisheries in Medieval Yorkshire, 1066–1300*. University of York, Borthwick Institute of Historical Research, Borthwick Papers, No. 60. York: University of York, 1981.

McEvoy, Arthur F. "Toward an interactive theory of nature and culture: ecology, production, and cognition in the California fishing industry," *Environmental Review* 11 (1987), 289–305.

McGuire, Brian P. *The Cistercians in Denmark: Their Attitudes, Roles, and Functions in Medieval Society*. Kalamazoo: Cistercian Publications, 1982.

McNeill, John R. *The Mountains of the Mediterranean World: An Environmental History*. Cambridge: Cambridge University Press, 1992.

*Something New under the Sun: An Environmental History of the Twentieth-Century World*. New York: Norton, 2000.

Meier, Dirk. "From nature to culture: landscape and settlement history of the North-Sea coast of Schleswig-Holstein, Germany." In Erik Thoen et al., eds., *Landscapes or Seascapes? The History of the Coastal Environment in the North Sea Area Reconsidered*, 85–110. Turnhout: Brepols, 2013

Melo, Arnaldo Sousa. "Women and work in the household economy: the social and linguistic evidence from Porto, *c.* 1340–1450," In C. Beattie, A. Maslakovic, and S. R. Jones, eds., *The Medieval Household in Christian Europe, c.850–c.1550: Managing Power, Wealth, and the Body*, 249–269. Turnhout: Brepols, 2003.

*Le Ménagier de Paris*, ed. Georgine E. Brereton and Janet M. Ferrier. Oxford: Clarendon, 1981.

Ménagier de Paris. *The Good Wife's Guide: A Medieval Household Book*, tr. Gina L. Greco and Christine M. Rose. Ithaca and London: Cornell University Press, 2009. (While valuable for the larger cultural context of the manuscript, the translation lacks technical precision on matters of material culture, so the section on fishes, pp. 201–210, requires caution.)

Menant, F. *Campagnes lombardes au Moyen Âge: l'économie et la société rurales dans la région de Bergame, de Crémone et de Brescia du X^e au XIII^e siècle*. Bibliothéque de l'École française de Rome, fasc. 281. Rome: Bibliothéque de l'École française de Rome, 1993.

Menjot, Denis, ed. *Manger et boire au Moyen Age*. 2 vols. Colloque de Nice, 1982, Publications de la Faculté des lettres et sciences humaines de Nice, nos. 27–28, 1^ère série. Paris: Belles lettres, 1984.

"Notes sur le marché de l'alimentation et la consommation alimentaire a Murcie à la fin du moyen age." In Menjot, ed., *Manger et boire*, I, 199–210. 1984.

Mensing, Scott A., Irene Tunno, Leonardo Sagnotti, Fabio Florindo, Paula Noble, Claire Archer, Susan Zimmerman, Francisco Javier Pavon-Carrasco, Gabriele Cifani, Susanna Passigli, and Gianluca Piovesan. "2700 years of Mediterranean environmental change in central Italy: a synthesis of sedimentary and cultural records to interpret past impacts of climate on society," *Quaternary Science Reviews* 116 (2015), 72–94.

Mensing, Scott A., Irene Tunno, Gabriele Cifani, Susanna Passigli, Paula Noble, Claire Archer, and Gianluca Piovesan. "Human and climatically induced environmental change in the Mediterranean during the Medieval Climate Anomaly and Little Ice Age: a case from central Italy," *Anthropocene* 15 (2016), 49–59. http://dx.doi.org/10.1016/j.ancene.2016.01.003.

Merchant, Carolyn. "The theoretical structure of ecological revolutions," *Environmental Review* 11 (1987), 265–274.

Mercier, Jean. *La seigneurie de Jaucourt en Champagne méridionale (approche socio-économique)*. Bar-sur-Aube: n.p., 1993.

Merg, Marie-Line, Olivier Dézerald, Karl Kreutzenberger, Samuel Demski, Yorick Reyjol, Philippe Usseglio-Polatera, and Jérôme Belliard. "Modeling diadromous fish loss from historical data: identification of anthropogenic drivers and testing of mitigation scenarios," *PLOS ONE* 15(7) (2020), e0236575. https://doi.org/10.1371/ journal.pone.0236575.

MGH Diplomata. *Die Urkunden der deutschen Könige und Kaiser. Diplomata regum et imperatorum Germaniae.* 19 vols. Hannover, Munich, and Vienna: various publishers, 1979–2016.

Mieck, Ilja. "Die Anfänge der Umweltschutzgesetzgebung in Frankreich," *Francia* 9 (1981), 331–368.

Mieris, Frans van, ed., *Groot charterboek der graaven van Holland, van Zeeland en Heeren van Vriesland.* 4 vols. Leyden: van der Eyck, 1753–1756.

Míka, Alois. "České rybnikářství a problém počatků původní akumulace kapitálu v českých zemích," *Československý Časopis Historický* 2 (1954), 262–272.

Mills, A. D. "Some late Middle English fish names," *Notes and Queries*, n.s. 11 [whole number 209] (1964), 170–171.

Minniti, Claudia. "Meat consumption as identity of status and profession during the Middle Ages: a review of the zooarchaeological evidence from Rome and Latium (Italy)." In Ben Jervis, Lee G. Broderick, and Idoia Grau Sologestoa, eds., *Objects, Environment, and Everyday Life in Medieval Europe*, 137–156. Studies in the History of Daily Life (800–1600), vol. 3. Turnhout: Brepols, 2016.

Mira, Giuseppi. *La pesca nel medioevo nelle acque interne italiane*. Milan: A. Giuffrè, 1937.

Moirez, Pauline. "L'approvisionnement de l'Hôtel du roi de France à la fin du XIVe siècle d'après les sources comptables." In Fabienne Ravoire and Anne Dietrich, eds., *La cuisine et la table dans la France de la fin du Moyen Age. Contenus et contenants du XIV au XVI siècle*, 79–85. Caen: Publications du CRAHM, 2009.

Molkenthin, R. "Der Rhein – Aller Wasser König." In U. Oster, ed., *Flüsse in Deutschland*, 35–52. Darmstadt: Primus Verlag, 2007.

Mollat, Michel. *Le Commerce maritime Normand à la fin du moyen age: Étude d'histoire économique et sociale*. Paris: Plon, 1952.

*La pêche a Dieppe au XVe siècle*. Rouen: Imprimerie Laine, 1939.

Mollat du Jourdin, Michel. *Europe and the Sea*, tr. Teresa L. Fagan. Oxford: Blackwell, 1993.

Mone, Franz J. "Ueber die Flussfischerei und den Vogelfang vom 14. bis 16. Jahrhundert," *Zeitschrift für die Geschichte des Oberrheins* 4 (1853), 67–97.

Montanari, Massimo. *L'alimentazione contadina nell'alto medioevo*. Naples: Liguori, 1979.

*The Culture of Food*, tr. Carl Ipsen. Oxford, 1994. [English tr. of his *Fame e l'abbondanza* (Rome, 1993).]

Mooney, Linne R. "A Middle English text on the seven liberal arts," *Speculum*, 68 (1993), 1027–1052.

Moorhouse, S. "Medieval fishponds: some thoughts." In Aston, ed., *Medieval Fish*, part ii, pp. 475–484. BAR British Series 182 (ii). 1988.

Morales-Muñiz, Dolores Carmen, Eufrasia Rosello Izquierdo, and Arturo Morales-Muñiz. "Pesquerías medievales hispanas: Las evidencias arqueofaunísticas." In *La Pesca en la Edad Media*, 145–165. Madrid: Sociedad Española de Estudios Medievales, 2009.

Morales-Muñiz, Arturo, Eduardo González-Gómez de Agüero, Carlos Fernández-Rodríguez, Fran Saborido Rey, Laura Llorente-Rodríguez, Begoña López-Arias, Eufrasia Roselló-Izquierdo, "Hindcasting to forecast: an archaeobiological approach to the European hake (*Merluccius merluccius*, Linnaeus 1758) fishery: Iberia and beyond," *Regional Studies in Marine Science* 21 (2018), 21–29. https://doi.org/10.1016/j.rsma.2017.11.001.

Morales-Muñiz, Arturo, Laura Llorente-Rodriguez, Lief Jonsson, D. Albertini, and Eufrasia Roselló-Izquierdo. "Looking for needles in haystacks: The early archaeological record of the Hake (*Merluccius merluccius* Linnaeus, 1758) in European waters," *Documenta Archaeobiologiae*, 14. In press.

Morales-Muñiz, Arturo, D. C. Morales, and E. Roselló. "Sobre la presencia del bacalao (*Gadus morhua*) en la Cartuja Sevillana de Santa María de las Cuevas (siglos xv–xvi)," *Arquivo Hispalense* 226 (1991), 17–24

Moreno, A., M. Morellón, C. Martín-Puertas, J. Frigola, M. Canals, I. Cacho, J. P. Corella, A. Pérez, Á. Belmonte, T. Vegas-Vilarrúbia, P. González-Rouco, L. Fernandez-Donado, C. C. Raible, and E. Zorita. "Hydrological pattern in the Iberian Peninsula region," *PAGES news* 19:1 (March 2011), 16–17.

Moreno-Chamarro, Eduardo, Davide Zanchettin, Katja Lohmann, Jürg Luterbacher, and Johann H. Jungclaus. "Winter amplification of the European Little Ice Age cooling by the subpolar gyre," *Scientific Reports* 7:1 (August 2017). DOI:10.1038/s41598-017-07969-0.

Mosler, Hans, ed. *Urkundenbuch der Abtei Altenberg*. 2 vols. Vol. 1: *1138–1400*. Bonn, 1912. Vol. 2: *1400–1893*. Urkundenbücher der geistlichen Stiftungen des Niederrheins, III/1–2. Bonn: Hanstein, 1912, and Düsseldorf: Lintz, 1955.

Moulinier, Laurence. "L'abbesse et les poissons: un aspect de la zoologie de Hildegarde de Bingen." In Jean Desse and Frédérique Audoin-Rouzeau, eds., *Exploitation des animaux sauvages à travers les temps*, XIIIe Recontres Internationales d'Archéologie de l'Histoire d'Antibes. IVe Colloque International de l'Homme et l'Animal, 461–472. Juan-les-Pins: APDCA, 1993.

Mousnier, Mireille. *L'Abbaye de Grandselve et sa place dans la société et l'économie méridionales XIIe début XIVe siècles*. Thèese présentée pour l'obtention du Doctorat de IIIe cycle, Université de Toulouse le Mirail. Toulouse: CNRS, 1982 and 2006.

Moyle, Peter B., and Joseph H. Cech, Jr. *Fishes: An Introduction to Ichthyology*. 5th ed. rev. Upper Saddle River, NJ: Pearson Prentice Hall, 2004.

Mück, Wolfgang. "Überlegungen zum Beginn der Teichwirtschaft in Aischgrund," *Jahrbuch des historischen Vereins für Mittelfranken* 95 (1990–1991), 13–18.

Muendel, John. "The distribution of mills in the Florentine countryside during the late Middle Ages." In J. Ambrose Raftis, ed., *Pathways to Medieval Peasants*, 83–115. Papers in Medieval Studies. Toronto: Pontifical Institute of Medieval Studies, 1981.

"The grain mills of Pistoia in 1350," *Bollettino Storico Pistoiese* 74 (1972), 39–64.

Muggenthaler, Hans. *Kolonisatorische und wirtschaftliche Tätigkeit eines deutschen Zisterzienserklosters im XII. und XIII. Jahrhundert.* Munich: H. Schmidt, 1924.

Müldner, G. H. "Investigating medieval diet and society by stable isotope analysis of human bone." In R. Gilchrist and A. Reynolds, eds., *Reflections: 50 Years of Medieval Archaeology, 1957–2007*, 327–346. Leeds: Maney, 2009.

"Marine fish consumption in medieval Britain: The isotope perspective from human skeletal remains." In Barrett and Orton, eds., *Cod and Herring*, 239–249. 2016.

Müldner, G. H., J. Montgomery, G., Cook, R. Ellam, A. Gledhill, and C. Lowe. "Isotopes and individuals: diet and mobility among the medieval Bishops of Whithorn," *Antiquity* 83:322 (2009), 1119–1133.

Müldner, G. H. and M. P. Richards, "Diet and diversity at later medieval Fishergate: the isotopic evidence," *American Journal of Physical Anthropology* 134:2 (2007), 162–174.

"Diet in medieval England: the evidence from stable isotopes." In C. Woolgar, D. Serjeantson and T. Waldron, eds., *Food in Medieval England: History and Archaeology*, 228–238. Oxford: Oxford University Press, 2006.

"Fast or feast: reconstructing diet in later medieval England by stable isotope analysis," *Journal of Archaeological Science* 32 (2005), 39–48.

"Stable isotope evidence for 1500 years of human diet at the city of York, UK," *American Journal of Physical Anthropology* 133:1 (2007), 682–697.

Müller, Hanns-Hermann. "Das Tierknochenmaterial aus den frühgeschichtlichen Siedlungen von Tornow, Kr. Calau." In J. Herrmann, ed., *Die germanischen und slawischen Siedlungen und das mittelalterliche Dorf von Tornow, Kr. Calau*, 267–310. Berlin: Akademiie-Verlag, 1973.

"Die Tierreste aus der slawischen Burganlage von Arkona auf der Insel Rügen," *Zeitschrift für Archäologie* 8 (1974), 255–295.

"Zoological and historical interpretation of bones from food and sacrifices in early medieval times." In Caroline Grigson and Juliet Clutton-Brock, eds., *Animals and Archaeology: 4. Husbandry in Europe*, 187–193. BAR International Series, 227. Oxford: BAR, 1983.

Müller, Miriam. "Food, hierarchy and class conflict." In Richard Goddard, John Langdon, and Miriam Müller, eds., *Survival and Discord in Medieval Society: Essays in Honour of Christopher Dyer*, 231–248. The Medieval Countryside, 4. Turnhout: Brepols, 2010.

Müller, Walter, ed. *Die Rechtsquellen der Abtei St. Gallen: 2. Reihe, 1 Bd., Die allgemeinen Rechtsquellen der alten Landschaft.* Aarau: Verlag Sauerländer, 1974.

Mulon, Marianne, ed. "Deux traités inédits d'art culinaire médiéval," *Bulletin Philologique et Historique (jusqu'à 1610) du Comité des Travaux Historiques et Scientifiques, année 1968*, 369–435. Paris, 1971.

Munro, John H. "Patterns of trade, money, and credit." In Thomas A. Brady, Jr., Heiko A. Oberman, and James D. Tracy, eds., *Handbook of European History 1400–1600. Late Middle Ages, Renaissance and Reformation*. Vol. I: *Structures and Assertions*, 147–196. Leiden: Brill, 1994.

Murray, James M. *Bruges, Cradle of Capitalism, 1280–1390*. Cambridge: Cambridge University Press, 2005.

Mutgé i Vives, Josefa. "L'abastament de peix i carn a Barcelona, en el primer terç del segle XIV." In *Alimentacio i Societat a la Catalunya Medieval* (Annuario de Estudio Medievales, Annex 20), 109–136. Barcelona: Consell Superior d'Investigacions Científiques, Institució Milà i Fontanals, Unitat d'Investigació d'Estudis Medievals, 1988.

*La ciudad de Barcelona durante el reinado de Alfonso el Benigno (1327–1336)*. Madrid and Barcelona: Consejo Superior de Investigaciones Científicas (Espanya), 1987.

Myers, A. R., ed. *The Household of Edward IV: The Black Book and the Ordinance of 1478*. Manchester: Manchester University Press,1959.

Myers, Ransom A., and Boris Worm. "Extinction, survival, or recovery of large predatory fishes," *Philosophical Transactions of the Royal Society B* 360 (2005), 13–20.

Myers, Ransom A., N. J. Barrowman, J. A. Hutchings and A. A. Rosenberg. "Population dynamics of exploited fish stocks at low population levels," *Science*, n.s., 269:5227 (25 Aug. 1995), 1106–1108.

Mylona, Dimitra, and Rebecca Nicholson, eds. *The Bountiful Sea: Fish Processing and Consumption in Mediterranean Antiquity*. Proceedings of the International Conference Held at Oxford, 6–8 September 2017, Journal of Maritime Archaeology Special Issue, 13.3. New York: Springer, 2018.

Nada Patrone, Anna Maria. *Il cibo del ricco ed il cibo del povero. Contributo alla storia qualitativa dell'alimentazione. L'area pedemontana negli ultimi secoli del Medio Evo*. Turin: Centro studi piemontesi,, 1981.

Nash, R. D. M., and M. Dickey-Collas. "The influence of life history dynamics and environment on the determination of year class strength in North Sea herring (*Clupea harengus*). *Fisheries Oceanography* 14 (2005), 279–291.

Naso, Irma. "Il pesce che nuoce, il pesce che sana. I prodotti ittici nel discorso medico fra medioevo e prima età moderna." In Duccio Balestracci and Paolo Pasini, eds., *Pesca e pescatori dal tardo medioevo alla prima età moderna*, 95–111. Milan: Leonardo Arte, 2001.

Nauwerck, Arnold. "Der Lachsfang in der Kinzig: In biologischer Beleuchtung. Nach Akten des Fürstlich Fürstenbergischen Rentamtes Wolfach aus dem 18. Jahrhundert," *Ortenau* 66 (1986), 499–525.

Neckham, Alexander. *De natura rerum libri duo. With the Poem of the Same Author De laudibus divinae sapientiae*, ed. Thomas Wright. Chronicles and Memorials of Great Britain and Ireland during the Middle Ages [Rolls Series], 34. London: Longman, 1863.

Nedkvitne, Arnved. "The development of the Norwegian long-distance stockfish trade." In Barrett and Orton, eds., *Cod and Herring*, 50–59. 2016.

"Fishing, whaling, and seal hunting." In Phillip Pulsiano et al., eds., *Medieval Scandinavia. An Encyclopedia*, 194–196. New York: Garland, 1993.

*The German Hansa and Bergen 1100–1600*. Quellen und Darstellungen zur Hansischen Geschichte, N.F. Bd. LXX. Cologne–Weimar–Vienna: Böhlau, 2014.

"Trade." In Phillip Pulsiano et al., eds., *Medieval Scandinavia: An Encyclopedia*, 649–653. New York: Garland, 1993.

Neis, Barbara. "Familial and social patriarchy in the Newfoundland fishing industry." In Newell and Ommer, eds., *Fishing Places, Fishing People*, 32–54. 1999.

Neumann, Augustin. *Prameny k dějinám duchovenstva v době předhusitské a Husovsě*. Olomouc: Nakl. Matice cyrilometodějské, 1926.

Newell, Dianne, and Rosemary E. Ommer, eds. *Fishing Places, Fishing People. Traditions and Issues in Canadian Small-Scale Fisheries*. Toronto: University of Toronto Press, 1999.

Newfield, Timothy P., and I. Labuhn. "Realizing consilience in studies of pre-instrumental climate and pre-laboratory disease." *Journal of Interdisciplinary History* 48:2 (2017), 211–240.

Newman, Barbara. "'Sibyl of the Rhine': Hildegard's life and times." In Barbara Newman, ed., *Voice of the Living Light: Hildegard of Bingen and Her World*, 1–29. Berkeley and London: University of California Press, 1998.

Nicholas, David. *The Metamorphosis of a Medieval City: Ghent in the Age of the Arteveldes, 1302–1390*. Lincoln: University of Nebraska Press, 1987.

Nicholson, Rebecca A. "The fish remains." In A. Hicks, ed., *Destined to Serve: Use of the Outer Grounds of St. Augustine's Abbey, Canterbury, before, during and after the Time of the Monks*, 285–291. Canterbury: Canterbury Archaeological Trust, 2017.

Nicoud, Marilyn. *Les régimes de santé au moyen âge. Naissance et diffusion d'une écriture médicale (XIII^e–XV^e siècle)*. 2 vols. Rome: École Française de Rome, 2007.

Nielssen, Alf R. "Early commercial fisheries and the interplay among farm, fishing station and fishing village in North Norway." In Barrett and Orton, eds., *Cod and Herring*, 42–49. 2016.

Nigro, Giampiero. "Mangiare di grasso, mangiari di magro: il consumo di carni e pesci tra Medioevo ed Età Moderna." In Simonetta Cavaciocchi, ed. *Alimentazione e Nutrizione Secc. XIII–XVIII*, 113–146. Instituto Internazionale di Storia Economica "F. Datini." Serie II – Atti delle "Settimane di Studi" e altri Convegni, 28. Prato [Firenze]: Le Monnier, 1996.

Nikulina. Elena A., and Ulrich Schmölcke. "Reconstruction of the historical distribution of sturgeons (Acipenseridae) in the eastern North Atlantic based on ancient DNA and bone morphology of archaeological remains: implications for conservation and restoration programmes," *Diversity and Distributions* 22:10 (October 2016), 1036–1044. DOI: 101111 / ddi 12461.

Nitz, H.-J. "Feudal woodland colonization as a strategy of the Carolingian empire in the conquest of Saxony: reconstruction of the spatial patterns of expansion and colonist settlement morphology in the Leine–Weser Region." In B. K. Roberts and R. E. Glasscock, eds., *Villages, Fields and Frontiers. Studies in European Rural Settlement in the Medieval and Early Modern Periods*, 171–184. BAR International Series, 185. Oxford: BAR, 1983.

Noble, R. A., I. G. Cowx, I. G. D. Goffaux, and P. Kestemont. "Assessing the health of European rivers using functional ecological guilds of fish communities: standardising species classification and approaches to metric selection," *Fisheries Management and Ecology*, 14:6 (2007), 381–392.

Noël de la Morinière, S. B. J. *Histoire général des pêches anciennes et modernes, dans les mers et les fleuves des deux continens*. Paris: de l'imprimerie royale, 1815.

Noichl, Elisabeth, ed. *Codex Falkensteinensis. Die Rechtsaufzeichnungen der Grafen von Falkenstein*. Quellen und Erorterungen zur Bayerischen Geschichte, N.F., vol. 29. Munich: Beck, 1978.

Nowakowski. Cezary. "Rybackie narzędzia kolne na Pomorzu Zachodnim w świetle zbiorów Muzeum Narodowie w Szczecinie," *Materiały Zachodniopomorskie* 27 (1981), 261–280.

Nyrek, Aleksander. *Gospodarka rybna na Górnym śląsku od połowy XVI do połowy XIX wieku*. Wrocław: Zakład Narodowy im. Ossolińskich, 1966.

Nyström, Urban, ed. *Poèmes français sur les biens d'un ménage depuis l'Oustillement au villain du XIIIe siècle jusqu'aux Controverses de Gratien du Pont*. Helsinki: Suomal. Tiedeakat, 1940.

O'Connor, T. "What shall we have for dinner? Food remains from urban sites." In D. Serjeantson and T. Waldron, eds., *Diet and Crafts in Towns: The Evidence of Animal Remains from the Roman to the Post-medieval Periods*, 13–23. BAR British Series, 199. Oxford: BAR, 1989.

O'Donnell, James J. *Cassiodorus*. Berkeley: University of California Press, 1979.

O'Sullivan, Aidan. *Foragers, Farmers and Fishers in a Coastal Landscape: An Intertidal Archaeological Survey of the Shannon Estuary*. Dublin: Royal Irish Academy, 2001.

"Place, memory and identity among estuarine fishing communities: interpreting the archaeology of early medieval fish weirs." *World Archaeology* 35:3 (2003), 449–468.

Occhipinti, Elisa. "Fortuna e crisi di un patrimonio monastico: Morimondo e le sue grangie fra XII e XIV secolo," *Studi storici* 26 (1985), 315–363.

Ogilvie, Astrid, and Graham Farmer. "Documenting the medieval climate." In Michael Hulm and Elaine Barrow, eds., *Climates of the British Isles: Past, Present, Future*, 112–133. London and New York: Routledge, 1997

Ogilvie, Astrid E. J., and Thomas H. McGovern. "Sagas and science: climate and human impacts in the North Atlantic." In William W. Fitzhugh and Elizabeth. Ward, eds. *Vikings: The North Atlantic Saga*, 385–393. Washington, DC: Smithsonian Institution Press, 2000.

Ojaveer, Henn, and Brian R. MacKenzie. "Historical development of fisheries in northern Europe: reconstructing chronology of interactions between nature and man," *Fisheries Research* 87 (2007), 102–105.

Ólafsdóttir, Guðbjörg Ásta, et al. "Historical DNA reveals the demographic history of Atlantic Cod (Gadus Morhua) in medieval and early modern Iceland," *Proceedings of the Royal Society of London B: Biological Sciences* 281 (2014): 20132976, with corrections *ibid.*, 7 September 2014 (1790): 20141279.

Oldfield, F., and R. L. Clark, "Environmental history – the environmental evidence." In Peter Brimblecombe and Christian Pfister, eds., *The Silent Countdown. Essays in European Environmental History*, 137–161. Berlin and Heidelberg: Springer, 1990.

Oliveira Marques, A. H. de. *Daily Life in Portugal during the Later Middle Ages.* Madison: University of Wisconsin Press, 1971.

*Hansa e Portugal na Idade Média*, 2d ed. rev. Lisbon: Editoral Presença, 1993.

Oliveira Marques, A. H. de, and João Pedro Ferro. "L'alimentation au Portugal du moyen âge au XVIIIe siècle." In Martin Aurell, Olivier Dumoulin, and Françoise Thélamon, eds., *La sociabilité à table: Commensalité et convivialité à travers les âges*, 283–291. Actes du Colloque de Rouen, 14–17 novembre 1990. Publications de l'Université de Rouen, 178. Rouen: Université de Rouen, 1992.

Olson, Sherri. *A Chronicle of All That Happens: Voices from the Village Court in Medieval England.* Studies and Texts, 124. Toronto: PIMS, 1996.

Olsson, Monica and Sven Isaksson. "Molecular and isotopic traces of cooking and consumption of fish at an Early Medieval manor site in eastern middle Sweden," *Journal of Archaeological Science* 35 (2008), 773–780.

Onori, Alberto Maria. *L'abbazia di San Salvatore a Sesto e il Lago di Bientina: Una signoria ecclesiastica 1250/1300.* Quaderni di Storia Urbana e Rurale, 3. Florence: Salimbene, 1984.

Oram, Richard. "Estuarine environments and resource exploitation in eastern Scotland c.1125 to c.1400. A comparative study of the Forth and Tay Estuaries." In Erik Thoen et al., eds., *Landscapes or Seascapes?. The History of the Coastal Environment in the North Sea Area Reconsidered*, 353–378. Turnhout: Brepols, 2013.

Origo, Iris. *The Merchant of Prato.* The Bedford Historical Series, 16. London: Jonathan Cape, 1960.

Ortega, Pablo, Flavio Lehner, Didier Swingedouw, Valerie Masson-Delmotte, Christoph C. Raible and Mathieu Casado. "A model-tested North Atlantic Oscillation reconstruction for the past millennium," *Nature* 523 (2015), 71–74.

Ortolani, Franco, and Silvana Pagliuca. "Cyclical climatic-environmental changes in the Mediterranean Area (2500 BP – present day)," *PAGES Past Global Changes News* 11:1 (April 2003), 15–17.

Orton, David C. "Archaeology as a tool for understanding past marine resource use and its impact." In Kathleen Schwerdtner Máñez and Bo Poulsen, eds. *Perspectives on Oceans Past. A Handbook of Marine Environmental History*, 47–69. Dordrecht: Springer, 2016.

Orton, David C., Alison Locker, James Morris, and James H. Barrett. "Fish for London." In Barrett and Orton, eds., *Cod and Herring*, 205–214. 2016.

Orton, David C., D. Makowiecki, T. de Roo, C. Johnstone, J. Harland, et al. "Stable isotope evidence for late medieval (14th–15th C) origins of the Eastern Baltic Cod (Gadus morhua) fishery. *PLoS ONE* 6:11(2011), e27568. doi:10.1371/journal.pone.0027568.

Orton, David C., James Morris, Alison Locker, and James H. Barrett. "Fish for the city: meta-analysis of archaeological cod remains and the growth of London's northern trade," *Antiquity* 88 (2014), 516–530.

Orton, David, James Morris, and Alan Pipe. "Catch per unit research effort: sampling intensity, chronological uncertainty, and the onset of marine fish consumption in historic London." *Open Quaternary* 3:1 (2017) 1–20, https://doi.org/10.5334/oq.29.

Osborn, Timothy J., and Keith R. Briffa. "The spatial extent of 20th-century warmth in the context of the past 1200 years," *Science* 311:5762 (10 February 2006), 841–844. DOI: 10.1126/science.1120514.

Ottenthal, E. von, ed. "Die ältesten Rechnungsbücher der Herren von Schlandersberg," *Mitteilungen des Instituts für Österreichische Geschichtsforschung* 2 (1881), 553–614.

Overland, James E., Juergen Alheit, Andrew Bakun, James W. Hurrell, David L. Mackas, and Arthur J. Miller. "Climate controls on marine ecosystems and fish populations," *Journal of Marine Systems* 79 (2010), 305–15

Øye, Ingvild, ed. *Bergen and the German Hansa*. Bergen: Bryggens Museum, 1994.

Page, William, ed. *Three Early Assize Rolls for the County of Northumberland, saec. xiii*. Publications of the Surtees Society, vol. 88. Durham: Published for the Society by Andrews & Co., 1891.

Palacín Gálvez, Maria del Carmen, and Luis Martínez-García, eds. *Documentacion del Hospital del Rey de Burgos 1136–1277*. Burgos: J.M. Garrido Garrido, 1990.

Papadopoli, Nicolò. *Le Moneta di Venezia descritte ed illustrate*, vol. 1. Facsimile reprint of Venezia-Milano, 1893. Bologna: Forni, 1970.

Pasda, Kerstin. "Tierhaltung und Jagd auf der mittelalterlichen Burg Sulzbach." In Mathias Hensch, *Burg Sulzbach in der Oberpfalz: archäologisch-historische Forschungen zur Entwicklung eines Herrschaftszentrums des 8. bis 14. Jahrhunderts in Nordbayern*, vol. 2, pp. 247–260. 3 vols., ed. Bayerisches Landesamt für Denkmalpflege. Materialien zur Archäologie in der Oberpfalz, Bd. 3. Büchenbach: Faustus, 2005.

*Tierknochen als Spiegel sozialer Verhältnisse im 8.–15. Jahrhundert in Bayern*. Praehistorika Monographien 1. Erlangen: Praehistorika archäologischer Verlag, 2004.

"Passio Quirini Tegernseensis," ed. Bruno Krutsch. *MGH SrM*, vol. 3, 8–20. Hannover: Hahn, 1896.

Passmore, D. G., and M. G. Macklin. "Geoarchaeology of the Tyne Basin: Holocene river environments and the archaeological record." In C. Tolan-Smith, ed., *Landscape Archaeology in Tynedale*, 11–28. Tyne–Solway Ancient and Historic Landscapes Research Monograph 1. Newcastle-upon-Tyne: University of Newcastle-upon-Tyne, 1997.

Pastor y Lluís, F. "La pesca de la saboga en el Ebro (Notas históricas)," *La Zuda* [Tortosa] num. 40, año 4 (1916), 108–109.

Pastre, Jean-François, et al. "Variations paléoenvironmentales et paléohydrologiques durant les 15 dernier millénaires: les réponses morphosédimentaires des vallées du Bassin Parisien (France)." In Bravard and Magny, eds., *Les fleuves ont une histoire*, 29–44. 2002.

Paul, A. "Knochenfunde aus dem mittelalterlich-neuzeitlichen Lübeck (Grabung Königstrasse 59–63)," *Lübecker Schriften zur Archäologie und Kulturgeschichte* 2 (1980), 7–104.

Pauly, Daniel. "Postscript. Anecdotes and the shifting baseline syndrome of fisheries," *Trends in Ecology and Evolution* 10:10 (1995), 430.

Pauly, Daniel, V. Christensen, J. Dalsgaard, J. R. Froese and F. Torres. "Fishing down marine food webs," *Science*, 279 (1998), 860–863.

Pavesi, Pietro, ed. *Ordini e statuti del paratico dei pescatori di Pavia*. Pavia: Fusi, 1894.[originally in *Bollettino Storico Pavese* anno 1:3–4 and anno 2:1–2].

Peck, Myron, Patricia Reglero, Motomitsu Takahashi, and Ignacio A. Catalán. "Life cycle ecophysiology of small pelagic fish and climate-driven changes in populations," *Progress in Oceanography* 116 (2013), 220–245.

Pederin, Ivan. "Commercio, economia, pesca, arti e mestieri in Arbe nel Quattrocentro." *Archivio Storico Italiano* 147:2:540 (1989), 215–249.

Pegolotti, Francesco Balducci. *La Pratica della Mercatura*, ed. Allan Evans. Cambridge, MA: Medieval Academy of America, 1936.

Pelc, Juljan. *Ceny w Krakowie w latach 1369–1600*. Badania z dziejów społecznych i gospodarczych, no. 14. Lwów: Skl. gl., Kasa im. J. Mianowskiego, 1935.

Percevaux, Paul. "Essai sur l'origine des étangs de Dombes," *Bulletin de la Société des Naturalists et Archéologues de l'Ain* (1962), 81–90.

"Structures et relations économiques en Dombes à la fin du Moyen Age," *Cahiers d'histoire* 12 (1967), 339–357.

Perdikaris, Sophia. "From chiefly provisioning to commercial fishery: long term economic change in Arctic Norway," *World Archaeology* 30:3 (1999), 388–402. [Special issue *Arctic Archaeology*, ed. Peter Rowley-Conwy.]

"Scaly heads and tales: Detecting commercialization in early fisheries." *Archaeofauna* 5 (1996) (special issue *Ichthyoarchaeology: Fish and the Archaeological Record: Proceedings of the Eighth Meeting of the ICAZ Fish Remains Working Group – Madrid 3–11 October 1995)*, 21–33.

Perdikaris, Sophia, and Thomas H. McGovern. "Codfish and kings, seals and subsistence: Norse marine resource use in the North Atlantic." In Torben C. Rick and Jon M. Erlandson, eds. *Human Impacts on Ancient Marine Ecosystems: A Global Perspective*, 187–214. Berkeley: University of California Press, 2008.

"Viking Age economics and the origins of commercial cod fisheries in the North Atlantic." In Louis Sicking and Darlene Abreu-Ferreira, eds., *Beyond the Catch: Fisheries of the North Atlantic, the North Sea and the Baltic, 900–1850*, 61–90. Leiden and Boston: Brill, 2009.

Pereira, Olegário Nelson Azevedo. "Pesca marítima na época medieval: o caso do Porto (Portugal) / Sea fishing in medieval epoch: the case of Porto (Portugal)," Antiteses *[Londrina]* 14:27 (August 2021), 55–83. DOI: 10.5433/1984-3356.2021v14n27p055

Perez-Embid Wamba, Javier. *El Cister en Castilla y Leon. Monacato y dominios rurales (Siglos XII–XV)*. [Salamanca?]: Junta de Castilla y Leon, Consejería de Educación y Cultura, 1986.

Perger, Richard. "Klosterneuburg im Mittelalter." In Floridus Röhrig, Gustav Otruba, and Michael Duscher, eds., for the Stadtgemeinde Klosterneuburg, *Klosterneuburg Geschichte und Kultur*. Bd. I. *Die Stadt*, 139–208. Klosterneuburg: Mayer & Co., n.d. [1992?]

Perrin, Charles-Edmond. *Recherches sur la Seigneurie rurale en Lorraine, d'après les plus anciens censiers (IXe–XIIe siècles)*. Publications de la Faculté des Lettres de l'Université de Strasbourg, fasc. 71. Paris: Belles lettres, 1935.

Pertz, G. H., ed. "Vita Iohannis abbatis Gorziensis,." In *MGH, SS*, IV, 335–377. Hannover: Hahn, 1841.

*La Pesca en la Edad Media*. Monographias de la Sociedad Española de Estudios Medievales, 1. Madrid: Sociedad Española de Estudios Medievales, 2009.

Petřík, Jan, Michal Hlavica, Libor Petr, Tomáš Chmela, Zdeněk Schenk, Hana Lukšíková, Peter Milo, Radim Vrla, Petr Odehnal, Zdeněk Petrůj, Martin Petrůj, and Petr Kočár. "Rybník Jako Součást Hospodářství Vrchnostenského Panství a Indikátor Podoby Krajiny Jižního Valašska v 15. až 17. Století." *Archaeologia historica* 42:2 (2017), 789–817.

Petrin, Silvia. "Das Archiv der Tullner Fischerzeche," *Mitteilungen aus dem Niederösterreichischen Landesarchiv* 3 (1979), 29–34.

Petts, Geoffrey. "Forested river corridors, a lost resource." In Denis Cosgrove, ed., *Water, Engineering, and Landscape: Water Control and Landscape Transformation in the Modern Period*, 12–24. London and New York: Belhaven Press, 1990.

Petts, Geoffrey E., Heino Moller, and A. L. Roux, eds. *Historical Change of Large Alluvial Rivers: Western Europe*. New York: Wiley, 1989.

Pfeifer, Wolfgang, et al., eds. *Etymologisches Wörterbuch des Deutschen*, 2d ed. rev. 3 vols. Berlin: Akademie-Verlag, 1993.

Pfister, Christian, G. Schwarz-Zanetti, and M. Wegmann. "Winter severity in Europe: the fourteenth century," *Climatic Change* 34 (1996), 91–108.

Philippe de Mézières. *Le Songe d'un vieil pèlerin*, ed. and tr. G. W. Coopland. 2 vols. Cambridge: Cambridge University Press, 1969.

*Songe du viel pelerin*, ed. Joël Blanchard with Antoine Calvet and Didier Kahn. 2 vols. Textes Littéraires Français. Geneva: Droz, 2015.

Phillips, Carla Rahn. "The long-term profitability of the tuna fisheries in southwestern Spain." In Simonetta Cavaciocchi, ed., *Ricchezza del mare – Ricchezza dal mare*, 2: 1121–1136 and 1146–1152. Atti delle "Settimane di Studi" e altri Convegni, Ser. II, 37, 2 vols. Prato: Istituto Internazionale di Storia Economica "F. Datini," 2006.

"Who owns the fish in the sea? The Dukes of Medina Sidonia and Spain's tuna fisheries." In Alan Karras and Laura J. Mitchell, eds., *Encounters Old and New in World History: Essays Inspired by Jerry H. Bentley*, 78–90. Perspectives on the Global Past. Honolulu: University of Hawai'i Press, 2017.

Piccolomini, Aeneas Silvius [Pope Pius II].*The Commentaries of Pius II*, tr. F. Gragg and L. Gabel. Abridged ed. New York: Capricorn, 1962.

*Commentarii rerum memorabilium que temporibus suis contigerunt*, ed. Adriano van Heck. 2 vols. Studi e testi, 312–313. Vatican City: Biblioteca apostolica Vaticana, 1984.

Pichot, Daniel, and Dominique Marguerie. "Approche pluridisciplinaire sur l'aménagement des petits cours d'eau par les sociétés médiévales dans l'ouest de la France (VII–XIIe siècle)." In Joëlle Burnouf and Phillipe

Leveau, eds., *Fleuves et marais, une histoire au croisement de la nature et de la culture. Sociétés préindustrielles et milieux fluviaux, lacustres et palustres: pratiques sociales et hydrosystèmes*, 113–125. Paris: CTHS, 2004.

Piekosiński, Franciszek, ed. *Kodeks dyplomatyczne miasta Krakowa (1257–1506)*, 2 vols., Monumenta medii aevi historica res gesta Poloniae illustrantia, vols. 5 and 7. Kraków: Nakł. Akad. Umiejętnośc 1879–1882. Reprinted New York, 1965. [NB: Charters and pages are numbered consecutively across the two volumes.]

ed. *Rachunki dworu króla Władysława Jagiełły i królowej Jadwigi z lat 1388 do 1420*. Monumenta medii aevi historica res gestas Poloniae illustrantia, vol. 15. Kraków: Nakł. Akad. Umiejętnośc, 1896.

Piekosiński, Franciszek, and Józef Szujski, eds. *Najstarsze księgi i rachunki miasta Krakowa od r. 1300 do 1400*. Monumenta medii aevi historica res gestas Poloniae illustrantia, vol. 4. Kraków: Nakł. Akad. Umiejętnośc, 1877–1878.

Pieters, Marnix. "The archaeology of fishery, trade, and piracy. the material environment of Walraversijde and other late medieval and early modern fishing communities along the southern North Sea." In Marnix Pieters, Frans Verhaeghe and Glann Gevaert, eds., *Fishery, Trade and Piracy: Fishermen and Fishermen's Settlements in and around the North Sea Area in the Middle Ages and Later*, 41–61. Brussels: Vlaams Instituut voor het Onroerend Erfgoed, 2006.

Pigière, Fabienne, Ides Boone, Mircea Udrescu, Wim Van Neer, and Sofie Vanpoucke. "Status as reflected in food refuse of late medieval noble and urban households at Namur (Belgium), in Sharyn Jones O'Day, Wim van Neer, and Anton Ervynck, eds., *Behaviour behind Bones. The Zooarchaeology of Ritual, Religion, Status and Identity*, 233–243. Proceedings of the 9th Conference of the International Council of Archaeozoology, Durham, August 2002. Oxford: Oxbow Books, 2004.

Pini, Antonio I. "Pesce, pescivendoli e mercanti de pesce in Bologna medievale," *Il Carrobio* 1 (1975), 328–349.

Pinnegar, John, and George H. Engelhard. "The 'shifting baseline' phenomenon: a global perspective," *Reviews in Fish Biology and Fisheries* 18:1 (April 2008), 1–16.

Pinto, Giuliano. "Incolti, fiumi, paludi. Alcune considerazioni sulle risorse naturali nella Toscana medievale e moderna." In Alberto Malvolti and Giuliano Pinto, eds., *Incolti, fiumi, paludi. Utilizzazione delle risorse naturali nella Toscana medievale e moderna*, 1–16. Firenze: Leo S. Olschki, 2003.

*Toscana medievale. Paesaggi e realtà sociali*. Firenze: Le Lettere, 1993.

Planche, A. "La Table comme signe de la classe. Le Témoignage du Roman du Comte d'Anjou (1316)." In Menjot, ed., *Manger et boire*, I:239–260. 1984.

Platina, Bartolomeo [aka Sacchi, Bartolomeo]. *De honesta voluptate et valetudine*. Rome:[n.p.] 1474, ed. Emilio Faccioli in his *Arte della cucina. Libri di recetti, testi sopra lo scalco, il trinciante e i vini dal XIV al XIX secolo*, I: 205–233. 2 vols. Milan: Edizioni Il Polifilio, 1966. Italian version is *Il piacere onesto e la buona salute* [Venice: [n.p.], 1487], ed. Emilio Faccioli. Turin: Giulio Einaudi, 1985.

Pleji, Herman. *Dreaming of Cockaigne: Medieval Fantasies of the Perfect Life*, tr. Diane Webb. New York: Columbia University Press, 2001.

Pliny [the Elder]. *Natural History*, ed. and tr. H. Rackham. 10 vols. London and Cambridge, MA, Harvard University Press, 1942–1984.

Plouchard, Pierre. "La Scarpe et les gens de rivière aux XIVe et XVe siècles," *Revue du Nord* 79:322 (1997), 841–858.

Plouvier, Liliane. "La gastronomie dans le 'Viandier de Taillevent' et le 'Ménagier de Paris'." In Menjot, ed., *Manger et boire*, vol. 2, pp. 149–159. 1984.

Plumettaz, Nicole. "Une pêcherie du XIIe siècle à Marin-Pré de la Mottaz (Neuchâtel, Suisse),." In Benoît et al., eds., *Pêche et pisciculture en eau douce*. 2004.

Podwińska, Zofia. "Rozmieszczenie wodnych młyno zbożowych w Małopolsce w XV wieku," *Kwartalnik Historii Kultury Materialnej* 18 (1970), 373–402.

Polet, Caroline and M. Anne Katzenberg. "Comportements alimentaires de trois populations médiévales belge: apports de la biogéochimie isotopique," *Revue belge de philologie et d'histoire* 80:4 (2002), 1371–1390.

Polet, C., and M. A. Katzenberg. "Reconstruction of the diet in a mediaeval monastic community from the coast of Belgium," *Journal of Archaeological Science* 30 (2001), 525–533.

Pollard, A. M., P. Ditchfield, E. Piva, S. Wallis, C. Falys, and S. Ford. "Sprouting like cockle amongst the wheat: the St Brice's Day Massacre and the isotopic analysis of human bones from St John's College, Oxford," *Oxford Journal of Archaeology* 31:1 (2012), 83–102.

Pöltsam, Inna. "Essen und Trinken in den livländischen Städten im Spätmittelalter." In Jüri Kivimäe and Juhan Kreem, eds., *Quotidianum Estonicum. Aspects of Daily Life in Medieval Estonia*, 118–127. Medium Aevum Quotidianum, Sonderband 5. Krems: Gesellschaft zur Erforschung der materiellen Kultur des Mittelalters, 1996.

Pontecorvo, Giulio, ed. *Fisheries Conflicts in the North Atlantic: Problems of Management and Jurisdiction*. Cambridge: Ballinger Publishing Company, 1974.

Pope, Peter. "Early estimates: assessment of catches in the Newfoundland cod fishery, 1660–1690." In Daniel Vickers, ed., *Marine Resources and Human Societies in the North Atlantic since 1500. Papers presented at the conference … October 20–22, 1995*. ISER Conference Paper Number 5. St. John's: Institute of Social and Economic Research, Memorial University of Newfoundland, 1995.

*The Many Landfalls of John Cabot*. Toronto: University of Toronto Press, 1997.

"Transformation of the maritime cultural landscape of Atlantic Canada by migratory European fishermen, 1500–1800." In Sicking and Abreu-Ferreira, eds., *Beyond the Catch*, 124–132. 2009.

Posthumus, N. W. *Inquiry into the History of Prices in Holland*. 2 vols. Leiden: Brill, 1946 and 1964.

Poulsen, Bjørn. "The herring fisheries off Heligoland, *c*1330–1550." In Poul Holm and David J. Starkey, eds., *North Atlantic Fisheries:*

*Markets and Modernization*, 7–38. Ejsberg: Fisheri- og Søfartsmuseets Forlag, 1998.

"Late medieval and early modern peasants of West Jutland and their markets." In Mette Goldberg, Poul Holm, and Per K. Madsen, eds., *Facing the North Sea: West Jutland and the World; Proceedings of the Ribe conference, April 6–8, 1992*, 37–52, Fiskeri- og Seefartsmuseets studieserie, 2. Esbjerg: Fiskeri- og Seefartsmuseets Forlag, 1993.

Poulsen, Bo. *Dutch Herring: An Environmental History, c.1600–1860*. Aksant: Amsterdam, 2008.

"The variability of fisheries and fish populations prior to industrialized fishing: an appraisal of the historical evidence," *Journal of Marine Systems* 79 (2010) 327–332.

Powell, Owen, ed. and tr. *Galen. On the Properties of Foodstuffs (De alimentorum facultatibus)*. Cambridge: Cambridge University Press, 2003.

Powers, James F., ed. *The Code of Cuenca: Municipal Law on the Twelfth-Century Castilian Frontier*. Philadelphia: University of Pennsylvania Press, 2000.

Prat, N. F., M. A. Puig, G. Gonzalez, M. F. Tort, and M. Estrada, "Llobregat." In B. A. Whitton, ed., *Ecology of European Rivers*, 527–552. Oxford: Blackwell, 1984.

Preston, Todd. "Fact and fish tales in Ælfric's Colloquy," *Mediaevalia* 40 (2019), 1–25.

Prestwich, Michael, ed. and tr. *York Civic Ordinances, 1301*. Borthwick Papers, no. 49. York: University of York. Borthwick Institute of Historical Research, 1976.

Pribram, Alfred F. *Materialien zur Geschichte der Preise und Löhne in Österreich*, vol. 1. Vienna: C. Ueberreuter, 1938.

Pribyl, Kathleen. *Farming, Famine, and Plague: The Impact of Climate in Late Medieval England*. Cham: Springer, 2017.

Prilloff, Ralf-Jürgen. *Tierknochen aus dem mittelalterlichen Konstanz. Eine archäologische Studie zur Ernährungswirtschaft und zu Handwerk im Hoch- und Spätmittelalter. Materialhefte zur Archäologie in Baden-Württemberg, Heft 50*. Landesdenkmalamt Baden-Württemberg. Stuttgart: Konrad Theiss Verlag, 2000.

Prior, M. *Fisher Row: Fishermen, Bargemen, and Canal Boatmen in Oxford, 1500–1900*. Oxford,: Clarendon,1982.

Privat, Karen L., Tamsin C. O'Connell, and Michael P. Richards. "Stable isotope analysis of human and faunal remains from the Anglo-Saxon cemetery at Berinsfield, Oxfordshire: dietary and social implications," *Journal of Archaeological Science* 29 (2002), 779–790.

Pross, A. Paul, and Susan McCorquodale. *Economic Resurgence and the Constitutional Agenda: The Case of the East Coast Fisheries*. Kingston, Ontario: Institute of Intergovernmental Relations, 1987.

Provansal, Mireille, Georges Pichard, and Edward J. Anthony. "Geomorphic changes in the Rhône Delta during the LIA: input from the analysis of ancient maps." In Mohamed Maanan and Marc Robin, eds., *Sediment Fluxes in Coastal Areas*, 47–72. Coastal Research Library, 10. Dordrecht: Springer Netherlands, 2015.

Ptaśnik, Jan, ed. *Cracovia artificum 1300–1500.* Kraków: Polska Akademia Umiejętności, 1917.

Pucci Donati, Francesca. "Il mercato del pesce. Regolamentazione dei prezzi a Bologna nel Medioevo (secc. XIII–XV)." In Giampiero Nigro, ed., *I Prezzi delle Cose nell'età Preindustriale / The Prices of Things in Pre-industrial Times*, pp. 35–56. Istituto Francesco Datini, XLVIII Settimana di Studi, 8–11 maggio 2016. Firenze: Firenze University Press, 2017.

Pudek, N. "Untersuchungen an Tierknochen des 13. bis 20. Jahrhunderts aus dem Heiligengeist Hospital in Lübeck," *Lübecker Schriften zur Archäologie und Kulturgeschichte* 2 (1980), 107–201.

Puig, Carole. "Les ressources de l'étang et de la mer, dans la partie occidentale du Golfe du Lion (du XII^e^ au XIV^e^ siècle)." In Marie-Claude Marandet, ed., *L'homme et l'animal dans les sociétés méditerranéennes*, 93–121. 4^e^ journée d'études du Centre de recherches historiques sur les sociétés méditerranéennes et du Pôle universitaire européen de Montpellier et du Languedoc–Roussillon. Université de Perpignan, 20 novembre 1998. Perpignan: Presses Universitaires de Perpignan, 2000.

Puñal Fernández, Tomás. *El Mercado en Madrid en la baja edad media. Estructura y sistemas de abastecimiento de un concejo medieval castellano (s. xv).* Madrid: Caja de Madrid, 1992.

Purkarthofer, Heinrich. "Die Teichwirtschaft der Herrschaft Herberstein," *Zeitschrift des Historischer Vereins für Steiermark* 56 (1965), 97–106.

Putelat, Olivier, and Thierry Logel, with collaboration of Hervé Bocherens, Eva-Maria Geigl, Silvia de Lima Guimaraes, and Bernhard Metz. "Une chasse aristocratique dans le Ried Centre-Alsace au premier Moyen Âge: l'apport de l'archéozoologie à la connaissance du site d'Ostheim *Birgelsgaerten* (Haut-Rhin, France)." In Édith Peytremann, ed., *Des fleuves et des hommes à l'époque mérovingienne. Territoire fluvial et société au premier moyen âge (V^e^–XII^e^ siècle).* Actes des 33^e^ journées internationales d'archéologie mérovingienne. Strasbourg, 28–30 septembre 2012, pp. 255–270. 42^ème^ supplément à la *Revue Archéologique de l'Est.* Dijon : S.A.E., 2016.

Querrien, Armelle. "Pêche et consommation du poisson en Berry au Moyen Age," *Bibliothèque de l'Ecole des Chartes* 161:2 (2003), 409–435.

"Pêche et consommation du poisson en Berry au Moyen Age,." In Benoît et al., eds., *Pêche et pisciculture en eau douce.* 2004.

Quinn, David B. *North America from Earliest Discovery to First Settlements: The Norse Voyages to 1612.* New York: Harper & Row, 1977.

Rackham, Oliver. *The History of the Countryside.* London: Dent, 1986. Reprinted London: Dent, 1993, and thereafter.

*Trees and Woodland in the British Landscape*, rev. ed. London: Dent, 1990. Reprinted London: Phoenix Press, 2001.

Rácz, Lajos. *The Steppe to Europe. An Environmental History of Hungary in the Traditional Age.* Cambridge: White Horse Press, 2013.

Radke, Christian. "Bemerkungen zum mittelalterliche Fischfang mit Heringszäunen in der Schlei," *Berichte über die Ausgrabungen in Haithabu* 10 (1977), 123–140.

Rasmussen, Kaare Lund, Lilian Skytte, Anne Juul Jensen, and Jesper Lier Boldsen. "Comparison of mercury and lead levels in the bones of rural and urban populations in southern Denmark and northern Germany during the Middle Ages," *Journal of Archaeological Science: Reports* 3 (2015), 358–370. doi:10.1016/j.jasrep.2015.06.021.

Rassart-Eeckhout, Emmanuelle. "Pratiques alimentaires à la fin du moyen âge. L'apport de la langue imagée." In Emmanuelle Rassart-Eeckhout, et al., eds., *La vie matérielle au moyen âge. L'apport des sources litteraires, normatives et de la pratique*. Actes du Colloque International de Louvain-la-Neuve 3–5 octobre 1996, 147–177. Louvain-la-Neuve: Université Catholique de Louvain, Publications de l'Institut d'Études Médiévales: Textes, Études, Congrès, vol. 18, 1997.

Ravoire, Fabienne, and Anne Dietrich, eds. *La cuisine et la table dans la France de la fin du Moyen Age. Contenus et contenants du XIV au XVI siècle*. Caen: Publications du CRAHM, 2009.

Real Academia Española. *Diccionario de Autoridades* [facsimile of the 1726 first edition]. Madrid: Real Academia Española, 1962.

Reese, David S. "The faunal remains from three cisterns," with a contribution by Alwyne Wheeler entitled "The fish remains." In *Excavations at Carthage*, Conducted by the University of Michigan, Institut National d'Archéologie et d'Art, and American Schools of Oriental Research, vol. 6, pp. 231–241. Ann Arbor: University of Michigan, 1981.

Regier, Henry J., Robin L. Welcomme, Robert J. Steedman, and H. Francis Henderson. "Rehabilitation of degraded river ecosystems." In Douglas P. Dodge, ed., *Proceedings of the International Large River Symposium*, 86–97. Canadian Special Publications in Fisheries and Aquatic Science, no. 106. Ottawa: Queen's Printer, 1989.

Reginald of Durham. *Libellus de vita et miraculis S. Godrici, heremitae de Finchale*, ed. Joseph Stevenson. Surtees Society, vol. 20. London: Surtees Society, 1847.

Reichstein, Hans. "Jagd und Fischfang." In G. Kossack, E.-E. Behtre and P. Schmid, eds., *Archäologische und Naturwissenschaftliche Untersuchungen an Siedlungen im deutschen Küstengebiet*. Bd. 1: *Ländliche Siedlungen*, 275–276. Weinheim: Verlag Chemie, 1984.

*Untersuchungen an Tierknochen von der Isenburg bei Hattingen/Ruhr: ein Beitrag zur Nahrungswirtschaft auf hochmittelalterlichen Burgen*. Hattingen: H. Eversberg für den Verein zur Erhaltung der Isenburg u. den Heimat- und Geschichtsverein Hattingen, 1981.

Reitsema, Laurie, J., D. E. Crews, and M. Polcyn. "Preliminary evidence for medieval Polish diet from carbon and nitrogen stable isotopes," *Journal of Archaeological Science* 37:7 (2010), 1413–23.

Reitsema, Laurie, J., T. Kozłowski, and D. Makowiecki. "Human–environmental interactions in medieval Poland: a perspective from the analysis of faunal stable isotope ratios," *Journal of Archaeological Science* 40 (2013), 3636–3646.

Reitsema, Laurie, J., and Giuseppe Vercellotti. "Stable isotope evidence for sex- and status-based variations in diet and life history at medieval Trino Vercellese, Italy," *American Journal of Physical Anthropology* 148:4 (August 2012), 589–600.

Renwick, R., ed. *Extracts from the Records of the Royal Burgh of Stirling, A.D. 1519–1666 with Appendix A.D. 1295–1666*. Glasgow: for the Glasgow Stirlingshire and Sons of the Rock Society, 1887.

Requate, H. "Die Jagdtier in den Nahrungssytem einiges frühgeschichtliches Siedlung in Schleswig-Holstein," *Schriften des Naturwissenschaftliches Vereins für Schleswig-Holstein* 28 (1956), 21–41.

Retsö, Dag, and Johan Söderberg. "Winter severity in medieval Sweden: the documentary evidence." In Andrea Kiss and Kathleen Pribyl, eds., *The Dance of Death in Late Medieval and Renaissance Europe: Environmental Stress, Mortality and Social Response*, 25–45. Routledge Environmental Humanities Series. London and New York: Routledge, 2020.

Reynolds, Rebecca. "The social complexities of early marine fish consumption: new evidence from southeast England." In Barrett and Orton, eds., *Cod and Herring*, 215–220. 2016.

Reynolds, Terry S. *Stronger Than a Hundred Men. A History of the Vertical Water Wheel*. Baltimore: Johns Hopkins University Press,1983.

Ribi, Adolf. *Die Fischbenennungen des Unterseegebietes; mit dem erstmaligen Abdruck des Fischbüchleins von Gregor Mangolt, nach dem Original-Handschrift, 1557*. Rüschlikon: Baublatt A.G., 1942.

Richard, Jean. "Le commerce du poisson en Bourgogne et les étangs de la région Autunoise," *Mémoires de la Société Éduenne* 44 (1983), 181–197.

"Les Etangs et le commerce du poisson en Bourgogne." In *L'eau en Bourgogne*, 97–101. Cahiers du Centre d'Etudes Régionales de Bourgogne, no. 3. Dijon: Centre d'Etudes Régionales de Bourgogne, 1986.

"Transports par eau et péages de Chalon à Avignon: à propos des fournitures de poisson à la cour des papes," *Actes du 112ᵉ Congrès national des Sociétés savantes, Lyon, 1987: Section d'histoire médiévale et de philologie*, 37–44. Paris, 1989.

Richards, John F. *The Unending Frontier. An Environmental History of the Early Modern World*. Berkeley: University of California Press, 2003.

Richardson, David M., and Anthony Ricciardi, "Misleading criticisms of invasion science: a field guide," *Diversity and Distribution* 19 (2013), 1461–1467.

Richental, Ulrich [aka von Richental]. *Konstanzer Konzilschronik* [facsimile], ed. Otto Feger, *Ulrich Richental: Das Konzil zu Konstanz, 1414–1418*. 2 vols. Starnberg: Josef Keller; Konstanz: Jan Thorbecke, 1964. Translated in Louise R. Loomis, *The Council of Constance: The Unification of the Church*, ed. and annotated by John Hine Mundy and Kennerly M. Woody, 84–199. New York: Columbia University Press, 1961.

Rick, Torben C., and Jon M. Erlandson, eds. *Human Impacts on Ancient Marine Ecosystems. A Global Perspective*. Berkeley: University of California Press, 2008.

Riddler, Ian, and Nicola Trzaska-Nartowski. "From Dover to New Romney: medieval fishing in south-east Kent, England." In Daniel Makowiecki et al., eds., *Fishes– Culture – Environment through Archaeoichthyology, Ethnography & History*, 35–37. The 15th Meeting of the International Council for Archaeozoology (ICAZ) Fish Remains Working Group (FRWG),

September 3–9, 2009, in Poznań & Toruń, Poland. Poznań: Bogucki Wydawnictwo Naukowe, 2009.

Riepertinger, Rainhard. "Typologie der Unruhen in Herzogtum Bayern, 1525," *Zeitschrift für bayerische Landesgeschichte* 51 (1988), 329–386.

Riera Melis, Antoni. "La pesca en el Mediterráneo Noroccidental durante la Baja Edad Media." In *Pesca en la Edad Media,* 121-144. Madrid: Sociedad Espanola de Estudios Medievales, 2009.

"Sistemes alimentaris i estructura social a la Catalunya de l'alta edat mitjana." In *Alimentació i societat a la Catalunya medieval*, Anuario de Estudios Medievales, Anex 20, 1-26. Barcelona: Instituto de Historia Medieval de España, 1988.

Rijnsdorp, Adriaan D., Myron A. Peck, Georg H. Engelhard, Christian Möllmann, and John K. Pinnegar. "Resolving the effect of climate change on fish populations." *ICES Journal of Marine Science* 66: (2009), 1570–1583.

Riley, Henry T., ed. and tr. *Liber Albus: The White Book of the City of London compiled A.D. 1419 by John Carpenter, Common Clerk, and Richard Whitington, Mayor.* London: R. Griffin, 1861.

Rippmann, Dorethee. "Un aliment sain dans un corps sain: santé et systèmes culinaires au Moyen Âge." In F. Audoin-Rouzeau and F. Sabban, eds., *Un aliment sain dans un corps sain. Perspectives historiques.* 2e colloque de l'Institut européen histoire et des cultures de alimentation, Tours, décembre 2002, 39–64. Tours: Presses Universitaires François Rabelais, 2007.

Rippmann, Dorethee, and Brigitta Neumeister-Taroni, eds., *Les manguers de l'an 1000. Archéologie et alimentation.* Vevey: Foundation Alimentarium, 2000.

Rippon, Stephen. "Making the most of a bad situation? Glastonbury Abbey, Meare, and the medieval exploitation of wetland resources at the Somerset Levels," *Medieval Archaeology* 48 (2004), 91–130

*The Transformation of Coastal Wetlands: Exploitation and Management of Marshland Landscapes in North West Europe during the Roman and Medieval Periods.* Oxford and New York: Oxford University Press, 2000.

Robbert, Louise B. "Twelfth-century Italian prices," *Social Science History* 7 (1983), 381–404.

Roberts, Brian K. "Fish culture in sixteenth-century Poland," *Agricultural History Review* 16 (1968), 161–163.

Roberts, Callum. *The Unnatural History of the Sea. The Past and Future of Man and Fishing.* London: Gaia, 2007.

Roberts, Edward. "The Bishop of Winchester's fishponds in Hampshire, 1150–1400: their development, function and management," *Proceedings of the Hampshire Field Club and Archaeological Society* 42 (1986), 125–138.

Robinson, Robb, and David J. Starkey. "The sea fisheries of the British Isles, 1376–1976: a preliminary survey." In Holm, Starkey, and Thor, eds., *The North Atlantic Fisheries, 1100–1976,* 121–144. Esbjerg: Fiskerei- og Søfartsmuseets Forlag, 1996.

Roca Cabau, Guillem. "Provision and consumption of fish in a Catalan inland city during the fourteenth and fifteenth centuries: the Case of Lleida," *Imago Temporis. Medium Aevum* 15 (2021), 273–306

Rodrigo Estevan, María L. "Fresco, frescal, salado, seco, remojado: abasto y mercado de pescado en Aragón (siglos XII–XV)." In Beatriz Arizaga Bolumburu and Jesús A. Solorzano Telechea, eds., *Alimentar la ciudad en la edad media*, 547–577. Actas de los V encuentros internacionales del medievo, celebrados del 22 al 25 de julio de 2008 en Nájera (La Rioja). Logroño: Instituto de Estudios Rojanos; Nájera: Ayuntamiento de Nájera, 2009.

Rogers, Charles, ed. *Rental Book of the Cistercian Abbey of Cupar-Angus, with the Breviary of the Register*. 2 vols. London: Printed for the Grampian Club, 1879–1880. [Cupar *sic*]

Rogers, James E. Thorold. *A History of Agriculture and Prices in England*. 7 vols. in 8. Oxford: Clarendon, 1866–1902.

Rogers, Raymond A. *The Oceans Are Emptying: Fish Wars and Sustainability*. Montreal: Black Rose Books, 1995.

Rohde, Klaus. "Latitudinal diversity gradients: equilibrium and nonequilibrium explanations." In his *The Balance of Nature and Human Impacts*, 155–167. Cambridge: Cambridge University Press, 2013.

Rohling, Eelco, Ramedan Abu-Zied, James Cusford, Angela Hayes, and Babette Hoogakker. "The marine environment: present and past." In Jamie Woodward, ed., *The Physical Geography of the Mediterranean*, 33–68. Oxford: Oxford University Press, 2009.

Röhrig, Floridus. "Die materiellen Kultur des Chorherrenstiftes Klosterneuburg unter besonderer Berücksichtigung der Aussage von Rechnungsbücher,." In *Klösterliche Sachkultur des Spätmittelalters*, 217–224. Veröffentlichungen des Instituts für mittelalterliche Realienkunde Österreichs, 3; Sitzungsberichte der Österreichischen Akademie der Wissenschaften, phil.-hist. Klasse 367. Vienna: ÖAW, 1980.

Romano, Dennis. *Markets and Marketplaces in Medieval Italy, c.1100 to c.1450*. New Haven: Yale University Press, 2015.

Rombai, Leonardo. "Le acque interne in Toscana tra Medioevo ed età moderna, Il Caso delle maremme." In Alberto Malvolti and Giuliano Pinto, eds., *Incolti, fiumi, paludi. Utilizzazione delle risorse naturali nella Toscana medievale e moderna*, 17–42 Firenze: Leo S. Olschki, 2003.

Rome [city]. *Statuta Almae Urbis Romae*. Rome: In aedibus Popoli Romani, 1580.

Rosa Salva, Paolo, and Sergio Sartori. *Laguna e pesca. Storia, tradizioni e prospettive*. Quaderni di Materiali veneti, n. 6, 9–20. Materiali veneti: 10–11 (Supplement). Venezia: Arsenale Cooperativa Editore, 1979.

Rose, George A. *Cod. The Ecological History of the North Atlantic Fisheries*. St. John's.: Breakwater Books, 2007.

Rosello Izquierdo, Eufrasia, Arturo Morales Muñiz, and D. C. Morales, "La Cartuja/Spain: Anthropogenic ichthyocenosis of culinary nature in a paleocultural context," *Offa* 51 (1994): 323–331

Rösener, Werner. *Bauern im Mittelalter*. Munich: C. H. Beck Verlag, 1985.

*Grundherrschaft im Wandel: Untersuchungen zur Entwicklung geistlicher Grundherrschaften im südwestdeutschen Raum vom 9. bis 14. Jahrhundert*. Veröffentlichungen des Max-Planck-Instituts für Geschichte, 102. Göttingen 1991.

*Reichsabtei Salem. Verfassungs- und Wirtschaftsgeschichte des Zisterzienserklosters von der Gründung bis zur Mitte des 14. Jh.* Sigmaringen: Thorbecke, 1974.

Rossiaud, Jacques. *Le Rhône au Moyen Âge. Histoire et répresentations d'un fleuve européen.* Paris: Aubier, 2007.

Rostafiński, Józef. *Średniowieczna historia naturalna w Polsce. Symbola ad historia naturalna medii aevi*, 2 vols. Munera saecularis universitatis Cracoviensis, vols. 7–8. Kraków: Nakł. Uniwersytetu, 1900.

Rouché, Michel. "La faim à l'époque carolingienne: essai sur quelques types de rations alimentaires," *Revue historique* 250 (1973), 295– 320.

Rouillard, Joséphine. "La législation royale de la pêche en eau douce du XIIIe au début du XVe siècle." In Benoît et al. eds., *Pêche et pisciculture en eau douce.* 2004.

Rouillard, Joséphine, and Véronique Maupoume. "Les étangs royaux sous Philippe le Bel d'après les comptes royaux." In Benoît et al., eds., *Pêche et pisciculture en eau douce.* 2004.

Rucquoi, A. "Alimentation des riches, alimentation des pauvres dans une ville castillane au XVe siecle." In Menjot, ed., *Manger et boire*, I, 297–312. Paris: Belles lettres, 1984.

Ruiz, Juan [Archpriest of Hita]. *Libro de Buen Amor*, ed. Anthony H. Zahares, tr. Saraly R. Daly. University Park, PA: Pennsylvania State University Press, 1991.

Rulewicz, Marian. *Rybołówstwo Gdańska na tle Ośrodków Miejskich Pomorza od IX do XIII Wieku.* Gdańsk Wczesnośredniowieczny, vol. 10. Gdańskie Towarzystwo Naukowe, Wydział i Nauk Społecznych i Humanistycznych: Prace Komisji Archeologicznej, nr. 11. Wrocław: Zakład Narodowy im. Ossolińskich, 1994.

*Ruodlieb: Faksimile-Ausgabe des Codex Latinus Monacensis 19486 der bayerischen Staatsbibliothek München und der Fragmente von Sankt Florian*, ed. W. Haug and Benedikt Vollmann, 2 vols. 2/1: *Kritischer Text.* Wiesbaden: Ludwig Reichert, 1974 and 1985.

Rybarski, Roman. *Gospodarstwo Księstwa Oświęcimskiego w XVI wieku.* Polska Akademia Umiejętności: Rozprawy Wydziału Historyczno-Filozoficznego, Serja II. T. 43, Nr. 2. Kraków: Polska Akademia Umiejętności, 1931.

Rynne, Colin. "Waterpower in medieval Ireland." In Squatriti, ed., *Working with Water in Medieval Europe*, 1–50. 2000.

Sacks, David H. *The Widening Gate: Bristol and the Atlantic Economy, 1450–1700.* Berkeley: University of California Press, 1991.

Salamon, M., A. Copps, M. McCormick, M. Rubini, R. Vargiu, and N. Tuross. "The consilience of historical and isotopic approaches in reconstructing the medieval Mediterranean diet," *Journal of Archaeological Science* 35 (2008), 1667–1672.

Salicru i Llunch, Roser. "En torno al comercio de pescado atlántico ibérico en el mediterráneo catalanoaragonés del siglo XV." In *Pesca en la Edad Media*, 167–180. Madrid: Sociedad Espaňnola de Estudios Medievales, 2009.

Salisbury, C. R. "Primitive British fishweirs." In G. L. Good et al., eds., *Waterfront Archaeology: Proceedings of the 3rd International Conference on*

*Waterfront Archaeology held at Bristol, 1988*, CBA research reports 74, 76–87. London: Council for British Archaeology, 1991.

Salvadori, Frank. "La pesca nel Medioevo: le evidenze della cultura material." In Jacopo De Grossi Mazzorin, D. Saccà, and C. Tozzi, eds. *Atti del 6° Convegno Nazionale di Archeozoologia* (Parco dell'Orecchiella, San Romano in Garfagnana – Lucca, 21–24 maggio 2009), 297–305. Lecce: Associazione Italiana di Archeozoologia, 2012.

"The transition from late antiquity to early Middle Ages in Italy: a zooarchaeological perspective," *Quaternary International* 499 (January 2019), 35–48.

*Salzburger Urkundenbuch* Bd. I: *Traditionscodices*, ed. Willibald Hauthaler. Salzburg: Selbstverlag der Gesellschaft für Salzburger Landeskunde. 1910; reprinted Aalen: Scientia, 1987.

Sánchez-Garduño, Faustino, Pedro Miramontes, and Tatiana T. Marquez-Lago. "Role reversal in a predator–prey interaction," *Royal Society Open Science*, 1:2 (1 October 2014), 140186. doi: 10.1098/rsos.140186.

Sánchez Quiñones, Julián. *Pesca y Comercio en el Reino de Castilla Durante la Edad Media: Los valles del Guadiana, Júcar y Tajo (siglos XII y XVI)*. Serie Histórica, 6. Madrid: Ediciones de La Ergástula, 2014.

"Los Precios del Pescado en Guadalajara en el Siglo XV: Problemas y Factores de Influencia." In *Pesca en la Edad Media*, 181–192. Monographias de la Sociedad Española de Estudios Medievales, 1. Madrid : Sociedad Española de Estudios Medievales, 2009.

Sanfaçon, Roland. *Défrichements, peuplement et institutions seigneuriales en Haut-Poitou du X au XIIIe siècle*. Quebec: les Presses de l'Universite Laval, 1967.

Sanfilippo, Isa Lori. *La Roma dei romani: arti, mestieri e professioni nella Roma del Trecento*. Rome: Istituto storico italiano per il Medio Evo, 2001.

Santich, Barbara. "Les éléments distinctifs de la cuisine médiévale méditerranéenne." In Carole Lambert, ed., *Du manuscrits à la table: Essais sur la cuisine au Moyen Âge et répertoire des manuscrits médiévaux contenant des recettes culinaires*, 133–139. Montréal: Champion-Slatkine, les Presses de l'Université de Montréal, 1992.

Santucci, M. "Nourritures et symboles dans le Banquet du Faisan et dans Jehan de Saintré." In Menjot, ed., *Manger et boire*, I, pp. 429–440. Paris: Belles lettres, 1984.

Sardi, Cesare. *Le Contrattazioni agrarie de medio evo studiate nei documenti lucchesi*. Sardi: Recti Lucca Giusti, 1914.

Sarnowsky, Jürgen. *Die Wirtschaftsführung des Deutschen Ordens in Preußen (1382–1454)*. Veröffentlichungen aus den Archiven Preussischer Kulturbesitz, Bd. 34. Cologne: Böhlau, 1993.

Sarti, R. *Europe at Home: Family and Material Culture 1500–1800*. tr. Allan Cameron. New Haven and London: Yale University Press, 2002.

Sasse, Barbara. *Die Sozialstruktur Böhmens in der Frühzeit. Historisch-archäologische Untersuchungen zum 9.–12. Jahrhundert*. Berliner Historischen Studien, Bd. 7; Germanica Slavica, vol. 4. Berlin: Duncker & Humblot, 1982.

Sauerländer, Willibald. "Architecture and the figurative arts: the north." In Robert L. Benson and Giles Constable, eds., *Renaissance and Renewal in the Twelfth Century*, 671–710. Cambridge, MA: Harvard University Press, 1982.

Saul, A. "Great Yarmouth and the Hundred Years War in the fourteenth century," *Bulletin of the Institute of Historical Research* 52 (1979), 105–115

"The herring industry at Great Yarmouth c. 1280–1400," *Norfolk Archaeology* 38 (1981), 33–43.

Sawicki, Kazimierz. "Jana Dubrawiusza, Biskupa Czeskiego, Nauka Niwelacji," *Kwartalnik Historii Nauki i Techniki* 15:4 (1970), 759–774.

Saxo Grammaticus. *Gesta Danorum / The History of the Danes*, ed. Karsten Friis-Jensen, tr. Peter Fisher. 2 vols. Oxford: Clarendon Press, 2015.

Sayers, W. "Some fishy etymologies: Eng. *Cod*, Norse *Þorskr*, Du. *Kabelhauw*, Sp. *Bacalao*," *NOWELE: North-West European Language Evolution* 41 (2002), 17–30.

Schäffer, J. and A. von den Driesch. "Tierknochenfunde aus fünf frühmittelalterlichen Siedlungen Altbayerns," *Documenta naturae* [Munich] 15 (1983), 1–78.

Schalk, Randall F. "The structure of an anadromous fish resource." In Lewis R. Binford, ed., *For Theory Building in Archaeology*, 207–249. New York, San Francisco, and London: Academic Press, 1977.

Scheffer, Marten. *Critical Transitions in Nature and Society*. Princeton: Princeton University Press, 2009.

Scheiber, Artur Maria. *Zur Geschichte der Fischerei in Oberösterreich, insbesondere der Traunfischerei. Sonderabdruck aus den "Heimatgauen": Zeitschrift für oberösterreichische Geschichte, Landes- und Volkskunde*. Linz: Verlag R. Pirngruber, 1930.

Schenk, Gerrit J. "'... prima ci fu la cagione de la mala provedenza de'Fiorentini.' Disaster and life world reactions in the commune of Florence to the flood of November 1333," *Medieval History Journal* 10 (2007), 355–386.

Schlosser, Julius Ritter von. "Ein veronesisches Bilderbuch und die höfische Kunst des XIV. Jahrhunderts," *Jahrbuch der Kunsthistorischen Sammlungen der Allerhöchste Kaiserhause* 16 (1895), 144–230.

Schmauderer, Eberhard. *Studien zur Geschichte der Lebensmittelwissenschaft*, 2 vols. Vierteljahrsschrift für Sozial- und Wirtschaftsgeschichte, Beiheft 62. Wiesbaden: Steiner, 1975.

Schmidt, Ulrich. "Himmelkron. Grundbesitz und wirtschatliche Verhältnisse des ehemaligen Zisterzeinserinnenklosters," *Archiv für Geschichte von Oberfranken*, 48 (1968), 7–70.

Schmidt, Volker ed. *Lieps. Eine slawische Siedlungskammer am Südende des Tollensesees*. Beiträge zur Ur- und Frühgeschichte der Bezirke Rostock, Schwerin und Neubrandenburg, Bd. 16. Berlin: Deutscher Verlag der Wissenschaften,1984.

Schmitt, Jean-Claude. *The Holy Greyhound: Guinefort, Healer of Children since the Thirteenth Century*, tr. Martin Thom., Cambridge Studies in Oral and Literate Culture, 6. Cambridge: Cambridge University Press, 1983.

Schmölcke, Ulrich, and Dirk Heinrich. "Die Tierknochen aus dem Hafen von Haithabu – Schlämmfunde." In C. Radtke, ed., *Untersuchungen an Skelettresten von Tieren aus dem Hafen von Haithabu*, 195–239. Berichte über die Ausgrabungen in Haithabu, Bericht 35. Neumünster: Wachholz, 2006.

Schneider, Reinhard. "Lebensverhältnsse bei den Zisterziensern im Spätmittelalter." In *Klösterliche Sachkultur des Spätmittelalters*, 43–71. Veröffentlichungen des Instituts für mittelalterliche Realienkunde Österreichs, 3. Sitzungsberiechte der Österreichischen Akademie der Wissenschaften, phil.-hist. Klasse, 367. Vienna: ÖAW. 1980.

Schofield, John, and Alan Vince. *Medieval Towns: The Archaeology of British Towns in Their European Setting*, 2d ed. London: Continuum, 2003.

Schreiber, Dorothee, and Camilla Brattland, eds. *Salmon Cultures. Indigenous Peoples and the Aquaculture Industry*. RCC Perspectives, 2012/4. Munich: Rachel Carson Center, 2012.

Schreiner, Klaus. "Zisterziensisches Möchtum und soziale Umwelt: Wirtschaftlicher und sozialer Strukturwandel in hoch- und spätmittelalterlichen Zisterzienserkonventen." In K. Elm, ed., *Die Zisterzienser. Ordensleben zwischen Ideal und Wirklichkeit, Erganzungsband*, 79–136. Schriften des Rheinische Museumsamtes, 18. Cologne: Rheinland-Verlag: Wienand Verlag, 1982.

Schubert, Ernst. *Essen und Trinken im Mittelalter*, 3ed ed. (Original ed. 2006.) Darmstadt: Wissenschaftliche Buchgesellschaft, 2016.

Schuhmann, Günther, and Gerhard Hirschmann, eds. *Urkundenregesten des Zisterzienserklosters Heilsbronn*. Erster Teil, *1132–1287*. Veröffentlichungen der Gesellschaft für fränkische Geschichte, ser. 3, vol. 3. Würzburg: Kommissionsverlag F. Schöningh, 1957.

Schulte, Aloys. "Die Urkunde Walafrid Strabos von 843. Eine Fälschung," Zeitschrift für die Geschichte des Oberrheins, vol. 42, NF 3 (1888), 345–353.

Schultze, Johannes. "Ein mittelalterlicher Fischkenner," *Archiv für Fischereigeschichte* 2 (1914), 133–137.

"Ein Strassburger Handschrift des 16. Jahrhunderts," *Archiv für Fischereigeschichte* 3 (1914), 228–231.

Schulz, Anna. *Essen und Trinken im Mittelalter (1000–1300). Literarische, kunsthistorische und archäologische Quellen*. Ergänzungsbände zum Reallexikon der Germanischen Altertumskunde, vol. 74. Berlin: De Gruyter, 2011.

Schwab, Ingo, ed. *Das Prümer Urbar*. Rheinische Urbare, vol. 5. Publikationen der Gesellschaft für Rheinische Geschichtskunde, vol. 20. Düsseldorf: Droste Verlag, 1983.

Schwarz, Dietrich. "Die Küchenordnung des Abtes Ulrich VIII. Rösch von St. Gallen für den Hof in Wil von etwa 1480." In *Klösterliche Sachkultur des Spätmittelalters*, 268–277. Veröffentlichungen des Instituts für mittelalterliche Realienkunde Österreichs, 3; Sitzungsberichte der Österreichischen Akademie der Wissenschaften, phil.-hist. Klasse 367. Vienna: ÖAW, 1980.

Schwarz, Klaus. "Nochmals: Der Lachs und die Dienstboten," *Bremisches Jahrbuch* 77 (1998), 277–283. Online at http://brema.suub.uni-bremen.de/periodical/titleinfo/23118.

"Der Weserlachs und die bremischen Dienstboten: Zur Geschichte des Fischverbrauchs in Norddeutschland," *Bremisches Jahrbuch* 74/75 (1995/6), 134–173. Available online at http://brema.suub.uni-bremen.de/periodical/titleinfo/23118.

Schwarz, Werner M. "Umweltstrafen. Fragen nach den historischen Bewertungskriterien von Umweltdelikten." In Gerhard Jaritz and Verena Winiwarter, eds., *Umweltbewältigung. Die historische Perspektive*, 77–98. Bielefeld: Verlag für Regionalgeschichte, 1994.

Schwerdtner Máñez, Kathleen, and Bo Poulsen, eds. *Perspectives on Oceans Past: A Handbook of Marine Environmental History*. Dordrecht: Springer, 2016.

Scialoja, Antonio. "Statuta et ordinamento artis piscium civitatis Perusii (1296–1376)," *Bollettino della Regia Deputazione di Storia Patria per l'Umbria*, 1910, 813–873.

Sclafert, Thérèse. *Le Haut-Dauphiné au Moyen Age*. Paris: Société anonyme du Recueil Sirey, 1926.

Scott, Anthony D. "Catch quotas and shares in the fishstock as property rights." In Edward Miles, Robert Pealy, and Robert Stokes, eds., *Natural Resources Economics and Policy Applications. Essays in Honor of James A. Crutchfield*, 72–80. Seattle: Institute for Marine Studies, Graduate School of Public Affairs, and the Institute for Public Policy and Management of the University of Washington, 1986.

"Conceptual origins of rights-based fishing." In Philip A. Neher, Ragnar Arnason, and Nina Mollett, eds., *Rights Based Fishing*, 11–38. NATO ASI Series E: Applied Science, no. 169. Dordrecht: Kluwer, 1989.

Scottish Burgh Records Society, ed. *Ancient Laws and Customs of the Burghs of Scotland*. 2 vols. [vol 1, 1124–1424; vol 2, 1424–1707]. Edinburgh: Scottish Burgh Records Society, 1868 and 1910.

Scully, Terence. *The Art of Cookery in the Middle Ages*. Woodbridge, UK: Boydell & Brewer, 1995.

ed. and tr. *Chiquart's On Cookery. A Fifteenth-Century Savoyard Culinary Treatise*. New York and Bern, 1986.

ed. "*Du fait de cuysine* par Maistre Chiquart 1420 (Ms. S 103 de la bibliothèque Supersaxo, à la Bibliothèque cantonale du Valais, à Sion)," *Vallesia* 40 (1985), 101–231.

"The *Opusculum de saporibus* of Magninus Mediolanensis," *Medium Aevum*, 54 (1985), 178–207.

"Tempering medieval food." In Melitta Weiss Adamson, ed., *Food in the Middle Ages: A Book of Essays*, 3–24. New York and London: Garland, 1995.

ed. and tr. *The Viandier of Taillevent: An Edition of All Extant Manuscripts*. Ottawa: University of Ottawa Press, 1988.

ed. and tr., with Rudolf Grewe. *Cuoco Napoletano: The Neapolitan Recipe Collection (New York Pierpont Morgan Library, MS Bühler, 19)*. Ann Arbor: University of Michigan Press, 2000.

Seabourne, Gwen. *Royal Regulation of Loans and Sales in Medieval England: "Monkish Superstition and Civil Tyranny."* Woodbridge, UK, and Rochester, NY: Boydell and Brewer, 2003.

Seager, R., and R. Burgmann, "Medieval hydroclimate revisited," *PAGES news* 19:1 (March 2011), 10–12.

Seddon, Alistair W. R. "What do we mean by regime shift? Distinguishing between extrinsic and intrinsic forcing in palaeoecological data," *Past Global Changes Magazine* 25:2 (2017), 94–95.

Seeman, M. "Monnickendam – the faunal remains." In H. A. Heidinga and H. H. van Regteren Altena, *Medemblik and Monnickendam: Aspects of Medieval Urbanization in Northern Holland,* 125–134. Cingula, 11. Amsterdam: Universiteit van Amsterdam, Albert Egges van Giffen Instituut voor Prae- en Protohistorie, 1989.

Segurado, P., P. Branco, A. P. Avelar and M. T. Ferreira. "Historical changes in the functional connectivity of rivers based on spatial network analysis and the past occurrences of diadromous species in Portugal," *Aquatic Sciences* 77:3 (2015), 427–440.

Seligo, Arthur. "Zur Geschichte der Fischerei in Westpreussen," *Mitteilungen des westpreussischen Fischereivereins* 14 (1902), 2–72.

Serjeantson, Dale, and Pam Crabtree, with Jacqui Mulville, Kathy Ayres, Claire Ingrem and Alison Locker. "How pious? How wealthy? The status of Eynsham and St Albans Abbeys between the 8th to the 12th centuries re-examined in the light of their food consumption." In Ben Jervis, ed., *The Middle Ages Revisited: Studies in the Archaeology and History of Medieval Southern England Presented to Professor David A. Hinton,* 115–140. Oxford: Archaeopress, 2018.

Serjeantson, D., and C. M. Woolgar. "Fish consumption in medieval England." In C. M. Woolgar, D. Serjeantson, and T. Waldron, eds., *Food in Medieval England: Diet and Nutrition,* 131–147. Oxford: Oxford University Press, 2006.

Serrano Larráyoz, Fernando. *La mesa del rey. Cocina y régimen alimentario en la corte de Carlos III el noble de Navarra (1411–1425).* Pamplona: Gobierno de Navarra, 2002.

Sicard, Germain. *Les moulins de Toulouse au Moyen Age.* Paris: Colin, 1953.

Sickel, T., ed. *Die Urkunden Otto des II. und Otto des III.* 2 vols. MGH, Diplomata, vol. 2, parts 1–2. Hannover: Hahn, 1888–1893.

Sicking, Louis. *Neptune and the Netherlands: State, Economy, and War at Sea in the Renaissance.* History of Warfare, 23. Leiden and Boston: Brill, 2004. [A revised and updated translation of his *Zeemacht en onmacht: maritieme politiek in de Nederlanden, 1488–1558.* Bijdragen tot de Nederlandse marinegeschiedenis, 7 Amsterdam: De Bataafse Leeuw, 1998.]

Sicking, Louis, and Darlene Abreu-Ferreira, eds. *Beyond the Catch: Fisheries of the North Atlantic, the North Sea and the Baltic, 900–1850.* Leiden and Boston: Brill, 2009.

Sicking, Louis, and Adri P. van Vliet. "'Our triumph of Holland'. War, violence, and the herring fishery of the Low Countries, c. 1400–1650." In Sicking and Abreu-Ferreira, eds., *Beyond the Catch,* 337–364. 2009.

Sidonius Apollinaris [C. Sollius Apollinaris Sidonius]. *Sidonius Poems and Letters*, ed. and tr. W. B. Anderson. 2 vols. Cambridge, MA: Loeb Classical Library, 1936 and 1965.

*Sidonie Apollinare*, ed. and tr. André Loyen. Tome 1: *Poems*; Tomes 2–3: *Lettres*. Paris, 1960 and 1970.

Sidrys, Raymond V. "Fish names in the Eastern Baltic: etymology, ecology, economy," *Istorija [Vilnus]* 41 (1999), 3–23 and appendix.

Sigebert of Gembloux. "Vita Deodorici episcopi Mettensis," ed. G. H. Pertz. *MGH, SS*, 4: 462–483. Hannover: Hahn, 1841.

Sigebert and Godescalc of Gembloux. "Gesta abbatum Gemblacensium," ed. G. H. Pertz. *MGH, SS*, 8: 523–557. Hannover: Hahn, 1848.

Simmons, Ian G. *Global Environmental History 10,000 BC to AD 2000*. Edinburgh: Edinburgh University Press. 2008.

Simpson, Ian A., James H. Barrett, and Karen B. Milek. "Interpreting the Viking Age to Medieval Period transition in Norse Orkney through cultural soil and sediment analyses," *Geoarchaeology: An International Journal* 20 (2005), 355–377.

Singer, David A. "Threshold of affordability: assessing fish remains for socio-economics." In Suzanne M. Spencer-Wood, ed., *Consumer Choice in Historical Archaeology*, 85–99. London and New York, 1987.

"The use of fish remains as a socio-economic measure: an example from 19th century New England," *Historical Archaeologist* 19 (1985), 110–113.

Sivery, G. *Structures agraires et vie rurale dans le Hainaut à la fin du moyen age*. Lille: Université de Lille III, 1977.

Slicher van Bath, B. H. *The Agrarian History of Western Europe A.D. 500–1850*, tr. Olive Ordish. New York: St. Martin's Press, 1963.

Smail, Daniel L. "Common violence: vengeance and inquisition in fourteenth-century Marseille," *Past & Present* 151 (May 1996), 28–59.

Šmelhaus, Vratislav. "Vývoj nížinného rybničního hospodářství ve východních čechách, zejména na Chlumecku a Bydžovsku." In Československá Akademie Zemědělská, ed., *Současnost, perspektivy rozvoje a tradicečeskoslovenského rybářství*, 75–82. Prague: [nákl. vl.], 1978.

Smil, Vaclav. *Harvesting the Biosphere: What We Have Taken from Nature*. Cambridge, MA, and London: MIT Press, 2013.

Society of Antiquaries. *A collection of Ordinances and Regulations for the Government of the Royal Household, made in divers reigns. Also Receipts in ancient Cookery*. London: Society of Antiquaries, 1790.

Soens, Tim. "Capitalisme, institutions et conflits hydrauliques autour de la Mer du Nord (XIIIe–XVIIIe siècle)." In Patrick Fournier and Sandrine Lavaud, eds., *Eaux et conflits dans l'Europe médiévale et moderne: Actes des XXXIIe journées internationales d'histoire de l'Abbaye de Flaran, 8 et 9 octobre 2010*, 149–171. Flaran, Toulouse: Presses Universitaires du Mirail, 2012.

"The origins of the Western Scheldt: environmental transformation, storm surges and human agency in the Flemish coastal plain (1250–1600)." In Erik Thoen et al., eds., *Landscapes or Seascapes? The History of the Coastal Environment in the North Sea Area Reconsidered*, 287–312. Turnhout: Brepols, 2013.

Sommé, Monique. "L'Alimentation quotidienne à la Cour de Bourgogne au milieu du XVe siècle,." In *Bulletin Philologique et Historique (jusqu'à 1610) du Comité des Travaux Historiques et Scientifiques*, 99ème Congrès des Sociétés savantes, Tours, 1968, vol. 1: *Les Problèmes de l'alimentation*, 103–117. Paris, 1971.

Sonnlechner, Christoph. "The establishment of new units of production in Carolingian times: making early medieval sources relevant for environmental history," *Viator* 35 (2004), 21–48.

Sørheim, Helge. "The birth of commercial fisheries and the trade of stockfish in the Borgundfjord, Norway," In Barrett and Orton, eds., *Cod and Herring*, 60–70. 2016.

Sowina, Urszula. *Water, Towns and People: Polish Lands against a European Background until the Mid-16th Century*, tr. Justyna Woldańska. Polish Studies: Transdisciplinary Perspectives, vol. 17. Frankfurt a.M.: Peter Lang, 2016. [Translates a projected 2d ed. rev. of Urszula Sowina, *Woda i Ludzie w mieście późnośredniowiecznym i wczesnonowożytnym. Ziemie polskie z Europa w tle*. Warsaw: Instytut Archeologii i Etnologii PAN, 2009.]

Sparholt, Henrik. "Fish species interactions in the Baltic Sea," *Dana* 10 (1994), 131–162.

Spicciani, Amleto. "Il Padule di Fucecchio nell'alto Medievale." In Alberto Malvolti and Giuliano Pinto, eds., *Incolti, fiumi, paludi. Utilizzazione delle risorse naturali nella Toscana medievale e moderna*, 57– 72. Firenze: Leo S. Olschki, 2003.

Spitzenberger, F. "Die Tierknochenfunde des Hausbergs zu Gaiselberg, einer Wehranlage des 12.–16. Jahrhunderts in Niederösterreich," *Zeitschrift für Archäologie des Mittelalters* 11 (1983), 121–161.

Sportiello, Anne. *Les pêcheurs du vieux-port: fêtes et traditions de la communauté des pêcheurs de Saint-Jean*. Marseille: J. Laffitte, 1981.

Squatriti, Paolo. "'Advent and conquests' of the water mill in Italy." In Elizabeth B. Smith and Michael Wolfe, eds., *Technology and Resource Use in Medieval Europe: Cathedrals, Mills and Mines*, 125–139. Aldershot: Ashgate, 1997.

"Marshes and mentalities in early medieval Ravenna," *Viator* 23 (1992), 1–16.

"Riverains et rivaux dans l'Italie du haut moyen âge." In Patrick Fournier and Sandrine Lavaud, eds., *Eaux et conflits dans l'Europe médievale et moderne: Actes des XXXIIe journées internationales d'histoire de l'Abbaye de Flaran, 8 et 9 octobre 2010*, 137–148. Flaran, Toulouse: Presses Universitaires du Mirail, 2012

*Water and Society in Early Medieval Italy*. Cambridge: Cambridge University Press, 1998.

ed. *Working with Water in Medieval Europe: Technology and Resource Use*. Leiden: Brill, 2000.

Squatriti, Paolo, and Betsy Meredith, tr. *Duccio Balestracci, the renaissance in the fields: family memoirs of a fifteenth-century Tuscan peasant*. University Park: Pennsylvania State University Press, 1999.

Staiger, X. *Salem oder Salmansweiler, ehemaliges Reichskloster Cisterzienser-Ordens*. Constanz [*sic*]: Stadler, 1863.

Star, Bastiaan, Sanne Boessenkool, Agata T. Gondek, Elena A. Nikulin, Anne Karin Hufthammer, Christophe Pampoulie, Halvor Knutsen, Carl André, Heidi M. Nistelberger, Jan Dierking, Christoph Petereit, Dirk Heinrich, Kjetill S. Jakobsen, Nils Chr. Stenseth, Sissel Jentoft, and James H. Barrett, "Ancient DNA reveals the Arctic origin of Viking Age cod from Haithabu, Germany," *PNAS* 114:34 (2017), 9152–9157. www.pnas.org/cgi/doi/10.1073/pnas.1710186114.

Starkel, Leszek, Roman Soja, and Danuta J. Michczyńska. "Past hydrological events reflected in Holocene history of Polish rivers," *Catena* 66 (2006), 24–33.

Starkey, David J., Jón Th. Thór, and Ingo Heidbrink, eds. *A History of the North Atlantic Fisheries*, vol. 1: *From Early Times to the Mid-Nineteenth Century*. Bremen: Verlag Hauschild, 2009.

Stauner, Valérie. "Les étangs de l'est du Berry du Moyen Âge à l'époque moderne: formation d'un paysage ou mise en valeur d'un milieu naturel?" In Jean-Michel Derex, ed., *La production des étangs du Moyen Âge à l'époque contemporaine: Actes de la journée d'étude 2005*, 23–39. Groupe d'Histoire des Zones Humides: Journées d'Étude. Paris: Groupe d'Histoire des Zones Humides, 2006.

Steane, John M. "The royal fishponds of medieval England." In Aston, ed., *Medieval Fish*, 39–68. BAR British Series 182 (i–ii). 1988.

Steane. John M., and M. Foreman. "The archaeology of medieval fishing tackle." In G. L. Good et al., eds., *Waterfront Archaeology: Proceedings of the 3rd International Conference on Waterfront Archaeology held at Bristol, 1988*, 88–101. CBA Research Reports, 74. London: Council for British Archaeology, 1991.

"Medieval fishing tackle." In Aston, ed., *Medieval Fish*, 137–186. BAR British Series 182 (i–ii). 1988.

Steinberg, Theodore. *Industrialization and the Waters of New England*. New York: Cambridge University Press,1991.

Steinhilber, F., and J. Beer. "Solar activity – the past 1200 years," *PAGES news* 19:1 (March 2011), 5–6.

Steinmeyer, Elias, and Eduard Sievers, eds. *Die althochdeutschen Glossen*. 5 vols. Berlin: Prussia: Ministerium für Wissenschaft, Kunst und Volksbildung; Deutsche Akademie der Wissenschaften zu Berlin, 1879–1922.

Sten, Sabine. "Trading with fish in medieval Sweden: some examples from archaeological bone finds," *Archaeofauna* [Madrid] 4 (1995), 65–69.

Sternberg, Myriam. "L'approvisionnement de Paris en poisson au 16ème siècle: que disent les sources bibliographiques? Que peut on espérer des données ichtyofauniques des Jardins du Carrousel?" *Anthropozoologica* 16 (1992), 127–130.

"Le rôle des fleuves dans la pêche du Ier au VIe siècle: état des connaissances." In Joëlle Burnouf and Phillipe Leveau, eds., *Fleuves et marais, une histoire au croisement de la nature et de la culture. Sociétés préindustrielles et milieux fluviaux, lacustres et palustres: pratiques sociales et hydrosystèmes*, 185–199. Paris: CTHS, 2004.

"Une spécificité de la cuisine monastique: l'abondancedes restes de poisson (XIIe–XVIe siècle)." In Benjamin Saint-Jean Vitus, ed., *Pas de fumet sans feu:*

*Cuisine et vie quotidienne auprès des moines de Tournus (fin du IXe–XVIe siècle)*, 83–94. Supplément, Revue Archéologique de l'Est, 35. Dijon: Revue Archéologique de l'Est, 2014.

Stolz, Otto. *Der geschichtliche Inhalt der Rechnungsbücher der Tiroler Landesfürsten von 1288–1350.* Schlern-Schriften, vol. 175. Innsbruck: Wagner, 1957.

*Geschichtskunde der Gewässer Tirols.* Schlern-Schriften, vol. 32. Innsbruck: Universitäts-Verlag Wagner, 1936.

Stouff, Louis. *Ravitaillement et alimentation en Provence aux XIVe et XVe siècles.* Paris: Mouton, 1970.

Straffin Eric C., and Michael D. Blum. "Late and post-glacial fluvial dynamics of the Lire River, Burgundy, France." In Jean-Paul Bravard and Michel Magny. eds. *Les fleuves ont une histoire. Paléo-environnement des rivières et des lacs français depuis 15000 ans*, 85–99. Paris: Errance, 2002.

Stromeyer, Hans. *Zur Geschichte der Badischen Fischerzünfte. Heidelberger volkswirtschaftliche Abhandlungen*, 1:3. Karlsruhe: G. Braun, 1910.

Strumieński, Olbrycht. *O sprawie, sypaniu, wymierzaniu i rybnieniu stawów.* Kraków, 1573; 2d ed. Kraków: F. Kucharzewski, 1897.

Susłowska, W., and K. Urbanowicz. "Szczątki kostne ryb z wczesnośredniowiecznego Gdańska (X–XIII w.)," *Gdańsk wczesnośredniowieczny* 6 (1967), 53–65.

Šusta, Josef. *Fünf Jahrhunderte der Teichwirtschaft zu Wittingau. Ein Beitrag zur Geschichte der Fischzucht mit besonderer Berücksichtigung der Gegenwart.* Stettin: Herrcke & Lebeling, 1898.

*Purkrabské účty panství Novohradského z let 1390–91.* Historické archiv, vol. 35. Prague: Nákl. České Akademie Císaře Františka Josefa pro Vědy, Slovesnost a Umění,, 1909.

Sutton, Anne F.,and P. W. Hammond, eds. *The Coronation of Richard III: The Extant Documents.* Gloucester: Sutton, 1983.

Svanberg, Ingvar, and Stanislaw Cios. "Petrus Magni and the history of freshwater aquaculture in the later Middle Ages," *Archives of Natural History* 41 (2014), 124–130.

Sykes, Naomi J. *The Norman Conquest: A Zooarchaeological Perspective.* BAR International Series 1656. Oxford: Archaeopress, 2007.

Szczygielski, Wojciech. *Gospodarka stawowa na ziemach połodniowo-zachodniej Rzeczypospolitej w XVI–XVIII w.: Studium z dziejów postępu w dawnym gospodarstwie wiejskim.* Prace Instytutu Historycznego Uniwersytetu Łódzkiego, nr. 11. Łódź: Instytutu Historycznego Uniwersytetu Łódzkiego, 1965.

*Z dziejów gospodarki rybnej w Polsce w XVI–XVIII wieku*, Instytut Historii Kultury Materialnej Polskiej Akademii Nauk, Studia z Dziejów Gospodarstwa Wiejskiego, vol. IX, no. 2. Warsaw: Państwowe wydawnictwo naukowe, 1967.

Szende, Katalin. "The Sopron fish market." In Dóra Mérai, Ágnes Drosztmér, Kyra Lyublyanovics, Judith Rasson, Zsuzsanna Papp Reed, András Vadas, Csilla Zatykó, eds., *Genius loci. Laszlovszky 60*, 159–164. Budapest: Archaeolingua Alapítvány, 2018.

"Stadt und Naturlandschaft im ungarischen Donauraum des Mittelalters." In Ferdinand Opll and Christoph Sonnlechner, eds., *Europäische Städte im*

*Mittelalter*, 365–397. Forschungen und Beiträge zur Wiener Stadtgeschichte Bd. 52. Innsbruck, Vienna, and Bozen: Studien Verlag, 2010.

Sznura, Franek. "Veleni e '*nobilissimi pesci*'. Appunti sulla legislazione fiorentina in tema di pesca nelle acque interne (secolo XV)." In Franek Sznura, ed., *Fiumi e laghi toscani fra passato e presente: pesca, memorie, regole*, 270–282. Atti del Convegno di Studi (Firenze, 11–12.XII.2006). Firenze: Aska, 2010.

Tabacco, Giovanni. *The Struggle for Power in Medieval Italy: Structures of Political Rule*, tr. Rosalind B. Jensen. Cambridge: Cambridge University Press, 1989.

Takács, Károly. "Medieval hydraulic systems in Hungary: written sources, archaeology and interpretation." In Jószef Laszlovszky and Péter Szabó, eds., *People and Nature in Historical Perspective*, 289–311. Budapest: Central European University, Department of Medieval Studies, and Archaeolingua, 2003.

Tavares, Alice. "Estancia e imagen de Portugal, según el viajero alemán – Jerónimo Münzer – en su periplo por la Península Ibérica (1494–1495). El caso de Lisboa," *Viaggiatori. Circolazioni, scambi ed Esilio (Secoli XII–XX)* [Naples] 2:2 (2018), 461–492.

"Pescados, pinnípedos, cetáceos, crustáceos, aves acuáticas y otros animales marinos en los archipiélagos de Portugal – Azores y Madeira – en los siglos XV–XVI. Una breve panorámica." *Viaggiatori. Circolazioni, scambi ed Esilio (Secoli XII–XX)* [Naples] 2:1 (2018), 225–258.

Taverner, John. *Certain Experiments Concerning Fish and Frvite: Practised by Iohn Taverner Gentleman, and by him published for the benefit of others.* London: William Ponsonby, 1600. Facsimile reprint The English Experience, no. 75. Amsterdam and New York, 1968.

Taylor, Jeffrey N., Walter R. Courtenay, Jr., and James A. McCann. "Known impacts of exotic fishes in the continental United States." In Walter R. Courtenay, Jr., and Jay R. Stauffer, Jr., eds., *Distribution, Biology, and Management of Exotic Fishes*, 322–373. Baltimore: Johns Hopkins University Press, 1984.

TeBrake, William H. *Medieval Frontier: Culture and Ecology in Rijnland.* College Station: Texas A&M University Press, 1985.

Teplý, František. "Obraz rybnikářství na něterých panstvích v čechach," *Věstník československého zemědelského musea* 8 (1935), 114–120.

"O rybnikářstvi v čechách do husitských válek," *Věstnik československého Zemědelského Musea* 8(1935) 10–20.

*Příspěvky k dějinám českého rybnikářství.* Publikace ministerstva zemědělství, Ročník 1937. Prague: Nákladem Ministerstva zemědělství Republiky Československé, 1937.

"Stavba rybníků a jejich cena," *Vsětník československého zemědelského musea* 8 (1935), 160–166.

"Význam Vítkoviců a Viléma z Pernštejna pro české rybnikářství," *Věstník československého zemědelského musea* 8 (1935), 62–68.

Teuteberg, Hans J. "Agenda for comparative European history of diet." In his *European Food History: A Research Review*, 1–18. Leicester: Leicester University Press,1992.

Theurot, Jacky. "Approche de la pêche et du ravitaillement en poisson dans les villes de la vallée du Doubs aux XIVe et XVe siècles." In Benoît et al., eds., *Pêche et pisciculture en eau douce*. 2004.

Thibault, Max, and Anne-Françoise Garçon. "Un problème d'écohistoire: le saumon dans les contrats de louage, une origine médiévale?" In Benoît et al., eds., *Pêche et pisciculture en eau douce*. 2004.

Thienemann, A. *Verbreitungsgeschichte der Süsswassertierwelt Europas*. Die Binnengewässer, Bd. 18. Stuttgart: Schweitzerbart, 1950.

Thieren, E., A. Ervynck, D. Brinkhuizen, A. Locker, and W. Van Neer. "The Holocene occurrence of *Acipenser* spp. in the southern North Sea: the archaeological record," *Journal of Fish Biology* 89:4 (2016), 1958–1973. doi:10.1111/jfb.13094.

Thoen, Erik, Guus J. Borger, Adriaan M. J. de Kraker, Tim Soens, Dries Tys, Lies Vervaet, and Henk J. T. Weerts, eds. *Landscapes or Seascapes?: The History of the Coastal Environment in the North Sea Area Reconsidered*. CORN Publication series, 13. Turnhout: Brepols, 2013.

Thomas Aquinas. *Summa Theologica. Latin Text and English Translation* [Blackfriars]. London and New York: Blackfriars and McGraw-Hill, 1964.

Thomas of Cantimpré [Thomas Cantimpratensis]. *Liber de natura rerum. Editio princeps secundum codices manuscriptos*. Teil I: *Text*, ed. H. Boese. Berlin: de Gruyter, 1973.

Thomas of Wratislavia [of Wrocław, of Breslau]. *Regimen Sanitatis*, in Jerzy Burchardt, ed. *Higiena wedłe Tomasza z Wrocławia*. Studia Copernicana, v., 36, 75–90. Warsaw: Wydawnicta IHN PAN, 1997.

Thomas, Françoise. "Hygiène, approvisionnement en eau et gestion hydrographique à Namur au XVème siècle," *Annales de la Société archéologique de Namur* 68 (1994), 235–305.

Thomazi, A. *Histoire de la pêche des âges de la pierre à nos jours*. Paris: Payot, 1947.

Thompson, D'Arcy W. *A Glossary of Greek Fishes*. London: Oxford University Press, 1947.

Thompson, Paul. "Women in the fishing: the roots of power between the sexes," *Comparative Studies in Society and History* 27 (1985), 3–32.

Thompson, Pauline. "The case of the salmon," *Scintilla* 1 (1984), 63–75.

Thorhallsson, Baldur, and Þorsteinn Kristinsson. "Iceland's external affairs from 1400 to the Reformation: Anglo-German economic and societal shelter in a Danish political vacuum," *Icelandic Review of Politics and Administration [Stjórnmál & stjórnsýsla]* 9:1 (2013), 113–137. DOI: 10.13177/irpa.a.2013.9.1.6

Thorndike, Lynn. "A mediaeval sauce-book," *Speculum* 9 (1934), 183–190.

"Three tracts on food in Basel manuscripts," *Bulletin of the History of Medicine* 8 (1940), 355–369.

Thorpe, John E., and L. Stradmeyer. "Management examples: the Atlantic salmon." In John E. Thorpe, Graham Gall, James Lannan, and Colin Nash, eds., *Conservation of Fish and Shellfish Resources: Managing Diversity*, 79–114. London: Academic Press, 1995.

Thrupp, Sylvia. *The Merchant Class of Medieval London [1300–1500]*. Ann Arbor: University of Michigan Press, 1948.

Tiedemann, Ralph, K. Moll, K. Paulus, M. Scheer, P. Williot, R. Bartel, J. Gessner, and F. Kirschbaum. "Atlantic sturgeons (*Acipenser sturio, Acipenser oxyrinchus*): American females successful in Europe," *Naturwissenschaften* 94 (2007), 213–217.

Tischler, J. "Fische: Sprachliches." In Heinrich Beck et al., eds., *Reallexikon der Germanischen Altertumskunde*, 2d ed. rev., vol. 9, pp. 120–126. Berlin and New York: de Gruyter, 1994.

"Fischerei und Fischereimethoden: Fischfang in den verschiedenen Texten." In Heinrich Beck et al., eds., *Reallexikon der Germanischen Altertumskunde*, vol. 9, pp. 138–140. Berlin and New York: de Gruyter, 1994.

Tock, Benoît-Michel, ed. *Les Chartes des évêques d'Arras (1093–1203)*. Collection de documents inédits sur l'histoire de France, Section d'histoire médiévales et de philologie, série in-8o, vol. 20 [title page says 20; original printed spine and cover say 21]. Paris: Comité des Travaux historiques et scientifiques, 1991.

Tockner, K., U. Uehlinger, and C. T. Robinson, eds. *Rivers of Europe*. Oxford: Academic Press, 2008.

Topolski, Jerzy. "Rybołówstwo i gospodarstwo rybne w XVI–XVIII w. na terenie latyfundium arcybiskupstwa gnieżnieńskiego." In Janina Leskiewiczowa, ed., *Studia z dziejów gospodarstwa wiejskiego*, vol. I, 179–197. Studia i materiały z historii kultury materialnej IHKM PAN, vol. 2. Wrocław: Zakład Im. Ossolińskich, Wydawnictwo Polskiej Akademii Nauk, 1957.

Toubert, Pierre. *Les Structures du Latium médiéval: Le Latium méridional et la Sabine du IX^e siècle à la fin du XII^e siècle*. Bibliothèque des Ecoles françaises d'Athènes et de Rome, 221. Paris and Rome: Ecole française de Rome, 1973.

Touchard, Henri. *Le Commerce maritime breton à la fin du moyen âge*. Paris: Belles lettres, 1967.

Toussaint-Samat, Maguelonne. "Gastronomie et fastes culinaires à la cour Sforzesque et chez les bourgeois de Milan au milieu du XV^e siècle." In Menjot, ed., *Manger et boire*, vol. 2, pp. 105–119. Paris: Belles Lettres, 1984.

Towle, Jerry C. "Authored ecosystems: Livingston Stone and the transformation of California fisheries," *Environmental History* 5 (2000), 54–74.

Trakadas, L. A. "'Exhausted by fishermen's nets': Roman sea fisheries and their management," *Journal of Mediterranean Studies* 16 (2006), 259–272.

Tranchant, Mathias. "Pêches et pêcheurs des villes-ports du Centre-Ouest français au Moyen Age." In Jesús Ángel Solórzano Telechea, Michel Bochaca, and Amélia Aguiar Andrade, eds., *Gentes de mar en la ciudad atlántica medieval*, 83–94. Ciencas históricas 24. Logroño: Instituto de Estudios Riojanos, 2012.

Trevisa, John. *On the Properties of Things. John Trevisa's Translation of Bartholomeus Anglicus De Proprietatibus Rerum. A Critical Text*, ed. M. C. Seymour *et al.* 2 vols. Oxford: Oxford University Press, 1975–1988,

Trexler, Richard. "Measures against water pollution in fifteenth century Florence," *Viator* 5 (1974), 455–467.

Troubat, Olivier. "Pêcheries fixes du I^er au VII^e siècle dans le lit du Cher à Saint-Victor et Vaux (Allier)." In Édith Peytremann, ed., *Des fleuves et des hommes à l'époque mérovingienne. Territoire fluvial et société au premier moyen âge (Ve–*

*XIIe siècle)*, 119–130. Actes des 33e journées internationales d'archéologie mérovingienne. Strasbourg, 28–30 septembre 2012, pp. 119–130. 42ème supplément à la *Revue Archéologique de l'Est*. Dijon, S.A.E., 2016.

Troubleyn, Liesbeth, Frank Kinnaer, Anton Ervynck, Luk Beeckmans, Danielle Caluwé, Brigitte Cooremans, Frans De Buyser, Koen Deforce, Konjev Desender, An Lentacker, Jan Moens, Gaston Van Bulck, Maarten Van Dijck, Wim Van Neer and Werner Wouters. "Consumption patterns and living conditions inside *Het Steen*, the late medieval prison of Malines (Mechelen, Belgium)" *Journal of Archaeology in the Low Countries* 1–2 (November 2009), 5–47.

Trouet, Valerie, Jan Esper, Nicholas E. Graham, Andy Baker, James D. Scourse, and David C. Frank, "Persistent positive North Atlantic Oscillation mode dominated the Medieval Climatic Anomaly," *Science* 324 (3 April 2009), 78–80.

Trusen, Winfried. "'Insula in flumine nata.' Ein kanonischer Zivilprozeß aus den Jahren 1357 bis 1363/64 um ein Insel im Mittelrhein," *Zeitschrift der Savigny-Stiftung für Rechtsgeschichte: Kanonistische Abteilung* 68 (1982), 294–338.

Tudge, Colin. *The Time before History: 5 Million Years of Human Impact*. New York: Touchstone, 1996.

Tumiati, Simone, et al. "The ancient mines of Servette (Saint-Marcel, Val d'Aosta, Italian Alps): a minerological, metallurgical, and charcoal analysis of furnace slags," *Archaeometry* 47:2 (2005), 317–340.

Turner, S. J., S. F. Thrush, J. E. Hewitt, V. J. Cummings, and G. Funnell. "Fishing impacts and the degradation or loss of habitat structure," *Fisheries Management and Ecology* 6 (1999), 401–420.

Turner, Thomas H., ed. *Manners and Household Expenses of England in the Thirteenth and Fifteenth Centuries Illustrated by Original Records*. Roxburghe Club. London: William Nichol, 1841.

Turville-Petre, Thorlac. "The earliest English manorial survey," *Speculum* 73 (1998), 58–79.

Tylmann, Wojciech, and Martin Grosjean. "Climate variability and human impacts in Central and Eastern Europe during the last two millennia." *PAGES Magazine* 23:2 (December 2015), 80.

Tys, Dries, and Marnix Pieters. "Understanding a medieval fishing settlement along the Southern North Sea: Walraversijde, c. 1200–1630." In Louis Sicking and Darlene Abreu-Ferreira, eds., *Beyond the Catch: Fisheries of the North Atlantic, the North Sea and the Baltic, 900–1850*, 91–122. Leiden and Boston: Brill, 2009.

Ulanowski, Bolesław, ed. *Księgi sądowe wiejskie*. 2 vols. Starodawne prawa polskiego pomniki, tom XI–XII. Kraków: Nakładem Polskiei Akademji Umiejętności, 1921.

Ulrich of Cluny [of Zell]. "Consuetudines Cluniacensis," In Migne *PL* 149, cols. 633–778. Paris: Garnier, 1882.

UNESCO Permanent Forum on Indigenous Issues, Ninth session, New York, 19–30 April 2010, *Report on Indigenous Fishing Rights in the Seas with Case Studies from Australia and Norway*. www.un.org/esa/socdev/unpfii/documents/E.C.19.2010.2EN.pdf. (accessed 27 October 2014).

Unger, Richard W. "Dutch herring, technology, and international trade in the seventeenth century," *Journal of Economic History* 40 (1980), 253–280.

"The Netherlands herring fishery in the late Middle Ages: the false legend of William Beukels of Biervliet," *Viator* 9 (1978), 335–356.

Unger, W. S. *De tol van Iersekeroord. Documenten en rekeningen 1321–1572*. The Hague: Martinus Nijhoff, 1939.

Unwin, George. *The Gilds and Companies of London*. London: George Allen & Unwin, 1908. Fourth [reprint] ed., London: Frank Cass, 1963.

Urbańczyk, Przemysław. *Medieval Arctic Norway*. Warsaw: Institute of Archaeology and Ethnology, 1992.

Ureña y Smenjaud, Rafael de, ed., *Fuero de Cuenca*. Madrid: Tipografia de Archivos, 1935.

Uytven, Raymond van. "L' approvisionnement des villes des anciens Pays-Bas au moyen âge." In his *Production and Consumption in the Low Countries, 13th–16th Centuries*, pp. XI:75–116. Aldershot: Ashgate, 2001. [Originally published in *Flaran* 5 (1983).]

*De Zinnelijke Middeleeuwen*. Leuven: Davidsfonds, 1998.

Vahrenhold, Wilhelm. *Kloster Marienfeld. Besitz- und Wirtschaftsgeschichte des Zisterzienzerklosters Marienfeld in Westfalen (1185–1456)*. Quellen und Forschungen zur Geschichte der Stadt Warendorf, vol. 4. Warendorf: Heimatverl. Schnell, 1966.

Val Valdivieso, Maria I. del. *Agua y poder en la Castilla bajomedieval: el papel del agua en el ejercicio del poder concejil a fines de la Edad Media*. Valladolid: Junta de Castilla y León, Consejería de Cultura y Turismo, 2003.

ed. *Vivir del agua en las ciudades medievales*. Valladolid: Universidad de Valladolid, Secretariado de Publicaciones e Intercambio Editorial, 2006.

Valiela, Ivan, Jennifer L. Bowen, and Joanna K. York, "Mangrove forests: one of the world's threatened major tropical environments," *BioScience* 51:10 (October 2001 ), 807– 815.

Válka, Josef. *Le grand domaine feodal en Bohème et Moravie du 16e au 18e siècle*. Hospodářské dějiny – Economic History, vol. 10. Prague: Institute of Czechoslavak and World History of the Czechoslovak Academy of Sciences, 1982.

Van Damme, Dirk, Nina Bogutskaya, Richard C. Hoffmann, and Carl Smith, "The introduction of the European Bitterling (*Rhodeus amarus*) to west and central Europe," *Fish and Fisheries* 8:2 (2007), 79–106

Van Denderen, P. Daniel, and Tobias Van Kooten. "Size-based species interactions shape herring and cod population dynamics in the face of exploitation," *Ecosphere* 4:1 (October 2013), 1–15.

van der Lingen, Carl D., and 13 co-authors. "Trophic dynamics." In Checkley et al., eds., *Climate Change and Small Pelagic Fish*, 112–157. 2009.

Van der Wee, Herman. *The Growth of the Antwerp Market and the European Economy (Fourteenth–Sixteenth Centuries)*. 3 vols. The Hague: M. Nijhoff, 1963.

Van Drival, E., ed. *Cartulaire de l'abbaye de Saint-Vaast d'Arras rédigé au XIIe siècle par Guimann*. Arras: A. Courtin, 1875.

van Houts, Elisabeth M. C., ed. and tr. *The "Gesta Normannorum ducum" of William of Jumièges, Orderic Vitalis, and Robert of Torigni, 1: Introduction and Books I–IV:2; 2: Books V–VIII*. Oxford Medieval Texts. Oxford: Clarendon, 1992–1995.

Van Molle, Leen, and Yves Segers, eds. *The Agro-Food Market: Production, Distribution, and Consumption*. Rural Economy and Society in North-Western Europe, 500–2000. Turnhout: Brepols, 2013.

Van Neer, Wim, ed., *Fish Exploitation in the Past. Proceedings of the 7th Meeting of the I.C.A.Z. Fish Remains Working Group*. Annales du Musée Royal de l'Afrique Central, Sciences Zoologiques, vol. 274. Tervuren: Musée Royal,1994.

Van Neer, Wim, and Anton Ervynck. "Apport de l'archéologie à la connaissance de l'exploitation de l'eau douce au Moyen Age et à l'Epoque moderne en Belgique." In Benoît et al., eds., *Pêche et pisciculture en eau douce*. 2004.

*Archeologie en Vis*. Herlevend Verleden, 1. Zellik: Instituut voor het Archeologisch Patrimonium, 1993.

"Food rules and status: patterns of fish consumption in a monastic community (Ename, Belgium)." *Archaeofauna* 5 (1996), 155–164.

"New data on fish remains from Belgian archaeological sites." In Wim Van Neer, ed., *Fish Exploitation in the Past. Proceedings of the 7th Meeting of the I.C.A.Z. Fish Remains Working Group*, 217–230. Annales du Musée Royal de l'Afrique Central, Sciences Zoologiques, vol. 274. Tervuren: Musée Royal, 1994.

"Remains of traded fish in archaeological sites: indicators of status, or bulk food?" In Sharyn Jones O'Day, Wim van Neer, and Anton Ervynck, eds., *Behaviour behind Bones. The Zooarchaeology of Ritual, Religion, Status and Identity*, 203–214. Proceedings of the 9th Conference of the International Council of Archaeozoology, Durham, August 2002. Oxford: Oxbow Books, 2004.

"The rise of sea fish consumption in inland Flanders, Belgium." In Barrett and Orton, eds., *Cod and Herring*, 156–171. 2016.

"The zooarchaeological reconstruction of the development of the exploitation of the sea: a *status questionis* for Flanders." In Marnix Pieters, Frans Verhaeghe and Glann Gevaert, eds., *Fishery, Trade and Piracy: Fishermen and Fishermen's Settlements in and around the North Sea Area in the Middle Ages and Later*, 95–115. Brussels: Vlaams Instituut voor het Onroerend Erfgoed, 2006.

Van Neer, Wim, Anton Ervynck, Benjamin T. Fuller, Patrick Degryse, and Wim Wouters. "Freshwater fisheries in Belgium during medieval and postmedieval times: looking for markers for the onset of overfishing and pollution." In Daniel Makowiecki et al., eds., *Fishes – Culture – Environment through Archaeoichthyology, Ethnography & History*, 31–34. The 15th Meeting of the International Council for Archaeozoology (ICAZ) Fish Remains Working Group (FRWG) September 3–9, 2009 in Poznań & Toruń, Poland. Poznań: Bogucki Wydawnictwo Naukowe, 2009.

VanderZwaag [*sic*], David L. *The Fish Feud: The U.S. and Canadian Boundary Dispute*. Toronto: Lexington Books, 1983.

Varano, Sara, Flavio De Angelis, Andrea Battistini, Luca Brancazi, Walter Pantano, Paola Ricci, Marco Romboni1, Paola Catalano, Valentina Gazzaniga, Carmine Lubritto, Riccardo Santangeli Valenzani, Cristina Martínez-Labarga, and Olga Rickards. "The edge of the Empire: diet characterization of medieval Rome through stable isotope analysis," *Archaeological and Anthropological Sciences* 12:196 (2020). https://doi.org/10.1007/s12520-020-01158-3.

Varley, Margaret. *British Freshwater Fishes: Factors Affecting Their Distribution.* London: Fishing News Books, 1967.

Vatin, Nicolas. "A propos de l'exotisme dans les Vâḳi'ât-i Sulṭân Cem: Le regard porté sur l'Europe occidentale à la fin du XVe siècle par un Turc Ottoman," *Journal Asiatique* 272 (1984), 237–248.

"Pratiques agricoles en Limousin à la fin du XVe siècle d'après une source ottomane," *Histoire et sociétés rurales* 3 (1995), 261–270.

Vendittelli, Marco. "Diritti ed impianti di pesca degli enti ecclesiastici Romani tra X e XIII secolo," Melanges de l'ecole française de Rome. *Moyen Age* 104 (1992), 387–430.

"La pesca nelle acque interne del territorio ninfesino. Tecniche di sfruttamento ed interessi di gestione." In Luigi Fiorani, ed., *Ninfa, una città, un giardino,* 113–137. Atti del Colloquio della Fondazione Camillo Caetani, Roma, Sermoneta, Ninfa, 7–9 ottobre 1988. Rome: L'Erma di Bretschneider, 1990.

Verdon, Jean. "Recherches sur la pêche et la pisciculture en occident durant le haut moyen age." *Actes du 102e Congrés national des sociétés savantes (Limoges, 1977). Section d'archéologie et d'histoire de l'art. Le Limousin. Études Archéologiques.* 337–349. Paris, 1979.

Verlinden, Charles, et al. *Dokumenten voor de Geschiedenis von Prijzen en Lonen in Vlaanderen en Brabant.* 4 parts in 5 vols. Rijksuniversiteit te Gent. Werken uitgegeven door de Faculteit van de Letteren en Wijsbegeerte, Aflevering 125, 136–137, 153, and 156. Brugge: De Tempel, 1959–1973.

Verriest, Léo, ed. *Corpus des records de coutumes et des lois de chefs-lieux de l'ancien Comté de Hainaut,* Sources de l'Histoire du droit rurale. Mons: Frameries, 1947.

ed. *Le polyptyque illustré dit "Veil Rentier" de messire Jehan de Pamele-Audenarde (vers 1275).* Brussels: Presses des Établie Duculot, 1950 [a facsimile transcript].

*Le Régime seigneurial dans le Comté de Hainaut du XIe siècle à la Révolution,* 2d ed. Louvain: P. Smeesters, 1956.

Vésteinsson, Orri. "Commercial fishing and the political economy of medieval Iceland." In Barrett and Orton, eds., *Cod and Herring,* 71–79. Oxford and Philadelphia: Oxbow, 2016.

Vickers, Danny. "'The price of fish': a price index for cod 1505–1892," *Acadiensis* 25:2 (1996), 92–104.

Vigne, Jean-Denis, Salvadir Bailon, Jacques Cuisin, and Nathalie Desse-Berset. "Sensibilité des microvertébrés aux fluctuations hydrométriques tardiglaciaires et holocènes: deux séquences en zon méditerranéenne humide

(Grítulu et Monte di Tuda, Haute-Corse)." In Bravard and Magny, eds. *Les fleuves ont une histoire*, 269–277. 2002.

Vigneras, L. A. "New light on the 1497 Cabot voyage to America," *Hispanic American Historical Review* 36 (1956), 503–509.

Vincent of Beauvais [Vincentius Bellovacensis]. *Speculum naturale*. Vol. 1 of his *Speculum quadruplex sive speculum maius*. 4 vols. Duaci: Baltazaris Belleri, 1624, reprinted Graz: Akademische Verlag, 1964–1965.

Vincent, Bernard. "Consommation alimentaire en Andalousie orientale (Les Achats de l'Hôpital Royal de Guadix, 1581–1582)," *Annales e. s. c.* 30:2–3 (1975), 445–453.

Vincenti, Valentina. "La tutela ambientale del Lago Trasimeno in eta medioevale / Environmental Protection of Lake Trasimene in the Middle Ages." In Thomas M. Robinson and Laura Westra, eds., *Thinking about the Environment: Our Debt to the Classical and Medieval Past*, 130–140. Lanham, MD: Lexington Books, 2002.

Vinther, B. M. "The Medieval Climate Anomaly in Greenland ice core data," *PAGES news* 19:1 (March 2011), 27–28.

Vinyoles i Vidal, Teresa M. "El rebost, la taula i la cuina dels frares barcelonins al 1400." In *Alimentació i societat a la Catalunya medieval*, Anuario de Estudios Medievales, Anex 20, 137–166. Barcelona: Instituto de Historia Medieval de España, 1988.

*Vita brevior sancti Maioli* in *Bibliotheca Cluniacensis*, ed. Martin Marrier and André Duchesne, cols. 1763–1782. Paris: n.p., 1614, reprinted Mâcon: Protat, 1915.

*Vita Hugberti episcopi Traiectensis*. MGH, SrM, vol. 6., pp. 471–476. Hannover: Hahn, 1913.

Vita-Finzi, C. *The Mediterranean Valleys: Geological Change in Historical Times*. Cambridge; Cambridge University Press, 1969.

Vogler, Werner. "Wirtschafts- und Finanzpolitik Abt Ulrich Röschs nach einer Zusammenstellung aus dem Ende des 15. Jahrhunderts." In Werner Vogler, ed., *Ulrich Rösch St. Galler Fürstabt und Landesherr. Beiträge zu seinem Wirken und zu seinem Zeit*, 131–150. St. Gallen: Stiftsarchiv, 1987.

Vogt, J. "Aspects of historical soil erosion in Western Europe." In Peter Brimblecombe and Christian Pfister, eds., *The Silent Countdown: Essays in European Environmental History*, 83–91. Berlin, Heidelberg, and New York, 1990.

Volk, Otto. *Wirtschaft und Gesellschaft am Mittelrhein vom 12. bis zum 16. Jahrhundert*. Veröffentlichungen des Historischen Kommission für Nassau, Bd. 63. Wiesbaden: Historische Komm. für Nassau, 1998.

Vollmann, Benedikt K, ed. and tr. "Ruodlieb" and "Stellenkommentar zum "Ruodlieb." In Walter Haug and Benedikt Vollmann, eds., *Frühe deutsche Literatur und lateinische Literatur in Deutschland, 800–1150*, pp. 388–551 and 1306–1406. Bibliothek des Mittelalters, 1. Frankfurt a.M.: Deutscher Klassiker, 1991.

von der Ohe, P. C., A. Prüß, R. B. Schäfer, M. Liess, E. De Dekeere, and W. Brack. "Water quality indices across Europe: a comparison of the good ecological status of five river basins," *Journal of Environmental Monitoring* 9:9 (2007), 970–978.

Vries, Jan de, and Ad van der Woude. *The First Modern Economy: Success, Failure, and Perseverance of the Dutch Economy, 1500–1815.* Cambridge: Cambridge University Press, 1997.

Wace. *Maistre Wace's Roman de Rou et des ducs de Normandie*, ed. Hugo Andresen. 2 vols. Heilbronn and Paris: Gebr. Henninger, 1877–1879.

Wagner, Norbert. "Zur Datierung des `Summarium Heinrici'," *Zeitschrift für deutsches Altertum und deutsche Literatur* 104 (1975), 118–126.

Waitz, G., ed. "Handschriften in Englischer Bibliotheken," *Neues Archiv der Gesellschaft für ältere Deutsche Geschichtskunde* 4 (Hannover 1879), 323–393.

Wakefield. *Court Rolls of the Manor of Wakefield*, vol. 3: *1313 to 1316, and 1286*, ed. John Lister. Yorkshire Archaeological Society Record Series, vol. 57. Cambridge: Cambridge University Press, 2013.

Walter, Robert C., and Dorothy J. Merritts. "Natural streams and the legacy of water-powered mills." *Science* 319:5861 (2008), 299–304.

Walton, Steven A., ed. *Wind and Water in the Middle Ages: Fluid Technologies from Antiquity to the Renaissance.* Tempe: ACMRS, 2006.

Ward-Perkins, B. "Informazioni sull'economia medievale e postmedievale di Pavia: le ossa dello scavo," *Archeologia Medievale* 5 (1978), 249–266.

Warner, George, ed. *The Libelle of Englyshe Polycye: A Poem on the Use of Sea-Power: 1436.* Oxford: Clarendon Press 1926. Electronic replica in English Poetry Full-Text Database as *Anon., ca. 1436: The libelle of Englyshe polycye (1926)*. Cambridge: Chadwyck-Healey, 1992.

Wartburg, Walther von. *Französisches Etymologisches Wörterbuch.* Basel: Zbinden, 1934, reprinted Basel: Zbinden,1971.

Weckström, Kaarina E., Jonathan P. Lewis, Elinor Andrén, Marianne Ellegaard, Peter Rasmussen, David B. Ryves, and Richard Telford. "Palaeoenvironmental history of the Baltic Sea: one of the largest brackish-water ecosystems in the world." In K. Weckström, K. Saunders, P. Gell, and C. Skilbeck, eds., *Applications of Paleoenvironmental Techniques in Estuarine Studies*, 615–662 Developments in Paleoenvironmental Research, vol 20. Dordrecht: Springer, 2017.

Weiss, Hildegard. *Die Zisterzienserabtei Ebrach: Eine Untersuchung zur Grundherrschaft, Gerichtsherrschaft und Dorfgemeinde im Frankischen Raum.* Stuttgart: G. Fischer, 1962.

Weiss, Stefan. *Die Versorgung des päpstlichen Hofes in Avignon mit Lebensmitteln (1316–1378): Studien zur Social- und Wirtschaftsgeschichte eines mittelalterliches Hofes.* Berlin: Akademie, 2002.

Weiss-Amer, Melitta. "Die 'Physica' Hildegards von Bingen als Quelle für das 'Kochbuch Meister Eberhards'," *Sudhoffs Archiv. Zeitschrift für Wissenschaftesgeschichte* 76 (1992), 85–96.

"The role of medieval physicians in the diffusion of culinary recipes and cooking practices." In Carole Lambert, ed., *Du manuscrits à la table: Essais sur la cuisine au Moyen Âge et répertoires des manuscrits médiévaux contenant des recettes culinaires*. 69–80. Montréal: Champion-Slatkine, les Presses de l'Université de Montréal, 1992.

Weißensteiner, Johann. *Tegernsee, die Bayern und Österreich. Studien zu Tegernseer Geschichtsquellen und der bayerischen Stammessage. Mit einer Edition der Passio secunda s. Quirini*. Archiv für Österreichische Geschichte, Bd. 133. Vienna: ÖAW, 1983.

Welch, Evelyn. *Shopping in the Renaissance: Consumer Cultures in Italy 1400–1600*. New Haven and London: Yale University Press, 2005.

Welcomme, R. L. "International transfers of inland fish species." In Walter R. Courtenay, Jr., and Jay R. Stauffer, Jr., eds., *Distribution, Biology, and Management of Exotic Fishes*, 22–40. Baltimore: Johns Hopkins University Press, 1984.

Went, A. E. J. "Ancient Irish fishing weirs for salmon." In B. Gunda, ed., *Fishing Culture*, vol 1, 455–468. Budapest: Akadémiai Kiadó, 1984.

Werther, Lukas. "Histoire(s) des vallées: études comparatives sur le développement des paysages fluviaux haut-médiévaux en Bavière entant que lieux d`habitation, d'économie et de communication." In Édith Peytremann, ed., *Des fleuves et des hommes à l'époque mérovingienne. Territoire fluvial et société au premier moyen âge (V^e^–XII^e^ siècle)*, 35–48. Actes des 33^e^ journées internationales d'archéologie mérovingienne. Strasbourg, 28–30 septembre 2012. 42^ème^ supplément à la Revue Archéologique de l'Est. Dijon, S.A.E., 2016.

Wheeler, Alwyne. *The Fishes of the British Isles and North-West Europe*. East Lansing: Michigan State University Press, 1969.

Wheeler, Alwyne C, and Andrew K. G. Jones. *Fishes*. Cambridge Manuals in Archeology. Cambridge: Cambridge University Press, 1989.

Whitton, B. A., ed. *Ecology of European Rivers*. Oxford: Blackwell, 1984.

Wicken, William C. *Mi'kmaq Treaties on Trial: History, Land and Donald Marshall Junior*. Toronto: University of Toronto Press, 2002.

Wickersheimer, Ernest. "Zur spätmittelalterlichen Fischdiätetik. Deutsche Texte aus dem 15. Jahrhundert," *Sudhoffs Archiv* 47 (1963), 411–416.

Wickham, Chris. "European forests in the early Middle Ages: landscape and land clearance," *Settimane di studio* 37:2 (1990), 479–545. [Reprinted in his *Land and Power: Studies in Italian and European Social History 400–1200*, pp. 155–199. London: British School at Rome, 1994.]

Wiedemann, F. B. and H. Bocherens. "Spurenelement- und Isotopen-analyse in archäologischem Knochen im Vergleich (am Beispiel der mittelalterlichen Skelettserie von Weingarten, Deutschland)," *Anthropologischer Anzeiger* 55:1 (1997), 147–154.

Wiessner, Edmund. *Kommentar zu Heinrich Wittenwilers Ring*. Deutsche Literatur: Reihe IV, Realistik des Spätmittelalters, Ergänzungsband. Leipzig: Philipp Reclam, 1936; reprinted Darmstadt: Wissenschaftliche Buchgesellschaft, 1977.

Wilkinson, M. "The fish remains." In Mark Maltby, *Faunal Studies on Urban Sites: The Animal Bones from Exeter 1971–1975*, 74–81. Exeter Archaeological

Reports, vol. 2. Sheffield: Department of Prehistory and Archeology, University of Sheffield, 1979.

Willam, Horst A. "Die Fischerei des Deutschen Ordens in Preussen bis zu Dietrich von Altenburg," *Jahrbuch der Albertus-Universität zu Königsberg/Pr.* 11 (1961), 64–152.

William of Malmesbury. *De gestis regum Anglorum*, ed. W. Stubbs, 2 vols. Rolls Series, vol. 90. London: HMSO, 1887–1889.

Williams, David H. *The Welsh Cistercians: Aspects of Their Economic History*. Pontypool: Griffin Press, 1969. 2d ed. rev. 2 vols. Caldey Island and Tenby, 1984.

Williamson, James A., ed. *The Cabot Voyages and Bristol Discovery under Henry VII*. Hakluyt Society, 2d series, vol. 120. Cambridge: The Hakluyt Society, 1962.

Williamson, Tom. *Shaping Medieval Landscapes: Settlement, Society, Environment*. Macclesfield: Windgather Press, 2003.

Williot, Patrick, Eric Rochard, Nathalie Desse-Berset, Frank Kirschbaum, and Jörn Gessner, eds., *Biology and Conservation of the European Sturgeon Acipenser sturio L. 1758: The Reunion of the European and Atlantic Sturgeons*. Berlin and Heidelberg: Springer-Verlag 2011.

Wilson, E. O. *Consilience: The Unity of Knowledge*. New York: Knopf, 1998.

Winchester, Angus J. L. *Landscape and Society in Medieval Cumbria*. Edinburgh: John Donald, 1987.

Winiwarter, Verena, and Martin Knoll. *Umweltgeschichte. Eine Einführung*. Cologne–Weimar–Vienna: Böhlau, 2007.

Winter, Gustav, ed. *Niederösterreichische Weistümer*. 4 vols. Vienna: Braumüller 1886–1913.

Winter, Johanna Maria van. "Fasting and abstinence in Christianity." In *Johanna Maria van Winter, Spices and Comfits: Collected Papers on Medieval Food*, 267–270. Totnes: Prospect Books, 2007.

"Fish recipes in late medieval and early modern cookbooks." In Johanna Maria van Winter, *Spices and Comfits: Collected Papers on Medieval Food*, 271–292. Totnes: Prospect Books, 2007.

"The Low Countries in the fifteenth and sixteenth centuries." In Melitta W. Adamson, ed., *Regional Cuisines in Medieval Europe: A Book of Essays*, 187–214. New York and London: Routledge, 2002.

"Nahrung auf dem Lobither Zollhaus auf Grund der Zollrechnungen aus den Jahren 1426–27, 1427–28 und 1428–29." In T. J. Hoekstra, H. L. Janssen, and I. W. L. Moerman, eds., *Liber Castellorum. 40 variaties op het thema kasteel*, 338–348. Zutphen: de Walburg Pers, 1981.

"Obligatory fasts and voluntary asceticism in the Middle Ages." In *Johanna Maria van Winter, Spices and Comfits: Collected Papers on Medieval Food*, 259–266. Totnes: Prospect Books, 2007.

Wittenweiler, Heinrich. *Der Ring*. Ed. Edmund Wiessner as *Heinrich Wittenwilers Ring*. Deutsche Literatur: Reihe IV, Realistik des Spätmittelalters, Bd. III. Leipzig: Philip Reclam, 1931. Tr. George F. Jones in his *Wittenwiler's Ring and the Anonymous Scots Poem "Colkelbie Sow": Two Comic-Didactic Works from the Fifteenth Century*. Chapel Hill: University of North Carolina Press, 1956.

Wolfram von Eschenbach. *Titurel*, ed. Walter J Schröder and Gisela Hollandt. Darmstadt, Wissenschaftliche Buchgesellschaft, 1971.

Wolter, Christian. "Historic catches, abundance, and decline of Atlantic salmon *Salmo salar* in the River Elbe," *Aquatic Sciences* 77:3 (July 2015), 367–380.

Wood, Diana. *Medieval Economic Thought*. Cambridge: Cambridge University Press, 2002.

Woolgar, C. M. *The Culture of Food in England, 1200–1500*. New Haven: Yale University Press, 2016.

"Fast and feast: conspicuous consumption and the diet of the nobility in the fifteenth century." In Michael Hicks, ed., *Revolution and Consumption in Late Medieval England*, 7–25. Woodbridge, UK, and Rochester, NY: Boydell and Brewer, 2001.

"Feasting and fasting: food and taste in Europe in the Middle Ages." In Paul Freedman, ed., *Food: The History of Taste*, 135–197. Berkeley and Los Angeles: University of California Press and London: Thames and Hudson, 2007.

*The Great Household in Late Medieval England*. New Haven: Yale University Press, 1999.

Woolgar, Christopher, D. Serjeantson, and T. Waldron, eds. *Food in Medieval England: History and Archaeology*. Oxford: Oxford University Press, 2006.

Worcester, William. *Itineraries*. edited from the unique MS. Corpus Christi College, Cambridge, 210 [and translated] by John H. Harvey. Oxford: Clarendon Press, 1969.

Worm, Boris, and Trevor A. Branch. "The future of fish," *Trends in Ecology and Evolution* 27:11 (November 2012), 594–599.

Worster, Donald. *The Ends of the Earth: Perspectives on Modern Environmental History*. Cambridge: Cambridge University Press,1988.

*Shrinking the Earth: The Rise and Decline of National Abundance*. Oxford: Oxford University Press, 2018.

Wright, Laura. *Sources of London English: Medieval Thames Vocabulary*. Oxford: Clarendon Press, 1996.

Wubs-Mrozewicz, Justyna. "Fish, stock and barrel. changes in the stockfish trade in northern Europe, c. 1360–1560." In Louis Sicking and Darlene Abreu-Ferreira, eds., *Beyond the Catch: Fisheries of the North Atlantic, the North Sea and the Baltic, 900–1850*, 187–208. Leiden: Brill, 2009.

Wundsch, H. H. "Noch einiges zum Gebrauch der 'Aalspeere' an der Ostseeküste der Deutschen Demokratischen Republik," *Deutsche Fischerei-Zeitung* 9 (1962), 206–209.

Wywra, Andrzej M., and Daniel Makowiecki. "Fish in the menu of the Cistercians from Łekno and Bierzwnik (Poland): an historical and archaeoichthyological consideration,." In Daniel Makowiecki et al., eds., *Fishes – Culture – Environment through Archaeoichthyology, Ethnography & History*, 63–68. The 15th Meeting of the International Council for Archaeozoology (ICAZ) Fish Remains Working Group (FRWG) September 3–9, 2009 in Poznań & Toruń, Poland. Poznań: Bogucki Wydawnictwo Naukowe, 2009.

Yoder, Cassady. "Diet in medieval Denmark: a regional and temporal comparison," *Journal of Archaeological Science* 37 (2010), 2224–2236.

Zago, R. *I Nicolotti. Storia di una comunità di pescatori a Venezia nell'età moderna*. Materiali e ricerche, 10. Padua: Francisci, 1982.

Zahlten, Johannes. *Creatio mundi: Darstellungen der sechs Schöpfungstage und naturwissenschaftliches Weltbild im Mittelalter*. Stuttgarter Beiträge zur Geschichte und Politik, vol. 13. Stuttgart: Klett-Cotta, 1979.

Zakrzewski, Ignacy, and Franciszek Piekosiński, eds. *Codex diplomaticus Majoris Poloniae. Kodeks dyplomatyczny Wielkopolska* [CdMP]. 5 vols. Poznań: Kraszewski and Bibliothecae Kornicensis, 1877–1908.

Zambernardi, Ambra. "Les derniers 'paysans de la mer'. La pêche au thon rouge par la madrague en mer Méditerranée." In Buti et al., eds. *Moissonner la mer*, 85–118. 2018.

Zatykó, Csilla. "Aspects of fishing in medieval Hungary." In J. Klapste and P. Sommer, eds. *Processing, Storage, Distribution of Food – Food in the Medieval Rural Environment*, 399–408. Ruralia VIII, 7–12 September 2009, Lorca, Spain. Turnhout: Brepols, 2011.

Zauner, Alois. "Die Beschwerden der oberösterrreichischen Bauern 1511/12 und 1525," *Mitteilungen des Oberösterreichischen Landesarchivs* 14 (1984), 95–122.

Zaunick, Rudolph. "Die Fischerei-Tollköder in Europa von Altertum bis zur Neuzeit; geschichtliche Studien zur angewandten Naturwissenschaft." In A. Thienemann, ed., *Archiv für Hydrobiologie, Supplement-Band, IV*, pp. 526–736. Stuttgart: Schweizerbart, 1928.

Zbierski, Andrezej. "Ichthyological studies on fishing in Gdańsk in the 9th–11th centuries based on archaeological materials from Pomerania," *Archaeologia Polona* 17 (1976), 247–255.

"Rybołowstwo, łowiectwo i zajęcia rolniczo-hodowlane." In Edmund Cieślak, ed., *Historia Gdańska (tom I do roku 1454)*, 173–193. Gdańsk: Wydawn. Morskie, 1978.

Zdekauer, Lodovico, ed. *Statutum potestatis comunis Pistorii anni MCCLXXXXVI*. Milan: Apud Ulricum Hoepli, 1888.

Zechini, Mariana E., Kristina Killgrove, Claudia M. Melisch, Bethany L. Turner, and Benjamin J. Schaefer. "Diachronic changes in diet in medieval Berlin: comparison of dietary isotopes from pre- and post-Black Death adults," *Journal of Archaeological Science: Reports* 38 (August 2021), p.103064 [sic].

Zeheter, Michael. *Die Ordnung der Fischer. Nachhaltigkeit und Fischerei am Bodensee (1350–1900)*. Umwelthistorische Forschungen. Cologne–Weimar–Vienna: Böhlau, 2014.

"To prevent a disorder in the lake: regulating the Lake Constance fisheries, ca. 1350–1800." In Philippe Béarez, Sandrine Grouard, and Benoît Clavel, eds., *Archéologie du poisson. 30 ans d'archéo-ichtyologogie au CNRS. Hommage aux travaus de Jean Desse et Nathalie Desse-Berset*, Actes de recontres 18–20 octobre 2007, 83–90. Antibes: Éditions APDCA, 2008.

Zeibig, Hartmann, ed. *Urkundenbuch des Stiftes Klosterneuburg bis zum Ende des vierzehnten Jahrhunderts*. 2 vols. Fontes rerum Austriacarum, II: Diplomataria et Acta, vols. 27–28. Vienna: Böhlau, 1867–1868.

Zillén, Lovisa, Daniel J. Conley, Thomas Andrén, Elinor Andrén, and Svante Björck. "Past occurrences of hypoxia in the Baltic Sea and the role of climate variability, environmental change and human impact," *Earth Science Reviews* 91 (2008), 77–92.

Zimmermann, G. *Ordensleben und Lebensstandard. Die Cura Corporis in den Ordensvorschriften des abendländischen Hochmittelalters. Beiträge zur Geschichte des alten Mönchtums und des Benediktinerordens*, 32. Münster: Aschendorff 1973.

Zimmermann, T. C. Price. *Paolo Giovio: The Historian and the Crisis of Sixteenth-Century Italy.* Princeton: Princeton University Press, 1995.

Ziolkowski, Jan M. *Fairy Tales from before Fairy Tales: The Medieval Latin Past of Wonderful Lies.* Ann Arbor: University of Michigan Press, 2007.

Zug Tucci, Hannelore. "Il mondo medievale dei pesci tra realta' e immaginazione." In *L'uomo di fronte al mondo animale nell'alto Medio Evo*, pp. 291–360 and discussion pp. 361–372. XXXI Settimane del Centro Italiano di Studi sull'alto Medio Evo. Spoleto: Presso la Sede del Centro, 1985.

"Pesca e caccia in Laguna." In L. Cracco Ruggini, M. Pavan, G. Cracco, and G. Ortalli, eds., *Storia di Venezia*, vol. 1, pp. 492–514. Rome: Istituto della Enciclopedia italiana, 1992.

Županović, Šime. *Ribarstvo Dalmacije u 18. Stoleću s Posebnim Osvrtom na Rezultate Suvremenih Istraživanja.* Zavod za Hrvatsku Poviejest Filozofskog Fakulteta Sveućilišta u Zagrebu. Split: Književni Krug, 1993.

# Index

**Index is to text. Except for a few substantial remarks in the notes, indicated by page numbers in *italics*, neither notes nor supplement are indexed.**

www.ingramcontent.com/pod-product-compliance
Lightning Source LLC
LaVergne TN
LVHW102012210926
848483LV00011B/277
*9781108845465*